# Fundamentals of Fluid–Solid Interactions

## Analytical and Computational Approaches

# MONOGRAPH SERIES ON NONLINEAR SCIENCE AND COMPLEXITY

# Fundamentals of Fluid–Solid Interactions

## Analytical and Computational Approaches

XIAODONG (SHELDON) WANG

*Department of Mathematical Sciences*
*New Jersey Institute of Technology*
*323 Martin Luther King, Jr. Blvd.*
*Newark, NJ 07102, USA*

ELSEVIER

AMSTERDAM • BOSTON • HEIDELBERG • LONDON • NEW YORK • OXFORD
PARIS • SAN DIEGO • SAN FRANCISCO • SINGAPORE • SYDNEY • TOKYO

Elsevier
Radarweg 29, PO Box 211, 1000 AE Amsterdam, The Netherlands
Linacre House, Jordan Hill, Oxford OX2 8DP, UK

First edition 2008



**Library of Congress Cataloging-in-Publication Data**
A catalog record for this book is available from the Library of Congress

**British Library Cataloguing in Publication Data**
A catalogue record for this book is available from the British Library

ISBN: 978-0-444-52807-0
ISSN: 1574-6917



Printed and bound in Hungary

08 09 10 11 12 10 9 8 7 6 5 4 3 2 1

*To my parents,*
*my wife, and*
*my children*

# Preface

The rapid development of computational mechanics and computer technology during the last two decades have brought immensely powerful computational tools to the disposal of scientists and engineers. With the recent advent of teraflop and parallel computers, simulation-based engineering processes are gaining more acceptance and are rapidly becoming viable design alternatives. The traditional classification of various disciplines of mechanics has been blurred, and interaction problems involving two or more disciplines are often encountered in ever more challenging engineering practices. One of these interacting disciplines is fluid–structure or fluid–solid interaction (FSI), which lies at the intersection of fluid mechanics, solid mechanics, and dynamics. In fact, the study of FSI systems is becoming a bridge between complex dynamical systems and computational mechanics.

The new research frontiers in quantitative understanding of biological systems compel researchers to grasp fundamentals with depth in the fields of physics, chemistry, and mathematics. In addition, complex system dynamics derived from interacting systems is essential for the understanding of phenomena pertaining to multiple temporal and spatial scales. The key question is how to acquire, within a reasonable period of time, a knowledge base in these disciplines with both depth and breadth. It is obviously not feasible for students to wade through the traditional curricula of applied mathematics, solid mechanics, continuum mechanics, theory of elasticity, structural mechanics, fluid mechanics, vibration, and dynamics.

In this book, we endeavor to present a unified exposition of fundamental principles and solution techniques to problems involving fluid–solid interactions. Through the publication of this work, we hope to start the transition of mechanics education to a more integrated model. The benefit of this approach is that it will provide students with a convenient entry into mechanics and applied mathematics and help them to quickly acquire sufficient background knowledge in preparation for subsequent focus on special areas which are often interdisciplinary in nature.

It is the intention of this book to serve as the basis of a course for senior undergraduate or first-year graduate students. The comprehensive presentation of the fundamentals of fluid–solid systems will also provide applied scientists and mathematicians with recent references and an overview of the subject.

In this book, we extract the unique physical and mathematical aspects of various FSI problems, lay a foundation of computational approaches for steady or stable transient analysis, explore numerical approaches for static and dynamic instability analysis, and identify feasible algorithms for studying chaotic and complex dynamical systems. Although many examples derived from various engineering fields are introduced for didactic purposes, it is not our intention to focus on problems in any specific industry. We would like to point out that this book is not all-encompassing and certainly not written as an encyclopedia of mechanics and mathematics. As a matter of fact, not all applications and subfields of fluid–solid interactions are fully addressed in this book.

I would like to acknowledge a few individuals without whom I would not have had the knowledge or time to prepare this book. Foremost is Professor Klaus Jürgen Bathe, my Ph.D. advisor at MIT. I met Jürgen when I was a callow graduate student. I had modeled the dynamical behaviors of the magnetic head flying over Winchester hard disks as my undergraduate project, working closely with Professor Binglin Tan at SJTU. This problem involves a rigid body suspended from a flexible structure interacting with fluids governed by the Reynolds or Boltzmann equation depending on the clearance. I was genuinely intrigued by the complexity of this type of FSI problems, so I told Jürgen that I would like to do research related to fluid–solid interactions. Jürgen actually gave me a complicated acoustoelastic problem with slosh to test my capability. I was lucky enough to pass his tough test and the solution to this type of problems is presented in Chapter 6 of this book. Jürgen took me on as his Ph.D. student and supported me financially. I was very fortunate to have Jürgen, a world-class pioneer in computational modeling of FSI systems, as my mentor at MIT. I believe it was those few years of studying and working closely with him and many other MIT faculty members that prepared me to work in this multidisciplinary area. In fact, many of the materials presented in this book came directly from courses and research projects I completed at MIT.

After I got my Ph.D. in 1995, I became a non-tenure track faculty member at the Institute of Paper Science and Technology, where I was exposed to various FSI problems encountered in paper making processes. While I was at the paper institute, I got a chance to work with Professor Jack Hale in the School of Mathematics at GaTech. During my year-long interaction with Jack, I learned many interesting aspects of dynamical systems which could emerge from fluid–solid interaction problems. Our collaboration also resulted in a nice paper on numerical procedures for dynamical systems.

In 1999, I relocated to Polytechnic University in Brooklyn, NY to be with my wife, who at the time was teaching at Bucknell University. My wife was teaching in the Department of Mathematics and I was teaching in the Department of Mechanical Engineering. During this time Polytechnic University was trying to redefine itself amid a sea of change in high education systems. I am very grateful to the Department of Mechanical Engineering for providing me with wonderful support during these few years. Many of the results presented in this book were obtained and constructed during this period. It was also during this period that I had the opportunity to sit in on Professor Charlie Peskin's course on immersed boundary methods at the Courant Institute and to attend Professor Van Mow's research meetings at Columbia University. I was searching for an effective way of handling large structural motions within viscous fluid and Charlie' immersed boundary methods provide just that. My interaction with Van exposed me to many biphasic and triphasic modeling problems in biological systems. These interacting systems are so complex and intricate that I believe many life times will be required before we achieve true understanding of the multiphysics and multiphysics phenomena.

It was very natural for me to connect immersed boundary methods with nonlinear finite element formulations for both fluids and solids and generalize immersed boundary methods to immersed continuum methods, which brings me to another person I would like to acknowledge in my academic career, Professor Wing Kam Liu. I have known of Wing Kam's work in fluid–solid interactions since I was a Ph.D. student at MIT. In this joint effort to combine immersed boundary methods with finite element formulations, I got a chance to work closely with him. Wing Kam is very well informed and always has keen physical insight regarding research problems surfacing from various science and engineering fields. Of course, for magical reasons, he always has many talented and hardworking students, many of whom I know and have worked with have already become faculty members in various universities.

In 2004, I took a faculty job in the Department of Mathematical Sciences at New Jersey Institute of Technology, an emerging national research university. This mathematical science department was recently expanded to national prominence with many first class researchers in applied mathematics under the leaderships of Professors Daljit Ahluwalia, Gregory Kriegsmann, and Robert Miura. I feel truly at home with so many wonderful colleagues in this department. It is really a blessing to have worked with so many friends over the course of preparing this book. I am eternally grateful for the help and friendship of Professors Tomasz Wierzbicki, Koichi Masubuchi, Frank A. McClintock, Anthony T. Patera, and Dick K.P. Yue at Massachusetts Institute of Technology, Professor Fred Bloom at Southern Illinois University, Professor Larry Forney at Georgia Institute of Technology, Professor Weizhu Bao at National University of Singapore, Professors Steve Arnold, Erwin Lutwak, and Zhong-Ping Jiang at Polytechnic University, Dr. David McQueen at New York University, Dr. Robert Kung at ABIOMED, Professor Lucy Zhang at Rensselaer Polytechnic Institute, Professor Yaling Liu at University of Texas at Arlington, and Professors John Tavantzis and Bruce Bukiet, and two wonderful graduate students Ye Yang and Tao Wu at New Jersey Institute of Technology. The financial support from National Science Foundation (NSF) is also gratefully acknowledged. I also would like to thank Rosemary Culver for her special efforts in editing the manuscript. Finally I would like to dedicate this book to my patient and loving family and dear friends, without whom this book would not be possible.

Xiaodong (Sheldon) Wang

# Contents

*Chapter 1*

# Introduction

Three centuries after Sir Isaac Newton, the advance of modern computers that brought a new perspective to classical mechanics. For the past four decades, computational mechanics has pushed the understanding of mechanics to a new horizon. As a consequence, many complex engineering problems, previously inconceivable or heavily dependent on experimental approaches, can now be solved numerically with ease and confidence.

With the development of applied mathematics and computational mechanics, the traditional classification of mechanics disciplines has been blurred, and coupled systems involving two or more disciplines are often encountered in engineering practice. Problems involving both fluid and solid media are often referred to as fluid-structure or fluid–solid interactions (FSI). Stemming from the studies of aircraft and missile structures, the FSI field has been expanding dramatically. Other names introduced for this field of study include flow-induced vibration, aero- and hydro-elasticity. FSI systems exist in mechanical, nuclear, ocean, aerospace, and aeronautic industries [31,185,208,269,270]. Recently, many such coupled systems have emerged in micro-electro-mechanical systems (MEMS), and increasing applications have been found in biomechanics areas, such as flexible blood vessels, blood cells, or heart valves interacting with viscous blood streams.

As illustrated in Figure 1.1, in this triangle discipline among fluid mechanics, solid mechanics, and dynamics, the interesting and unique aspects are the complex physical phenomena, often dynamical and nonlinear. One of the objectives of this book is to attempt to provide a systematic treatment of their fundamental aspects and the myriad of practical FSI problems. In particular, we wish to bridge the gap between nonlinear dynamics and computational mechanics through the study of FSI problems.

## 1.1. Historical background

The search for understanding of the mechanical universe dates back almost to original human curiosity about nature. It is well accepted that Sir Issac Newton (1642–1727) introduced the systematic and mathematical approaches to the

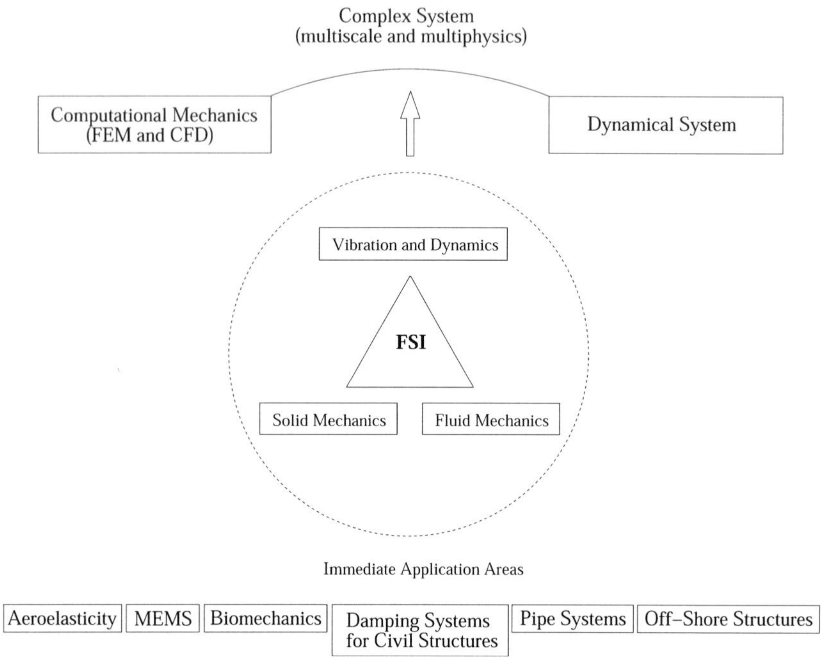


Figure 1.1. Studies of FSI systems and beyond.

study of mechanics. In many aspects, the early work of Galileo Galilei (1564–1642) and the three laws of Johannes Kepler (1571–1630), which are based on the astronomical observations of Tycho Brahe (1546–1601), along with the calculus co-invented by Gottfried Wilhelm von Leibniz (1646–1716) provided Newton with the necessary foundations and tools. Although as continuous media fluids and solids share similar properties, traditionally, due to different response characteristics and solution strategies, distinct disciplines have been assigned.

FSI discipline originates from aero- and hydro-elasticity, which are often related to aeronautics and aerospace as well as nuclear industries. FSI systems, in particular, those with highly deformable immersed structures/solids, still pose unique challenges to applied mathematics and computational mechanics communities. The original work has been well documented by Fung, Bisplinghoff, Ashley, and Halfman [233,234,316]. More recent accomplishments, following the tradition of analytical and experimental approaches, have been presented in Dowell [68–70], Haszpra [101], Blevins [228], and Paidoussis [191,192]. In practice, within the scope of nuclear, civil, aerospace, ocean, chemical, and mechanical engineering, there are many terminologies involved, i.e., flow-induced vibration, aero-elasticity, hydro-elasticity, fluid-structure interaction, and fluid–solid interaction. Typical problems include structure interaction with surface and sound waves, and vibrations and stabilities of cables, pipes, plates, and shells.

Each subject is by itself very extensive. For example, the aero-elasticity area also includes the sound vibration discussed in Junger and Feit [179], Cremer and Heckl [47], and Howe [197]. Reviews on various finite element formulations for FSI systems are available in Morand and Ohayon [105] and Bathe [145]. This area is quickly expanding and it is very likely that new breakthroughs may appear while this book is being prepared for publication. Nevertheless, we hope that the discussion of analytical and computational approaches in this book may draw attention to new possibilities in the study of FSI problems. We feel strongly that effective numerical algorithms for large FSI systems are essential to instability analysis of linear non-autonomous systems; the computation of limit sets, which include stationary points, periodic orbits, quasi-periodic orbits, and strange attractors; and the computation of turning points, which include super- and sub-critical saddle-node, pitchfork, and Hopf bifurcations in parameter space.

## 1.2. Motivation

This book focuses on the common physical and mathematical aspects of various FSI problems along with some generic case studies. Following the lead of Fung, Bispinghoff, Ashley, Blevins, Moon, Dowell et al., interesting chaotic oscillations along with aeroelasticity and flow-induced vibration will be introduced. In addition, following the seminal work of Paidoussis, instability analysis related to internal and concentric flows will be addressed. Finally, following the lead of Morand and Ohayon, this book will include recent advances in computational algorithms associated with FSI problems, such as immersed boundary/continuum methods, arbitrary Lagrangian–Eulerian (ALE) formulations, mixed formulations, upwind techniques, and effective iterative solvers. Some general FSI problems will be identified and solved by general purpose computational mechanics software. Considering the fact that most FSI problems are transient and nonlinear, we emphasize that the potential challenge of coupled large systems is going to be the stability analysis, which ties in with the future area of adopting nonlinear dynamics and chaotic oscillations in computational mechanics for large systems.

In practice, FSI systems can be associated with all four types of fluid models, i.e., ideal potential flow, acoustic fluid, incompressible viscous flow, and compressible flow, along with most solid material models such as linear and nonlinear elastic materials or plates and shells. Three steps are commonly used in the analysis of FSI systems. The first step is the identification of the physical system under consideration and the construction of a proper mathematical model, which includes assumptions on geometry, time evolution domain, kinematics, material law, loading, and boundary condition. In many cases, we must also predict what results might be derived from the analysis of the mathematical models. The second step is the formulation of governing equations for the mathematical model and the selection of appropriate solution methods. The third step is the interpretation of the solution and the comparison with the expectations and physics. Of course, iterations among these three steps are commonly needed.

## 1.3. Outline

This book contains seven chapters, including this introductory chapter. Chapter 2 provides a brief summary of the essence of mechanics and the associated fundamentals of mathematics. Notations introduced in Chapter 2 will be used throughout the book. Chapter 3 is devoted to the understanding of stability issues and chaos with respect to various FSI problems. Chapter 4 concerns one of the original FSI problems: flow-induced vibration. Although this particular set of problems has been well understood, complex engineering problems involving coupled systems with a large number of degrees of freedom still require further investigation, in particular, further development of computational approaches. In Chapter 5 we present a unified illustration of Green's function and boundary element methods associated with potential flow, low Reynolds number flow, and acoustic field. Chapter 6 addresses the computational approaches for linear FSI systems such as acoustoelastic FSI systems. Chapter 7 addresses nonlinear structural responses due to various steady or transient fluid flows. Because it is impossible to ignore oscillations, which are often chaotic, numerical analysis of nonlinear dynamical behaviors of large systems, which still pose a challenge to applied scientists and mathematicians, will be briefly discussed.

*Chapter 2*

# Aspects of Mathematics and Mechanics

Although Newton's laws retain their importance throughout the modern study of mechanics, the mathematical tools required in various disciplines of mechanics have evolved into something far more complicated than original calculus.

The analysis of FSI problems requires the idealization of both fluid and solid domains in a form that can be solved systematically and repetitively. In order to coherently present the formulation of such mathematical models as well as the solution procedures, we aim in this chapter to establish a mathematical foundation for kinematic and dynamic descriptions of fluids and solids.

In Section 2.1, we present basic mathematical concepts, notations employed in this book, and some fundamental concepts of linear algebra, complex analysis, and advanced calculus. In Section 2.2, we introduce kinematics for mass points, rigid bodies, and continuous media. Also in Section 2.2, we discuss Reynolds transport theorems and variation principles, from which both integral and differential equilibria of mass, momentum, and energy conservation laws derive. Some useful mathematical tools are discussed in Section 2.3. We expect that Chapter 2 not only streamlines the presentation of this book, but also enables the reader to quickly pick up fundamental concepts crucial for building up necessary tools to tackle FSI problems without wading through numerous traditional books on mathematics and mechanics.

## 2.1. Preliminary concepts and notations

The purpose of this section is to form a foundation of elementary mathematical concepts and to present notations employed throughout this book.

### *2.1.1. Logic; set, group, and field; Boolean algebra*

*Logic*

Mathematical tools are based on logistical principles. Since we also present a few proofs in this book, we will briefly summarize key concepts of logic, and more importantly, the mathematical notations. The logical concepts and symbols stipulated in this chapter will be used throughout this book.

Let $p$ and $q$ represent propositions which could be true or false, but not both true and false. We introduce the following symbolic languages:

$p \wedge q$ and $p \vee q$ are the conjunction and disjunction of $p$ and $q$.

Implication $p \to q$ means antecedent $p$ implies consequent $q$.

$p \Leftrightarrow q$ is the equivalence relation, and $\sim p$ is the negation of $p$.

$\exists p$ means there exists a $p$ and $\forall p$ means for all $p$.

Thus, it can be easily deduced that

$$\sim(p \wedge q) \Leftrightarrow \big[(\sim p) \vee (\sim q)\big] \quad \text{and} \quad \sim(p \vee q) \Leftrightarrow \big[(\sim p) \wedge (\sim q)\big].$$

Moreover, for a given implication, $p \to q$, we have the converse proposition $q \to p$, the inverse proposition $\sim p \to (\sim q)$, and the contrapositive proposition $\sim q \to (\sim p)$. Finally, a tautology is a proposition formed by combining other propositions, e.g., $p, q, r, \ldots$, which is true regardless of the truth or falsehood of $p, q, r, \ldots$. For example, $\sim(\sim p) \Leftrightarrow p$, $(p \Leftrightarrow q) \Leftrightarrow [(p \to q) \wedge (q \to p)]$, $\sim[p \wedge (\sim p)]$, $[(p \to q) \wedge (q \to r)] \to (p \to r)$, and $(p \to q) \Leftrightarrow [(\sim q) \to (\sim p)]$ are tautologies, which are useful for both direct and indirect proofs.

*Set, group, and field*

A set is defined as a collection of objects sharing common characteristics, such as numbers, points, lines, etc., and is denoted in this book with capital calligraphic letters. If a set contains no elements, it is called the null set, denoted as $\emptyset$.

Two sets are said to be equivalent when they can be placed in one-to-one correspondence, and two sets are said to be identical if every element of one set is an element of the other. Moreover, if every element of a set $\mathcal{A}$ is an element of a set $\mathcal{B}$, $\mathcal{A}$ is called a subset of $\mathcal{B}$, or $\mathcal{A}$ is included in $\mathcal{B}$, denoted as $\mathcal{A} \subseteq \mathcal{B}$. In addition, if at least one element of $\mathcal{B}$ is not an element of $\mathcal{A}$, it is called a proper subset of $\mathcal{B}$, denoted as $\mathcal{A} \subset \mathcal{B}$. In this book, we denote a set of natural numbers as $\bar{\mathcal{N}}$, a set of integers as $\mathcal{Z}$, a set of nonnegative integers as $\mathcal{N}$, a set of rational numbers as $\mathcal{Q}$, a set of real numbers as $\mathcal{R}$, and a set of complex numbers as $\mathcal{C}$.

For an abstract mathematical system, operations within or between different sets must be defined. For example, logic operations such as $\wedge$ and $\vee$ and arithmetic operations such as addition $(+)$, subtraction $(-)$, multiplication $(\times)$, and division ($\div$ or $/$) are often used. As a side remark, if the difference of two integers $a$ and $b$ is divisible by an integer $m$, $a$ and $b$ are said to be congruent with respect to the modulus $m$, or denoted as $b = a \bmod m$, which is often named as the fifth arithmetic operation. Furthermore, distributivity, commutativity, and associativity are in general introduced in the basic laws of such systems, along with a zero element, an identity element, and an inverse element.

One such mathematical system, the idea of a group is introduced by Evariste Galois (1811–1832). Assume undefined elements $a, b, c, \ldots,$ belonging to a set $\mathcal{S}$, and an undefined operator $\circ$, used to pair two elements as $a \circ b$, the structure of a group is given by the following axioms:

**Closure:** For each pair, $a$ and $b \in \mathcal{S}$, the combination $c = a \circ b$ is a unique element of $\mathcal{S}$.

**Associativity:** For each triple, $a$, $b$, and $c \in \mathcal{S}$,

$$(a \circ b) \circ c = a \circ (b \circ c).$$

**Identity:** There exists a unique element $i \in \mathcal{S}$, called identity, having the property that for every element $a \in \mathcal{S}$,

$$a \circ i = i \circ a = a.$$

**Inverse:** Corresponding to each element $a \in \mathcal{S}$, there is a unique element $a'$, called the inverse of $a$, having the property that for every element $a \in \mathcal{S}$,

$$a \circ a' = a' \circ a = i.$$

More specifically, any set with a closed operation is called a groupoid, any groupoid which satisfies associativity is called a semi-group, any semi-group with an identity element is called a monoid, and any monoid with an inverse is called a group. In addition, if a group satisfies the commutative axiom, i.e., for every $a$ and $b \in \mathcal{S}$,

$$a \circ b = b \circ a,$$

it is called a commutative group or an Abelian group, named after Niels Henrik Abel (1802–1829).

Thus, a few commonly used results about groups can be easily verified:

If $a = b$, then $a \circ c = b \circ c$ and if $a \circ c = b \circ c$, then $a = b$.

There exists a unique element $x$, such that $a \circ x = b$ and $(a')' = a$.

As a group is concerned with a single operation, abstract systems may involve two operations. Assume undefined elements $a, b, c, \ldots,$ belonging to a set $\mathcal{S}$, and two undefined operations $\circ$ and $\bullet$, a field is an abstract system satisfying the following axioms:

**Closure:** For each pair, $a$ and $b \in \mathcal{S}$, the combinations $a \circ b$ and $a \bullet b$ are unique elements of $\mathcal{S}$.

**Associativity:** For each triple, $a$, $b$, and $c \in \mathcal{S}$,

$$(a \circ b) \circ c = a \circ (b \circ c) \quad \text{and} \quad (a \bullet b) \bullet c = a \bullet (b \bullet c).$$

**Zero:** There exists a unique element $0 \in \mathcal{S}$, called zero or additive identity, having the property that for every element $a \in \mathcal{S}$,

$$a \circ 0 = 0 \circ a = a.$$

**Unit:** There exists a unique element $1 \in \mathcal{S}$, called unit element, such that for every element $a \in \mathcal{S}$,

$$a \bullet 1 = 1 \bullet a = a.$$

**Additive inverse:** Corresponding to each element $a \in \mathcal{S}$, there is a unique element $-a \in \mathcal{S}$, called the additive inverse of $a$, such that

$$a \circ -a = -a \circ a = 0.$$

**Multiplicative inverse:** Corresponding to each element $a \in \mathcal{S}$, with $a \neq 0$, there is a unique element $1/a \in \mathcal{S}$, called the multiplicative inverse of $a$, such that

$$a \bullet 1/a = 1/a \bullet a = 1.$$

**Commutativity:** For each pair, $a$ and $b \in \mathcal{S}$,

$$a \circ b = b \circ a \quad \text{and} \quad a \bullet b = b \bullet a.$$

**Distributivity:** For every triple, $a$, $b$, and $c \in \mathcal{S}$,

$$a \bullet (b \circ c) = (a \bullet b) \circ (a \bullet c).$$

Therefore, if we suppose the two operations $\circ$ and $\bullet$ represent addition $(+)$ and multiplication $(\times)$, respectively, and the set $\mathcal{S}$ is a set of real numbers, we have the following results:

$$\begin{aligned}
&\big[(a = b) \wedge (c = d)\big] \to (a + c = b + d),\\
&\big[(a = b) \wedge (c = d)\big] \to (a \times c = b \times d);\\
&(a + x = b) \to \big[x = (-a) + b\big];\\
&\big[(a \times x = b) \wedge (a \neq 0)\big] \to \big[x = (1/a) \times b\big];\\
&\big[(a + c) = (b + c)\big] \to (a = b);\\
&\big[(a \times c) = (b \times c) \wedge (c \neq 0)\big] \to (a = b).
\end{aligned}$$

Moreover, if we denote the difference $a + (-b)$ as $a - b$, and the quotient $a \times (1/b)$ as $a/b$, we have

$$\begin{aligned}
&\big[(a = b) \wedge (c = d)\big] \to (a - c = b - d),\\
&\big[(a = b) \wedge (c = d) \wedge (c \neq 0)\big] \to (a/c = b/d);\\
&-(-a) = a, \qquad 1/(1/a) = a, \quad \text{when } a \neq 0;
\end{aligned}$$

$$-(a+b) = -a - b, \qquad 1/(a \times b) = 1/a \times 1/b,$$
$$\text{when } a \neq 0 \text{ and } b \neq 0;$$
$$-(a-b) = -a + b, \qquad 1/(a \times 1/b) = b/a, \quad \text{when } a \neq 0 \text{ and } b \neq 0;$$

and in this book, we denote $a \times b$ by $ab$.

*Boolean algebra*

Based on the similar concepts proposed for groups and fields, Boolean algebra, named after George Boole (1815–1864), can be easily introduced as the algebra for sets. The union and intersection operations of two sets, $\mathcal{A}$ and $\mathcal{B}$, are denoted as $\mathcal{A} \cup \mathcal{B}$ and $\mathcal{A} \cap \mathcal{B}$, respectively. The union of $\mathcal{A}$ and $\mathcal{B}$ is the set of elements which belong either to $\mathcal{A}$ or to $\mathcal{B}$ or to both $\mathcal{A}$ and $\mathcal{B}$; the intersection of $\mathcal{A}$ and $\mathcal{B}$ is the set of elements which belong to both $\mathcal{A}$ and $\mathcal{B}$. Moreover, if all the sets considered are subsets of a fixed set $\mathcal{W}$, the complement of a set $\mathcal{A}$ relative to $\mathcal{W}$ is the set of elements of $\mathcal{W}$ which are not elements of $\mathcal{A}$, denoted as $\mathcal{A}'$, or $\mathcal{W} \setminus A$.

Almost identical to the symbols and axioms of groups and fields, Boolean algebra gives us the following de Morgan's laws named after Augustus de Morgan (1806–1871):

| | |
|---|---|
| *Closure*: | $\mathcal{A} \cup \mathcal{B} \subseteq \mathcal{W}, \qquad \mathcal{A} \cap \mathcal{B} \subseteq \mathcal{W};$ |
| *Associativity*: | $\mathcal{A} \cup (\mathcal{B} \cup \mathcal{C}) = (\mathcal{A} \cup \mathcal{B}) \cup \mathcal{C},$ |
| | $\mathcal{A} \cap (\mathcal{B} \cap \mathcal{C}) = (\mathcal{A} \cap \mathcal{B}) \cap \mathcal{C};$ |
| *Commutativity*: | $\mathcal{A} \cup \mathcal{B} = \mathcal{B} \cup \mathcal{A}, \qquad \mathcal{A} \cap \mathcal{B} = \mathcal{B} \cap \mathcal{A};$ |
| *Zero and identity*: | $\mathcal{A} \cup \emptyset = \mathcal{A}, \qquad \mathcal{A} \cap \mathcal{I} = \mathcal{A};$ |
| *Complement*: | $\mathcal{A} \cup \mathcal{A}' = \mathcal{W}, \qquad \mathcal{A} \cap \mathcal{A}' = \emptyset;$ |
| *Distributivity*: | $\mathcal{A} \cap (\mathcal{B} \cup \mathcal{C}) = (\mathcal{A} \cap \mathcal{B}) \cup (\mathcal{A} \cap \mathcal{C}),$ |
| | $\mathcal{A} \cup (\mathcal{B} \cap \mathcal{C}) = (\mathcal{A} \cup \mathcal{B}) \cap (\mathcal{A} \cup \mathcal{C});$ |
| *de Morgan*: | $\mathcal{W} \setminus (\mathcal{A} \cup \mathcal{B}) = (\mathcal{W} \setminus \mathcal{A}) \cap (\mathcal{W} \setminus \mathcal{B}),$ |
| | $\mathcal{W} \setminus (\mathcal{A} \cap \mathcal{B}) = (\mathcal{W} \setminus \mathcal{A}) \cup (\mathcal{W} \setminus \mathcal{B});$ |
| *Duality*: | $\mathcal{W} \setminus \bigcup_{k=1}^{n} \mathcal{A}_k = \bigcap_{k=1}^{n} (\mathcal{W} \setminus \mathcal{A}_k),$ |
| | $\mathcal{W} \setminus \bigcap_{k=1}^{n} \mathcal{A}_k = \bigcup_{k=1}^{n} (\mathcal{W} \setminus \mathcal{A}_k).$ |

### 2.1.2. Mapping and linear space; grad, div, and curl

*Mapping and linear space*

A mapping (function, transformation) is a connection from a set $\mathcal{X}$ to a set $\mathcal{Y}$, denoted as $f : \mathcal{X} \to \mathcal{Y}$, which assigns for every $x \in \mathcal{X}$, a unique element $y \in \mathcal{Y}$, and is usually written as $y = f(x)$ or $(x, y)$. A mapping $f$ is surjective (onto), if and only if (iff) every $y \in \mathcal{Y}$ is the image of some element $x \in \mathcal{X}$, injective (one-to-one), iff every $y \in \mathcal{Y}$ is the image of at most one element of $\mathcal{X}$, and bijective (one-to-one and onto), iff every $y \in \mathcal{Y}$ is the unique image of some element

$x \in \mathcal{X}$, i.e., both surjective and injective. Moreover, $\mathcal{X}$ is called the domain of $f$, denoted as $\mathcal{D}(f)$, and the set of elements in $\mathcal{Y}$ which are images of elements in $\mathcal{X}$ is called the range of $f$, denoted as $\mathcal{R}(f)$. In the case where the domain and range are the same, $f$ is also called an operator. As discussed in Section 2.1, we introduce an operation $\circ$ between two mappings $f : \mathcal{X} \to \mathcal{Y}$ and $g : \mathcal{Y} \to \mathcal{Z}$, such that $\forall x \in \mathcal{X}$, $g \circ f(x) = g(f(x))$. In addition, the identity mapping $i_x$ and $i_y$ are defined as $i_x : \mathcal{X} \to \mathcal{X}$ with $i_x(x) = x$, $\forall x \in \mathcal{X}$, and $i_y : \mathcal{Y} \to \mathcal{Y}$ with $i_y(y) = y$, $\forall y \in \mathcal{Y}$, respectively. Thus, a mapping $f$ is left invertible, if there exists a mapping $g : \mathcal{Y} \to \mathcal{X}$ such that $g \circ f = i_x$, and right invertible, if $f \circ g = i_y$. Of course, a mapping is invertible if it is both left and right invertible, and customarily denoted as $f^{-1}$. It is clear that the invertible mapping is bijective and its inverse mapping is unique.

Furthermore, a mapping $f : \mathcal{X} \to \mathcal{Y}$ is called a linear mapping, if for $x_1, x_2 \in \mathcal{X}$ and $\lambda \in \mathcal{R}$,

$$\begin{aligned} f(x_1 + x_2) &= f(x_1) + f(x_2), \\ f(\lambda x_1) &= \lambda f(x_1). \end{aligned} \tag{1.1}$$

There are quantities in physics characterized by magnitude only, such as mass, length, and temperature. Such quantities are often called scalars. Apart from their physical units, scalars are just real numbers, and are denoted in this book by upper or lower case ordinary letters. To describe physical quantities characterized by both magnitude and direction, such as force, velocity, and heat flux, we introduce the concept of vectors. The geometric representation of a vector is a directed line segment. Since the translation of a vector in space will not change its magnitude or direction, we can always assign the origin $o$ and a spatial point $p$ as the initial point and the terminal point of a vector, respectively, and a vector always corresponds to a position vector $\vec{op}$. As in number systems, we also introduce rules, so-called linear algebra, for the manipulation of vectors.

To represent three-dimensional space in nature, we often introduce a rectangular coordinate frame $oxyz$: the Cartesian coordinate system, named after René Descartes (1596–1650), with three real numbers $x_1$, $x_2$, and $x_3$. Such a space is denoted as $\mathcal{R}^3$, and called the Euclidean space, named after Euclid (~300 BC). Thus, a spatial point $p$ can be represented by $(x_1, x_2, x_3)$ where $x_1$, $x_2$, and $x_3$ stand for the distances relative to the origin $o(0, 0, 0)$ along the $x_1$-, $x_2$-, and $x_3$-axes. Likewise, we can introduce a real $n$-dimensional Cartesian space (hyperspace, or manifold) $\mathcal{R}^n$, as a set of all $n$-tuples of real numbers. A typical element in $\mathcal{R}^n$ is called a point (or vector) and denoted as $\mathbf{r}$, $\mathbf{x}$, or $(x_1, \ldots, x_n)$, where $x_i$ is the $i$th coordinate of that point, or the $i$th component of that vector. Conventionally, $(x_1, \ldots, x_n)$ refers a row vector. However, a typical vector $\mathbf{x}$ refers to a column vector. To delineate $(x_1, \ldots, x_n)$ from the column vector $\mathbf{x}$, we introduce $\mathbf{x} = \langle x_1, \ldots, x_n \rangle$. Naturally, for two elements $\mathbf{x}_1$ and $\mathbf{x}_2$ in $\mathcal{R}^n$ and $\lambda \in \mathcal{R}$, we have $\mathbf{x}_1 + \mathbf{x}_2 = \langle x_{11} + x_{21}, \ldots, x_{1n} + x_{2n}\rangle$ and $\lambda\mathbf{x}_1 = \langle \lambda x_{11}, \ldots, \lambda x_{1n}\rangle$. Furthermore, the magnitude of vector $\mathbf{r}$ is defined as $|\mathbf{r}| = \sqrt{x_i x_i}$. Note that, without specification, in indicial notations, we adopt both the Einstein summation convention, i.e., a dummy index (superscript or subscript) repeated in any expression is summed throughout its range, and the range convention, i.e., a dummy index unrepeated in a term takes all values in its range. In this book, vectors are denoted as boldface lowercase letters. A null or zero vector with a magnitude of 0, and a unit vector with a magnitude of 1, are denoted as $\mathbf{0}$ and $\mathbf{e}$, respectively.

Denoting the unit vectors along the $x_1, \ldots, x_n$-axes as $\mathbf{e}_1, \ldots, \mathbf{e}_n$, e.g., $\mathbf{e}_i$ is a vector with 1 as the $i$th coordinate and 0 elsewhere. We can then express a vector $\mathbf{r}$ as $\mathbf{r} = x_i\mathbf{e}_i$. The basic laws of vector algebra are similar to those of group and field definitions in Section 2.1, including commutativity, associativity, and distributivity:

$$\begin{aligned}
&\mathbf{x}_1 + \mathbf{x}_2 = \mathbf{x}_2 + \mathbf{x}_1,\\
&\mathbf{x}_1 + (\mathbf{x}_2 + \mathbf{x}_3) = (\mathbf{x}_1 + \mathbf{x}_2) + \mathbf{x}_3,\\
&\lambda_1(\lambda_2\mathbf{x}) = (\lambda_1\lambda_2)\mathbf{x} = \lambda_2(\lambda_1\mathbf{x}),\\
&(\lambda_1 + \lambda_2)\mathbf{x} = \lambda_1\mathbf{x} + \lambda_2\mathbf{x},\\
&\lambda(\mathbf{x}_1 + \mathbf{x}_2) = \lambda\mathbf{x}_1 + \lambda\mathbf{x}_2.
\end{aligned} \tag{1.2}$$

Denoting the Kronecker delta $\delta_{ij}$, named after Leopold Kronecker (1823–1891), as 1, if $i = j$, or 0, if $i \neq j$. As a consequence, the following laws are valid:

$$\begin{aligned}
&\mathbf{e}_i \cdot \mathbf{e}_j = \delta_{ij}, \qquad \mathbf{x}_1 \cdot \mathbf{x}_2 = \mathbf{x}_2 \cdot \mathbf{x}_1 = x_{1i}x_{2i},\\
&\mathbf{x}_1 \cdot (\mathbf{x}_2 + \mathbf{x}_3) = \mathbf{x}_1 \cdot \mathbf{x}_2 + \mathbf{x}_1 \cdot \mathbf{x}_3,\\
&\lambda(\mathbf{x}_1 \cdot \mathbf{x}_2) = (\lambda\mathbf{x}_1) \cdot \mathbf{x}_2 = \mathbf{x}_1 \cdot (\lambda\mathbf{x}_2) = (\mathbf{x}_1 \cdot \mathbf{x}_2)\lambda,\\
&\mathbf{x}_1 \text{ is perpendicular to } \mathbf{x}_2, \text{ if } \mathbf{x}_1 \cdot \mathbf{x}_2 = 0, \text{ with } \mathbf{x}_1, \mathbf{x}_2 \neq \mathbf{0}.
\end{aligned} \tag{1.3}$$

Furthermore, we define the dot or scalar product of two vectors $\mathbf{x}_1$ and $\mathbf{x}_2$ as

$$\mathbf{x}_1 \cdot \mathbf{x}_2 = x_{1i}x_{2i} = |\mathbf{x}_1||\mathbf{x}_2|\cos\theta, \quad 0 \leqslant \theta \leqslant \pi, \tag{1.4}$$

where $\theta$ is the angle between vectors.

The angle $\theta$ between two nonzero vectors is defined as the ordinary interior angle $0 \leqslant \theta \leqslant \pi$ between these two vectors positioned tail to tail. The physical significance of the dot product is the projection of one vector onto the other. It is obvious that the magnitude of a vector is in fact the square root of the dot product of the vector with itself, and the vector from a point $\mathbf{x}_1$ to a point $\mathbf{x}_2$ can be expressed as $\mathbf{x}_2 - \mathbf{x}_1$.

A set of $n$ vectors $\mathbf{x}_i$, $i = 1, \ldots, k \leqslant n$, is called linearly dependent if there exists a set of scalar coefficients $c_1, \ldots, c_k$, not all zero, such that $c_i\mathbf{x}_i = \mathbf{0}$. Otherwise the set of vectors is called linearly independent. In fact, for an $n$-dimensional space, a set of linearly independent vectors consists of at most $n$ vectors. Moreover, such a set of linearly independent vectors $\mathbf{x}_1, \ldots, \mathbf{x}_n \in \mathcal{S}$ is said to span $\mathcal{S}$ and form a basis of $\mathcal{S}$ if every vector in $\mathcal{S}$ is a linear combination of $\mathbf{x}_1, \ldots, \mathbf{x}_n$. Succinctly, the dimension of a vector space, denoted as $\dim(\mathcal{S})$, is the number of elements in a basis, e.g., $\dim(\mathcal{R}^n) = n$. In addition, the corresponding coefficients are called the coordinates with respect to the basis. One example of the basis is the so-called standard basis $\mathbf{e}_i$. It is obvious that by adjoining to a set of $k < n$ independent vectors suitable $n - k$ vectors, one can form a basis.

Now we introduce a matrix, denoted in this book with boldface capital letters, as an array of numbers formed in rows and columns. For example, matrix $\mathbf{A} = [A_{ij}]$ with $m$ rows and $n$ columns can be expressed as

$$\mathbf{A} = [A_{ij}] = \begin{bmatrix} A_{11} & A_{12} & \ldots & A_{1n} \\ A_{21} & A_{22} & \ldots & A_{2n} \\ \vdots & \vdots & \ddots & \vdots \\ A_{m1} & A_{m2} & \ldots & A_{mn} \end{bmatrix}. \tag{1.5}$$

In fact, each $m \times n$ matrix $\mathbf{A}$ is associated with a linear mapping $\mathbf{A} : \mathcal{R}^n \to \mathcal{R}^m$, where $m$ and $n$ are the row and column numbers, respectively. In addition, the transpose of this matrix, denoted as $\mathbf{A}^T = [A_{ji}]$, corresponds to a linear mapping $\mathbf{A}^T : \mathcal{R}^m \to \mathcal{R}^n$. Of course, an $n$-dimensional vector can be viewed as an $n \times 1$ matrix, and is traditionally denoted as a column, i.e., $\mathbf{x} = \langle x_1, \ldots, x_n \rangle = [x_i] = [x_1 \quad \ldots \quad x_n]^T$. Furthermore, let $\mathbf{B}$ be an $n \times l$ matrix, the composite map $\mathbf{AB}$ corresponds to an $m \times l$ matrix, $\mathbf{C} = [C_{ij}]$, with $C_{ij} = A_{ik} B_{kj}$. Defining $\mathbf{I}$ as an identity matrix (unit matrix), i.e., $I_{ij} = \delta_{ij}$, if there exists an $n \times m$ matrix $\mathbf{B}$ such that $\mathbf{BA} = \mathbf{I}_n$, where $\mathbf{I}_n$ is an $n \times n$ identity matrix, the matrix $\mathbf{B}$ is called the left inverse of the left invertible matrix $\mathbf{A}$. If there exists another $n \times m$ matrix $\mathbf{B}$ such that $\mathbf{AB} = \mathbf{I}_m$, where $\mathbf{I}_m$ is an $m \times m$ identity matrix, the matrix $\mathbf{B}$ is called the right inverse of the right invertible matrix $\mathbf{A}$. Of course, for a linear operator $\mathbf{A}$ on $\mathcal{R}^n$, if there exists an operator $\mathbf{B}$, such that $\mathbf{AB} = \mathbf{BA} = \mathbf{I}$, $\mathbf{A}$ is called invertible, i.e., both left and right invertible, hence, $\mathbf{B} = \mathbf{A}^{-1}$ and $\mathbf{A} = \mathbf{B}^{-1}$.

A nonempty subset $\mathcal{S} \subseteq \mathcal{R}^n$ is called a linear subspace (vector space, linear space), if $\mathcal{S}$ is closed under the operations of addition and scalar multiplication in $\mathcal{R}^n$. If $\mathcal{S}$ contains only $\mathbf{0}$, $\mathcal{S}$ is called the trivial subspace, and if $\mathcal{S} \neq \mathcal{R}^n$, it is called a proper subspace of $\mathcal{R}^n$. Moreover, if both $\mathcal{S}_1$ and $\mathcal{S}_2$ are subspaces of $\mathcal{R}^n$ and $\mathcal{S}_1 \subset \mathcal{S}_2$, $\mathcal{S}_1$ is a subspace of $\mathcal{S}_2$.

In $\mathcal{R}^3$ we also define the cross or vector product of two vectors $\mathbf{f}_1$ and $\mathbf{f}_2$ as

$$\mathbf{f}_1 \times \mathbf{f}_2 = |\mathbf{f}_1||\mathbf{f}_2| \sin\theta \mathbf{n}, \quad 0 \leqslant \theta \leqslant \pi, \tag{1.6}$$

where $\theta$ is the angle between vectors, $\mathbf{n}$ is a unit vector indicating the direction of the resulting vector $\mathbf{f}_3 = \mathbf{f}_1 \times \mathbf{f}_2$ which is perpendicular to the plane of $\mathbf{f}_1$ and $\mathbf{f}_2$, and $\mathbf{f}_1$, $\mathbf{f}_2$, and $\mathbf{f}_3$ form a right-handed system.

### 2.1.3. Tensors and curvilinear coordinates; complex variables

*Tensors and curvilinear coordinates*

In practice, a scalar is called a tensor of order zero, a vector is a tensor of order one, and a tensor of order two can be expressed as a $3 \times 3$ matrix, also denoted in this book by a boldface uppercase letter or a diadic product of two vectors, e.g., $\mathbf{F} = \mathbf{f}_1 \odot \mathbf{f}_2 = f_{1i} f_{2j} \mathbf{e}_i \mathbf{e}_j$. In this book, without specification, tensors are referred to tensors of second order. A proper introduction of tensors must begin with some basic concepts of linear mapping, coordinate transformations, and contravariant and covariant vectors.

Consider a linear mapping represented by an $m \times n$ matrix $\mathbf{A} : \mathcal{R}^n \rightarrow \mathcal{R}^m$, where $\mathcal{R}^n$ and $\mathcal{R}^m$ are $n$- and $m$-dimensional vector spaces, respectively. The kernel (null space) of $\mathbf{A}$ is defined as

$$\ker(\mathbf{A}) = \{\mathbf{x} \in \mathcal{R}^n \mid \mathbf{A}\mathbf{x} = \mathbf{0}\}, \tag{1.18}$$

and the image (column space) of $\mathbf{A}$ is defined as

$$\operatorname{im}(\mathbf{A}) = \{\mathbf{y} \in \mathcal{R}^m \mid \mathbf{A}\mathbf{x} = \mathbf{y} \text{ for some } \mathbf{x} \in \mathcal{R}^n\}. \tag{1.19}$$

Of course, $\ker(\mathbf{A})$ and $\operatorname{im}(\mathbf{A})$ are subspaces of $\mathcal{R}^n$ and $\mathcal{R}^m$, respectively, and

$$\dim\bigl(\operatorname{im}(\mathbf{A})\bigr) + \dim\bigl(\ker(\mathbf{A})\bigr) = n. \tag{1.20}$$

Likewise, the transpose of the matrix $\mathbf{A}$ will denote a linear mapping $\mathbf{A}^T : \mathcal{R}^m \rightarrow \mathcal{R}^n$. The kernel (null space) of $\mathbf{A}^T$ or the left null space of $\mathbf{A}$ is defined as

$$\ker(\mathbf{A}^T) = \{\mathbf{y} \in \mathcal{R}^m \mid \mathbf{A}^T\mathbf{y} = \mathbf{0}\}, \tag{1.21}$$

and the image of $\mathbf{A}^T$ or the row space of $\mathbf{A}$ is defined as

$$\operatorname{im}(\mathbf{A}^T) = \{\mathbf{x} \in \mathcal{R}^n \mid \mathbf{A}^T\mathbf{y} = \mathbf{x} \text{ for some } \mathbf{y} \in \mathcal{R}^m\}. \tag{1.22}$$

Again, $\ker(\mathbf{A}^T)$ and $\operatorname{im}(\mathbf{A}^T)$ are subspaces of $\mathcal{R}^m$ and $\mathcal{R}^n$, respectively, and

$$\dim\bigl(\operatorname{im}(\mathbf{A}^T)\bigr) + \dim\bigl(\ker(\mathbf{A}^T)\bigr) = m. \tag{1.23}$$

It is now clear that the matrix $\mathbf{A}$ is characterized by its four fundamental subspaces, namely, the null space $\ker(\mathbf{A})$ and the row space $\operatorname{im}(\mathbf{A}^T)$ within $\mathcal{R}^n$ and the left null space $\ker(\mathbf{A}^T)$ and the column space $\operatorname{im}(\mathbf{A})$ within $\mathcal{R}^m$. Based on the definition, all the vectors within the null space must be perpendicular to all the vectors within the row space, while all the vectors within the left null space must be perpendicular to all the vectors within the column space. Therefore, within $\mathcal{R}^n$, the subspace $\ker(\mathbf{A})$ is orthogonal to the subspace $\operatorname{im}(\mathbf{A}^T)$; while the subspace $\ker(\mathbf{A}^T)$ is orthogonal to the subspace $\operatorname{im}(\mathbf{A})$. In fact the fundamental theorem of linear algebra or Fredholm alternative, named after Erik Ivar Fredholm (1866–1927), is stated in the context of the four fundamental subspaces.

THEOREM 2.1.1 *(Fundamental theorem of linear algebra). For any matrix* $\mathbf{A}$ *and vector* $\mathbf{b}$*, one and only one of the following systems has a solution* (1) $\mathbf{Ax} = \mathbf{b}$ *or* (2) $\mathbf{A}^T\mathbf{y} = \mathbf{0}$, $\mathbf{y}^T\mathbf{b} \neq \mathbf{0}$*. The understanding of this alternative can be simply viewed as either the vector* $\mathbf{b}$ *is in the column space of* $\mathbf{A}$ *as in* (1) *or the vector* $\mathbf{b}$ *is not in the column space of* $\mathbf{A}$ *as in* (2).

Furthermore, $\mathbf{A}$ is a one-to-one mapping iff $\ker(\mathbf{A}) = \{\mathbf{0}\}$, and the kernel of a linear mapping $\mathbf{A}: \mathcal{R}^n \to \mathcal{R}^m$ corresponds to the solution space of the linear algebraic equation $\mathbf{Ax} = \mathbf{0}$ with $\mathbf{x} \in \mathcal{R}^n$. Obviously, solving the system $\mathbf{Ax} = \mathbf{0}$ is equivalent to finding a basis for $\ker(\mathbf{A})$. If the mapping $\mathbf{A}: \mathcal{R}^n \to \mathcal{R}^m$ is bijective, $\mathbf{A}$ is called an isomorphism and $\mathcal{R}^n$ and $\mathcal{R}^m$ are called isomorphic vector spaces with $n = m$. Hence, suppose $n = m$, we have the following equivalent logical statements:

$$\ker(\mathbf{A}) = \mathbf{0} \;\Leftrightarrow\; \mathrm{im}(\mathbf{A}) = \mathcal{R}^m \;\Leftrightarrow\; \mathbf{A} \text{ is an isomorphism} \;\Leftrightarrow\; \mathbf{A} \text{ is invertible.}$$

A crucial fact of the four fundamental subspaces is that the dimension of the column space must be equal to the dimension of the row space, which is easy to establish through Gauss elimination or row reduction procedures, named after Johann Carl Friedrich Gauss (1777–1855). Thus, we define the rank of the matrix $\mathbf{A}$ as $r$ with $\dim(\mathrm{im}(\mathbf{A}^T)) = \dim(\mathrm{im}(\mathbf{A})) = r$. Obviously, we should have $r \leqslant m$ and $r \leqslant n$. Moreover, if $r = m$, the matrix $\mathbf{A}$ has full row rank, and the right inverse of the matrix $\mathbf{A}$ is denoted as $\mathbf{A}^T(\mathbf{AA}^T)^{-1}$. Likewise, if $r = n$, the matrix $\mathbf{A}$ has full column rank, and the left inverse of the matrix $\mathbf{A}$ is denoted as $(\mathbf{A}^T\mathbf{A})^{-1}\mathbf{A}^T$.

Consider two sets of base vectors, $\mathbf{e}_i$ and $\hat{\mathbf{e}}_i \in \mathcal{R}^n$, with the corresponding coordinates $x_i$ and $\hat{x}_i$. That means that the same vector $\mathbf{x}$ can also be expressed as $x_i\mathbf{e}_i$ or $\hat{x}_i\mathbf{e}_i$. Since base vector $\hat{\mathbf{e}}_i$ must be a linear combination of base vector $\mathbf{e}_i$, and we can define an $n \times n$ transformation matrix $\mathbf{T}$, called the transformation matrix for the basis, such that $\hat{\mathbf{e}}_i = T_{ij}\mathbf{e}_j$, with $T_{ij} = \hat{\mathbf{e}}_i \cdot \mathbf{e}_j$. Then any vector, including the base vectors, represented in the system $\mathbf{e}_i$ as $\mathbf{f}$, can be expressed in the system $\hat{\mathbf{e}}_i$ as $\hat{\mathbf{f}}$, with the following associations:

$$\hat{f}_i = T_{ij} f_j \quad \text{or} \quad \hat{\mathbf{f}} = \mathbf{Tf}. \tag{1.24}$$

It is easy to verify that there exists another $n \times n$ matrix $\mathbf{S}$, named as the transformation matrix for the coordinates, such that the coordinate change from one base vector to another is defined by $\hat{x}_i = S_{ij}x_j$ and $T_{ki}S_{kj} = \delta_{ij}$, i.e., $\mathbf{T}^T\mathbf{S} = \mathbf{I}$. Obviously, these invertible matrices themselves can be viewed as linear mappings.

Consider linear mapping $\mathbf{A}: \mathcal{R}^n \to \mathcal{R}^n$, corresponding to one basis $\mathbf{e}_i$, i.e., there is $\mathbf{y} = \mathbf{Ax} \in \mathcal{R}^n$, $\forall \mathbf{x} \in \mathcal{R}^n$. Suppose the same linear mapping can be expressed as $\mathbf{B}: \mathcal{R}^n \to \mathcal{R}^n$ in another basis $\hat{\mathbf{e}}_i$ and $\hat{\mathbf{e}}_i = T_{ij}\mathbf{e}_j$, we then have

$$\mathbf{B} = (\mathbf{T}^T)^{-1}\mathbf{A}(\mathbf{T}^T) = \mathbf{SAS}^{-1}, \tag{1.25}$$

where $\mathbf{A}$ and $\mathbf{B}$ are called similar matrices.

More specifically, in $\mathcal{R}^3$ we have sets of linearly independent vectors consisting of at most three vectors, and any vector can be expressed as a unique linear combination of a set of three linearly independent vectors, often called a set of base vectors, or basis. In a Cartesian coordinate system, $\mathbf{e}_i$ form a set of base vectors, and are orthonormal, i.e., a triad of mutually orthogonal unit vectors. Of course, in general, a set of base vectors does not have to be unit vectors or orthogonal to each other, and vectors in one basis must be a linear combination of the vectors of the other basis.

Let us now introduce the curvilinear coordinate systems. We first provide some definitions. Suppose that the rectangular coordinates $x_1, x_2$, and $x_3$ are expressed in terms of new coordinates $u_1, u_2$, and $u_3$ with $x_i = x_i(u_1, u_2, u_3)$, or in abbreviated form $x_i = x_i(\mathbf{u})$, and that, conversely, these relations can be inverted to express $u_1, u_2$, and $u_3$ in terms of $x_1, x_2$, and $x_3$, i.e., $u_i = u_i(x_1, x_2, x_3)$, or in abbreviated form $u_i = u_i(\mathbf{x})$, with suitable restrictions. Then, in some regions, any point $(x_1, x_2, x_3)$ corresponds to a point $(u_1, u_2, u_3)$ in the new coordinate system. Moreover, we define the $u_1$-curve as the curve generated in space with $u_2$ and $u_3$ held constant, and $u_1$-surface as the surface with $u_1$ held constant. In general, the coordinate curves are not necessarily mutually perpendicular at every point, and such coordinates are called curvilinear coordinates.

One simple way of describing contravariant and covariant vectors is to introduce the following chain rules

$$du_i = \frac{\partial u_i}{\partial x_k} dx_k, \tag{1.26}$$

$$\frac{\partial \phi}{\partial u_i} = \frac{\partial \phi}{\partial x_k} \frac{\partial x_k}{\partial u_i}, \tag{1.27}$$

where $du_i$ and $dx_i$ represent the differential elements of coordinates $u_i$ and $x_i$ and $\partial \phi/\partial u_i$ and $\partial \phi/\partial x_i$ stand for the gradient of an invariant function $\phi$ with respect to coordinates $u_i$ and $x_i$.

According to convention, we call $dx_i$ as the components of a contravariant vector and $\partial \phi/\partial x_i$ as the components of a covariant vector. In general, a set of quantities $t^i$, associated with a point $p$, are said to be the components of a contravariant vector, if the corresponding quantities in the transformed coordinate system can be written as

$$\hat{t}^i = t^k \frac{\partial u_i}{\partial x_k}. \tag{1.28}$$

Analogously, a set of quantities $t_i$, associated with a point $p$, are said to be the components of a covariant vector, if the corresponding quantities in the transformed coordinate system can be written as

$$\hat{t}_i = t_k \frac{\partial x_k}{\partial u_i}. \tag{1.29}$$

Clearly, for the degenerate case in which the new coordinates $u_i$ are linearly related to the original coordinates $x_i$, $\partial x_k/\partial u_i$ and $\partial u_i/\partial x_k$ are in fact the transformation matrices. Likewise, a set of quantities $T^{ij}$ is said to be the components of a contravariant tensor of order two, if it transforms according to the following equation:

$$\hat{T}^{ij} = T^{mn}\frac{\partial u_i}{\partial x_m}\frac{\partial u_j}{\partial x_n}. \tag{1.30}$$

A set of quantities $T_{ij}$ is said to be components of a covariant tensor of order two, if it transforms according to the following equation:

$$\hat{T}_{ij} = T_{mn}\frac{\partial x_m}{\partial u_i}\frac{\partial x_n}{\partial u_j}. \tag{1.31}$$

In addition, a mixed tensor of order two corresponds to the transformation

$$\hat{T}^i_j = T^m_n\frac{\partial u_i}{\partial x_m}\frac{\partial x_n}{\partial u_j}. \tag{1.32}$$

Suppose we have a set of contravariant vectors $\mathbf{g}^i$ and a set of covariant vectors $\mathbf{g}_i$ as two sets of base vectors, and in addition,

$$\mathbf{g}_i \cdot \mathbf{g}^j = \delta_i^j, \tag{1.33}$$

with the mixed Kronecker delta $\delta_i^j = \delta_{ij}$, which implies that the contravariant vectors form the dual basis, or reciprocal basis, of the covariant vectors.

Hence, using Eq. (1.33), we can derive

$$f_i = \mathbf{f} \cdot \mathbf{g}_i \quad \text{and} \quad f^i = \mathbf{f} \cdot \mathbf{g}^i, \tag{1.34}$$

where any vector $\mathbf{f}$ can be expressed as $f^i\mathbf{g}_i$ or $f_i\mathbf{g}^i$.

If we introduce another set of covariant vectors $\hat{\mathbf{g}}_i$ as another basis, analogous to the linear transformation in (1.24), we can certainly express $\hat{\mathbf{g}}_i$ as a linear combination of $\mathbf{g}_i$, and vice versa:

$$\hat{\mathbf{g}}_i = A_{ij}\mathbf{g}_j \quad \text{and} \quad \mathbf{g}_i = \hat{A}_{ij}\hat{\mathbf{g}}_j. \tag{1.35}$$

Note that base vectors are not necessarily unit vectors or mutually orthogonal to each other. Therefore, unlike the merely rotation of an orthonormal basis, in general, we have $\hat{\mathbf{A}} = \mathbf{A}^{-1} \neq \mathbf{A}^T$.

Therefore, it can be shown that if we express the vector $\mathbf{f}$ with the basis $\hat{\mathbf{g}}_i$, the corresponding contravariant components must be written as

$$\hat{f}^i = \hat{A}_{ji}f^j \quad \text{and} \quad f^i = A_{ji}\hat{f}^j, \tag{1.36}$$

which, compared with Eq. (1.35), uses a backward matrix in a forward transformation, or vice versa, and originates the name *contravariant* for the contravariant components $f^i$.

Of course, similar coordinate transformations in the form of Eq. (1.35) can be also operated on contravariant bases $\mathbf{g}_i$ and $\hat{\mathbf{g}}_i$. Hence, if we express the vector $\mathbf{f}$ with the basis $\hat{\mathbf{g}}^i$, we can easily obtain

$$\hat{f}_i = A_{ij} f_j \quad \text{and} \quad f_i = \hat{A}_{ij} \hat{f}_j, \tag{1.37}$$

which, compared with Eq. (1.35), uses a forward matrix in a forward transformation, or vice versa, and originates the name *covariant*.

Moreover, if we denote $B_{ij} = \mathbf{g}_i \cdot \mathbf{g}_j$ and $B^{ij} = \mathbf{g}^i \cdot \mathbf{g}^j$, we have

$$f_i = B_{ij} f^j \quad \text{and} \quad f^i = B^{ij} f_j, \tag{1.38}$$

which are used in the process of the so-called raising or lowering indices.

Therefore, without much difficulty, we can show that $B_{ij}$ is the linear transformation from contravariant components or vectors to covariant components or vectors, and $B^{ij}$ is the linear transformation from covariant components or vectors to covariant components or vectors. Thus the matrices $B_{ij}$ and $B^{ij}$ must be inverse, i.e., $[B_{ij}] = [B^{ij}]^{-1}$. In fact, $B^{ij}$ and $B_{ij}$ are called the contravariant and covariant components of a unit tensor $\mathbf{I}$, i.e., $I_{ij} = \delta_{ij}$. A typical example is the so-called metric tensor illustrated with the following chain rules:

$$\begin{aligned} dx_k\, dx_k &= \frac{\partial x_k}{\partial u_i} \frac{\partial x_k}{\partial u_j} du_i\, du_j = G_{ij}\, du_i\, du_j, \\ du_k\, du_k &= \frac{\partial u_k}{\partial x_i} \frac{\partial u_k}{\partial x_j} dx_i\, dx_j = G^{ij}\, dx_i\, dx_j. \end{aligned} \tag{1.39}$$

From the definitions of tensors, we can associate one vector, through a tensor, with another vector,

$$\mathbf{a} = \mathbf{T}\mathbf{b}, \tag{1.40}$$

and more specifically, analogous to Eq. (1.38) we can have

$$a^i = T^{ij} b_j, \qquad a_i = T_{ij} b^j, \qquad a_i = T_i^j b_j. \tag{1.41}$$

Using the same coordinate transformation in (1.35), we can easily derive

$$\hat{T}^{ij} = \hat{A}_{mi} T^{mn} \hat{A}_{nj}, \qquad \hat{T}_{ij} = A_{im} T_{mn} A_{jn}, \qquad \hat{T}_i^j = A_{im} T_m^n \hat{A}_{nj}. \tag{1.42}$$

Here we introduce both superscript and subscript to delineate contravariant and covariant vectors. In practice, we often prefer the subscript and introduce the superscript only if the subscript is adopted for another purpose. In the curvilinear system, the tangential vector to the $u_i$-curve at a point $p$ is defined as

$$\mathbf{u}_i = \frac{\partial \mathbf{r}}{\partial u_i} = \frac{\partial \mathbf{r}}{\partial s_i} \frac{\partial s_i}{\partial u_i} = h_i \mathbf{e}_u^i \quad \text{(without summation)}, \tag{1.43}$$

where $s_i$ is the arc length along the $u_i$-curve, $\mathbf{e}_u^i$ denotes the unit vector $\partial \mathbf{r}/\partial s_i$, and $h_i$ represents the length of $\mathbf{u}_i$.

It is clear from the definition of the unit vector $\mathbf{e}_u^i$ that we have

$$\mathbf{e}_u^i \cdot \mathbf{e}_u^j = \delta_{ij}, \tag{1.44}$$

implying the coordinate curves are mutually orthogonal at every point.

Eq. (1.43) can also be converted to the scalar differential form

$$ds_i = h_i \, du_i \quad \text{(without summation)}. \tag{1.45}$$

Therefore, the arc length along any curve can be represented as

$$ds^2 = h_i^2 \, du_i^2, \tag{1.46}$$

from which we can easily identify that $h_i^2 = G_{ii}$ as defined in (1.39).

Since, from the chain rule, we have $\mathbf{u}_i = \partial \mathbf{r}/\partial x_k \partial x_k/\partial u_i$, according to Eq. (1.29), $\mathbf{u}_i$ are covariant vectors and can be written as $\mathbf{g}_i$. Conversely,

$$\mathbf{e}_i = \frac{\partial \mathbf{r}}{\partial x_i} = \frac{\partial \mathbf{r}}{\partial u_k} \frac{\partial u_k}{\partial x_i} = \frac{\partial u_k}{\partial x_i} \mathbf{g}_k, \tag{1.47}$$

and $\mathbf{e}_i$ is in fact $\hat{\mathbf{g}}_i$ defined with the transformation $A_{ij} = \partial u_i/\partial x_j$ and $\hat{A}_{ij} = \partial x_i/\partial u_j$.

In the curvilinear coordinate system, we define the normal vector to the $u_i$-surface at a point $p$ as

$$\mathbf{n}_i = \frac{\partial u_i}{\partial x_j} \mathbf{e}_j, \tag{1.48}$$

which can be viewed as contravariant vectors and written as $\mathbf{g}^i$. Then, according to Eq. (1.28), we have,

$$\hat{\mathbf{g}}^i = \mathbf{g}^k \frac{\partial x_i}{\partial u_k} = \frac{\partial u_k}{\partial x_j} \mathbf{e}_j \frac{\partial x_i}{\partial u_k} = \mathbf{e}_i. \tag{1.49}$$

Eq. (1.49) in fact suggests that in the Cartesian system there is no distinction between contravariant and covariant vectors, namely, $\mathbf{e}_i = \mathbf{e}^i$. On the $u_1$-surface, the differential element of surface area $d\mathbf{s}_1$ is given by the vector product $\mathbf{u}_2 \, du_2 \times \mathbf{u}_3 \, du_3$ following the assumed right-hand rule, and hence with $\mathbf{e}_u^1 = \mathbf{e}_u^2 \times \mathbf{e}_u^3$, we obtain

$$d\mathbf{s}_1 = h_2 h_3 \, du_2 \, du_3 \, \mathbf{n}_1. \tag{1.50}$$

Analogously, we have for the differential element of volume $d\Omega$

$$d\Omega = (\mathbf{u}_1 \, du_1 \times \mathbf{u}_2 \, du_2) \cdot \mathbf{u}_3 \, du_3 = h_1 h_2 h_3 \, du_1 \, du_2 \, du_3. \tag{1.51}$$

REMARK 2.1.1. In general, we refer to curvilinear coordinate systems as orthogonal curvilinear coordinate systems, with typical examples such as the cylindrical coordinate system and the spherical coordinate system. In these cases, the unit vectors corresponding to contravariant and covariant vectors form the identical triad of mutually orthogonal unit vectors. However, we must point out that in the definition of nonlinear strain and stress components, the contravariant and covariant base vectors are no longer orthogonal to each other.

We now proceed to determine the expressions for the gradient, divergence, curl, and the Laplacian operator $\nabla^2$ in the curvilinear coordinate system. To determine the gradient of a scalar function $f$, we begin from the relation

$$df = \nabla f \cdot d\mathbf{r}. \tag{1.52}$$

In terms of the coordinates $u_1$, $u_2$, and $u_3$, we have

$$df = \frac{\partial f}{\partial u_i}\, du_i, \qquad d\mathbf{r} = \mathbf{u}_i\, du_i, \qquad \nabla f = \lambda_i \mathbf{e}_u^i, \tag{1.53}$$

where $\lambda_i$ is expressed as

$$\lambda_i = \frac{1}{h_i}\frac{\partial f}{\partial u_i} \quad \text{(without summation)}. \tag{1.54}$$

Thus, the gradient of a scalar function $f$ in orthogonal curvilinear coordinate systems can be written as

$$\nabla f = \frac{\mathbf{e}_u^1}{h_1}\frac{\partial f}{\partial u_1} + \frac{\mathbf{e}_u^2}{h_2}\frac{\partial f}{\partial u_2} + \frac{\mathbf{e}_u^3}{h_3}\frac{\partial f}{\partial u_3}. \tag{1.55}$$

Consequently, we obtain

$$\nabla u_i = \frac{\mathbf{e}_u^i}{h_i} \quad \text{(without summation)}, \tag{1.56}$$

and following Eq. (1.17), we have

$$\nabla \times \frac{\mathbf{e}_u^i}{h_i} = \mathbf{0} \quad \text{(without summation)}. \tag{1.57}$$

Furthermore, according to the mutual orthogonality in (1.44), we also obtain

$$\frac{\mathbf{e}_u^1}{h_2 h_3} = \frac{\mathbf{e}_u^2}{h_2} \times \frac{\mathbf{e}_u^3}{h_3} = (\nabla u_2) \times (\nabla u_3), \tag{1.58}$$

and following Eq. (1.17), we have

$$\nabla \cdot \frac{\mathbf{e}_u^1}{h_2 h_3} = \nabla \cdot \frac{\mathbf{e}_u^2}{h_3 h_1} = \nabla \cdot \frac{\mathbf{e}_u^3}{h_1 h_2} = 0. \tag{1.59}$$

Therefore, for a vector $\mathbf{f} = f_i \mathbf{e}_u^i$, using Eqs. (1.17), (1.55), and (1.59), we have

$$\nabla \cdot \left(f_1 \mathbf{e}_u^1\right) = \frac{\mathbf{e}_u^1}{h_2 h_3} \cdot \nabla (h_2 h_3 f_1) = \frac{1}{h_1 h_2 h_3} \frac{\partial}{\partial u_1} (h_2 h_3 f_1). \tag{1.60}$$

Hence, treating the other terms in a similar fashion, we obtain

$$\nabla \cdot \mathbf{f} = \frac{1}{h_1 h_2 h_3} \left[ \frac{\partial}{\partial u_1} (h_2 h_3 f_1) + \frac{\partial}{\partial u_2} (h_3 h_1 f_2) + \frac{\partial}{\partial u_3} (h_1 h_2 f_3) \right]. \tag{1.61}$$

Likewise, combining Eqs. (1.55) and (1.61), the expression $\nabla^2$, or $\nabla \cdot \nabla$ can be written as

$$\begin{aligned} \nabla^2 = \frac{1}{h_1 h_2 h_3} & \left[ \frac{\partial}{\partial u_1} \left( \frac{h_2 h_3}{h_1} \frac{\partial}{\partial u_1} \right) + \frac{\partial}{\partial u_2} \left( \frac{h_3 h_1}{h_2} \frac{\partial}{\partial u_2} \right) \right. \\ & \left. + \frac{\partial}{\partial u_3} \left( \frac{h_1 h_2}{h_3} \frac{\partial}{\partial u_3} \right) \right]. \end{aligned} \tag{1.62}$$

Furthermore, using Eqs. (1.17), (1.55), and (1.57), we also derive

$$\begin{aligned} \nabla \times (f_1 \mathbf{e}_u^1) &= \nabla (h_1 f_1) \times \frac{\mathbf{e}_u^1}{h_1} \\ &= -\frac{\mathbf{e}_u^3}{h_1 h_2} \frac{\partial}{\partial u_2} (h_1 f_1) + \frac{\mathbf{e}_u^2}{h_3 h_1} \frac{\partial}{\partial u_3} (h_1 f_1). \end{aligned} \tag{1.63}$$

Again, treating the other terms in a similar fashion, we have

$$\nabla \times \mathbf{f} = \frac{1}{h_1 h_2 h_3} \begin{vmatrix} h_1 \mathbf{e}_u^1 & h_2 \mathbf{e}_u^2 & h_3 \mathbf{e}_u^3 \\ \partial/\partial u_1 & \partial/\partial u_2 & \partial/\partial u_3 \\ h_1 f_1 & h_2 f_2 & h_3 f_3 \end{vmatrix}. \tag{1.64}$$

*Complex variables*

Complex analysis was mainly founded in the 19th century by Augustin-Louis Cauchy (1789–1857), Bernhard Riemann (1826–1866), Karl Weierstrass (1815–1897), and Johann Carl Friedrich Gauss (1777–1855). The concepts of complex variables are important in a number of areas, such as linear algebra and contour integral. In this section, we briefly introduce some fundamentals. The concepts of contour integral and conformal mapping will be discussed in Section 2.3.2 and the detailed application will be integrated in respective chapters.

A complex variable $z$ is denoted as $z = x + iy$, where $i = \sqrt{-1}$ is the imaginary unit, and $x$ and $y$ are real numbers, denoted as the real part of $z$, i.e., $x = \mathrm{Re}(z)$, and the imaginary part of $z$, i.e., $y = \mathrm{Im}(z)$, respectively. Geometrically, each complex number $z$ corresponds to the point $(x, y)$ in a Cartesian plane, a so-called complex plane, with $x$ and $y$ as abscissa and ordinate, respectively. Therefore, the distance between two complex numbers $z_1$ and $z_2$ is defined as the distance between two points $(x_1, y_1)$ and $(x_2, y_2)$ in the complex plane, i.e.,

$$|z_1 - z_2| = \sqrt{(x_1 - x_2)^2 + (y_1 - y_2)^2}. \tag{1.65}$$

Naturally, the absolute value (modulus) of $z$ is defined as

$$|z| = \sqrt{x^2 + y^2}. \tag{1.66}$$

The complex conjugate of a complex number corresponds to a complex number with the same real part and the imaginary part with an opposite sign, denoted as $\bar{z} = x - iy$. Furthermore, two complex numbers are the same if and only if both the imaginary and real parts are equal. Thus, the complex number is zero if and only if the corresponding real and imaginary parts are equal. The complex number $z$ can also be written in the polar coordinate frame as

$$z = r\cos\theta + ir\sin\theta, \tag{1.67}$$

with $x = r\cos\theta$ and $y = r\sin\theta$, where $r$ is called the modulus of $z$ (amplitude), and $\theta$ is the argument of $z$, denoted as $\arg z$.

It is obvious that $\theta$ has infinitely many values, and it can be expressed as

$$\theta = \theta_p + 2n\pi, \quad n \in \mathcal{Z}, \tag{1.68}$$

where $\theta_p$ is the principal value of $\theta$ with $r\cos\theta_p = x$, $r\sin\theta_p = y$, and $r = \sqrt{x^2 + y^2}$; and of course, any interval of length $2\pi$, e.g., $[0, 2\pi)$ or $[-\pi, \pi)$ can be used for $\theta_p$.

The simplest function of a complex variable is the integral power function $f(z) = z^n$, with $n \in \mathcal{N}$. Analogous to the power series of a real variable, a complex variable function can be expressed as a complex power series about $z = 0$ or $z = z_o$

$$f(z) = \sum_{n=0}^{N} a_n z^n \quad \text{or} \quad f(z) = \sum_{n=0}^{N} a_n (z - z_o)^n, \tag{1.69}$$

with $a_n \in \mathcal{C}$ and $n \in \mathcal{N}$, and if $|z - z_o| \lim_{n\to+\infty} |a_{n+1}/a_n| < 1$ with $z_o \in \mathcal{C}$, the infinite series $\sum_{i=0}^{+\infty} a_n (z - z_o)^n$ converges.

Hence, by adopting the work of Karl Weierstrass, the exponential function of a complex variable $f(z) = e^z$ can be expressed as

$$e^z = \sum_{n=0}^{+\infty} \frac{z^n}{n!}. \tag{1.70}$$

As a consequence, $e^{z_1+z_2} = e^{z_1}e^{z_2}$ and if $n \in \mathcal{N}$, $(e^z)^n = e^{nz}$. Furthermore, similar to the trigonometric function of a real variable, we define trigonometric (circular) functions of a complex variable as

$$\begin{aligned} \sin z &= \frac{e^{iz} - e^{-iz}}{2i} = \sum_{n=0}^{+\infty} (-1)^n \frac{z^{2n+1}}{(2n+1)!}, \\ \cos z &= \frac{e^{iz} + e^{-iz}}{2} = \sum_{n=0}^{+\infty} (-1)^n \frac{z^{2n}}{(2n)!}. \end{aligned} \tag{1.71}$$

We then have Euler's formula, named after Leonhard Euler (1707–1783):

$$e^{iz} = \cos z + i \sin z \quad \text{and} \quad z = r(\cos\theta + i \sin\theta) = re^{i\theta}. \tag{1.72}$$

Furthermore, from Eq. (1.72), we can deduce de Moivre's theorem, named after Abraham de Moivre (1667–1754):

$$\begin{aligned} &(\cos\theta + i \sin\theta)^n = \cos n\theta + i \sin n\theta \quad \text{and} \\ &z^n = r^n(\cos n\theta + i \sin n\theta). \end{aligned} \tag{1.73}$$

Analogous to Eq. (1.71), hyperbolic functions of a complex variable are defined as

$$\begin{aligned} \sinh z &= \frac{e^{z} - e^{-z}}{2} = \sum_{n=0}^{+\infty} \frac{z^{2n+1}}{(2n+1)!}, \\ \cosh z &= \frac{e^{z} + e^{-z}}{2} = \sum_{n=0}^{+\infty} \frac{z^{2n}}{(2n)!}. \end{aligned} \tag{1.74}$$

By comparing Eq. (1.74) with Eq. (1.71), we can relate hyperbolic functions with trigonometric functions:

$$\begin{aligned} &\sinh iz = i \sin z, \qquad \sin iz = i \sinh z, \\ &\cosh iz = \cos z, \qquad \cos iz = \cosh z; \end{aligned} \tag{1.75}$$

and eventually derive the following formulas for trigonometric and hyperbolic functions of a complex variable $z$:

$$\begin{aligned} e^z &= e^{x+iy} = e^x \cos y + i e^x \sin y, \\ \sin z &= \sin(x+iy) = \sin x \cosh y + i \cos x \sinh y, \\ \cos z &= \cos(x+iy) = \cos x \cosh y - i \sin x \sinh y, \\ \sinh z &= \sinh(x+iy) = \sinh x \cos y + i \cosh x \sin y, \\ \cosh z &= \cosh(x+iy) = \cosh x \cos y + i \sinh x \sin y. \end{aligned} \tag{1.76}$$

THEOREM 2.1.2 *(Fundamental theorem of algebra). The algebraic equation*

$$a_0 + a_1 x + a_2 x^2 + a_3 x^3 + \cdots + a_n x^n = 0, \quad a_n \neq 0 \tag{1.77}$$

*with complex coefficients* $a_i$, $n \in \bar{\mathcal{N}}$, *and* $n \geqslant 1$, *always has a complex number as a solution.*

Although Theorem 2.1.2 indicates that the set of complex numbers is algebraically closed, we must note that for other purposes, mathematicians have developed systems of hypercomplex, such as quaternions and Cayley numbers.

### 2.1.4. Eigenvalues and eigenvectors; singular value decomposition

#### Eigenvalues and eigenvectors

One of the most important aspects in linear algebra is the study of eigenvalues and eigenvectors, which is directly related with dynamics, stability analysis, and numerical methods. Let $\mathbf{A}$ be an $n \times n$ matrix. Denote $\mathbf{A}_{ij}$ as the $(n-1) \times (n-1)$ matrix obtained by deleting the $i$th row and $j$th column. The determinant of the matrix $\mathbf{A}_{ij}$ is called a minor and is denoted as $\det(\mathbf{A}_{ij})$, or $|\mathbf{A}_{ij}|$, and if the determinant of a $1 \times 1$ matrix $[a]$ is defined as $a$, the determinant of the $n \times n$ matrix is then defined inductively as $\det(\mathbf{A}) = |\mathbf{A}| = \sum_{j=1}^{n} (-1)^{i+j} a_{ij} \det(\mathbf{A}_{ij})$, where $i$ is a fixed integer with $1 \leqslant i \leqslant n$. The basic properties of the determinant function are as follows:

$$\begin{aligned} &\det(\mathbf{I}) = 1, \qquad \det(\mathbf{AB}) = \det(\mathbf{A}) \det(\mathbf{B}), \\ &\det(\mathbf{A}) \neq 0 \quad \Leftrightarrow \quad \mathbf{A} \text{ invertible} \quad \Leftrightarrow \quad \ker(\mathbf{A}) = \{\mathbf{0}\}. \end{aligned} \tag{1.78}$$

Furthermore, if the kernel of the operator $\mathbf{A} - \alpha\mathbf{I} : \mathcal{S} \subseteq \mathcal{R}^n \to \mathcal{S}$ is nontrivial, it is called the $\alpha$-eigenspace of $\mathbf{A}$, which consists of all eigenvectors belonging to $\alpha$, the so-called eigenvalue of $\mathbf{A}$. The solution of the eigenvalue is equivalent to the solution of

$$\det(\mathbf{A} - \alpha\mathbf{I}) = 0, \tag{1.79}$$

the result of which is a polynomial $p(\alpha)$, often called the characteristic polynomial of $\mathbf{A}$.

Thus, let $\lambda_1, \ldots, \lambda_r$ be the distinct roots of $p(\lambda)$ and the integer $n_k \geqslant 1$, with $k = 1, \ldots, r$, stands for the multiplicity of $\lambda_k$ with $\sum_{k=1}^{r} n_k = \dim(\mathcal{S})$. We can express the characteristic polynomial as

$$p(\alpha) = \prod_{k=1}^{r} (\alpha - \lambda_k)^{n_k}. \tag{1.80}$$

Similar to the eigenspace

$$\ker(\mathbf{A} - \lambda_k\mathbf{I}) \subset \mathcal{S}, \tag{1.81}$$

the generalized eigenspace of $\mathbf{A}$ associated with $\lambda_k$ is a subspace of $\mathcal{S}$, denoted as

$$\ker(\mathbf{A} - \lambda_k\mathbf{I})^{n_k} \subset \mathcal{S}. \tag{1.82}$$

If we change the basis of the vector space from $\mathbf{e}_i$ to $\hat{\mathbf{e}}_i$, with $\hat{\mathbf{e}}_i = \mathbf{T}\mathbf{e}_i$, the linear operator $\mathbf{A}$ is transformed as $\mathbf{B} = (\mathbf{T}^T)^{-1}\mathbf{A}(\mathbf{T}^T)$ and consequently, the characteristic polynomial of $\mathbf{B}$ can be written as

$$\det(\mathbf{B} - \alpha\mathbf{I}) = \det(\mathbf{T}^T)^{-1}\det(\mathbf{A} - \alpha\mathbf{I})\det(\mathbf{T}^T) = \det(\mathbf{A} - \alpha\mathbf{I}) = 0, \quad (1.83)$$

which means that the characteristic polynomial $p(\alpha)$ is independent of the basis, i.e., invariant.

Another invariant is the trace of a matrix $\mathbf{A} = [A_{ij}]$, denoted as $\mathrm{tr}(\mathbf{A}) = \sum_{i=1}^{n} A_{ii}$, which has the following properties

$$\mathrm{tr}(\mathbf{AB}) = \mathrm{tr}(\mathbf{A})\,\mathrm{tr}(\mathbf{B}) \quad \text{and} \quad \mathrm{tr}(\mathbf{A} + \mathbf{B}) = \mathrm{tr}(\mathbf{A}) + \mathrm{tr}(\mathbf{B}). \quad (1.84)$$

Moreover, supposing the vector space $\mathcal{S}$ has a set of base vectors $\hat{\mathbf{e}}_i$ as the eigenvectors of $\mathbf{A}$, then $\mathbf{A}$ is considered diagonalizable and the corresponding eigenvalues are called simple, i.e., there exists an invertible matrix $\mathbf{S}$ such that

$$\mathbf{SAS}^{-1} = \mathrm{diag}\{\lambda_1, \ldots, \lambda_n\} = \begin{bmatrix} \lambda_1 & 0 & \ldots & 0 \\ 0 & \lambda_2 & \ldots & 0 \\ \vdots & \vdots & \ddots & \vdots \\ 0 & 0 & \ldots & \lambda_n \end{bmatrix}. \quad (1.85)$$

So far, we have discussed linear operators in real spaces. However, the elementary properties can be directly extended to complex spaces, which consist of complex elements. This is of particular importance when dealing with FSI systems. Just like the real Cartesian space, the complex Cartesian space $\mathcal{C}^n$ has a set of $n$-tuples complex number, denoted as $(z_1, z_2, \ldots, z_n)$. In particular, a linear operator $\mathbf{A}$ in a complex space is diagonalizable if the characteristic polynomial has distinct roots. To proceed further, we have to introduce the concept of the direct sum of subspaces. Consider a set of subspaces $\mathcal{S}_1, \mathcal{S}_2, \ldots, \mathcal{S}_n \subset \mathcal{S}$. If every vector $\mathbf{x} \in \mathcal{S}$ can be expressed uniquely with the vectors $\mathbf{x}_i \in \mathcal{S}_i$ as $[\mathbf{x}_1^T \quad \ldots \quad \mathbf{x}_n^T]^T$, then the vector space $\mathcal{S}$ is called a direct sum of $\mathcal{S}_i$ and denoted as

$$\mathcal{S} = \mathcal{S}_1 \oplus \cdots \oplus \mathcal{S}_n = \bigoplus_{i=1}^{n} \mathcal{S}_i. \quad (1.86)$$

Moreover, if each $\mathcal{S}_i$ is invariant under a linear operator $\mathbf{A}_i$, i.e., $\mathbf{A}_i : \mathcal{S}_i \to \mathcal{S}_i$, then the linear operator $\mathbf{A} : \mathcal{S} \to \mathcal{S}$ is the direct sum of $\mathbf{A}_i$, denoted as

$$\mathbf{A} = \mathbf{A}_1 \oplus \cdots \oplus \mathbf{A}_n = \bigoplus_{i=1}^{n} \mathbf{A}_i, \quad (1.87)$$

where each $\mathcal{S}_i$ is invariant under $\mathbf{A}$, i.e., $\mathbf{A}(\mathcal{S}_i) \subset \mathcal{S}_i$ or $\mathbf{Ax} = \mathbf{A}_i\mathbf{x}_i$ with $\mathbf{x} = [\mathbf{0}^T \quad \ldots \quad \mathbf{x}_i^T \quad \ldots \quad \mathbf{0}^T]^T$, denoted as $\mathbf{x} \in \mathcal{S}_i$. Obviously, if the basis of $\mathbf{A}$ is the union of the basis elements of the $\mathcal{S}_i$, we have

$$\mathbf{A} = \text{diag}\{\mathbf{A}_1, \ldots, \mathbf{A}_n\} = \begin{bmatrix} \mathbf{A}_1 & \mathbf{0} & \ldots & \mathbf{0} \\ \mathbf{0} & \mathbf{A}_2 & \ldots & \mathbf{0} \\ \vdots & \vdots & \ddots & \vdots \\ \mathbf{0} & \mathbf{0} & \ldots & \mathbf{A}_n \end{bmatrix}. \tag{1.88}$$

It can be also verified that

$$\det(\mathbf{A}) = \det(\mathbf{A}_1) \cdot \ldots \cdot \det(\mathbf{A}_n) = \prod_{i=1}^{n} \det(\mathbf{A}_i). \tag{1.89}$$

THEOREM 2.1.3 *(Eigenspace of real matrix). If* $\mathbf{A} : \mathcal{S} \to \mathcal{S}$ *is a real operator, then the set of its eigenvalues is preserved under complex conjugation, i.e., if* $\zeta = \xi + i\eta$ *is an eigenvalue, so is its complex conjugate* $\bar{\zeta} = \xi - i\eta$. *Therefore, we can write the eigenvalues of* $\mathbf{A}$ *as:* ($r$ *all real eigenvalues*) $\lambda_1, \ldots, \lambda_r$ *and* ($2s$ *all complex eigenvalues*) $\zeta_1, \bar{\zeta}_1, \ldots, \zeta_s, \bar{\zeta}_s$. *Moreover, supposing* $\mathbf{A}$ *has distinct eigenvalues, then both* $\mathcal{S}$ *and* $\mathbf{A}$ *have a direct sum decomposition:*

$$\mathcal{S} = \mathcal{S}_1 \oplus \mathcal{S}_2 \quad \textit{and} \quad \mathbf{A} = \mathbf{A}_1 \oplus \mathbf{A}_2, \tag{1.90}$$

*with*

$$\mathbf{A}_1 : \mathcal{S}_1 \to \mathcal{S}_1 \quad \textit{and} \quad \mathbf{A}_2 : \mathcal{S}_2 \to \mathcal{S}_2, \tag{1.91}$$

*where* $\mathbf{A}_1$ *has real eigenvalues and* $\mathbf{A}_2$ *has complex eigenvalues.*

Consider a linear mapping $\mathbf{A} : \mathcal{S} \to \mathcal{S}$ be an operator on a real vector space with distinct complex eigenvalues $\zeta_1, \bar{\zeta}_1, \ldots, \zeta_s, \bar{\zeta}_s$. Then, as a direct consequence of Theorem 2.1.3, there is an invariant direct sum decomposition for $\mathcal{S}$ and a corresponding direct sum decomposition for $\mathcal{A}$, such that

$$\mathcal{S} = \bigoplus_{i=1}^{r} \mathcal{S}_i \quad \text{and} \quad \mathbf{A} = \bigoplus_{i=1}^{r} \mathbf{A}_i, \tag{1.92}$$

where $\mathcal{S}_i$ and $\mathbf{A}_i$ are two-dimensional and $\mathbf{A}_i$ has eigenvalues $\zeta_i$ and $\bar{\zeta}_i$.

Therefore, consider a linear operator $\mathbf{A}$ on a two-dimensional vector space $\mathcal{S} \subset \mathcal{R}^2$ with nonreal eigenvalues $\zeta$ and $\bar{\zeta}$, i.e., $\zeta = \xi + i\eta$ and $\eta \neq 0$. It can be shown that we have two eigenvectors $\boldsymbol{\varphi} = \boldsymbol{\phi} + i\boldsymbol{\psi}$ and $\bar{\boldsymbol{\varphi}} = \boldsymbol{\phi} - i\boldsymbol{\psi}$ corresponding to $\zeta$ and $\bar{\zeta}$, respectively. Hence,

$$\begin{aligned} \mathbf{A}\boldsymbol{\varphi} &= \mathbf{A}(\boldsymbol{\phi} + i\boldsymbol{\psi}) = \zeta\boldsymbol{\varphi} = (\xi + i\eta)(\boldsymbol{\phi} + i\boldsymbol{\psi}), \\ \mathbf{A}\bar{\boldsymbol{\varphi}} &= \mathbf{A}(\boldsymbol{\phi} - i\boldsymbol{\psi}) = \bar{\zeta}\bar{\boldsymbol{\varphi}} = (\xi - i\eta)(\boldsymbol{\phi} - i\boldsymbol{\psi}), \end{aligned} \tag{1.93}$$

we have

$$\mathbf{A}\boldsymbol{\phi} = \xi\boldsymbol{\phi} - \eta\boldsymbol{\psi}, \qquad \mathbf{A}\boldsymbol{\psi} = \eta\boldsymbol{\phi} + \xi\boldsymbol{\psi}. \tag{1.94}$$

Note that for complex vectors and matrices, the transpose or the Hermitian, named after Charles Hermite (1822–1901), corresponds to a transpose for real vectors and matrices along with a complex conjugate operation, denoted as $(\bar{\mathbf{A}})^T$ or $\mathbf{A}^H$, for a typical matrix $\mathbf{A}$. If $\boldsymbol{\phi}$ and $\boldsymbol{\psi}$ are normalized such that $\boldsymbol{\phi}^T\boldsymbol{\phi} = \boldsymbol{\psi}^T\boldsymbol{\psi} = 1$ and $\boldsymbol{\phi}^T\boldsymbol{\psi} = 0$, we then have

$$\begin{aligned} \bar{\varphi}^T\mathbf{A}\varphi &= \boldsymbol{\phi}^T\mathbf{A}\boldsymbol{\phi} + \boldsymbol{\psi}^T\mathbf{A}\boldsymbol{\psi} - i\boldsymbol{\psi}^T\mathbf{A}\boldsymbol{\phi} + i\boldsymbol{\phi}^T\mathbf{A}\boldsymbol{\psi} = 2\xi + 2i\eta, \\ \varphi^T\mathbf{A}\varphi &= \boldsymbol{\phi}^T\mathbf{A}\boldsymbol{\phi} - \boldsymbol{\psi}^T\mathbf{A}\boldsymbol{\psi} + i\boldsymbol{\psi}^T\mathbf{A}\boldsymbol{\phi} + i\boldsymbol{\phi}^T\mathbf{A}\boldsymbol{\psi} = 0. \end{aligned} \tag{1.95}$$

From Eq. (1.95), we can easily derive

$$\begin{aligned} \boldsymbol{\phi}^T\mathbf{A}\boldsymbol{\phi} &= \xi, \qquad & \boldsymbol{\psi}^T\mathbf{A}\boldsymbol{\psi} &= \xi \\ \boldsymbol{\psi}^T\mathbf{A}\boldsymbol{\phi} &= -\eta, \qquad & \boldsymbol{\phi}^T\mathbf{A}\boldsymbol{\psi} &= \eta. \end{aligned} \tag{1.96}$$

It is obvious that $\boldsymbol{\phi}$ and $\boldsymbol{\psi}$ are independent vectors, and if we assign $\mathbf{S}^{-1} = [\boldsymbol{\psi} \quad \boldsymbol{\phi}]$, we have

$$\mathbf{B} = \mathbf{S}\mathbf{A}\mathbf{S}^{-1} = \begin{bmatrix} \xi & -\eta \\ \eta & \xi \end{bmatrix}. \tag{1.97}$$

Compared with the linear transformation (1.25), the matrix $\mathbf{T} = (\mathbf{S}^{-1})^T$ corresponds to the transformation matrix from the standard basis $\mathbf{e}_1$ and $\mathbf{e}_2$ to the basis $\boldsymbol{\psi}$ and $\boldsymbol{\phi}$. Eq. (1.97) demonstrates why the operator on a complex space having distinct complex eigenvalues is called semi-simple, while the operator on a real space having distinct real eigenvalues is called simple.

As a natural extension of Taylor's expansion for scalar functions, e.g., Eq. (1.70), exponentials of operators can also be expanded in the same form, i.e.,

$$e^{\mathbf{A}} = \sum_{k=0}^{+\infty} \frac{\mathbf{A}^k}{k!}. \tag{1.98}$$

Moreover, if $\mathbf{AB} = \mathbf{BA}$, we have $e^{\mathbf{A}+\mathbf{B}} = e^{\mathbf{A}}e^{\mathbf{B}}$; if $\mathbf{A} = \left[\begin{smallmatrix} a & -b \\ b & a \end{smallmatrix}\right]$, we have

$$e^{\mathbf{A}} = e^a \begin{bmatrix} \cos b & -\sin b \\ \sin b & \cos b \end{bmatrix};$$

and if $\mathbf{B} = \mathbf{S}\mathbf{A}\mathbf{S}^{-1}$, then $e^{\mathbf{B}} = \mathbf{S}e^{\mathbf{A}}\mathbf{S}^{-1}$.

THEOREM 2.1.4 *(Decomposition). Let* $\mathbf{A}$ *be an operator on* $\mathcal{S}$*, where* $\mathcal{S}$ *is a complex vector space, or else* $\mathcal{S}$ *is real and* $\mathbf{A}$ *has real eigenvalues. Then* $\mathcal{S}$ *is the* *direct sum of the generalized eigenspaces of* $\mathbf{A}$. *The dimension of each generalized eigenspace equals the multiplicity of the corresponding eigenvalue.* $\mathbf{A}$ *can be uniquely decomposed into* $\mathbf{N}$ *and* $\mathbf{D}$, *where* $\mathbf{N}$ *is a so-called nilpotent matrix, i.e., there exists a natural number* $m$ *such that* $\mathbf{N}^m = \mathbf{0}$, *and* $\mathbf{D}$ *is diagonalizable.*

Suppose for an $n_i \times n_i$ matrix $\mathbf{A}_i : \mathcal{S}_i \to \mathcal{S}_i$, there is only one eigenvalue $\lambda_i$ with a multiplicity of $n_i = \dim(\mathcal{S}_i)$. Based on Theorem 2.1.4, if we consider $\mathbf{A}_i = \mathbf{N}_i + \mathbf{D}_i$, with $\mathbf{N}_i = \mathbf{A}_i - \lambda_i \mathbf{I}_i$ and $\mathbf{D}_i = \lambda_i \mathbf{I}_i$, where $\mathbf{I}_i$ is an $n_i \times n_i$ unity matrix, since $\mathbf{N}_i \mathbf{D}_i = \mathbf{D}_i \mathbf{N}_i$, we have $e^{\mathbf{A}_i} = e^{\mathbf{N}_i} e^{\mathbf{D}_i}$. Furthermore, based on the fact that $\dim(\mathbf{N}_i^n) = \dim(\mathcal{S}_i)$, i.e., $\ker(\mathbf{N}_i^n) = \mathcal{S}_i$, we know $\mathbf{N}_i$ is nilpotent and $\mathbf{N}_i^{n_i} = \mathbf{0}$, hence, $e^{\mathbf{A}_i} = e^{\lambda_i} \sum_{k=0}^{n_i - 1} (\mathbf{N}_i^k / k!)$. This in fact changes the infinite series of the exponential operator $e^{\mathbf{A}}$ to a finite series.

More importantly, if we decompose an $n \times n$ matrix $\mathbf{A}$ into $\mathbf{A} = \mathbf{A}_1 \oplus \cdots \oplus \mathbf{A}_r$, where each matrix $\mathbf{A}_i$ has only one eigenvalue with multiplicity $n_i = \dim(\mathbf{A}_i) \geqslant 1$ and $\sum_{i=1}^{r} n_i = \dim(\mathcal{S})$. Apply the previous results, with $\mathbf{D}_i = \lambda_i \mathbf{I}_i$ and $\mathbf{N}_i = \mathbf{A}_i - \mathbf{D}_i$, $\mathbf{N}_i$ is a nilpotent matrix of order $n_i$, and if we assign

$$\mathbf{D} = \mathbf{D}_1 \oplus \cdots \oplus \mathbf{D}_r \quad \text{and} \quad \mathbf{N} = \mathbf{N}_1 \oplus \cdots \oplus \mathbf{N}_r, \tag{1.99}$$

it is obvious that $\mathbf{D}$ is diagonalizable, $\mathbf{N}$ is nilpotent of $m = \max(n_1, \ldots, n_r)$, and $\mathbf{DN} = \mathbf{ND}$.

THEOREM 2.1.5 *(Cayley–Hamilton). Consider* $\mathbf{A}$ *be an operator on* $\mathcal{S}$ *(real or complex). If the characteristic polynomial is denoted as* $p(\alpha) = \sum_{k=0}^{n} a_k \alpha^k$, *then* $p(\mathbf{A}) = \mathbf{0}$, *i.e.,*

$$\sum_{k=0}^{n} a_k \mathbf{A}^k \mathbf{x} = \mathbf{0}, \quad \forall \mathbf{x} \in \mathcal{S}. \tag{1.100}$$

This theorem is named after Arthur Cayley (1821–1895) and William R. Hamilton (1805–1865). Introducing a matrix

$$\mathbf{N} = \begin{bmatrix} 0 & 1 & 0 & \ldots & 0 & 0 \\ 0 & 0 & 1 & \ldots & 0 & 0 \\ \vdots & \vdots & \ddots & \ddots & \vdots & \vdots \\ 0 & 0 & 0 & \ldots & 0 & 1 \\ 0 & 0 & 0 & \ldots & 0 & 0 \end{bmatrix}, \tag{1.101}$$

since $\mathbf{N}\mathbf{e}_1 = \mathbf{0}$ and $\mathbf{N}\mathbf{e}_i = \mathbf{e}_{i-1}$ with $2 \leqslant i \leqslant n$, we have $\mathbf{N}^n \mathbf{e}_k = 0$ for $1 \leqslant k \leqslant n$, i.e., $\mathbf{N}^n = \mathbf{0}$.

Thus, $\mathbf{N}$ is nilpotent of order $n$ and is called the elementary nilpotent matrix with $\dim(\ker(\mathbf{N})) = 1$ and $\dim(\mathbf{N}) = n - 1$. In addition, we also call the matrix

$\lambda \mathbf{I} + \mathbf{N}$ the elementary Jordan matrix, denoted as

$$\mathbf{J} = \begin{bmatrix} \lambda & 1 & 0 & \dots & 0 & 0 \\ 0 & \lambda & 1 & \dots & 0 & 0 \\ \vdots & \vdots & \ddots & \ddots & \vdots & \vdots \\ 0 & 0 & 0 & \dots & \lambda & 1 \\ 0 & 0 & 0 & \dots & 0 & \lambda \end{bmatrix}. \tag{1.102}$$

Note that in this book, we use the upper nilpotent matrix and upper Jordan matrix. The properties are the same for lower nilpotent matrix and lower Jordan matrix and they are similar matrices.

THEOREM 2.1.6 *(Jordan form). Let* $\mathbf{A}$ *be an operator on* $\mathcal{S}$*, where* $\mathcal{S}$ *is a complex vector space, or else* $\mathcal{S}$ *is real and* $\mathbf{A}$ *has real eigenvalues. Then* $\mathbf{A}$ *can be transposed into* $\mathbf{N} + \lambda\mathbf{I}$ *with* $\mathbf{N}$ *nilpotent. Moreover,* $\mathbf{N}$ *can be uniquely decomposed into* $\mathbf{N} = \operatorname{diag}\{\mathbf{N}_1, \dots, \mathbf{N}_r\}$*, where* $\mathbf{N}_i$ *is the elementary nilpotent matrix whose size is a nonincreasing function of i with* $1 \leqslant i \leqslant r$*, with* $\sum_{i=1}^{r} n_i = \dim(\mathbf{A})$ *and* $r = \dim(\ker(\mathbf{A}))$.

This theorem is named after Marie Ennemond Camille Jordan (1838–1922).

Finally, consider a linear operator $\mathbf{A}$, which can be decomposed into a diagonal matrix and a nilpotent matrix, i.e., $\mathbf{A} = \lambda \mathbf{I} + \mathbf{N}$, the corresponding exponential operator can be expressed as,

$$e^{t\mathbf{A}} = e^{t\lambda} \sum_{k=0}^{n-1} \frac{t^k \mathbf{N}^k}{k!} = e^{t\lambda} \begin{bmatrix} 1 & t & \frac{t^2}{2!} & \dots & \frac{t^{n-2}}{(n-2)!} & \frac{t^{n-1}}{(n-1)!} \\ 0 & 1 & t & \dots & \frac{t^{n-3}}{(n-3)!} & \frac{t^{n-2}}{(n-2)!} \\ \vdots & \vdots & \ddots & \ddots & \vdots & \vdots \\ 0 & 0 & 0 & \dots & 1 & t \\ 0 & 0 & 0 & \dots & 0 & 1 \end{bmatrix}. \tag{1.103}$$

*Singular value decomposition*

Consider an $m \times m$ matrix $\mathbf{A}$. Because of the definition of the determinant, the characteristic polynomial of the matrix $\mathbf{A}$ is the same as that of its transpose. Suppose the eigenvector of the matrix $\mathbf{A}$ corresponding to the eigenvalue $\lambda_i$ is $\mathbf{u}_i$. Since $\lambda_i$ must be one of the eigenvalues of the matrix $\mathbf{A}^T$ and the corresponding eigenvector is $\mathbf{v}_i$. In practice, we often call $\mathbf{u}_i$ the right eigenvector and $\mathbf{v}_i$ the left eigenvector. It is clear that the right eigenvectors span the row space whereas the left eigenvectors make the column space. Now supposing there exist two distinct eigenvalues $\lambda_i$ and $\lambda_j$, we have

$$\mathbf{v}_j^T \mathbf{A} \mathbf{u}_i = \lambda_i \mathbf{v}_j^T \mathbf{u}_i \quad \text{and} \quad \mathbf{u}_i^T \mathbf{A}^T \mathbf{v}_j = \lambda_j \mathbf{u}_i^T \mathbf{v}_j, \tag{1.104}$$

which implies that $\mathbf{u}_i$ is orthogonal to $\mathbf{v}_j$ if $\lambda_i \neq \lambda_j$.

Consider an $m \times n$ matrix $\mathbf{A}$. We can easily construct two symmetric matrices $\mathbf{A}^T\mathbf{A}$ and $\mathbf{A}\mathbf{A}^T$. Suppose the rank of the matrix $\mathbf{A}$ is $r$, there will be $r$ nonzero eigenvalues of these two symmetric matrices. In fact, these $r$ eigenvalues are positive. However, although it can be shown that the positive eigenvalues of the $n \times n$ matrix $\mathbf{A}^T\mathbf{A}$ are the same as those of the $m \times m$ matrix $\mathbf{A}\mathbf{A}^T$, the number of zero eigenvalues will be different. For the symmetric matrix $\mathbf{A}^T\mathbf{A}$, the dimension of the null space will be $n - r$, which means there are $n - r$ null space vectors. It can be easily shown that these $n - r$ null space vectors of the matrix $\mathbf{A}^T\mathbf{A}$ are also the null space vectors of the matrix $\mathbf{A}$. Likewise, there will be $m - r$ null space vectors of the matrix $\mathbf{A}\mathbf{A}^T$ and they are the same as the left null space vector of the matrix $\mathbf{A}$.

Consider the four fundamental spaces of the matrix elaborated in Section 2.1. It is easy to confirm that the orthogonal subspace of the null space must be the row space while the orthogonal subspace of the left null space is the column space. As a consequence, the eigenvectors of the matrix $\mathbf{A}^T\mathbf{A}$ denoted as $\mathbf{v}_i$ span the row space of the matrix $\mathbf{A}$ and the eigenvectors of the matrix $\mathbf{A}\mathbf{A}^T$ represented with $\mathbf{u}_i$ construct the column space of the matrix $\mathbf{A}$. In fact, $\mathbf{u}_i$ and $\mathbf{v}_i$ are similar to the right and left eigenvectors of the square matrix.

Denote a matrix $\mathbf{U}$, the first $r$ columns of which are the eigenvectors of the symmetric matrix $\mathbf{A}^T\mathbf{A}$, and the last $m - r$ columns are constructed with the left null space vectors. In addition, introduce a matrix $\mathbf{V}$, the first $r$ columns of which are the eigenvectors of the symmetric matrix $\mathbf{A}\mathbf{A}^T$, and the last $n - r$ columns are made of the null space vectors. The singular value decomposition (SVD) of the matrix $\mathbf{A}$ is then depicted as

$$\mathbf{A} = \mathbf{U}\Sigma\mathbf{V}^T, \tag{1.105}$$

where the first $r$ diagonal entities of the $m \times n$ matrix $\Sigma$ are the square root of the nonzero eigenvalues of the matrix $\mathbf{A}^T\mathbf{A}$ or $\mathbf{A}\mathbf{A}^T$ or the singular values of the matrix $\mathbf{A}$.

This singular value decomposition could be considered the true makeup of any matrix $\mathbf{A}$ and specifically illustrates the four fundamental subspaces and the eigenspaces of the row and column spaces. This decomposition will be abundantly used in the model reduction procedures as well as the principle component analysis.

## 2.2. Kinematical descriptions and conservation laws

Kinematical descriptions play an important role in the study of FSI systems. In this section, we present the so-called arbitrary Lagrangian–Eulerian (ALE) kinematic description as well as the derivation of basic equations in mechanics.

### *2.2.1. Mass point and rigid body; continuum*

To study the mechanical behavior of objects in nature, we introduce kinematics to mathematically describe the geometrical locations of material particles or mesh points. Kinematical considerations center on the behavior in time of displacements, velocities, and accelerations, or the parametric family of configurations, i.e., mappings and deformations. By introducing a real number system $t$ to represent time, the habitual three-dimensional representation of space and time can be viewed as a four-dimensional continuum.

Two types of kinematical descriptions are commonly used, namely, the Lagrangian description and the Eulerian description. In the Lagrangian description, named after Joseph-Louis Lagrange (1736–1813), state variables are associated with material particles identified with their initial positions. In the Eulerian description, named after Leonhard Euler (1707–1783), state variables are described at spatial locations without regard to specific material particles. Therefore, it is more convenient to use the Eulerian description with velocities as the primitive variables in fluid mechanics, and the Lagrangian description with displacements as the primitive variable in solid mechanics. Furthermore, to represent the motion of solids in computational mechanics, we commonly use the Lagrangian mesh, the motion of which coincides with the material particle motion. Likewise, an Eulerian mesh, fixed in space, is used to represent the motion of fluids. In numerical approaches for FSI problems, if we are to follow the moving boundaries, in order to avoid excessive mesh distortions we often need to arbitrarily select proper mesh motions using the so-called arbitrary Lagrangian–Eulerian (ALE) kinematic description for continuum media, in which the mesh position for the description of the state variables is neither fixed in space nor attached with material particles.

*Mass point and rigid body*

A point mass is a single mass point whose spatial dimension is negligibly small in comparison with the distances involved in the problem under consideration. A rigid body is an idealization of a continuum medium of which the internal deformation is neglected in comparison with the motion of the body as a whole. Naturally, the dynamical behavior of a point mass is governed by the dynamical motion of that point, often along a trajectory or curve. In this case, the Lagrangian description must be used. Since a rigid body can also be considered as a distribution of mass rigidly fixed to a coordinate frame, rotations of such a rigid frame must also be considered in the description of the dynamical behavior of the rigid body.

Let us start with the parametric representation of a curve,

$$\mathbf{x}(s) = x_i(s)\mathbf{e}_i, \tag{2.106}$$

with the real parameter $s$.

Thus, a simple closed curve corresponds to a close curve such that $s_1 \neq s_2$ implies that $\mathbf{x}(s_1) \neq \mathbf{x}(s_2)$. If the parameter $s$ stands for the arc length, i.e., $ds = \sqrt{dx_i\, dx_i}$, the unit tangent vector can be written as

$$\mathbf{t} = \frac{dx_i(s)}{ds}\mathbf{e}_i. \tag{2.107}$$

Moreover, differentiating $\mathbf{t} \cdot \mathbf{t} = 1$ with respect to $s$, we have

$$\frac{d\mathbf{t}}{ds} = \kappa\mathbf{n} = \frac{1}{\rho}\mathbf{n}, \tag{2.108}$$

where $\mathbf{n}$ is the unit normal vector, $\kappa$ is the curvature, and $\rho$ is the radius of curvature.

Denoting the unit vector $\mathbf{m}$ normal to the osculating plane of $\mathbf{t}$ and $\mathbf{n}$ as the so-called binormal vector, we then form a triad of orthogonal unit vectors ($\mathbf{n}$, $\mathbf{t}$, $\mathbf{m}$), as shown in Figure 2.1. From the right-hand rule we have $\mathbf{m} = \mathbf{n}\times\mathbf{t}$ and $\mathbf{n} = \mathbf{t}\times\mathbf{m}$. Hence using Eq. (1.17), we can easily verify

$$\frac{d\mathbf{m}}{ds} = \nu\mathbf{n}, \tag{2.109}$$

$$\frac{d\mathbf{n}}{ds} = -\nu\mathbf{m} - \kappa\mathbf{t}, \tag{2.110}$$

where $\nu$ is called the torsion of the curve.

Eqs. (2.108), (2.109), and (2.110) are often called the Serret–Frenet formula, named after Joseph Alfred Serret (1819–1885) and Fredari–Jean Frenet (1816–1900). Two of the most widely used orthogonal curvilinear coordinate systems are cylindrical and spherical coordinate systems, as shown in Figure 2.2. Although we have mentioned them in Section 2.1, we feel that this section is the place to present them in detail.

The common denotations for these two coordinate frames are cylindrical system $(r, \theta, z)$: $x = r\cos\theta$, $y = r\sin\theta$, $z = z$ with $r \geqslant 0$, $0 \leqslant \theta < 2\pi$, and $z \in \mathcal{R}$; and spherical system $(r, \phi, \theta)$: $x = r\sin\phi\cos\theta$, $y = r\sin\phi\sin\theta$, and $z = r\cos\phi$, with $r \geqslant 0$, $0 \leqslant \phi \leqslant \pi$, and $0 \leqslant \theta < 2\pi$.

Hence, we can express the unit vectors in the cylindrical coordinate system as

$$\begin{aligned} \mathbf{e}_r &= \cos\theta\mathbf{e}_1 + \sin\theta\mathbf{e}_2, \\ \mathbf{e}_\theta &= -\sin\theta\mathbf{e}_1 + \cos\theta\mathbf{e}_2, \\ \mathbf{e}_z &= \mathbf{e}_3, \end{aligned} \tag{2.111}$$

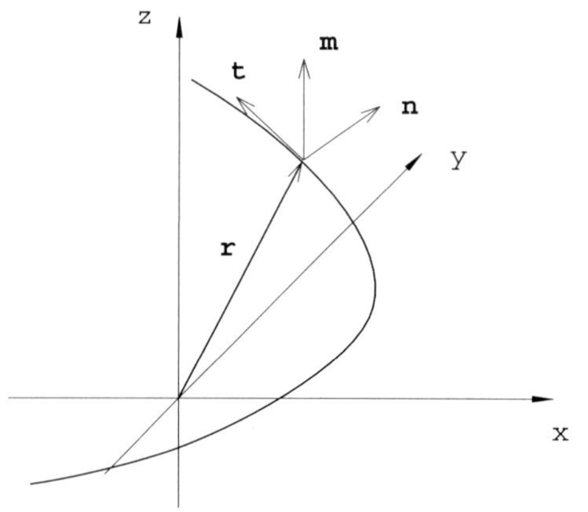


Figure 2.1. Normal, tangent, and binormal unit vectors.

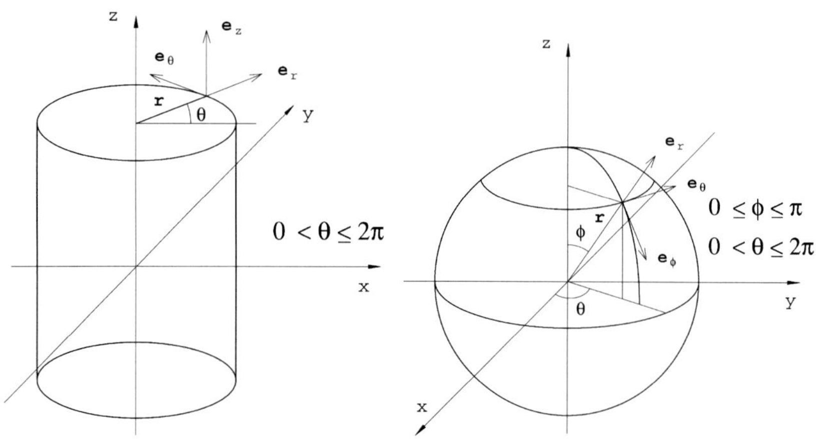


Figure 2.2. Cylindrical and spherical coordinate systems.

and consequently,

$$\begin{aligned} &\frac{\partial \mathbf{e}_r}{\partial r} = \mathbf{0}, \qquad \frac{\partial \mathbf{e}_r}{\partial \theta} = \mathbf{e}_\theta, \qquad \frac{\partial \mathbf{e}_r}{\partial z} = \mathbf{0}; \\ &\frac{\partial \mathbf{e}_\theta}{\partial r} = \mathbf{0}, \qquad \frac{\partial \mathbf{e}_\theta}{\partial \theta} = -\mathbf{e}_r, \qquad \frac{\partial \mathbf{e}_\theta}{\partial z} = \mathbf{0}; \\ &\frac{\partial \mathbf{e}_z}{\partial r} = \mathbf{0}, \qquad \frac{\partial \mathbf{e}_z}{\partial \theta} = \mathbf{0}, \qquad \frac{\partial \mathbf{e}_z}{\partial z} = \mathbf{0}. \end{aligned} \tag{2.112}$$

Similarly, the spherical coordinate system can be represented as

$$\begin{aligned}\mathbf{e}_r &= \sin\phi\cos\theta\mathbf{e}_1 + \sin\phi\sin\theta\mathbf{e}_2 + \cos\phi\mathbf{e}_3,\\ \mathbf{e}_\phi &= \cos\phi\cos\theta\mathbf{e}_1 + \cos\phi\sin\theta\mathbf{e}_2 - \sin\phi\mathbf{e}_3,\\ \mathbf{e}_\theta &= -\sin\theta\mathbf{e}_1 + \cos\theta\mathbf{e}_2,\end{aligned} \tag{2.113}$$

and consequently,

$$\begin{aligned}&\frac{\partial\mathbf{e}_r}{\partial r} = \mathbf{0}, && \frac{\partial\mathbf{e}_r}{\partial\phi} = \mathbf{e}_\phi, && \frac{\partial\mathbf{e}_r}{\partial\theta} = \sin\phi\mathbf{e}_\theta;\\ &\frac{\partial\mathbf{e}_\phi}{\partial r} = \mathbf{0}, && \frac{\partial\mathbf{e}_\phi}{\partial\phi} = -\mathbf{e}_r, && \frac{\partial\mathbf{e}_\phi}{\partial\theta} = \cos\phi\mathbf{e}_\theta;\\ &\frac{\partial\mathbf{e}_\theta}{\partial r} = \mathbf{0}, && \frac{\partial\mathbf{e}_\theta}{\partial\phi} = \mathbf{0}, && \frac{\partial\mathbf{e}_\theta}{\partial\theta} = -\sin\phi\mathbf{e}_r - \cos\phi\mathbf{e}_\phi.\end{aligned} \tag{2.114}$$

In essence, for any vector $\mathbf{a}$ of fixed length, i.e., $\frac{d}{dt}(\mathbf{a}\cdot\mathbf{a}) = 2\frac{d\mathbf{a}}{dt}\cdot\mathbf{a} = 0$, we can introduce an angular velocity $\boldsymbol{\omega}$ from the expression

$$\frac{d\mathbf{a}}{dt} = \boldsymbol{\omega}\times\mathbf{a}. \tag{2.115}$$

Thus, if $\mathcal{B}$ is a rigid body, $\boldsymbol{\omega}$ must represent the three rotations in space, and it is easy to verify that if $\mathcal{B}$ undergoes an angular velocity $\dot{\theta}(t)$ and is expressed in cylindrical coordinate system, we have $\boldsymbol{\omega} = \dot{\theta}\mathbf{e}_z$ and

$$\frac{d\mathbf{e}_r}{dt} = \boldsymbol{\omega}\times\mathbf{e}_r = \dot{\theta}\mathbf{e}_\theta. \tag{2.116}$$

In fact, if we use Eq. (2.112) and chain differentiation, we have

$$\frac{d\mathbf{e}_r}{dt} = \frac{\partial\mathbf{e}_r}{\partial r}\frac{dr}{dt} + \frac{\partial\mathbf{e}_r}{\partial\theta}\frac{d\theta}{dt} + \frac{\partial\mathbf{e}_r}{\partial z}\frac{dz}{dt} = \dot{\theta}\frac{\partial\mathbf{e}_r}{\partial\theta} = \dot{\theta}\mathbf{e}_\theta. \tag{2.117}$$

Now, if we present the curve as

$$\mathbf{r} = r\mathbf{e}_r + z\mathbf{e}_z, \tag{2.118}$$

the velocity and acceleration are represented as

$$\begin{aligned}\mathbf{v} &= \dot{r}\mathbf{e}_r + r\dot{\theta}\mathbf{e}_\theta + \dot{z}\mathbf{e}_z,\\ \mathbf{a} &= (\ddot{r} - r\dot{\theta}^2)\mathbf{e}_r + (r\ddot{\theta} + 2\dot{r}\dot{\theta})\mathbf{e}_\theta + \ddot{z}\mathbf{e}_z.\end{aligned} \tag{2.119}$$

Likewise, if we employ spherical coordinate system, we obtain similar expressions:

$$\begin{aligned}\mathbf{r} &= r\mathbf{e}_r,\\ \mathbf{v} &= \dot{r}\mathbf{e}_r + r\dot{\phi}\mathbf{e}_\phi + r\dot{\theta}\sin\phi\mathbf{e}_\theta,\\ \mathbf{a} &= (\ddot{r} - r\dot{\phi}^2 - r\dot{\theta}^2\sin^2\phi)\mathbf{e}_r + (r\ddot{\phi} + 2\dot{r}\dot{\phi} - r\dot{\theta}^2\sin\phi\cos\phi)\mathbf{e}_\phi\\ &\quad + (r\ddot{\theta}\sin\phi + 2\dot{r}\dot{\phi}\sin\phi + 2r\dot{\theta}\dot{\phi}\cos\phi)\mathbf{e}_\theta.\end{aligned} \tag{2.120}$$

*Continuum*

Let a body $\mathcal{B}$ occupy a region $\mathcal{R}_x$ in the Euclidean space $\mathcal{R}^3$ at time $t_o$. The motion of any material particle of $\mathcal{B}$ is defined by a mapping or function $\mathbf{y} : \mathcal{R}_x \times t \to \mathcal{R}^3$. The region occupied by the body $\mathcal{B}$ at time $t$ or the image of $\mathcal{R}_x$ at time $t$ is denoted by $\mathcal{R}_y$. Thus the position at time $t$ of a material particle which at time $t_o$ occupied the position $\mathbf{x}$ or the image of $\mathbf{x} \in \mathcal{R}_x$ at time $t$ is denoted by

$$\mathbf{y} = \mathbf{y}(\mathbf{x}, t). \tag{2.121}$$

We then define another domain, the referential domain $\mathcal{R}_z$, e.g., the finite element mesh domain, and another mapping or function $\hat{\mathbf{y}} : \mathcal{R}_z \times t \to \mathcal{R}^3$ whose range is also $\mathcal{R}_y$. This mapping $\hat{\mathbf{y}}$ defines the motion of the mesh. The motion of any mesh point or the image of $\mathbf{z} \in \mathcal{R}_z$ at time $t$ is defined by

$$\mathbf{y} = \hat{\mathbf{y}}(\mathbf{z}, t). \tag{2.122}$$

The choice of $\mathcal{R}_z \neq \mathcal{R}_x$ only serves to underscore the difference between the mesh motion and the material motion. In practice, the mesh configuration has to coincide with the material configuration, i.e., $\mathcal{R}_z = \mathcal{R}_x$. Nevertheless, although in general, $\hat{\mathbf{y}} \neq \mathbf{y}$, it is worth mentioning that if $\mathcal{R}_z = \mathcal{R}_x$ and $\hat{\mathbf{y}} = \mathbf{y}$ for all $t$ the mesh-referential description is then identical to the Lagrangian description and the mesh motion is identical to the material motion. Likewise, if $\mathcal{R}_z = \mathcal{R}_y$ and $\hat{\mathbf{y}} = \mathbf{1}$ (identity mapping) the mesh-referential description specializes the Eulerian description. The mappings $\mathbf{y}$ and $\hat{\mathbf{y}}$ induce another mapping $\tilde{\mathbf{y}} : \mathcal{R}_x \times t \to \mathcal{R}_z$ and we can write

$$\mathbf{z} = \tilde{\mathbf{y}}(\mathbf{x}, t) = \hat{\mathbf{y}}^{-1}\big(\mathbf{y}(\mathbf{x}, t), t\big). \tag{2.123}$$

A material particle which occupied $\mathbf{x}$ at time $t_o$ is in general, at different times $t_1$ and $t_2$, represented by different point $\mathbf{z}_1$ and $\mathbf{z}_2$ in the referential domain. The mapping $\tilde{\mathbf{y}}$ defines such a relation. Consider a physical quantity $f$ (e.g., density, particle velocity components). To express this quantity in the Eulerian description we write

$$f = f(\mathbf{y}, t), \tag{2.124}$$

while in the Lagrangian description we write

$$f = f\big(\mathbf{y}(\mathbf{x}, t), t\big) = \tilde{f}(\mathbf{x}, t). \tag{2.125}$$

We then introduce the "mesh-referential description", in which we associate the value of $f$ with $\mathbf{z} \in \mathcal{R}_z$ via the mesh motion $\hat{\mathbf{y}}(\mathbf{z}, t)$, i.e.

$$f = f\big(\hat{\mathbf{y}}(\mathbf{z}, t), t\big) = \hat{f}(\mathbf{z}, t), \tag{2.126}$$

which means that we attach the variable $f$ to mesh points.

In the Eulerian description, the time rate of change is the rate observed at a fixed point in space. The so-called spatial time derivative is defined as

$$f' = \frac{\partial}{\partial t} f(\mathbf{y}, t), \tag{2.127}$$

whereas the time rate of change of $f$ with respect to a fixed material particle, the so-called material time derivative, is defined as

$$\dot{f} = \frac{\partial}{\partial t} \tilde{f}(\mathbf{x}, t). \tag{2.128}$$

Finally the time rate of change of $f$ with respect to a given reference point $\mathbf{z} \in \mathcal{R}_z$ (a given mesh point), the so-called mesh-referential time derivative, is defined as

$$\check{f} = f^* = \frac{\partial}{\partial t} \hat{f}(\mathbf{z}, t). \tag{2.129}$$

The spatial time derivative and the material time derivative are related by

$$\dot{f} = \frac{\partial f(\mathbf{y}, t)}{\partial t} + \frac{\partial f(\mathbf{y}, t)}{\partial \mathbf{y}} \frac{\partial \mathbf{y}}{\partial t}\bigg|_{\mathbf{y}=\mathbf{y}(\mathbf{x},t)} = f' + \mathbf{v} \cdot \nabla f. \tag{2.130}$$

The convective term $\mathbf{v} \cdot \nabla f$ arises from the fact that a fixed point in space $\mathbf{y}$ does not always represent the same material particle. Likewise, using the mesh-referential time derivative and the spatial time derivative we obtain

$$\check{f} = f^* = \frac{\partial f(\mathbf{y}, t)}{\partial t} + \frac{\partial f(\mathbf{y}, t)}{\partial \mathbf{y}} \cdot \frac{\partial \mathbf{y}}{\partial t}\bigg|_{\mathbf{y}=\hat{\mathbf{y}}(\mathbf{z},t)} = f' + \hat{\mathbf{v}} \cdot \nabla f, \tag{2.131}$$

where $\hat{\mathbf{v}} = \check{\mathbf{y}} = \partial \hat{\mathbf{y}} / \partial t$ is the velocity of a mesh point.

Note that if we denote the mesh displacement as $\hat{\mathbf{u}} = \hat{\mathbf{u}}(\mathbf{z}, t) = \hat{\mathbf{y}}(\mathbf{z}, t) - \hat{\mathbf{y}}(\mathbf{z}, t_o)$, the mesh velocity can be also represented as $\hat{\mathbf{u}}^*$. We also note that if the mesh motion coincides with the material particle motion, i.e., $\hat{\mathbf{v}} = \mathbf{v}$, the mesh-referential time derivative is the material derivative, also denoted as $d/dt$. On the other hand, if the mesh is fixed in space, i.e., $\hat{\mathbf{v}} = \mathbf{0}$, then the mesh-referential time derivative and the spatial time derivative are identical.

### 2.2.2. *Vector calculus; transport theorems and conservation laws; variational principles*

#### *Vector calculus*

We begin with a brief introduction to Green's theorem, named after George Green (1793–1841), Stokes' theorem, named after George Gabriel Stokes (1819–1903), and Gauss' theorem, named after Johann Carl Friedrich Gauss (1777–1855). These theorems are commonly used in the derivation of transport theorems and variational principles. In this book, a curve is called a closed curve, if the end points coincide; if not, it is called an arc. A curve is simple if it does not intersect itself, and a closed curve is called a simple closed curve if it does not intersect itself. A curve is smooth if it poses a tangent vector that varies continuously along the length, and is piecewise smooth if it is comprised of finite number of smooth segments.

Let $\mathbf{f}$ be a continuous vector field defined on a smooth curve $C$ given by a vector function $\mathbf{r}(t)$ with $t \in [a, b]$ or $\mathbf{r}(s)$ with $s \in [0, L]$ where $L$ is the total arc length defined as $L = \int_C ds$. Then the line integral of $\mathbf{f}$ along $C$ is

$$\int_C \mathbf{f} \cdot d\mathbf{r} = \int_0^L \mathbf{f} \cdot \mathbf{t}\, ds = \int_a^b \mathbf{f}\big(\mathbf{r}(t)\big) \cdot \mathbf{r}'(t)\, dt, \tag{2.132}$$

with the unit tangent vector $\mathbf{t}$ defined as $d\mathbf{r}/ds$.

One important result of the line contour integral is for a differentiable function $f(\mathbf{r})$, whose gradient $\nabla f$, often called a conservative vector field, is continuous along $C$. Hence, employing the Fundamental Theorem of Calculus, i.e., $\int_a^b f'(x)\, dx = f(b) - f(a)$, we have

$$\int_C \nabla f \cdot d\mathbf{r} = \int_C df = f\big(\mathbf{r}(b)\big) - f\big(\mathbf{r}(a)\big). \tag{2.133}$$

Eq. (2.133) clearly shows that the contour integral for the conservative vector field is path independent.

THEOREM 2.2.1 *(Green's theorem). Consider a continuous differentiable or $C^1$ vector field $\mathbf{f} = f_1\mathbf{e}_1 + f_2\mathbf{e}_2$ defined in a two-dimensional region $S$ bounded by a piecewise smooth simple closed curve $C$, positively oriented namely counterclockwise. Then*

$$\int_C \mathbf{f} \cdot d\mathbf{r} = \int_C (f_1\, dx + f_2\, dy) = \int_S \left(\frac{\partial f_2}{\partial x_1} - \frac{\partial f_1}{\partial x_2}\right) dS. \tag{2.134}$$

Consider a smooth surface $S$ represented by $\mathbf{r}(u, v)$ with two parameters $u$ and $v$ and $(u, v) \in D$. The surface area is defined as

$$A = \int_D |\mathbf{r}_u \times \mathbf{r}_v|\, du\, dv, \tag{2.135}$$

with $\mathbf{r}_u = \partial\mathbf{r}/\partial u$ and $\mathbf{r}_v = \partial\mathbf{r}/\partial v$.

The surface integral of a continuous function $f(\mathbf{r})$ can be expressed as

$$\int_S f(\mathbf{r})\, dS = \int_D f\big(\mathbf{r}(u, v)\big)|\mathbf{r}_u \times \mathbf{r}_v|\, du\, dv. \tag{2.136}$$

For a continuous vector field $\mathbf{f}(\mathbf{r})$ defined on an oriented surface $S$ with the unit normal vector $\mathbf{n}$ often defined as $\mathbf{r}_u \times \mathbf{r}_v / |\mathbf{r}_u \times \mathbf{r}_v|$, the surface integral is

$$\int_S \mathbf{f}(\mathbf{r}) \cdot \mathbf{n}\, dS = \int_D \mathbf{f}\big(\mathbf{r}(u, v)\big) \cdot (\mathbf{r}_u \times \mathbf{r}_v)\, du\, dv. \tag{2.137}$$

THEOREM 2.2.2 *(Stokes' theorem). Let $S$ be a simple surface enclosed by the boundary curve $C$ given with positive, namely counterclockwise, orientation. Let $\mathbf{f}$ be a $C^1$ vector field defined in $S$. Let $\mathbf{t}$ be the unit tangential vector along the curve $C$ chosen so that the vector $\mathbf{n} \times \mathbf{t}$ points into $S$ from $V$. Then*

$$\int_S \nabla \times \mathbf{f} \cdot \mathbf{n}\, dS = \int_C \mathbf{f} \cdot \mathbf{t}\, ds. \tag{2.138}$$

In fact, Green's theorem is the simple two-dimensional representation of Stokes' theorem. The geometric divergence of a vector function at a point $p$ is a scalar function of position and represents the outflow per unit volume at the point $p$. In fact, the other definition of the divergence can be written as

$$\operatorname{div} \mathbf{f} = \lim_{\Delta V \to 0} \frac{1}{\Delta V} \int_S \mathbf{f} \cdot \mathbf{n}\, dS. \tag{2.139}$$

Of course, the divergence is uniquely defined with no dependence of any particular coordinate system. The detail physical account is discussed in Ref. [108]. We can have the divergence theorem for $n$-dimensional systems as proven in Ref. [135].

THEOREM 2.2.3 *(Divergence or Gauss' theorem). Let $V$ be a simple solid region enclosed by the boundary surface $S$ given with positive, namely outward, orientation. Let $\mathbf{f}$ be a continuous differentiable vector field and $\mathbf{n}$ be the unit normal vector on the boundary surface $S$ that points outwards from $V$. Then*

$$\int_V \nabla \cdot \mathbf{f}\, dV = \int_S \mathbf{f} \cdot \mathbf{n}\, dS. \tag{2.140}$$

Thus, it can be easily established that

$$\int_V \nabla^2 f\, dV = \int_S \nabla f \cdot \mathbf{n}\, dS, \tag{2.141}$$

$$\int_V \nabla \times \mathbf{f}\, dV = \int_S \mathbf{n} \times \mathbf{f}\, dS. \tag{2.142}$$

In particular, employing the permutation symbol, we can easily prove that Eq. (2.142) can be rewritten in the following form:

$$-\int_V \nabla \times (\mathbf{r} f)\, dV = \int_S (\mathbf{r} \times \mathbf{n}) f\, dS. \tag{2.143}$$

Consequently, from the divergence theorem it is easy to derive two forms of Green's theorem.

*The first form of Green's theorem*:

$$\int_S \mathbf{n} \cdot (f_1 \nabla f_2)\, dS = \int_V \left[f_1 \nabla^2 f_2 + (\nabla f_1) \cdot (\nabla f_2)\right] dV. \tag{2.144}$$

*The second form of Green's theorem*:

$$\int_S \mathbf{n} \cdot (f_1 \nabla f_2 - f_2 \nabla f_1)\, dS = \int_V \left(f_1 \nabla^2 f_2 - f_2 \nabla^2 f_1\right) dV. \tag{2.145}$$

REMARK 2.2.1. To simplify the notations, we replace the original volumetric integral sign $\iiint$ with $\int_V$, and the surface integral sign $\iint$ with $\int_S$, where $V$ and $S$ denote the volume and surface, respectively.

*Transport theorems and conservation laws*

To express the fundamental laws of mechanics, i.e., the laws of mass, linear (angular) momentum, and energy conservation in integral form, we define a material volume as a volume which always contains the same material particles. In other words, a material volume $\tilde{V}(t)$ at some time $t$ is bounded by a closed surface $\tilde{S}(t)$ in which every point moves with the material velocity $\tilde{\mathbf{v}}$. We define a control volume as a volume arbitrarily chosen for the convenience of the observer. Unlike the material volume which is governed by the motions of the same material particles, a proper definition of a control volume $V(t)$ with the bounding surface $S(t)$ must include the velocity $\mathbf{v}(\mathbf{x}, t)$ for all time $t$.

Let $f(\mathbf{x}, t)$ be a continuous property per unit volume, and consider integral $I(t) = \int_{V(t)} f(\mathbf{x}, t)\, dV$ with $\mathbf{v}(\mathbf{x}, t)$. Thus, the derivative with respect to $t$ is expressed as $dI(t)/dt = \lim_{\Delta t \to 0} [I(t + \Delta t) - I(t)]/\Delta t$, and according to the Taylor's expansion, named after Brook Taylor (1685–1731), we have

$$f(\mathbf{x}, t + \Delta t) = f(\mathbf{x}, t) + \frac{\partial}{\partial t} f(\mathbf{x}, t)\Delta t + O(\Delta t^2). \tag{2.146}$$

Moreover,

$$\frac{d}{dt} I(t) = \int_{V(t)} \frac{\partial f(\mathbf{x},t)}{\partial t} dV + \lim_{\Delta t \to 0} \Big[ \int_{V(t+\Delta t)} f\, dV - \int_{V(t)} f\, dV \Big] / \Delta t. \tag{2.147}$$

From the geometric arguments, analogous to the Leibniz rule for one-dimensional integral, named after Gottfried Wilhelm von Leibniz (1646–1716), we obtain the following variation of the integral:

$$\int_{V(t+\Delta t)} f(\mathbf{x},t)\, dV = \int_{V(t)} f\, dV + \int_{S(t)} f \mathbf{v} \cdot \mathbf{n} \Delta t\, dS. \tag{2.148}$$

Hence, derived from Eq. (2.147), the *first Reynolds transport theorem* (kinematic transport theorem), named after Osborne Reynolds (1842–1912), can be expressed as

$$\frac{d}{dt} \int_{V(t)} f(\mathbf{x},t)\, dV = \int_{V(t)} \Big[ \frac{\partial f}{\partial t} + \nabla \cdot (f \mathbf{v}) \Big] dV. \tag{2.149}$$

Of course, considering the material volume and the control volume coinciding at time $t$, namely at time $t$, $V(t) = \tilde{V}(t)$ and $S(t) = \tilde{S}(t)$, the first kinematic transport theorem can then be written as

$$\frac{d}{dt} \int_{\tilde{V}(t)} f(\mathbf{x},t)\, dV = \int_{V(t)} \frac{\partial f}{\partial t} dV + \int_{S(t)} f \tilde{\mathbf{v}} \cdot \mathbf{n}\, dS, \tag{2.150}$$

or

$$\frac{d}{dt} \int_{\tilde{V}(t)} f(\mathbf{x},t)\, dV = \frac{d}{dt} \int_{V(t)} f\, dV + \int_{\tilde{S}(t)} f(\tilde{\mathbf{v}} - \mathbf{v}) \cdot \mathbf{n}\, dS, \tag{2.151}$$

with the unit normal vector $\mathbf{n}$ pointing outward from the volume.

Note that in Eq. (2.151), the time derivative on the right-hand side follows the time evolution of the control volume $V$ instead of the material volume $\tilde{V}$. Eq. (2.150) does not use the control volume velocity $\mathbf{v}(\mathbf{x},t)$ and is convenient for steady problems. Eq. (2.151) only requires the relative material velocity $\tilde{\mathbf{v}}(\mathbf{x},t) - \mathbf{v}(\mathbf{x},t)$ and is particularly useful for unsteady problems such as rocket buster analysis. If $f(\mathbf{x},t)$ stands for the density $\rho(\mathbf{x},t)$ and $\tilde{\mathbf{v}}(\mathbf{x},t) = \mathbf{v}(\mathbf{x},t)$, i.e., the selected control volume coincides with the material volume at all time, according to the law of mass conservation, we have

$$\int_{\Omega} \Big[ \frac{\partial \rho}{\partial t} + \nabla \cdot (\rho \mathbf{v}) \Big] dV = 0. \tag{2.152}$$

Since Eq. (2.152) holds for arbitrary material volumes, the law of mass conservation can then be written as

$$\frac{\partial \rho}{\partial t} + \nabla \cdot (\rho \mathbf{v}) = 0. \tag{2.153}$$

Moreover, because Eq. (2.153) holds for any material volume, we can use Eq. (1.17) to get

$$\frac{d\rho}{dt} + \rho \nabla \cdot \mathbf{v} = 0. \tag{2.154}$$

In general, $\rho$ is a function of the position $\mathbf{x}$, pressure $p$, temperature $T$, etc. For incompressible materials, i.e., $d\rho/dt = 0$, we simply have

$$\nabla \cdot \mathbf{v} = 0. \tag{2.155}$$

If in the first kinematic transport theorem, we replace $f$ in Eq. (2.149) with $\rho f$, we get

$$\frac{d}{dt} \int_{V(t)} \rho f \, dV = \int_{V(t)} \left[ \frac{\partial}{\partial t}(\rho f) + \nabla \cdot (\rho f \mathbf{v}) \right] dV. \tag{2.156}$$

Similarly, with the chain rule and in Eq. (1.17), we obtain

$$\frac{\partial}{\partial t}(\rho f) + \nabla \cdot (\rho f \mathbf{v}) = \rho \frac{df}{dt} + f \left[ \frac{\partial \rho}{\partial t} + \nabla \cdot (\rho \mathbf{v}) \right]. \tag{2.157}$$

Following the law of mass conservation, we obtain the *second Reynolds transport theorem*,

$$\frac{d}{dt} \int_{V(t)} \rho f \, dV = \int_{V(t)} \rho \frac{df}{dt} \, dV. \tag{2.158}$$

In the ALE finite element formulation needed for the study of FSI systems, the mesh velocity $\hat{\mathbf{v}}$ is specified by some updating rules to the material velocity $\mathbf{v}$. Fundamentally, there are two ways of relating the mesh and material particle motions. The first is used for problems with closed domains or complete prescribed boundaries. In this ALE description, the mesh velocity can be assigned arbitrarily as long as the mesh boundary coincides with the material boundary. The second is for problems with unknown material boundaries, such as large amplitude free surfaces and FSI interfaces. In these cases, the mesh configurations cannot be assigned arbitrarily, due to the fact that the mesh boundary has to move with the material boundary. Their positions in general must be solved for any time $t$.

Obviously, in both cases, choices for the mesh velocities are not unique. The kinematic relations developed in the study of continuum mechanics can be used with mesh motions by reinterpreting the particle motion as the motion of a mesh point. Define mesh deformation gradient as $\hat{\mathbf{D}} = \partial \mathbf{y}/\partial \mathbf{z}$, mesh Jacobian determinant as $\hat{J} = \det(\hat{\mathbf{D}})$, mesh velocity gradient as $\hat{\mathbf{L}} = \partial \hat{\mathbf{v}}/\partial \mathbf{y}$, spatial gradient operator as $\nabla = \partial/\partial \mathbf{y}$, and referential gradient operator as $\hat{\nabla} = \partial/\partial \mathbf{z}$. Relationships between the referential and spatial derivatives both in time and space for a scalar $f$ and a vector $\mathbf{a}$ can be derived as follows:

$$\begin{aligned} &\check{f} = f' + \nabla f \cdot \hat{\mathbf{v}}, && \check{\mathbf{a}} = \mathbf{a}' + (\nabla \mathbf{a})\hat{\mathbf{v}}, \\ &\hat{\nabla} f = \hat{\mathbf{D}}^T \nabla f, && \hat{\nabla}\mathbf{a} = (\nabla \mathbf{a})\hat{\mathbf{D}}, \\ &\hat{\mathbf{L}} = \hat{\mathbf{D}}^* \hat{\mathbf{D}}^{-1}, && (\hat{J})^* = \hat{J}\,\mathrm{tr}(\hat{\mathbf{L}}) = \hat{J} \nabla \cdot \hat{\mathbf{v}}, \\ &(\nabla f)^* = -\hat{\mathbf{L}}^T \nabla f + \nabla \check{f}, && (\nabla \mathbf{a})^* = -(\nabla \mathbf{a})\hat{\mathbf{L}} + \nabla \check{\mathbf{a}}. \end{aligned} \tag{2.159}$$

Consider a subregion $V \subset \mathcal{R}_y$ and a surface $S \subset \mathcal{R}_y$ that represent the same mesh region and the same mesh surface at all time, using

$$(\nabla \cdot \mathbf{a})^* = -(\nabla \mathbf{a}) \cdot \hat{\mathbf{L}} + \nabla \cdot \check{\mathbf{a}}, \tag{2.160}$$

we have

$$\begin{aligned} \left( \int_V f\, dV \right)^* &= \int_V \left[ \check{f} + f\,\mathrm{tr}(\hat{\mathbf{L}}) \right] dV, \\ \left( \int_S \mathbf{a} \cdot \mathbf{n}\, dS \right)^* &= \int_S \left[ \check{\mathbf{a}} + \mathbf{a}\,\mathrm{tr}(\hat{\mathbf{L}}) - \hat{\mathbf{L}}\mathbf{a} \right] \cdot \mathbf{n}\, dS, \end{aligned} \tag{2.161}$$

where $\mathbf{n}$ is the surface normal vector corresponding to $S$.

The above relations will be elaborated upon and used extensively in the linearization of nonlinear governing equations in Chapter 6. Of course, the proof of the above relations require the use of Eq. (1.17) and the concepts illustrated in this chapter.

We call a particular collection of continuum medium in question a system. A closed system means that the system does not exchange matter with its surroundings. An isolated system means that no interactions between the system and its surroundings occur. In addition, any process within the system which is thermally insulated is called adiabatic. Therefore, the conservation of mass simply states that the mass of the same system stays constant, which has been adequately explained during the discussion of the second Reynolds transport theorem. Newton's law of motion (the law of linear momentum conservation) equates the rate of change of linear momentum in the material volume to the total external force exerted on the volume, of which the $i$th component can be expressed as $\int_V f_i^b \, dV + \int_S f_i^s \, dS$, where $f_i^b$ and $f_i^s$ stand for the $i$th component of the body force and the surface traction vectors, respectively. If $f(\mathbf{x}, t)$ in Eq. (2.158) represents the $i$th velocity component, with the second Reynolds transport theorem, we then have

$$\int_V f_i^b dV + \int_S f_i^s\, dS = \frac{d}{dt}\int_V \rho v_i\, dV = \int_V \rho\left(\frac{\partial v_i}{\partial t} + \mathbf{v}\cdot\nabla v_i\right) dV. \tag{2.162}$$

Denoting the stress tensor as $\sigma_{ij}$, the surface traction can be expressed as $f_i^s = \sigma_{ij}n_j$. Therefore, employing Gauss' theorem and the second Reynolds transport theorem, we have

$$\sigma_{ij,j} + f_i^b = \rho\frac{dv_i}{dt}. \tag{2.163}$$

In addition to the conservation of mass and linear momentum, we must also discuss the conservation of energy. To properly present the concepts, we start with the laws of thermodynamics. The first law of thermodynamics (the law of energy conservation) states that the rate of increase of the total energy in the material volume $U$, which includes the potential, kinetic, and internal energy associated with the thermal state of the material, to the sum of the net external heat input $Q$ and the net external work of the volume $W$,

$$dQ = dU - dW. \tag{2.164}$$

Naturally, the first law of thermodynamics implies that if a thermally insulated system can be taken from one state to another by alternative paths, the work done on the system $W$ has the same value for every adiabatic path. The second law of thermodynamics for a homogeneous system states that there exist two single-valued functions of state, namely, the absolute temperature $T$ and the entropy $S$. The commonly used unit for the temperature $T$ is Kelvin, named after William Thomson Kelvin (1824–1907). With this unit, the temperature is always a positive function. In general, the energy of the system consists of the kinetic energy, the potential energy, e.g., the gravitational energy, and the internal energy. The key aspect of energy conservation is to understand the concept of internal energy. For example, the internal energy of a gas is a function of its state. It has been shown in various experiments that only two of the three states, or measurable properties, namely, pressure $p$, density $\rho$, and temperature $T$, are independent. In practice, we often define the specific quantities with respect to the unit mass of gas. For example, the specific volume $v$ is equivalent to $V/m$ or $1/\rho$ where $V$ represent the volume occupied by the gas and $m$ the mass of the gas. For a perfect gas where molecules are far enough apart so that intermolecular forces are negligible and occupied volume is much larger than the molecular volume itself, we have

$$pV = mRT, \quad \text{or} \quad p = \rho RT, \quad \text{or} \quad pv = RT, \tag{2.165}$$

where $R$ is a constant for the gas.

Moreover, the increment of the entropy $S$ can be further decomposed as

$$dS = d_e S + d_i S, \tag{2.166}$$

where $d_e S$ denotes the part of the increase due to interaction with the surroundings and can be expressed as $d_e S = dQ/T$, and $d_i S$ denotes the part of the increase due to changes taking place inside the system.

Note that $d_i S$ is always nonnegative. In particular, if $d_i S$ is zero, the process is said to be reversible, and if $d_i S$ is positive, the process is said to be irreversible. In general, the internal energy per unit mass of gas $e$ can be expressed as

$$de = \left.\frac{\partial e}{\partial v}\right|_T dv + \left.\frac{\partial e}{\partial T}\right|_v dT. \tag{2.167}$$

Now, if heat is added to the gas and the temperature rise is observed according to the first law of thermodynamics, we have

$$dq = pdv + \left.\frac{\partial e}{\partial T}\right|_v dT + \left.\frac{\partial e}{\partial v}\right|_T dv, \tag{2.168}$$

where $q$ stands for the heat input per unit mass of gas.

According to Joule's experiments, the internal energy of a perfect gas is contained entirely within the molecule itself, as energy of translation and rotation, and the intermolecular forces are not important. Therefore, for a perfect gas, we have

$$\left.\frac{\partial e}{\partial v}\right|_T = 0. \tag{2.169}$$

Employing Eq. (2.169), directly from Eq. (2.168), we define the specific heat at constant volume $C_v$ as

$$C_v = \left.\frac{\partial q}{\partial T}\right|_v = \left.\frac{\partial e}{\partial T}\right|_v. \tag{2.170}$$

Hence, integrating Eq. (2.170), assuming that the internal energy vanishes at zero temperature, which is in accordance with the third law of thermodynamics, we derive

$$e = C_v T. \tag{2.171}$$

If Eq. (2.171) holds, the gas is called polytropic gas which means that the internal energy is simply proportional to the temperature. Moreover, applying Eq. (2.169), the first law of thermodynamics can be expressed as

$$dq = d(pv) - v\,dp + C_v\,dT. \tag{2.172}$$

Using Eq. (2.165), from Eq. (2.172), we can express the specific heat at constant pressure $C_p$ as follows:

$$C_p = \left.\frac{\partial q}{\partial T}\right|_p = \frac{d(pv)}{dT} + C_v = R + C_v. \tag{2.173}$$

Furthermore, using the definition for the entropy $S$, in reversible process, the entropy per unit mass $s$ can be written as

$$ds = C_p \frac{dT}{T} - R\frac{dp}{p}, \tag{2.174}$$

or rather

$$s - s_o = C_p \ln \frac{T}{T_o} - R \ln \frac{p}{p_o}, \tag{2.175}$$

where $s_o$, $T_o$, and $p_o$ represent the initial entropy per unit mass, temperature, and pressure.

Defining the ratio of the specific heat at constant volume and the specific heat at constant pressure or the adiabatic exponent $\gamma$ as

$$\gamma = \frac{C_p}{C_v}, \tag{2.176}$$

we have $C_v = R/(\gamma - 1)$.

Hence, we can easily derive from Eqs. (2.165), (2.173), and (2.174), for isentropic process, i.e., the entropy remains a constant,

$$pv^\gamma = C, \tag{2.177}$$

where $C$ is a constant.

According to the basic principle in classical kinetic theory, the internal energy of each degree of freedom of the individual molecules of a gas can be expressed as $KT/2$, where $K$ represents the Boltzmann constant. Denoting the number of molecules of a gas per unit mass as $n$, we have $R = nK$. Thus, if there are $N$ degrees of freedom, the internal energy per unit mass can be calculated as

$$e = \frac{NRT}{2}. \tag{2.178}$$

Combining Eqs. (2.171) and (2.178), we have

$$C_p = \frac{NR}{2} + R = \left(\frac{N}{2} + 1\right)R, \tag{2.179}$$

and consequently,

$$\gamma = \frac{N+2}{N}. \tag{2.180}$$

For a monatomic gas which has three degrees of freedom in translation, we have $\gamma = 1.67$; for a diatomic gas which has five degrees of freedom (three translations and two rotations), we have $\gamma = 1.4$. For air, which is a mixture of diatomic molecules such as Nitrogen (78.06%), Oxygen (21%), and Argon (0.94%), we also have $\gamma = 1.4$.

REMARK 2.2.2. Internal energy provides a thermal dynamic quantification of the molecular motion in microscopic scale, whereas kinetic energy provides a macroscopic representation of a motion of continuum. In Eq. (2.168), although the molecule of gas is obviously not stationary, the continuum is stationary. Therefore, there is no kinetic energy term other than the internal energy.

In engineering practice, instead of the internal energy per unit mass $e$, we introduce the enthalpy per unit mass $h$ as

$$h = pv + e. \tag{2.181}$$

Hence, from Eq. (2.181), employing the first law of thermal dynamics, we have

$$dh = v\,dp + p\,dv + de = v\,dp + dq, \tag{2.182}$$

which implies that the change of enthalpy is identical to the heat added to the gas at a constant pressure.

In an ideal gas, like internal energy $e$, $h$ is evidently a function of the temperature. Furthermore, for polytropic gas, the specific enthalpy can also be expressed as

$$h = \frac{\gamma RT}{\gamma - 1} = \frac{\gamma pv}{\gamma - 1}. \tag{2.183}$$

*Variational principles*

Based on the physical fact the heat is transferred from high temperature to low temperature, in engineering practice we introduce the heat flux vector $\mathbf{h}$ calculated as $-k\nabla T$, where $k$ is the so-called heat conductivity. Consequently, we have the rate of heat change expressed as

$$\frac{dQ}{dt} = \int_V \rho \frac{dq}{dt}\,dV = -\int_S h_i n_i\,dS, \tag{2.184}$$

where $\mathbf{n}$ is the surface normal unit vector normally pointing outwards and $q$ is the specific heat transfer to the system.

Obviously, the rate of work done on the system with the material volume $V$ from the surrounding can be expressed as $\int_V f_i^b v_i\,dV + \int_S f_i^s v_i\,dS$, where the kinetic energy is simply $\int_V \rho v_i v_i/2\,dV$. Moreover, the internal energy is defined

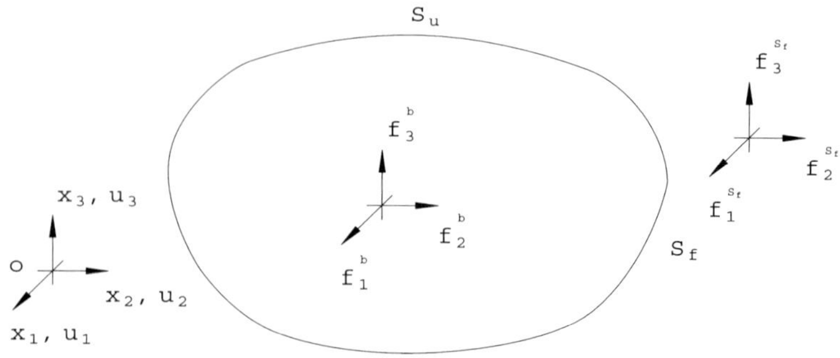


Figure 2.3. A typical three-dimensional continuous body.

as $\int_V \rho e\, dV$, where $e$ is the internal energy per unit mass. By using the second Reynolds transport theorem, we can easily derive

$$\int_V f_i^b v_i\, dV + \int_S f_i^s v_i\, dS = \int_V \rho\left(\frac{de}{dt} + v_i \frac{dv_i}{dt}\right) dV - \int_V \rho \frac{dq}{dt}\, dV. \tag{2.185}$$

Applying Eq. (2.163), for small deformations we can easily derive that

$$\rho \frac{de}{dt} = \frac{dq}{dt} + \sigma_{ij} \frac{d\epsilon_{ij}}{dt}, \tag{2.186}$$

or employ Eq. (2.184),

$$\rho \frac{de}{dt} = -\frac{\partial h_i}{\partial x_i} + \sigma_{ij} \frac{d\epsilon_{ij}}{dt}. \tag{2.187}$$

Consider a general three-dimensional body $V$ with the surface area $S = \partial V$, as shown in Figure 2.3. Assume $\mathbf{u}_p$ as the prescribed displacement on the Dirichlet boundary $S_u \subset S$, named after Johann Peter Gustav Lejeune Dirichlet (1805–1859) and $\mathbf{f}^{S_f}$, as the prescribed surface traction on the Neumann boundary $S_f \subset S$, named after Carl Neumann (1832–1925). In addition, we must have $S_u \cup S_f = S$ and $S_u \cap S_f = \emptyset$. For simplicity, we represent stress and strain tensors with the corresponding vector forms:

$$\begin{aligned} \boldsymbol{\sigma} &= \langle \sigma_{11} \quad \sigma_{22} \quad \sigma_{33} \quad \sigma_{12} \quad \sigma_{23} \quad \sigma_{31} \rangle, \\ \boldsymbol{\epsilon} &= \langle \epsilon_{11} \quad \epsilon_{22} \quad \epsilon_{33} \quad \gamma_{12} \quad \gamma_{23} \quad \gamma_{31} \rangle, \end{aligned} \tag{2.188}$$

with $\gamma_{ij} = 2\epsilon_{ij}$ for $i \neq j$.

If the material is linear elastic, the stress-strain relation can be denoted as

$$\boldsymbol{\sigma} = \mathbf{C}\boldsymbol{\epsilon}, \tag{2.189}$$

or

$$\boldsymbol{\sigma} = \mathbf{C}\boldsymbol{\epsilon} + \boldsymbol{\sigma}^I, \tag{2.190}$$

if there exists a stress vector $\boldsymbol{\sigma}^I$ representing the initial stress such as the thermal stress.

On the other hand, the stress vector $\boldsymbol{\sigma}$ and the strain vector $\boldsymbol{\epsilon}$ in Eq. (2.189) can be rewritten as tensorial quantities expressed by the following $3 \times 3$ matrices:

$$\boldsymbol{\sigma} = \begin{bmatrix} \sigma_{11} & \sigma_{12} & \sigma_{13} \\ \sigma_{21} & \sigma_{22} & \sigma_{23} \\ \sigma_{31} & \sigma_{32} & \sigma_{33} \end{bmatrix} \quad \text{and} \quad \boldsymbol{\epsilon} = \begin{bmatrix} \epsilon_{11} & \epsilon_{12} & \epsilon_{13} \\ \epsilon_{21} & \epsilon_{22} & \epsilon_{23} \\ \epsilon_{31} & \epsilon_{32} & \epsilon_{33} \end{bmatrix}. \tag{2.191}$$

Thus, the relation between stress and strain tensors can be represented as

$$\begin{aligned} \sigma_{ij} &= C_{ijrs}\epsilon_{rs} = \left[\lambda\delta_{ij}\delta_{rs} + \mu(\delta_{ir}\delta_{js} + \delta_{is}\delta_{jr})\right]\epsilon_{rs} \\ &= \lambda\delta_{ij}\epsilon_{rr} + \mu\epsilon_{ij} + \mu\epsilon_{ji}, \end{aligned} \tag{2.192}$$

with the Lamé constants, named after Gabriel Lamé (1795–1870), $\lambda$ and $\mu$ defined as

$$\lambda = \frac{E(1-\nu)}{(1+\nu)(1-2\nu)} \quad \text{and} \quad \mu = \frac{E}{2(1+\nu)},$$

where $E$ and $\nu$ are Young's modulus and Poisson's ratio, named after Thomas Young (1773–1829) and Simé-Denis Poisson (1781–1840), respectively.

In addition, in Eq. (2.189), we can write the constant material matrix as

$$\mathbf{C} = \lambda \begin{bmatrix} 1 & \nu/(1-\nu) & \nu/(1-\nu) & 0 & 0 & 0 \\ \nu/(1-\nu) & 1 & \nu/(1-\nu) & 0 & 0 & 0 \\ \nu/(1-\nu) & \nu/(1-\nu) & 1 & 0 & 0 & 0 \\ 0 & 0 & 0 & G/\lambda & 0 & 0 \\ 0 & 0 & 0 & 0 & G/\lambda & 0 \\ 0 & 0 & 0 & 0 & 0 & G/\lambda \end{bmatrix}.$$

If we assign the volumetric strain as $\epsilon_v = \epsilon_{rr}$, and the deviatoric strain as

$$\epsilon'_{ij} = \epsilon_{ij} - \epsilon_v\delta_{ij}/3, \tag{2.193}$$

Eq. (2.192) can be expressed as

$$\sigma_{ij} = \left(\lambda + \frac{2\mu}{3}\right)\epsilon_v\delta_{ij} + 2\mu\epsilon'_{ij}. \tag{2.194}$$

Furthermore, if we denote the pressure as

$$p = -\sigma_{rr}/3 = -\kappa\epsilon_v, \tag{2.195}$$

and the deviatoric stress as

$$\mathbf{s} = \boldsymbol{\sigma} + p\delta_{ij}, \tag{2.196}$$

with $\kappa = \lambda + 2\mu/3 = E/3(1-2\nu)$, Eq. (2.192) can also be rewritten as

$$\sigma_{ij} = -p\delta_{ij} + 2G\epsilon'_{ij}, \tag{2.197}$$

with $G = \mu$ as the shear modulus.

For small displacements, the linear strain-displacement relation (so-called compatibility condition) can be expressed as

$$\boldsymbol{\epsilon} = \boldsymbol{\partial}_\epsilon \mathbf{u} \tag{2.198}$$

with

$$\boldsymbol{\partial}_\epsilon = \begin{bmatrix} \partial/\partial x_1 & 0 & 0 \\ 0 & \partial/\partial x_2 & 0 \\ 0 & 0 & \partial/\partial x_3 \\ \partial/\partial x_2 & \partial/\partial x_1 & 0 \\ 0 & \partial/\partial x_3 & \partial/\partial x_2 \\ \partial/\partial x_3 & 0 & \partial/\partial x_1 \end{bmatrix}.$$

The principle of virtual displacements (the principle of virtual work) states that the equilibrium of the body $V$ requires that for any compatible small virtual displacements $\bar{\mathbf{u}}$, i.e., $\bar{\mathbf{u}} = \mathbf{0}$ on $S_u$ and $\bar{\boldsymbol{\epsilon}} = \boldsymbol{\partial}_\epsilon \bar{\mathbf{u}}$, the total internal virtual work is equal to the total external virtual work:

$$\int_V \bar{\boldsymbol{\epsilon}}^T \boldsymbol{\sigma}\, dV = \int_V \bar{\mathbf{u}}^T \mathbf{f}^b\, dV + \int_{S_f} \bar{\mathbf{u}}^{S_f^T} \mathbf{f}^{S_f}\, dS, \tag{2.199}$$

where $\bar{\boldsymbol{\epsilon}}$ represents the virtual strain vector associated with the virtual displacement $\bar{\mathbf{u}}$.

Here we introduce the variational indicator (so-called energy form):

$$\Pi_o = \int_V \frac{1}{2}\boldsymbol{\epsilon}^T \mathbf{C}\boldsymbol{\epsilon}\, dV - \int_V \mathbf{u}^T \mathbf{f}^b\, dV - \int_{S_f} \mathbf{u}^{S_f^T} \mathbf{f}^{S_f}\, dS, \tag{2.200}$$

with the displacement boundary condition $\mathbf{u}^{S_u} = \mathbf{u}_p$ on $S_u$.

Invoking the stationarity of $\Pi_o$, i.e., evaluating $\delta\Pi_o = 0$ with respect to the displacement $\mathbf{u}$ which also appears in the strain we obtain:

$$\int_V \delta\boldsymbol{\epsilon}^T \mathbf{C}\boldsymbol{\epsilon}\, dV - \int_V \delta\mathbf{u}^T \mathbf{f}^b\, dV - \int_{S_f} \delta\mathbf{u}^{S_f^T} \mathbf{f}^{S_f}\, dS = 0. \tag{2.201}$$

It is obvious that $\delta \mathbf{u}^{S_u} = \delta \mathbf{u}_p = \mathbf{0}$ on $S_u$. By using Eq. (2.189), we can easily identify that $\bar{\boldsymbol{\epsilon}} = \delta \boldsymbol{\epsilon}$, $\bar{\mathbf{u}} = \delta \mathbf{u}$, and $\bar{\mathbf{u}}^{S_f} = \delta \mathbf{u}^{S_f}$. In fact, the principle of virtual displacements corresponding to the minimum of the energy form in Eq. (2.200) and is often called the minimum energy principle. Of course, we assume that the displacement boundary condition and the compatibility condition are satisfied. To arrive at more general forms of variational principles, we first relax the compatibility condition and introduce the following variation indicator:

$$\Pi_1 = \Pi_o - \int_V \boldsymbol{\lambda}_\epsilon^T (\boldsymbol{\epsilon} - \boldsymbol{\partial}_\epsilon \mathbf{u})\, dV, \tag{2.202}$$

where $\boldsymbol{\lambda}_\epsilon$ is a Lagrangian multiplier.

Invoking the stationarity of $\Pi_1$, i.e., evaluating $\delta \Pi_1 = 0$ with respect to $\mathbf{u}$, $\boldsymbol{\epsilon}$, and $\boldsymbol{\lambda}_\epsilon$, we can easily verify that $\boldsymbol{\lambda}_\epsilon = \boldsymbol{\sigma} = \mathbf{C}\boldsymbol{\epsilon}$ and recover the compatibility condition. Since $\boldsymbol{\lambda}_\epsilon^T (\boldsymbol{\partial}_\epsilon \delta \mathbf{u}) = \sigma_{ij} \delta u_{i,j} = (\sigma_{ij} \delta u_j)_{,j} - \sigma_{ij,j} \delta u_i$, using the divergence theorem, we also recover the force equilibrium equation (2.163).

The variational principle associated with $\Pi_1$ is often called Hellinger–Reissner principle, named after Ernst Hellinger (1883–1950) and Eric Reissner (1913–1996). Furthermore, to relax the compatibility condition as well as the prescribed displacements on $S_u$, we introduce another variational indicator

$$\Pi_2 = \Pi_o - \int_V \boldsymbol{\lambda}_\epsilon^T (\boldsymbol{\epsilon} - \boldsymbol{\partial}_\epsilon \mathbf{u})\, dV - \int_{S_u} \boldsymbol{\lambda}_u^T \left(\mathbf{u}^{S_u} - \mathbf{u}_p\right) dS, \tag{2.203}$$

where $\boldsymbol{\lambda}_\epsilon$ and $\boldsymbol{\lambda}_u$ are Lagrangian multipliers.

Invoking the stationarity of $\Pi_2$, i.e., evaluating $\delta \Pi_2 = 0$ with respect to $\mathbf{u}$, $\boldsymbol{\epsilon}$, $\boldsymbol{\lambda}_\epsilon$, and $\boldsymbol{\lambda}_u$, we can easily identify $\boldsymbol{\lambda}_\epsilon = \boldsymbol{\sigma}$ and $\boldsymbol{\lambda}_u = \mathbf{f}^{S_u}$ and recover the equilibrium condition, the compatibility condition, and the displacement boundary condition, along with $f_i^{S_u} = \sigma_{ij} n_j$ on $S_u$ and $f_i^{S_f} = \sigma_{ij} n_j$ on $S_f$. The variational principle associated with $\Pi_2$ is often called Hu–Washizu principle.

REMARK 2.2.3. The fifteen equations governing six stress components, six strain components, and three displacements can be directly derived from Hu–Washizu principle as well as the geometrical (essential, Dirichlet) and force (natural, Neumann) boundary conditions.

In this section, we only provide a foundation of continuum mechanics approaches. The detailed discussion on geometrical nonlinearity with large displacements and deformations and material nonlinearity will be presented in Chapter 7.

## 2.3. Some mathematical tools

In this section, we briefly review some mathematical tools pertaining to the analysis of fluid–solid systems. Some concepts are related to the analytical solutions of fluid–solid systems and others provide foundations for various numerical approaches.

### 2.3.1. *Fourier series and transform; Laplace transform*

*Fourier series and transform*

One of the most important concepts in engineering mathematics is the use of Fourier Series and Transforms, named after Jean Baptiste Joseph Fourier (1768–1830). Based on the following integrals, for $m$ and $n \in \mathcal{Z}$,

$$\frac{2}{T}\int_0^T \cos\frac{2\pi mt}{T}\cos\frac{2\pi nt}{T}\,dt = \begin{cases} 1, & m = n, \\ 0, & m \neq n; \end{cases}$$
$$\frac{2}{T}\int_0^T \sin\frac{2\pi mt}{T}\sin\frac{2\pi nt}{T}\,dt = \begin{cases} 1, & m = n, \\ 0, & m \neq n; \end{cases} \tag{3.204}$$

and

$$\int_0^T \sin\frac{2\pi mt}{T}\cos\frac{2\pi nt}{T}\,dt = \int_0^T \sin\frac{2\pi mt}{T}\,dt = \int_0^T \cos\frac{2\pi nt}{T}\,dt = 0, \tag{3.205}$$

we can express any $T$-periodic function $f(t)$ with $t \in (-\infty, +\infty)$ by a Fourier series:

$$f(t) = a_o + \sum_{n=1}^{+\infty}\left(a_n\cos\frac{2\pi nt}{T} + b_n\sin\frac{2\pi nt}{T}\right), \tag{3.206}$$

with

$$a_o = \frac{1}{T}\int_0^T f(t)\,dt,$$

$$a_n = \frac{2}{T}\int_0^T f(t)\cos\frac{2\pi nt}{T}\,dt,$$

$$b_n = \frac{2}{T} \int_0^T f(t) \sin \frac{2\pi nt}{T} \, dt.$$

Furthermore, using the following integral with respect to a series of even functions, for $m$ and $n \in \mathcal{Z}$,

$$\int_0^{T/2} \cos \frac{2\pi mt}{T} \cos \frac{2\pi nt}{T} \, dt = \begin{cases} T/2, & m = n = 0, \\ T/4, & m = n \neq 0, \\ 0, & m \neq n; \end{cases} \tag{3.207}$$

any even $T$-periodic function $f_e(t)$ with $t \in (-\infty, +\infty)$ can be expressed as

$$f_e(t) = A_o + \sum_{n=1}^{+\infty} A_n \cos \frac{2\pi nt}{T}, \tag{3.208}$$

with

$$A_o = \frac{2}{T} \int_0^{T/2} f_e(t) \, dt,$$

$$A_n = \frac{4}{T} \int_0^{T/2} f_e(t) \cos \frac{2\pi nt}{T} \, dt.$$

Analogously, using the following integral with respect to a series of odd functions, for $m$ and $n \in \mathcal{Z}$,

$$\int_0^{T/2} \sin \frac{2\pi mt}{T} \sin \frac{2\pi nt}{T} \, dt = \begin{cases} T/4, & m = n, \\ 0, & m \neq n; \end{cases} \tag{3.209}$$

any odd $T$-periodic function $f_o(t)$ with $t \in (-\infty, +\infty)$ can be expressed as

$$f_o(t) = \sum_{n=1}^{+\infty} B_n \sin \frac{2\pi nt}{T}, \tag{3.210}$$

with

$$B_n = \frac{4}{T} \int_0^{T/2} f_o(t) \sin \frac{2\pi nt}{T} \, dt.$$

In addition, it is easy to verify that for a $T$-periodic function $f(t)$, we have:

$$\begin{aligned}\int_{T/2}^{T} f(t)\sin\frac{2\pi nt}{T}\,dt &= \int_{0}^{T/2} -f(-t)\sin\frac{2\pi nt}{T}\,dt,\\ \int_{T/2}^{T} f(t)\cos\frac{2\pi nt}{T}\,dt &= \int_{0}^{T/2} f(-t)\cos\frac{2\pi nt}{T}\,dt.\end{aligned}\tag{3.211}$$

Using Eq. (3.211) and the definitions of odd and even functions, we can show that odd functions have sine expansions whereas even functions have cosine expansions. Moreover, the coefficients of the sine and cosine expansions only involve half period integrations. Since any $T$-periodic function $f(t)$ with $t \in (-\infty, +\infty)$ can be written as a combination of a $T$-periodic odd function $f_o(t) = [f(t) - f(-t)]/2$ and a $T$-periodic even function $f_e(t) = [f(t) + f(-t)]/2$, using Eq. (3.211), the Fourier series of $f(t)$ written in the form of (3.206) can be viewed as the combination of forms (3.208) and (3.210) with

$$a_o = \frac{2}{T}\int_{0}^{T/2} \frac{[f(t)+f(-t)]}{2}\,dt = \frac{1}{T}\int_{0}^{T} f(t)\,dt = A_o,$$

$$a_n = \frac{4}{T}\int_{0}^{T/2} \frac{[f(t)+f(-t)]}{2}\cos\frac{2\pi nt}{T}\,dt = \frac{2}{T}\int_{0}^{T} f(t)\cos\frac{2\pi nt}{T}\,dt = A_n,$$

$$b_n = \frac{4}{T}\int_{0}^{T/2} \frac{[f(t)-f(-t)]}{2}\sin\frac{2\pi nt}{T}\,dt = \frac{2}{T}\int_{0}^{T} f(t)\sin\frac{2\pi nt}{T}\,dt = B_n.$$

Furthermore, employing the complex variable representations in Eq. (1.76), we can rewrite Eq. (3.206) in the following complex exponential form:

$$\begin{aligned}f(t) &= a_o + \sum_{n=1}^{+\infty}\left(a_n\frac{e^{i2\pi nt/T}+e^{-i2\pi nt/T}}{2} + b_n\frac{e^{i2\pi nt/T}-e^{-i2\pi nt/T}}{2i}\right)\\ &= a_o + \sum_{n=1}^{+\infty}\left(\frac{a_n - ib_n}{2}e^{i2\pi nt/T} + \frac{a_n+ib_n}{2}e^{-i2\pi nt/T}\right)\\ &= \sum_{k=-\infty}^{+\infty} c_k e^{i2\pi kt/T},\end{aligned}\tag{3.212}$$

with $c_k = (a_k - ib_k)/2, k \in \mathcal{Z}$, and

$$c_k = \frac{1}{T} \int_0^T f(t) e^{-i2\pi kt/T} \, dt. \tag{3.213}$$

At this point, we are ready to discuss the convergence issue of this Fourier series expansion. By using integration by parts, we also have

$$c_k = \frac{i}{2\pi k} \big[ f(T) - f(0) \big] + \frac{T}{(2\pi k)^2} \big[ f'(T) - f'(0) \big] - \frac{T}{(2\pi k)^2} \int_0^T f''(t) e^{-i2\pi kt/T} \, dt. \tag{3.214}$$

From Eq. (3.214) we can obtain the rate of convergence of the Fourier coefficients as $k \to \pm\infty$. For instance, if $f(T) \neq f(0)$, the Fourier coefficients $c_k$ converge as $O(1/k)$, and if $f(T) = f(0)$ and $f'(T) \neq f'(0)$, the Fourier coefficients $c_k$ converge as $O(1/k^2)$. Furthermore, any function $f(t)$ with $t \in (-\infty, +\infty)$ can be considered as a $T$-periodic function with $T \to +\infty$. It can be proven (refer to [79]) that for the even function $f_e(t)$, we have

$$\begin{aligned} f_e(t) &= \int_0^{+\infty} A(\omega) \cos \omega t \, d\omega, \\ A(\omega) &= \frac{2}{\pi} \int_0^{+\infty} f_e(t) \cos \omega t \, dt; \end{aligned} \tag{3.215}$$

whereas for the odd function $f_o(t)$

$$\begin{aligned} f_o(t) &= \int_0^{+\infty} B(\omega) \sin \omega t \, d\omega, \\ B(\omega) &= \frac{2}{\pi} \int_0^{+\infty} f_o(t) \sin \omega t \, dt. \end{aligned} \tag{3.216}$$

Again, since any function can be viewed as a combination of an odd function with an even function, analogous to the Fourier series expansion, based on

Eqs. (3.216) and (3.215), we derive the following Fourier transform:

$$
\begin{aligned}
f(t) &= \frac{1}{2\pi} \int_{-\infty}^{+\infty} F(\omega) e^{i\omega t}\, d\omega, \\
F(\omega) &= \int_{-\infty}^{+\infty} f(t) e^{-i\omega t}\, dt.
\end{aligned} \tag{3.217}
$$

In some cases, the Fourier transform is also written as

$$
\begin{aligned}
f(t) &= \int_{-\infty}^{+\infty} F(f) e^{i2\pi f t}\, df, \\
F(f) &= \int_{-\infty}^{+\infty} f(t) e^{-i2\pi f t}\, dt.
\end{aligned} \tag{3.218}
$$

The important prerequisite for the existence of the Fourier series or transform is that $\int_{-\infty}^{+\infty} |f(t)|\, dt$ exists, which means $c_k$ or $F(f) \to 0$ as $k$ or $f \to \pm\infty$.

An interesting aspect of the Fourier series is the Gibbs' phenomenon, named after Josiah Willard Gibbs (1839–1903). Consider the $\delta$ function or Dirac function, named after Paul Adrien Maurice Dirac (1902–1984):

$$
f(t) = \delta(t) = \begin{cases} 1, & t = 0, \\ 0, & \text{elsewhere}; \end{cases} \tag{3.219}
$$

with its Fourier transform $F(f) = 1$.

If we truncate $F(f) = 1$ to a symmetric square pulse, we get the approximation for $\delta(t)$ as

$$
\tilde{f}(t) = \int_{-k}^{k} F(f) e^{i2\pi f t}\, df = \frac{\sin kt}{\pi t}, \tag{3.220}
$$

a so-called sinc function, where $t = n\pi/k, n \in \mathcal{Z}$ corresponds to $\tilde{f}(t) = 0$.

Now, suppose we have a Heaviside function (or step function), named after Oliver Heaviside (1850–1925):

$$
f(t) = H(t) = \int_{-\infty}^{t} \delta(s)\, ds. \tag{3.221}
$$

Its approximate Fourier transform can be expressed as

$$\tilde{F}(t) = \int_{-\infty}^{t} \frac{\sin ks}{\pi s}\, ds = \frac{1}{2} + \int_{0}^{t} \frac{\sin ks}{\pi s}\, ds, \tag{3.222}$$

with

$$\int_{-\infty}^{0} \frac{\sin x}{x}\, dx = \int_{0}^{+\infty} \frac{\sin x}{x}\, dx = \frac{1}{2}\int_{-\infty}^{+\infty} \frac{\sin x}{x}\, dx = \frac{\pi}{2}. \tag{3.223}$$

Eq. (3.223) can be easily proven using a complex contour integral $\oint_C \frac{e^{iz}}{z}\, dz$, to be discussed in Section 2.3.2. The important fact is that the maximum of $\tilde{F}(f) = \tilde{F}(f)|_{t=\pi/k} = \frac{1}{2} + \frac{1}{\pi}\int_0^{\pi} \frac{\sin t}{t}\, dt$, which corresponds to $\frac{d}{dt}\int_0^t \frac{\sin ks}{\pi s}\, ds = \frac{\sin kt}{\pi t} = 0$ and is independent of $k$. Thus, if we have a jump (or discontinuity) of $f(t)$ at $t_o$, we can always define a new value $f(t_o) = [f_+(t_o) + f_-(t_o)]/2$, where $f_+(t_o) = \lim_{h\to 0} f(t_o+h)$ and $f_-(t_o) = \lim_{h\to 0} f(t_o-h)$, with $h > 0$. Nevertheless, since the overshoot is independent of $k$, i.e., the number of the Fourier series taken, the overshoot will always exist. This is called Gibbs' phenomenon.

*Laplace transform*

We are now ready to introduce the Laplace transform, named after Pierre-Simon Laplace (1749–1827) and often denoted as $L$ of a real variable function $f(t)$,

$$F(s) = \int_{0}^{+\infty} e^{-st} f(t)\, dt, \tag{3.224}$$

with the inverse Laplace transform,

$$f(t) = \lim_{b\to+\infty} \frac{1}{2\pi i} \int_{a-ib}^{a+ib} e^{st} F(s)\, ds, \tag{3.225}$$

where $f(t)$ is of exponential order, i.e., there exist real constants $M$, $t_m$, and $c$, such that $|f(t)| \leqslant M e^{ct_m}$, $\forall t > t_m$; $a$ is a real constant, and $s$ is a complex number with $\mathrm{Re}(s) = a > c$.

Consider the following real variable function:

$$x(t) = \begin{cases} e^{-at} f(t), & t \geqslant 0, \\ 0, & t < 0. \end{cases} \tag{3.226}$$

Since $f(t)$ is of exponential order and $a > c$, $\int_{-\infty}^{+\infty} |f(t)|\, dt$ exists, by using the Fourier transform of Eq. (3.217), we have

$$\begin{aligned} X(\omega) &= \int_{0}^{+\infty} f(t)e^{-at}e^{-i\omega t}\, dt, \\ f(t)e^{-at} &= \frac{1}{2\pi}\int_{-\infty}^{+\infty} X(\omega)e^{i\omega t}\, d\omega. \end{aligned} \tag{3.227}$$

Thus, if we consider $s = a + iy$, with $a > c$, i.e. $ds = i\, dy$, the inverse Laplace transform becomes

$$\begin{aligned} \lim_{b\to+\infty}\frac{1}{2\pi i}\int_{a-ib}^{a+ib} e^{st}F(s)\, ds &= \lim_{b\to+\infty}\frac{1}{2\pi i}\int_{a-ib}^{a+ib}\int_{0}^{+\infty} e^{st-su}f(u)\, du\, ds \\ &= \lim_{b\to+\infty}\frac{1}{2\pi}e^{at}\int_{-b}^{b} e^{iyt}\, dy\int_{0}^{+\infty} e^{-iyu}\big[e^{-au}f(u)\big]\, du, \end{aligned}$$

and using the Fourier transform of Eq. (3.227),

$$\lim_{b\to+\infty}\frac{1}{2\pi i}\int_{a-ib}^{a+ib} e^{st}F(s)\, ds = \frac{1}{2\pi}e^{at}2\pi f(t)e^{-at} = f(t),$$

which proves that the inverse Laplace transform is indeed valid.

Moreover, let $f(t)$ have a period $T > 0$, i.e., $f(t) = f(t+T)$. It can be proven that

$$L\big[f(t)\big] = F(s) = \int_{0}^{T} e^{-st}f(t)\, dt/(1 - e^{-sT}). \tag{3.228}$$

The evaluation of the inverse Laplace transform often requires the use of contour integrals to be discussed in Section 2.3.2.

### 2.3.2. *Contour integral; conformal mapping*

#### *Contour integral*

To better understand the inverse transformations, we must introduce the concepts of contour integral and conformal mapping in the complex plane. These concepts are of tremendous value in the solutions of problems of heat transfer, potential theory, fluid mechanics, elasticity, and many science and engineering disciplines.

Further discussions relevant to the solution of radiation and scattering problems are available in Chapter 5.

The key ingredient of the complex contour integral and conformal mapping involves the differentiability (analyticity) of a complex function under consideration. Although, parallel to real variable theories, the limit and continuity of a complex variable function are defined the same as those of a real variable function, the differentiability is quite different. In fact, it is clear from the definition of the derivative of $f(z)$ at $z$, $f'(z) = \lim_{\Delta z \to 0}(f(z + \Delta z) - f(z))/\Delta z$, $f'(z)$ can have different values based on different paths of $\Delta z$.

THEOREM 2.3.1 *(Cauchy–Riemann). Suppose $f(z) = \phi(x, y) + i\psi(x, y)$ is defined in a domain $\mathcal{S}$. The necessary condition of $f(z)$ being differentiable at $z$ (analytic, regular, or homomorphic) is that the following Cauchy–Riemann equations are satisfied*:

$$\frac{\partial \phi}{\partial x} = \frac{\partial \psi}{\partial y} \quad \textit{and} \quad \frac{\partial \phi}{\partial y} = -\frac{\partial \psi}{\partial x}; \tag{3.229}$$

*and the sufficient condition corresponds to an additional requirement of both $\phi$ and $\psi$ being $C^1$ at $z$ in $\mathcal{S}$.*

Eq. (3.229), so-called Cauchy–Riemann equations, named after August Louis Cauchy (1789–1857) and Georg Friedrich Bernhard Riemann (1826–1866), can also be expressed in the polar coordinate system as

$$\frac{\partial \phi}{\partial r} = \frac{1}{r}\frac{\partial \psi}{\partial \theta} \quad \text{and} \quad \frac{1}{r}\frac{\partial \phi}{\partial \theta} = -\frac{\partial \psi}{\partial r}. \tag{3.230}$$

In fact, assuming $f(z)$ is analytic, i.e., $f'(z)$ is uniquely defined, we have for the Cartesian coordinate system,

$$\frac{\partial f(x, y)}{\partial x} = f'(z)\frac{\partial (x + iy)}{\partial x} \quad \text{and} \quad \frac{\partial f(x, y)}{\partial y} = f'(z)\frac{\partial (x + iy)}{\partial y}; \tag{3.231}$$

and for the polar coordinate system

$$\frac{\partial f(r, \theta)}{\partial r} = f'(z)\frac{\partial (re^{i\theta})}{\partial r} \quad \text{and} \quad \frac{\partial f(r, \theta)}{\partial \theta} = f'(z)\frac{\partial (re^{i\theta})}{\partial \theta}. \tag{3.232}$$

It is then clear that by matching the real and imaginary parts of the equations in (3.231), we can easily confirm the Cauchy–Riemann equation in the Cartesian coordinate system (3.229). Likewise, from the equations in (3.232), the same relationship in the polar coordinate system can be derived as (3.230).

Obviously, if both $\phi$ and $\psi$ are $C^1$ and $f(z)$ is analytic in $\mathcal{S}$, then both $\phi$ and $\psi$ satisfy Laplace's equation, i.e., $\nabla^2\phi = \nabla^2\psi = 0$ in $\mathcal{S}$, and are called harmonic functions, or harmonic conjugate functions. Note that the complex conjugate function $f(z) = \bar{z}$ is not analytic. To make multiple-valued functions single-valued, we set up an artificial barrier, a so-called branch cut, which often starts from a branch point to the infinity or between branch points. Hence, each branch of a multiple-valued function is represented by a complex sheet and the whole complex plane can be viewed as those sheets superimposed on each other and joined together at the branch cut. Thus, the Riemann surface of a function, named after Bernhard Riemann (1826–1866), is a collection of those sheets. Obviously, $\ln z$ has infinitely many sheets within its Riemann surface, whereas the complex function

$$z^{1/3} = r^{1/3}e^{i\theta/3} = r^{1/3}e^{i\theta_p/3+i2\pi m/3}, \tag{3.233}$$

with $m \in \mathcal{Z}$, has three sheets in its Riemann surface corresponding to its three branches of solutions, i.e., $r^{1/3}e^{i\theta_p/3+i2\pi k}$, $r^{1/3}e^{i\theta_p/3+i2\pi(3k+1)/3}$, and $r^{1/3}e^{i\theta_p/3+i2\pi(3k+2)/3}$ with $k \in \mathcal{Z}$.

Furthermore, if $f(z)$ is analytic in a region, so are $f'(z)$, $f''(z), \ldots, f^{(n)}(z)$ with $n \in \bar{\mathcal{N}}$ and $n \geqslant 3$ in the same region, i.e., all higher derivatives exist in that region. A point at which $f(z)$ is not analytic is a singular point or singularity of $f(z)$. More specifically, if we can find a circle $|z - z_o| = \delta > 0$ which encloses no singular point other than $z_o$, the point $z_o$ is called an isolated singularity; if we can find $n \in \bar{\mathcal{N}}$ such that $\lim_{z\to z_o}(z - z_o)^n f(z)$ exists, then the singularity point $z_o$ is called an $n$th order pole (the first-order pole is often called a simple pole). The singularity $z_o$ is also called a removable singularity if $\lim_{z\to z_o} f(z)$ exists. A singularity of a multiple-valued function is also called a branch point. Finally, a singularity which is not a pole, a removable singularity, or a branch point is called an essential singularity.

We define the complex logarithmic function by mimicking the real variable formula as

$$f(z) = \ln z = \ln(r) + i\theta, \quad \text{with } r = |z|, \tag{3.234}$$

where the argument $\theta$ has infinitely many angles differing by integral multiplies of $2\pi$ which renders $\ln z$ multiple-valued.

We begin the discussion on the complex contour integral by defining a simply-connected domain as a domain in which any simple closed curve can be shrunk to a point without leaving the domain. The opposite of a simply-connected domain is a multiply-connected domain, which can be divided into a simply-connected domain with some simple closed continuous curves called Jordan curves, named after Marie Ennemond Camille Jordan (1838–1922), enclosing infinitesimally small area.

Analogous to the integral $\int_a^b f(x)\,dx$ for a real variable function, the complex integral is defined as $\int_C f(z)\,dz$, where the path $C$ can generally be a piecewise smooth simple curve (an arc or a closed curve). In fact, if $f(z)$ is analytic in a simply-connected domain and $z_o$ is any point within the domain, $\int_{z_o}^{z} f(\zeta)\,d\zeta = F(z) - F(z_o)$ is also analytic in the same domain with $F'(z) = f(z)$. According

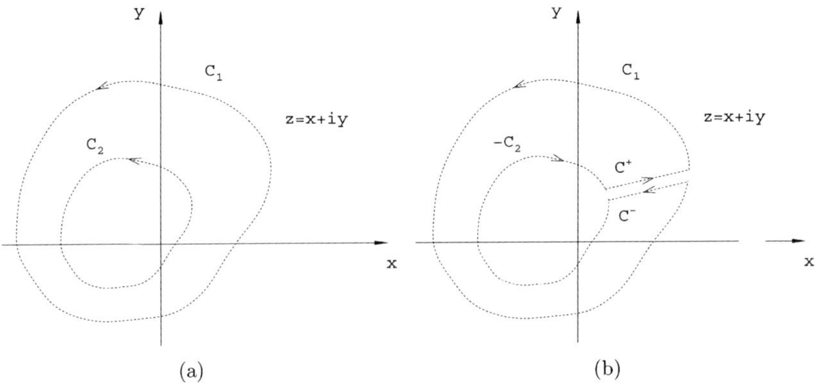


Figure 2.4. Path independence of complex contour integrals.

to convention, the region of integration must lie to the left of curve $C$. Hence, if $C$ is a closed curve, with the contour integral expressed as $\oint_C f(z)\,dz$, $C$ is in the counterclockwise direction.

THEOREM 2.3.2 *(Cauchy's theorem). If $f(z)$ is analytic in a simply-connected domain enclosed by a piecewise smooth simple closed curve $C$, then*

$$\oint_C f(z)\,dz = 0. \tag{3.235}$$

Thus, if $f(z)$ is analytic in a simply-connected domain, then $\int_C f(z)\,dz$ is independent of path $C$ from point $p_1$ to point $p_2$ within the domain, i.e., $\int_{C_1} f(z)\,dz = \int_{C_2} f(z)\,dz$. Furthermore, $\oint_C f(z)\,dz$ is also independent of closed path $C$ within the domain enclosing a point, which means if $f(z)$ is analytic on and between $C_1$ and $C_2$, as shown in Figure 2.4, $\oint_{C_1} f(z)\,dz = \oint_{C_2} f(z)\,dz$. This can be easily proven by using a Jordan curve to cut the multiply-connected domain between $C_1$ and $C_2$ into a simply-connected domain and applying Cauchy's theorem.

THEOREM 2.3.3 *(Cauchy integral theorem). Let $f(z)$ be analytic in a simply-connected domain enclosed by a piecewise counterclockwise smooth simple closed curve $C$ and $z_o$ be any point within the domain, then*

$$\oint_C \frac{f(z)}{z - z_o}\,dz = 2\pi i f(z_o). \tag{3.236}$$

Although the Taylor series for an analytic function $f(z)$ in a domain $\mathcal{D}$ is the same as for a real variable function, i.e., $f(z) = \sum_{n=0}^{+\infty}(f^{(n)}(a)/n!)(z-a)^n$, with the disk $|z - a|$ lying entirely within $\mathcal{D}$, a more general series expansion in complex analysis is the Laurent series, named after Pierre Alphonse Laurent (1813–1854),

$$f(z) = \sum_{n=-\infty}^{+\infty} c_n (z - z_o)^n, \tag{3.237}$$

employing Cauchy Integral the coefficient $c_n$ can be easily derived as $c_n = \frac{1}{2\pi i} \oint_C (f(\zeta)/(\zeta - z_o)^{n+1})\, d\zeta$, with any piecewise smooth simple closed counterclockwise path $C$ in a domain enclosed by two concentric circles centered at $z = z_o$.

THEOREM 2.3.4 *(Generalized Cauchy Integral theorem). Let $f(z)$ be analytic in a simply-connected domain enclosed by a piecewise smooth simple closed curve $C$ oriented counterclockwise and let $z_o$ be any point within the domain. Then*

$$\oint_C \frac{f(z)}{(z - z_o)^{n+1}}\, dz = \frac{2\pi i}{n!} f^{(n)}(z_o), \quad \textit{with } n \in \bar{\mathcal{N}}. \tag{3.238}$$

In addition, as an extension of Cauchy Integral theorem and Generalized Cauchy Integral theorem, we can easily prove the following Residue theorem, by employing the Laurent series.

THEOREM 2.3.5 *(Residue theorem). Let $f(z)$ be an analytic function inside the domain enclosed by a piecewise smooth simple closed curve $C$ oriented counterclockwise, and $f(z)$ is meromorphic, i.e., analytic everywhere except at a finite number of $n_j$th poles $z_j$, with $j = 1, \ldots, k$, where $k$ is the number of poles and $n_j$ is the order of the $j$th pole. Then*

$$\oint_C f(z)\, dz = 2\pi i \sum_{j=1}^{k} c_{-1}^{(j)}, \tag{3.239}$$

*with*

$$c_{-1}^{(j)} = \frac{1}{(n_j - 1)!} \lim_{z \to z_j} \frac{d^{n_j - 1}}{dz^{n_j - 1}} \big[ (z - z_j)^{n_j} f(z) \big]. \tag{3.240}$$

Cauchy's theorem is used to prove an important theorem named after Joseph Liouville (1809–1882).

THEOREM 2.3.6 *(Liouville theorem). If $f(z)$ is analytic in the entire complex plane and $f(z)$ is bounded, then $f(z)$ must be a constant.*

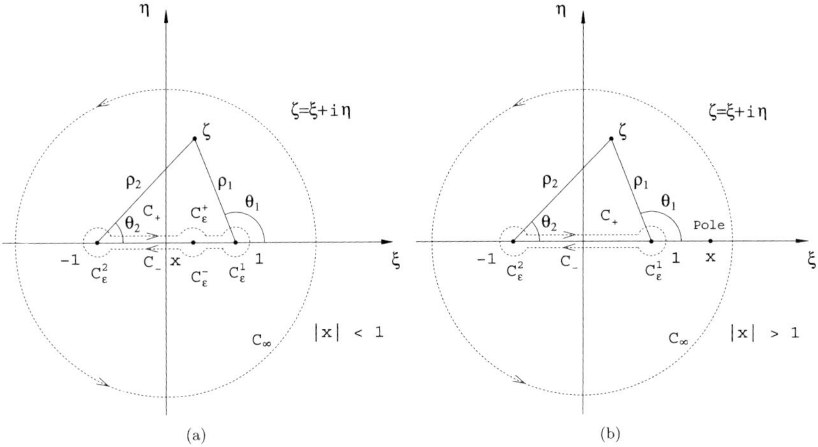


Figure 2.5. Complex contours of Example 2.1.

For further illustrations, we use the following few examples to show that the complex contour integral is of great importance in evaluating the inverse Fourier and Laplace transforms.

EXAMPLE 2.1. Evaluate the following integral

$$I(x) = \int_{-1}^{1} \frac{du}{(x-u)\sqrt{1-u^2}}, \tag{3.241}$$

where $x$ and $u$ are real numbers with $|x| \neq 1$.

Note that $\fint_{t_1}^{t_2} f(t)\,dt$ stands for the Cauchy principal value of an improper real integral, as $\int_{t_1}^{t_2} f(t)\,dt$ stands for the proper real integral. A detailed account is available in Ref. [79]. Consider first the case of $|x| < 1$. If we construct a complex contour integral, $\oint_C d\zeta/(x-\zeta)\sqrt{1-\zeta^2}$, we have three branch points, namely, $-1$, $x$, and 1. The natural branch cuts are selected as the line segments among these three points, as shown in Figure 2.5. Thus, $\sqrt{1-\zeta^2} = e^{i\pi/2}\sqrt{\zeta-1}\sqrt{\zeta+1} = e^{i(\pi/2+\theta_1/2+\theta_2/2)}\sqrt{\rho_1}\sqrt{\rho_2}$. Of course, we use the branch of the Riemann surface with $0 \leqslant \theta_1, \theta_2 \leqslant \pi$ for $\mathrm{Im}(\zeta) \geqslant 0$, and $\pi \leqslant \theta_1 \leqslant 2\pi$, $-\pi \leqslant \theta_2 \leqslant 0$ for $\mathrm{Im}(\zeta) \leqslant 0$. Therefore, the integrand is analytic in the domain enclosed by $C_\infty$, $C_\epsilon^1$, $C_+$, $C_\epsilon^+$, $C_\epsilon^2$, $C_-$, and $C_\epsilon^-$. Assigning $\zeta - x = \epsilon e^{i\theta}$, we have

$$\int_{C_\epsilon^+} \frac{d\zeta}{(x-\zeta)\sqrt{1-\zeta^2}} = \int_0^{\pi} \frac{\epsilon e^{i\theta} i\, d\theta}{-\epsilon e^{i\theta} e^{i\pi/2}\sqrt{1-x^2}e^{i\pi/2}} = \frac{i\pi}{\sqrt{1-x^2}},$$

$$\int_{C_\epsilon^-} \frac{d\zeta}{(x-\zeta)\sqrt{1-\zeta^2}} = \int_{\pi}^{2\pi} \frac{\epsilon e^{i\theta} i\, d\theta}{-\epsilon e^{i\theta} e^{i\pi/2}\sqrt{1-x^2}e^{i3\pi/2}} = -\frac{i\pi}{\sqrt{1-x^2}}.$$

Thus,

$$\int_{C_\epsilon} \frac{d\zeta}{(x-\zeta)\sqrt{1-\zeta^2}} = \int_{C_\epsilon^+} \frac{d\zeta}{(x-\zeta)\sqrt{1-\zeta^2}} + \int_{C_\epsilon^-} \frac{d\zeta}{(x-\zeta)\sqrt{1-\zeta^2}} = 0.$$

Similarly, assigning $\zeta - 1 = \epsilon e^{i\theta}$ and $\zeta + 1 = \epsilon e^{i\theta}$, respectively, we have, as $\epsilon \to 0$,

$$\int_{C_\epsilon^1} \frac{d\zeta}{(x-\zeta)\sqrt{1-\zeta^2}} = \int_{-\pi}^{\pi} \frac{\epsilon e^{i\theta} i\, d\theta}{(x-1-\epsilon e^{i\theta})e^{i\pi/2}\sqrt{\epsilon}e^{i\theta/2}\sqrt{2+\epsilon e^{i\theta}}} = 0,$$

$$\int_{C_\epsilon^2} \frac{d\zeta}{(x-\zeta)\sqrt{1-\zeta^2}} = \int_{0}^{2\pi} \frac{\epsilon e^{i\theta} i\, d\theta}{(x+1-\epsilon e^{i\theta})e^{i\pi/2}\sqrt{\epsilon}e^{i\theta/2}\sqrt{\epsilon e^{i\theta}-2}} = 0.$$

Moreover, letting $C_\infty$ as a circle with the radius $R \to +\infty$. Denoting $\zeta = Re^{i\theta}$, we have, as $R \to +\infty$,

$$\int_{C_\infty} \frac{d\zeta}{(x-\zeta)\sqrt{1-\zeta^2}} = \int_{0}^{2\pi} \frac{Re^{i\theta} i\, d\theta}{(x-Re^{i\theta})e^{i\pi/2}\sqrt{R^2e^{i2\theta}-1}} = 0. \tag{3.242}$$

Finally, on the top and bottom of the line segments of $(-1, x)$ and $(x, 1)$, i.e., $C_+$ and $C_-$ we have

$$\int_{C_+} \frac{d\zeta}{(x-\zeta)\sqrt{1-\zeta^2}} = \int_{C_-} \frac{d\zeta}{(x-\zeta)\sqrt{1-\zeta^2}} = -\int_{-1}^{1} \frac{d\xi}{(x-\xi)\sqrt{1-\xi^2}}.$$

Hence, from Cauchy's theorem, we have

$$\oint_{C_1} \frac{d\zeta}{(x-\zeta)\sqrt{1-\zeta^2}} + \oint_{C_2} \frac{d\zeta}{(x-\zeta)\sqrt{1-\zeta^2}} = 0, \tag{3.243}$$

with $C_1 = C_\infty$ and $C_2 = C_\epsilon^1 + C_+ + C_\epsilon^+ + C_\epsilon^2 + C_- + C_\epsilon^-$, and without much difficulty, we obtain

$$\int_{-1}^{1} \frac{du}{(x-u)\sqrt{1-u^2}} = 0. \tag{3.244}$$

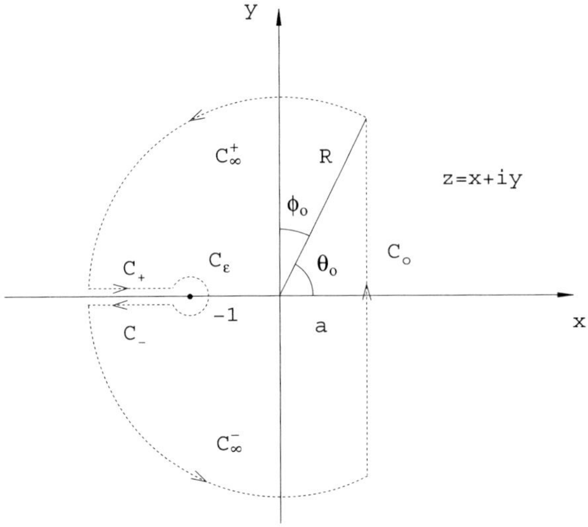


Figure 2.6. Complex contours of Example 2.2.

The conclusion in Eq. (3.244) will be used in Chapter 5. Consider the case of $|x| > 1$. We can construct a same complex contour integral, $\oint_C d\zeta/(x-\zeta)\times\sqrt{1-\zeta^2}$, which includes two branch points $-1$ and $+1$; $x$ is a simple pole outside the domain $(-1, 1)$. In this case, the branch cut is just the line segment between $\zeta = -1$ and $\zeta = 1$, as shown in Figure 2.5. Using the Residue theorem, we obtain

$$\oint_{C_1} \frac{d\zeta}{(x-\zeta)\sqrt{1-\zeta^2}} + \oint_{C_2} \frac{d\zeta}{(x-\zeta)\sqrt{1-\zeta^2}} = 2\pi i \frac{1}{i\sqrt{x^2-1}}, \tag{3.245}$$

with $C_1 = C_\infty$ and $C_2 = C_\epsilon^1 + C_+ + C_\epsilon^2 + C_-$, and further we obtain

$$\int_{-1}^{1} \frac{du}{(x-u)\sqrt{1-u^2}} = \frac{\pi}{\sqrt{x^2-1}}. \tag{3.246}$$

EXAMPLE 2.2. Evaluate the inverse Laplace transform $L^{-1}[1/\sqrt{s+1}]$.

The inverse Laplace transform is written as $\frac{1}{2\pi i}\int_{a-i\infty}^{a+i\infty}(e^{st}/\sqrt{s+1})\,ds$, with $a > 0$ corresponding to the complex contour integral $\oint_C (e^{zt}/\sqrt{z+1})\,dz$. Since $z = -1$ is the only branch point of the integrand, we consider the contour integral as shown in Figure 2.6 with $C = C_\infty^+ + C_+ + C_\epsilon + C_- + C_\infty^- + C_o$. Let us consider $C_\infty^+$ first. Assign $z = Re^{i\theta}$. Since $\phi_o = \pi/2 - \theta_o$ and $\theta_o \leqslant \theta < \pi/2$, we have $\lim_{R\to+\infty}\phi_o = \sin\phi_o = a/R$, and moreover, assigning $\phi = \pi/2 - \theta$, we have $\lim_{R\to+\infty}\phi = \sin\phi$. Thus, from $|1/\sqrt{z+1}| \leqslant M/\sqrt{R}$, with $M > 0$, we obtain, as $R \to +\infty$,

$$\left| \int_{\theta_o}^{\pi/2} \frac{e^{zt}}{\sqrt{z+1}} \, dz \right| \leqslant \frac{M}{\sqrt{R}} \int_0^{\phi_o} Re^{(R\sin\phi)t} \, d\phi = \frac{M}{t\sqrt{R}}(e^a - 1).$$

Therefore, as $R \to +\infty$, $\int_{\theta_o}^{\pi/2} (e^{zt}/\sqrt{z+1}) \, dz \to 0$. Analogously, if we assign $z = Re^{i\theta}$ with $\pi/2 \leqslant \theta \leqslant \pi$ and $\phi = \theta - \pi/2$, we have

$$\left| \int_{\pi/2}^{\pi} \frac{e^{zt}}{\sqrt{z+1}} \, dz \right| \leqslant \frac{M}{\sqrt{R}} \int_{\pi/2}^{\pi} Re^{(R\cos\theta)t} \, d\theta = \frac{M}{\sqrt{R}} \int_0^{\pi/2} Re^{-(R\sin\phi)t} \, d\phi.$$

Since $\sin\phi \geqslant 2\phi/\pi$ for $0 \leqslant \phi \leqslant \pi/2$, we have

$$\frac{M}{\sqrt{R}} \int_0^{\pi/2} Re^{-(R\sin\phi)t} \, d\phi \leqslant \frac{M}{\sqrt{R}} \int_0^{\pi/2} Re^{-R2\phi t/\pi} \, d\phi \leqslant \frac{M}{\sqrt{R}} \frac{\pi}{2t}(1 - e^{-Rt}),$$

and as a consequence, as $R \to +\infty$, $\int_{\pi/2}^{\pi} (e^{zt}/\sqrt{z+1}) \, dz \to 0$.

Hence, as $R \to +\infty$:

$$\int_{C_\infty^+} \frac{e^{zt}}{\sqrt{z+1}} \, dz = 0.$$

Likewise, we also have, as $R \to +\infty$:

$$\int_{C_\infty^-} \frac{e^{zt}}{\sqrt{z+1}} \, dz = 0.$$

In addition, if we assign $z = -1 + \epsilon e^{i\theta}$, we have, as $\epsilon \to 0$:

$$\int_{C_\epsilon} \frac{e^{zt}}{\sqrt{z+1}} \, dz = \int_{\pi}^{-\pi} \frac{e^{(-1+\epsilon e^{i\theta})t}}{\sqrt{\epsilon} e^{i\theta/2}} \epsilon i e^{i\theta} \, d\theta = 0.$$

Finally, employing

$$\begin{aligned} \int_{C_+} \frac{e^{zt}}{\sqrt{z+1}} \, dz &= \int_{-R}^{-\epsilon-1} \frac{e^{xt}}{\sqrt{-1-x} e^{i\pi/2}} \, dx = \int_{R-1}^{\epsilon} -\frac{e^{-(v+1)t}}{i\sqrt{v}} \, dv, \\ \int_{C_-} \frac{e^{zt}}{\sqrt{z+1}} \, dz &= \int_{-\epsilon-1}^{-R} \frac{e^{xt}}{\sqrt{-1-x} e^{-i\pi/2}} \, dx = \int_{\epsilon}^{R-1} \frac{e^{-(v+1)t}}{i\sqrt{v}} \, dv, \end{aligned} \tag{3.247}$$

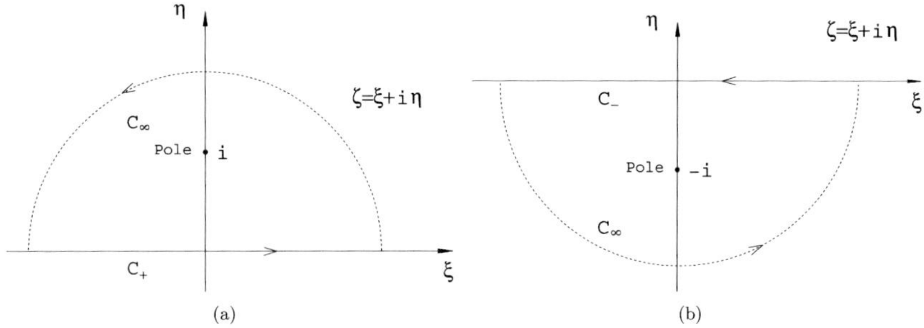


Figure 2.7. Complex contours of Example 2.3.

and using Cauchy's theorem, we have $\oint_C (e^{zt}/\sqrt{z+1})\,dz = 0$,

$$\frac{1}{2\pi i}\int_{a-i\infty}^{a+i\infty} \frac{e^{st}}{\sqrt{s+1}}\,ds = \frac{1}{2\pi i}\int_{C_o} \frac{e^{zt}}{\sqrt{z+1}}\,dz = \frac{1}{\pi i}\int_0^{+\infty} \frac{i}{\sqrt{v}} e^{-(v+1)t}\,dv$$

$$= \frac{2e^{-t}}{\pi\sqrt{t}}\int_0^{+\infty} e^{-p^2}\,dp = \frac{e^{-t}}{\sqrt{\pi t}}.$$

EXAMPLE 2.3. Evaluate the inverse Fourier transform $f(x) = \frac{1}{2\pi}\int_{-\infty}^{+\infty}(e^{ix\omega}/(\omega^2+1))\,d\omega$.

For $x \geqslant 0$, consider a complex contour integral $\oint_C (e^{ix\zeta}/(\zeta^2+1))\,d\zeta$ with $C = C_+ + C_\infty$ as shown in Figure 2.7. According to the Residue theorem, we have $\oint_C (e^{ix\zeta}/(\zeta^2+1))\,d\zeta = 2\pi i(e^{ixi}/(i+i)) = \pi e^{-x}$. We also have $\int_{C_\infty}(e^{ix\zeta}/(\zeta^2+1))\,d\zeta = 0$ as $R \to +\infty$. Therefore,

$$\frac{1}{2\pi}\int_{-\infty}^{+\infty} \frac{e^{ix\omega}}{\omega^2+1}\,d\omega = \frac{1}{2\pi}\int_{C_+} \frac{e^{ix\zeta}}{\zeta^2+1}\,d\zeta = \frac{1}{2\pi}\oint_C \frac{e^{ix\zeta}}{\zeta^2+1}\,d\zeta = \frac{1}{2}e^{-x}.$$

For $x < 0$, analogously, if we select the contour with $C = C_- + C_\infty$ as shown in Figure 2.7 and use the Residue theorem, we have $\oint_C (e^{ix\zeta}/(\zeta^2+1))\,d\zeta = 2\pi i(e^{ix(-i)}/(-i-i)) = -\pi e^{x}$, and obtain

$$\frac{1}{2\pi}\int_{-\infty}^{+\infty} \frac{e^{ix\omega}}{\omega^2+1}\,d\omega = -\frac{1}{2\pi}\int_{C_-} \frac{e^{ix\zeta}}{\zeta^2+1}\,d\zeta = -\frac{1}{2\pi}\oint_C \frac{e^{ix\zeta}}{\zeta^2+1}\,d\zeta = \frac{1}{2}e^{x}.$$

Therefore, combining both cases, we have

$$\frac{1}{2\pi}\int_{-\infty}^{+\infty}\frac{e^{ix\omega}}{\omega^2+1}\,d\omega=\frac{1}{2}e^{-|x|}.$$

*Conformal mapping*

Consider a complex variable function $\zeta = f(z)$ with $\zeta = \xi + i\eta$ and $z = x + iy$ as a mapping from a complex $z$ plane to a complex $\zeta$ plane. To ensure a one-to-one correspondence between the domain $\mathcal{D}$ in the $z$ plane and the domain $\bar{\mathcal{D}}$ in the $\zeta$ plane, we assume that $\zeta = \zeta(x, y)$ and $\eta = \eta(x, y)$ are single-valued, and so are the inverse functions $x = x(\xi, \eta)$ and $y = y(\xi, \eta)$. Notice, however, from the Cauchy–Riemann equations, we have $\xi_x = \eta_y$ and $\xi_y = -\eta_x$. Naturally, we must then have the Jacobian

$$\frac{\partial(\xi,\eta)}{\partial(x,y)}=\begin{vmatrix}\xi_x & \xi_y\\ \eta_x & \eta_y\end{vmatrix}=\xi_x\eta_y-\xi_y\eta_x=\xi_x^2+\xi_y^2=\left|f'(z)\right|^2\neq 0.$$

THEOREM 2.3.7. *Consider an analytic function $\zeta = f(z) = \xi(x, y) + i\eta(x, y)$ in a domain $\mathcal{D}$ with $|f'(z)| \neq 0$. Assuming the mapping $\zeta = f(z)$ from the domain $\mathcal{D}$ in the $z$ plane to the domain $\bar{\mathcal{D}}$ in the $\zeta$ plane is bijective, if $\phi(x, y)$ is harmonic in $\mathcal{D}$, then $\phi(x(\xi, \eta), y(\xi, \eta)) = \hat{\phi}(\xi, \eta)$ is harmonic in $\hat{\mathcal{D}}$.*

Consider two smooth curves $C_1$ and $C_2$ intersecting at $z_o$ in the complex $z$ plane. Assume a mapping $\zeta = f(z)$ with $f'(z_o) \neq 0$, and the corresponding two smooth curves $\hat{C}_1$ and $\hat{C}_2$ intersecting at $\zeta_o = f(z_o)$. If we parameterize $C_1$ and $C_2$ as $z_1(s) = x_1(s) + iy_1(s)$ and $z_2(s) = x_2(s) + iy_2(s)$ with the arc length $s$, the corresponding parameterized forms of $\hat{C}_1$ and $\hat{C}_2$ are also denoted as $\zeta_1(s) = \xi_1(s) + i\eta_1(s)$ and $\zeta_2(s) = \xi_2(s) + i\eta_2(s)$, and the angle between the two curves $\hat{C}_1$ and $\hat{C}_2$ at $\zeta_o$ can be expressed as

$$\begin{aligned}\arg\dot{\zeta}_2(0)-\arg\dot{\zeta}_1(0)&=\arg f'(z_o)+\arg\dot{z}_2(0)-\arg f'(z_o)-\arg\dot{z}_1(0)\\&=\arg\dot{z}_2(0)-\arg\dot{z}_1(0),\end{aligned}$$

which is equal to the angle between the two curves $C_1$ and $C_2$ at $z_o$ in the complex $z$ plane.

Hence the mapping $\zeta = f(z)$ is called the conformal mapping. In fact, any simply-connected domain $D \subset \mathcal{R}^2$ can be mapped one-to-one and conformally onto any other simply-connected domain $D' \subset \mathcal{R}^2$. In this section, we illustrated four typical examples of conformal mappings in Figures 2.8–2.11. In Figure 2.8, an upper plane is mapped into the interior of a unit circle through a bilinear transformation. If $z_1$, $z_2$, $z_3$, and $z_4$ are distinct, $((z_4 - z_1)(z_2 - z_3))/((z_2 - z_1)(z_4 - z_3))$ is invariant under the general bilinear

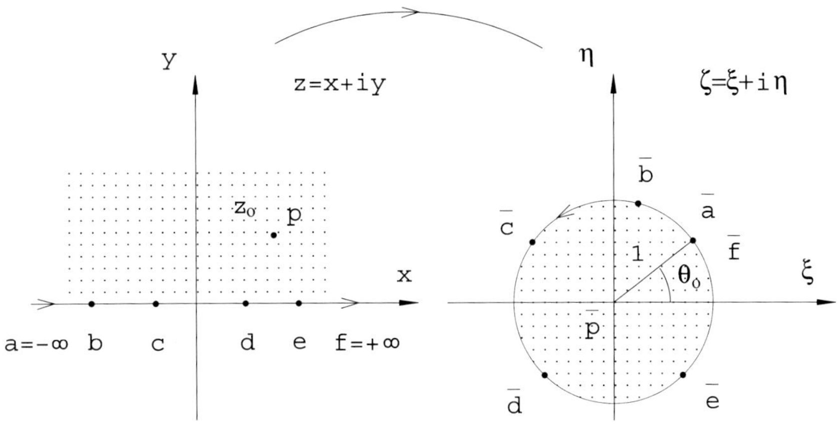


Figure 2.8. The conformal mapping $\zeta = e^{i\theta_o}(z - z_o)/(z - \bar{z}_o)$.

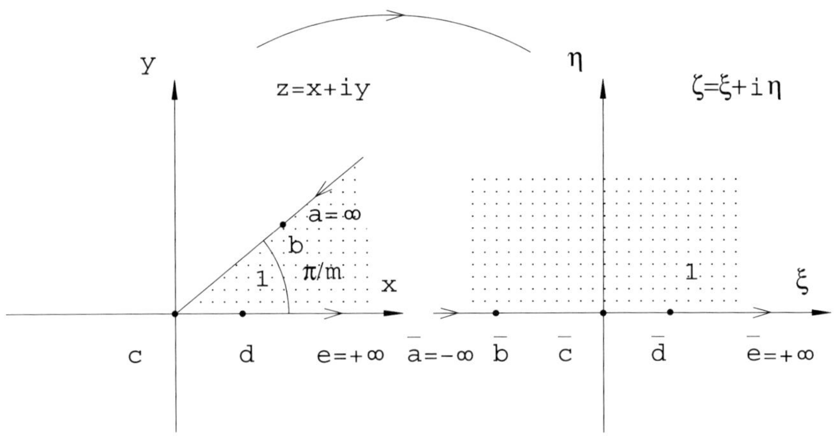


Figure 2.9. The conformal mapping $\zeta = z^m$ with $m \geqslant 1/2$.

transformation $\zeta = (c_1 z + c_2)/(c_3 z + c_4)$, where $c_i$ are complex constants with $c_1 c_4 - c_2 c_3 \neq 0$. This bilinear transformation includes four basic operations: scaling and rotation ($\zeta = cz$), translation ($\zeta = z + c$), and inversion ($\zeta = 1/z$), and can map any circle or straight line (circle with an infinite radius) to a circle or a straight line. In Figure 2.9, an interior of a corner $\pi/m$ is expanded to the entire upper plane with a power transformation. In Figure 2.10, the exterior of a semi-circle is mapped to the entire upper plane with a so-called Joukowski transformation. In Figure 2.11, a semi-infinite strip is transformed into the entire upper plane with a hyperbolic function.

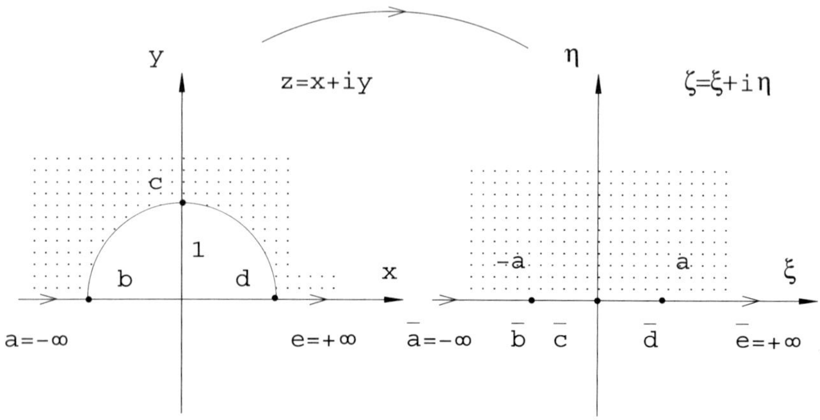


Figure 2.10. The conformal mapping $\zeta = a/2(z + 1/z)$.

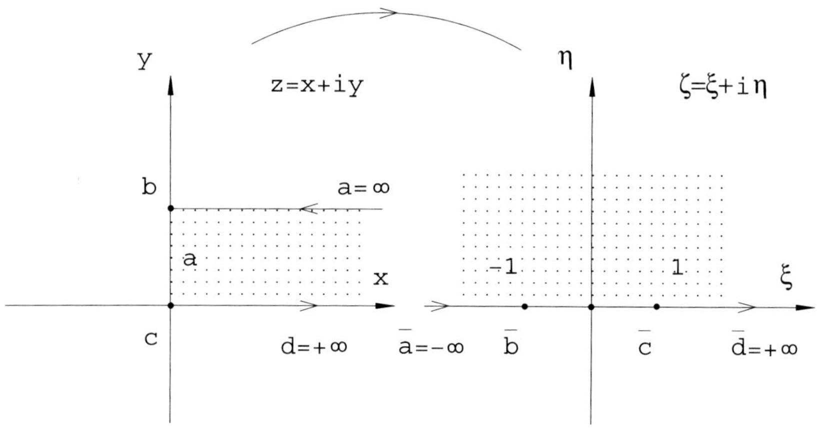


Figure 2.11. The conformal mapping $\zeta = \cosh \pi z/a$.

More discussion on specific transformation is presented in Chapter 5. For a comprehensive study of this subject, the reader is referred to Ref. [79]. Finally, we present a useful theorem for conformal mapping named after Karl Herman Amandus Schwarz (1843–1921) and Elwin Bruno Christoffel (1829–1900).

THEOREM 2.3.8 *(Schwarz–Christoffel transformation). The conformal mapping which maps the upper complex z plane to a polygon in the complex $\zeta$ plane is often expressed as $\zeta = c_1 \int \prod_{i=1}^{n} (z - x_i)^{\alpha_i/\pi - 1}\, dz + c_2$. The complex constants $c_1$ and $c_2$ determine the size, orientation, and position of the polygon, and $\alpha_i$ corresponds to the interior angles. Moreover, the points $x_i$ along the real axis of the z plane* *are mapped to the vertices of the polygon in the $\zeta$ plane, and the order from $-\infty$ to $+\infty$ in the $z$ plane corresponds to the counterclockwise order in the $\zeta$ plane. In addition, any three of the points $x_i$ can be chosen at will, including the infinity.*

### 2.3.3. Sturm–Liouville problems; Frobenius' methods and special functions

*Sturm–Liouville problems*

To generalize the concept of orthogonal functions beyond trigonometric functions, we define two generic orthogonal functions, $\phi_m(x)$ and $\phi_n(x)$, on an interval $[a, b]$, such that

$$\int_a^b \phi_m(x)\phi_n(x)\,dx = 0 \quad \text{or} \quad \int_a^b r(x)\phi_m(x)\phi_n(x)\,dx = 0, \tag{3.248}$$

where $r(x)$ is a weighting function.

Consider the linear homogeneous second-order differential equation

$$\frac{d}{dx}\left[p(x)\frac{dy}{dx}\right] + \left[q(x) + \lambda r(x)\right]y = 0, \tag{3.249}$$

which in general, represents

$$a_o(x)\frac{d^2y}{dx^2} + a_1(x)\frac{dy}{dx} + \left[a_2(x) + \lambda a_3(x)\right]y = 0, \tag{3.250}$$

with $p(x) = e^{\int a_1/a_o\,dx}$, $q(x) = a_2 p/a_o$, and $r(x) = a_3 p/a_o$.

Suppose $\lambda_m$ and $\lambda_n$ are any two different characteristic values of Eq. (3.249) corresponding to characteristic functions $\phi_m(x)$ and $\phi_n(x)$, we can then derive from the governing equation (3.249)

$$\begin{aligned} &\frac{d}{dx}\left(p\frac{d\phi_m}{dx}\right) + (q + \lambda_m r)\phi_m = 0, \\ &\frac{d}{dx}\left(p\frac{d\phi_n}{dx}\right) + (q + \lambda_n r)\phi_n = 0. \end{aligned} \tag{3.251}$$

Therefore, combining both equations in (3.251), we have

$$\phi_n\frac{d}{dx}\left(p\frac{d\phi_m}{dx}\right) - \phi_m e\frac{d}{dx}\left(p\frac{d\phi_n}{dx}\right) + (\lambda_m - \lambda_n)r\phi_m\phi_n = 0. \tag{3.252}$$

Using integration by parts, we obtain

$$\begin{aligned} (\lambda_n - \lambda_m)\int_a^b r\phi_m\phi_n\,dx &= \int_a^b \left[\phi_n\frac{d}{dx}\left(p\frac{\phi_m}{dx}\right) - \phi_m\frac{d}{dx}\left(p\frac{\phi_n}{dx}\right)\right]dx \\ &= p\left[\phi_n\frac{d\phi_m}{dx} - \phi_m\frac{d\phi_n}{dx}\right]\Bigg|_a^b. \end{aligned} \tag{3.253}$$

Thus, if we have the kind of boundary conditions at $x = a$ and $x = b$, e.g., $y = 0$, $dy/dx = 0$, or $y + \alpha dy/dx = 0$, with $\alpha$ a constant, expression (3.253) will be zero, and since $\lambda_m \neq \lambda_n$, we then have the orthogonal relation of form (3.248). Such problems are called Sturm–Liouville problems, named after Charles–François Sturm (1803–1855) and Joseph Liouville (1809–1882).

For a proper Sturm–Liouville problem, all characteristic values are real and nonnegative, and the corresponding characteristic functions are real (or can be made so by rejecting a possible complex constant multiplicative factor). In addition, there are infinitely many discretely distributed characteristic values. In most applications, the functions $p(x)$ and $r(x)$ are positive in $[a, b]$, while the function $q(x)$ is nonpositive in $[a, b]$. Therefore, to ensure positive characteristic values, when $y + \alpha\, dy/dx$ boundary conditions are used, $\alpha$ must be negative at $x = a$, and positive at $x = b$, which can be easily verified through the expansion of $\lambda_1 \int_a^b r\phi_1^2\, dx$ by employing the governing equation (3.249) and integration by parts.

Furthermore, as with the Fourier series, we can expand any function in a series of orthogonal characteristic functions as

$$f(x) = \sum_{n=1}^{\infty} A_n \phi_n(x), \tag{3.254}$$

with

$$A_n = \int_a^b r(x) f(x) \phi_n(x)\, dx \Big/ \int_a^b r(x) \big[\phi_n(x)\big]^2\, dx. \tag{3.255}$$

Hence, solutions of the nonhomogeneous boundary-value problems

$$\frac{d}{dx}\left[p(x)\frac{dy}{dx}\right] + \big[q(x) + \Lambda r(x)\big] y = h(x), \tag{3.256}$$

can be expanded with characteristic functions of the corresponding homogeneous equation (3.249), denoted as

$$y(x) = \sum_{n=1}^{\infty} a_n \phi_n(x). \tag{3.257}$$

By substituting Eq. (3.257) into Eq. (3.256) and employing the characteristic equations, we can have a form of Eq. (3.254), with $A_n = a_n(\Lambda - \lambda_n)$ and $f(x) = h(x)/r(x)$. Hence $a_n$ is evaluated as

$$a_n = \frac{\int_a^b h(x)\phi_n(x)\, dx}{(\Lambda - \lambda_n) \int_a^b r(x)[\phi_n(x)]^2\, dx}. \tag{3.258}$$

REMARK 2.3.1. Because of the growth of factor $\Lambda - \lambda_n$ in the denominator as $n \to \infty$, $a_n$ converges faster than $A_n$. In fact, the sequence of $a_n$ is convergent, even when $A_n$ is not.

We must point out that there are many types of orthogonal functions in addition to Fourier series. In particular, Bessel functions, named after Friedrich Wilhelm Bessel (1784–1846), and Legendre functions, named after Adrien–Marie Legendre (1752–1833), are very useful for fluid–solid systems expressed in cylindrical and spherical coordinate systems.

*Frobenius' methods and special functions*

In order to discuss special functions such as Bessels functions, we must first consider power series expansion near a point $x_o$:

$$\lim_{N\to\infty} \sum_{n=0}^{N} A_n (x - x_o)^n.$$

This series converges if $\lim_{n\to\infty} |A_{n+1}/A_n||x - x_o| < 1$. Naturally, the interval of convergence near the point $x_o$ can be expressed as

$$\left( x_o - \lim_{n\to\infty} \left| \frac{A_n}{A_{n+1}} \right|, x_o + \lim_{n\to\infty} \left| \frac{A_n}{A_{n+1}} \right| \right).$$

It is also clear that for a standard homogeneous second-order linear differential equation written as

$$\frac{d^2 y}{dx^2} + a_1(x)\frac{dy}{dx} + a_2(x)y = 0, \tag{3.259}$$

the behavior of solutions near a point $x_o$ is dependent on the behavior of the coefficients $a_1(x)$ and $a_2(x)$ near $x_o$.

In particular, the point $x_o$ is called an ordinary point of the differential equation if both $a_1(x)$ and $a_2(x)$ can be expanded in power series in an interval including $x_o$. Otherwise, it is called a singular point of the differential equation. In addition, it is called a regular singular point, if both $(x - x_o)a_1(x)$ and $(x - x_o)^2 a_2(x)$ are regular at $x_o$. For all the other scenarios of the singular point, it is called an irregular singular point. In this section, for simplicity, we restrict attention to solutions in the neighborhood of the point $x = 0$. Naturally, solutions near a more general point $x = x_o$ can be obtained in an analogous way or through a change of variable from $x$ to $x - x_o$. These power series expansion techniques are often called Frobenius' methods, named after Ferdinand Georg Frobenius (1849–1917).

Consider the differential equation in the following form:

$$R(x)\frac{d^2 y}{dx^2} + \frac{1}{x}P(x)\frac{dy}{dx} + \frac{1}{x^2}Q(x)y = 0, \tag{3.260}$$

where $R(x)$, $P(x)$, and $Q(x)$ are regular at $x = 0$.

Eq. (3.260) is often abbreviated as $L[y] = 0$, where $L$ represents the linear operator with respect to this second order equation. Furthermore, with $R(0) \neq 0$, if we rewrite Eq. (3.260) in the form of Eq. (3.259), it is obvious that $xa_1(x) = P(x)/R(x)$ and $x^2 a_2(x) = Q(x)/R(x)$ are regular at $x = 0$. Without loss of generality, we expand $R(x)$, $P(x)$, and $Q(x)$ as follows:

$$\begin{aligned} R(x) &= 1 + R_1 x + R_2 x^2 + \cdots, \\ P(x) &= P_o + P_1 x + P_2 x^2 + \cdots, \\ Q(x) &= Q_o + Q_1 x + Q_2 x^2 + \cdots. \end{aligned} \tag{3.261}$$

Let us attempt to express the nontrivial solution in the power series

$$y(x) = x^s \sum_{k=0}^{\infty} A_k x^k, \tag{3.262}$$

with $s$ to be determined.

If we substitute Eq. (3.262) into Eq. (3.260), we must have all the coefficients of the powers of $x$ vanish to satisfy Eq. (3.260):

$$f(s) = 0, \tag{3.263}$$

$$f(s+k)A_k = -\sum_{n=1}^{k} g_n(s+k)A_{k-n}, \quad \text{with } k \geqslant 1, \tag{3.264}$$

with

$$f(s) = s^2 + (P_o - 1)s + Q_o, \tag{3.265}$$

$$g_n(s) = R_n(s-n)^2 + (P_n - R_n)(s-n) + Q_n. \tag{3.266}$$

Eq. (3.263) is the so-called indicial equation, from which we determine $s$, whereas Eq. (3.264) is the so-called recurrence formula. Moreover, to obtain unique solutions $A_k$ as functions of preceding $A_{k-n}$ with $1 \leqslant n \leqslant k$, we must also have $f(s+k) \neq 0$.

Supposing $s_1$ and $s_2$ are solutions of Eq. (3.263), we have $f(s_1 + k) = k[k + (s_1 - s_2)]$ and $f(s_2 + k) = k[k - (s_1 - s_2)]$. Now, if $s_1$ and $s_2$ are complex conjugates, both $f(s_1 + k)$ and $f(s_2 + k)$ are nonzero, which means that we have two distinct power series solutions with respect to $s_1$ and $s_2$, respectively, the general solution of Eq. (3.263) at the neighborhood of $x = 0$ can be written as

$$y(x) = c_1 y_1(x) + c_2 y_2(x). \tag{3.267}$$

Note that the coefficient $A_o$ of the power series solutions is absorbed in $c_1$ and $c_2$. Moreover, supposing both $s_1$ and $s_2$ are real with $s_1 > s_2$, and $s_1 - s_2$ is not a positive integer, we have two distinct solutions and the general solution is in the same form as Eq. (3.267). However, if $s_1$ and $s_2$ are real and $p = s_1 - s_2$ is a positive integer, we have two constants to be determined for the solution in the form (3.262), i.e., $A_o$ and $A_p$. Note that since the coefficient $f(s_2 + p)$ of $A_p$ in Eq. (3.264) is zero, $A_p$ can be arbitrary.

To illustrate the power series expansion procedure, we consider an example with $R(x) = 1$, $P(x) = 1+x$, and $Q(x) = -1$. From Eq. (3.263), we have $s_1 = 1$ and $s_2 = -1$, and consequently $p = 2$. Employing $s = s_1 = 1$, from the recurrence formula (3.264) and Eq. (3.266), we have for $k \geqslant 1$, $A_k = -A_{k-1}/(k+2)$ and from Eq. (3.262), the power series expansion

$$y(x) = x^1 A_o\left(1 - \frac{x}{3} + \frac{x^2}{12} + \cdots\right) = 2A_o \frac{e^{-x} - 1 + x}{x}. \tag{3.268}$$

On the other hand, employing $s = s_2 = -1$, the recurrence formula (3.264), and Eq. (3.266), we have $A_1 = A_o$, and $A_k = -A_{k-1}/k$ for $k \geqslant 3$. Note that $A_2$ is arbitrary. For convenience, we choose $A_2$ and consequently from the recurrence formula $A_k = 0$ for $k \geqslant 3$. Therefore, the general solution becomes

$$y(x) = c_1 \frac{e^{-1} - 1 + x}{x} + c_2 \frac{1-x}{x}, \quad \text{or} \quad y(x) = c_1 \frac{e^{-x}}{x} + c_2 \frac{1-x}{x}. \tag{3.269}$$

In fact, employing L'Hosptial's rule, it is clear the only solution regular at $x = 0$ is expressed as

$$y(x) = C \frac{e^{-x} - 1 + x}{x}. \tag{3.270}$$

In essence, Eq. (3.262) can be expressed as

$$y(x) = y(x, s)\big|_{s=s_1,s_2} = x^s \sum_{k=0}^{\infty} A_k(s) x^k \Bigg|_{s=s_1,s_2}. \tag{3.271}$$

As a consequence, with the satisfaction of the recurrence formula (3.264), we have

$$L\big[y(x, s)\big] = A_o f(s) x^{s-2}. \tag{3.272}$$

For the case $s_1 = s_2$, we have

$$\frac{\partial}{\partial s} L\big[y(x, s)\big] = A_o\big[2(s - s_1) + (s - s_1)^2 \ln x\big] x^{s-2}, \tag{3.273}$$

whereas for the case $s_1 - s_2 = p$, we have

$$\begin{aligned} &\frac{\partial}{\partial s} L\big[(s - s_2) y(x, s)\big] \\ &\quad = A_o\big[2(s - s_1)(s - s_2) + (s - s_1)(s - s_2)^2 \ln x\big] x^{s-2}. \end{aligned} \tag{3.274}$$

Therefore, since the operator $L$ and $\partial/\partial s$ are commutative, if $s_1 = s_2 = (1 - P_o)/2$ with $(1 - P_o)^2 = 4Q_o$, we have $f(s_1 + k) = f(s_2 + k) = k^2 \neq 0$, which means that only one solution of the form (3.262) can be obtained. From the solution written in the form (3.271), the second solution can then be expressed as $\frac{\partial y(x,s)}{\partial s}|_{s=s_1}$. Similarly, if $s_1 - s_2 = p$, we in fact can derive the second solution as $\frac{\partial}{\partial s}[(s - s_2)y(x, s)]|_{s=s_2}$.

Many special functions stem from a particular class of equation which is written as follows:

$$\left(1 + R_m x^m\right)\frac{d^2 y}{dx^2} + \frac{1}{x}\left(P_o + P_m x^m\right)\frac{dy}{dx} + \frac{1}{x^2}\left(Q_o + Q_m x^m\right)y = 0. \tag{3.275}$$

For example, the following characteristic equations for the special functions can be all written in the form of Eq. (3.275).

*Bessel's equation* ($m = 2$):

$$x^2\frac{d^2 y}{dx^2} + x\frac{dy}{dx} + (x^2 - p^2)y = 0. \tag{3.276}$$

*Legendre's equation* ($m = 2$):

$$(1 - x^2)\frac{d^2 y}{dx^2} - 2x\frac{dy}{dx} + p(p + 1)y = 0. \tag{3.277}$$

*Gauss' equation* ($m = 1$):

$$(x - x^2)\frac{d^2 y}{dx^2} + \left[\gamma - (\alpha + \beta + 1)x\right]\frac{dy}{dx} - \alpha\beta y = 0. \tag{3.278}$$

*Hermite's equation* ($m = 2$):

$$\frac{d^2 y}{dx^2} - 2x\frac{dy}{dx} + 2ny = 0. \tag{3.279}$$

*Chebyshev's equation* ($m = 2$):

$$(1 - x^2)\frac{d^2 y}{dx^2} - x\frac{dy}{dx} + n^2 y = 0. \tag{3.280}$$

*Jacobi's equation* ($m = 1$):

$$(x - x^2)\frac{d^2 y}{dx^2} + \left[a - (1 + b)x\right]\frac{dy}{dx} + n(b + n)y = 0. \tag{3.281}$$

Substituting the power series expansion (3.262) into Eq. (3.275), from the indicial equation (3.262) and the recurrence formula (3.266), we have $A_1 = A_2 = \cdots = A_{M-1} = 0$. Consequently, all coefficients not an integral multiple of $m$ are zero. Hence, the power series expansion can be rewritten as

$$y(x, s) = \sum_{k=0}^{\infty} B_k x^{mk+s}. \tag{3.282}$$

Let us first consider Bessel's equation, from which Bessel function is derived:

$$x^2 \frac{d^2 y}{dx^2} + x \frac{dy}{dx} + (x^2 - p^2) y = 0. \tag{3.283}$$

From Frobenius' methods, we derive the solution in the form

$$y(x, s) = \sum_{k=0}^{\infty} B_k x^{2k+s}. \tag{3.284}$$

Since the two roots of the indicial equation (3.262) are $s_1 = p$ and $s_2 = -p$, the recurrence formula can then be expressed as

$$(s + 2k + p)(s + 2k - p) B_k = -B_{k-1}, \quad \text{with } k \geqslant 1. \tag{3.285}$$

Thus, with respect to the exponent $s_1 = p$, we have

$$B_k(p) = (-1)^k \frac{1}{(1 + p)(2 + p) \cdots (k + p)} \frac{B_o}{2^{2k} k!}, \quad \text{with } k \geqslant 1. \tag{3.286}$$

Define first a Gamma function:

$$\Gamma(p) = \int_0^{\infty} e^{-x} x^{p-1} \, dx, \quad \text{with } p > 0, \tag{3.287}$$

where $p$ is not necessarily an integer.

It is easy to verify, based on Eq. (3.287), that the Laplace transform of $t^p$ can be expressed as a Gamma function:

$$L[t^p] = \frac{\Gamma(p + 1)}{s^{p+1}}, \quad \text{with } p > -1. \tag{3.288}$$

In addition, employing integration by parts, we also have

$$\Gamma(p + 1) = p\Gamma(p), \qquad \Gamma(1) = 0! = 1, \quad \text{and} \quad \Gamma(n + 1) = n! \tag{3.289}$$

Therefore, according to Eq. (3.289), the solution with respect to $s_1 = p$ can be written as

$$B_o 2^p \Gamma(p + 1) \sum_{k=0}^{\infty} \frac{(-1)^k (x/2)^{2k+p}}{(k + p)! k!} = B_o 2^p \Gamma(p + 1) J_p(x), \tag{3.290}$$

and likewise, the solution corresponding to $s_2 = -p$ can be written as

$$B_o 2^p \Gamma(-p+1) \sum_{k=0}^{\infty} \frac{(-1)^k (x/2)^{2k-p}}{(k-p)!k!} = B_o 2^{-p} \Gamma(-p+1) J_{-p}(x). \tag{3.291}$$

$J_p(x)$ is the so-called Bessel function of the first kind of order $p$. Therefore, if $p$ is not zero or a positive integer, the complete solution of Bessel's equation can be expressed as

$$y(x) = c_1 J_p(x) + c_2 J_{-p}(x). \tag{3.292}$$

Now, supposing $p = 0$, two solutions in the form of (3.277) become identical, therefore, the second solution can be derived according to (3.273):

$$\left.\frac{\partial y(x,s)}{\partial s}\right|_{s=0} = B_o \left[ J_o(x) \ln x + \sum_{k=1}^{\infty} (-1)^{k+1} \phi(k) \frac{(x/2)^{2k}}{(k!)^2} \right], \tag{3.293}$$

where $\phi(0) = 0$ and $\phi(k) = -\sum_{m=1}^{k} 1/m$ for $k \geqslant 1$.

The coefficient of $B_o$ in Eq. (3.293) represents the second solution of Bessel's equation with $p = 0$ and is also called Bessel function of the second kind of order zero. In particular, the expression of the second equation (3.293) is called Neumann's form and expressed as $Y^{(0)}(x)$. Since the complete solution of Bessel's equation is a combination of these two solutions, the second solution can be also written in Weber's form as

$$Y_o(x) = \frac{2}{\pi} \left[ Y^{(0)}(x) + (\gamma - \ln 2) J_o(x) \right], \tag{3.294}$$

where $\gamma$ is Euler's constant defined as $\gamma = \lim_{k\to\infty}[\phi(k) - \ln k]$.

Therefore the complete solution of Bessel's function for $p = 0$ is expressed as

$$y(x) = c_1 J_o(x) + c_2 Y_o(x). \tag{3.295}$$

For the other degenerate case with $p = n$, where $n$ is a positive integer, according to (3.274), the second solution can be evaluated as

$$\begin{aligned} Y_n(x) = \frac{2}{\pi} \Bigg[ \left( \ln \frac{x}{2} + \gamma \right) J_n(x) - \frac{1}{2} \sum_{k=0}^{n-1} \frac{(n-k-1)!(x/2)^{2k-n}}{k!} \\ + \frac{1}{2} \sum_{k=0}^{\infty} (-1)^{k+1} \left[ \phi(k) + \phi(k+n) \right] \frac{(x/2)^{2k+n}}{k!(n+k)!} \Bigg], \end{aligned} \tag{3.296}$$

where $Y_n$ is called Weber's form of the Bessel function of the second kind of order $n$.

Notice that if $p$ is not zero or a positive integer, we also have

$$Y_p(x) = \frac{(\cos p\pi) J_p(x) - J_{-p}(x)}{\sin p\pi}. \tag{3.297}$$

Of course, in this case, we can directly use $J_p(x)$ and $J_{-p}(x)$ instead of $J_p(x)$ and $Y_p(x)$ as two independent solutions.

In order to understand the asymptotic behaviors of Bessel's function, we perform a transformation with $u(x) = y(x)\sqrt{x}$. Hence, Eq. (3.276) can be expressed as

$$\frac{d^2u}{dx^2} + \left(1 - \frac{p^2 - 1/4}{x^2}\right) u = 0. \tag{3.298}$$

Obviously, for sufficiently large $x$, the solution can be simply expressed as

$$u(x) = A\cos(x - \theta). \tag{3.299}$$

By comparing with the expression of $J_p(x)$ and $Y_p(x)$, we can derive for $x \to \infty$,

$$\begin{aligned} J_p(x) &\sim \sqrt{\frac{2}{\pi x}} \cos(x - \theta_p), \\ Y_p(x) &\sim \sqrt{\frac{2}{\pi x}} \sin(x - \theta_p), \end{aligned} \tag{3.300}$$

with $\theta_p = (2p + 1)\pi/4$.

Therefore, following the exponential expression of the complex variables in Section 2.1, we have for $x \to \infty$

$$\begin{aligned} J_p(x) + iY_p(x) &\sim \sqrt{\frac{2}{\pi x}} e^{i(x-\theta_p)}, \\ J_p(x) - iY_p(x) &\sim \sqrt{\frac{2}{\pi x}} e^{-i(x-\theta_p)}. \end{aligned} \tag{3.301}$$

In general, we define the Hankel functions of the first and second kinds, or Bessel functions of the third kind, as

$$\begin{aligned} H_p^{(1)}(x) &= J_p(x) + iY_p(x), \\ H_p^{(2)}(x) &= J_p(x) - iY_p(x). \end{aligned} \tag{3.302}$$

Consider now a modified equation:

$$x^2\frac{d^2y}{dx^2} + x\frac{dy}{dx} - (x^2 + p^2)y = 0. \tag{3.303}$$

Substituting $t$ for $ix$, we arrive at Bessel's equation with respect to $t$. Thus, to reiterate, if $p$ is not zero or a positive integer, the general solution is of the form

$$y(x) = c_1 J_p(ix) + c_2 J_{-p}(ix). \tag{3.304}$$

To combine the cases of $p = 0$ and a positive integer with Weber's forms, we can also express the general solution as

$$y(x) = c_1 J_n(ix) + c_2 Y_n(ix), \quad \text{with } n \in \mathcal{N}. \tag{3.305}$$

From the definitions of (3.290) and (3.291), we have

$$\begin{aligned} J_p(ix) &= \sum_{k=0}^{\infty} \frac{(-1)^k i^{2k+p} (x/2)^{2k+p}}{k!(k+p)!} \\ &= i^p \sum_{k=0}^{\infty} \frac{(x/2)^{2k+p}}{k!(k+p)!} = i^p I_p(x), \end{aligned} \tag{3.306}$$

where $I_p(x)$ is called the modified Bessel function of the first kind of order $p$.

Consequently, if $p$ is not zero or a positive integer, the complete solution of Eq. (3.303) can be expressed as

$$y(x) = c_1 I_p(x) + c_2 I_{-p}(x). \tag{3.307}$$

Likewise, in the case of $p = n$ where $n$ is zero or a positive integer, it is convenient to define the following modified Bessel function of the second kind of order $n$ as

$$K_n(x) = \frac{\pi}{2} i^{n+1} \big[ J_n(ix) + i Y_n(ix) \big] = \frac{\pi}{2} i^{n+1} H_n^{(1)}(ix). \tag{3.308}$$

Consequently, the general solution is given as

$$y(x) = c_1 I_n(x) + c_2 K_n(x). \tag{3.309}$$

Similar to the case of Weber's form of the Bessel function of the second kind in Eq. (3.297), if $p$ is not zero or a positive integer, the function $K_p(x)$ is defined by the following:

$$K_p(x) = \frac{\pi}{2} \frac{I_{-p}(x) - I_p(x)}{\sin p\pi}. \tag{3.310}$$

Naturally, from the asymptotic relation (3.301), for large values of $x$ the modified functions have the following asymptotic behavior:

$$\begin{aligned} I_p(x) &\sim \frac{e^x}{\sqrt{2\pi x}}, \\ K_p(x) &\sim \frac{e^{-x}}{\sqrt{(2/\pi)x}}. \end{aligned} \tag{3.311}$$

In practice, for large values of $x$, we have the additional asymptotic behavior

$$\begin{aligned} J_p(x) &\sim \sqrt{\frac{2}{\pi x}} \cos(x - \theta_p), \\ Y_p(x) &\sim \sqrt{\frac{2}{\pi x}} \sin(x - \theta_p); \end{aligned} \tag{3.312}$$

whereas for small values of $x$, we have

$$
\begin{aligned}
J_p(x) &\sim \frac{1}{2^p\Gamma(p+1)}x^p,\\
J_{-p}(x) &\sim \frac{2^p}{\Gamma(-p+1)}x^{-p}, \quad \text{with } p \neq n,\\
Y_p(x) &\sim -\frac{2^p\Gamma(p)}{\pi}x^{-p}, \quad \text{with } p \neq 0,\\
Y_o(x) &\sim \frac{2}{\pi}\ln x,\\
I_p(x) &\sim \frac{1}{2^p\Gamma(p+1)}x^p,\\
I_{-p}(x) &\sim \frac{2^p}{\Gamma(-p+1)}x^{-p}, \quad \text{with } p \neq n,\\
K_p(x) &\sim 2^{p-1}\Gamma(p)x^{-p}, \quad \text{with } p \neq 0,\\
K_o(x) &\sim -\ln x.
\end{aligned}
\tag{3.313}
$$

Moreover, the following derivative formulas are also used frequently:

$$
\frac{d}{dx}\left[x^p y_p(\alpha x)\right] = \begin{cases} \alpha x^p y_{p-1}(\alpha x), & \text{with } y = J, Y, I, H,\\ -\alpha x^p y_{p-1}(\alpha x), & \text{with } y = K; \end{cases}
\tag{3.314}
$$

$$
\frac{d}{dx}\left[x^{-p} y_p(\alpha x)\right] = \begin{cases} -\alpha x^{-p} y_{p+1}(\alpha x), & \text{with } y = J, Y, K, H,\\ \alpha x^{-p} y_{p+1}(\alpha x), & \text{with } y = I; \end{cases}
\tag{3.315}
$$

$$
\frac{d}{dx} y_p(\alpha x) = \begin{cases} \alpha y_{p-1}(\alpha x) - \frac{p}{x} y_p(\alpha x), & \text{with } y = J, Y, I, H,\\ -\alpha y_{p-1}(\alpha x) - \frac{p}{x} y_p(\alpha x), & \text{with } y = K; \end{cases}
\tag{3.316}
$$

$$
\frac{d}{dx} y_p(\alpha x) = \begin{cases} -\alpha y_{p+1}(\alpha x) + \frac{p}{x} y_p(\alpha x), & \text{with } y = J, Y, K, H,\\ \alpha y_{p+1}(\alpha x) + \frac{p}{x} y_p(\alpha x), & \text{with } y = I. \end{cases}
\tag{3.317}
$$

The schematics for typical Bessel functions are depicted in Figure 2.12. The expression for $J_o(x)$, $J_1(x)$, $I_o(x)$, $K_o(x)$, $Y_o(x)$, and $K_o(x)$ can be expressed as

$$
J_o(x) = \sum_{k=0}^{\infty} \frac{(-1)^k (x/2)^{2k}}{(k!)^2},
\tag{3.318}
$$

$$
J_1(x) = \sum_{k=0}^{\infty} \frac{(-1)^k (x/2)^{2k+1}}{k!(k+1)!},
\tag{3.319}
$$

$$
I_o(x) = \sum_{k=0}^{\infty} \frac{(x/2)^{2k}}{(k!)^2},
\tag{3.320}
$$

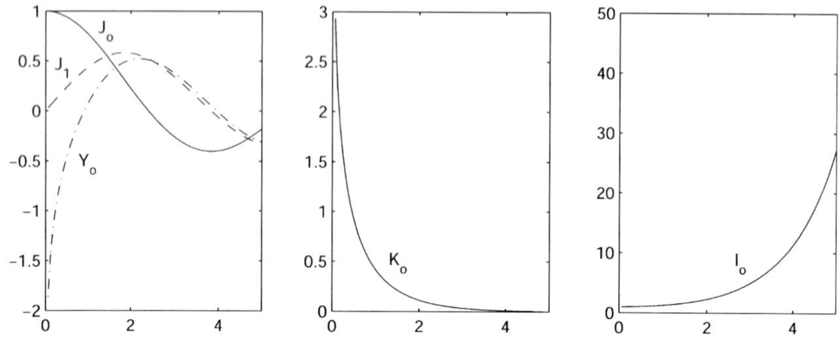


Figure 2.12. The schematic behaviors of Bessel functions.

$$Y_o(x) = \frac{2}{\pi}\big(\ln(x/2) + \gamma\big)J_o(x) + \frac{2}{\pi}\sum_{k=0}^{\infty}\frac{(-1)^{k+1}\phi(k)(x/2)^{2k}}{(k!)^2}, \tag{3.321}$$

$$K_o(x) = -\big(\ln(x/2) + \gamma\big)I_o(x) + \sum_{k=0}^{\infty}\frac{\phi(k)(x/2)^{2k}}{(k!)^2}. \tag{3.322}$$

Bessel functions $J_p(x)$ and $Y_p(x)$ are oscillatory in nature. The amplitude of oscillation around a zero value tends to decrease with $\sqrt{2/(\pi x)}$ and the distance between successive zeros of the function decreasing towards $\pi$. On the other hand, Bessel functions $I_p(x)$ and $K_p(x)$ are not oscillatory. As depicted in Figure 2.12, the former function essentially increases exponentially with $x$, whereas the second essentially decreases exponentially.

Let us return to the previous discussion of Sturm–Liouville problems. One of the most important applications of Bessel functions is the Fourier–Bessel series. Consider the following general form of Bessel's equation:

$$x^2\frac{d^2y}{dx^2} + x\frac{dy}{dx} + (\mu^2x^2 - p^2)y = 0. \tag{3.323}$$

Based on the previous discussion, we can write the general solution in the following form:

$$y(x) = \begin{cases} c_1J_p(\mu x) + c_2J_{-p}(\mu x), & (p \text{ not an integer}), \\ c_1J_p(\mu x) + c_2Y_p(\mu x), & (p \text{ an integer}). \end{cases} \tag{3.324}$$

Furthermore, by comparing Eqs. (3.323) and (3.249), we can easily identify $p(x) = x$, $q(x) = -p^2/x$, $\lambda = \mu^2$, and $r(x) = x$. This implies that we can expand the solution of the boundary value problem as a series of Bessel functions, just as we use Fourier series. Notice that although all Bessel functions satisfy the governing equations, the choice of Bessel functions depends on the boundary conditions. Suppose at $x = 0$, $y$ is finite and $x\,dy/dx$ is zero, and at $x = l$, $y(l)$

is zero. From the asymptotic behavior of $J_{-p}$ or $Y_p$ near $x = 0$, we must have $c_2 = 0$. To satisfy the boundary condition at $x = l$, we also have $J_p(\mu_n l) = 0$, where $\mu_n l$ is the $n$th dimensionless root of $J_p(x) = 0$. Of course, for a more general boundary condition at $x = l$, $y + \alpha x\, dy/dx = 0$, the $n$th dimensionless root of $J_p(\mu_n l) + \alpha \mu_n l J_p'(\mu_n l) = 0$. The rest of the orthogonal property is simply the same as Sturm–Liouville problems, replacing $\phi_n(x)$ with $J_p(\mu_n x)$.

*Chapter 3*

# Dynamical Systems

For the past two decades, paralleling the development of computational mechanics, tremendous understanding and progress have been achieved in the field of nonlinear dynamics and chaos. The works of Lorenz, Smale, Feigenbaum, Crutchfield, Hénon, Shaw, Mandelbrot, Ruelle, Takens, Holmes, Chow, Hale, Guckenheimer, May, and Yorke have inspired researchers in virtually all scientific communities. A good survey of and introduction to this subject is presented by Gleick [92]. One of the most interesting recent developments of this area, also a paragon of practical application of nonlinear dynamics, is the use of dynamical systems to model turbulence [25,109]. Although our first concern is the structural chaotic oscillations of FSI systems, we will also present some of the common tools associated with hydroinstabilities and dynamical system approaches to turbulence. These topics are also discussed in depth in Refs. [15,132,203,283].

Since the work on chaotic oscillations of FSI problems by Holmes and Marsden [110,111], significant progress has been made in areas such as flow-induced vibration (FIV) of pipes and cables, flutter of plates and shells, and various biomechanical subjects. A comprehensive review is available in Refs. [80,194].

A practical computational approach to dealing with FSI problems is to use two general-purpose computational softwares with staggered interactions. This means that the true Jacobian matrices are not formulated and local bifurcation phenomena are ignored. Such approaches are effective, particularly for static coupled problems, in solving various complex coupled problems by employing solid codes and fluid codes interactively. Nevertheless, because we seldom know *a priori* stability issues hidden behind millions of degrees of freedoms from the discretizations of both fluid and solid domains, it is very difficult for practicing engineers to find a convergent solution for FSI problems with trial and error methods, especially for nonlinear dynamical coupling between structures and flows. In actuality, simple convergence problems between fluids and solids, associated with stabilities, cannot be solved by refinement of meshes.

Because of the fact that FSI problems are in general nonlinear and dynamical, the understanding of stabilities, chaos or nonlinear dynamical behavior is crucial to the success of many FSI computations. In general, after the traditional semi-discretization procedures with finite element, finite difference, or finite volume methods, we obtain systems of ordinary differential equations. Nevertheless, although low-dimensional dynamical systems have been studied abundantly, often with analytical tools and the aid of geometrical and topological approaches, complicated dynamical aspects presented by FSI problems, remain unsolved in high-dimensional cases. The stability issues of fully coupled FSI problems still challenge researchers in this field. In essence, dominant dynamical modes have to be factored out from large governing matrices. We believe that in the near future, FSI problems will give rise to numerical approaches for stability and nonlinear dynamical analysis of large complex systems. This subject will be worthy of exploitation, and applied scientists and mathematicians will devote much more attention to this field.

## 3.1. Single- and multi-degree of freedom systems

In this section, we summarize some key aspects of dynamical systems such as limit sets and bifurcations. In addition, Lagrangian dynamics and typical nonlinear oscillations are also depicted.

### *3.1.1. Basic concepts; limit sets*

*Basic concepts*

A dynamical system is a $C^1$ mapping $\boldsymbol{\phi} : \mathcal{M} \times \mathcal{R} \to \mathcal{M}$, where $\mathcal{M}$ is a smooth manifold in $\mathcal{R}^n$ and $\boldsymbol{\phi}_t(\mathbf{x}) = \boldsymbol{\phi}(\mathbf{x}, t)$ satisfies

(1) $\boldsymbol{\phi}_o(\mathbf{x}) = \boldsymbol{\phi}(\mathbf{x}, 0) = \mathbf{x}$, i.e., $\boldsymbol{\phi}_o : \mathcal{M} \to \mathcal{M}$ is the identity mapping;
(2) $\boldsymbol{\phi}_{t_1+t_2} = \boldsymbol{\phi}_{t_1} \circ \boldsymbol{\phi}_{t_2}, \forall t_1, t_2 \in \mathcal{R}$, i.e., $\boldsymbol{\phi}(\boldsymbol{\phi}(\mathbf{x}, t_1), t_2) = \boldsymbol{\phi}(\mathbf{x}, t_1 + t_2)$.

The definition of a dynamical system implies that $\boldsymbol{\phi}(\mathbf{x}, t)$ is a diffeomorphism and has a $C^1$ inverse $\boldsymbol{\phi}^{-1}(\mathbf{x}, t) = \boldsymbol{\phi}(\mathbf{x}, -t)$. On the other hand, $\boldsymbol{\phi}(\mathbf{x}, t)$ is called a homeomorphism if it has a continuous inverse $\boldsymbol{\phi}(\mathbf{x}, -t)$. Furthermore, the mapping $\boldsymbol{\phi}$ is also called a flow in a continuous system with $t, t_o \in \mathcal{R}$, or a map in a discrete system with $t, t_o \in \mathcal{Z}$. Thus, a continuous system can be written as a differential equation, a so-called autonomous system,

$$\dot{\mathbf{x}} = \mathbf{f}(\mathbf{x}), \qquad \mathbf{x}(t_o) = \mathbf{x}_o, \tag{1.1}$$

or

$$\dot{\mathbf{x}} = \mathbf{f}(\mathbf{x}, t), \qquad \mathbf{x}(t_o) = \mathbf{x}_o, \tag{1.2}$$

a so-called nonautonomous system, while a discrete system can be represented by a difference equation:

$$\mathbf{x}_{k+1} = \mathbf{m}(\mathbf{x}_k), \quad k \in \mathcal{N}. \tag{1.3}$$

Of course, an $n$-dimensional nonautonomous system can be changed to an $(n+1)$-dimensional autonomous system by assigning an additional variable $x_{n+1} = t$. For example, the general $n$th order scalar differential equation $\frac{d^n x}{dt^n} = f(x, \frac{dx}{dt}, \ldots, \frac{d^{n-1}x}{dt^{n-1}}, t)$ with $x(t_o) = c_1$, $\frac{dx}{dt}(t_o) = c_2, \ldots, \frac{d^{n-1}x}{dt^{n-1}}(t_o) = c_n$ can be written as an $n$-dimensional nonautonomous system by introducing $x_k = \frac{d^{k-1}x}{dt^{k-1}}$, with $k = 1, \ldots, n$ and $\mathbf{x}_o = \langle c_1, \ldots, c_n \rangle$ or as an $(n+1)$-dimensional autonomous system by introducing $x_k = \frac{d^{k-1}x}{dt^{k-1}}$, with $k = 1, \ldots, n$, $x_{n+1} = t$, and $\mathbf{x}_o = \langle c_1, \ldots, c_n, t_o \rangle$. To emphasize the dependence of a solution $\mathbf{x}(t)$ through $\mathbf{x}_o$ at $t = t_o$, we often represent the solution of the autonomous system (1.1) as $\mathbf{x}(t) = \boldsymbol{\phi}_t(\mathbf{x}_o)$ and the solution of the nonautonomous system (1.2) as $\mathbf{x}(t) = \boldsymbol{\phi}_t(\mathbf{x}_o, t_o)$, i.e., $\boldsymbol{\phi}_{t_o}(\mathbf{x}_o) = \boldsymbol{\phi}_{t_o}(\mathbf{x}_o, t_o) = \mathbf{x}_o$. It is then obvious that for autonomous systems, since the vector field $\mathbf{f}(\mathbf{x})$ is independent of time $t$, we can always set $t_o = 0$. This is not the case for nonautonomous systems. In general, autonomous and nonautonomous systems bear different characteristics and consequently require different solution procedures.

The dynamical systems (1.1) and (1.2) are linear if the vector fields $\mathbf{f}(\mathbf{x})$ and $\mathbf{f}(\mathbf{x}, t)$ are linear with respect to $\mathbf{x}$. To discuss the existence, uniqueness, and continuity of the solutions of both continuous dynamical systems, we first introduce Lipschitz continuity and Grönwall inequality, named after Rudolf Otto Sigismund Lipschitz (1832–1903) and Thomas Hakon Grönwall (1877–1932). A vector function $\mathbf{f}(\mathbf{x})$ with $\mathbf{f}: \mathcal{F} \to \mathcal{R}^n$, where $\mathcal{F}$ is a normed vector space, is said to be Lipschitz continuous on $\mathcal{F}$ if there exists a constant $L_{\mathcal{F}}$ such that

$$\left\| \mathbf{f}(\mathbf{x}_1) - \mathbf{f}(\mathbf{x}_2) \right\| \leqslant L_{\mathcal{F}} \|\mathbf{x}_1 - \mathbf{x}_2\|, \quad \forall \mathbf{x}_1, \mathbf{x}_2 \in \mathcal{F}.$$

We call $L_{\mathcal{F}}$ the Lipschitz constant for $\mathbf{f}$ on $\mathcal{F}$. It is obvious that continuous differentiability ($C^1$) implies the Lipschitz continuity. However, although Lipschitz continuity implies continuity, it does not imply uniform continuity in both variables $\mathbf{x}$ and $t$. This explains the possibility of chaotic motion. Consider continuous nonnegative functions $u, v: [t_o, t_1] \to \mathcal{R}$. If

$$u(t) \leqslant C_1 + C_2 \int_{t_o}^{t} v(s)u(s)\, ds, \quad \forall t \in [t_o, t_1],$$

with $C_1, C_2 \geqslant 0$, we have the following Grönwall inequality which states that

$$u(t) \leqslant C_1 \exp\left( C_2 \int_{t_o}^{t} v(s)\, ds \right).$$

THEOREM 3.1.1 *(Existence and uniqueness). Suppose* $\dot{\mathbf{x}} = \mathbf{f}(\mathbf{x}, t)$ *with* $\mathbf{x}(t) = \mathbf{x}_o$ *and a* $C^1$ *mapping* $\mathbf{f}: \mathcal{A} \times \mathcal{B} \to \mathcal{R}^n$, *where* $\mathcal{A} = \{\mathbf{x} \mid \|\mathbf{x} - \mathbf{x}_o\| \leqslant b\}$, $\mathcal{B} =$

*$[t_o - a, t_o + a]$, and $a$ and $b$ are positive constants, then there exists a positive number $\epsilon = \min(a, \frac{b}{c})$, with $c = \sup_{\mathcal{A}\times\mathcal{B}} \|\mathbf{f}(\mathbf{x}, t)\|$, such that a solution $\mathbf{x}(t)$ exists and is unique for $|t - t_o| \leqslant \epsilon$.*

THEOREM 3.1.2 *(Continuity of solutions). Let $\mathbf{x}_1(t)$ and $\mathbf{x}_2(t)$ be solutions to $\dot{\mathbf{x}} = \mathbf{f}(\mathbf{x})$ with $\mathbf{f}\colon \mathcal{F} \to \mathcal{R}^n$, and $L_{\mathcal{F}}$ is the Lipschitz constant on $\mathcal{F} \subseteq \mathcal{R}^n$. Then, we have*

$$\left\|\mathbf{x}_1(t) - \mathbf{x}_2(t)\right\| \leqslant \left\|\mathbf{x}_1(t_o) - \mathbf{x}_2(t_o)\right\| e^{L_{\mathcal{F}}(t-t_o)}, \quad \forall t \in [t_o, t_1].$$

We define the trajectory through $\mathbf{x}_o$ of the dynamical system (1.1) or (1.2) as the set of solution points $\mathbf{x}(t)$. In other words, the trajectory is a subset of $\mathbf{x}$-$t$ space, the so-called state space or solution space, represented by $(x_1, \ldots, x_n, t)$, $\forall t \in \mathcal{R}$, where every point of the trajectory satisfies (1.1) or (1.2). From the definition of a dynamical system, for any time $t$, we have $\boldsymbol{\phi}_t(\mathbf{x}_1) = \boldsymbol{\phi}_t(\mathbf{x}_2)$ if and only if $\mathbf{x}_1 = \mathbf{x}_2$, while for any time $t$ and $t_o$, $\boldsymbol{\phi}_t(\mathbf{x}_1, t_o) = \boldsymbol{\phi}_t(\mathbf{x}_2, t_o)$ if and only if $\mathbf{x}_1 = \mathbf{x}_2$. A trajectory of an autonomous system is uniquely specified by its initial condition $\mathbf{x}_o$ and distinct trajectories cannot intersect, whereas, although the solution of a nonautonomous system is uniquely determined by its initial condition $t_o$ and $\mathbf{x}_o$, there could exist a time $t_1$ such that $\boldsymbol{\phi}_{t_1}(\mathbf{x}_o, t_o) = \boldsymbol{\phi}_{t_1}(\hat{\mathbf{x}}_o, \hat{t}_o)$ along with a time $t_2$ such that $\boldsymbol{\phi}_{t_2}(\mathbf{x}_o, t_o) \neq \boldsymbol{\phi}_{t_2}(\hat{\mathbf{x}}_o, \hat{t}_o)$, which indicates that distinct solutions of nonautonomous systems can intersect.

Furthermore, the projection of the trajectory through which goes $\mathbf{x}_o$ on $x_i$-$t$ plane is tangent to the direction $f_i(\mathbf{x})$ or $f_i(\mathbf{x}, t)$ at every point through which the trajectory passes on that plane. Thus, the slope with respect to $t$ at every point of the trajectory through $\mathbf{x}_o$ is $\mathbf{f}(\mathbf{x})$ or $\mathbf{f}(\mathbf{x}, t)$. In addition, phase space of a differential equation is defined as an $n$-dimensional space $\mathcal{R}^n$ parameterized by time. Hence, the orbit through $\mathbf{x}_o$ is defined as the projection of the trajectory through $\mathbf{x}_o$ on the phase space. For an autonomous system (1.1), since $\mathbf{f}(\mathbf{x})$ is independent of $t$, on any line parallel to the $t$ axis, the direction of the trajectory is the same. As a consequence, we need only to consider its orbit. Unlike linear dynamical systems, nonlinear dynamical systems do not generally have closed-form solutions, nor does the principle of superposition hold. Thus, qualitative studies of the solution behavior as $t \to +\infty$, or $t \to -\infty$, become important.

*Limit sets*

In phase space, a stationary point $\mathbf{x}_o$ (or fixed point, or equilibrium) of an autonomous system, corresponding to the flow $\boldsymbol{\phi}_t(\mathbf{x}_o) = \mathbf{x}_o$, $\forall t \in \mathcal{R}$, is represented by a single point. In general, at a stationary point, the vector field $\mathbf{f}$ vanishes, i.e., $\mathbf{f}(\mathbf{x}_o) = \mathbf{0}$. Obviously, there is no corresponding stationary point for nonautonomous systems.

For autonomous systems, a periodic solution is a trajectory such that $\boldsymbol{\phi}_{t+T}(\mathbf{x}_o) = \boldsymbol{\phi}_t(\mathbf{x}_o)$, $\forall t \in \mathcal{R}$, and $\boldsymbol{\phi}_{t+s}(\mathbf{x}_o) \neq \boldsymbol{\phi}_t(\mathbf{x}_o)$, $\forall s \in (0, T)$. We denote $T > 0$ as the minimum period and in phase space the trajectory is a closed curve, or periodic orbit. Notice that for autonomous systems, $\mathbf{x}_o$ is not uniquely determined by a periodic solution and changing $\mathbf{x}_o$ corresponds to changing $t_o$. In fact, any point within the periodic solution can be $\mathbf{x}_o$. Moreover, if at an isolated periodic orbit, there exists a neighborhood containing no other periodic solution, we called it a limit cycle. Of course, the spectrum of each component of a periodic solution contains spikes at the fundamental frequency $1/T$ and its harmonics (integer multiplies of the fundamental frequency) $k/T$, with $k \in \mathcal{Z}$. A reliable way of identifying a limit cycle from the spectrum is to compare both the first nonzero frequency spike and the spacing of the remaining frequency spikes.

Consider an nonautonomous system as a time periodic nonautonomous system with a period $T_o$, i.e., $\mathbf{f}(\mathbf{x}, t) = \mathbf{f}(\mathbf{x}, t + T_o)$. With the transformation $\theta = 2\pi t/T_o \bmod 2\pi$, we can represent the solution space with the cylindrical space $\mathcal{R}^n \times \mathcal{T}$ with $\mathcal{T} = [0, 2\pi)$ rather than the Cartesian space $\mathcal{R}^{n+1}$, and the solution can be expressed as

$$\begin{bmatrix} \mathbf{x}(t) \\ \theta(t) \end{bmatrix} = \begin{bmatrix} \boldsymbol{\phi}_t(t_o, \mathbf{x}_o) \\ 2\pi t/T_o \bmod 2\pi \end{bmatrix}. \tag{1.4}$$

The vector field is a periodic function for a time periodic nonautonomous system. The minimum period of a periodic solution of a nonautonomous equation must be some integer multiple of $T_o$, such as $T = mT_o$, with $m \in \overline{\mathcal{N}}$, and

$$\boldsymbol{\phi}_t(\mathbf{x}_o, t_o) = \boldsymbol{\phi}_{t+T}(\mathbf{x}_o, t_o). \tag{1.5}$$

The solution $\boldsymbol{\phi}_t(\mathbf{x}_o, t_o)$ is often called a period-$m$ solution, or an $m$th-order subharmonic. Notice that for a periodic-1 solution, i.e., a fundamental solution, $\mathbf{x}_o$ is uniquely defined with a fixed $t_o$. For a periodic-$m$ solution, given $t_o$, according to Eq. (1.4), there exist $m - 1$ other values of $\mathbf{x}_o$ corresponding to the same periodic solution $\boldsymbol{\phi}_t(t_o, \mathbf{x}_o)$, i.e., $\mathbf{x}_o^i = \boldsymbol{\phi}_{t_o+iT_o}(\mathbf{x}_o, t_o)$, with $i = 1, \ldots, m - 1$. We must point out that such subharmonics do not exist for autonomous systems and are different from period-doubling bifurcation due to the change of parameters inherent in dynamical systems instead of the initial conditions.

A quasi-periodic solution can be expressed as a sum of finite number of periodic functions,

$$\mathbf{x}(t) = \sum_{i=1}^{k} \mathbf{x}_{T_i}(t), \tag{1.6}$$

where $T_i$ is the minimum period of the periodic function $\mathbf{x}_{T_i}(t)$ and $f_i = 1/T_i$, with $i = 1, \ldots, k$, forms a series of rationally independent frequencies, i.e., incommensurable frequencies.

More precisely, consider that there exist a set of base frequencies $\hat{f}_i$, with $i = 1, \dots, m$ and $m \leqslant k$, i.e., a set of rationally independent frequencies: that is, there does not exist a set of integers $k_i$, not all zero, such that $k_i \hat{f}_i = 0$. If every $f_i$ can be expressed as $f_i = k_{ij} \hat{f}_j$ with $k_{ij} \in \mathcal{Z}$, a quasi-periodic solution in Eq. (1.6) is $m$-periodic. In phase space, an $m$-periodic quasi-periodic solution corresponds to an $m$-torus.

Obviously, stationary points, periodic, or quasi-periodic orbits are part of an invariant set $\mathcal{M}$, such that $\boldsymbol{\phi}(\mathbf{x}, t) \in \mathcal{M}$, $\forall \mathbf{x} \in \mathcal{M}$ and $t \in \mathcal{R}$. In addition, set $\mathcal{M}$ is forward invariant if $\boldsymbol{\phi}(\mathbf{x}, t) \in \mathcal{M}$, $\forall \mathbf{x} \in \mathcal{M}$ and $t > 0$, and is backward invariant if $\boldsymbol{\phi}(\mathbf{x}, t) \in \mathcal{M}$, $\forall \mathbf{x} \in \mathcal{M}$ and $t < 0$. If we define the trajectory through $\mathbf{x}$ as the set $\gamma(\mathbf{x}) = \bigcup_{t \in \mathcal{R}} \boldsymbol{\phi}(\mathbf{x}, t)$ and the positive and negative semi-trajectories as $\gamma^+(\mathbf{x}) = \bigcup_{t \geqslant 0} \boldsymbol{\phi}(\mathbf{x}, t)$ and $\gamma^-(\mathbf{x}) = \bigcup_{t \leqslant 0} \boldsymbol{\phi}(\mathbf{x}, t)$, respectively, the following lemma can be easily verified.

LEMMA 3.1.1. *$\mathcal{M}$ is invariant iff $\gamma(\mathbf{x}) \subset \mathcal{M}$, $\forall \mathbf{x} \in \mathcal{M}$.*

*$\mathcal{M}$ is invariant iff $\mathcal{R}^n \setminus \mathcal{M}$ is invariant.*

*For a countable collection of invariant sets $\mathcal{M}_i$, $\bigcup_i \mathcal{M}_i$ and $\bigcap_i \mathcal{M}_i$ are also invariant.*

Furthermore, the $\omega$-limit set $\mathcal{O}(\mathbf{x})$ is the set of points which the trajectories through $\mathbf{x}$ tend to as $t \to +\infty$, i.e., the limit points of $\gamma^+(\mathbf{x})$ noted as $\mathcal{O}(\mathbf{x}) = \bigcap_{\mathbf{y} \in \gamma(\mathbf{x})} \mathrm{cl}(\gamma^+(\mathbf{y}))$. The $\alpha$-limit set $\mathcal{A}(\mathbf{x})$ is the set of points which the trajectories through $\mathbf{x}$ tend to as $t \to -\infty$, i.e., the limit points of $\gamma^-(\mathbf{x})$ noted as $\mathcal{A}(\mathbf{x}) = \bigcap_{\mathbf{y} \in \gamma(\mathbf{x})} \mathrm{cl}(\gamma^-(\mathbf{y}))$. It is easy to prove that $\mathcal{O}(\mathbf{x})$ and $\mathcal{A}(\mathbf{x})$ are invariant, and that nonempty compact (closed and bounded), if $\gamma^+(\mathbf{x})$ and $\gamma^-(\mathbf{x})$ are bounded [130].

At this point, we provide a commonly accepted definition of chaos as a bounded steady-state behavior: $t \to +\infty$, that is not a stationary point, not periodic, and not quasi-periodic. In phase space, the geometrical object to which chaotic orbits are attracted is called a strange attractor. In general, we refer to stationary points, periodic orbits, quasi-periodic orbits, and strange attractors as limit sets. The most important aspect of dynamical systems is the stability of such limit sets.

A limit set $\mathcal{L}$ is stable or Lyapunov stable, named after Aleksandr Mikhailovich Lyapunov (1857–1918), if for every open neighborhood $\mathcal{S}_1(\mathcal{L})$, there exists an open neighborhood $\mathcal{S}_2(\mathcal{L})$ such that $\forall \mathbf{x} \in \mathcal{S}_2(\mathcal{L})$ and $t > 0$, $\boldsymbol{\phi}(\mathbf{x}, t) \in \mathcal{S}_1(\mathcal{L})$. A limit set is asymptotically stable if there exists an open neighborhood $\mathcal{S}(\mathcal{L})$ such that the $\omega$-limit set of every point in $\mathcal{S}(\mathcal{L})$ is $\mathcal{L}$. A limit set is unstable if there exists an open neighborhood $\mathcal{S}(\mathcal{L})$ such that the $\alpha$-limit set of every point in $\mathcal{S}(\mathcal{L})$ is $\mathcal{L}$. A limit set is non-stable if every neighborhood $\mathcal{S}(\mathcal{L})$ contains at least one point not in $\mathcal{L}$ whose $\omega$-limit set is $\mathcal{L}$ and at least one point not in $\mathcal{L}$ whose $\alpha$-limit set is $\mathcal{L}$. Obviously, an asymptotically stable limit set is also stable, while an unstable limit set is asymptotically stable in reverse time. Moreover, non-stable limit sets remain non-stable in reverse time.

Consider $\mathbf{x}_o$ as a stationary point of a differential equation $\dot{\mathbf{x}} = \mathbf{f}(\mathbf{x})$. $\mathbf{x}_o$ is asymptotically stable iff all eigenvalues of $\mathbf{Df}(\mathbf{x}_o)$ have negative real parts. In addition, the stationary point $\mathbf{x}_o$ is called hyperbolic if $\mathbf{Df}(\mathbf{x}_o)$ has no eigenvalue with real part zero. Furthermore, suppose every eigenvalue of $\mathbf{Df}(\mathbf{x}_o)$ has real part less than $-c$, with $c > 0$. There is then a neighborhood $\mathcal{S}(\mathbf{x}_o)$, or basin of attraction, such that $\|\boldsymbol{\phi}(\mathbf{x}, t) - \mathbf{x}_o\| \leqslant e^{-ct}\|\mathbf{x} - \mathbf{x}_o\|$, $\forall \mathbf{x} \in \mathcal{S}(\mathbf{x}_o)$ and $t \geqslant 0$. Further, $\mathbf{x}_o$ is globally asymptotically stable if $\mathcal{S}(\mathbf{x}_o) = \mathcal{R}^n$.

Let $\mathbf{x}_o \in \mathcal{M} \subseteq \mathcal{R}^n$ be a stationary point and $V : \mathcal{S}(\mathbf{x}_o) \to \mathcal{R}$ be a continuous function, defined on a neighborhood $\mathcal{S}(\mathbf{x}_o) \subset \mathcal{M}$, differentiable on $\mathcal{S}(\mathbf{x}_o) \setminus \{\mathbf{x}_o\}$, such that $V(\mathbf{x}_o) = 0$, and $V(\mathbf{x}) > 0$ and $\dot{V}(\mathbf{x}) \leqslant 0$, $\forall \mathbf{x} \in \mathcal{S}(\mathbf{x}_o) \setminus \{\mathbf{x}_o\}$. $\mathbf{x}_o$ is then stable and the function $V(\mathbf{x})$ is called the Lyapunov function. It is then obvious that the fixed point $\mathbf{x}_o$ is asymptotically stable, if $\dot{V}(\mathbf{x}) < 0$, $\forall \mathbf{x} \in \mathcal{S}(\mathbf{x}_o) \setminus \{\mathbf{x}_o\}$.

THEOREM 3.1.3 *(La Salle's invariance). Suppose $\mathbf{x}_o$ is a stationary point of a dynamical system $\dot{\mathbf{x}} = \mathbf{f}(\mathbf{x})$ and $V(\mathbf{x})$ is a Lyapunov function on some neighborhood $\mathcal{S}(\mathbf{x}_o)$. If there exist $\mathcal{O}(\mathbf{x}) = \bigcap_{\mathbf{y} \in \gamma(\mathbf{x})} \mathrm{cl}(\gamma^+(\mathbf{y})) \subseteq \mathcal{S}(\mathbf{x}_o)$ and $\mathcal{M}$ is the largest invariant subset of $\{\mathbf{x} \mid \mathbf{x} \in \mathcal{S}(\mathbf{x}_o)$ and $\dot{V}(\mathbf{x}) = 0\}$, then $\boldsymbol{\phi}(\mathbf{x}, t) \to \mathcal{M}$ as $t \to +\infty$.*

We should point out that although Lyapunov function is difficult to construct in general, for dynamical systems derived from mechanical and electrical systems, energy is often a Lyapunov function.

To study the dynamical behavior near the stationary point $\mathbf{x}_o$, we first consider the linearized system

$$\dot{\mathbf{x}} = \mathbf{A}\mathbf{x}, \tag{1.7}$$

with $\mathbf{A} = \mathbf{Df}(\mathbf{x}_o)$ and $\mathbf{f}(\mathbf{x}_o) = \mathbf{0}$.

Of course, $\mathbf{x}$ in Eq. (1.7) stands for the coordinate deviation from the stationary point, i.e., $\mathbf{x} - \mathbf{x}_o$, and the solution of Eq. (1.7) can be simply written as $\mathbf{x}(t) = e^{\mathbf{A}t}\hat{\mathbf{x}}_o$, where $\hat{\mathbf{x}}_o$ is given as the initial position relative to the stationary point $\mathbf{x}_o$ and the matrix exponent $e^{\mathbf{A}t}$ as discussed in Section 3.2.

We can also introduce the concept of a funnel, where solutions approach each other in the phase diagram as if they form a tunnel. The same behavior as $t \to -\infty$ is called an antifunnel. Consider a function $\dot{\mathbf{x}} = \mathbf{f}(t, \mathbf{x})$, $\forall t \in \mathcal{D} \subset \mathcal{R}$. A fence is a $C^1$ function $\hat{\boldsymbol{\phi}}(t)$, which channels the solution $\boldsymbol{\phi}(t)$ in the direction of the slope field. More specifically, a fence is called a lower fence, denoted as $\hat{\boldsymbol{\phi}}(t)^l$, if $\frac{\partial}{\partial t}\hat{\boldsymbol{\phi}}(t) \leqslant \mathbf{f}(t, \hat{\boldsymbol{\phi}}(t))$, $\forall t \in \mathcal{D}$, or an upper fence, denoted as $\hat{\boldsymbol{\phi}}(t)^u$, if $\frac{\partial}{\partial t}\hat{\boldsymbol{\phi}}(t) \geqslant \mathbf{f}(t, \hat{\boldsymbol{\phi}}(t))$, $\forall t \in \mathcal{D}$. Therefore, the precise definition of funnel is the set of points $(t, \mathbf{x})$, $\forall t$, $\hat{\boldsymbol{\phi}}(t)^l \leqslant \mathbf{x} \leqslant \hat{\boldsymbol{\phi}}(t)^u$.

Consider the dynamical system $\dot{\mathbf{x}} = \mathbf{f}(\mathbf{x}) = -\boldsymbol{\nabla} F(\mathbf{x})$, the so-called gradient system, where $F$ is the potential function of $\mathbf{f}(\mathbf{x})$. It is evident that the stationary point $\mathbf{x}_o$, with $\mathbf{f}(\mathbf{x}_o) = \mathbf{0}$ is the extreme point of the potential function $F$. Since $\frac{d}{dt}F(\mathbf{x}(t)) = \nabla F(\mathbf{x}(t))\dot{\mathbf{x}}(t) = -|\mathbf{f}(\mathbf{x})|^2 \leqslant 0$, $F$ is a decreasing function with respect to $t$.

Most of the notation used in this book is standard. For example, the partial derivative of a function $\phi$ with respect to time $t$ is denoted by $\phi_{,t}$, $\phi'$, or $\partial\phi/\partial t$. In ordinary differential equations, the differentiation with respect to time is denoted as $\dot{\phi}$.

THEOREM 3.1.4 *(Peixoto). Let* $\mathbf{f}: \mathcal{R}^2 \to \mathcal{R}^2$ *be a twice differentiable vector field and let* $\mathcal{D}$ *be a compact, connected subset of* $\mathcal{R}^2$ *bounded by the simple closed curve* $\partial\mathcal{D}$ *with outward normal* $\mathbf{n}$*. Suppose that* $\mathbf{f}\cdot\mathbf{n} \neq 0$ *on* $\partial\mathcal{D}$*.* $\mathbf{f}$ *is structurally stable on* $\mathcal{D}$ *iff*

(1) *all stationary points are hyperbolic*;
(2) *all periodic orbits are hyperbolic*;
(3) *if* $\mathbf{x}$ *and* $\mathbf{y}$ *are hyperbolic saddles (possibly,* $\mathbf{x} = \mathbf{y}$*), then stable manifold* $W^s(\mathbf{x}) \cap$ *unstable manifold* $W^u(\mathbf{y}) = \emptyset$.

This theorem was named after Maurício Matos Peixoto (1921–). Poincaré Index is an invariant of closed curves (not necessarily orbits) in phase space which can be used to rule out the possibility of periodic orbits in certain systems.

We start with the study of second-order autonomous systems in the form of Eq. (1.7). Such a simple dynamical system provides basic ideas, specially the geometrical and topological representation of orbits in phase space. The locus in phase space ($x_1$-$x_2$ plane in this case) of the solution $\mathbf{x}(t) = \langle x_1(t), x_2(t)\rangle$ is a curve passing through the starting (or initial) point $\hat{\mathbf{x}}_o = \langle \hat{x}_{01}, \hat{x}_{02}\rangle$. The tangent vector of such an orbit is $\dot{\mathbf{x}}(t) = \langle \dot{x}_1(t), \dot{x}_2(t)\rangle = \langle f_1(\mathbf{x}), f_2(\mathbf{x})\rangle$. Moreover, the contour of the vector field with a constant slope of the curve, i.e., $f_2/f_1 = c$, is called an isocline. Thus, the tangent of the orbit crossing the isocline is given by $c$.

According to Section 2.1, for a $2\times 2$ real matrix $\mathbf{A}$, there exists a real nonsingular matrix $\mathbf{S}$ (a transformation matrix), such that $\mathbf{A}$ can be converted to a canonical real Jordan form $\mathbf{B}$, with $\mathbf{B} = \mathbf{SAS}^{-1}$ expressed in the following three forms:

$$\begin{bmatrix} \lambda_1 & 0 \\ 0 & \lambda_2 \end{bmatrix}, \quad \begin{bmatrix} \lambda & 1 \\ 0 & \lambda \end{bmatrix}, \quad \text{and} \quad \begin{bmatrix} \xi & -\eta \\ \eta & \xi \end{bmatrix},$$

where the eigenvalues $\lambda_1$ and $\lambda_2$ are real (distinct or identical), $\lambda$ is the eigenvalue with a multiplicity 2, and the complex eigenvalue pair $\zeta = \xi + i\eta$ and $\bar{\zeta} = \xi - i\eta$ with $\eta \neq 0$.

Thus, the solution for Eq. (1.7) with the initial condition $\mathbf{x}(0) = \hat{\mathbf{x}}_o$ can be represented as $\mathbf{x}(t) = \mathbf{S}^{-1}e^{\mathbf{B}t}\mathbf{S}\hat{\mathbf{x}}_o$.

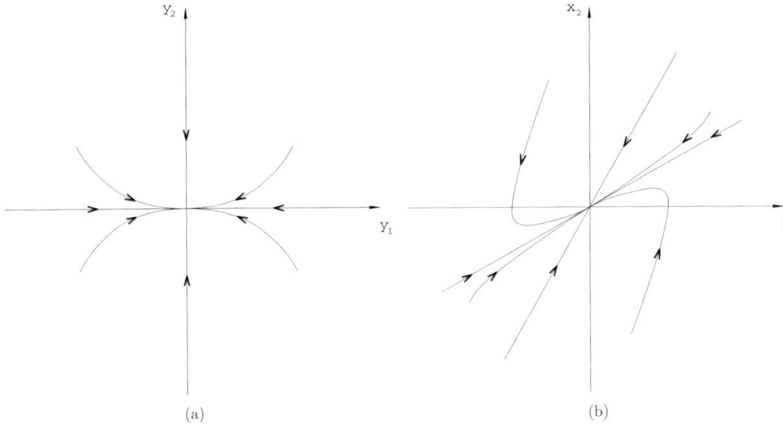


Figure 3.1. A stable node.

Consider $\mathbf{B} = \left[\begin{smallmatrix}\lambda_1 & 0\\ 0 & \lambda_2\end{smallmatrix}\right]$. We obviously have two distinct real eigenvectors $\boldsymbol{\phi}_1$ and $\boldsymbol{\phi}_2$ and $\mathbf{S}^{-1} = [\boldsymbol{\phi}_1 \quad \boldsymbol{\phi}_2]$. Hence, Eq. (1.7) can be transformed into the following two decoupled equations:

$$\begin{aligned}\dot{y}_1 &= \lambda_1 y_1,\\ \dot{y}_2 &= \lambda_2 y_2,\end{aligned} \tag{1.8}$$

and the solution can be written as $y_1(t) = \hat{y}_{01}e^{\lambda_1 t}$ and $y_2(t) = \hat{y}_{02}e^{\lambda_2 t}$, with $\hat{\mathbf{y}}_o = \mathbf{S}\hat{\mathbf{x}}_o$.

Without loss of generality, suppose $\lambda_2 < \lambda_1 < 0$. All orbits approach $\mathbf{y} = \mathbf{0}$ (or $\mathbf{x} = \mathbf{0}$), which means that all points in the vicinity of the stationary point $\mathbf{x}_o$ of the manifold of Eqs. (1.1) or (1.7) approach $\mathbf{x}_o$, and the stationary point $\mathbf{x}_o$ is called a stable node. As shown in Figure 3.1, the slope of the orbit, $c = \lambda_2 y_2/\lambda_1 y_1$, approaches 0 as $y_2 \to 0$ (or $y_1 \to \infty$) and approaches $\infty$ as $y_2 \to \infty$ (or $y_1 \to 0$). Conversely, for $\lambda_2 > \lambda_1 > 0$, $\mathbf{x}_o$ is called an unstable node as shown in Figure 3.2. In addition, the node corresponding to $\lambda_1 = \lambda_2 < 0$ is called a stable star as shown in Figure 3.3, and the node corresponding to $\lambda_1 = \lambda_2 > 0$ is called an unstable star as shown in Figure 3.4. If the eigenvalues have opposite signs, i.e., $\lambda_1\lambda_2 < 0$, $\mathbf{x}_o$ is called a saddle as shown in Figure 3.5.

When $\lambda_1\lambda_2 = 0$, the phase portrait is in some sense degenerate, i.e., $\ker(\mathbf{A}) \neq \{\mathbf{0}\}$. Supposing $\lambda_1 = 0$ and $\lambda_2 < 0$, we have $y_1(t) = \hat{y}_{01}$ and $y_2(t) = \hat{y}_{02}e^{\lambda_2 t}$. Clearly, the $y_2$-coordinate of all points converges from $\hat{y}_{02}$ to 0. Conversely, for the case $\lambda_1 = 0$ and $\lambda_2 > 0$, the $y_2$-coordinate of all points diverges from $\hat{y}_{02}$. The orbits of such stable and unstable degenerate nodes are shown in Figures 3.6 and 3.7, respectively.

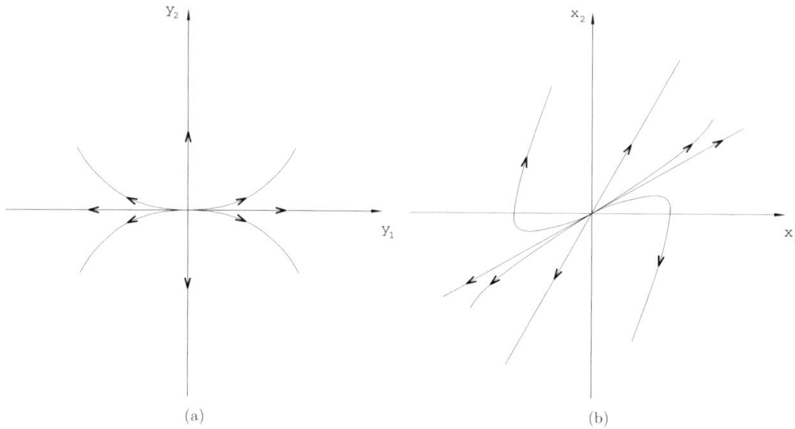


Figure 3.2. An unstable node.

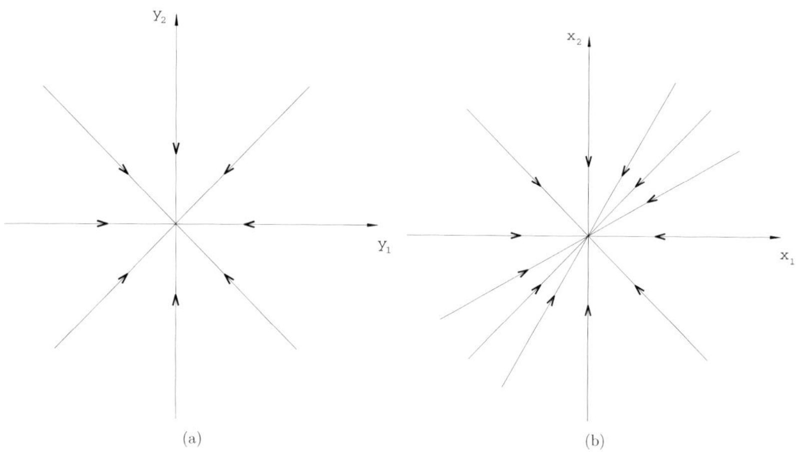


Figure 3.3. A stable star.

For the trivial case $\lambda_1 = \lambda_2 = 0$, it is clear that all points in phase plane are equilibrium points for Eq. (1.7).

Consider $\mathbf{B} = \left[\begin{smallmatrix} \lambda & 1 \\ 0 & \lambda \end{smallmatrix}\right]$. The transformed system can be expressed as

$$\begin{aligned} \dot{y}_1 &= \lambda y_1 + y_2, \\ \dot{y}_2 &= \lambda y_2. \end{aligned} \tag{1.9}$$

Thus, the solution can be written as $y_1(t) = (\hat{y}_{01} + \hat{y}_{02}t)e^{\lambda t}$ and $y_2(t) = \hat{y}_{02}e^{\lambda t}$. The slope of orbit, $c = \lambda y_2/(\lambda y_1 + y_2)$, approaches $\lambda$ as $y_1 \to 0$ (or $y_2 \to \infty$).

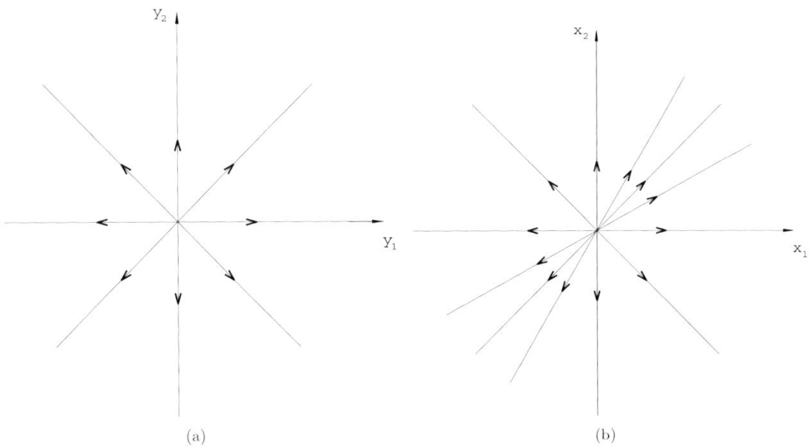


Figure 3.4. An unstable star.

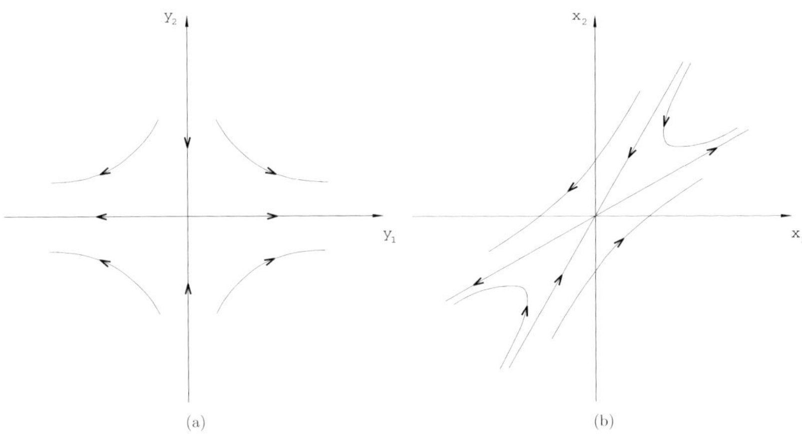


Figure 3.5. A saddle.

As shown in Figure 3.8, if $\lambda < 0$, all points approach $\mathbf{y} = \mathbf{0}$ (or $\mathbf{x} = \mathbf{0}$), and the stationary point is called a stable improper node. Analogously, as shown in Figure 3.9, if $\lambda > 0$, the stationary point is called an unstable improper node.

The corresponding degenerate case is $\lambda = 0$, where the solution can be written as $y_1(t) = \hat{y}_{01} + \hat{y}_{02}t$ and $y_2(t) = \hat{y}_{02}$, as shown in Figure 3.10. Obviously, the orbit moves parallel to $y_1$-axis and the direction of the change of the $y_2$-coordinate depends on the sign of the $y_2$-coordinate of the starting point. In particular, the whole $y_1$-axis is the equilibrium subspace.

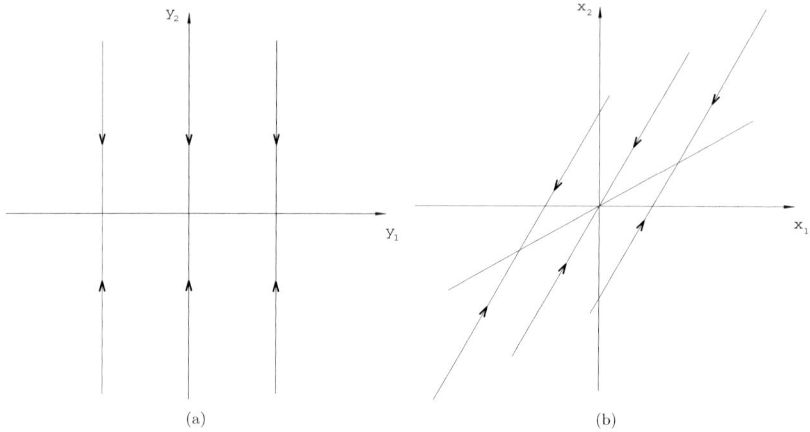


Figure 3.6. A stable degenerate node.

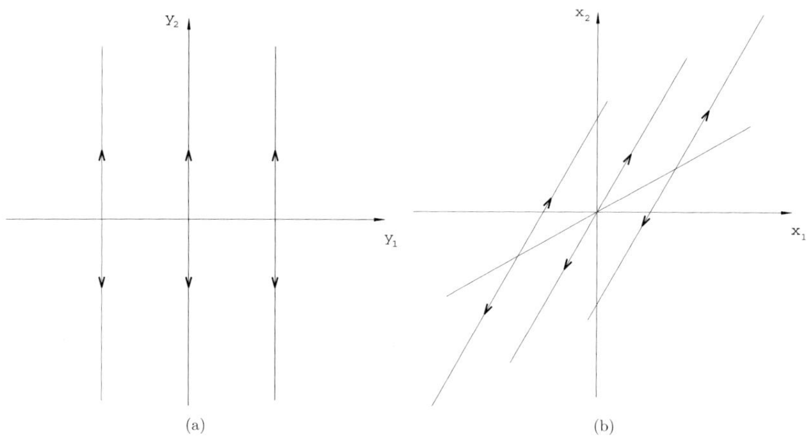


Figure 3.7. An unstable degenerate node.

Finally, consider $\mathbf{B} = \left[\begin{smallmatrix} \xi & -\eta \\ \eta & \xi \end{smallmatrix}\right]$. The transformed equation can be written as

$$\begin{aligned} \dot{y}_1 &= \xi y_1 - \eta y_2, \\ \dot{y}_2 &= \eta y_1 + \xi y_2, \end{aligned} \tag{1.10}$$

or in polar coordinate system $y_1 = r\cos\theta$ and $y_2 = r\sin\theta$,

$$\begin{aligned} \dot{r} &= \xi r, \\ \dot{\theta} &= \eta. \end{aligned} \tag{1.11}$$

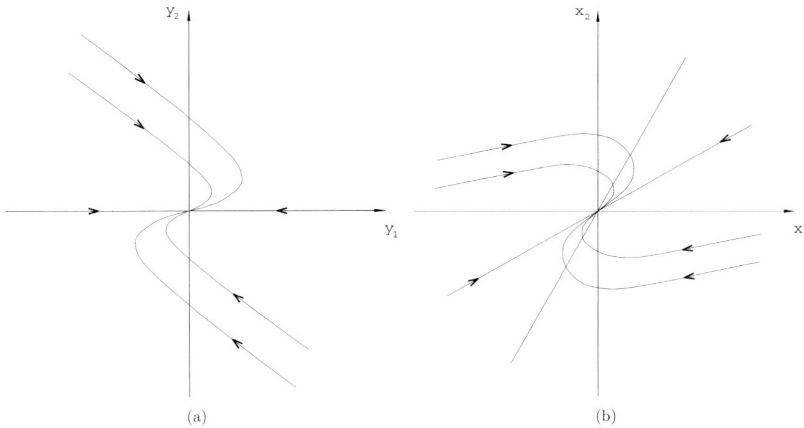


Figure 3.8. A stable improper node.

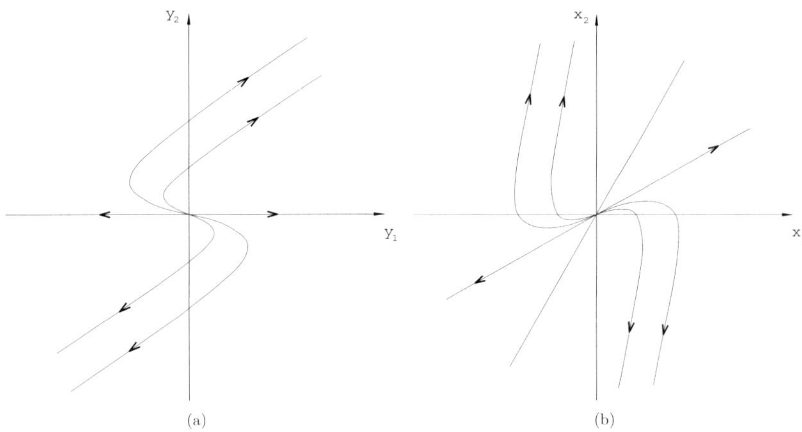


Figure 3.9. An unstable improper node.

The solution of Eq. (1.11) can be easily derived as $r(t) = r_o e^{\xi t}$ and $\theta(t) = \theta_o + \eta t$ with $r_o = \sqrt{\hat{y}_{01}^2 + \hat{y}_{02}^2}$ and $\theta_o = \tan^{-1}(\hat{y}_{02}/\hat{y}_{01})$. If $\xi < 0$, the orbits in phase plane are converging logarithmic spirals towards $\mathbf{y} = \mathbf{0}$ (or $\mathbf{x} = \mathbf{0}$) as shown in Figure 3.11, in which case the stationary point is called a stable focus. Notice that the rotational direction of the spiral illustrated in Figure 3.11 indicates that $\eta < 0$. Conversely, if $\xi > 0$, the orbits in phase plane are diverging logarithmic spirals away from $\mathbf{y} = \mathbf{0}$ (or $\mathbf{x} = \mathbf{0}$) as shown in Figure 3.12 and the stationary point is called an unstable focus. Here, the rotational direction of

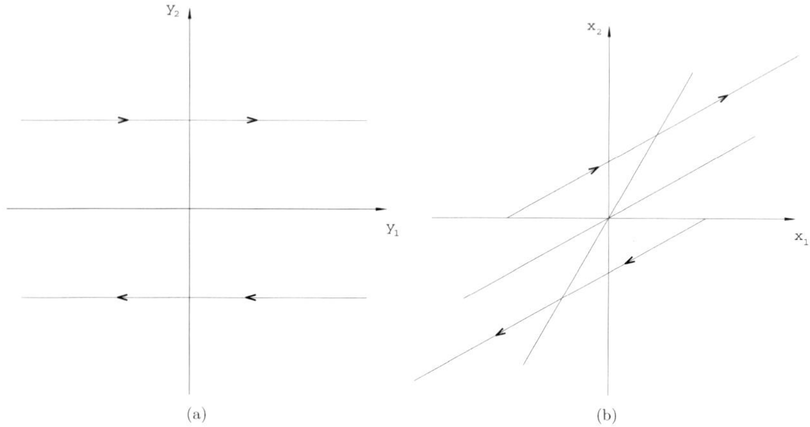


Figure 3.10. A degenerate improper node.

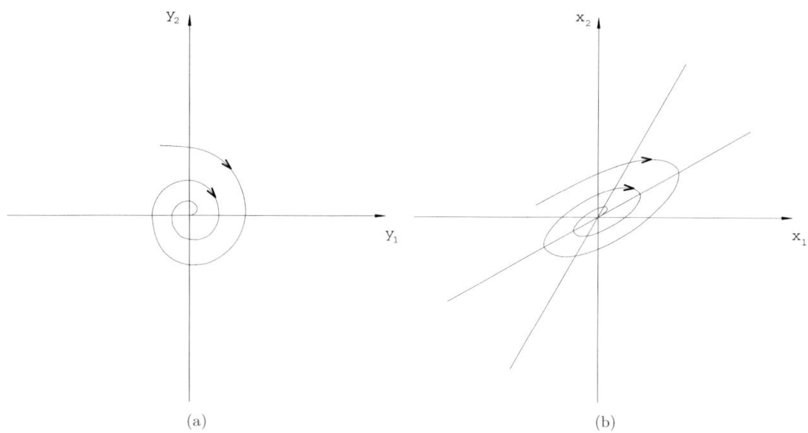


Figure 3.11. A stable focus.

the spiral illustrated in Figure 3.12 indicates that $\eta > 0$. Of course, when $\xi = 0$, the trajectories are concentric circles around $\mathbf{y} = \mathbf{0}$ (or $\mathbf{x} = \mathbf{0}$), as shown in Figure 3.13, and the stationary point is called a center.

In summary, a stationary point corresponds to a single point $\mathbf{y} = \mathbf{0}$ in phase space ($y_1$-$y_2$ plane), or $\mathbf{x} = \mathbf{x}_o$ in the original $x_1$-$x_2$ plane. In general, since we study the relative distance to the point $\mathbf{x} = \mathbf{x}_o$, a coordinate translation is used to replace $\mathbf{x} = \mathbf{x}_o$ with $\mathbf{x} = \mathbf{0}$ in the current $x_1$-$x_2$ plane. If the stationary point is a center, it is Lyapunov stable. Moreover, the stationary point is called a source (unstable) if the eigenvalues have positive real parts, and a sink (asymptotically

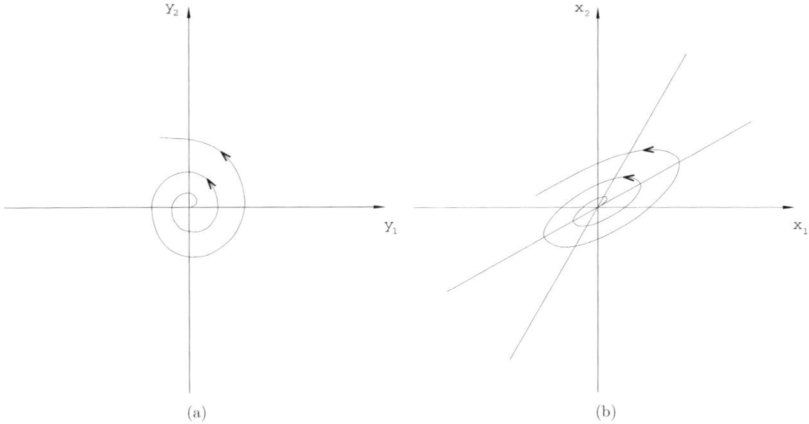


Figure 3.12. An unstable focus.

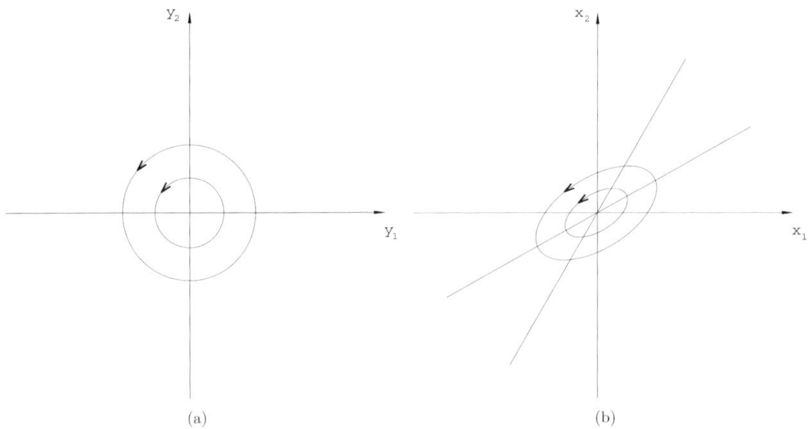


Figure 3.13. A center.

stable) if the eigenvalues have negative signs. Naturally, saddles are non-stable, and stable and unstable nodes, improper nodes, stars, and foci are various manifestations of sinks and sources, respectively.

In particular, if the matrix **A** is diagonalizable and the two eigenvalues are both negative, the sink is called a stable node, and if the eigenvalues of the diagonalizable matrix have the same negative real number, the sink is called a stable star. Conversely, if the matrix **A** is diagonalizable and the two eigenvalues are positive, the source is called an unstable node. If the eigenvalues of the diagonalizable matrix have the same positive real number, the source is called an unstable star. If, however, the matrix $\mathbf{A}$ has an eigenvalue of multiplicity of two and is not diagonalizable, the sink is called an improper stable node and, conversely, the source is called an improper unstable node. Source and sink are called spiral source and spiral sink if the eigenvalues are simple complex conjugates.

In practice, a simple $2 \times 2$ dynamical system as depicted in Eq. (1.7) represents the following linear vibration model of one mass point:

$$\ddot{x} + 2\xi\omega_o\dot{x} + \omega_o^2 x = 0, \tag{1.12}$$

with initial boundary conditions $x(0) = x_o$ and $\dot{x}(0) = \dot{x}_o$, where $\omega_o$ and $\xi$ are called the natural frequency and the damping ratio, respectively.

Thus, if we introduce a state variable $\mathbf{x} = \langle x, \dot{x} \rangle$, Eq. (1.12) can be easily rewritten as Eq. (1.7), with

$$\mathbf{A} = \begin{bmatrix} 0 & 1 \\ -\omega_o^2 & -2\xi\omega_o \end{bmatrix}. \tag{1.13}$$

Notice that in this case, so-called image phase space $(x, \dot{x})$ is in fact phase space. Obviously, $\xi < 0$ corresponds to a source (physically it is a case of negative damping), $\xi = 0$ corresponds to a center (physically it is a case of no damping), $\xi > 0$ corresponds to a sink (physically it is a case of positive damping), and in particular, $\xi = 1$ corresponds to an improper stable node (physically, it is a case of critical damping).

### *3.1.2. Bifurcations; Lagrangian dynamics*

#### *Bifurcations*

Although there exist no general solutions for nonlinear dynamical systems, a great deal of qualitative and quantitative information about the local behavior of the solution has been studied, particularly the dependence of the structure of orbits or flows on variable parameters. For FSI systems, we often have to address the question of the qualitative behavior as we alter some system parameters in both fluid and solid domains. Generally, such systems can be written as $\dot{\mathbf{x}} = \mathbf{f}(\mathbf{x}, \mathbf{c})$, where $\mathbf{c} \in \mathcal{R}^s$ designates physical parameters. In this section, we summarize a few basic concepts of bifurcations near the vicinity of fixed points and their dependence on variable physical parameters. We begin with an important theorem.

THEOREM 3.1.5 *(Implicit function theorem). Suppose that $f : \mathcal{R} \times \mathcal{R}^k \to \mathcal{R}$, and $(x, \mathbf{c}) \mapsto f(x, \mathbf{c})$, is a $C^1$ function satisfying*

$$f(0, \mathbf{0}) = 0 \quad \text{and} \quad \frac{\partial f}{\partial x}(0, \mathbf{0}) \neq 0. \tag{1.14}$$

*There are then constants $\delta_1 > 0$ and $\delta_2 > 0$, and a $C^1$ function:*

$$g : \{\mathbf{c}\colon \|\mathbf{c}\| < \delta_1\} \to \mathcal{R},$$

*such that*

$$g(\mathbf{0}) = 0 \quad \text{and} \quad f\big(g(\mathbf{c}), \mathbf{c}\big) = 0 \quad \text{for } \|\mathbf{c}\| < \delta_1.$$

*Moreover, if there is a* $(x_o, \mathbf{c}_o) \in \mathcal{R} \times \mathcal{R}^k$ *such that* $\|\mathbf{c}_o\| < \delta_1$ *and* $|x_o| < \delta_2$, *and* $f(x_o, \mathbf{c}_o) = 0$, *then* $x_o = g(\mathbf{c}_o)$.

In general, one condition of all variable parameters corresponds to a surface, or a hypersurface in parameter space $\mathcal{R}^k$ with $c_1, \ldots, c_k$ as its coordinates. Of course, at bifurcation points, or critical points, where a differential equation must have multiple equilibrium points and often different stability conditions, one condition of $\mathbf{c}$ can be derived from

$$\begin{aligned} &\mathbf{f}(\mathbf{x}, \mathbf{c}) = 0, \\ &\frac{\partial \mathbf{f}(\mathbf{x}, \mathbf{c})}{\partial \mathbf{x}} = \mathbf{0}. \end{aligned} \tag{1.15}$$

Thus, bifurcations are called codimension-1 if it is sufficient to set one condition of all variable parameters so that bifurcations can take place if such a condition is satisfied. The geometrical representation of codimension-1 bifurcation is a line or a simple curve intersecting the surface or hypersurface. Similarly, if it is sufficient to set two conditions of all variable parameters so that bifurcations can take place if these conditions are satisfied, such bifurcations are called codimension-2. The geometrical representation of codimension-2 bifurcation is a surface or hypersurface passing an intersection (a line or a simple curve) of another two surfaces or hypersurfaces. Naturally, the simplest bifurcation scenario is the so-called codimension-1 problem, which implies that it is sufficient to move along a simple curve in parameter space in order to encounter bifurcation at a point.

Furthermore, supercritical bifurcation (or normal bifurcation) corresponds to the cases in which higher order terms produce opposite effects on the stability compared with lower order terms, whereas subcritical bifurcation (or inverse bifurcation) corresponds to the cases in which higher order terms produce the same effects on the stability as lower order terms. To illustrate elementary bifurcations, we use a few codimension-1 examples.

Consider a simple linear differential equation:

$$\dot{x} = c - x, \tag{1.16}$$

where $c$ is a variable parameter, and Eq. (1.16) can be viewed as a perturbation of the equation $\dot{x} = -x$.

Since the Jacobian of the function $f(x, c) = c - x$ is $-1$, the fixed point $x = c$ is always hyperbolic and asymptotically stable. We use this particular example to illustrate that structure of orbits can be insensitive to certain variable parameters.

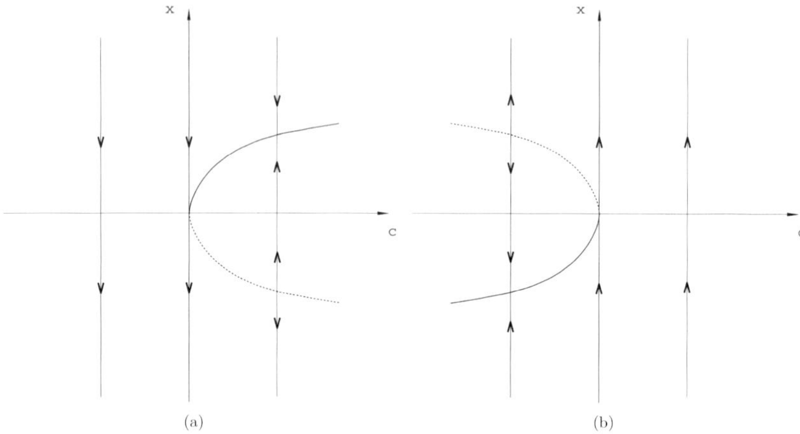


Figure 3.14. Supercritical and subcritical saddle-node bifurcations.

Of course, the dynamical behavior of Eq. (1.16) can be explained by a vertical translation of the $x$-axis.

Consider the equation

$$\dot{x} = c - x^2, \tag{1.17}$$

where $c$ is a variable parameter, and Eq. (1.17) can be viewed as a perturbation of the equation $\dot{x} = -x^2$.

The dynamical behavior of fixed points ($x = \pm\sqrt{c}$ with $c \geqslant 0$) is shown in Figure 3.14. We represent stable equilibria with solid curves and unstable equilibria with dotted curves. The bifurcation, with respect to parameter $c$, is called the supercritical saddle-node bifurcation. Analogously, the subcritical saddle-node bifurcation corresponds to the following equation, as shown in Figure 3.14:

$$\dot{x} = c + x^2. \tag{1.18}$$

Consider the equation

$$\dot{x} = cx - x^2, \tag{1.19}$$

where $c$ is a variable parameter, and Eq. (1.19) can be viewed as a perturbation of the equation $\dot{x} = -x^2$.

The bifurcation diagram is shown in Figure 3.15, and the bifurcation is called the supercritical transcritical bifurcation. Analogously, the subcritical transcritical bifurcation as shown in Figure 3.15 corresponds to

$$\dot{x} = cx + x^2. \tag{1.20}$$

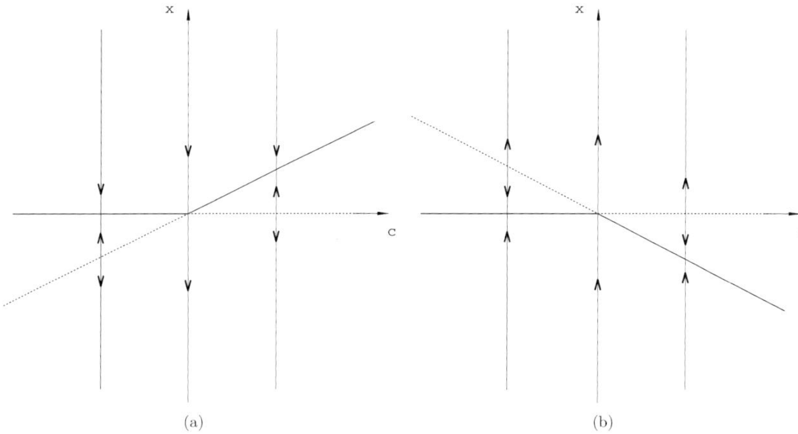


Figure 3.15. Supercritical and subcritical transcritical bifurcations.

Consider the equation

$$\dot{x} = cx - x^3, \tag{1.21}$$

where $c$ is a variable parameter, and Eq. (1.21) can be viewed as a perturbation of the equation $\dot{x} = -x^3$.

The associated bifurcation, as shown in Figure 3.16, is called the supercritical pitchfork bifurcation, and the subcritical pitchfork bifurcation corresponds to

$$\dot{x} = cx + x^3. \tag{1.22}$$

Consider the equation

$$\dot{z} = cz - z|z|^2, \tag{1.23}$$

where $c$ is a complex variable parameter, and Eq. (1.23) can be viewed as a perturbation of the equation $\dot{z} = -z|z|^2$.

Unlike the previous three cases, $z$ and $c$ are complex numbers ($z = x + iy$, $c = \mathrm{Re}(c) + i\,\mathrm{Im}(c)$) and the bifurcation as shown in Figure 3.17 is called the supercritical Hopf bifurcation, named after Heinz Hopf (1894–1971).

Notice that Eq. (1.21) can be better understood in polar coordinate system, written thusly:

$$\begin{aligned} \dot{r} &= \mathrm{Re}(c)r - r^3, \\ \dot{\theta} &= \mathrm{Im}(c), \end{aligned} \tag{1.24}$$

with $x = r\cos\theta$ and $y = r\sin\theta$.

Obviously, as far as the modulus $r$ is concerned, the Hopf bifurcation is similar to the pitchfork bifurcation. In addition, the bifurcation is driven solely by the real

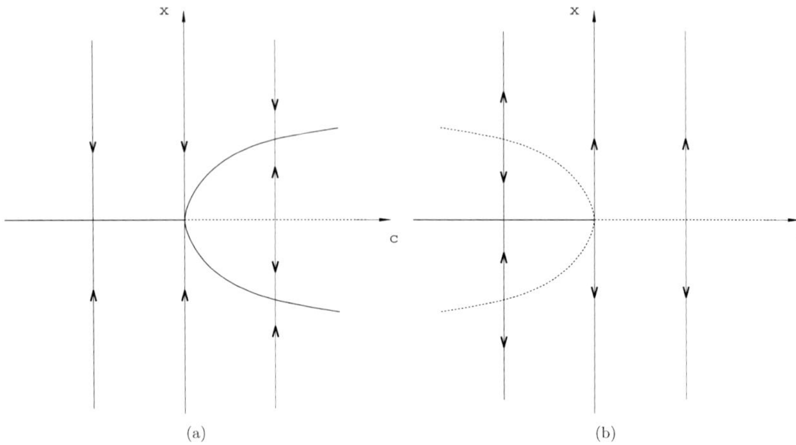


Figure 3.16. Supercritical and subcritical pitchfork bifurcations.

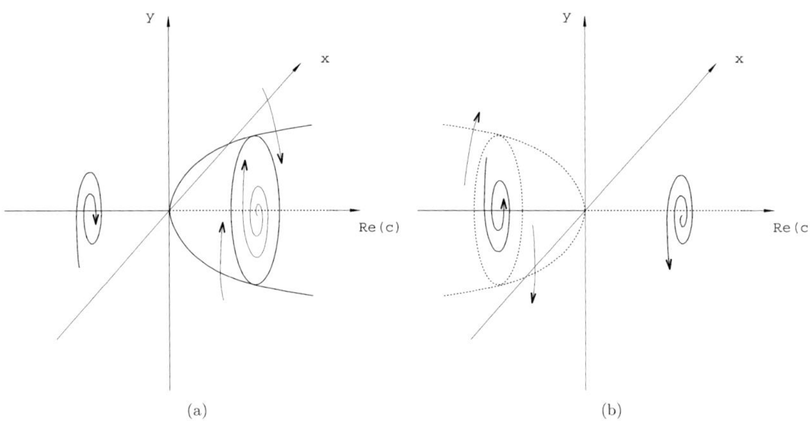


Figure 3.17. Supercritical and subcritical Hopf bifurcations.

part of the complex variable parameter $c$, while the imaginary part of the complex variable parameter $c$ controls the rotational direction and speed of the complex variable $z$.

Of course, the bifurcation for the case

$$\dot{z} = cz + z|z|^2, \tag{1.25}$$

is called the subcritical Hopf bifurcation, as shown in Figure 3.17.

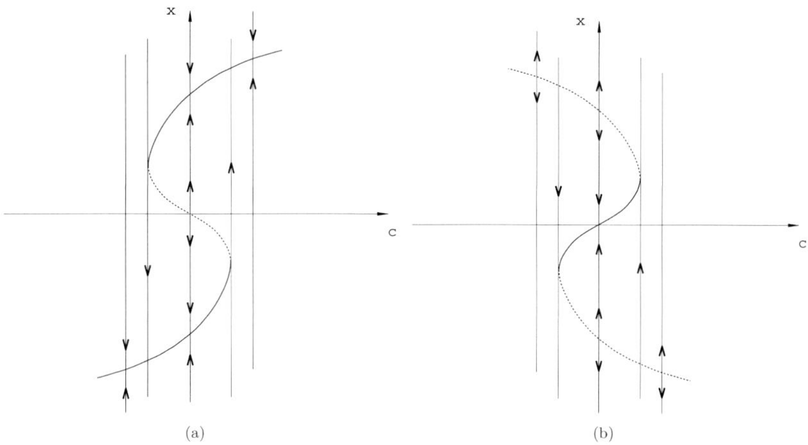


Figure 3.18. Supercritical and subcritical hysteresis bifurcations.

Consider the equation

$$\dot{x} = c + x - x^3, \tag{1.26}$$

where $c$ is a variable parameter, and Eq. (1.26) can be viewed as a perturbation of the equation $\dot{x} = x - x^3$.

The associated bifurcation is shown in Figure 3.18, which is called the supercritical hysteresis loop, and the subcritical hysteresis bifurcation corresponds to

$$\dot{x} = c - x + x^3. \tag{1.27}$$

We must point out that for the following equation

$$\dot{x} = c + x + x^3, \tag{1.28}$$

the dynamical behavior is trivial, because the Jacobian $df(x, c)/dx = 1 + 3x^2$ is always positive, which indicates that all fixed points must be inherently unstable.

Finally, we consider a codimension-2 case with the following governing equation:

$$\dot{x} = c_1 + c_2 x - x^3, \tag{1.29}$$

with two variable parameters, $c_1$ and $c_2$.

The bifurcation diagram is three-dimensional as shown in Figure 3.19, and is called the supercritical fold or cusp. Interestingly, Eq. (1.15) corresponds to

$$\begin{aligned} c_1 &= -2x^3, \\ c_2 &= 3x^2. \end{aligned} \tag{1.30}$$

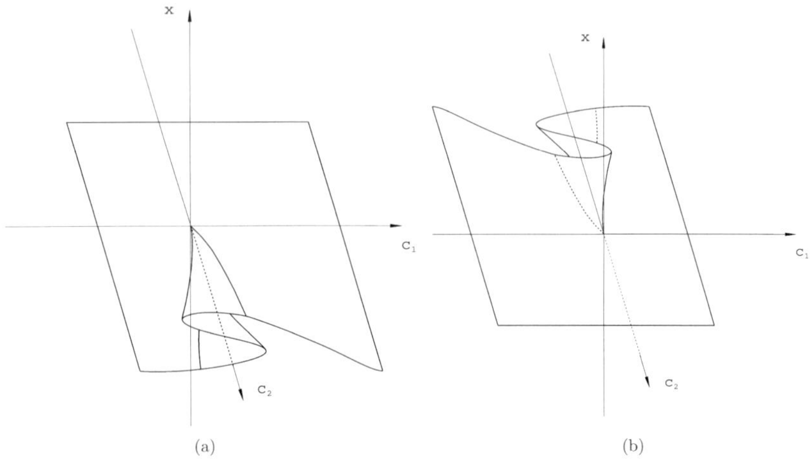


Figure 3.19. Supercritical and subcritical fold or cusp bifurcations.

By eliminating $x$, we have one bifurcation curve in parameter space, rather than a bifurcation point in codimension-1 cases, represented by

$$\left(\frac{c_1}{2}\right)^2 = \left(\frac{c_2}{3}\right)^3. \tag{1.31}$$

The geometrical representation of Eq. (1.31) is shown in Figure 3.20. It is then obvious that Eq. (1.29) has two variable parameters and the bifurcation line corresponds to a simple curve in $c_1$-$c_2$ plane, as shown in Figure 3.20. In fact, Figure 3.20 is the projection of the locus of points in the manifold in the bifurcation diagram, as shown in Figure 3.19, which have vertical tangency. Analogously, the subcritical fold or cusp bifurcation corresponds to

$$\dot{x} = c_1 + c_2 x + x^3. \tag{1.32}$$

We also want to point out that the fold or cusp bifurcation consists of both the hysteresis loop and pitchfork bifurcations, including their variations. For example, if we choose $c_1 = 1$ and $c_2 = c$, the equation

$$\dot{x} = 1 + cx - x^3 \tag{1.33}$$

represents a modified supercritical pitchfork bifurcation as shown in Figure 3.21.

*Lagrangian dynamics*

Variational principles discussed in Section 3.2 can also be viewed as dynamic variational principles if we incorporate the inertia force $-\rho\, d^2\mathbf{u}/dt^2$, or

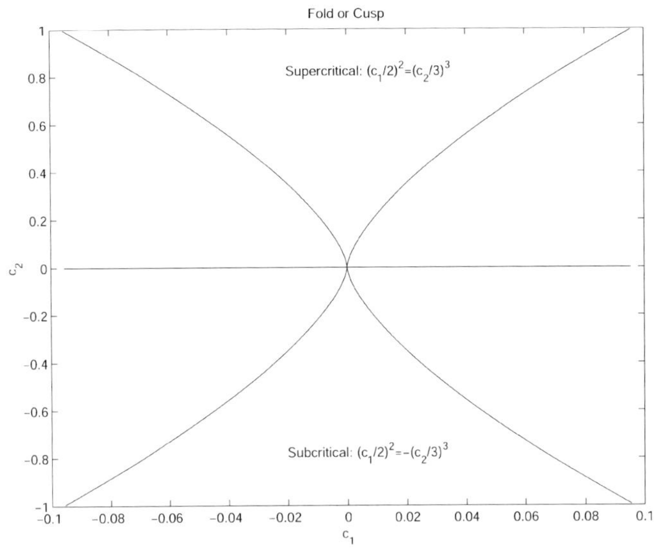


Figure 3.20. The locus of bifurcations in parameter space.

d'Alembert's force, named after Jean Le Rond d'Alembert (1717–1783), in the expression of the body forces $\mathbf{f}^b$. However, in the study of the dynamical behaviors of rigid bodies or mass points, we generally resort to the tools of Lagrangian dynamics.

We start by defining the kinetic energy $T$ and the kinetic coenergy $\bar{T}$ for a mass point $m$ with a linear momentum $\mathbf{p}$:

$$T = \int_o^{\mathbf{p}} \mathbf{v} \cdot d\mathbf{p} \quad \text{and} \quad \bar{T} = \int_o^{\mathbf{v}} \mathbf{p} \cdot d\mathbf{v} = \mathbf{v} \cdot \mathbf{p} - T. \tag{1.34}$$

If we consider the special theory of relativity, denote $p = |\mathbf{p}|$ and $v = |\mathbf{v}|$, from

$$p = \frac{mv}{\sqrt{1 - v^2/c^2}} \quad \text{and} \quad v = \frac{p/m}{\sqrt{1 + p^2/m^2c^2}}, \tag{1.35}$$

where $m$ is called the rest mass and $c$ is the speed of light, we have

$$T = mc^2\left(\sqrt{1 + p^2/m^2c^2} - 1\right) \quad \text{and} \quad \bar{T} = mc^2\left(1 - \sqrt{1 - v^2/c^2}\right). \tag{1.36}$$

In this book we consider the Newtonian mechanical system, and will not distinct between the kinetic coenergy and the kinetic energy. Therefore, the kinetic energy of a mass point $m$ is simply written as $T = mv^2/2$.

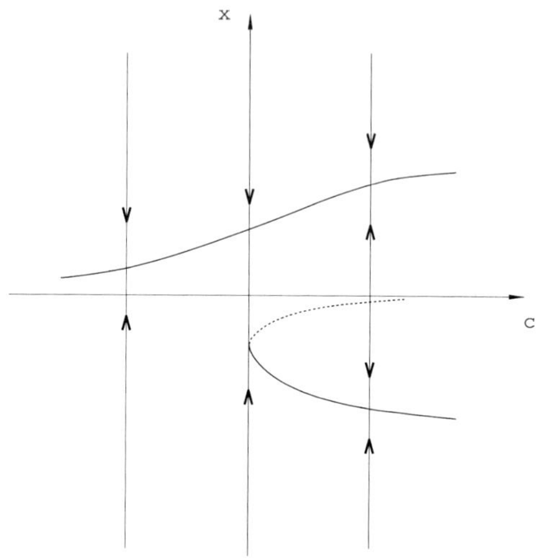


Figure 3.21. The modified supercritical pitchfork bifurcation.

A geometric constraint which can be expressed analytically as an equation relating generalized coordinates and time is said to be holonomic. In particular, it displays a one-to-one correspondence between a restriction on a generalized coordinate and a restriction on the infinitesimal variation. If all the constraints in a system are holonomic, implying that the number of independent generalized coordinates in a complete set is the same as the number of degrees of freedom, the system is holonomic. Consider a holonomic system of $N$ mass point, $m_1, \ldots, m_N$ acted upon by forces $\mathbf{f}_1, \ldots, \mathbf{f}_N$, respectively. The locations of the mass points are represented as $\mathbf{r}_i = x_{ik}\mathbf{e}_k$, with $i = 1, \ldots, N$. Thus, from Newton's law $\mathbf{f}_i = m_i\ddot{\mathbf{r}}_i$, for any linear virtual displacement $\delta\mathbf{r}_i$, we must have

$$\sum_{i=1}^{N} m_i\ddot{\mathbf{r}}_i \cdot \delta\mathbf{r}_i = \sum_{i=1}^{N} \mathbf{f}_i \cdot \delta\mathbf{r}_i. \tag{1.37}$$

Eq. (1.37) is called d'Alembert's equation. Hence, if we consider a set of $N$ generalized coordinates $q_1, \ldots, q_N$, Eq. (1.37) can be also written as the form of Lagrange's equation:

$$\frac{d}{dt}\left(\frac{\partial T}{\partial \dot{q}_k}\right) - \frac{\partial T}{\partial q_k} = f_{q_k}, \tag{1.38}$$

with the kinetic energy $T = \sum_{i=1}^{N} m_i|\dot{\mathbf{r}}_i|^2/2$, and the generalized force $f_{q_k} = \mathbf{f}_i \cdot \partial\mathbf{r}_i/\partial q_k$.

A force field (a vector field) is conservative if we can find a scalar function $U$ such that $\nabla U = -\mathbf{f}$. The negative sign is inherited from classical mechanics tradition. Since $\nabla \times (\nabla U) = \mathbf{0}$, using Stokes' theorem in Eq. (3.138) in Chapter 2, we have

$$\int_S (\nabla \times \mathbf{f}) \cdot \mathbf{n}\, dS = \oint_C \mathbf{f} \cdot d\mathbf{r} = 0, \tag{1.39}$$

where the surface $S$ is bounded by a closed curve $C$ and $\mathbf{n}$ is the normal vector of the surface $S$.

Hence, if we consider the total force acting on the mass point $m_i$ as a combination of the conservative force $-\nabla U$ and the nonconservative force $\mathbf{f}_i$, Lagrange's equation (1.38) can also be expressed as

$$\frac{d}{dt}\left(\frac{\partial T}{\partial \dot{q}_k}\right) - \frac{\partial (T-U)}{\partial q_k} = f_{q_k}, \tag{1.40}$$

where $-\partial U/\partial q_k$ is the generalized conservative force and $f_{q_k}$ is the generalized nonconservative force.

Moreover, if we assign the Lagrangian function $L = T - U$, Eq. (1.40) can be written as

$$\frac{d}{dt}\left(\frac{\partial L}{\partial \dot{q}_k}\right) - \frac{\partial L}{\partial q_k} = f_{q_k}. \tag{1.41}$$

In fact, Eq. (1.41) is equivalent to Hamilton's principle: an admissible motion of the system between specified configurations at $t_1$ and $t_2$ is a natural motion if, and only if, the following equation holds for arbitrary admissible variations $\delta q_k$ which vanish at $t = t_1$ and $t = t_2$:

$$\int_{t_1}^{t_2} \left[ \delta(T-U) + \sum_{i=1}^{N} f_i \delta q_i \right] dt = 0. \tag{1.42}$$

For conservative systems, i.e., the nonconservative force $\mathbf{f}_i = \mathbf{0}$, Lagrange's equation becomes

$$\frac{d}{dt}\left(\frac{\partial T}{\partial \dot{q}_k}\right) - \frac{\partial (T-U)}{\partial q_k} = 0. \tag{1.43}$$

For a mass point $m$ with a linear momentum $\mathbf{p}$, according to Newton's law, we have

$$\mathbf{f} = \frac{d\mathbf{p}}{dt} \quad \text{and} \quad \mathbf{r} \times \mathbf{f} = \frac{d}{dt}(\mathbf{r} \times \mathbf{p}), \tag{1.44}$$

where $\mathbf{r}$, $\mathbf{f}$, $\mathbf{r} \times \mathbf{f}$, and $\mathbf{r} \times \mathbf{p}$ stand for the position vector, the total force corresponding to the mass point, the torque, and the angular momentum with respect to the fixed coordinate center $o$.

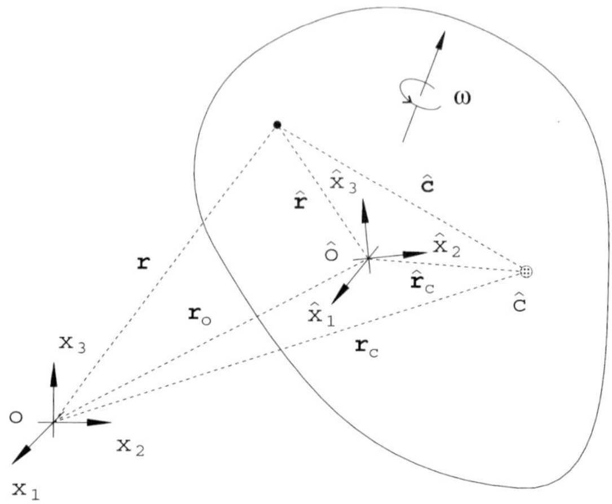


Figure 3.22. Moving coordinate frame $\hat{o}\hat{x}\hat{y}\hat{z}$ and fixed coordinate frame $oxyz$.

Suppose a rigid body, to which an intermediate or moving coordinate frame is attached, locates at a position $\mathbf{r}_o$, translates at a velocity $\mathbf{v}_o = d\mathbf{r}_o/dt$, and rotates at an angular velocity $\boldsymbol{\omega}$ with respect to the fixed Cartesian coordinate frame as shown in Figure 3.22. Let $\hat{\mathbf{f}}$ be any vector represented in the moving coordinate frame, i.e., $\hat{\mathbf{f}} = \hat{f}_i \hat{\mathbf{e}}_i$. According to Eq. (2.115) in Chapter 2, we have $d\hat{\mathbf{e}}_i/dt = \boldsymbol{\omega} \times \hat{\mathbf{e}}_i$ and consequently,

$$\frac{d\hat{\mathbf{f}}}{dt} = \frac{d\hat{f}_i}{dt}\hat{\mathbf{e}}_i + \boldsymbol{\omega} \times \hat{\mathbf{f}}. \tag{1.45}$$

Thus, it is easy to verify that

$$\begin{aligned}
\mathbf{r} &= \mathbf{r}_o + \hat{\mathbf{r}}, \\
\mathbf{v} &= \mathbf{v}_o + \hat{\mathbf{v}} + \boldsymbol{\omega} \times \hat{\mathbf{r}}, \\
\mathbf{a} &= \dot{\mathbf{v}}_o + \hat{\mathbf{a}} + 2\boldsymbol{\omega} \times \hat{\mathbf{v}} + \dot{\boldsymbol{\omega}} \times \hat{\mathbf{r}} + \boldsymbol{\omega} \times (\boldsymbol{\omega} \times \hat{\mathbf{r}}),
\end{aligned} \tag{1.46}$$

where $\hat{\mathbf{r}}$, $\hat{\mathbf{v}}$, and $\hat{\mathbf{a}}$ represent the displacement, velocity, and acceleration vectors with respect to the moving coordinate frame, respectively.

The term $2\boldsymbol{\omega} \times \hat{\mathbf{v}}$ is called the Coriolis acceleration, named after Gustave Gaspard de Coriolis (1792–1843), and the term $\boldsymbol{\omega} \times (\boldsymbol{\omega} \times \hat{\mathbf{r}})$ is called the centripetal acceleration.

Define the mass center, or centroid, as

$$\mathbf{r}_c = \int_\Omega \rho \mathbf{r}\, d\Omega \Big/ \int_\Omega \rho\, d\Omega \quad \text{and} \quad \hat{\mathbf{r}}_c = \int_\Omega \rho \hat{\mathbf{r}}\, d\Omega \Big/ \int_\Omega \rho\, d\Omega, \tag{1.47}$$

where $\mathbf{r}_c$ and $\hat{\mathbf{r}}_c$ represent the position vectors of the centroid with respect to the origins of the fixed and the moving coordinate systems, respectively.

Since $\mathbf{v}_c = d\mathbf{r}_c/dt$, based on Eq. (1.46), we have, with $\mathbf{r} = \mathbf{r}_c + \hat{\mathbf{c}}$,

$$\mathbf{v} = \mathbf{v}_c + \boldsymbol{\omega} \times \hat{\mathbf{c}}, \tag{1.48}$$

where $\hat{\mathbf{c}}$ stands for the position vector in the moving coordinate frame with respect to the centroid $\hat{c}$.

Obviously, Eq. (1.47) also gives us

$$\int_{\Omega} \rho \hat{\mathbf{c}} \, d\Omega = \mathbf{0}. \tag{1.49}$$

Thus, the total linear momentum of the rigid body can be expressed as

$$\mathbf{p} = \int_{\Omega} \rho \mathbf{v} \, d\Omega = \int_{\Omega} \rho (\mathbf{v}_c + \boldsymbol{\omega} \times \hat{\mathbf{c}}) \, d\Omega = M \mathbf{v}_c, \tag{1.50}$$

with $M = \int_{\Omega} \rho \, d\Omega$.

More importantly, the total angular momentum with respect to $o$, $\hat{o}$, and $\hat{c}$ can be represented as

$$\mathbf{l}_o = \int_{\Omega} \rho \mathbf{r} \times \mathbf{v} \, d\Omega, \qquad \mathbf{l}_{\hat{o}} = \int_{\Omega} \rho \hat{\mathbf{r}} \times \mathbf{v} \, d\Omega, \quad \text{and} \quad \mathbf{l}_{\hat{c}} = \int_{\Omega} \rho \hat{\mathbf{c}} \times \mathbf{v} \, d\Omega. \tag{1.51}$$

In particular, using Eq. (1.47), we have

$$\begin{aligned} \mathbf{l}_{\hat{c}} &= \int_{\Omega} \rho \hat{\mathbf{c}} \times \mathbf{v} \, d\Omega = \int_{\Omega} \rho \hat{\mathbf{c}} \times (\mathbf{v}_c + \boldsymbol{\omega} \times \hat{\mathbf{c}}) \, d\Omega \\ &= \int_{\Omega} \rho \hat{\mathbf{c}} \, d\Omega \times \mathbf{v}_c + \int_{\Omega} \rho \hat{\mathbf{c}} \times (\boldsymbol{\omega} \times \hat{\mathbf{c}}) \, d\Omega = \int_{\Omega} \rho \hat{\mathbf{c}} \times (\boldsymbol{\omega} \times \hat{\mathbf{c}}) \, d\Omega. \end{aligned} \tag{1.52}$$

Therefore, the angular momentum corresponding to other centers can be expressed as

$$\mathbf{l}_{\hat{o}} = \int_{\Omega} \rho \hat{\mathbf{r}} \times \mathbf{v} \, d\Omega = \int_{\Omega} \rho (\hat{\mathbf{r}}_c + \hat{\mathbf{c}}) \times \mathbf{v} \, d\Omega = \hat{\mathbf{r}}_c \times \mathbf{p} + \mathbf{l}_{\hat{c}}, \tag{1.53}$$

$$\mathbf{l}_o = \int_{\Omega} \rho \mathbf{r} \times \mathbf{v} \, d\Omega = \int_{\Omega} \rho (\mathbf{r}_o + \hat{\mathbf{r}}) \times \mathbf{v} \, d\Omega = \mathbf{r}_o \times \mathbf{p} + \mathbf{l}_{\hat{o}}. \tag{1.54}$$

Assume $\boldsymbol{\omega} = \langle \omega_1, \omega_2, \omega_3 \rangle$, and move $\hat{o}$ to $\hat{c}$, i.e., assign $\hat{\mathbf{r}}_c = \mathbf{0}$, and using Eq. (1.11) in Chapter 2, it can be easily verified that

$$\int_{\Omega} \rho \hat{\mathbf{c}} \times (\boldsymbol{\omega} \times \hat{\mathbf{c}}) \, d\Omega = \mathbf{I} \boldsymbol{\omega}, \tag{1.55}$$

with

$$\mathbf{I} = \begin{bmatrix} I_{11} & I_{12} & I_{13} \\ I_{21} & I_{22} & I_{23} \\ I_{31} & I_{32} & I_{33} \end{bmatrix},$$

and

$$I_{11} = \int_{\Omega} \rho\left(\hat{x}_2^2 + \hat{x}_3^2\right) d\Omega, \qquad I_{22} = \int_{\Omega} \rho\left(\hat{x}_3^2 + \hat{x}_1^2\right) d\Omega;$$

$$I_{33} = \int_{\Omega} \rho\left(\hat{x}_1^2 + \hat{x}_2^2\right) d\Omega, \qquad I_{12} = -\int_{\Omega} \rho(\hat{x}_1 \hat{x}_2)\, d\Omega;$$

$$I_{23} = -\int_{\Omega} \rho(\hat{x}_2 \hat{x}_3)\, d\Omega, \qquad I_{31} = -\int_{\Omega} \rho(\hat{x}_3 \hat{x}_1)\, d\Omega.$$

Notice that **I** is the so-called inertia matrix with respect to a coordinate system originating at the centroid $\hat{c}$, and is strictly dependent on the geometry of the rigid body which includes the location of the centroid. Therefore, the total kinetic energy within the rigid body can be expressed as

$$\begin{aligned} T &= \frac{1}{2} \int_{\Omega} \rho \mathbf{v} \cdot \mathbf{v}\, d\Omega = \frac{1}{2} \int_{\Omega} \rho(\mathbf{v}_c + \boldsymbol{\omega} \times \hat{\mathbf{c}}) \cdot (\mathbf{v}_c + \boldsymbol{\omega} \times \hat{\mathbf{c}})\, d\Omega \\ &= \frac{1}{2} \int_{\Omega} \rho\left[\mathbf{v}_c \cdot \mathbf{v}_c + 2\mathbf{v}_c \cdot (\boldsymbol{\omega} \times \hat{\mathbf{c}}) + (\boldsymbol{\omega} \times \hat{\mathbf{c}}) \cdot (\boldsymbol{\omega} \times \hat{\mathbf{c}})\right] d\Omega. \end{aligned} \tag{1.56}$$

Using Eqs. (1.11) in Chapter 2 and (1.49), we have

$$T = \frac{1}{2} M \mathbf{v}_c^T \mathbf{v}_c + \frac{1}{2} \boldsymbol{\omega}^T \mathbf{I} \boldsymbol{\omega}. \tag{1.57}$$

We must address two important points here: Firstly, if we are dealing with a system of mass points, the angular momentum derivations still hold, provided we replace the integration of the volume with the summation over the mass points. Secondly, the angular velocity $\boldsymbol{\omega}$, the angular momentum $\mathbf{l}$, and the inertia matrix **I** are subject to the rules of coordinate transformations as are all vectors and tensors.

### *3.1.3. Van der Pol; perturbation method; Duffing*

#### *Van der Pol*

In this section, we introduce a few low-dimensional dynamical system examples to illustrate and confirm the concepts introduced in previous sections.

First, we discuss van der Pol's equation:

$$\ddot{x} + \epsilon(x^2 - 1)\dot{x} + x = 0, \quad 0 < \epsilon \ll 1, \tag{1.58}$$

with the initial conditions $x(0) = x_o$ and $\dot{x}(0) = \dot{x}_o$.

Eq. (1.58) represents the so-called self-excited oscillation which could exist in both mechanical and electrical systems. Before we introduce the numerical approaches to dynamical systems, let us first apply two classical methods for non-linear oscillations.

Using the averaging technique, we assume the solution in the following form:

$$x(t) = a(t)\cos(t + \phi(t)), \quad a(t) \geqslant 0. \tag{1.59}$$

It is obvious that the trivial case $a(t) = 0$ represents the stationary point of this dynamical system. The stability of this point will be discussed along with the dynamical system approach.

Therefore, we have

$$\dot{x}(t) = -a(t)\sin(t + \phi(t))(1 + \dot{\phi}(t)) + \dot{a}(t)\cos(t + \phi(t)). \tag{1.60}$$

In order to have an analogous periodic solution as the harmonic solution for the linear free-oscillation, assuming that there is one, we impose a consistency condition:

$$\dot{a}(t)\cos(t + \phi(t)) = \dot{\phi}(t)a(t)\sin(t + \phi(t)), \tag{1.61}$$

therefore

$$\dot{x}(t) = -a(t)\sin(t + \phi(t)). \tag{1.62}$$

Furthermore, from Eq. (1.58), we derive

$$\begin{aligned} &\dot{a}(t)\sin(t + \phi(t)) + a(t)\dot{\phi}(t)\cos(t + \phi(t)) \\ &\quad = \epsilon\left[1 - a^2(t)\cos^2(t + \phi(t))\right]a(t)\sin(t + \phi(t)). \end{aligned} \tag{1.63}$$

As a consequence of Eqs. (1.61) and (1.63), we obtain

$$\dot{a} = \epsilon a(1 - a^2\cos^2\theta)\sin^2\theta, \tag{1.64}$$

$$\dot{\phi} = \epsilon(1 - a^2\cos^2\theta)\sin\theta\cos\theta, \tag{1.65}$$

with $\theta(t) = t + \phi(t)$.

Obviously, if $\epsilon = 0$, we can recover the harmonic solution $x(t) = a\cos(t + \phi)$ where both $a$ and $\phi$ are constants with $x_o = a\cos\phi$ and $\dot{x}_o = -a\sin\phi$. Now, since $0 < \epsilon \ll 1$, we expect both $a$ and $\phi$ to vary very little during a period of oscillation, and we may average out the fast oscillation by introducing

$$\dot{a} = \frac{\epsilon a}{2\pi}\int_0^{2\pi}(1 - a^2\cos^2\theta)\sin^2\theta\, d\theta = \epsilon\left(\frac{a}{2} - \frac{a^3}{8}\right). \tag{1.66}$$

This procedure is confirmed by

$$\dot{\phi} = \frac{\epsilon}{2\pi} \int_0^{2\pi} \left(1 - a^2 \cos^2 \theta\right) \sin\theta \cos\theta \, d\theta = 0. \tag{1.67}$$

The fact that $\dot{a} = 0$ at $a = 2$ indicates the existence of a limit cycle. Moreover, since $\dot{a} < 0$ for $a > 2$ and $\dot{a} > 0$ for $a < 2$, both solutions of $a < 2$ and $a > 2$ tend to limit cycle $a = 2$. This confirms that the limit cycle is stable. To be more specific, we derive

$$a(t) = 2/\sqrt{1 + Ce^{-\epsilon t}}, \tag{1.68}$$

from Eq. (1.66), with $C = (4 - a^2(0))/a^2(0)$.

Naturally, the larger $\epsilon$ is, the greater the approaching speed becomes.

*Perturbation method*

Using the method of multiple scales, we recognize two time scales: a fast time scale $T_1 = t$, where solutions oscillate, and a slow time scale $T_2 = \epsilon t$, where amplitude and phase evolve. Thus, a proper asymptotic expansion must reflect both time scales. If we express the solution of Eq. (1.58) as

$$x(t) = x_o(T_1, T_2) + \epsilon x_1(T_1, T_2) + O\left(\epsilon^2\right), \tag{1.69}$$

and treat $T_1$ and $T_2$ as two independent variables, i.e., $d/dt = \partial/\partial T_1 + \epsilon \partial/\partial T_2$, we have

$$\dot{x}(t) = \frac{\partial x_o}{\partial T_1} + \epsilon \frac{\partial x_o}{\partial T_2} + \epsilon \frac{\partial x_1}{\partial T_1} + O\left(\epsilon^2\right), \tag{1.70}$$

$$\ddot{x}(t) = \frac{\partial^2 x_o}{\partial T_1^2} + 2\epsilon \frac{\partial^2 x_o}{\partial T_1 \partial T_2} + \epsilon \frac{\partial^2 x_1}{\partial T_1^2} + O\left(\epsilon^2\right). \tag{1.71}$$

Collecting the coefficients of different orders of $\epsilon$ in Eq. (1.58), we obtain

$$O(1): \quad \frac{\partial^2 x_o}{\partial T_1^2} + x_o = 0, \tag{1.72}$$

with $x_o(0, 0) = x_o$ and $\frac{\partial x_o}{\partial T_1}(0, 0) = \dot{x}_o$, and

$$O(\epsilon): \quad \frac{\partial^2 x_1}{\partial T_1^2} + x_1 = -2\frac{\partial^2 x_o}{\partial T_1 \partial T_2} - \frac{\partial x_o}{\partial T_1}\left(x_o^2 - 1\right), \tag{1.73}$$

with $x_1(0, 0) = 0$ and $\frac{\partial x_1}{\partial T_1}(0, 0) = 0$.

The solution of Eq. (1.72) is easily obtained and can be expressed as

$$x_o(T_1, T_2) = a(T_2) \cos\left(T_1 + \phi(T_2)\right). \tag{1.74}$$

Substituting $x_o(T_1, T_2)$ into Eq. (1.73), we have

$$\frac{\partial^2 x_1}{\partial T_1^2} + x_1 = \left(2\dot{a}(T_2) + \frac{a^3(T_2)}{4} - a(T_2)\right)\sin\bigl(T_1 + \phi(T_2)\bigr) + a(T_2)\dot{\phi}(T_2)\cos\bigl(T_1 + \phi(T_2)\bigr) + \frac{a^3(T_2)}{4}\sin\bigl(3T_1 + 3\phi(T_2)\bigr). \tag{1.75}$$

Therefore, to avoid resonance in Eq. (1.75), we must have coefficients of the harmonical terms with the same frequency as the natural frequency vanish, i.e.,

$$\dot{a}(T_2) = \frac{a(T_2)}{2} - \frac{a^3(T_2)}{8}, \tag{1.76}$$

$$\dot{\phi}(T_2) = 0. \tag{1.77}$$

Note that $T_2 = \epsilon t$, Eqs. (1.76) and (1.77) are the same set of equations as Eqs. (1.64) and (1.65). The same solutions as Eq. (1.68) can be obtained and the averaging technique is confirmed to be physically correct. However, to accommodate more general conditions, e.g., $\epsilon = 2 > 1$, or high-dimensional systems, it is clear that classical perturbation methods are limited and numerical approaches for dynamical systems must be used. In this section, we employ the second-order Runge–Kutta (RK2) scheme, named after Carl David Tolmé Runge (1856–1927) and Martin Wilhelm Kutta (1867–1944), with sufficiently small time steps. Further discussion on the selection of critical time steps as well as various numerical schemes will follow in this chapter.

As a foundation for various numerical approaches, we introduce the following theorem, named after Charles Émile Picard (1856–1941) and Ernst Leonard Lindelöf (1870–1946).

THEOREM 3.1.6 *(Picard–Lindelöf iteration). Consider a dynamical system* $\dot{\mathbf{x}} = \mathbf{f}(\mathbf{x}, t)$ *with* $\mathbf{x}(t_o) = \mathbf{x}_o$ *and a* $C^1$ *mapping* $\mathbf{f}$. *If* $\mathbf{x}^{k+1}(t) = \mathbf{x}_o + \int_{t_o}^{t} \mathbf{f}(\mathbf{x}^k(t), s)\,ds$, *then the sequence* $\mathbf{x}^k$ *converges uniformly to the solution* $\mathbf{x}$.

As discussed in the previous section, Eq. (1.58) can also be written as

$$\begin{aligned} \dot{x}_1 &= x_2, \\ \dot{x}_2 &= -x_1 - \epsilon\bigl(x_1^2 - 1\bigr)x_2, \end{aligned} \tag{1.78}$$

with the initial conditions corresponding to $\mathbf{x}_o = \langle x_o, \dot{x}_o\rangle$.

Thus, the stationary point $\mathbf{x}_o$ of Eq. (1.78) is equal to $\mathbf{0}$, the trivial solution. Moreover, we have

$$\mathbf{Df}(\mathbf{x}_o) = \begin{bmatrix} 0 & 1 \\ -1 & \epsilon \end{bmatrix}, \tag{1.79}$$

where $f_1 = x_2$ and $f_2 = -x_1 - \epsilon(x_1^2 - 1)x_2$.

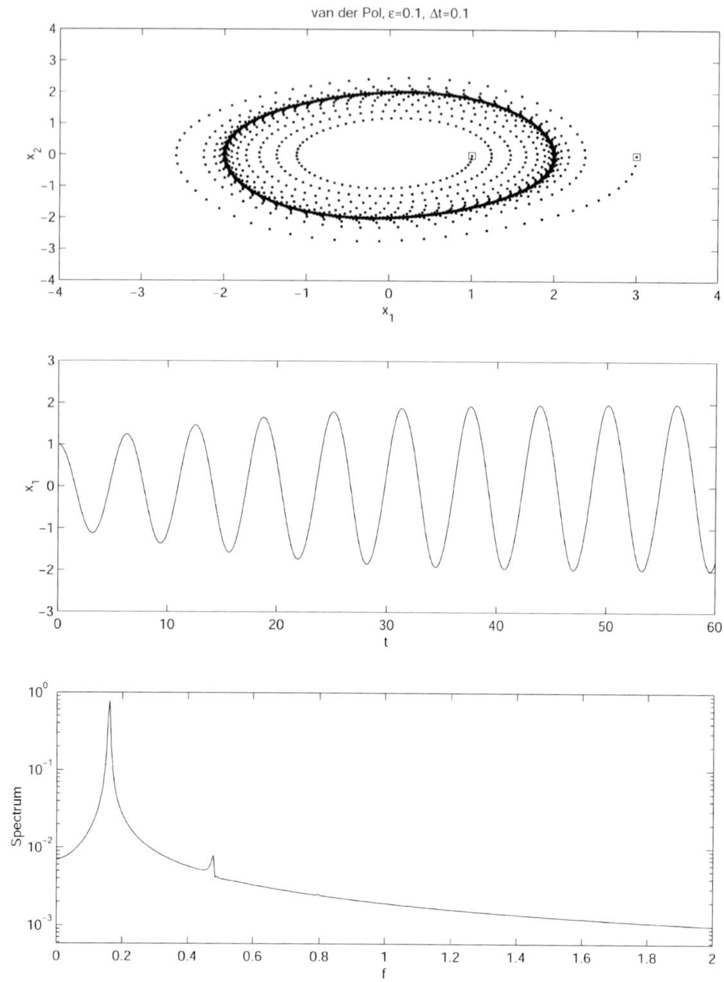


Figure 3.23. A periodic solution of van der Pol's equation.

The eigenvalues of $\mathbf{Df}(\mathbf{x}_o)$ can be expressed as $\lambda_1 = \epsilon/2 + i\sqrt{1-(\epsilon/2)^2}$ and $\lambda_2 = \epsilon/2 - i\sqrt{1-(\epsilon/2)^2}$. It is clear that for $0 < \epsilon \ll 1$, the real part of the conjugate eigenvalue pair is positive and the stationary point is unstable. In fact, for any real value of $\epsilon$, the real part of the eigenvalues is positive. The van der Pol limit cycle is stable and can be approached from both inside and outside, as shown in Figures 3.23 and 3.24. The basin of attraction is the whole phase space and the convergence is monotonic. In addition, as suggested in Eq. (1.68), $\epsilon = 2$ corresponds to higher approaching speed compared with $\epsilon = 0.1$. Of course, in-

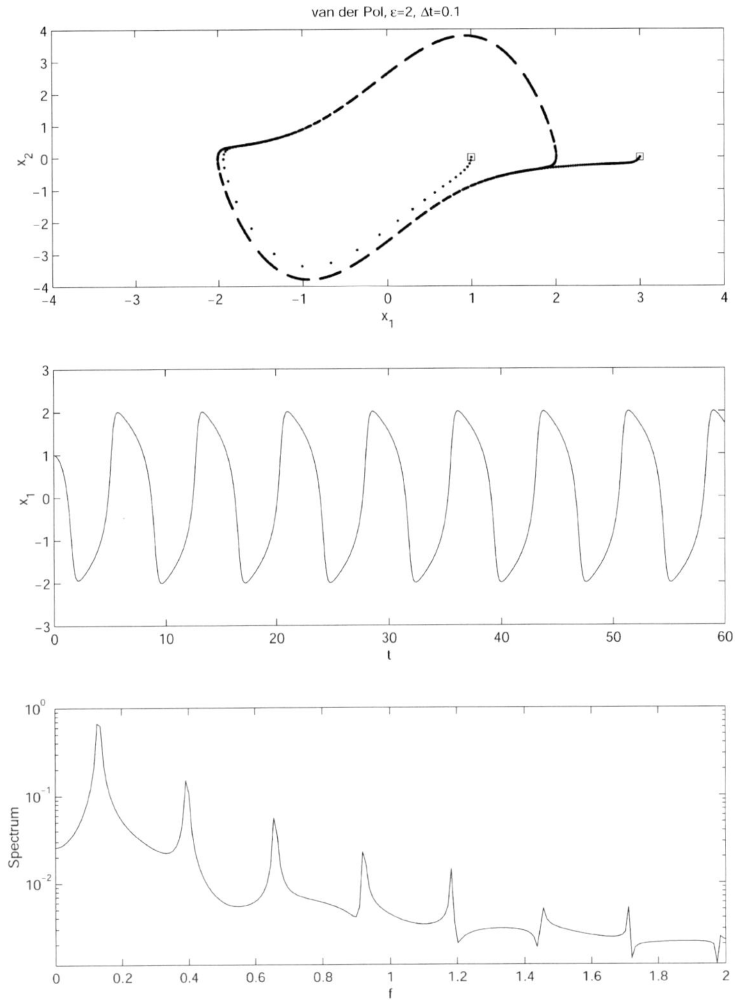


Figure 3.24. A periodic solution of van der Pol's equation.

dependent of the initial conditions, odd harmonics are dominant in Figures 3.23 and 3.24, due to the symmetry close to the sine wave. Likewise, if the time waveform has symmetry close to the cosine wave, even harmonics must be dominant. Moreover, as depicted in Figure 3.25, a quasi-periodic solution could exist for the following forced van der Pol's equation:

$$\ddot{x} + \epsilon\left(x^2 - 1\right)\dot{x} + x = \gamma \cos \omega_o t, \tag{1.80}$$

with the initial conditions $x(0) = x_o$ and $\dot{x}(0) = \dot{x}_o$.

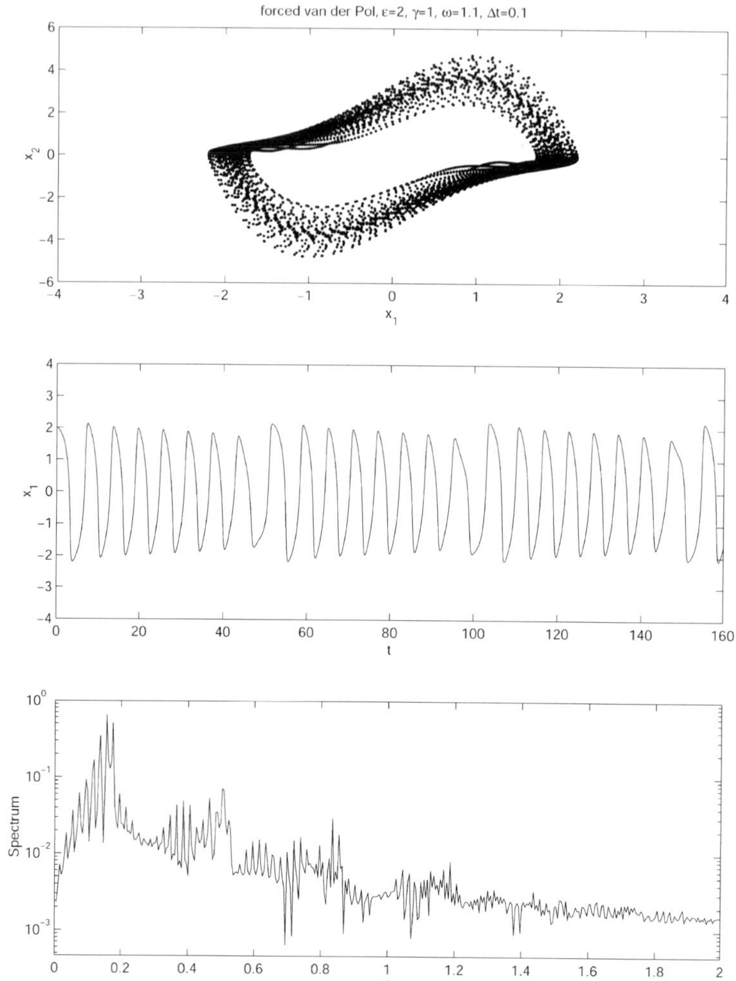


Figure 3.25. A quasi-periodic solution of forced van der Pol's equation.

*Duffing*

Consider another paradigm of nonlinear equations widely encountered in engineering [2]: Duffing's equation,

$$\ddot{x} + \epsilon \dot{x} + \alpha x + \beta x^3 = 0, \tag{1.81}$$

with the initial conditions $x(0) = x_o$ and $\dot{x}(0) = \dot{x}_o$.

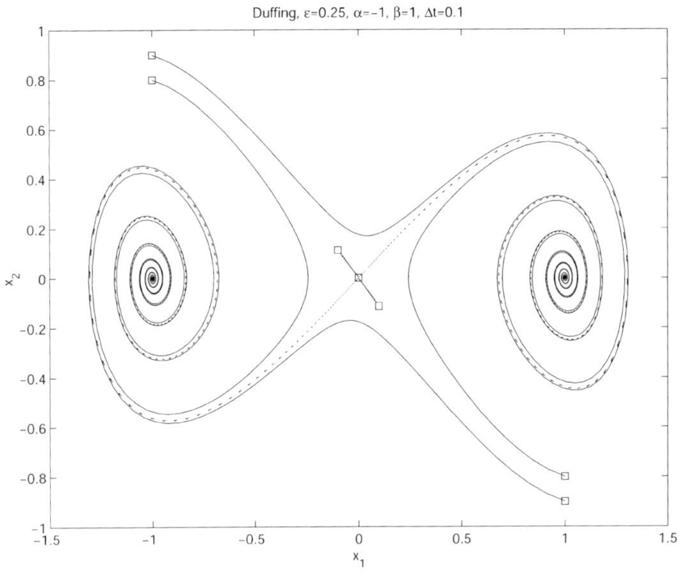


Figure 3.26. Solutions of Duffing's equation.

Similarly, Eq. (1.81) can be also written as

$$\begin{aligned} \dot{x}_1 &= x_2, \\ \dot{x}_2 &= -\alpha x_1 - \beta x_1^3 - \epsilon x_2, \end{aligned} \tag{1.82}$$

and the initial conditions correspond to $\mathbf{x}_o = \langle x_o \quad \dot{x}_o \rangle$.

Thus, if $\alpha = -1$ and $\beta = 1$, i.e., $f_1 = x_2$ and $f_2 = x_1 - x_1^3 - \epsilon x_2$, the stationary point $\mathbf{x}_o$ of Eq. (1.82) corresponds to (0, 0), (1, 0), and (−1, 0).

When $\mathbf{x}_o = (0, 0)$, we have

$$\mathbf{Df}(\mathbf{x}_o) = \begin{bmatrix} 0 & 1 \\ 1 & -\epsilon \end{bmatrix}. \tag{1.83}$$

Thus, the eigenvalues of $\mathbf{Df}(\mathbf{x}_o)$ can be expressed as $\lambda_1 = -\epsilon/2 + \sqrt{1 + (\epsilon/2)^2}$ and $\lambda_2 = -\epsilon/2 - \sqrt{1 + (\epsilon/2)^2}$. Obviously, the fixed point (0, 0) is a saddle point.

When $\mathbf{x}_o = (1, 0)$, or $(-1, 0)$, we have

$$\mathbf{Df}(\mathbf{x}_o) = \begin{bmatrix} 0 & 1 \\ -2 & -\epsilon \end{bmatrix}. \tag{1.84}$$

Thus, the eigenvalues of $\mathbf{Df}(\mathbf{x}_o)$ can be expressed as $\lambda_1 = -\epsilon/2 + i\sqrt{2 - (\epsilon/2)^2}$ and $\lambda_2 = -\epsilon/2 - i\sqrt{2 - (\epsilon/2)^2}$. The fixed points (1, 0) and (−1, 0) are asymptotically stable points, i.e., sinks. All three fixed points are illustrated in Figure 3.26, where solutions stemming from various initial conditions are presented.

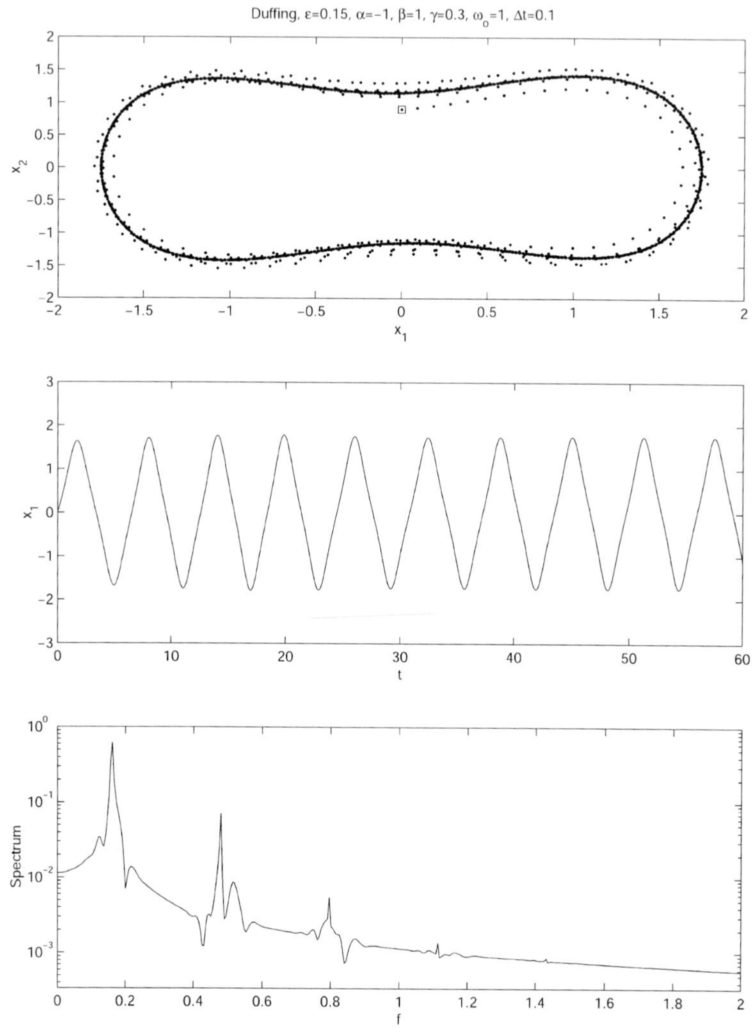


Figure 3.27. A periodic-1 solution of Duffing's equation.

Again, the solutions become more interesting when we consider the following forced Duffing's equation:

$$\ddot{x} + \epsilon\dot{x} + \alpha x + \beta x^3 = \gamma \cos \omega_o t, \tag{1.85}$$

with the initial conditions $x(0) = x_o$ and $\dot{x}(0) = \dot{x}_o$.

A periodic-1 solution is shown in Figure 3.27, and a periodic-3 solution in Figure 3.28. As suggested in Figure 3.27, the convergence to the periodic-1 limit

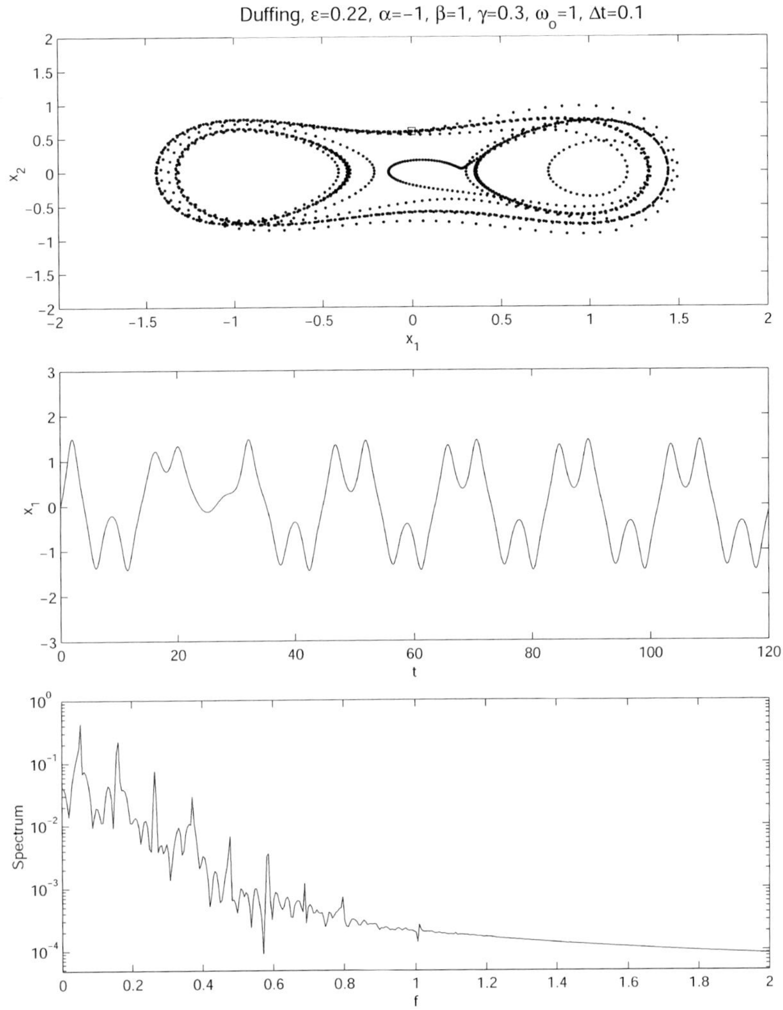


Figure 3.28. A periodic-3 solution of Duffing's equation.

cycle is oscillatory. Although subharmonics of the time-periodic nonautonomous system with minimum period $T_o$ bear resemblance to period-doubling cases, we must point out that subharmonics of different orders can exist in Duffing's equation for the same set of parameters. In fact, for the same set of parameters, $\epsilon = 0.15$, $\gamma = 0.3$, $\omega_o = 1$, we obtain a chaotic solution as shown in Figure 3.29 rather than a periodic-1 solution as shown in Figure 3.27 by selecting $x_o = 0$ and $\dot{x}_o = 0.9$ as opposed to $x_o = 0.9$ and $\dot{x}_o = 0$.

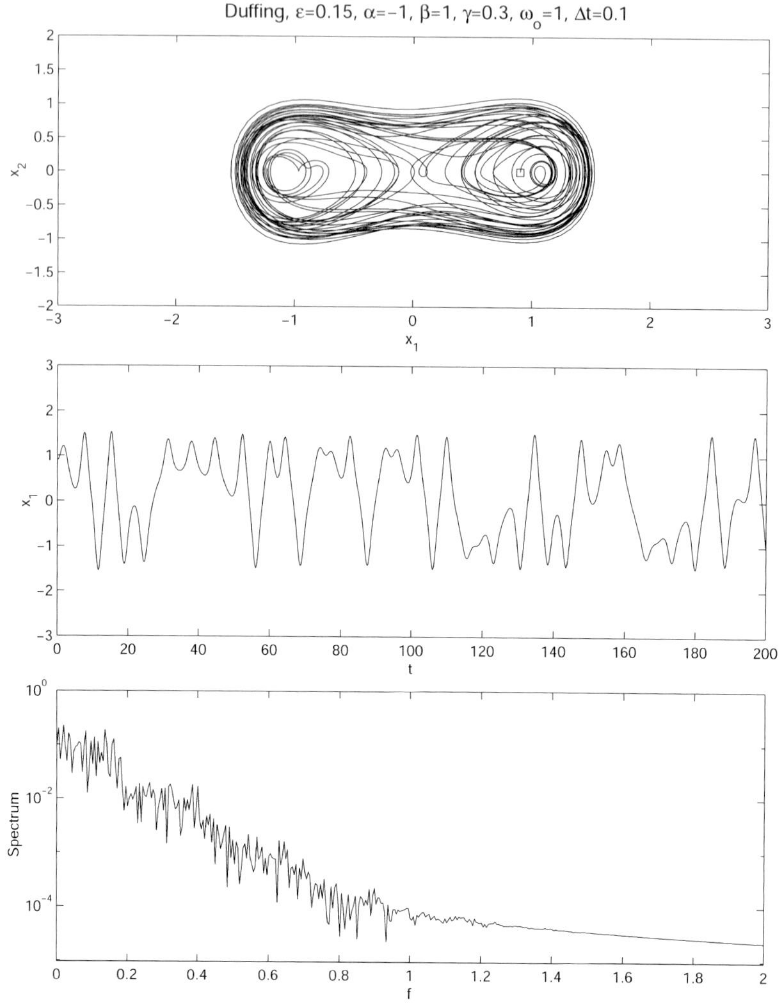


Figure 3.29. A chaotic solution of Duffing's equation.

To investigate the onset of chaos, we introduce the following Rössler's equation:

$$\begin{aligned} \dot{x}_1 &= -x_2 - x_3, \\ \dot{x}_2 &= x_1 + 0.2x_2, \\ \dot{x}_3 &= 0.2 + (x_1 - c)x_3, \end{aligned} \tag{1.86}$$

where $c$ is a variable parameter.

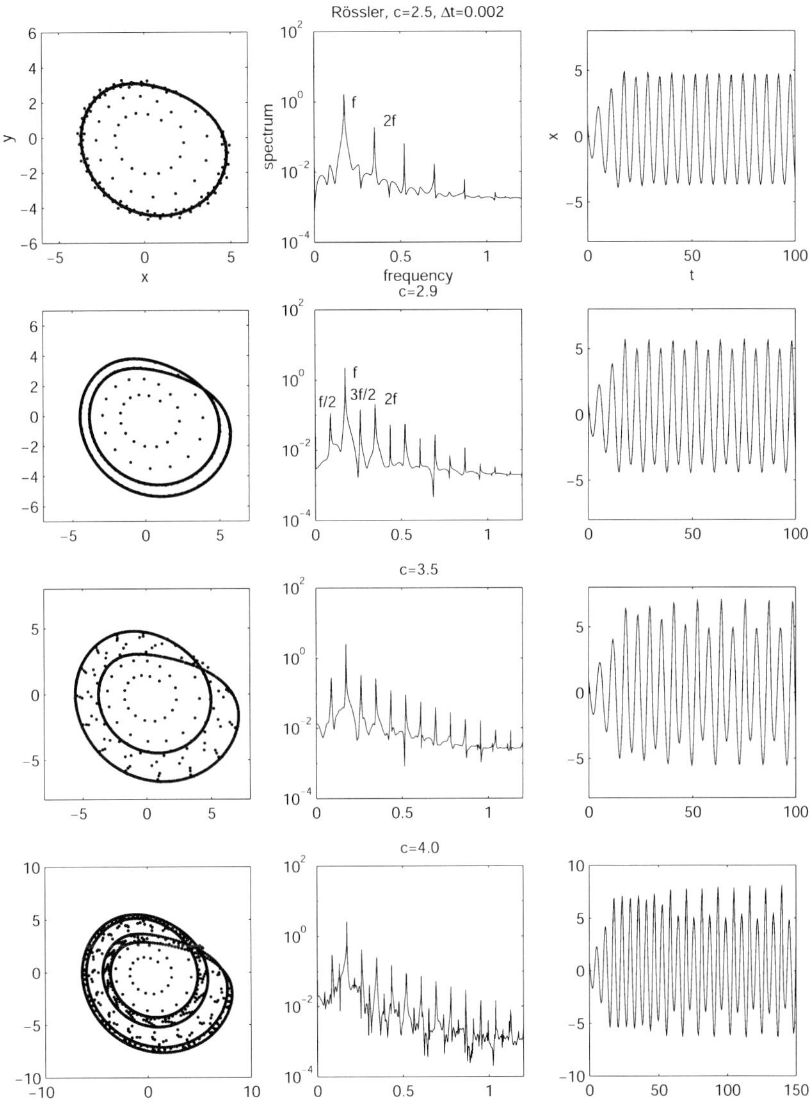


Figure 3.30. Period-doubling solutions of Rössler's equation.

Figure 3.30 shows the gradual process of periodic doubling as parameter $c$ increases and Figure 3.31 demonstrates the final chaotic solution at $c = 5.7$. Two other typical examples of chaos are Chua's equation and Lorenz's equation.

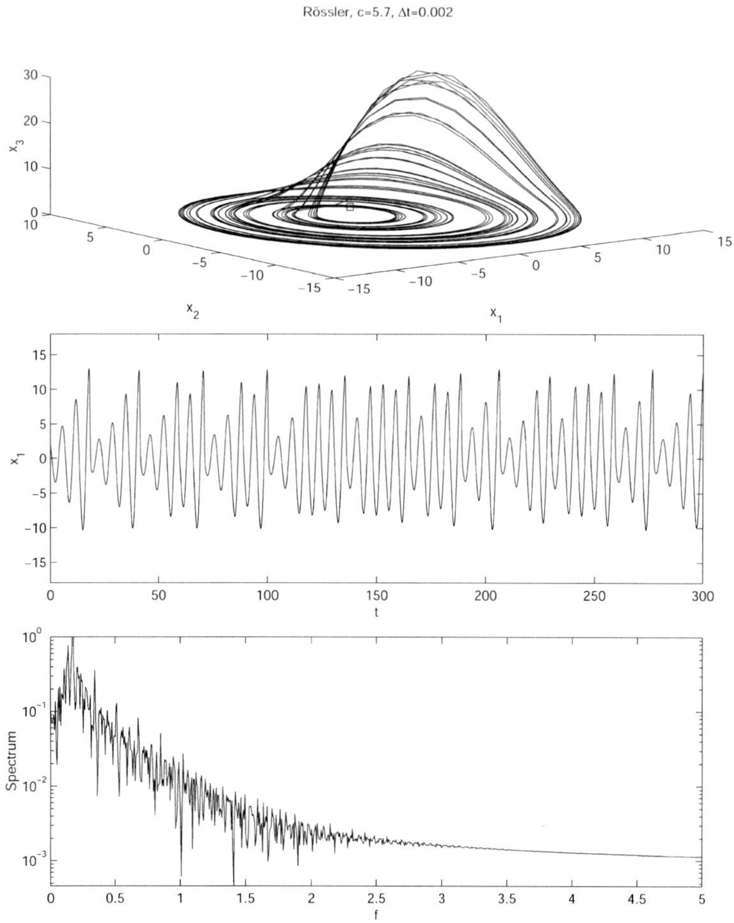


Figure 3.31. A chaotic solution of Rössler's equation.

Chua's equation is expressed as

$$\begin{aligned}\dot{x}_1 &= \alpha\big(x_2 - h(x_1)\big),\\ \dot{x}_2 &= x_1 - x_2 + x_3,\\ \dot{x}_3 &= -\beta x_2,\end{aligned} \tag{1.87}$$

with a piecewise-linear function $h(x_1)$,

$$h(x_1) = \begin{cases} m_1 x_1 + (m_0 - m_1), & x_1 \geqslant 1,\\ m_0 x_1, & |x_1| \leqslant 1,\\ m_1 x_1 - (m_0 - m_1), & x_1 \leqslant -1,\end{cases} \tag{1.88}$$

where $\alpha$, $\beta$, $m_0$, and $m_1$ are variable parameters.

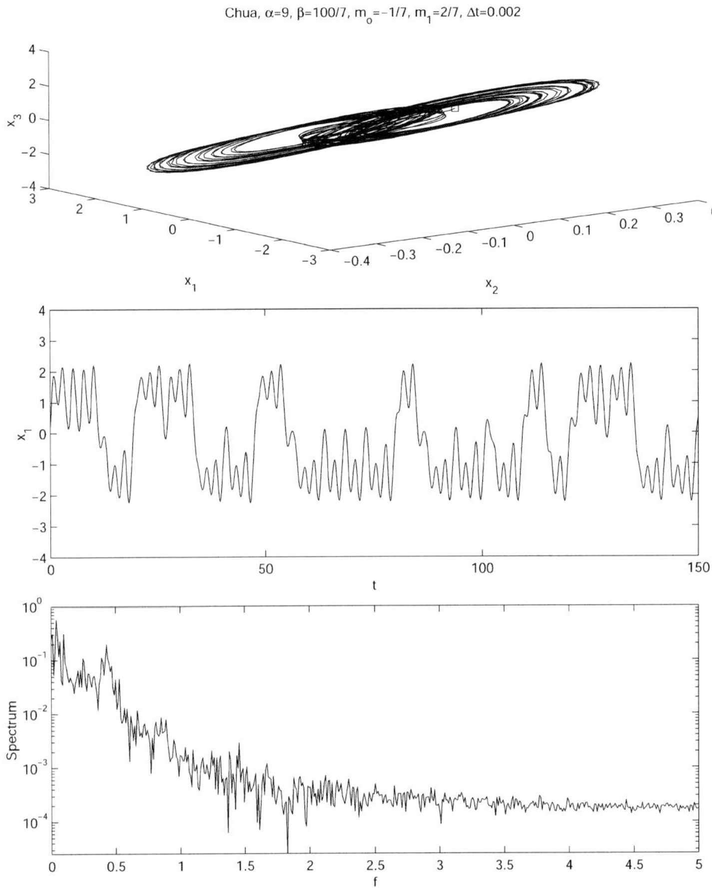


Figure 3.32. A chaotic solution of Chua's equation.

Lorenz' equation is often written as

$$\begin{aligned}\dot{x}_1 &= 10(x_2 - x_1),\\ \dot{x}_2 &= r x_1 - x_2 - x_1 x_3,\\ \dot{x}_3 &= -\frac{8}{3} x_3 + x_1 x_2,\end{aligned} \tag{1.89}$$

where $r$ is a variable parameter.

Figures 3.32 and 3.33 illustrate typical chaotical solutions. Essentially, the common feature of such states is the continuous distribution of their spectra. Chaos will be discussed further later in this chapter.

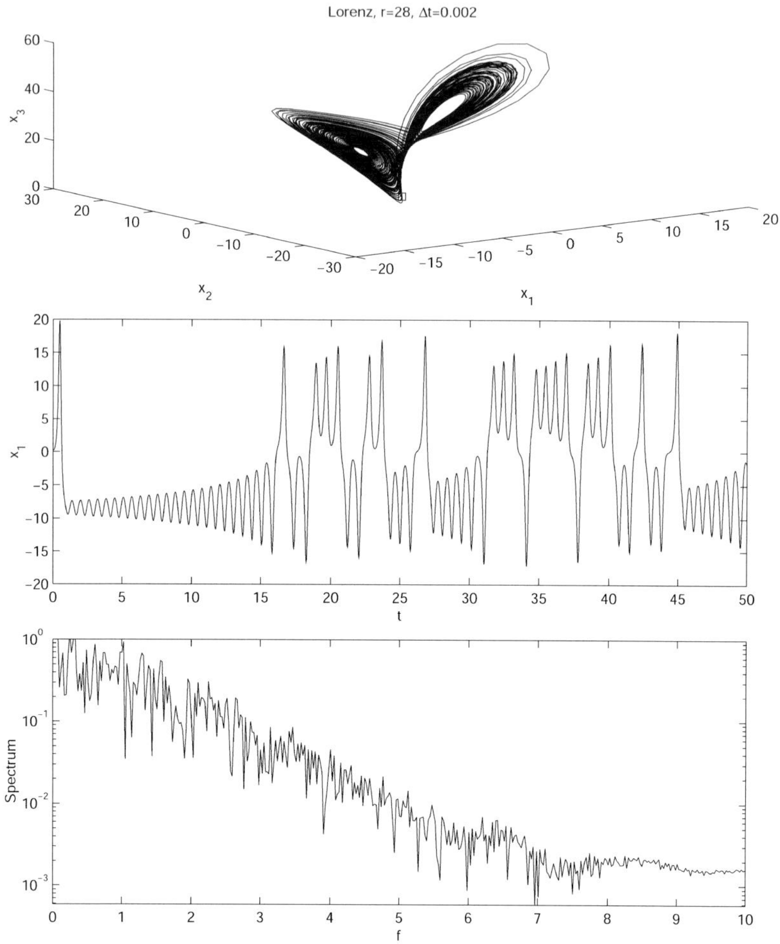


Figure 3.33. A chaotic solution of Lorenz's equation.

## 3.2. Lyapunov–Schmidt method

We now present a study on the critical time step for numerical integration based on the Runge–Kutta method of the monodromy matrix (the fundamental matrix solution) associated with a set of $n$ first-order linear ordinary differential equations with periodic coefficients. By applying the Lyapunov–Schmidt method, for any dimension $n$ and systems which are perturbations of autonomous systems, we obtain an approximation of the critical time step which involves the autonomous part as well as periodic perturbation.

In practice, in the absence of mathematical analysis of numerical errors or convergence, we can keep halving the step size $h$ until the results of successive computations over a given time interval $t$ show no perceptible difference. We must point out that for some models, such as $\dot{x} = t - x^2$, numerical chaos can be obtained with a sufficiently long time interval.

In many engineering applications, such as the vibration of pipes conveying pulsatile flows, column structures under periodic axial loading, and moving webs subjected to periodic excitation forces on boundaries, we encounter linear ordinary differential equations with periodic coefficients [193,252,292,296,299].

For low-dimensional systems, the traditional approach is the Galerkin–Ritz method, named after Boris Grigoryevich Galerkin (1871–1945) and Walther Ritz (1878–1909). Using one or two terms of the series expansions can be effective in determining the monodromy matrix (see, for example, [285]). For systems of very large dimensions, we often incorporate the direct time integration for computation of the monodromy matrix. Because we deduce the dynamic stability from the eigenvalues of the monodromy matrix [4,132,283], it is critical to understand the accuracy and stability of the numerical schemes used to derive the monodromy matrix. In practice, especially when dealing with large dynamical systems, the efficiency of the numerical algorithm depends on the choice of time step [296,299]. Although the critical time steps of various numerical schemes used for linear ordinary differential equations with constant coefficients are well understood [124], no available literature exists on *a priori* selection of the time step in the numerical integration of the monodromy matrix. This section pursues this topic and aims to provide the critical time step as a function of key parameters of dynamical systems with periodic perturbations.

Let us briefly describe the ideas contained in this section. Suppose that $\mathbf{A}(t)$, a continuous matrix $\in \mathcal{R}^{n\times n}$, is periodic in $t$ of period $T_o$, and consider the first order differential system

$$\dot{\mathbf{x}} = \mathbf{A}(t)\mathbf{x}. \tag{2.90}$$

If $\mathbf{X}(t)$, with $\mathbf{X}(0) = \mathbf{I}$, is an $n \times n$ matrix solution of Eq. (2.90), then the monodromy matrix is defined as $\mathbf{X}(T_o)$. The eigenvalues of this matrix are the Floquet multipliers of Eq. (2.90). If each Floquet multiplier has modulus less than one, then the origin is exponentially stable. If at least one multiplier has modulus greater than one, then the origin is unstable. A complex number $\rho = e^{\lambda T_o}$ is a Floquet multiplier of Eq. (2.90) if and only if there is a nonzero $n$-dimensional vector function $\mathbf{p}(t)$, periodic of period $T_o$, such that $\mathbf{x}(t) = e^{\lambda t}\mathbf{p}(t)$ is a solution of Eq. (2.90). In applications, the matrix in Eq. (2.90) is given as $\mathbf{A}(t, \mathbf{c})$, where $\mathbf{c} \in \mathcal{R}^s$ designates physical parameters. The problem is then to find the regions of stability in parameter space and, especially, to find the surfaces in parameter space that represent surfaces of transition from stability to instability. In this section, we consider dynamical systems with two parameters, i.e., $s = 2$.

For the determination of the approximate Floquet multipliers of the perturbed system, we often use the Lyapunov–Schmidt method, which works as follows: if the Floquet multipliers of the unperturbed system are simple, then the determination of the Floquet multipliers of the perturbed system reduces the determination of the eigenvalues of an $n \times n$ matrix to the solution of $n$ one-dimensional problems. Furthermore, each Floquet multiplier can be given to any desired degree of accuracy. In this section, we demonstrate the procedure with a second-order expansion. Although theoretically, based on the Lyapunov–Schmidt method, we can explicitly obtain the analytical approximations for the Floquet multipliers and the solutions of the perturbed system up to any desired order of accuracy, the evaluation of these algebraic expressions as functions of both $\mathbf{c}$ and $t$ is an insurmountable task, especially for problems with large dimensions. Therefore, we instead use numerical schemes to determine the monodromy matrix $\mathbf{X}(T_o)$. The monodromy matrix computation is also a very difficult and time-consuming task if the dimension $n$ of Eq. (2.90) is very large and the time step is very small. Moreover, to obtain dynamical stability regions within the parameter space of interest, we have to evaluate the monodromy matrices and their eigenvalues at various points within that parameter space. Consequently, one would like to take the largest time step possible which preserves stability of the numerical method and provides correct dynamical stability information. We refer to this number as the critical time step, defined as $\Delta t_c$.

In many applications, Eq. (2.90) is the perturbation of an autonomous one; that is, the perturbation of a linear system with constant coefficients. Knowing the complete behavior of the autonomous system should give some information about the critical step size that can be used when the perturbed system is considered. Of course, there is a critical step size for the autonomous problem, and it is not unreasonable to take this as the step size for the perturbed equation. However, such a step size uses no information whatsoever about the nature of the perturbation. It also assumes that the perturbation is very small relative to $\Delta t$ which, in general, is not true. On the other hand, if we can find a critical step size which incorporates some information about the perturbation, then it is to be expected that this numerical approach can lead to more efficient schemes and perhaps can be a guide to problems for which the perturbation is not so small.

The approach taken in this section is to obtain, using the Lyapunov–Schmidt method, the critical time step $\Delta t_c$ as a function of the perturbation terms, and, as a consequence, to select an optimum time step $\Delta t \leqslant \Delta t_c$ for the numerical integration of the monodromy matrix $\mathbf{X}(T_o)$. We note that the first few terms in the Taylor expansions of Floquet multipliers in terms of the perturbation are easy to obtain and yield reasonable approximations for the critical time step. As previously noted, it is generally felt that the step size chosen will be valid for a much wider range of perturbation. This point will be demonstrated in the third numerical example.

### 3.2.1. Floquet theory

Consider the following perturbed $n$-dimensional linear differential equations:

$$\dot{\mathbf{x}}(t) = \big(\mathbf{A}_o + \epsilon \mathbf{A}_1(t)\big)\mathbf{x}(t), \tag{2.91}$$

where $\mathbf{A}_o$ is a constant matrix $\in \mathcal{R}^{n\times n}$, $\mathbf{A}_1(t)$ is a $T_o$-periodic continuous coefficient matrix $\in \mathcal{R}^{n\times n}$, and the perturbation constant $\epsilon$ is small.

For simplicity, we assume that the monodromy matrix $e^{\mathbf{A}_o T_o}$ for Eq. (2.91) at $\epsilon = 0$ has simple eigenvalues; that is,

$$e^{(-\lambda_j+\lambda_k)T_o} \neq 1, \quad j \neq k,\ k \geqslant 1, \tag{2.92}$$

where $\lambda_m \in \mathcal{C}$, with $m = 1, \ldots, n$, represents the eigenvalues of the matrix $\mathbf{A}_o$.

If we introduce the modal matrix $\boldsymbol{\Phi} \in \mathcal{C}^{n\times n}$ and the generalized solution vector $\boldsymbol{\xi}(t) \in \mathcal{C}^n$, with

$$\boldsymbol{\Phi}^{-1}\mathbf{A}_o\boldsymbol{\Phi} = \boldsymbol{\Lambda} = \operatorname{diag}(\lambda_1, \ldots, \lambda_n), \tag{2.93}$$

and $\mathbf{x}(t) = \boldsymbol{\Phi}\boldsymbol{\xi}(t)$, we obtain

$$\dot{\boldsymbol{\xi}}(t) = \big(\boldsymbol{\Lambda} + \epsilon\bar{\mathbf{A}}_1(t)\big)\boldsymbol{\xi}(t), \tag{2.94}$$

with $\bar{\mathbf{A}}_1(t) = \boldsymbol{\Phi}^{-1}\mathbf{A}_1(t)\boldsymbol{\Phi}$.

If $\mathbf{X}_\epsilon(t)$, with $\mathbf{X}_\epsilon(0) = \mathbf{I}$, is a fundamental matrix solution of Eq. (2.91), then $\mathbf{X}_\epsilon(T_o)$ is the monodromy matrix. Moreover, $\boldsymbol{\Xi}(t) = \boldsymbol{\Phi}^{-1}\mathbf{X}_\epsilon(t)$ is a fundamental matrix solution of Eq. (2.94). Note that the monodromy matrix $\mathbf{X}_\epsilon(T_o)$ for Eq. (2.91) is an analytical function of $\epsilon$, and the eigenvalues of $\mathbf{X}_\epsilon(T_o)$ are the Floquet multipliers which we write as $e^{\mu_j(\epsilon)T_o}$. Since we are assuming that the Floquet multipliers of Eq. (2.91) for $\epsilon = 0$ are simple, it follows from Theorem 3.1.5 that the Floquet multipliers of Eq. (2.91) for $\epsilon \neq 0$ and sufficiently small are analytic functions of $\epsilon$. Therefore, we can choose the exponents $\mu_j(\epsilon)$ as analytic functions of $\epsilon$ where $\mu_j(0) = \lambda_j$, with $j = 1, \ldots, n$. As noted above, $e^{\mu_j(\epsilon)T_o}$ is a Floquet multiplier of Eq. (2.91) if and only if there is a nonzero solution $e^{\mu_j(\epsilon)T_o}\mathbf{p}_\epsilon(t)$ of Eq. (2.91) with $\mathbf{p}_\epsilon(t)$ a $T_o$-periodic $n$-dimensional vector function.

If we introduce the following transformation:

$$\boldsymbol{\xi}(t) = e^{\mu_j(\epsilon)t}\mathbf{w}(t), \tag{2.95}$$

we change the original problem to the determination of the constant $\mu_j(\epsilon) \in \mathcal{C}$, as a function of $\epsilon$, and the $T_o$-periodic vector function $\mathbf{w}(t) \in \mathcal{C}^n$. Substituting Eq. (2.95) into Eq. (2.94), we obtain

$$\dot{\mathbf{w}}(t) = \big(\boldsymbol{\Lambda} - \mu_j(\epsilon)\mathbf{I}\big)\mathbf{w}(t) + \epsilon\bar{\mathbf{A}}_1(t)\mathbf{w}(t), \tag{2.96}$$

with the identity matrix $\mathbf{I} \in \mathcal{R}^{n\times n}$.

For the application of the Lyapunov–Schmidt method, we need the following well-known results based on Fredholm alternative in Chapter 2.

THEOREM 3.2.1. *Suppose that* $\mathbf{B}_o(t)$ *is a* $T_o$*-periodic continuous* $n \times n$ *matrix, and* $\mathbf{f}(t)$ *is a continuous* $T_o$*-periodic* $n$*-dimensional vector function and consider the equation*

$$\dot{\mathbf{w}}(t) = \mathbf{B}_o(t)\mathbf{w}(t) + \mathbf{f}(t). \tag{2.97}$$

*There exists a* $T_o$*-periodic solution* $\mathbf{w}(t)$ *of Eq.* (2.97) *if and only if*

$$\int_0^{T_o} \hat{\mathbf{w}}(s)\mathbf{f}(s)\,ds = 0 \tag{2.98}$$

*for all* $T_o$*-periodic solutions of the adjoint equation*

$$\dot{\hat{\mathbf{w}}}(t) = -\hat{\mathbf{w}}(t)\mathbf{B}_o(t), \tag{2.99}$$

*where* $\hat{\mathbf{w}}(t)$ *is an* $n$*-dimensional row vector.*

*In addition, if there is a* $T_o$*-periodic solution* $\mathbf{w}^*(t)$ *of Eq.* (2.97), *and* $\mathbf{w}_1(t), \ldots, \mathbf{w}_k(t)$ *are the* $T_o$*-periodic solutions of the equation*

$$\dot{\mathbf{w}}(t) = \mathbf{B}_o(t)\mathbf{w}(t), \tag{2.100}$$

*then every* $T_o$*-periodic solution of Eq.* (2.97) *has the form* $\mathbf{w}(t) = \sum_{j=1}^{k} c_j \mathbf{w}_j(t) + \mathbf{w}^*(t)$.

THEOREM 3.2.2. *There exists a unique* $T_o$*-periodic solution of Eq.* (2.97) *for every* $\mathbf{f}(t)$ *if and only if there is no non-trivial* $T_o$*-periodic solution of Eq.* (2.100). *The resulting* $T_o$*-periodic solution* $\mathbf{w}^*(t, \mathbf{f})$ *is a continuous linear functional on* $\mathbf{f}(t)$; *that is, there is a constant* $K$ *such that* $\sup_t \|\mathbf{w}^*(t, \mathbf{f})\| \leqslant K \sup_t \|\mathbf{f}(t)\|$ *for any function* $\mathbf{f}(t)$.

The first part of this result is obvious from Theorem 3.2.1. The second part requires additional information about the $T_o$-periodic solution and is not completely trivial.

In preparation for the reduction from a multi-dimensional problem to a scalar one, let

$$\boldsymbol{\xi}(t) = \begin{bmatrix} \xi_1(t) \\ \boldsymbol{\eta}(t) \end{bmatrix}, \tag{2.101}$$

with $\xi_1(t) \in \mathcal{C}$ and $\boldsymbol{\eta}(t) \in \mathcal{C}^{n-1}$, and rewrite matrices $\bar{\mathbf{A}}_1(t)$ and $\boldsymbol{\Lambda}$ as

$$\bar{\mathbf{A}}_1(t) = \begin{bmatrix} \bar{A}_{11}(t) & \bar{\mathbf{A}}_{12}(t) \\ \bar{\mathbf{A}}_{21}(t) & \bar{\mathbf{A}}_{22}(t) \end{bmatrix}, \tag{2.102}$$

$$\boldsymbol{\Lambda} = \text{diag}(\lambda_1, \bar{\boldsymbol{\Lambda}}) \quad \text{and} \quad \bar{\boldsymbol{\Lambda}} = \text{diag}(\lambda_2, \ldots, \lambda_n). \tag{2.103}$$

Thus, the original system depicted with Eq. (2.94) can be decomposed into one scalar and one vector equation:

$$\dot{\xi}_1(t) = \lambda_1 \xi_1(t) + \epsilon \bar{A}_{11}(t)\xi_1(t) + \epsilon \bar{\mathbf{A}}_{12}(t)\boldsymbol{\eta}(t), \tag{2.104}$$

$$\dot{\boldsymbol{\eta}}(t) = \bar{\boldsymbol{\Lambda}}\boldsymbol{\eta}(t) + \epsilon \bar{\mathbf{A}}_{21}(t)\xi_1(t) + \epsilon \bar{\mathbf{A}}_{22}(t)\boldsymbol{\eta}(t). \tag{2.105}$$

Furthermore, we substitute in Eqs. (2.104) and (2.105) the following change of variables,

$$\boldsymbol{\xi}(t) = \begin{bmatrix} \xi_1(t) \\ \boldsymbol{\eta}(t) \end{bmatrix} = e^{\tau_1 t} \begin{bmatrix} \bar{\xi}_1(t) \\ \bar{\boldsymbol{\eta}}(t) \end{bmatrix}, \tag{2.106}$$

and obtain

$$\dot{\bar{\xi}}_1(t) = (\lambda_1 - \tau_1)\bar{\xi}_1(t) + \epsilon \bar{A}_{11}(t)\bar{\xi}_1(t) + \epsilon \bar{\mathbf{A}}_{12}(t)\bar{\boldsymbol{\eta}}(t), \tag{2.107}$$

$$\dot{\bar{\boldsymbol{\eta}}}(t) = (\bar{\boldsymbol{\Lambda}} - \tau_1 \mathbf{I}_{n-1})\bar{\boldsymbol{\eta}}(t) + \epsilon \bar{\mathbf{A}}_{21}(t)\bar{\xi}_1(t) + \epsilon \bar{\mathbf{A}}_{22}(t)\bar{\boldsymbol{\eta}}(t), \tag{2.108}$$

with the identity matrix $\mathbf{I}_{n-1} \in \mathcal{R}^{(n-1)\times(n-1)}$.

For the original problem (2.91) to have $e^{\tau_1 T_o}$ as a Floquet multiplier, we need to determine $\tau_1$ so that Eqs. (2.107) and (2.108) have a non-trivial $T_o$-periodic solution $(\bar{\xi}_1(t), \bar{\boldsymbol{\eta}}(t))$. Denote by $\mathcal{P}_{T_o}$ the space of continuous $T_o$-periodic functions. For any $\hat{\xi}_1 \in \mathcal{P}_{T_o}$ consider the equation

$$\begin{aligned}\dot{\bar{\boldsymbol{\eta}}}(t) &= \left[\bar{\boldsymbol{\Lambda}} - \tau_1 \mathbf{I}_{n-1} + \epsilon \bar{\mathbf{A}}_{22}(t)\right]\bar{\boldsymbol{\eta}}(t) + \epsilon \bar{\mathbf{A}}_{21}(t)\hat{\xi}_1(t) \\ &\stackrel{\text{def}}{=} \mathbf{B}_o(t, \epsilon, \tau_1)\bar{\boldsymbol{\eta}}(t) + \epsilon \bar{\mathbf{A}}_{21}(t)\hat{\xi}_1(t).\end{aligned} \tag{2.109}$$

For $\epsilon = 0$ and $\tau_1 = \lambda_1$, we have $\mathbf{B}_o(t, 0, \lambda_1) = \bar{\boldsymbol{\Lambda}} - \lambda_1 \mathbf{I}_{n-1}$. In addition, assumption (2.92) implies that there is no $T_o$-periodic solution of the equation

$$\dot{\mathbf{w}}(t) = (\bar{\boldsymbol{\Lambda}} - \lambda_1 \mathbf{I}_{n-1})\mathbf{w}(t). \tag{2.110}$$

That is, there is no Floquet multiplier equal to one.

Furthermore, Theorem 3.1.5 implies that the monodromy matrix of the equation

$$\dot{\mathbf{w}}(t) = \mathbf{B}_o(t, \epsilon, \tau_1)\mathbf{w}(t), \tag{2.111}$$

is continuous in $\epsilon$ and $\tau_1$, and that there is no Floquet multiplier of Eq. (2.111) equal to one if $|\epsilon|$ and $|\tau_1 - \lambda_1|$ are small. Therefore, Eq. (2.111) has no non-trivial $T_o$-periodic solution. Based on Theorems 3.1.5, 3.2.1, and 3.2.2, Eq. (2.109) has a unique $T_o$-periodic solution $\bar{\boldsymbol{\eta}}(\hat{\xi}_1, \epsilon, \tau_1)(t)$ which is linear and continuous in $\hat{\xi}_1$, and continuous and analytic in $\epsilon$ and $\tau_1$, with $\bar{\boldsymbol{\eta}}(\hat{\xi}_1, 0, \tau_1)(t) = \mathbf{0}$.

Consequently, the functions $\hat{\xi}_1(t)$ and $\bar{\boldsymbol{\eta}}(\hat{\xi}_1, \epsilon, \tau_1)(t)$ will be $T_o$-periodic solutions of Eqs. (2.107) and (2.108) if and only if $\hat{\xi}_1(t)$ is a $T_o$-periodic solution of the equation

$$\dot{\hat{\xi}}_1(t) = \big[\lambda_1 - \tau_1 + \epsilon \bar{A}_{11}(t)\big]\hat{\xi}_1(t) + \epsilon \bar{\mathbf{A}}_{12}(t)\bar{\boldsymbol{\eta}}(\hat{\xi}_1, \epsilon, \tau_1)(t)$$
$$\stackrel{\text{def}}{=} \alpha(\lambda_1 - \tau_1, \epsilon, t)\hat{\xi}_1(t) + \epsilon \bar{\mathbf{A}}_{12}(t)\bar{\boldsymbol{\eta}}(\hat{\xi}_1, \epsilon, \tau_1)(t). \qquad (2.112)$$

We must now determine $\tau_1 = \tau_1(\epsilon)$ in such a way that Eq. (2.112) has a $T_o$-periodic solution. Note that although Eq. (2.112) is not a differential equation, we can still analyze its properties.

Eq. (2.112) has a $T_o$-periodic solution $\hat{\xi}_1$ if and only if

$$\int_0^{T_o} \big[\alpha(\lambda_1 - \tau_1, \epsilon, s)\hat{\xi}_1(s) + \epsilon \bar{\mathbf{A}}_{12}(s)\bar{\boldsymbol{\eta}}(\hat{\xi}_1, \epsilon, \tau_1)(s)\big]\, ds = 0. \qquad (2.113)$$

Based on the previous discussions of $\bar{\boldsymbol{\eta}}(\hat{\xi}_1, \epsilon, \tau_1)(t)$, it is clear that Eq. (2.113) is linear in $\hat{\xi}_1$, and can be used to determine $\tau_1 = \tau_1(\epsilon)$ and $\hat{\xi}_1(t) = \hat{\xi}_1(\epsilon, t)$. Note that, based on Theorem 3.1.5, all functions are analytical in $\epsilon$ and we may theoretically determine the power series of these quantities to any desired accuracy. To reiterate, we want to determine $\tau_1 = \tau_1(\epsilon)$ so that $e^{\tau_1(\epsilon)T_o}$ satisfies $\tau_1(0) = \lambda_1$; that is, the Floquet multiplier is close to $e^{\lambda_1 T_o}$. As remarked earlier, $\tau_1(\epsilon)$ is also an analytical function of $\epsilon$ and can be written as

$$\tau_1(\epsilon) = \lambda_1 + \beta_1\epsilon + \beta_2\epsilon^2 + O(\epsilon^3), \qquad (2.114)$$

where $\beta_1$ and $\beta_2$ are constants to be determined.

Furthermore, at $\epsilon = 0$, Eq. (2.112) becomes $\dot{\hat{\xi}}_1(t) = 0$, which means that $\hat{\xi}_1$ is a constant. In order to normalize $\hat{\xi}_1(\epsilon, t)$ so that $\hat{\xi}_1(0, t) = 1$, we let

$$\hat{\xi}_1(\epsilon, t) = 1 + \epsilon\hat{\xi}_{11}(t) + \epsilon^2\hat{\xi}_{12}(t) + O(\epsilon^3), \qquad (2.115)$$

and deduce from Eq. (2.109) (refer to Hale [131] for details)

$$\bar{\boldsymbol{\eta}}\big(\hat{\xi}_1, \epsilon, \tau_1(\epsilon)\big)(t) = \epsilon\bar{\boldsymbol{\eta}}_1(t) + O(\epsilon^2), \qquad (2.116)$$

with

$$\bar{\boldsymbol{\eta}}_1(t) = e^{(\bar{\mathbf{A}} - \lambda_1\mathbf{I}_{n-1})t}\big[e^{-(\bar{\mathbf{A}} - \lambda_1\mathbf{I}_{n-1})T_o} - \mathbf{I}\big]^{-1} \int_t^{t+T_o} e^{-(\bar{\mathbf{A}} - \lambda_1\mathbf{I}_{n-1})s}\bar{\mathbf{A}}_{21}(s)\, ds.$$

Note that, if $\bar{\mathbf{A}}_{21}(t)$ is not a function of $t$, we can simply have

$$\bar{\boldsymbol{\eta}}\big(\hat{\xi}_1, \epsilon, \tau_1(\epsilon)\big)(t) = -\epsilon(\bar{\mathbf{A}} - \lambda_1\mathbf{I}_{n-1})^{-1}\bar{\mathbf{A}}_{21} + O(\epsilon^2). \qquad (2.117)$$

Furthermore, substituting Eq. (2.116) in Eq. (2.113), we obtain,

$$\beta_1 = \frac{1}{T_o}\int_0^{T_o} \bar{A}_{11}(s)\, ds, \qquad (2.118)$$

$$\beta_2 = \frac{1}{T_o}\int_0^{T_o} \bar{\mathbf{A}}_{12}(s)\bar{\boldsymbol{\eta}}_1(s)\,ds. \tag{2.119}$$

Therefore, we see that Eq. (2.112) up to terms of order $\epsilon^3$ is equivalent to the scalar ordinary differential equation

$$\dot{\bar{\xi}}_1(t) = \{\epsilon[\bar{A}_{11}(t) - \beta_1] + \epsilon^2[\bar{\mathbf{A}}_{12}(t)\bar{\boldsymbol{\eta}}_1(t) - \beta_2]\}\bar{\xi}_1(t) + O(\epsilon^3). \tag{2.120}$$

The first-order approximation is then given by

$$\dot{\xi}_1(t) = a(\epsilon, t)\xi_1(t) + O(\epsilon^2), \tag{2.121}$$

with

$$a(\epsilon, t) = \lambda_1 + \epsilon\bar{A}_{11}(t),$$

which is much easier to calculate than Eq. (2.120) for systems of high dimensions.

Note that the same approaches should be applied to all eigenvalues of the system. It is clear at this point that although we can derive approximations of the solutions of Eqs. (2.104) and (2.105) with any desired order of accuracy, the analytical evaluation of the matrices $\bar{\mathbf{A}}_{21}(t)$, $\bar{\mathbf{A}}_{12}(t)$, and $\bar{\mathbf{A}}_{22}(t)$ can be very challenging, if at all feasible, especially for high-dimensional problems.

### *3.2.2. Critical time step*

Let $\mathbf{N} = \mathbf{X}(T_o)$ be the monodromy matrix. With $\mathbf{X}(0) = \mathbf{I}$, we can determine the monodromy matrix $\mathbf{N}$ by numerically integrating Eq. (2.90) for $t \in [0, T_o]$. However, due to the presence of the time-varying coefficient matrix $\mathbf{A}(t) = \mathbf{A}_o + \epsilon\mathbf{A}_1(t)$, we cannot use full implicit numerical schemes. This limits us to explicit schemes which often require small time steps. In this section, we investigate the use of the second-order Runge–Kutta (RK2) scheme, depicted as follows:

$$\mathbf{x}^{k+1} = \mathbf{x}^k + \Delta t(\mathbf{k}_1 + \mathbf{k}_2)/2, \tag{2.122}$$

where

$$\mathbf{k}_1 = \mathbf{A}^k\mathbf{x}^k \quad \text{and} \quad \mathbf{k}_2 = \mathbf{A}^{k+1}(\mathbf{x}^k + \Delta t\mathbf{A}^k\mathbf{x}^k),$$

with $\mathbf{A}^k = \mathbf{A}(k\Delta t)$ and $0 \leqslant k \leqslant T_o/\Delta t$.

Note that the $m$th column of the matrix $\mathbf{N}$ corresponds to the numerical solution of Eq. (2.90) with the $m$th column of the identity matrix $\mathbf{I}$ as the initial condition. Therefore, the numerical integration illustrated in (2.122) has to be performed $n$ times to form a monodromy matrix. In general, the explicit nature of the Runge–Kutta scheme requires the use of excessively small time steps and the construction of the monodromy matrix can be very expensive. Furthermore, in order to obtain the dynamical stability regions, we have to compute the monodromy matrices and their eigenvalues at all parameter spatial grid points within a parameter space subdivided into parameter spatial divisions. Because we do not know *a priori* the structure of the matrix $\mathbf{N}$, and because a poorly constructed matrix $\mathbf{N}$ can lead to incorrect conclusions of the dynamical instability, we need to know the critical time step $\Delta t_c$ before the numerical integration, and try to avoid the costly trial and error process.

Because we obtain Eq. (2.94) from Eq. (2.91) through the transformation with the constant modal matrix $\boldsymbol{\Phi}$, the scheme presented in (2.122) can be applied equivalently to Eq. (2.94). Hence, for the unperturbed system $\mathbf{A}^k = \mathbf{A}_o$, the equivalent scheme is written as:

for $m = 1, \dots, n$

$$\xi_m^{k+1} = \xi_m^k + \Delta t(\tilde{k}_1 + \tilde{k}_2)/2, \tag{2.123}$$

where

$$\tilde{k}_1 = \lambda_m \xi_m^k,$$
$$\tilde{k}_2 = \lambda_m\left(\xi_m^k + \Delta t \lambda_m \xi_m^k\right),$$

or

$$\xi_m^{k+1} = G_m \xi_m^k, \tag{2.124}$$

with $G_m = 1 + \lambda_m \Delta t + (\lambda_m \Delta t)^2/2$.

Furthermore, the critical time step $\Delta t_c$, governed by the stability requirement of the RK2 scheme, corresponds to

$$|G_m| \leqslant 1, \quad \forall \lambda_m, \text{ with } m = 1, \dots, n. \tag{2.125}$$

Of course, we need to have $\mathrm{Re}(\lambda_m) < 0$, based on the stability of the unperturbed system. For the perturbed system, the matrix $\bar{\mathbf{A}}_1(t)$ is, in general, not diagonal, and a direct study of the scheme in the form of (2.122) becomes very difficult. Based on the discussion in previous sections, we can use the Lyapunov–Schmidt method to transform the original non-autonomous linear system in Eq. (2.94) to $n$ one-dimensional problems in the form of (2.121), the equivalent scalar non-autonomous equation as the first-order approximation of $\epsilon$. Again, because application of the scheme in (2.122) to Eq. (2.91) is equal to its application to Eqs. (2.94) and (2.121), we are prepared to discuss the numerical stability issues for the perturbed systems.

Introducing the RK2 scheme in (2.122) to Eq. (2.121), we obtain

$$\xi_1^{k+1} = G_1^k \xi_1^k + O\left(\epsilon^2\right), \tag{2.126}$$

where

$$G_1^k = 1 + \left[a^k(\epsilon) + a^{k+1}(\epsilon)\right]\Delta t/2 + a^k(\epsilon)a^{k+1}(\epsilon)\Delta t^2/2, \tag{2.127}$$

with $\bar{A}_{11}^k = \bar{A}_{11}(k\Delta t)$, $a^k(\epsilon) = a(\epsilon, k\Delta t) = \lambda_1 + \epsilon\bar{A}_{11}^k$, and $0 \leqslant k \leqslant T_o/\Delta t$.

We introduce the following supnorm $\|G_1\|_{\text{sup}}$, such that,

$$\|G_1\|_{\text{sup}} = \sup_{0\leqslant k\leqslant T_o/\Delta t} \left|G_1^k\right|, \tag{2.128}$$

and without loss of generality, if we apply the same approach to all the eigenvalues $\lambda_m$, with $m = 1, \ldots, n$, the critical time step $\Delta t_c$ satisfies

$$\|G_m\|_{\text{sup}} \leqslant 1, \quad \forall\lambda_m, \text{ with } m = 1, \ldots, n. \tag{2.129}$$

Notice that the derivation of $\Delta t_c$ is of $O(\epsilon^2)$. With $\epsilon = 0$, (2.129) is equivalent to (2.125). Hence, we denote $\Delta t_c(0)$ as the critical time step of the corresponding autonomous system, and $\Delta t_c(\epsilon)$ as the critical time step of the non-autonomous system with the perturbation $\epsilon$.

To confirm the proposed critical time step as a function of the autonomous part as well as the periodic perturbation, we present three paradigms of Mathieu–Hill equations, named after Émile Léonard Mathieu (1835–1890) and George William Hill (1838–1914). We begin by introducing the following second-order linear system:

$$\mathbf{M}_o\ddot{\mathbf{Y}}(t) + \mathbf{C}_o\dot{\mathbf{Y}}(t) + \mathbf{K}_o\mathbf{Y}(t) = \mathbf{0}, \tag{2.130}$$

with solution vector $\mathbf{Y}(t) \in \mathcal{R}^r$, and constant coefficient matrices $\mathbf{M}_o$, $\mathbf{C}_o$, and $\mathbf{K}_o \in \mathcal{R}^{r\times r}$.

If we assume a characteristic solution, $\mathbf{Y}(t) = e^{i\omega t}\widehat{\mathbf{Y}}$, with $i = \sqrt{-1}$, where $\widehat{\mathbf{Y}}$ represents the mode shape of the natural frequency $\omega = 2\pi f$, the stable system corresponds to $\text{Im}(\omega) \geqslant 0$ with $\text{Re}(\omega) \neq 0$. In engineering practice, we often define the buckling instability as $\text{Re}(\omega) \to 0$ with $\text{Im}(\omega) \geqslant 0$, and the flutter instability as $\text{Im}(\omega) < 0$ with $\text{Re}(\omega) \neq 0$. Moreover, having the set of $r$ second-order linear ordinary differential equations in Eq. (2.130), if we introduce a new solution vector, $\mathbf{x}(t) = \langle\mathbf{Y}(t), \dot{\mathbf{Y}}(t)\rangle \in \mathcal{R}^n$, with $n = 2r$, we can replace Eq. (2.130) with a system of $n$ first-order linear ordinary differential equations in the form of Eq. (2.91) with

$$\mathbf{A}(t) = \mathbf{A}_o = \begin{bmatrix} \mathbf{0} & \mathbf{I} \\ -\mathbf{M}_o^{-1}\mathbf{K}_o & -\mathbf{M}_o^{-1}\mathbf{C}_o \end{bmatrix}.$$

Now, let us consider a perturbation of Eq. (2.130),

$$\mathbf{M}_o\ddot{\mathbf{Y}}(t) + \big(\mathbf{C}_o + \epsilon\mathbf{C}_1(t)\big)\dot{\mathbf{Y}}(t) + \big(\mathbf{K}_o + \epsilon\mathbf{K}_1(t)\big)\mathbf{Y}(t) = \mathbf{0}, \tag{2.131}$$

where $\mathbf{C}_1(t)$ and $\mathbf{K}_1(t)$ are $T_o$-periodic coefficient matrices, and the perturbation constant $\epsilon$ is small.

Similarly, Eq. (2.131) can be written in the form of Eq. (2.91) with

$$\mathbf{A}(t) = \mathbf{A}_o + \epsilon \mathbf{A}_1(t) \quad \text{and} \quad \mathbf{A}_1(t) = \begin{bmatrix} \mathbf{0} & \mathbf{0} \\ -\mathbf{M}_o^{-1}\mathbf{K}_1(t) & -\mathbf{M}_o^{-1}\mathbf{C}_1(t) \end{bmatrix}.$$

### 3.2.3. *Mathieu–Hill system*

To implement the proposed critical time step presented in (2.129), we will study a few examples.

EXAMPLE 3.1.

$$\ddot{x} + 2\zeta\omega\dot{x} + \omega^2(1 + \epsilon \cos \omega_o t)x = 0, \tag{2.132}$$

with $T_o = 2\pi/\omega_o$, $\omega = \sqrt{2}$, $\zeta \in [0, 1]$, $\omega_o \in [2.0, 3.6]$, and $\epsilon \in [0, 1]$.

For simplicity, in order to compare with analytical solutions, we consider the damping ratio $\zeta = 0$. Hence, we have $\lambda_1 = i\omega$ and $\lambda_2 = -i\omega$. In general, we use $\lambda_{2m-1} = i\omega_m$ and $\lambda_{2m} = -i\omega_m$, with $m = 1, \ldots, r$. To determine the critical time step according to (2.129), we implement a simple program to evaluate $\|G_m\|_{\sup}$, $\forall \lambda_m$, with $m = 1, \ldots, n$, in which we compute the eigenvalues and eigenvectors of the constant matrix $\mathbf{A}_o$, and the corresponding $\bar{A}_{11}$. It is important to note that we only have to evaluate the eigensolutions of the matrix $\mathbf{A}_o$ once, and the computation effort is comparable to the determination of the eigensolutions of one monodromy matrix. Of course, prior to the expensive monodromy matrix computation for every parameter spatial grid point, using the simple program, we can conveniently determine the maximum time step satisfying (2.129) as a function of parameters. For instance, we can easily obtain $\lambda_2 = -i\omega$, $\bar{A}_{11}(t) = -i\omega \cos \omega_o t/2$, and the corresponding $a(t) = -i\omega(1 + \epsilon \cos \omega_o t/2)$. Using Eq. (2.127), it is straight-forward to derive $\Delta t_c(\epsilon)$ from (2.129) for various values of $\epsilon$ and $\omega_o$. Although we should check every eigensolution of the matrix $\mathbf{A}_o$ in (2.129), we intuitively understand that the numerical stability requirement is mainly governed by the perturbation $\epsilon$, the highest natural frequency $\omega_r$, or $\lambda_n$, and its corresponding $\bar{A}_{11}(t)$.

As shown in Figure 3.34, the critical time step $\Delta t_c(\epsilon)$ corresponding to $\|G_2\|_{\sup} = 1$ of the first example is a function of the perturbation $\epsilon$ and does not seem to be dependent on the period $T_o$ for periodic perturbations. Note that the critical time step is normalized with the critical time step for the corresponding autonomous system. If we take a time step $\Delta t = 0.66\Delta t_c(0)$ with $\Delta t_c(0) = 0.2114$, satisfying $\Delta t \leqslant \Delta t_c(\epsilon)$, $\forall \epsilon \in [0, 1]$, according to (2.129) and the discussion in the previous sections, the numerical integration scheme should be stable, and from the eigenvalues of monodromy matrices at various values of $\epsilon$ and $\omega_o$, we should obtain the correct dynamic stability regions within the parameter space

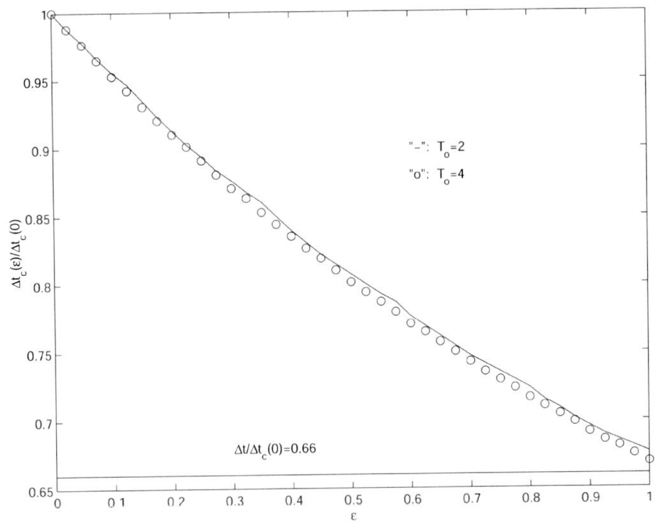


Figure 3.34. Critical time step $\Delta t_c(\epsilon)$ with $\|G_2\|_{\sup} = 1$ of Example 3.1.

of $\epsilon \in [0, 1]$ and $\omega_o \in [2.6, 3.6]$. Figure 3.35 confirms our predictions, and by comparing with the asymptotic solutions discussed in Ref. [97], it is clear that the time step selection based on (2.129) is appropriate and sufficient. As expected, the stability boundaries are smoothed and the instability regions are reduced with damping effects. Figure 3.35 also shows that when we select $\Delta t = \Delta t_c(0)$, a critical time step of the corresponding autonomous system, the dynamical stability results are very poor.

The regions of instability of the first example are presented with $\omega^2 = 2$, and the parameter space of $\epsilon \in [0, 1]$ and $\omega_o \in [2.0, 3.6]$ is obtained with $10 \times 10$ divisions. For the first example, with the correct answers in mind, we can afford to try out various time steps to empirically determine the convergence of the numerical integration of the monodromy matrix. In practice, unfortunately, we neither know *a priori* stability solutions nor can we afford the trial and error procedures for multi-dimensional problems. To further explore the main points of this section, we consider the second example.

EXAMPLE 3.2.

$$\mathbf{M}_o = \begin{bmatrix} 1 & 0 \\ 0 & 1 \end{bmatrix}, \quad \mathbf{C}_o = \begin{bmatrix} 0.1 & -0.2 \\ 0.2 & 0.2 \end{bmatrix}, \quad \mathbf{K}_o = \begin{bmatrix} 2 & 1 \\ 1 & 30 \end{bmatrix},$$
$$\mathbf{C}_1 = \begin{bmatrix} 0.1\cos\omega_o t & 0 \\ 0 & 0 \end{bmatrix}, \quad \mathbf{K}_1 = \begin{bmatrix} 2\cos\omega_o t & 0 \\ 0 & 30\cos\omega_o t \end{bmatrix}, \tag{2.133}$$

for $\epsilon \in [0, 1]$ and $\omega_o \in [1.0, 3.6]$.

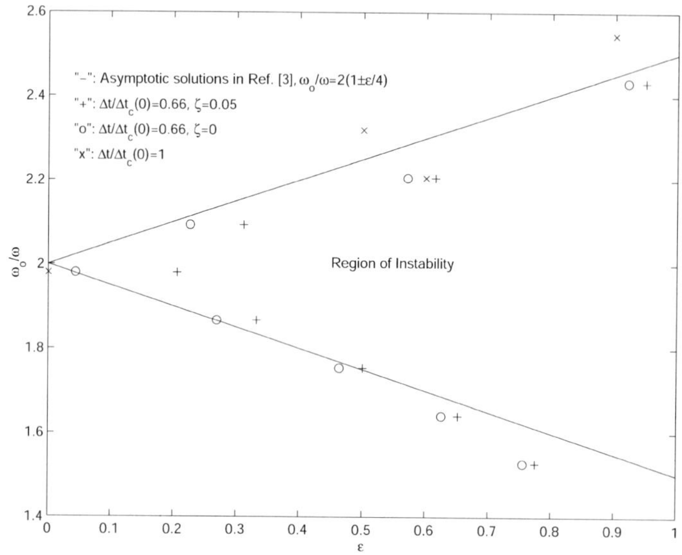


Figure 3.35. Instability region of Example 3.1.

From Eq. (2.130), based on the theorems on real operators with complex eigenvalues and eigenvectors [182,203], we can easily obtain the eigenvalues and eigenvectors of this dynamical system:

$$\boldsymbol{\Phi} = \begin{bmatrix} -0.002 & 0.626 & -0.000 & 0.008 \\ -0.006 & -0.022 & 0.111 & 0.103 \\ 0.876 & -0.028 & 0.044 & 0.000 \\ -0.031 & 0.010 & 0.554 & -0.617 \end{bmatrix} \begin{bmatrix} 1 & 1 & 0 & 0 \\ -i & i & 0 & 0 \\ 0 & 0 & 1 & 1 \\ 0 & 0 & -i & i \end{bmatrix},$$

$$\lambda_1 = -0.050 + 1.400i, \qquad \lambda_2 = -0.050 - 1.400i,$$

$$\lambda_3 = -0.100 + 5.483i, \qquad \lambda_4 = -0.100 - 5.483i.$$

It is then clear that the original system with the constant matrices defined in (2.133) is stable. Moreover, for $\lambda_4$, we have

$$\bar{A}_{11}(t) = (0.0002 - 2.7325i)\cos\omega_o t,$$

$$a(t) = (-0.100 - 5.483i) + \epsilon(0.0002 - 2.7325i)\cos\omega_o t.$$

As with the approach for the first example, with the aid of a simple program, we can easily obtain the critical time step as a function of the perturbation. As shown in Figure 3.36, we select a time step $\Delta t = 0.56\Delta t_c(0)$ with $\Delta t_c(0) = 0.1003$ prior to the monodromy matrix computation, such that $\Delta t < \Delta t_c(\epsilon)$, $\forall\epsilon \in [0, 1]$.

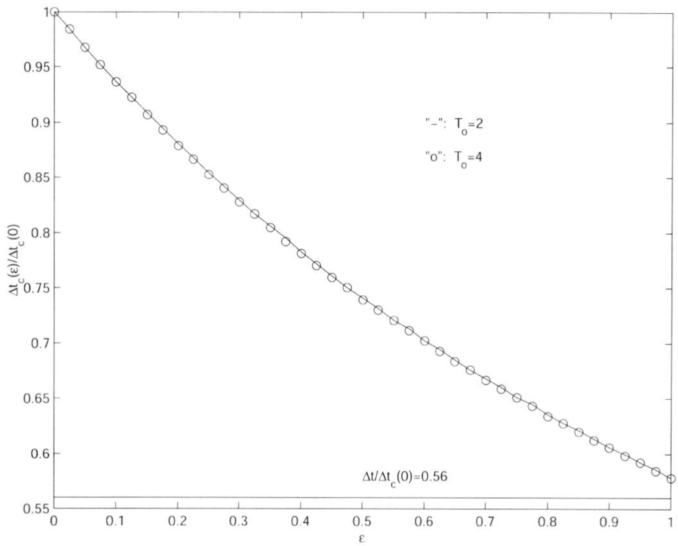


Figure 3.36. Critical time step $\Delta t_c(\epsilon)$ with $\|G_4\|_{\text{sup}} = 1$ of Example 3.2.

Figure 3.36 also indicates that the critical time step does not depend on the period of the perturbation. With the dynamical stability results obtained with a sufficiently small time step as a reference solution, it is demonstrated that the time step $\Delta t = \Delta t_c(0)$ does not provide an accurate solution, while the time step $\Delta t = 0.56\Delta t_c(0)$, judiciously chosen according to (2.129), yields much better results.

Figure 3.37 shows the regions of instability of the second example problem with $40 \times 40$ divisions in the parameter space of $\epsilon \in [0, 1]$ and $\omega_o \in [1.0, 3.6]$. In addition, the second example demonstrates that the maximum system eigenvalue $\lambda_n$ and its corresponding perturbation terms contribute significantly to the critical time step $\Delta t_c$. In general, if we are interested in the dynamical instability region around the lower harmonics and their resonances, because of the nature of the explicit scheme, the time step selection $\Delta t \leqslant \Delta t_c$ often satisfies the accuracy requirement of $\Delta t \leqslant T_o/20$.

As depicted in Figures 3.34 and 3.36, with small perturbations in the first two examples, the critical time step can be reduced to 50% of the critical time step of the corresponding autonomous system. In many engineering practices, the periodic perturbation could be significant. Because the numerical scheme illustrated in (2.122) does not limit itself to small perturbations, and is in fact applicable to general non-autonomous first-order linear ordinary differential equations. We introduce the third example in order to verify the possible extension of the approximation results of the critical time step.

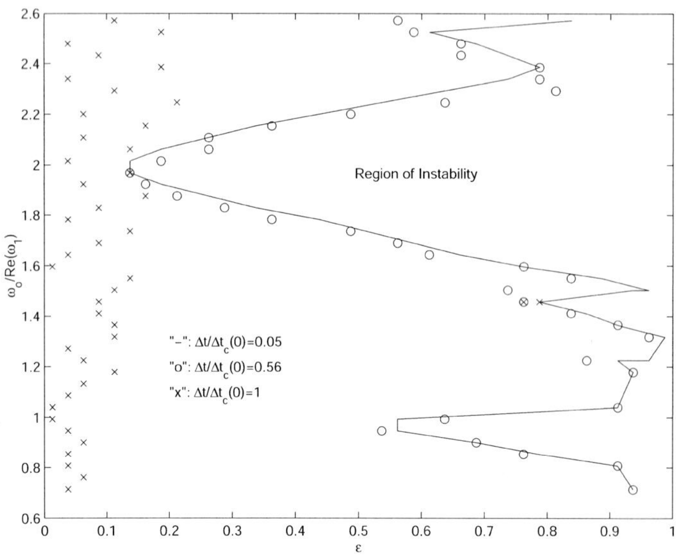


Figure 3.37. Instability region of Example 3.2.

EXAMPLE 3.3.

$$\ddot{x} + (a + b\cos t)x = 0, \tag{2.134}$$

with $a \in [0, 2]$ and $b \in [0, 4]$.

Consider $\lambda_2 = -i\sqrt{a}$. We obtain the corresponding $\bar{A}_{11}(t) = -i\sqrt{a}\cos t/2$ and $a(t) = -i\sqrt{a}(1 + b/2a\cos t)$. Figure 3.38 presents the critical time step $\Delta t_c$ as a function of the parameters $a$ and $b$ with $\Delta t_c(2, 0) = 0.2111$, $\Delta t_c(1, 0) = 0.2989$, and $\Delta t_c(0.5, 0) = 0.4233$. It is obvious that as $a$ increases, $\Delta t_c(a, b)$ decreases, and for a fixed value of $a$, with the increase of $b$, i.e., the perturbation $\epsilon = b/a$, the critical time step $\Delta t_c(a, b)$ can be reduced to a mere 10% of the critical time step of the corresponding autonomous system. As shown in Figure 3.38, the time step $\Delta t = 0.075$ satisfies (2.129), $\forall a \in [0, 2]$ and $b \in [0, 4]$, and should provide a sufficiently accurate dynamical stability result. With the reference of the result derived from a very small time step $\Delta t = 0.01\pi$ and the graph presented in Ref. [250], the results shown in Figure 3.39 with $50 \times 50$ divisions in the parameter space confirm our predictions. In addition, when we select $\Delta t = \Delta t_c(2, 0)$, the dynamical stability results are only accurate within the region of $a \in [0, 0.5]$ and $b \in [0, 1]$, where $\Delta t_c(2, 0)$ is smaller than $\Delta t_c(a, b)$. Therefore, we conjecture that (2.129) can be useful concerning problems with significant periodic perturbations.

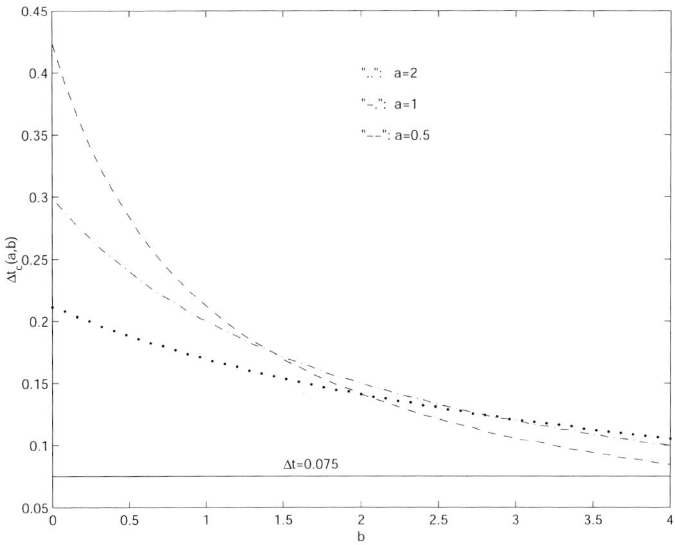


Figure 3.38. Critical time step $\Delta t_c(a, b)$ with $\|G_2\|_{\text{sup}} = 1$ of Example 3.3.

As a final remark, although the number of parameter spatial divisions does not directly affect the monodromy matrix computation, when the stability zones are very narrow, sufficiently refined parameter spatial divisions are required. Note that in order to obtain the dynamical stability information, we have to find the eigenvalues of the monodromy matrices at all parameter spatial grid points, which in fact further demonstrates the need to obtain an optimum time step *a priori*.

Although the critical time step is generally governed by the largest eigenvalue $\lambda_n$, and the corresponding perturbation term $\epsilon \bar{A}_{11}$, a simple program for the computation of $a(\epsilon, t)\Delta t = (\lambda_m + \epsilon \bar{A}_{11})\Delta t$ and $\|G_m\|_{\text{sup}}$, $\forall \lambda_m$, with $m = 1, \ldots, n$, can easily be implemented and applied to find an optimum time step $\Delta t \leqslant \Delta t_c$ prior to the monodromy matrix computation. Thus, the general procedure for the monodromy matrix computation contains two steps: first, we search for the critical time step as a function of parameters and select a time step smaller than the minimum value of the critical time step within the region of the parameter space of interest; second, we perform the numerical integration for $t \in [0, T_o]$ to obtain $\mathbf{X}(T_o)$, with $\mathbf{X}(0) = \mathbf{I}$, for all parameter spatial grid points. Of course, to derive the dynamical stability information, we need to compute the eigenvalues of the monodromy matrices.

Although we discuss extensively the second-order Runge–Kutta scheme, the same approaches can be directly applied to the other numerical integration schemes. For dynamical systems with small damping ratios, we should strive to

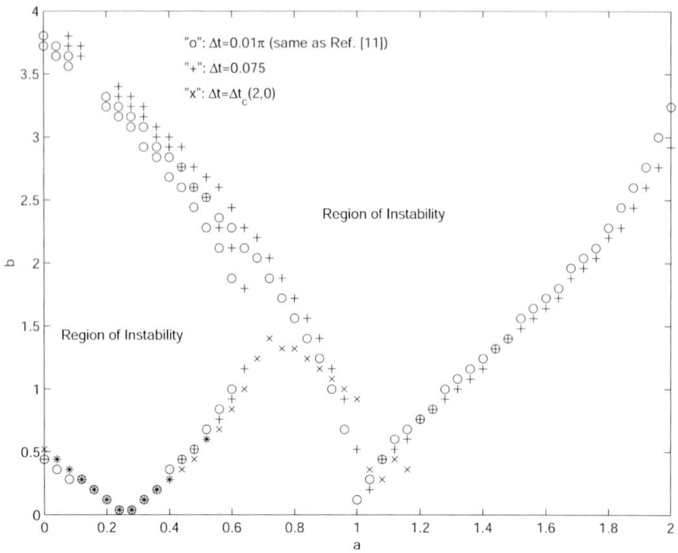


Figure 3.39. Instability region of Example 3.3.

choose the numerical scheme (explicit) which covers more areas of the imaginary axis in the complex $a(\epsilon, t)\Delta t$ plane.

## 3.3. Basin of attractions and control of chaos

### *3.3.1. Lyapunov methods*

In this section, we present an exploration of the global stability of a seemingly simple yet complex three-dimensional dynamical system with a strong nonlinear term. We start with some physical background of the dynamical system in question, then, based on its dynamical behaviors near the stationary point as well as some physical intuition, we discuss certain aspects of existence and uniqueness of its corresponding Lyapunov functions. The key contributions of this work are twofold: first, we introduce a straightforward numerical strategy to identify the stability areas within a finite distance from the stationary point. Second, we design a specific linear transformation to convert the original dynamical system, derived from control theory, to a dynamical system which belongs to a particular set of systems linked to the dynamics of mechanical systems containing a gyroscopic pendulum. The combination of both numerical and analytical approaches finally enables us to make some observations with respect to global stability and the prospect of numerical search algorithms for Lyapunov functions. In particular, with the guidance of the discovered theorem, we hope to be able to generate new controllers for global asymptotic stabilization.

Compared with local (near limit sets) stability and bifurcation issues, which have been studied abundantly [93,130,132], little is written or known with respect to global stability [244,304]. In this section, we focus on the global stability issues of a problem derived from the control of a linear model with the presence of a strong nonlinear disturbance. In fact, the system belongs to a particular set of dynamical systems involving a gyroscopic pendulum such as the monorail.

The following is the dynamical system in question:

$$\begin{aligned} \dot{x} &= -x + \lambda y^2, \\ \dot{y} &= -k_1 y - k_2 z + x, \\ \dot{z} &= y, \end{aligned} \tag{3.135}$$

where $k_1$ and $k_2$ are positive real numbers with $0 < \lambda < 1$.

This model is characterized by a second-order linear ordinary differential equation (refer to [127]):

$$\begin{aligned} \dot{y} &= u + x, \\ \dot{z} &= y, \end{aligned} \tag{3.136}$$

where $u$ is the control force, $(y, z)$ is the state, and $x$ is the disturbance driven by state $y$ as in the first equation of (3.135).

One of the fundamental questions in control theory is whether a feedback controller that stabilizes the nominal model (i.e., in the absence of disturbances) will also stabilize the real plant (i.e., in the presence of disturbances). For the considered control system, a direct application of linear control theory yields a linear control law $u$ that stabilizes the nominal model:

$$u = -k_1 y - k_2 z, \tag{3.137}$$

where $k_1$ and $k_2$ are positive real numbers and whose choice of feedback law leads us to system (3.135).

Clearly, by means of the Lyapunov first method or Lyapunov indirect method, the only stationary point $(0, 0, 0)$ is locally asymptotically stable for any pair of positive real numbers $(k_1, k_2)$. Furthermore, we can render the domain of attraction of system (3.135) as large as possible if we select large enough feedback gains $k_1$ and $k_2$ [8]. In other words, the size of the domain of attraction is proportional to the values of $k_1$ and $k_2$.

The interesting question is this: can we find a *fixed* pair of $(k_1, k_2)$ so that system (3.135) is asymptotically stable in the large, i.e., with $\mathcal{R}^3$ as its domain of attraction? It is obviously not a trivial task to directly address this question with a search for a well-chosen or -defined Lyapunov function for (3.135). Nevertheless, we will show that the Lyapunov second method or Lyapunov direct method can be still useful.

In this section, we report some interesting numerical observations which can be used as a guide, or at least that was our initial intention in the search of Lyapunov functions. The numerical integration is straightforward, RK2, RK4, etc. The basic scheme is this: we identify a domain of interest, say $x, y, z \in (-10, 10)$. Then, we subdivide the domain into different grid (or mesh) points. Denote a typical grid point as $\mathbf{x}_o = (x_o, y_o, z_o)$, assuming $\mathbf{x}_o \neq \mathbf{0}$, and use it as the initial position. Now if system (3.135) were globally asymptotically stable, after a finite number of time integrations, the trajectory should approach to the origin. The strategy we adopt is very simple. We assign a reasonable $\Delta t$ based on the linearized part of this system [299]. If, after a large number of time steps, say $M$ steps, the new position becomes $\mathbf{x}^M$ and $|\mathbf{x}^M|/|\mathbf{x}_o| \geqslant \delta$, where $\delta$ is a preassigned large positive number, we identify the point $\mathbf{x}_o$ as outside the global stability region and vice versa. For these particular three equations, a brutal implementation of this idea is not expensive. Naturally, a parallel algorithm is more economical for large systems.

Our numerical results clearly show that a well-defined global stability region exists for this system. Further numerical investigation which includes the convergence study (reducing time step size and grid size) confirms that such stability or instability patterns are characteristic of the system and are generic. This bring us to the following theorem:

THEOREM 3.3.1. *Consider a dynamical system $\dot{\mathbf{x}} = \mathbf{A}\mathbf{x} + \mathbf{g}(\mathbf{x})$, with $\|\mathbf{g}(\mathbf{x})\|/\|\mathbf{x}\| \to 0$, as $\mathbf{x} \to 0$. If all eigenvalues of the real matrix $\mathbf{A}$ have strictly negative real parts, then to every negative definite quadratic form $U(\mathbf{x})$, $\forall \mathbf{x} \in \mathcal{R}^n$, there corresponds one and only one quadratic form Lyapunov function $V$, which is positive definite such that*

$$\left(\frac{\partial V}{\partial \mathbf{x}}, \mathbf{A}\mathbf{x}\right) = U. \tag{3.138}$$

A detailed discussion of Theorem 3.3.1 is available in Refs. [164,237]. Since all eigenvalues of the linear part of our system satisfy the preassumptions of this theorem, i.e., real parts are negative, it should not be surprising that more than one form of appropriate Lyapunov function can exist. This seems to suggest that further analytical investigation of this system is needed.

To our best knowledge, there have not been any studies done or published for the system in question. Therefore, the search for Lyapunov functions could be a daunting task. One potential route is to use Theorem 3.3.1 and try to come up with some quadratic representation of Lyapunov functions either numerically or analytically, which we may pursue in a separate work. The other avenue is to invent a function. Interestingly, one thing leading to another, we choose the hard road in this work. We first notice that if we use the following transformation,

$$s = x + y - k_2 z, \tag{3.139}$$

where $x + y$ part is suggested by the first two equations of system (3.135), and we replace $s$ with $z$ and $z$ with $x$, we derive the following dynamical system in a familiar form:

$$\begin{aligned} \dot{x} &= y, \\ \dot{y} &= -ay + z, \\ \dot{z} &= -\phi(y) - f(x), \end{aligned} \tag{3.140}$$

with $\phi(y) = (k_1 + k_2)y - \lambda y^2$, $a = k_1 + 1$, and $f(x) = k_2 x$, which has been studied in Ref. [237].

THEOREM 3.3.2. *For system* (3.140): *let $\phi$ and $f$ be two functions from $\mathcal{R}$ to $\mathcal{R}$, $f$ of class $C^1$ with $f(0) = 0$, $\phi$ continuous with $\phi(0) = 0$, and let $a > 0$. Define $V(x, y, z) = aF(x) + f(x)y + \Phi(y) + z^2/2$, with $F(x) = \int_0^x f(\xi)\,d\xi$ and $\Phi(y) = \int_0^y \phi(\eta)\,d\eta$. If the following three conditions are satisfied, the origin is globally asymptotically stable*:

(i) $f(x)x > 0$ *for* $x \neq 0$,
(ii) $a\phi(y)y - y^2 f'(x) > 0$ *for any* $x$, *if* $y \neq 0$,
(iii) $aF(x) + f(x)y + \Phi(y) \to \infty$ *as* $x^2 + y^2 \to \infty$.

By using transformation (3.139), we have discovered that our system is indeed covered by this theorem. It is also clear that condition (i) is satisfied for any positive $k_2$. However, conditions (ii) and (iii) are not satisfied in general. Thus, based on Theorem 3.3.2, the origin is *unlikely* to be globally asymptotically stable, since conditions of Theorem 3.3.2 are not necessary conditions for global asymptotic stability, which is implied in Theorem 3.3.1 and confirmed in Figure 3.40. Furthermore, if the regions identified by conditions (ii) and (iii) were identical to our numerical simulation, which they were not, it would be a very strong proof for conditions of Theorem 3.3.2 to be necessary, i.e.:

from condition (ii), we have

$$-(k_1 + 1)\lambda y^3 + k_1(k_1 + 1)y^2 + k_1 k_2 y^2 > 0, \tag{3.141}$$

and from condition (iii), we derive

$$(k_1 + 1)k_2 \frac{x^2}{2} + k_2 xy + (k_1 + k_2)\frac{y^2}{2} - \frac{\lambda y^3}{3} + \frac{z^2}{2} > 0. \tag{3.142}$$

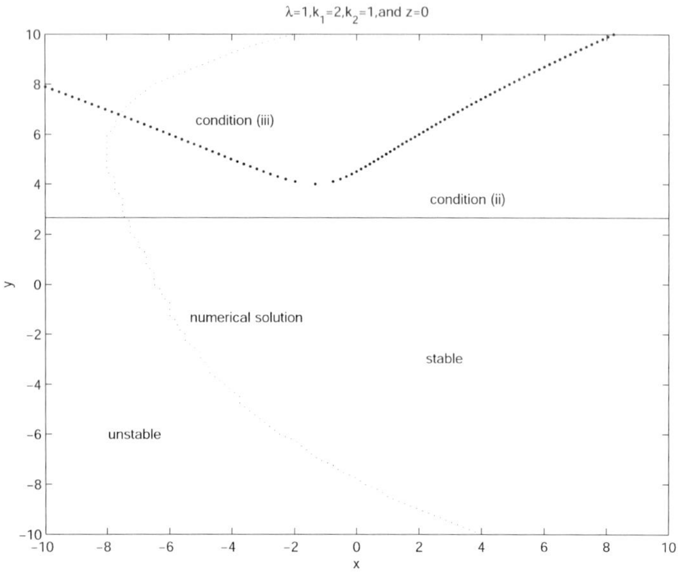


Figure 3.40. Stability region and prediction from conditions (ii) and (iii).

Notice here for condition (iii), because the function $aF(x) + f(x)y + \Phi(y)$ is homogeneous with respect to $x$ and $y$, we only have to require this function to be positive. At this stage, although we are unable to prove analytically that system (3.135) is not globally asymptotically stable, Theorem 3.3.2 pinpoints an interesting class of nonlinear systems for which globally asymptotic stability is guaranteed. In essence, if, in system (3.135), we replace $\lambda y^2$ with $\psi(y)$, which is a continuous function satisfying that $y\psi(y) \leqslant 0$ for all $y \in \mathcal{R}$, for example, $\psi(y) = -\lambda y^3, -\lambda y^5, \ldots$, the stationary point (0, 0, 0) is globally asymptotically stable provided $k_1, k_2 > 0$. This conclusion has also been confirmed in our numerical simulation.

This example demonstrates an interesting combination of numerical and analytical procedures for the global stability analysis of nonlinear dynamical systems. The key is the solution of a particular nonlinear control problem posed as system (3.135) through a transformation (3.139) to system (3.140). Using Theorem 3.3.2 for system (3.140), we hope we are able to generate new controllers and make the stationary point globally asymptotically stable. The implications of this discovery through our exercise of global stability can be significant for nonlinear control applications.

### 3.3.2. Robust control; sliding mode; adaptive control; FIV model

In the control algorithm considered in this section, a procedure similar to the back-stepping method in nonlinear control is used to select the control signal $u(t)$. In practice, it is difficult to precisely identify the fluid forces of the fluid–solid systems. To simplify, the disturbance (external flow-induced force) $d(t)$ is assumed to be bounded, i.e., $|d(t)| \leqslant \Delta$. For example, if we denote $d(t) = \rho U^2 D C_L e^{i\omega t}/2$, with $\omega = 2\pi f$, where $C_L$, $\rho$, $\omega$, and $f$ represent the lift coefficient, the fluid density, and the circular and natural frequencies of the fluid force, respectively, we have $\Delta = \rho U^2 D C_L/2$. By introducing a control signal $u(t)$, the governing equation can be rewritten as

$$m(x,t)\ddot{x} + c(x,t)\dot{x} + k(x,t)x = d(x,t) + u(t). \tag{3.143}$$

The control schemes presented in this section are in general applicable to various nonlinear systems with bounded disturbance and system parameters. To validate the controller selections, in addition to constant or randomly fluctuating system parameters, we also study a dynamical system with van der Pol damping, $\epsilon(x^2 - 1)\dot{x}$, and Duffing's stiffness, $\alpha x + \beta x^3$, where $\epsilon$, $\alpha$, and $\beta$ are constants with respect to damping and stiffness. Thus, Eq. (3.165) can be modified as

$$m\ddot{x} + \epsilon\left(x^2 - 1\right)\dot{x} + \left(\alpha + \beta x^2\right)x = d(x,t) + u(t). \tag{3.144}$$

The control problem to be addressed in the section is stated as follows: Given a desired reference (tracking) signal $x_{1d}(t)$ of class $C^2$, with bounded derivatives $\dot{x}_{1d}(t)$ and $\ddot{x}_{1d}(t)$, design a control law $u(t)$ to make the dynamic response $x_1(t)$ follow closely the desired trajectory $x_{1d}(t)$ for any given initial conditions and admissible system parameters. In more specific terms, we want to guarantee that the solutions of the closed-loop systems are (globally) uniformly ultimately bounded.

In order to put Eq. (3.143) into the standard state-space representation, let $x_1(t) = x(t)$ and $x_2(t) = \dot{x}(t)$, we obtain

$$\begin{aligned} \dot{x}_1 &= x_2, \\ \dot{x}_2 &= \frac{1}{m}(u - kx_1 - cx_2 + d). \end{aligned} \tag{3.145}$$

Likewise, Eq. (3.144) can be rewritten as

$$\begin{aligned} \dot{x}_1 &= x_2, \\ \dot{x}_2 &= \frac{1}{m}\left[u - \alpha x_1 - \beta x_1^3 - \epsilon\left(x_1^2 - 1\right)x_2 + d\right]. \end{aligned} \tag{3.146}$$

To solve our control problem, we will derive a suitable Lyapunov function candidate $V(\mathbf{x}, t)$, with $\mathbf{x} = \langle x_1, x_2\rangle$ and a control law $u(t)$ in such a way that $V$ is strictly decreasing, along the closed-loop solutions of (3.145) or (3.146), outside a neighborhood of the origin. Stability analysis is based upon the second method of Lyapunov [164,237].

Next, we introduce three different control approaches to solve the above-stated control problem. First of all, denote the change of variable $z_1 = x_1 - x_{1d}$, a stabilizing function $\alpha(z_1) = -k_1 z_1 + \dot{x}_{1d}$, with $k_1 > 0$, where $k_1$ is a positive design parameter, and a change of variable $z_2 = x_2 - \alpha(z_1)$. Consequently, we have $\dot{z}_1 = \dot{x}_1 - \dot{x}_{1d} = x_2 - \dot{x}_{1d}$ and $\dot{z}_2 = \dot{x}_2 + k_1\dot{z}_1 - \ddot{x}_{1d}$. Thus, the state-space formulation of Eq. (3.143) can be written in the new coordinates as follows:

$$\dot{z}_1 = -k_1 z_1 + z_2, \tag{3.147}$$

$$\dot{z}_2 = \frac{1}{m}(u - kx_1 - cx_2 + d) + k_1(x_2 - \dot{x}_{1d}) - \ddot{x}_{1d}. \tag{3.148}$$

The key is to construct a particular control law $u(t)$ and a suitable Lyapunov function candidate $V$ such that the system is globally asymptotically stable in the absence of nonvanishing disturbance $d$ and is globally uniformly ultimately bounded in the presence of $d$. By means of a discontinuous sliding mode controller, we also achieve the property of global asymptotic stability in $z$-coordinates.

*Robust control*

The first scheme is called robust control. Motivated by the backstepping methodology, we consider a quadratic Lyapunov function of the form $V = (z_1^2 + z_2^2)/2$. We then have

$$\dot{V} = -k_1 z_1^2 + z_1 z_2 + (u + d - kx_1 - cx_2)/m + k_1(x_2 - \dot{x}_{1d}) - \ddot{x}_{1d}. \tag{3.149}$$

Selecting $u = -k_2 z_2$, with $k_2$ a positive design parameter, we obtain

$$\dot{V} = -k_1 z_1^2 - k_2 z_2^2/m + z_1 z_2 + (d - kx_1 - cx_2)/m + k_1(x_2 - \dot{x}_{1d}) - \ddot{x}_{1d}. \tag{3.150}$$

It is often the case that in flow-induced vibration problems, system parameters $m$, $k$, and $c$ are not defined due to added mass and stiffness as well as viscous effects, however, in general, we know beforehand the upper and lower bounds of these parameters. Hence, the following assumptions are also needed on the system (3.143) or (3.144):

$$\begin{aligned} &\left|x_{1d}(t)\right| \leqslant \delta_0, \qquad \left|\dot{x}_{1d}(t)\right| \leqslant \delta_1, \quad \text{and} \quad \left|\ddot{x}_{1d}(t)\right| \leqslant \delta_2, \quad \forall t \geqslant 0, \\ &0 < m_1 \leqslant m \leqslant m_2, \\ &|kx_1 + cx_2| \leqslant p_1^*|x_1| + p_2^*|x_2|, \quad \forall t \geqslant 0, \end{aligned} \tag{3.151}$$

where $\delta_0$, $\delta_1$, $\delta_2$, $m_1$, $m_2$, $p_1^*$, and $p_2^*$, are predetermined (known) positive constants.

Notice that both $k_1$ and $k_2$ are to be determined later. Consequently, we will relax the knowledge of upper bounds $p_1^*$ and $p_2^*$ later by means of adaptive control. From our control scheme, it is clearly shown that nonlinear disturbances $m$, $k$, $c$, and $d$ can be treated as long as the above conditions (3.151) hold.

Under (3.151), Eq. (3.150) implies

$$\dot{V} \leqslant -k_1 z_1^2 - \frac{k_2}{m_2} z_2^2 + \frac{|z_2|}{m_1}\big[\big(m_2 + p_1^* + k_1 c_3\big)|z_1| + c_2|z_2| + c_4\big], \quad (3.152)$$

with $c_1 = c - mk_1$, $c_2 = p_1^* + m_2 k_1$, $c_3 = p_2^* + m_2 k_1$, and $c_4 = \Delta + p_1^*\delta_0 + p_2^*\delta_1 + m_2\delta_2$.

Obviously, given any $\lambda > 0$, we can always find $k_1, k_2 > 0$ such that

$$\dot{V} \leqslant -\lambda z_1^2 - \lambda z_2^2 + c_4^2. \quad (3.153)$$

From here, according to Theorem 3.3.2, which can be directly derived from Gronwall's inequality [107,164], we can easily show that a robust control scheme suppresses flow-induced vibration and the system is globally uniformly ultimately bounded.

*Sliding mode*

The use of a sliding mode controller achieves perfect tracking, instead of practical tracking or ultimate boundedness. We modify the above robust controller as follows:

$$u = -k_2 z_2 - c_4 \operatorname{sign}(z_2), \quad (3.154)$$

which leads to

$$\dot{V} \leqslant -k_1 z_1^2 - \frac{k_2}{m_2} z_2^2 + \frac{|z_2|}{m_1}\big[\big(m_2 + p_1^* + k_1 c_3\big)|z_1| + c_2|z_2|\big]. \quad (3.155)$$

From (3.155), it is clear that $k_1 > 0$, $k_2 > 0$ can be selected large enough to make the right-hand side of (3.155) negative definite using Cauchy's inequality. Therefore, according to Theorem 3.3.2, we can derive the property of global asymptotic stability for the closed-loop system (3.145) or (3.146).

*Adaptive control*

Here, we remove the assumption that the upper bounds $p_1^*$ and $p_2^*$ are known. To this end, an adaptive control is effectively used. If we denote $p_i = \max(p_i^*, p_i^{*^2})$, with $i = 1, 2$, the Lyapunov function is modified as

$$V = \frac{z_1^2}{2} + \frac{z_2^2}{2} + \frac{1}{2\gamma_1}(\hat{p}_1 - p_1)^2 + \frac{1}{2\gamma_2}(\hat{p}_2 - p_2)^2, \quad (3.156)$$

where $\gamma_1$ and $\gamma_2$ are two positive design parameters representing adaptive gains, and $\hat{p}_1$ and $\hat{p}_2$ are two additional variables introduced by adaptive control scheme.

Thus, the time derivative of the Lyapunov function is given by

$$\dot{V} = z_1\dot{z}_1 + z_2\dot{z}_2 + \frac{1}{\gamma_1}(\hat{p}_1 - p_1)\dot{\hat{p}}_1 + \frac{1}{\gamma_2}(\hat{p}_2 - p_2)\dot{\hat{p}}_2. \tag{3.157}$$

Using the initial assumption (3.151) as well as the following:

$$-z_2\ddot{x}_{1d} \leqslant \frac{z_2^2}{2} + \frac{\ddot{x}_{1d}^2}{2} \quad \text{and} \quad -z_1z_2\big(k_1^2 - 1\big) \leqslant \frac{(k_1^2-1)^2z_2^2}{2k_1} + \frac{k_1z_1^2}{2}, \tag{3.158}$$

we have

$$\begin{aligned}
\dot{V} &= -k_1z_1^2 + z_1z_2 + z_2\big(-k_1^2z_1 + k_1z_2\big) + \frac{z_2u}{m} - \frac{z_2}{m}(kx_1 + cx_2) \\
&\quad + \frac{z_2d}{m} - z_2\ddot{x}_{1d} + \frac{1}{\gamma_1}(\hat{p}_1 - p_1)\dot{\hat{p}}_1 + \frac{1}{\gamma_2}(\hat{p}_2 - p_2)\dot{\hat{p}}_2 \\
&\leqslant -k_1z_1^2 + \frac{(k_1^2-1)^2z_2^2}{2k_1} + \frac{k_1z_1^2}{2} + k_1z_2^2 + \frac{z_2u}{m} - \frac{z_2}{m}(kx_1 + cx_2) \\
&\quad + \frac{z_2^2}{2m} + \frac{\Delta^2}{2m} + \frac{z_2^2}{2} + \frac{\ddot{x}_{1d}^2}{2} + \frac{1}{\gamma_1}(\hat{p}_1 - p_1)\dot{\hat{p}}_1 + \frac{1}{\gamma_2}(\hat{p}_2 - p_2)\dot{\hat{p}}_2 \\
&\leqslant -\frac{k_1}{2}z_1^2 + \frac{z_2}{m}\left[u + \frac{z_2}{2} + m_2z_2\frac{(k_1^2-1)^2 + 2k_1^2 + k_1}{2k_1}\right] \\
&\quad + \frac{z_2}{m}(-kx_1 - cx_2) + \frac{1}{\gamma_1}(\hat{p}_1 - p_1)\dot{\hat{p}}_1 + \frac{1}{\gamma_2}(\hat{p}_2 - p_2)\dot{\hat{p}}_2 \\
&\quad + \frac{\Delta^2}{2m} + \frac{\ddot{x}_{1d}^2}{2},
\end{aligned} \tag{3.159}$$

the details of which can be referred to Ref. [36].

Furthermore, if we introduce the following inequality, which can also be derived from the assumption (3.151) and Cauchy's inequality,

$$\begin{aligned}
&-\frac{z_2}{m}(kx_1 + cx_2) \\
&\leqslant \frac{|z_2|}{m}|kx_1 + cx_2| \leqslant \frac{p_1^*|z_2||x_1|}{m} + \frac{p_2^*|z_2||x_2|}{m} \\
&\leqslant \frac{p_1z_2^2x_1^2}{4\epsilon_1} + \frac{\epsilon_1}{m^2} + \frac{p_2z_2^2x_2^2}{4\epsilon_2} + \frac{\epsilon_2}{m^2} \\
&= \frac{\hat{p}_1z_2^2x_1^2}{4\epsilon_1} + \frac{\epsilon_1}{m^2} + \frac{\hat{p}_2z_2^2x_2^2}{4\epsilon_2} + \frac{\epsilon_2}{m^2} + (p_1 - \hat{p}_1)\frac{z_2^2x_1^2}{4\epsilon_1} + (p_2 - \hat{p}_2)\frac{z_2^2x_2^2}{4\epsilon_2},
\end{aligned} \tag{3.160}$$

along with the new term $\sigma_i(\hat{p}_i - p_i)(\hat{p}_i - p_i^o)$, with $i = 1, 2$,

$$u(t) = -k_2 z_2 - z_2 \frac{k_1 + m_2[(k_1^2 - 1)^2 + 2k_1^2 + k_1]}{2k_1} - m_2 z_2 \left( \frac{\hat{p}_1 x_1^2}{4\epsilon_1} + \frac{\hat{p}_2 x_2^2}{4\epsilon_2} \right), \tag{3.161}$$

and the additional two equations

$$\begin{aligned} \dot{\hat{p}}_1 &= -\gamma_1 \sigma_1 (\hat{p}_1 - p_1^o) + \gamma_1 \frac{z_2^2 x_1^2}{4\epsilon_1}, \\ \dot{\hat{p}}_2 &= -\gamma_2 \sigma_2 (\hat{p}_2 - p_2^o) + \gamma_2 \frac{z_2^2 x_2^2}{4\epsilon_2}, \end{aligned} \tag{3.162}$$

where $\gamma_i$, $\sigma_i$, and $\epsilon_i$, with $i = 1, 2$ are positive design constants, we have,

$$\begin{aligned} \dot{V} &\leqslant \frac{z_2}{m} \left[ u + z_2 \frac{k_1 + m_2[(k_1^2 - 1)^2 + 2k_1^2 + k_1]}{2k_1} + z_2 \frac{m_2 \hat{p}_1 x_1^2}{4\epsilon_1} \right. \\ &\quad \left. + z_2 \frac{m_2 \hat{p}_2 x_2^2}{4\epsilon_2} \right] \\ &\quad + (\hat{p}_1 - p_1) \left[ \frac{\dot{\hat{p}}_1}{\gamma_1} - \frac{z_2^2 x_1^2}{4\epsilon_1} + \sigma_1 (\hat{p}_1 - p_1^o) \right] - \sigma_1 (\hat{p}_1 - p_1)(\hat{p}_1 - p_1^o) \\ &\quad + (\hat{p}_2 - p_2) \left[ \frac{\dot{\hat{p}}_2}{\gamma_2} - \frac{z_2^2 x_2^2}{4\epsilon_2} + \sigma_2 (\hat{p}_2 - p_2^o) \right] - \sigma_2 (\hat{p}_2 - p_2)(\hat{p}_2 - p_2^o) \\ &\quad + \ddot{x}_{1d}^2 2 + \frac{\Delta^2}{2m_1} + \frac{\epsilon_1}{m_1^2} + \frac{\epsilon_2}{m_1^2} - \frac{k_1}{2} z_1^2 \\ &\leqslant -\frac{k_1}{2} z_1^2 - \frac{k_2}{m_2} z_2^2 - \frac{1}{2} \sigma_1 (\hat{p}_1 - p_1)^2 - \frac{1}{2} \sigma_2 (\hat{p}_2 - p_2)^2 + \frac{\ddot{x}_{1d}^2}{2} + \frac{\Delta^2}{2m_1} \\ &\quad + \frac{\epsilon_1}{m_1^2} + \frac{\epsilon_2}{m_1^2} + \frac{1}{2} \sigma_1 (p_1 - p_1^o)^2 + \frac{1}{2} \sigma_2 (p_2 - p_2^o)^2 \\ &\leqslant -\lambda V + \mu, \end{aligned} \tag{3.163}$$

with $\lambda = \min(k_1, 2k_2/m_2, \gamma_1\sigma_1, \gamma_2\sigma_2)$ and $\mu = \ddot{x}_{1d}^2/2 + \Delta^2/2m_1 + \epsilon_1/m_1^2 + \epsilon_2/m_1^2 + \frac{\sigma_1}{2}(p_1 - p_1^o)^2 + \frac{\sigma_2}{2}(p_2 - p_2^o)^2$.

The inequality (3.163) indicates that the augmented system of $(z_1, z_2, \hat{p}_1, \hat{p}_2)$ is globally uniformally ultimately bounded. Note that again the ultimate bound for the tracking error $z_1 = x_1 - x_{1d}$ can be reduced to arbitrarily small by picking large enough feedback gains of $k_i$ and $\sigma_i$, with $i = 1, 2$.

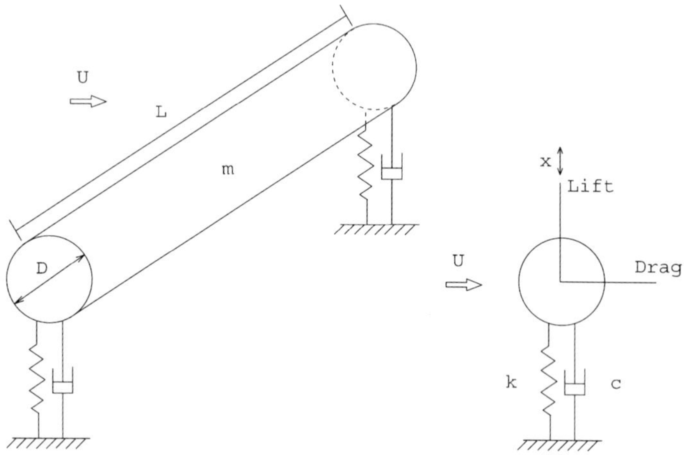


Figure 3.41. A flow-induced vibration model.

*FIV model*

In the late 1800's century, Strouhal and Rayleigh discovered that the Aerolian tones generated by a wire are proportional to the ratio of wind speed over the wire diameter [261], and the wire vibrates primarily normal to the wind direction [226]. In the early 1900's century, Bénard [16] and von Karman [284] found that periodic vortex shedding can occur in the wake of a bluff body, which suggests that the oscillatory pressure fields around the body may introduce or enhance the structural vibration at certain frequencies. These events can be seen as a conscious start of the research of flow-induced vibration (FIV) phenomena and the inception of a set of multidisciplinary problems, the so-called fluid-structure interaction (FSI) problems [68,101,228,316]. Almost a century later, tremendous understanding and progress have been achieved in the field of flow-induced (or rather vortex-induced) vibration of flexible structures [70]. Furthermore, similar flow-induced vibrations are found to be of great practical importance in the design of aeroplanes, off-shore structures, bridges, antennas, cables, heat exchangers, and information storage devices. Failure to recognize the importance of flow-induced vibration in engineering design often leads to catastrophe. One of the most famous failures is the collapse of Tacoma Narrow Bridge in 1940.

A representative model in this area is the flow passing an elastically supported rigid circular cylinder, which is discussed in detail in Ref. [316]. In general, the model involves an elastically supported rigid cylinder in a laminar (or turbulent) incompressible (or compressible) flow as shown in Figure 3.41. The forces acting on the cylinder are categorized as lift and drag forces, which are the results of the distributions of pressure and viscous shear. Due to the complex nature of this type of coupling problem, early work relied heavily on experimental means [228, 236]. Various experiments revealed that if the Reynolds number (*Re*) exceeds a critical value, stable unsymmetric vortex shedding arises. To be more specific, according to Ref. [163], for the Reynolds number to be in the range of 40 to 150, the Strouhal number (*St*), named after Vincenc Strouhal (1850–1922), which governs the vortex shedding frequency, can be expressed as

$$St = 0.195\left(1 - \frac{20.1}{Re}\right), \tag{3.164}$$

with $Re = UD/\nu$ and $St = fD/U$, where $U$, $D$, $\nu$, and $f$ represent the mean flow velocity, the rigid cylinder diameter, the flow kinematic viscosity, and the vortex shedding frequency.

Notice the one-to-one correspondence between *St* and *Re*. Moreover, if the amplitude is finite, additional fluid forces must be associated with the shifting of the flow separate points. As confirmed by the experimental measurements [236,316], in return, the cylinder motion (or sound) induced by vortex shedding may increase the strength of vortices and spanwise correlation, i.e., more three-dimensional effects, cause the shedding frequency to shift to the natural frequency of the cylinder vibration (frequency lock-in), increase the mean drag, and alter phase, sequence, and pattern of vortices. A detailed illustration of different modes of vortex shedding is documented in Ref. [186].

Recently, with the advance of computational tools, finite element methods (FEM) and computational fluid dynamics (CFD), certain flow-induced vibration problems can be accurately simulated [291,295,300]. In Refs. [144,206], an ALE finite element method is presented so that we may study the interaction between a rigid body and the surrounding moving viscous fluid. This ALE procedure, also employed in Ref. [293], can be efficiently used to avoid excessive mesh distortion due to structural motions. In addition, to completely circumvent moving boundaries, the similar flow-induced vibration problem can be solved with the immersed boundary method, in which the submerged elastic boundaries are replaced with a set of nodal force distributions [48,162]. Furthermore, in order to obtain more accurate prediction of lift and drag coefficients, direct numerical simulation based on spectral methods can be used [76,204].

In this section, we focus our attention on flow-induced vibration and the corresponding control schemes. To simplify the illustration, we employ the traditional model of a uniform flow passing an elastically supported rigid cylinder as illustrated in Figure 3.41. In this model, instead of predicting accurately the fluctuating fluid forces due to the fluid-structure interactions, which affect the system mass, stiffness, and damping, we assume that the flow-induced force and system parameters are bounded, which is a reasonable assumption considering the localization of fluid-structure interactions. Thus, being different from other active control of flow-induced vibration, three control schemes are designed for a generic system with bounded disturbance and variable system parameters.

As explained previously, the flow-induced vibration problem lends itself to an application object of advanced nonlinear control. Significant progress has been made by several researchers in recent literature regarding advanced feedback control of nonlinear systems. A detailed account of recent developments in nonlinear control system design and application can be found in Refs. [107,117,128,129, 159,177]. One of the new systematic nonlinear controller designs is the so-called backstepping technique [159], which allows us to address control problems for important and reasonably large classes of nonlinear systems, which are subject to strong nonlinearities. Performance improvements over traditional linear and nonlinear schemes are also demonstrated through the systematic use of this backstepping approach. In this section, we show how recent results in nonlinear control can be invoked to provide new solutions to the traditional flow-induced vibration problem. For the sake of simplicity, we focus on a simplified nonlinear model with bounded disturbance and system parameters. Evidently, because of Lyapunov-like arguments adopted in our control algorithm, the extension of the proposed results to a more general context of nonlinear, not necessarily bounded, disturbance is rather straightforward [128]. We will report results in this direction elsewhere, along with experimental verifications.

The governing equation of the mathematical model of flow-induced structure vibration as illustrated in Figure 3.41 is written as follows:

$$m(x,t)\ddot{x} + c(x,t)\dot{x} + k(x,t)x = d(x,t), \tag{3.165}$$

where $x$, $d$, $m$, $k$, and $c$ stand for the vertical position of the cylinder, the flow-induced force, the effective cylinder mass, the effective stiffness due to the elastic mount, and the effective damping, respectively.

Notice, however, that for flow-induced vibration problems we do not know beforehand the system parameters $m$, $k$, and $c$, and in some cases, these parameters may depend strongly on the position, velocity, and acceleration of the structure. For example, the added mass for the cylindrical structure is equivalent to the same structure volume multiplied by the fluid density. Detailed discussion of this subject can be found in Chapter 5. The system (3.165), with the effects of added mass and stiffness as well as viscous shear introduced in fluid-structure interactions is, in general, nonlinear.

To illustrate the characteristics of these three control schemes, we model a typical flow-induced vibration system with the following physical parameters: $0.5 \leqslant m \leqslant 1.5$, $|c| \leqslant 2$, $2 \leqslant k \leqslant 4$, $|d| \leqslant 4$, $k_1 = 15$, $k_2 = 20$, $\sigma_1 = 10$, $\sigma_2 = 10$, $\gamma_1 = 20$, $\gamma_2 = 20$, $\epsilon_1 = 20$, $\epsilon_2 = 20$, $p_1^o = p_1 = 20$, and $p_2^o = p_2 = 20$. Note that for simplification, all the parameters are non-dimensionalized, thus no units are involved.

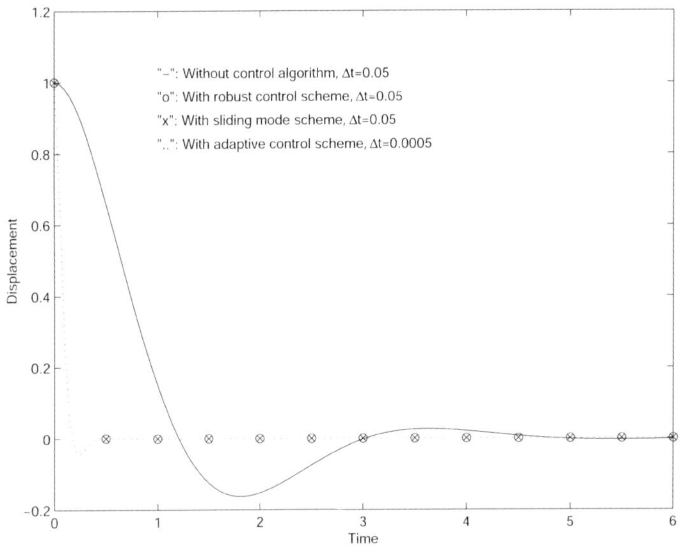


Figure 3.42. A free vibration test with various control schemes.

We employ the second-order Runge–Kutta (RK2) scheme to integrate the dynamical system of (3.145), (3.146), and (3.162). To evaluate the numerical scheme as well as the dynamical system in question, we first test a free vibration system with constant system parameters, i.e., $m = 1, c = 2, k = 4, x_1(0) = 1, x_2(0) = 0$, and $\Delta = 0$.

Results in Figure 3.42 with $m = 1$, $k = 4$, and $c = 2$ demonstrate that all three control schemes provide additional damping for the case with no external excitations. However, for the selected adaptive control parameters, the system with adaptive control scheme is very stiff. Consequently, to avoid numerical instability, we must take a very small time step: one hundredth of the time step needed for the other control schemes or the original system. This suggests the additional computation cost in using adaptive control for large systems.

Now we apply the same numerical scheme to flow-induced vibration with the excitation force given as $d = 4\sin(2t)$. To study the effects of various control schemes, we choose the worst scenario: zero damping, i.e., $c = 0$ at resonant frequency. In addition, for comparison, the initial conditions remain the same, i.e., $x_1(0) = 1$ and $x_2(0) = 0$.

Results at the resonant frequency in Figure 3.43 demonstrate that all three proposed control schemes work well with the flow-induced vibration case with so-called lock-in phenomena. However, adaptive control still requires an excessively small time step. In order to identify the performance of different control

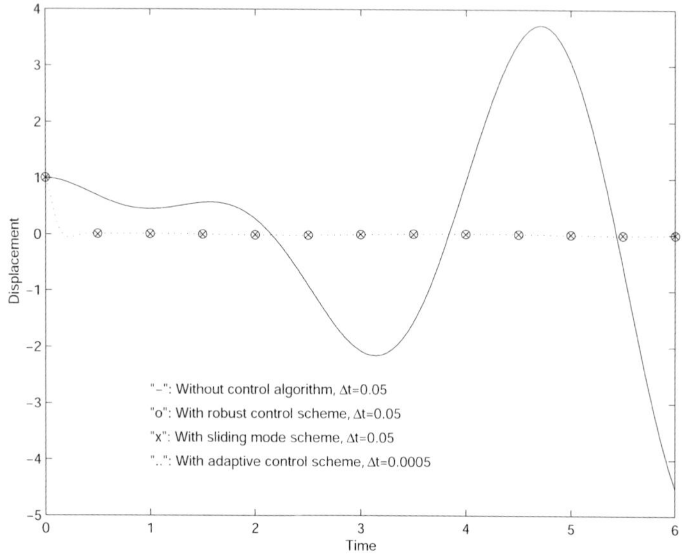


Figure 3.43. An FSI lock-in model with various control schemes.

schemes, we reset the flow-induced vibration test with $x_1(0) = 0$ and $x_2(0) = 0$. As shown in Figure 3.44, the residual signals of a robust control scheme and adaptive control at the resonant frequency, so-called lock-in phenomena, are almost identical, since both control schemes convert the dynamical system to uniformly ultimately bounded. However, Figure 3.44 also indicates that the residual of a sliding mode scheme is very much dependent on the time step of the numerical integration. Although a sliding mode scheme makes the dynamical system asymptotically stable, in the numerical implementation, due to the non-smooth nature of the *sign* function, there remains some residual with respect to the time step. Furthermore, Figure 3.45 confirms that for smooth control schemes, such as robust control and adaptive control when refining the time step, the residuals will not continue to diminish. Instead, they approach an asymptotical function.

As shown in Figure 3.46, the control signal $u(t)$ for sliding mode scheme is not as smooth as that of robust control and adaptive control schemes. It is, in fact, zigzagging, which suggests that in actual engineering implementation, unlike the smooth control signals of robust control and adaptive control schemes, the sliding model scheme has fast switching fluctuations (chattering). In addition, for a flow-induced vibration test at the resonant frequency, Figure 3.46 also demonstrates that the control signals for robust control and adaptive control schemes are equivalent to the additional signal with a $\pi$ phase shift from the disturbance, i.e., $u(t) = -4 \sin 2t$.

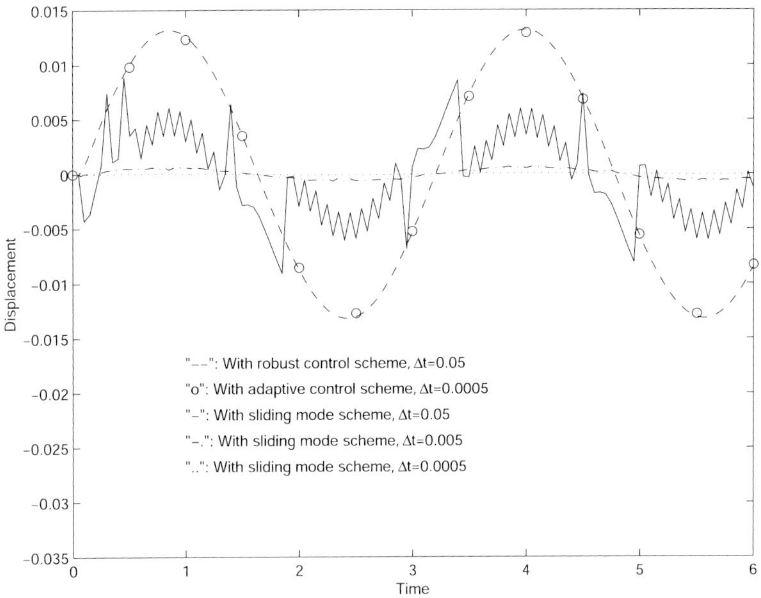


Figure 3.44. Residuals of an FSI model with various control schemes.

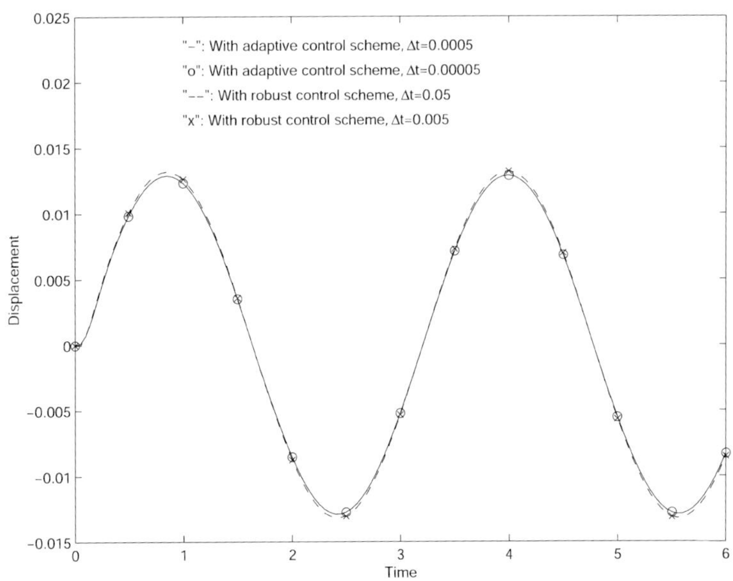


Figure 3.45. Residuals of an FSI model with smooth control schemes.

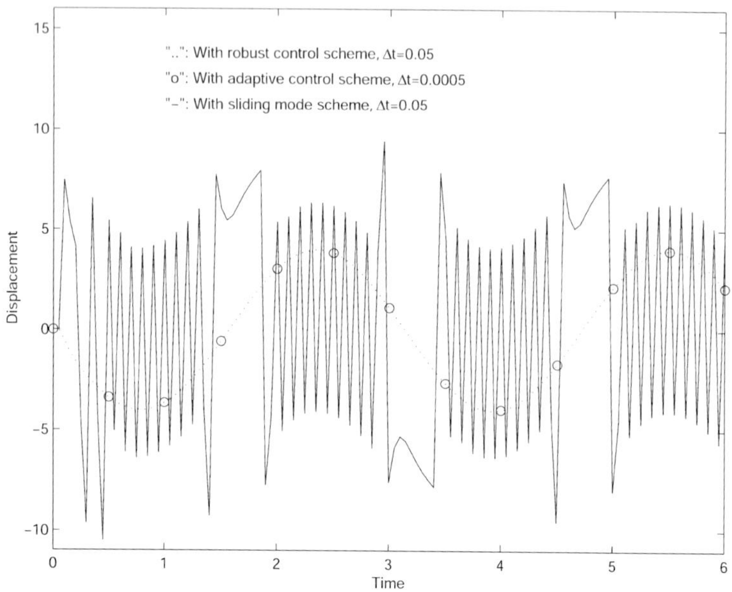


Figure 3.46. Control signals of an FSI model with various control schemes.

To validate the versatility of the adaptive control scheme, we introduce a more realistic flow-induced vibration case in which due to added mass and stiffness as well as fluid viscosity, the system parameters $m$, $k$, and $c$ are bounded time-dependent unknowns.

As shown in Figure 3.47, the system parameters vary with respect to time. So does the disturbance signal. There are characteristics of FSI problems. Interestingly, all control signals vary at a high frequency. In particular, the adaptive control signal is smaller than the robust control signal, whereas the sliding mode signal can be significantly reduced with a smaller integration time step or sampling step. All three schemes, as clearly shown in Figure 3.48, are very effective. Of course, in practice, high frequency signals are often filtered out.

For system (3.146) with the disturbance $\Delta \cos \omega_o t$, we select $\epsilon$, $\alpha$, and $\beta$ such that the assumptions in (3.151) are satisfied. As shown in Figure 3.49, with a properly selected adaptive control parameter, chaotic solutions of van der Pol and Duffing's equation with $m = 1$, $\Delta = 0.9$, $\epsilon = 0.01$, $\alpha = -1$, $\beta = 1$, and $\omega_o = 1$ can be effectively eliminated.

Finally, we apply all three control schemes to follow a reference signal or tracking trajectory to van der Pol and Duffing's equation. Given a reference signal, $x_{1d}(t) = 0.5 \cos 0.25t$, the results in Figure 3.50 with $m = 1$, $\Delta = 0.9$, $\epsilon = 0.01$, $\alpha = -1$, $\beta = 1$, and $\omega_o = 1$ again demonstrate the effectiveness of

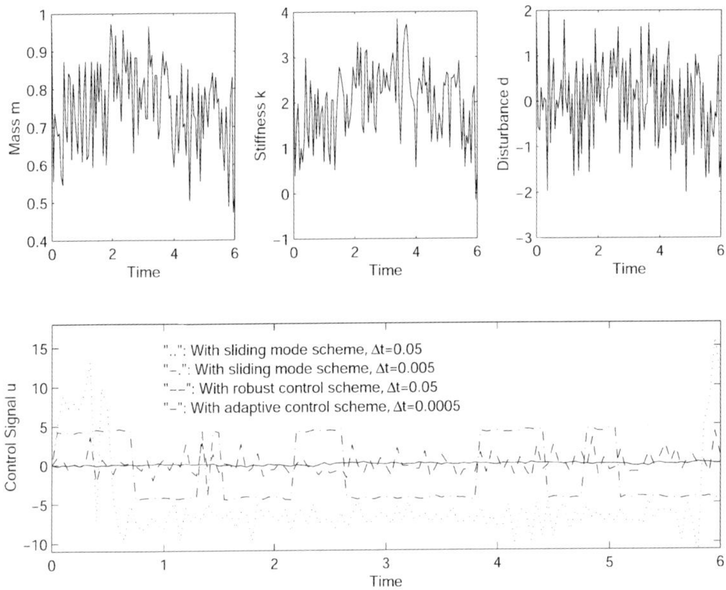


Figure 3.47. System parameters and disturbance for an FSI model with various control schemes.

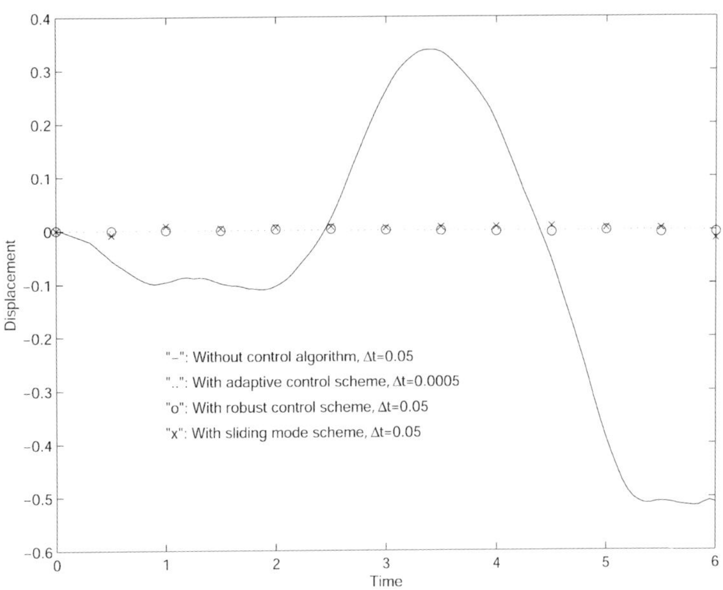


Figure 3.48. An FSI model with variable system parameters.

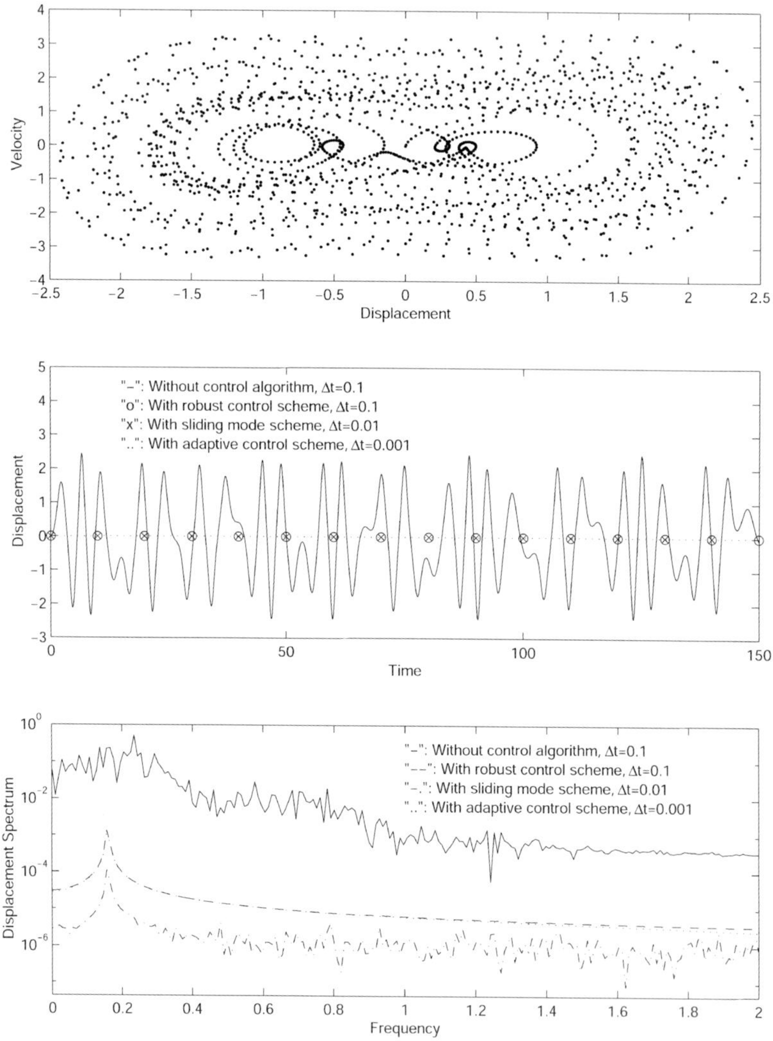


Figure 3.49. Chaotic solutions of van der Pol and Duffing.

these control schemes for tracking. To reiterate, the main aspect of this section is the formulation of a control algorithm with three different schemes for systems with bounded disturbance and variable parameters. These control schemes are particularly useful for flow-induced vibration problems, in which both disturbance and system parameter variations are bounded due to the localization of fluid-structure interactions. In particular, robust control and adaptive control

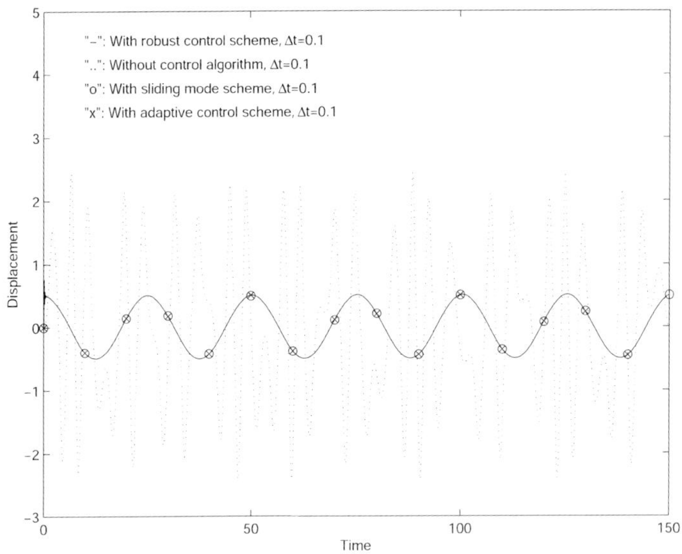


Figure 3.50. Tracking solutions of van der Pol and Duffing.

schemes are (globally) ultimately uniformly bounded, thus, as we reduce the integration time step, the residuals will approach to a bounded (finite) asymptotic function. On the other hand, sliding mode scheme is (globally) asymptotically stable; by reducing the integration time step, the residual will approach zero. Furthermore, due to self-tuning, the gain of the adaptive control scheme is relatively small, yet the computation cost is higher because of the excessively small time step requirement for numerical integration. With respect to the sliding mode scheme, the control signal is discontinuous due to the sign function. Consequently, the practical implementation has fast switching fluctuations (chattering).

*Chapter 4*

# Flow-Induced Vibrations

FSI problems are of great interest and complexity, and their solutions have practical applications in every branch of engineering. Fully coupled FSI systems, in particular, those involving three-dimensional large structural motions, still pose tremendous challenges to engineers and applied scientists. More importantly, some traditional approaches yield fairly good estimations of solutions and provide good comparisons or gauges for experimental and computational results. Therefore, there is still merit in applying traditional approaches to FIV problems. The objective of this chapter is to illustrate traditional analytical and computational approaches to FIV problems. Some recent advances in fully coupled computational fluid dynamics and finite element methods with respect to FSI systems will be discussed in Chapters 6 and 7. To provide a link to the further evolution of the book, this chapter is devoted to the study of FIV problems which until now have relied heavily on experimental approaches.

We focus on two FIV categories. In the aerospace industry, structures vibrating in steady flows can introduce oscillatory aerodynamic forces, which in turn induce so-called galloping and flutter [316]. In practice, flutter corresponds to an instability induced by aerodynamic forces which are comparable to the weight and flexural or torsional rigidity of the cross section and are significant enough to alter natural frequencies. Galloping is associated with an instability due to low structural damping combined with small oscillatory aerodynamic forces which do not shift natural frequencies. In practice, flutter often involves a combination of a torsion and a displacement mode, while galloping often happens to a single mode. In the nuclear industry, the vibration of pipes conveying both steady and pulsatile flows is also of practical concern [191,228]. The vibration mode of pipelines or pipe arrays can be seen as similar to problems with garden hoses and water hammers.

## 4.1. Attenuator and flutter suppression

With the increase in machine speed and higher precision requirements in paper, thin film, and textile manufacturing along with other technical processes in which flexible materials are pulled at high speeds over fixed points, two problems have received much attention. On the material supply side, it has become important to control the pressure pulsation within piping, whereas on the material transport side, it is crucial to control the transverse oscillations of moving materials. The design of current pressure pulsation attenuators can significantly reduce pressure variations. However, they have been found to be ineffective in dealing with low frequency pulsations (<1 Hz). The basic design concept of the attenuator used in the paper industry is the Helmholtz resonator. The characteristic property of such a resonator is its ability to effectively absorb pressure waves around a particular frequency, i.e., the so-called resonant frequency. As one of the fundamental acoustic devices, Helmholtz resonators have been studied extensively [115,168,216]. On the other hand, the existence and persistence of excessive oscillation introduced by the aerodynamic instability of the surrounding air, roll eccentricity, material basis weight, and tension variations will contribute to wrinkling, sheet breakage, and other runnability problems. To reduce vibration with minimal modification of existing machine designs, researchers have proposed to stabilize the moving structure at various locations with self-acting air bearings or frictionless guides [33,188]. Similar control of vibration by means of a viscous fluid has been discussed by Ingard and Akay [160]. In fact, such dynamic systems are very much the same as those associated with high-speed magnetic tapes and flexible disk drives interacting with their recording heads [121]. In the paper industry, the initial idea of using lubrication effects to help handle moving paper webs was proposed by Page [229] and Müftü and Benson [199]. In this section, we use attenuator and flutter suppression device as examples to illustrate analytical and computational modeling procedures for simplified FSI systems.

### 4.1.1. Helmholtz resonator; attenuator design

#### Helmholtz resonator

We model the Helmholtz resonator as depicted in Figure 4.1, with one-dimensional isentropic acoustic models, in which the air cavity is separated from the neck of water and the main pipe flow by a membrane. The membrane with the cross-sectional area $A_2$, thickness $d$, Young's modulus $E_o$, and density $\rho_o$ is modeled as a simple spring-mass system with stiffness $K_o$ and mass $M_o$. We also denote the cross-sectional area, length, displacement, pressure, density, bulk modulus, wave number, and wave speed as $A_i$, $L_i$, $w_i$, $p_i$, $\rho_i$, $\beta_i$, $k_i$, and $c_i$, respectively, with $i = 1$ and 2, representing the air and the water. In addition, we define the air cavity volume as $V_1 = A_1 L_1$ and the water neck volume as $V_2 = A_2 L_2$. From the constitutive relation $p_i = -\beta_i \partial w_i/\partial x$ and the linear momentum equation $\rho_i \partial^2 w_i/\partial t^2 = -\partial p_i/\partial x$, we obtain the following governing equations:

$$\frac{\partial^2 w_1}{\partial x^2} = \frac{1}{c_1^2}\frac{\partial^2 w_1}{\partial t^2}, \quad 0 \leqslant x \leqslant L_1, \tag{1.1}$$

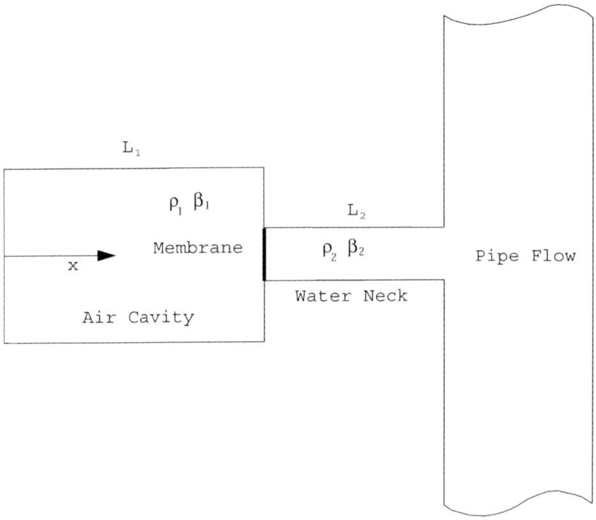


Figure 4.1. The Helmholtz resonator with air cavity and membrane.

$$\frac{\partial^2 w_2}{\partial x^2} = \frac{1}{c_2^2}\frac{\partial^2 w_2}{\partial t^2}, \qquad L_1 \leqslant x \leqslant L_1 + L_2. \tag{1.2}$$

Furthermore, the characteristic solutions for displacements $w_1$ and $w_2$ can be written as

$$w_1 = a_1 e^{-j(\omega t + k_1 x)} + b_1 e^{-j(\omega t - k_1 x)}, \qquad 0 \leqslant x \leqslant L_1, \tag{1.3}$$

$$w_2 = a_2 e^{-j(\omega t + k_2 x)} + b_2 e^{-j(\omega t - k_2 x)}, \qquad L_1 \leqslant x \leqslant L_1 + L_2, \tag{1.4}$$

where $\omega$ stands for the system natural frequency (angular frequency) and the four constants, $a_1$, $a_2$, $b_1$, and $b_2$ can be determined from the fixed boundary condition at $x = 0$, the zero pressure boundary condition at $x = L_1 + L_2$, and the kinematic and dynamic matching conditions at $x = L_1$, i.e.,

$$\begin{aligned}
&w_1 = 0, && \text{at } x = 0,\\
&A_1 \frac{\partial w_1}{\partial t} = A_2 \frac{\partial w_2}{\partial t}, && \text{at } x = L_1,\\
&A_2\left(-\beta_1 \frac{\partial w_1}{\partial x} + \beta_2 \frac{\partial w_2}{\partial x}\right) = K_o w_2 + M_o \frac{\partial^2 w_2}{\partial t^2}, && \text{at } x = L_1,\\
&\frac{\partial w_2}{\partial x} = 0, && \text{at } x = L_1 + L_2.
\end{aligned} \tag{1.5}$$

By introducing the dispersion relation $k_i = \omega/c_i$, with $c_i = \sqrt{\beta_i/\rho_i}$, we obtain the following characteristic equation:

$$\frac{\beta_1 A_2}{A_1}\frac{\omega}{c_1}\cot\left(\frac{\omega}{c_1}L_1\right) = \beta_2\frac{\omega}{c_2}\tan\left(\frac{\omega}{c_2}L_2\right) - \frac{(K_o - \omega^2 M_o)}{A_2}. \tag{1.6}$$

If we consider the same acoustic media (e.g., water) in both the cavity and the neck and ignore membrane effects, Eq. (1.6) becomes

$$\frac{A_2}{A_1}\cot\left(\frac{\omega}{c_2}L_1\right) = \tan\left(\frac{\omega}{c_2}L_2\right), \tag{1.7}$$

which is the same characteristic equation discussed in Ref. [241].

Since we are interested in the low resonant frequency, i.e., the coupled system natural frequency, we assume $k_i L_i = (\omega/c_i)L_i \ll 1$, which implies that the typical wave length is much larger than the cavity and orifice dimensions. Therefore, Eq. (1.6) is simplified as

$$\frac{\beta_1 A_2}{A_1}\frac{\omega}{c_1} = \frac{\omega}{c_1}L_1\beta_2\frac{\omega}{c_2}\frac{\omega}{c_2}L_2 - \frac{\omega}{c_1}L_1\frac{(K_o - \omega^2 M_o)}{A_2}, \tag{1.8}$$

from which we derive the analytical expression for the natural frequency $f$,

$$f = \frac{\omega}{2\pi} = \frac{1}{2\pi}\sqrt{\frac{\beta_1 A_2^2 + K_o V_1}{(\rho_2 V_2 + M_o)V_1}}. \tag{1.9}$$

Interestingly, we find that Eq. (1.9) can also be derived from the force equilibrium at $x = L_1$ with a similar approach introduced in Ref. [156]. Assuming the membrane displacement is $\xi$ or $w_2(L_1)$, the restoring force exerted on the membrane from the air cavity side is $-\xi\beta_1 A_2^2/V_1$, and the inertia force contributed by the water neck is $-\rho_2 V_2\ddot{\xi}$. When we add the spring-mass system of the membrane, we arrive at the following equation of motion:

$$(\rho_2 V_2 + M_o)\ddot{\xi} + \left(\frac{\beta_1 A_2^2}{V_1} + K_o\right)\xi = 0, \tag{1.10}$$

which gives us the natural frequency expression in Eq. (1.9).

Notice that the coupling between air and water does not introduce the complex wave number $k_i$ or natural frequency $\omega$, which indicates that no physical damping is introduced by the membrane. For the limiting case, if we neglect the existence of the membrane, i.e., $M_o = K_o = 0$, we obtain

$$f = \frac{c_1}{2\pi}\sqrt{\frac{\rho_1 A_2}{\rho_2 L_2 V_1}}. \tag{1.11}$$

Furthermore, if the air cavity is filled with water, we recover the classical resonant frequency for the Helmholtz resonator [227]:

$$f = \frac{c_2}{2\pi}\sqrt{\frac{A_2}{L_2 V_1}}. \tag{1.12}$$

*Attenuator design*

To determine the effects of various design variations on the resonance frequency, we test the example depicted in Figure 4.1 with the following parameters: $\rho_1 = 1.21\ \mathrm{kg/m^3}$; $\rho_2 = 1000\ \mathrm{kg/m^3}$; $\beta_1 = 1.41 \times 10^5$ Pa; $\beta_2 = 2.1 \times 10^9$ Pa; $L_1 = 2$ m; $L_2 = 0.8$ m; $A_1 = 1\ \mathrm{m^2}$; $A_2 = 0.14\ \mathrm{m^2}$; $\rho_o = 1260\ \mathrm{kg/m^3}$; $E_o = 1.41 \times$

Table 4.1 Resonant frequencies

| Resonant frequency (Hz) | Models |
|---|---|
| 4.02 | With the membrane effects |
| 0.56 | Neglect the membrane effects |
| 68.2 | Cavity filled with water |

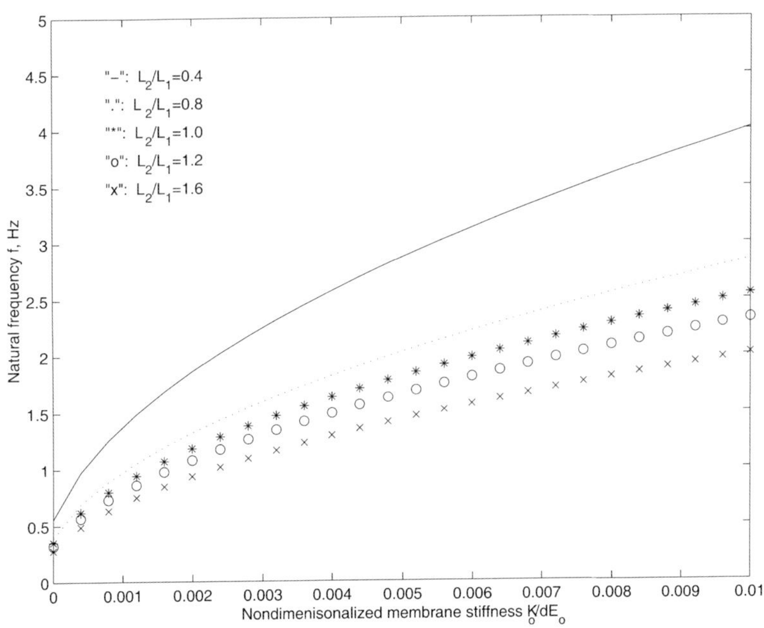


Figure 4.2. Resonant frequency vs. membrane stiffness with different length.

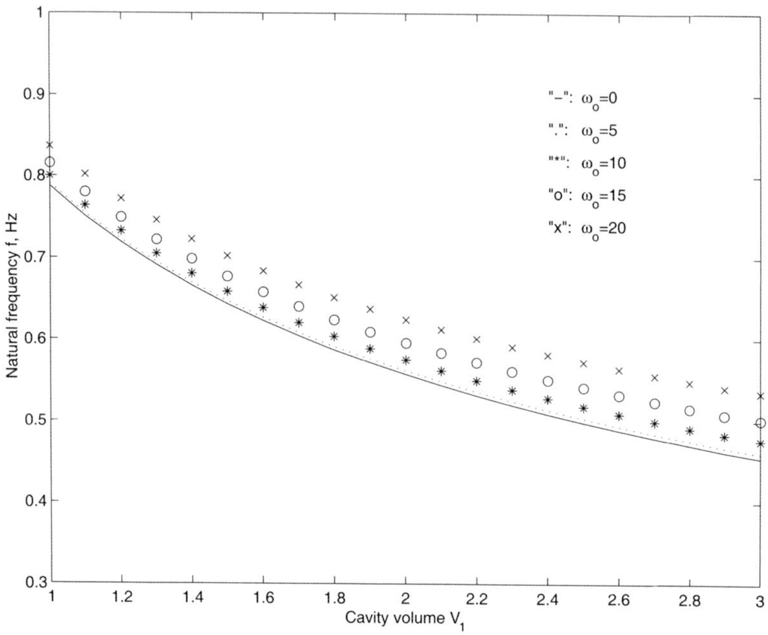


Figure 4.3. Resonant frequency vs. air cavity volume with different membrane stiffness.

$10^9$ Pa; and $d = 0.005$ m. For the particular configuration described above, we obtain the spring-mass system constants $K_o = 7.05 \times 10^4$ N/m $= 1\%\ dE_o$, $M_o = d\rho_o A_2 = 0.882$ kg and $\omega_o = \sqrt{K_o/M_o} = 282.72$ rad/s.

Notice that the resonant frequency of the traditional Helmholtz resonator filled with water, as illustrated in Table 4.1, is significantly higher than that of the modified version with air cavity and membrane. In addition, we find that the net effect of the membrane leads to a resonant frequency augmentation, which is further confirmed in Figure 4.3. Of course, in the actual design, we will never achieve the stable separation between air and water right at the location of $x = L_1$, and the case in which we neglect the membrane effects gives us a lower bound for the resonant frequency. As illustrated in Figures 4.2 to 4.4, the resonant frequency can be significantly reduced by minimizing the membrane stiffness or increasing the neck length or the air cavity volume. Figure 4.4 also shows that if the membrane stiffness is dominant, the increase of the cross-sectional area ratio $A_2/A_1$ will reduce the resonant frequency, which is in contradiction to the design intuitions of the Helmholtz attenuators without membrane effects as illustrated in Eqs. (1.11) and (1.12).

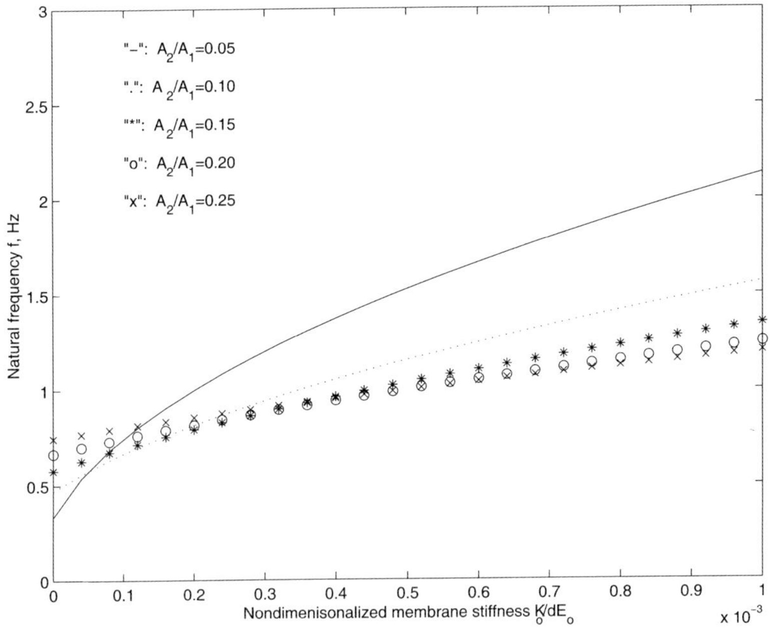


Figure 4.4. Resonant frequency vs. membrane stiffness with different cross-sectional area.

### 4.1.2. Reynolds equations; spatial and temporal discretizations; suppression results

#### Reynolds equations

The air bearings used to suppress web vibration consist of two rigid surfaces, positioned close to the axially moving web at a constant speed $V$. The gap formed between the moving web and two rigid surfaces is filled by a two-dimensional laminar and incompressible viscous air with the dynamic viscosity $\mu$ which is governed by the so-called Reynolds equation, named after Osborne Reynolds (1842–1912). All thermal effects as well as slip of the fluid are ignored. The air bearings shown in Figure 4.5 are located in the region $x \in [a_i, b_i]$, and the coupled governing equations are constructed as follows:

$$m\frac{\partial^2 w}{\partial t^2} + 2mV\frac{\partial^2 w}{\partial x \partial t} + (mV^2 - T)\frac{\partial^2 w}{\partial x^2} = f + p_2 - p_1, \quad x \in [0, L], \tag{1.13}$$

$$-\frac{\partial}{\partial x}\left(\frac{1}{12\mu}\frac{\partial p_1}{\partial x}h_1^3\right) + \frac{\partial h_1}{\partial t} + \frac{V}{2}\frac{\partial h_1}{\partial x} + v_d = 0, \quad x \in [a_i, b_i], \tag{1.14}$$

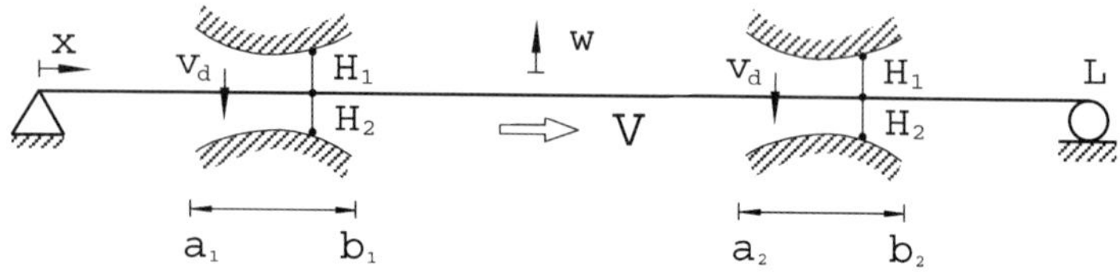


Figure 4.5. The coupled system of the moving web and two-sided suppression devices.

$$-\frac{\partial}{\partial x}\left(\frac{1}{12\mu}\frac{\partial p_2}{\partial x}h_2^3\right)+\frac{\partial h_2}{\partial t}+\frac{V}{2}\frac{\partial h_2}{\partial x}-v_d=0, \qquad x\in[a_i,b_i], \tag{1.15}$$

with the boundary conditions $w(0,t) = w(L,t) = 0$ and $p_1(x,t) = p_2(x,t) = 0$, $\forall x \in [0,L]/(a_i,b_i)$, where $w$, $f$, $v_d$, $h_j$, and $p_j$ $(j = 1,2)$ stand for the transverse web deflection, the external force, the net downward velocity of the diffusive flow through the porous web surface, the air bearing clearances, and pressures, respectively.

Notice that instead of the coupling with the full-fledged viscous fluid governed by the Navier–Stokes equation which will be elaborated upon in Chapters 6 and 7, the moving string is coupled with four air bearings governed by the Reynolds equations. For the two air bearing surface profiles $H_1$ and $H_2$, the upper and lower clearances are defined by

$$h_1 = H_1 - w \quad \text{and} \quad h_2 = H_2 + w. \tag{1.16}$$

Assuming that the web thickness $d$ is small, from the Darcy's law, named after Henry Philibert Gaspard Darcy (1803–1858), we obtain

$$v_d = \frac{k}{\mu}\frac{(p_1 - p_2)}{d}, \tag{1.17}$$

where $k$ is the paper permeability.

In the nonlinear coupled equations (1.13), (1.14), and (1.15), we have three unknowns $w$, $p_1$, and $p_2$. To introduce the surrounding air effects, we introduce $m_j = \rho\beta L$ $(j = 1,2,3)$ to the inertia, Coriolis, and centrifugal mass terms, respectively, where $\beta$ is a function of the web width.

*Spatial and temporal discretizations*

The variations of internal energy and external work are defined as $\delta\mathbf{W}$ and $\delta\mathbf{R}$. At any time $t$, the standard variational form of Eqs. (1.13), (1.14), and (1.15) can be written as

$$\delta^t\mathbf{W} = \delta^t\mathbf{R}, \tag{1.18}$$

with $\delta\mathbf{W} = \langle \delta W_w, \delta W_{p_1}, \delta W_{p_2} \rangle$, $\delta\mathbf{R} = \langle \delta R_w, 0, 0 \rangle$, and

$$\delta W_w = \int_0^L \left\{ \delta w \left( m \frac{\partial^2 w}{\partial t^2} + 2mV \frac{\partial^2 w}{\partial x \partial t} \right) - \frac{\partial \delta w}{\partial x} (mV^2 - T) \frac{\partial w}{\partial x} + \delta w (p_1 - p_2) \right\} dx, \tag{1.19}$$

$$\delta W_{p_1} = \int_{a_i}^{b_i} \left\{ \frac{\partial \delta p_1}{\partial x} \frac{1}{12\mu} (H_1 - w)^3 \frac{\partial p_1}{\partial x} - \delta p_1 \frac{\partial w}{\partial t} + \delta p_1 \frac{k}{\mu} \frac{(p_1 - p_2)}{d} + \delta p_1 \frac{V}{2} \frac{\partial H_1}{\partial x} - \delta p_1 \frac{V}{2} \frac{\partial w}{\partial x} \right\} dx, \tag{1.20}$$

$$\delta W_{p_2} = \int_{a_i}^{b_i} \left\{ \frac{\partial \delta p_2}{\partial x} \frac{1}{12\mu} (H_2 + w)^3 \frac{\partial p_2}{\partial x} + \delta p_2 \frac{\partial w}{\partial t} + \delta p_2 \frac{k}{\mu} \frac{(p_2 - p_1)}{d} + \delta p_2 \frac{V}{2} \frac{\partial H_2}{\partial x} + \delta p_2 \frac{V}{2} \frac{\partial w}{\partial x} \right\} dx, \tag{1.21}$$

$$\delta R_w = \int_0^L \delta w f \, dx. \tag{1.22}$$

Note that the displacement variation $\delta w$ is the arbitrary admissible out-of-plane displacement, i.e., $\delta w = 0$, at $x = 0, L$, and the pressure variations $\delta p_1$ and $\delta p_2$ are arbitrary within the interval $a_i < x < b_i$, but for which $\delta p_1 = \delta p_2 = 0$ at $x = a_i$ and $b_i$. From Taylor's expansion, we have

$$\delta \Delta^t \mathbf{W} = \delta^t \dot{\mathbf{W}} \Delta t \simeq \delta^{t+\Delta t} \mathbf{R} - \delta^t \mathbf{W}, \tag{1.23}$$

with

$$\delta \dot{W}_w \Delta t = \int_0^L \left\{ \delta w \left( m \frac{\partial^2 \Delta w}{\partial t^2} + 2mV \frac{\partial^2 \Delta w}{\partial x \partial t} \right) - \frac{\partial \delta w}{\partial x} (mV^2 - T) \frac{\partial \Delta w}{\partial x} + \delta w (\Delta p_1 - \Delta p_2) \right\} dx, \tag{1.24}$$

$$\delta \dot{W}_{p_1} \Delta t = \int_{a_i}^{b_i} \left\{ \frac{\partial \delta p_1}{\partial x} \frac{1}{12\mu} (H_1 - w)^3 \frac{\partial \Delta p_1}{\partial x} - \frac{\partial \delta p_1}{\partial x} \frac{1}{4\mu} (H_1 - w)^2 \Delta w \frac{\partial p_1}{\partial x} - \delta p_1 \frac{\partial \Delta w}{\partial t} + \delta p_1 \frac{k}{\mu} \frac{(\Delta p_1 - \Delta p_2)}{d} - \delta p_1 \frac{V}{2} \frac{\partial \Delta w}{\partial x} \right\} dx, \tag{1.25}$$

$$\delta \dot{W}_{p_2} \Delta t = \int_{a_i}^{b_i} \left\{ \frac{\partial \delta p_2}{\partial x} \frac{1}{12\mu} (H_2 + w)^3 \frac{\partial \Delta p_2}{\partial x} + \frac{\partial \delta p_2}{\partial x} \frac{1}{4\mu} (H_2 + w)^2 \Delta w \frac{\partial p_2}{\partial x} \right.$$

$$\left. + \delta p_2 \frac{\partial \Delta w}{\partial t} + \delta p_2 \frac{k}{\mu} \frac{(\Delta p_2 - \Delta p_1)}{d} + \delta p_2 \frac{V}{2} \frac{\partial \Delta w}{\partial x} \right\} dx. \quad (1.26)$$

It is clear that when Eqs. (1.19), (1.20), and (1.21) are expanded we retain those terms that are linear in the small increments of $\boldsymbol{\theta} = \langle w, p_1, p_2 \rangle$ with $\Delta \boldsymbol{\theta} = \dot{\boldsymbol{\theta}} \Delta t$. In other words, we neglect all squares, products, and higher powers of those small increments. Introducing the nodal unknowns $\boldsymbol{\Theta} = \langle \hat{\mathbf{W}}, \hat{\mathbf{P}}_1, \hat{\mathbf{P}}_2 \rangle$, and by applying the standard Galerkin finite element procedure, we obtain

$$\mathbf{M} \Delta \ddot{\boldsymbol{\Theta}} + \mathbf{C} \Delta \dot{\boldsymbol{\Theta}} + \mathbf{K} \Delta \boldsymbol{\Theta} = \Delta \mathbf{R}, \quad (1.27)$$

with

$$\mathbf{M} = \begin{bmatrix} \mathbf{M}_{ww} & \mathbf{0} & \mathbf{0} \\ \mathbf{0} & \mathbf{0} & \mathbf{0} \\ \mathbf{0} & \mathbf{0} & \mathbf{0} \end{bmatrix}, \qquad \mathbf{C} = \begin{bmatrix} \mathbf{C}_{ww} & \mathbf{0} & \mathbf{0} \\ \mathbf{C}_{p_1 w} & \mathbf{0} & \mathbf{0} \\ \mathbf{C}_{p_2 w} & \mathbf{0} & \mathbf{0} \end{bmatrix},$$

$$\mathbf{K} = \begin{bmatrix} \mathbf{K}_{ww} & \mathbf{K}_{wp_1} & \mathbf{K}_{wp_2} \\ \mathbf{K}_{p_1 w} & \mathbf{K}_{p_1 p_1} & \mathbf{K}_{p_1 p_2} \\ \mathbf{K}_{p_2 w} & \mathbf{K}_{p_2 p_1} & \mathbf{K}_{p_2 p_2} \end{bmatrix}, \qquad \Delta \mathbf{R} = \begin{bmatrix} \mathbf{R}_w \\ \mathbf{0} \\ \mathbf{0} \end{bmatrix} - \begin{bmatrix} \mathbf{F}_w \\ \mathbf{F}_{p_1} \\ \mathbf{F}_{p_2} \end{bmatrix},$$

$$\delta \mathbf{W}^T \mathbf{F}_w = \delta W_w, \qquad \delta \mathbf{P}_1^T \mathbf{F}_{p_1} = \delta W_{p_1},$$

$$\delta \mathbf{P}_2^T \mathbf{F}_{p_2} = \delta W_{p_2}, \qquad \delta \mathbf{W}^T \mathbf{R}_w = \delta R_w;$$

and

$$\delta \mathbf{W}^T \mathbf{K}_{wp_1} \Delta \hat{\mathbf{P}}_1 = \int_{a_i}^{b_i} \delta w \Delta p_1 \, dx,$$

$$\delta \mathbf{W}^T \mathbf{C}_{ww} \Delta \dot{\hat{\mathbf{W}}} = \int_0^L 2mV \delta w \frac{\partial^2 \Delta w}{\partial x \partial t} \, dx,$$

$$\delta \mathbf{P}_1^T \mathbf{K}_{p_1 p_2} \Delta \hat{\mathbf{P}}_2 = -\int_{a_i}^{b_i} \delta p_1 \frac{k}{\mu} \frac{\Delta p_2}{d} \, dx,$$

$$\delta \mathbf{W}^T \mathbf{M}_{ww} \Delta \ddot{\hat{\mathbf{W}}} = \int_0^L m \delta w \frac{\partial^2 \Delta w}{\partial t^2} \, dx,$$

$$\delta \mathbf{W}^T \mathbf{K}_{wp_2} \Delta \hat{\mathbf{P}}_2 = -\int_{a_i}^{b_i} \delta w \Delta p_2 \, dx,$$

$$\delta \mathbf{P}_1^T \mathbf{C}_{p_1 w} \Delta \dot{\hat{\mathbf{W}}} = -\int_{a_i}^{b_i} \delta p_1 \frac{\partial \Delta w}{\partial t} \, dx,$$

$$\delta \mathbf{P}_2^T \mathbf{K}_{p_2 p_1} \Delta \hat{\mathbf{P}}_1 = -\int_{a_i}^{b_i} \delta p_2 \frac{k}{\mu} \frac{\Delta p_1}{d} \, dx,$$

$$\delta \mathbf{P}_2^T \mathbf{C}_{p_2 w} \Delta \dot{\hat{\mathbf{W}}} = \int_{a_i}^{b_i} \delta p_2 \frac{\partial \Delta w}{\partial t} \, dx,$$

$$\delta \mathbf{W}^T \mathbf{K}_{ww} \Delta \hat{\mathbf{W}} = -\int_0^L \frac{\partial \delta w}{\partial x} (mV^2 - T) \frac{\partial \Delta w}{\partial x} \, dx,$$

$$\delta \mathbf{P}_1^T \mathbf{K}_{p_1 p_1} \Delta \hat{\mathbf{P}}_1 = \int_{a_i}^{b_i} \left\{ \frac{\partial \delta p_1}{\partial x} \frac{1}{12\mu} (H_1 - w)^3 \frac{\partial \Delta p_1}{\partial x} + \delta p_1 \frac{k}{\mu} \frac{\Delta p_1}{d} \right\} dx,$$

$$\delta \mathbf{P}_1^T \mathbf{K}_{p_1 w} \Delta \hat{\mathbf{W}} = -\int_{a_i}^{b_i} \left\{ \frac{\partial \delta p_1}{\partial x} \frac{1}{4\mu} (H_1 - w)^2 \Delta w \frac{\partial p_1}{\partial x} + \delta p_1 \frac{V}{2} \frac{\partial \Delta w}{\partial x} \right\} dx,$$

$$\delta \mathbf{P}_2^T \mathbf{K}_{p_2 p_2} \Delta \hat{\mathbf{P}}_2 = \int_{a_i}^{b_i} \left\{ \frac{\partial \delta p_2}{\partial x} \frac{1}{12\mu} (H_2 + w)^3 \frac{\partial \Delta p_2}{\partial x} + \delta p_2 \frac{k}{\mu} \frac{\Delta p_2}{d} \right\} dx,$$

$$\delta \mathbf{P}_2^T \mathbf{K}_{p_2 w} \Delta \hat{\mathbf{W}} = \int_{a_i}^{b_i} \left\{ \frac{\partial \delta p_2}{\partial x} \frac{1}{4\mu} (H_2 + w)^2 \Delta w \frac{\partial p_2}{\partial x} + \delta p_2 \frac{V}{2} \frac{\partial \Delta w}{\partial x} \right\} dx.$$

For the direct time integration along with the Newton–Raphson iteration, we apply the predictor–corrector procedure which will be discussed in Chapter 6. Note that because the convective terms $\frac{V}{2}\partial w/\partial x$ and $2mV\,\partial^2 w/\partial x \partial t$ introduce nonsymmetric **K** and **C** matrices, an effective nonsymmetric column solver is implemented [143]. However, it is found that these nonsymmetric terms do not introduce spatial oscillations that need to be treated with upwind discretizations in Chapter 7.

#### *4.1.2.1. Suppression results*

The proposed algorithm is implemented to analyze a typical moving paper web traveling at different velocities ($V = 0$, 4.85, and 9.7 m/s). A set of cylindrical surface air bearings as depicted in Figure 4.6 is located at $x = 0.5$ and $x = 1.9$

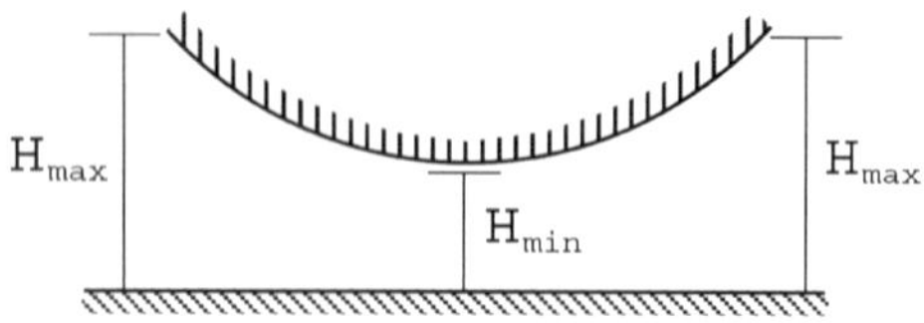


Figure 4.6. The cylindrical surface air bearing profile.

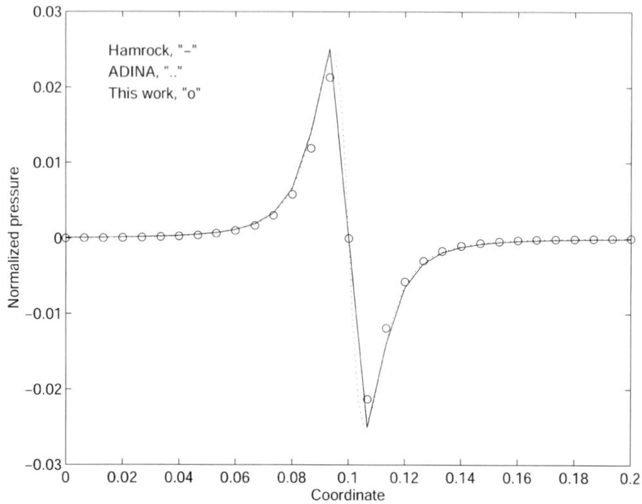


Figure 4.7. Pressure distributions of cylindrical surface bearing with the slide motion.

as suppression devices. The physical parameters are given as follows: $H_{\min} = 0.0002$ m; $H_{\max} = 0.02$ m; $m = 0.0355\ \mathrm{kg/m^2}$; $T = 280$ N/m; $k = 5.0 \times 10^{-13}\ \mathrm{m^2}$; $d = 5.0 \times 10^{-5}$ m; $\mu = 2.0 \times 10^{-5}$ Pa s; $\rho = 1.2\ \mathrm{kg/m^3}$; $l_j = b_j - a_j = 0.2$ m; and $L = 2.4$ m. The added mass coefficient $\beta$ is assumed to be 0.34 with the web length/width ratio $L/B = 5.0$.

For this FSI system, the lubrication effects are achieved by the slide and squeeze motions which correlate to the transverse and axial motions of the web. To verify the implementation of the proposed finite element algorithm, in Figures 4.7 and 4.8, for stationary webs, i.e., $V = 0$, the computed pressure distributions are compared with the analytical results published by Hamrock [21] as well as ADINA FSI solutions. The pressures are normalized by $\mu U R / H_{\min}^2$, where $R$ and $U$ stand for the cylindrical surface radius and the slide or squeeze velocity. As can be seen, the results are in agreement.

It is interesting to note that for this case, with zero web transporting velocity, ADINA FSI solutions match the Reynolds equation solutions. Nevertheless, if the air bearing inertia effects are significant, i.e., the Reynolds number $Re =$

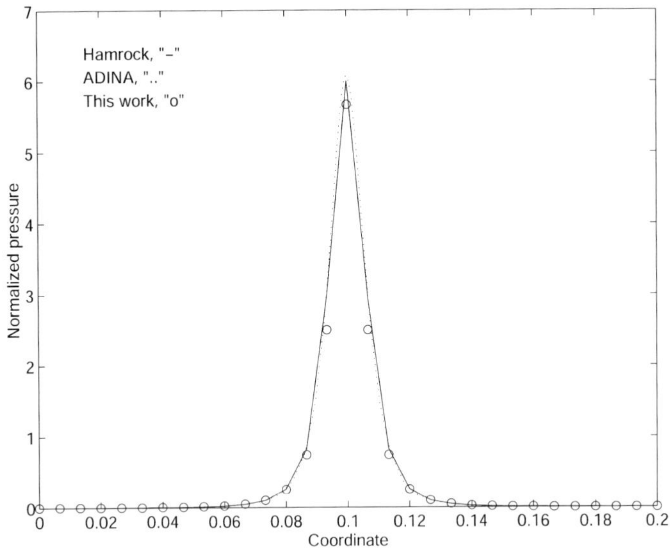


Figure 4.8. Pressure distributions of cylindrical surface bearing with the squeeze motion.

$UH^2_{\text{min}}/\nu l_j \sim O(1)$, the Reynolds equation can no longer accurately describe the lubrication zones, see Figure 4.9.

Because the purposes of using these suppression devices are the reduction of oscillations and the stabilization of the moving web, it is of interest to calculate the time history of the following measures of average positive and negative nodal displacements ($W_i^+$ and $W_i^-$) along with the mid-span nodal displacement:

$$\bar{W}^+ = \sum_{i=1}^{N^+} W_i^+/N^+ \quad \text{and} \quad \bar{W}^- = \sum_{i=1}^{N^-} W_i^-/N^-, \tag{1.28}$$

where $N^+$ and $N^-$ are the total numbers of positive and negative nodal displacements.

For the concentrated dynamic load applied at the mid-span node with the time functions shown in Figure 4.10, the salient features of the mid-span dynamic responses with and without suppression devices shown in Figures 4.11 to 4.14 demonstrate the stabilizing effects of such suppression devices. The average positive and negative displacement measures shown in Figure 4.15 confirm the conclusions drawn based on the mid-span displacement responses. Two points may be noticed: first, the suppression devices provide the damping needed to stabilize the traveling web; second, the suppression devices alter the steady response of the moving web. For the web traveling at velocity $V$, the stiffness of the steady response is reduced to $T - mV^2$. Figures 4.16 and 4.17 illustrate paper web poros-

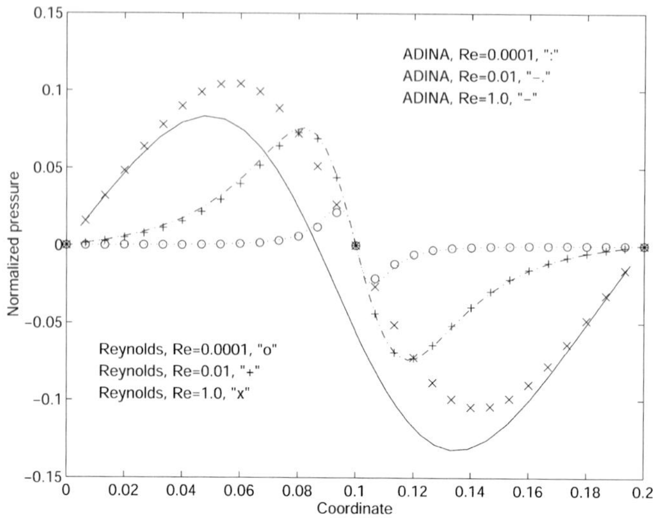


Figure 4.9. The Reynolds equation approximations.

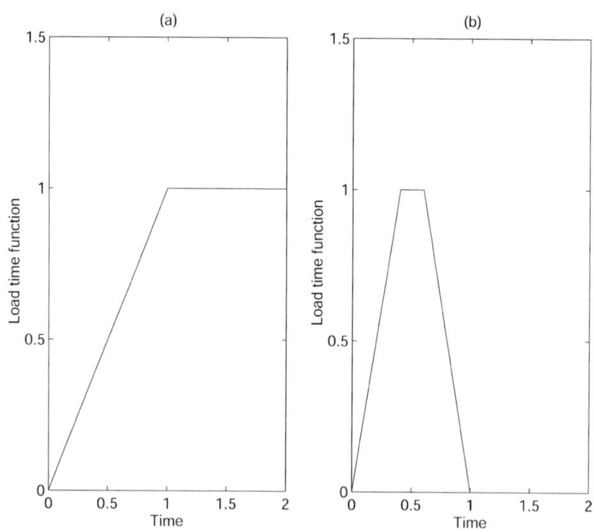


Figure 4.10. Load time functions.

ity effects on suppression devices. Because the moving paper web travels through the upper and lower air bearings, the porosity effects are minimized due to the fact that the air mass is always preserved within the bearing couples.

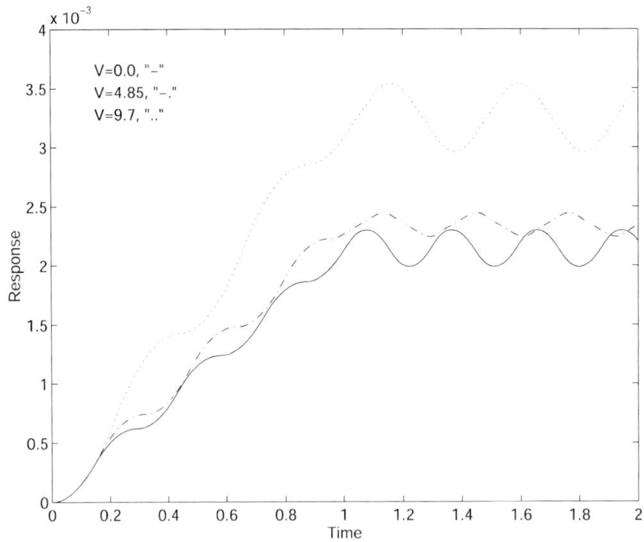


Figure 4.11. Response due to load (a) without suppression device.

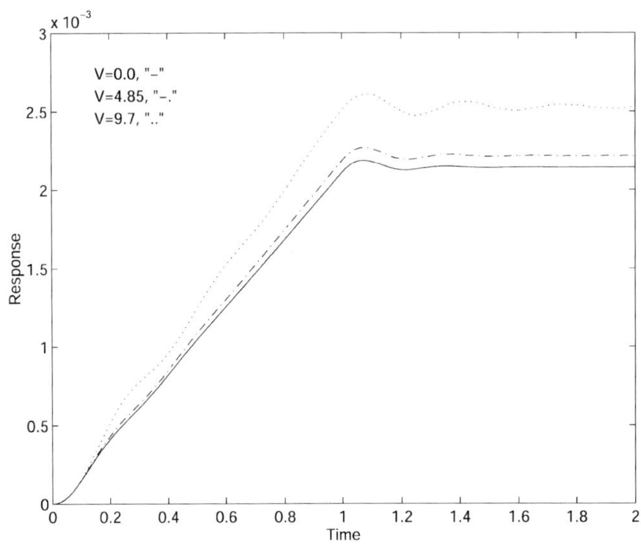


Figure 4.12. Response due to load (a) with suppression device.

Three modifications of suppression devices are made separately: first, the bearing location ($x = 0.7$ and $x = 1.7$); second, the bearing clearance ($H_{\min} = 0.0001$ m and $H_{\max} = 0.0199$ m); and third, the bearing profile ($R = 0.1$ m

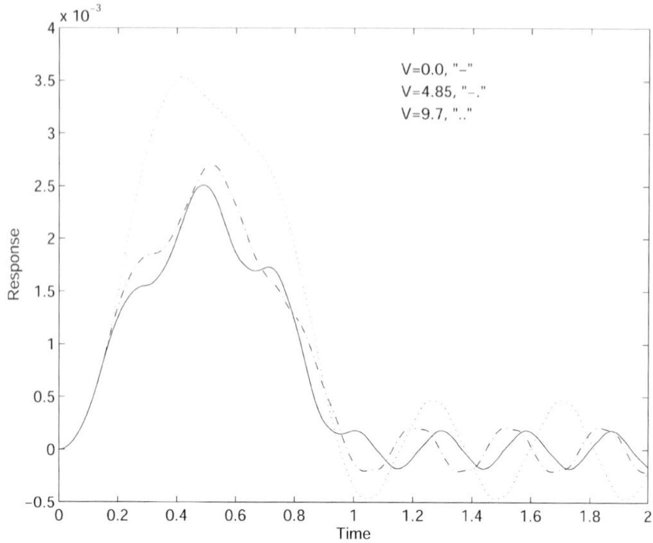


Figure 4.13. Response due to load (b) without suppression device.

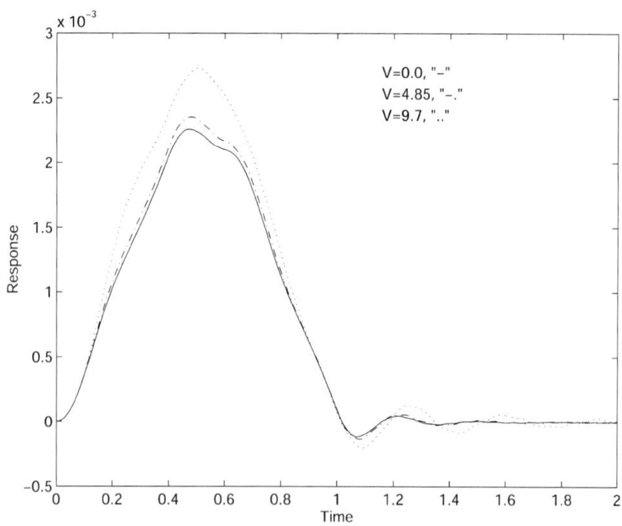


Figure 4.14. Response due to load (b) with suppression device.

and $l_j = 0.1$ m). The comparisons of their dynamic responses are shown in Figure 4.18. It is confirmed that both the steady response and damping effects can be modified by varying the bearing locations, clearances, and profiles. More specif-

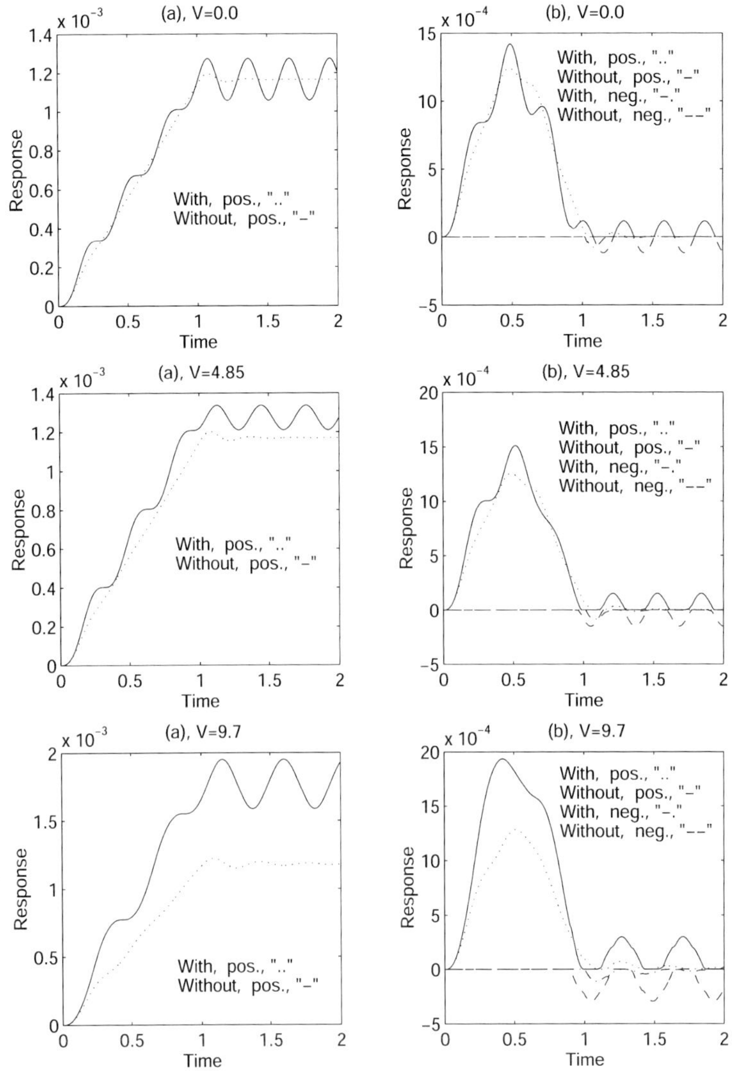


Figure 4.15. Average positive and negative displacements with and without suppression device.

ically, we observe that the reduction of clearances will increase the damping and stiffness of the air bearings; by shifting the bearings to the middle of the moving web, the mid-span deflection can be reduced significantly; and the reduction of bearing length will weaken the suppression effects.

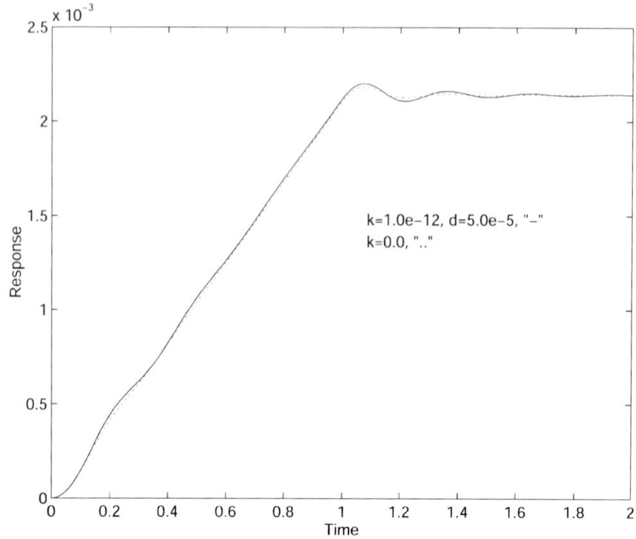


Figure 4.16. The web porosity effects with load (a).

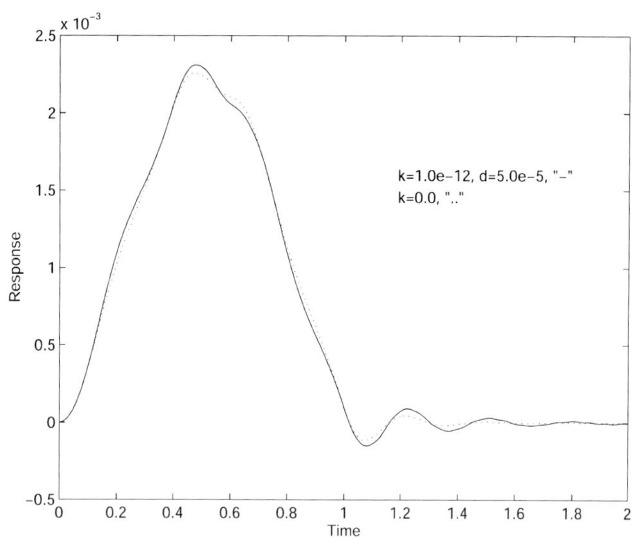


Figure 4.17. The web porosity effects with load (b).

For the tensioned web with zero traveling velocity ($V = 0$), it is possible to solve the four sets of the Navier–Stokes equations instead of the Reynolds equations coupled with the oscillating web. As shown in Figure 4.19, both the

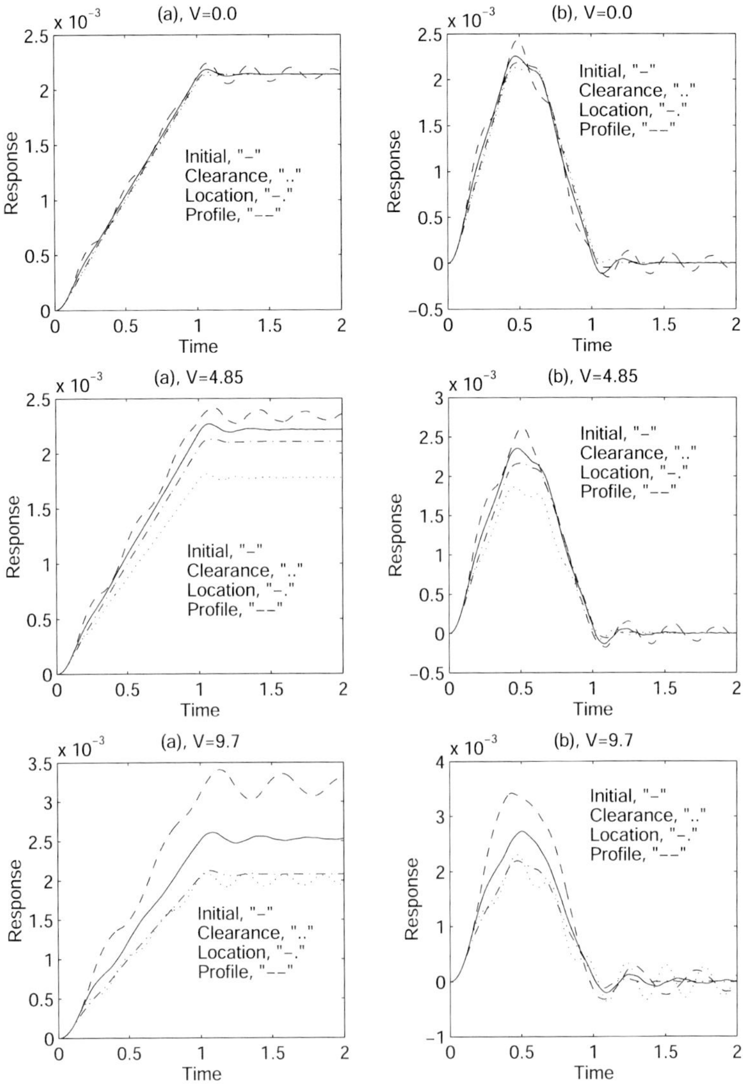


Figure 4.18. Modifications of bearing location, clearance, and profile.

displacement and pressure results match each other. In general, the optimized suppression device design is highly dependent on the web moving velocity. Therefore, an adjustable installation of such devices is beneficial when the web traveling velocity needs to be varied. Based on the results, we also confirm that for the cases with low Reynolds number, the Reynolds lubrication equation can be effectively

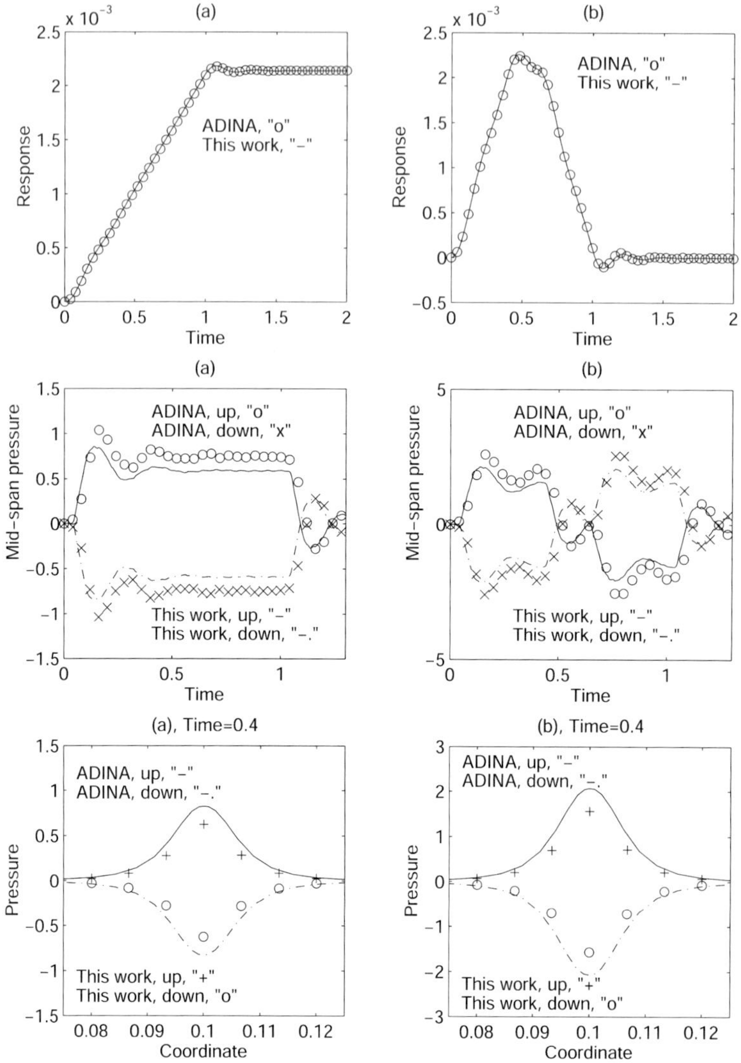


Figure 4.19. Comparison with ADINA for the zero traveling velocity cases.

used with good accuracy. However, when the lubricant inertia contribution is significant, full-fledged Navier–Stokes equations needed to couple with the moving web. It is also of interest to note that the web permeability barely affects the dynamic response with the double-sided suppression devices. In practice, the nips formed between suppression device surfaces also act as effective supports of the moving web and can significantly reduce the critical traveling speed by reducing the open draw length.

## 4.2. Stability issues of axial flow models

Fluid conveying pipes are widely used in engineering applications. One of the design challenges is to avoid pipe buckling and flutter under various operation conditions. Initial work on such flow-induced vibration analysis was reported by Ashley and Haviland [9], Benjamin [17,18], and Paidoussis [189,190]. A recent survey of the subject is available in Paidoussis [191]. The subject of flow through collapsible (deformable) tubes is also ubiquitous in cardiovascular systems [191,275,276]. Typical examples include ureters and urethras, intra-thoracic airways, veins, and arteries. The main objective of this section is to propose a simple mathematical model for stability behaviors of submerged flexible pipes containing viscous fluid flow.

With the development of numerical algorithms, it is now possible to simulate a two-dimensional coupled system of a very flexible tube with internal and external fluids. In this section, we focus on the simplified models in the study of static and dynamic instabilities. Specific examples of collapsible tubes for fully coupled nonlinear systems will be presented in Chapter 7.

### *4.2.1. Internal flow; concentric flow*

In this section, the governing equations of various pipe systems are derived under small deformation assumptions with consideration of fluid frictional forces and inertia effects.

#### *Internal flow*

Figure 4.20 shows the mathematical model of a flow-through collapsible tube system with the longitudinal direction being that of the $x$-axis. The mathematical model includes the inner pipe with a length $L$, an inner radius $R_i$, and an outer radius $R_o$. We denote the external hydrostatic pressure as $p_e$, and there is no shear stress acting on the external surface of the tube. Under the action of the gravitational force and fluid forces, i.e., pressure and frictional forces, the inner pipe will deform and oscillate.

Using small displacement and small strain assumptions, the body coordinates $(\xi, \eta, \zeta)$ corresponding to the deformed configuration can be projected onto the initial coordinates $(x, y, z)$. Moreover, in-plane ($x$-$y$ plane) and out-of-plane displacements are denoted as $y = y(x, t)$ and $z = z(x, t)$ as shown in Figure 4.21. Due to symmetry, we only consider $y(x, t)$. Assigning $\rho_i$ and $U_i$ as the fluid density and the spatially-averaged flow velocity for the pipe flow region, we initiate

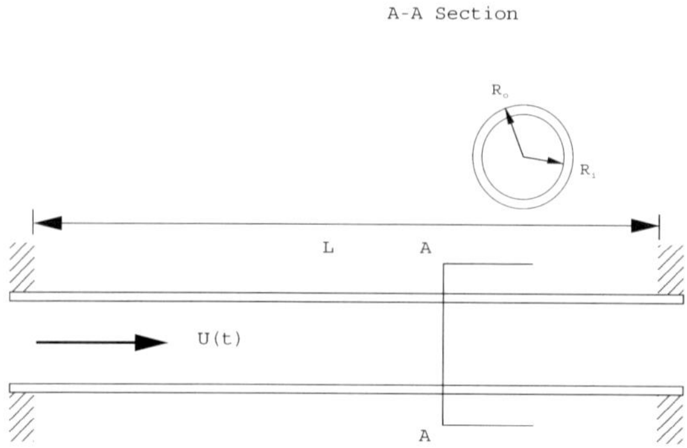


Figure 4.20. Schematic diagram of flow through a collapsible tube.

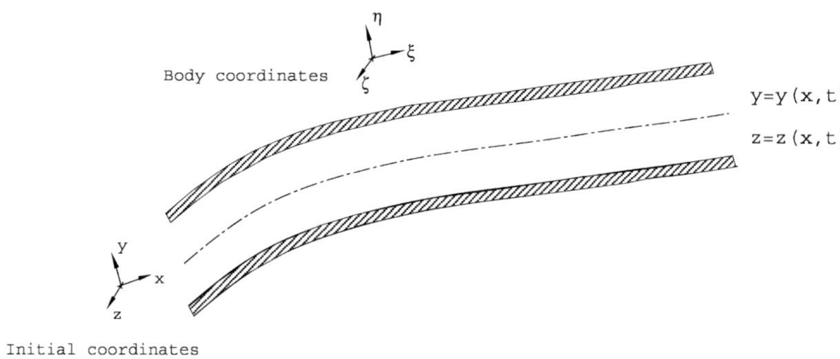


Figure 4.21. Deformed element of the tubular beam.

our derivation of the governing equations for $y(x, t)$ by writing down the force equilibrium for the tubular beam, which has a flexural rigidity $EI$, where $E$ is Young's modulus and $I$, expressed as $I = \pi(R_o^4 - R_i^4)/4$, is the moment of inertia of the tubular beam cross-sectional area. To include the inertia effects of the tubular beam we also denote $m$ as the mass per unit length. For the pulsatile pipe flow we have $U_i = \bar{U}_i(1 + \epsilon \cos \omega_o t)$, where $\bar{U}_i$, $\epsilon$, and $\omega_o$ represent the time-mean value of the spatially-averaged pipe flow velocity, the velocity perturbation magnitude, and the perturbation frequency, respectively. According to the discussion in Paidoussis and Issid [193], the inertia term $\rho A_i \, \partial U_i / \partial t$ needs to be added to the axial force equilibrium of the internal flow. By the chain rule, for the transverse inertia effects, we have

$$\left( \frac{\partial}{\partial t} + U_i \frac{\partial}{\partial x} \right)^2 = \frac{\partial^2}{\partial t^2} + 2U_i \frac{\partial^2}{\partial x \partial t} + U_i^2 \frac{\partial^2}{\partial x^2} + \frac{\partial U_i}{\partial t} \frac{\partial}{\partial x}. \tag{2.29}$$

Ignoring the axial inertia effect, the governing equation for the tubular beam can be expressed as

$$\frac{\partial}{\partial x}\left(T\frac{\partial y}{\partial x}\right) + EI\frac{\partial^4 y}{\partial x^4} + \left(F_\eta^i + F_\xi^i\frac{\partial y}{\partial x}\right) + \frac{\partial}{\partial x}\left(p_e A_o \frac{\partial y}{\partial x}\right)$$
$$= m\frac{\partial^2 y}{\partial t^2} + mg. \tag{2.30}$$

Furthermore, based on the force equilibrium for internal flow (see, for example, Hannoyer and Paidoussis [99], and Paidoussis and Pettigrew [195]), we express internal fluid forces as follows:

$$F_\eta^i + F_\xi^i\frac{\partial y}{\partial x} = -\frac{\partial}{\partial x}\left(p_i A_i \frac{\partial y}{\partial x}\right) - \rho_i A_i g - \rho_i A_i \left(\frac{\partial}{\partial t} + U_i\frac{\partial}{\partial x}\right)^2 y. \tag{2.31}$$

The key point in simplifying the governing equation is to obtain the explicit expression for the tension $T$, based on the assumption $p_e|_{x=l} = p_i|_{x=l} = p_2$. The mass per unit length $m$ can be expressed as $\rho\pi(R_o^2 - R_i^2)$. The gravitational effect is included as a distributed force derived from the gravitational constant $g$. Thus, in a manner similar to the derivation in Refs. [296,297], we find the following governing equation for $y(x,t)$:

$$c_1\frac{\partial^4 y}{\partial x^4} + c_2\frac{\partial^2 y}{\partial x^2} + c_3\frac{\partial^2 y}{\partial x \partial t} + c_4\frac{\partial^2 y}{\partial t^2} + c_5 = 0, \tag{2.32}$$

where the coefficients are given by

$$c_1 = EI, \qquad c_2 = \rho A_i U_i^2 - (T - p_i A_i + p_e A_o), \qquad c_3 = 2\rho A_i U_i,$$
$$c_4 = m + \rho A_i + \chi\rho A_o, \quad \text{and} \quad c_5 = (m + \rho A_i - \rho A_o)g.$$

REMARK 4.2.1. The inhomogeneous term due to gravity $c_5$ affects the dynamical behaviors of the tubular beam, but not the stability issues.

REMARK 4.2.2. The tension $T$ is imposed, and is uniform, based on linearization assumptions. However, the total effective axial force within the tubular beam depends on $\rho A_i\, \partial U_i/\partial t$, the internal pressure $p_i$, and the external pressure $p_e$

$$(T - p_i A_i + p_e A_o) = T - p_2 A + (x - l)\rho A_i\, \partial U_i/\partial t. \tag{2.33}$$

REMARK 4.2.3. Although $p_e$ is constant through the entire length of the flexible tube, due to the viscous effect, the internal pressure $p_i$ is a linear function of the spatial location $x$. Denoting the dynamical viscosity as $\mu$ and the flow rate as $Q$, we have

$$p_i = p_1 + (p_2 - p_1)x/l, \tag{2.34}$$

where $p_1$ is the internal fluid pressure at the starting point of the flexible tube with $(p_1 - p_2)/l = 8\mu Q/(\pi R_i^3)$.

Notice that for the case of clamped (or built-in) boundary condition at $x = 0$ and $x = l$, we have

$$\begin{aligned} y(0,t) &= 0, \qquad \frac{\partial y(0,t)}{\partial x} = 0, \\ y(l,t) &= 0, \qquad \frac{\partial y(l,t)}{\partial x} = 0. \end{aligned} \tag{2.35}$$

We also recognize that the coefficients $c_1$ to $c_5$ can be variables depending on position $x$. For pulsatile flow, only the coefficients $c_2$ and $c_3$ are periodic in time with the period $T_o = \omega_o/2\pi$. Considering the fact that we have third- and fourth-order spatial derivatives in Eq. (2.36), standard $C^0$ finite elements cannot be used. An elaborate discussion on the use of $C^1$ finite elements or mixed formulations is available in Refs. [143,207]. In this book, we employ the method of finite differences to replace the partial differential equation (2.32) with a set of ordinary differential equations with respect to time. We denote $Y^k(t)$ as the solution variable $y(x)$ evaluated at the $k$th spatial grid (or nodal) point depicted in Figure 4.22, its corresponding time derivative as $\dot{Y}^k(t)$. Using an equal spacing $h$ between finite difference grid points, we obtain the following finite difference approximations for various differentiations:

$$\begin{aligned} \left.\frac{\partial y}{\partial x}\right|_k &= \frac{Y^{k+1} - Y^{k-1}}{2h}, \\ \left.\frac{\partial^2 y}{\partial x^2}\right|_k &= \frac{Y^{k+1} - 2Y^k + Y^{k-1}}{h^2}, \\ \left.\frac{\partial^3 y}{\partial x^3}\right|_k &= \frac{Y^{k+2} - 2Y^{k+1} + 2Y^{k-1} - Y^{k-2}}{2h^3}, \\ \left.\frac{\partial^4 y}{\partial x^4}\right|_k &= \frac{Y^{k+2} - 4Y^{k+1} + 6Y^k - 4Y^{k-1} + Y^{k-2}}{h^4}. \end{aligned} \tag{2.36}$$

Equivalent difference schemes are also used for the boundary conditions in Eq. (2.35) and yield

$$Y^0 = 0, \qquad Y^{-1} = Y^1, \qquad Y^N = 0, \quad \text{and} \quad Y^{N-1} = Y^{N+1}. \tag{2.37}$$

Using Eq. (2.37), the partial differential equation (2.32) is replaced with a set of ordinary differential equations with respect to time. Equivalent difference schemes are also used for the boundary conditions in Eq. (2.35). We define the solution variable $y(x,t)$ at the spatial grid (or nodal) point $i$ as $Y^i(t)$ as depicted in Figure 4.22, and its corresponding time derivative as $\dot{Y}^i(t)$. As shown

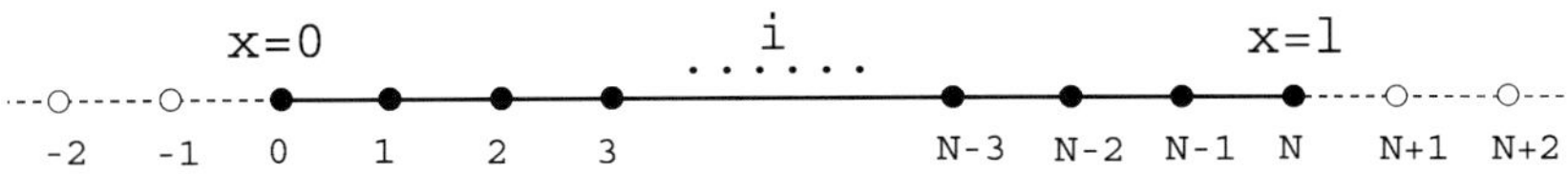


Figure 4.22. Finite difference stations on the tubular beam.

in Figure 4.22, using an equal spacing $h$ between finite difference stations, and employing the finite difference approximations, we obtain the discretized characteristic equation for node $i$ $(1 \leqslant i \leqslant N-1)$:

$$c_4^i \ddot{Y}^i + \frac{c_3^i}{2h}\dot{Y}^{i+1} - \frac{c_3^i}{2h}\dot{Y}^{i-1} + \frac{c_1^i}{h^4}Y^{i+2} + \left(\frac{c_2^i}{h^2} - \frac{4c_1^i}{h^4}\right)Y^{i+1}$$
$$+ \left(\frac{6c_1^i}{h^4} - \frac{2c_2^i}{h^2}\right)Y^i + \left(\frac{c_2^i}{h^2} - \frac{4c_1^i}{h^4}\right)Y^{i-1} + \frac{c_1^i}{h^4}Y^{i-2} = 0. \tag{2.38}$$

As the variable coefficients $c_1$ to $c_4$ could be functions of $x$, they are denoted as $c_1^i$ to $c_4^i$ at the nodal point $i$. Therefore, Eq. (2.38) becomes

$$\mathbf{M}\ddot{\mathbf{Y}} + \mathbf{C}\dot{\mathbf{Y}} + \mathbf{K}\mathbf{Y} = \mathbf{0}, \tag{2.39}$$

where $\mathbf{Y}$ is the solution vector, and $\mathbf{M}$, $\mathbf{C}$, and $\mathbf{K}$ stand for the mass, damping including gyroscopic terms, and stiffness algebraic coefficient matrices, respectively.

*Concentric flow*

In the paper industry, approach flow systems are generally used to dilute fiber stock with water. These systems consist of many pumps, screens, deaerators, and piping components. One of the key components in these systems is the so-called silo water mixing unit, a cylindrical water storage tank with a constant water level, as depicted in Figure 4.23. The inner pipe protruding into the fan pump inlet zone contains a higher consistency fiber stock, and the concentric outer pipe collects the recirculating diluted stock. At the inlet end, the outer pipe is welded on the silo side wall and the inner pipe is welded on the outer pipe wall and connected to a large elbow. In this book, we assume both the inner and outer concentric pipes are fixed at the inlet end. It has been discovered that the turbulent mixing of jets coming out of the concentric pipes before the fan pump contributes significantly to the smooth operation of impellers, the uniformity of stock consistency, and the minimization of pressure variations. In addition, the turbulent jets may introduce strong oscillations in the suspended pipes, which can cause structural damages such as fatigue or failure of pipe joints. Therefore, in a proper silo pipe system design, vibration problems associated with such pipes must be understood and resolved.

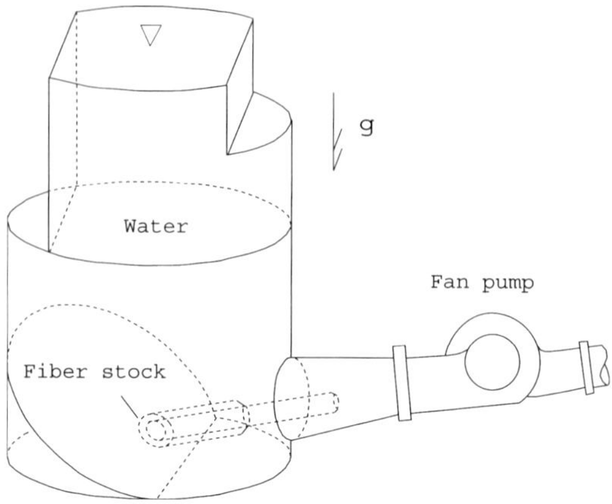


Figure 4.23. Location of the mixing pipe in the silo unit.

Considering the fact that the inner pipe carries thick fiber stock and has a much smaller diameter than the outer pipe, we will focus exclusively on the inner pipe vibration. In this mathematical model for inclined, submerged, concentric pipes with different pipe lengths, we consider the fluid forces exerted by both the confined and unconfined external flows. A schematic diagram of the location and general configuration of the mixing pipe arrangement within the silo is shown in Figure 4.23. As depicted in Figure 4.24, the mathematical model of the suspended concentric pipes includes the inner pipe with a length $l$ and the outer pipe with a length $L < l$. We note that all the pipes are submerged in silo water, and that continuous flow between the two concentric cylinders only occurs in the domain $0 \leqslant x \leqslant L$. Under the action of the gravitational force and fluid forces, i.e., pressure and frictional forces, the inner pipe will deform and oscillate.

Again, using small displacement and small strain assumptions, the body coordinates $(\xi, \eta, \zeta)$ corresponding to the deformed configuration are projected onto the initial coordinates $(x, y, z)$, as shown in Figure 4.21, with the following results:

$$\begin{aligned} &\xi_x \simeq \xi, \qquad \xi_y \simeq \xi\frac{\partial y}{\partial x}, \qquad \xi_z \simeq \xi\frac{\partial z}{\partial x}, \\ &\eta_x \simeq -\eta\frac{\partial y}{\partial x}, \qquad \eta_y \simeq \eta, \qquad \eta_z \simeq 0, \\ &\zeta_x \simeq -\zeta\frac{\partial z}{\partial x}, \qquad \zeta_y \simeq 0, \qquad \zeta_z \simeq \zeta, \end{aligned} \tag{2.40}$$

where the in-plane ($x$-$y$ plane) and out-of-plane displacements are denoted as $y = y(x, t)$ and $z = z(x, t)$.

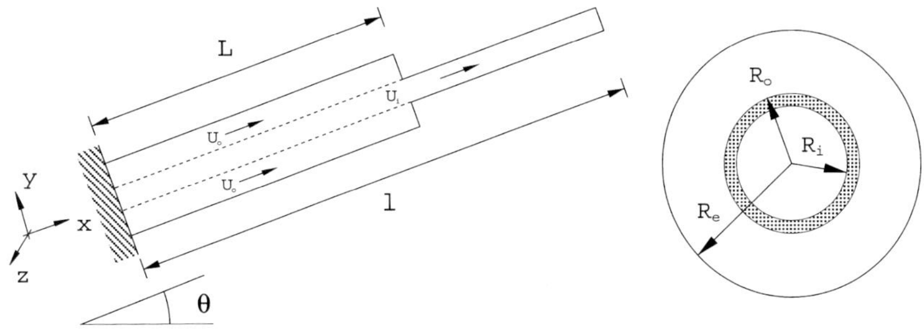


Figure 4.24. The concentric piping equilibrium configuration.

In order to retain certain nonlinear terms at the beginning of our derivation, we establish the force equilibrium on the deformed configurations. We consider that the fluid flows within the inner tube and the concentric region between the inner and outer pipes are fully developed turbulent flows, and express the forces exerted on the inner tubular beam from the internal and external fluid flows as $(F_\xi^i, F_\eta^i, F_\zeta^i)$ and $(F_\xi^e, F_\eta^e, F_\zeta^e)$, respectively. As a consequence of Eq. (2.40), we may project the internal fluid forces on the initial coordinates $(x, y, z)$ as follows:

$$
\begin{aligned}
&(F_\xi^i)_x \simeq F_\xi^i, && (F_\xi^i)_y \simeq F_\xi^i \frac{\partial y}{\partial x}, && (F_\xi^i)_z \simeq F_\xi^i \frac{\partial z}{\partial x},\\
&(F_\eta^i)_x \simeq -F_\eta^i \frac{\partial y}{\partial x}, && (F_\eta^i)_y \simeq F_\eta^i, && (F_\eta^i)_z \simeq 0,\\
&(F_\zeta^i)_x \simeq -F_\zeta^i \frac{\partial z}{\partial x}, && (F_\zeta^i)_y \simeq 0, && (F_\zeta^i)_z \simeq F_\zeta^i,
\end{aligned}
\tag{2.41}
$$

and similar projections can be obtained with respect to the external fluid forces.

It is clear that the internal fluid occupies a domain with cross-sectional area $A_i = \pi R_i^2$. In addition, we assign $\rho_i$ and $\rho_e$, $R_e$, $U_i$ and $U_e$ to be the internal and external fluid densities, inner radius of the external pipe, constant averaged turbulent flow velocities for internal and external regions. We will initiate our derivation of the governing equations for $y(x, t)$ and $z(x, t)$ by writing down the force equilibrium for the inner tubular beam. As depicted in Figure 4.25, a differential element $dx$ of the beam is acted upon by forces due to gravity $g$ with an inclination angle $\theta$, tension $T$, fluid forces $(F_\xi^i, F_\eta^i, F_\zeta^i)$ and $(F_\xi^e, F_\eta^e, F_\zeta^e)$, and transverse shear forces $(Q_y, Q_z)$. Moreover, we will, in the present analysis, discount the influence of any moments that may be exerted by internal and external flows. Also depicted in Figure 4.25 are the bending moments $M_y$ and $M_z$ in accord with $y(x, t)$ and $z(x, t)$. In the standard manner, if we ignore viscoelastic damping effects and consider the beam as an Euler–Bernoulli beam with constant cross-sectional area, the bending moments $M_y$ and $M_z$ and the shear forces $Q_y$

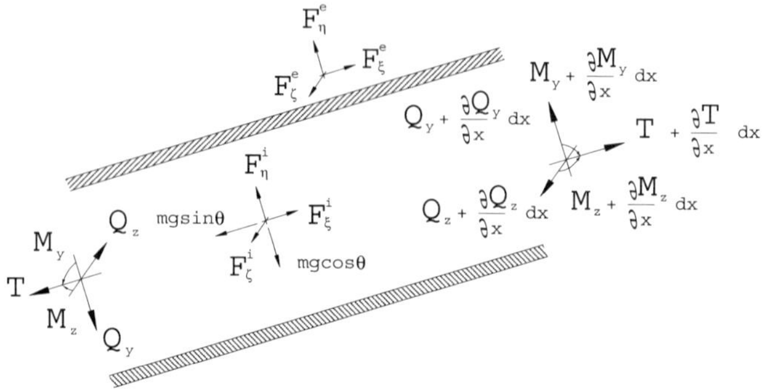


Figure 4.25. Forces acting on a differential element of the inner tubular beam.

and $Q_z$ assume the form

$$M_y = EI\frac{\partial^2 y}{\partial x^2}, \qquad Q_y = -EI\frac{\partial^3 y}{\partial x^3},$$
$$M_z = EI\frac{\partial^2 z}{\partial x^2}, \qquad Q_z = -EI\frac{\partial^3 z}{\partial x^3}. \tag{2.42}$$

Thus we ignore the axial inertia effect and derive the governing equations for the tubular beam as

$$\frac{\partial T}{\partial x} + \left(F_\xi^i - F_\eta^i\frac{\partial y}{\partial x} - F_\zeta^i\frac{\partial z}{\partial x}\right) + \left(F_\xi^e - F_\eta^e\frac{\partial y}{\partial x} - F_\zeta^e\frac{\partial z}{\partial x}\right)$$
$$- \frac{\partial}{\partial x}\left(Q_y\frac{\partial y}{\partial x}\right) - \frac{\partial}{\partial x}\left(Q_z\frac{\partial z}{\partial x}\right) - mg\sin\theta = 0, \tag{2.43}$$

$$\frac{\partial}{\partial x}\left(T\frac{\partial y}{\partial x}\right) + \frac{\partial Q_y}{\partial x} + \left(F_\eta^i + F_\xi^i\frac{\partial y}{\partial x}\right) + \left(F_\eta^e + F_\xi^e\frac{\partial y}{\partial x}\right)$$
$$= m\frac{\partial^2 y}{\partial t^2} + mg\cos\theta, \tag{2.44}$$

$$\frac{\partial}{\partial x}\left(T\frac{\partial z}{\partial x}\right) + \frac{\partial Q_z}{\partial x} + \left(F_\zeta^i + F_\xi^i\frac{\partial z}{\partial x}\right) + \left(F_\zeta^e + F_\xi^e\frac{\partial z}{\partial x}\right) = m\frac{\partial^2 z}{\partial t^2}, \tag{2.45}$$

where $m$ denotes the mass per unit length of the tubular beam.

Hence, from the force equilibrium for the internal flow, we can express the internal fluid forces as follows:

$$F_\xi^i - F_\eta^i\frac{\partial y}{\partial x} - F_\zeta^i\frac{\partial z}{\partial x} = -\frac{\partial}{\partial x}(p_i A_i) - \rho_i A_i g\sin\theta, \tag{2.46}$$

$$F_\eta^i + F_\xi^i\frac{\partial y}{\partial x} = -\frac{\partial}{\partial x}\left(p_i A_i\frac{\partial y}{\partial x}\right) - \rho_i A_i g\cos\theta - \rho_i A_i\left(\frac{\partial}{\partial t} + U_i\frac{\partial}{\partial x}\right)^2 y, \tag{2.47}$$

$$F_\zeta^i + F_\xi^i \frac{\partial z}{\partial x} = -\frac{\partial}{\partial x}\left(p_i A_i \frac{\partial z}{\partial x}\right) - \rho_i A_i \left(\frac{\partial}{\partial t} + U_i \frac{\partial}{\partial x}\right)^2 z. \tag{2.48}$$

Finally, we turn our attention to the contributions from the external fluid flow. We treat the external flow a little differently because of the fact that we can no longer directly replace the forces acting on the outer surface of the inner pipe with the equivalent flow pressure, inertia, and gravitational forces. We note that we have two different regions to consider, i.e., the confined region $0 \leqslant x < L$ and the unconfined region $L \leqslant x \leqslant l$. However, due to the outlet opening at the location $x = L$, the hydrostatic pressure is continuous in both regions, and for the tubular beam with a uniform cross-sectional area, the components of the net forces attributable to the external hydrostatic and hydrodynamic fluid pressures are given by

$$\begin{aligned} \mathcal{F}_x^1 &= 0, \\ \mathcal{F}_y^1 &= \rho_e A_o g \cos\theta + \frac{\partial}{\partial x}\left((p_e + p_o) A_o \frac{\partial y}{\partial x}\right), \\ \mathcal{F}_z^1 &= \frac{\partial}{\partial x}\left((p_e + p_o) A_o \frac{\partial z}{\partial x}\right), \end{aligned} \tag{2.49}$$

where $p_e$ and $p_o$ stand for the hydrostatic and hydrodynamic pressures, respectively, and the cross-sectional area $A_o$ is defined as $\pi R_o^2$.

By denoting the free surface level measured from the origin $(0, 0, 0)$ as $y_o$, such that the hydrostatic pressure at the tip of the submerged beam ($x = l$) is given by $\bar{p} = \rho_e g y_o - \rho_e g l \sin\theta$, we obtain the expression for the hydrostatic pressure of the external fluid,

$$p_e = (l - x)\rho_e g \sin\theta - y\rho_e g \cos\theta + \bar{p}. \tag{2.50}$$

Moreover, we have the following expression for the hydrodynamic pressure in the concentric flow region:

$$p_o A_o = \frac{1}{2}\rho_e D_o U_e^2 C_f h(x), \tag{2.51}$$

where the friction coefficient $C_f$ has different values in the confined and unconfined regions,

$$\begin{aligned} C_f &= \begin{cases} C_f^1, & 0 \leqslant x < L, \\ C_f^2, & L \leqslant x \leqslant l, \end{cases} \\ h(x) &= \begin{cases} D_o/(D_e - D_o) \cdot (L - x), & 0 \leqslant x < L, \\ 0, & L \leqslant x \leqslant l, \end{cases} \end{aligned} \tag{2.52}$$

along with

$$\frac{dh(x)}{dx} = \begin{cases} -D_o/(D_e - D_o), & 0 \leqslant x < L, \\ 0, & L \leqslant x \leqslant l. \end{cases} \tag{2.53}$$

Thus the external flow exerts upon the tubular beam the following viscous forces per unit length in both the normal ($\eta$, $\zeta$) and longitudinal ($\xi$) directions,

$$
\begin{aligned}
f_\xi^e &= \frac{1}{2}\rho_e D_o U_e^2 C_f,\\
f_\eta^e &= -\frac{1}{2}\rho_e D_o U_e C_f\left(\frac{\partial y}{\partial t} + U_e\frac{\partial y}{\partial x}\right),\\
f_\zeta^e &= -\frac{1}{2}\rho_e D_o U_e C_f\left(\frac{\partial z}{\partial t} + U_e\frac{\partial z}{\partial x}\right),
\end{aligned}
\tag{2.54}
$$

where $D_o$ is the outer diameter of the inner pipe.

Turning now to the fluid inertia forces, we take $x_o$ to be the entrance distance associated with the turbulent boundary layer, define the functions $\sigma = 1 + 0.4(x_o/L)C_f^1$ and $\alpha = 0.4C_f^1/\sigma$ for the confined external flow region, and introduce for the components of the inertia forces

$$
\begin{aligned}
\mathcal{F}_x^2 &= 0,\\
\mathcal{F}_y^2 &= -\chi\rho_e A_o\left(\frac{\partial}{\partial t} + \tilde{U}_e\frac{\partial}{\partial x}\right)\left(\frac{\partial}{\partial t} + U_e\frac{\partial}{\partial x}\right)y,\\
\mathcal{F}_z^2 &= -\chi\rho_e A_o\left(\frac{\partial}{\partial t} + \tilde{U}_e\frac{\partial}{\partial x}\right)\left(\frac{\partial}{\partial t} + U_e\frac{\partial}{\partial x}\right)z,
\end{aligned}
\tag{2.55}
$$

with

$$
\begin{aligned}
\chi &= \begin{cases} (R_e^2 + R_o^2)/(R_e^2 - R_o^2), & 0 \leqslant x < L,\\ 1, & L \leqslant x \leqslant l,\end{cases}\\
\tilde{U}_e &= \begin{cases} U_e(1 - \alpha(x/L)^2)/\sigma, & 0 \leqslant x < L,\\ U_e, & L \leqslant x \leqslant l.\end{cases}
\end{aligned}
\tag{2.56}
$$

Therefore, the overall external fluid forces are written as,

$$
F_\xi^e - F_\eta^e\frac{\partial y}{\partial x} - F_\zeta^e\frac{\partial z}{\partial x} = \mathcal{F}_x^1 + \left(f_\xi^e - f_\eta^e\frac{\partial y}{\partial x} - f_\zeta^e\frac{\partial z}{\partial x}\right) + \mathcal{F}_x^2, \tag{2.57}
$$

$$
F_\eta^e + F_\xi^e\frac{\partial y}{\partial x} = \mathcal{F}_y^1 + \left(f_\eta^e + f_\xi^e\frac{\partial y}{\partial x}\right) + \mathcal{F}_y^2, \tag{2.58}
$$

$$
F_\zeta^e + F_\xi^e\frac{\partial z}{\partial x} = \mathcal{F}_z^1 + \left(f_\zeta^e + f_\xi^e\frac{\partial z}{\partial x}\right) + \mathcal{F}_z^2. \tag{2.59}
$$

From this point on, we will ignore all nonlinear terms that appear in the three sets of force balance equations as consequences of small displacement and small strain assumptions. The key point in simplifying the governing equations (2.43) to (2.45) is to obtain the explicit expression for the tension $T$, based on the assumption $p_i|_{x=l} = \bar{p} - \frac{1}{2}\rho_e D_o^2 U_e^2 C_b/A$, and the axial force equilibrium at the tip of the tubular beam, i.e., $T|_{x=l} = -\bar{p}A + \frac{1}{2}\rho_e D_o^2 U_e^2 C_b$, where $A$ denotes the cross-sectional area of the tubular beam, which is given by $\pi(R_o^2 - R_i^2)$, and $C_b$ is the coefficient representing the base drag. Thus, if the tubular beam density is $\rho$, the mass per unit length $m$ can be expressed as $\rho A$. Abbreviating $-\bar{p}A_o + \frac{1}{2}\rho_e D_o^2 U_e^2 C_b A_o/A$ as $\mathcal{G}_0$, $(m + \rho_i A_i)g \sin\theta$ as $\mathcal{G}_1$, and $\frac{1}{2}\rho_e D_o U_e^2$ as $\mathcal{G}_2$, we obtain from Eqs. (2.43), (2.46), and (2.57),

$$\frac{\partial(T - p_i A_i)}{\partial x} = \mathcal{G}_1 - \mathcal{G}_2 C_f, \tag{2.60}$$

so that

$$T - p_i A_i = \mathcal{G}_0 + (x - l)\mathcal{G}_1 + \mathcal{G}_2 \begin{cases} C_f^1(L - x) + C_f^2(l - L), & 0 \leqslant x < L, \\ C_f^2(l - x), & L \leqslant x \leqslant l. \end{cases} \tag{2.61}$$

By neglecting nonlinear terms, the governing equations for $y(x,t)$ and $z(x,t)$ turn out to be decoupled. In addition, the difference in the inhomogeneous terms will not affect the linear dynamic analysis based on the characteristic equations. Therefore, we focus on the $y$ displacement. Using the sets of Eqs. (2.30), (2.31), and (2.58), it is now an easy task to derive the following governing linearized equation:

$$c_1 \frac{\partial^4 y}{\partial x^4} + c_2 \frac{\partial^2 y}{\partial x^2} + c_3 \frac{\partial^2 y}{\partial x \partial t} + c_4 \frac{\partial^2 y}{\partial t^2} + c_5 \frac{\partial y}{\partial x} + c_6 \frac{\partial y}{\partial t} + c_7 = 0, \tag{2.62}$$

where the set of coefficients are given by

$$\begin{aligned} c_1 &= EI, \\ c_2 &= \rho_i A_i U_i^2 - \left(T - p_i A_i + p_e A_o + \frac{1}{2}\rho_e D_o U_e^2 C_f h(x)\right) + \chi\rho_e A_o \tilde{U}_e U_e, \\ c_3 &= 2\rho_i A_i U_i + \chi\rho_e A_o(U_e + \tilde{U}_e), \\ c_4 &= m + \rho_i A_i + \chi\rho_e A_o, \\ c_5 &= -(m + \rho_i A_i - \rho_e A_o)g \sin\theta + \frac{1}{2}\rho_e D_o U_e^2 C_f \left(1 - \frac{dh(x)}{dx}\right), \\ c_6 &= \frac{1}{2}\rho_e D_o U_e C_f, \end{aligned} \tag{2.63}$$

along with the inhomogeneous term $c_7 = (m + \rho_i A_i - \rho_e A_o)g \cos\theta$.

REMARK 4.2.4. For the clamped (or built-in) boundary condition at $x = 0$, we have

$$y(0,t) = 0 \quad \text{and} \quad \frac{\partial y(0,t)}{\partial x} = 0, \tag{2.64}$$

while for the free end of the tubular beam at $x = l$, we have

$$\frac{\partial^2 y(l,t)}{\partial x^2} = 0 \quad \text{and}$$
$$EI\frac{\partial^3 y(l,t)}{\partial x^3} + \left(\bar{p}A - \frac{1}{2}\rho_e D_o^2 U_e^2 C_b\right)\frac{\partial y(l,t)}{\partial x} = 0. \tag{2.65}$$

REMARK 4.2.5. We recognize that the coefficients $c_2$ to $c_6$ are variables depending on the position $x$. To circumvent the discontinuity at the location $x = L$, where the confined and unconfined domains are separated, we prescribe a nodal point at that location.

REMARK 4.2.6. Because of small displacement and small strain assumptions, we may ignore the contribution of the term involving $y$ and rewrite the expression for the hydrostatic pressure of the external fluid equation (2.50) as

$$p_e \simeq (l - x)\rho_e g \sin\theta + \bar{p}. \tag{2.66}$$

Therefore, using the difference scheme illustrated in Eq. (2.36), the discretized characteristic equation based on the equilibrium equation (2.62) can be written as follows: node $i$ $(1 \leqslant i \leqslant N)$

$$c_4^i \ddot{Y}^i + \frac{c_3^i}{2h}\dot{Y}^{i+1} + c_6^i \dot{Y}^i - \frac{c_3^i}{2h}\dot{Y}^{i-1} + \frac{c_1^i}{h^4}Y^{i+2} + \left(\frac{c_2^i}{h^2} - \frac{4c_1^i}{h^4} + \frac{c_5^i}{2h}\right)Y^{i+1}$$
$$+ \left(\frac{6c_1^i}{h^4} - \frac{2c_2^i}{h^2}\right)Y^i + \left(\frac{c_2^i}{h^2} - \frac{4c_1^i}{h^4} - \frac{c_5^i}{2h}\right)Y^{i-1} + \frac{c_1^i}{h^4}Y^{i-2} = 0. \tag{2.67}$$

Employing the same finite difference procedure, we obtain from the boundary conditions in Eqs. (2.64) and (2.65)

$$Y^0 = 0, \qquad Y^{-1} = Y^1, \qquad Y^{N+1} = 2Y^N - Y^{N-1},$$
$$Y^{N+2} = \left[4 - \frac{(2\bar{p}A - \rho_e D_o^2 U_e^2 C_b)h^2}{EI}\right](Y^N - Y^{N-1}) + Y^{N-2}. \tag{2.68}$$

Note that the variable coefficients $c_1$ to $c_6$ in Eq. (2.63) are functions of $x$ and are denoted as $c_1^i$ to $c_6^i$ at the nodal point $i$. Moreover, substituting Eq. (2.68) in Eq. (2.67) yields the same form of governing equations (2.39).

### *4.2.2. Buckling and flutter; parametric instability and Bolotin method*

Based on mathematical models for submerged pipe systems with both unconfined and confined external flows, we can study both static and dynamic stability issues. We present ranges of both buckling and flutter instability for pipes conveying

Table 4.2 Resonance complex frequencies

| Coupling | First resonant frequency (Hz) | Second resonant frequency (Hz) |
|---|---|---|
| With | $34.15 + 0.08467i$ | $201.0 + 0.09622i$ |
| Without | $45.04 + 0i$ | $282.3 + 0i$ |

steady flows for a case study. For pipes conveying pulsatile fluids, we present two methods to determine the regions of dynamic instability.

*Buckling and flutter*

Once we have the set of second-order ordinary differential equations in Eq. (2.39), we can assume a characteristic solution $\mathbf{Y} = e^{i\omega t}\boldsymbol{\phi}$, where $\boldsymbol{\phi}$ is the mode shape corresponding to the natural frequency of the FSI system $\omega = 2\pi f$, and employ standard eigenvalue solution techniques. Of course, $\mathrm{Re}(f)$ corresponds to the true resonant frequency, and $\mathrm{Im}(f)$ represents the damping. For the steady flow case, i.e., $\epsilon = 0$, the stable system corresponds to $\mathrm{Im}(\omega) \geqslant 0$ with $\mathrm{Re}(\omega) \neq 0$. We define the buckling instability as $\mathrm{Re}(\omega) \to 0$ with $\mathrm{Im}(\omega) \geqslant 0$, and the flutter instability as $\mathrm{Im}(\omega) < 0$ with $\mathrm{Re}(\omega) \neq 0$. The analytical solutions for the natural frequencies of the cantilever beam, without coupling to the surrounding fluids, are expressed as $f_i = \lambda_i^2/2\pi L^2\sqrt{EI/m}$, $i = 1, 2, \ldots$, where $\lambda_i$ are constants listed in Blevins [227]. In addition to the introduction of damping effects, the interaction between the tubular beam and the internal and external fluids significantly reduce the natural frequencies.

To obtain critical design criteria concerning the dynamical behavior of the tubular beam in internal flow models, we need to vary the pipe length, flow velocity, axial, and pipe radius. In other words, if we reduce the inner pipe radius, the fluid flow velocity will increase accordingly, and vice versa. Table 4.2 shows the difference between the dry structural frequency and the wet frequency, namely, the coupled frequency with the effects of the surrounding fluid. Figures 4.26, 4.27, and 4.28 show the first three coupled frequencies (real part of the eigenvalues) of the FSI system for various pipe physical parameters with forty-one grid points and $\rho_s/\rho_f = 3$, $\mu = 0.01$ Pa, and $T = 1$ N. Comparing Figure 4.26 with Figure 4.27, we can clearly identify that the nondimensional frequency slightly increases and frequency loci on the right-hand side are also modified. Note that the frequency actually decreases with the beam length which is different from nondimensional frequency. In addition, comparing Figure 4.26 with Figure 4.27, we can confirm that the nondimensional critical buckling velocities increase slightly, which means that doubling the length does not reduce the buckling velocity by half, or that the beam length and the buckling velocity have a nonlinear relationship. Similarly, comparing Figure 4.26 with Figure 4.28, when the tube radius is reduced, the

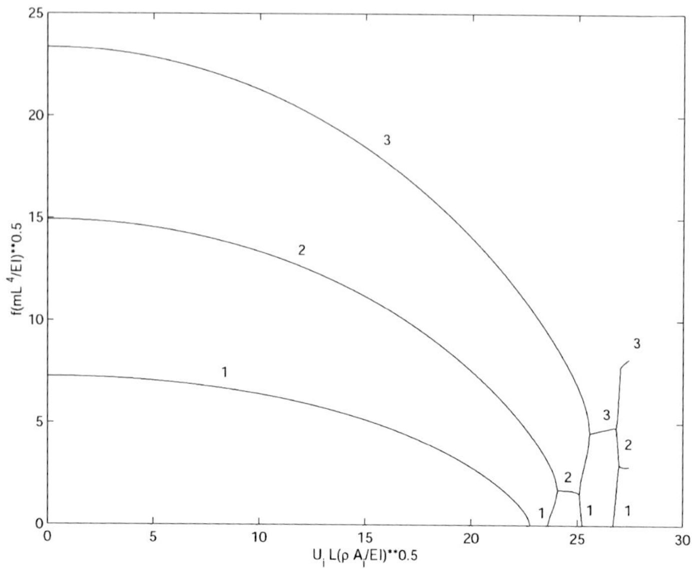


Figure 4.26. First three coupled system frequencies with $L = 0.2375$ m, $R_i = 4.763$ m, $R_o = 6.350$ m.

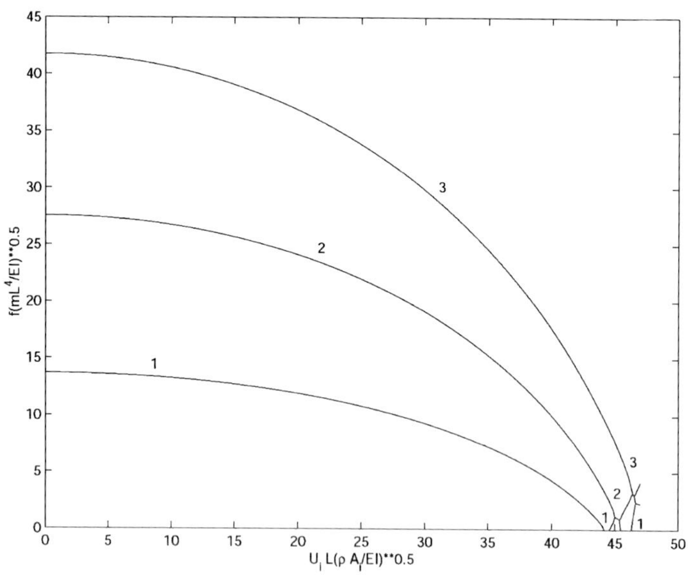


Figure 4.27. First three coupled system frequencies with $L = 0.4750$ m, $R_i = 4.763$ m, $R_o = 6.350$ m.

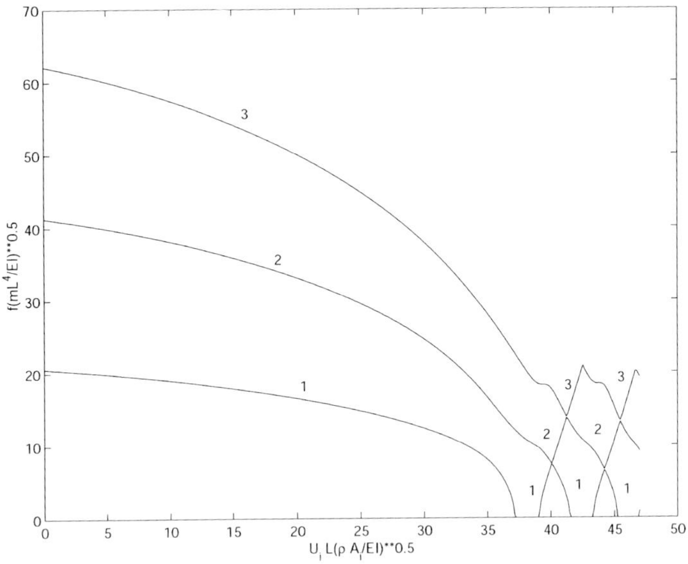


Figure 4.28. First three coupled system frequencies with $L = 0.2375$ m, $R_i = 2.382$ m, $R_o = 3.175$ m.

nondimensional coupled frequency also increases along with the corresponding nondimensional buckling velocity.

For concentric flow models, the gyroscopic term governed by $c_3$ is much more significant than the frictional damping term $c_6$, while $c_1$ is the predominant stiffness term. To find the frequency range of a particular pipe system design, we model a system currently used in the paper industry with the following physical parameters: $\rho_i = \rho_e = 1000\,\text{kg/m}^3$; $\rho = 7800\,\text{kg/m}^3$; $l = 2.392$ m; $L = 1.135$ m; $x_o = 2.4$ m; $y_o = 6.155$ m; $R_i = 0.1165$ m; $R_o = 0.1397$ m; $R_e = 0.2096$ m; $g = 9.8\,\text{m/s}^2$; $C_f^1 = 0.004\pi$; $C_f^2 = 0.5\pi R_i/l$; $C_b = 0.0125\pi$; $E = 200$ GPa; $U_i = 2$ m/s; and $U_e = 1$ m/s. In order to evaluate the friction coefficients, we need to calculate the Reynolds number defined as $Re = ua/\nu$, where $a$ is a characteristic length, such as one of the pipe diameters in this case; $u$ is a characteristic flow velocity, such as $U_i$ and $U_e$; and $\nu$ stands for the kinematic viscosity; in this work, we use $\nu = 1.13226 \times 10^{-6}\,\text{m}^2/\text{s}$. Notice that when $Re < 10^4$, the friction coefficients as well as the viscosity of the fiber water mixture can be significantly different from those of the pure water.

Because the first three natural frequencies of the inner tubular pipe are very small, they should be eliminated from the silo pipe system design if possible. We conjecture that such low frequency variations will show up as consistency or pressure fluctuations in the stock delivered to the paper forming area. To come

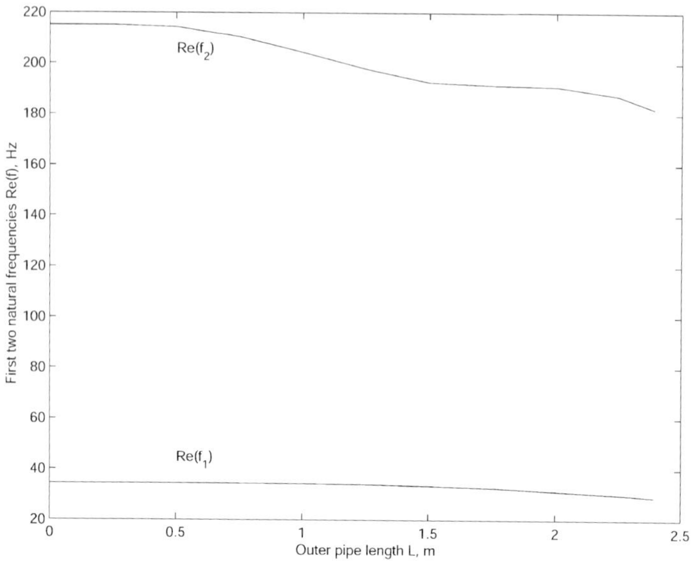


Figure 4.29. First two system frequencies vs. the outer pipe length $L$.

up with critical design criteria concerning the dynamical behavior of the tubular beam, we need to vary the pipe lengths, inclination angle, free surface level, flow velocities, and pipe radii. We will hold the volume flow rates constant to $8.5277 \times 10^{-2}$ $m^3/s$ for the internal flow, and to $7.6705 \times 10^{-2}$ $m^3/s$ for the external flow. In other words, if we reduce the inner pipe radius, the inner fluid flow velocity will increase accordingly, and vice versa. The same is true for the outer pipe. In addition, with a constant free surface level, the external hydrostatic pressure $p_e$ will vary in accord with the inner pipe length and the inclination angle. Finally, we vary $y_o$ which governs the depth of the immersed pipe system. However, we recognize that a design parameter such as $y_o$ is often related to the fan pump pressure drop and, in practice, cannot be modified easily.

Results in Figure 4.29 with forty-two grid points show the first two coupled frequencies of FSI systems Re($f_1$) and Re($f_2$) for various outer pipe lengths. As can be seen, the longer the outer pipe, the smaller are the natural frequencies of the inner pipe, i.e., the structure is more flexible. In Figure 4.30 with forty-two grid points, we observe that varying the inner pipe length has a similar, but much stronger effect on the coupled system frequencies. The physical explanation is that the inner pipe length directly determines the inner pipe structural stiffness and mass, and the outer pipe length contributes only to the fluid-structure coupling part. In fact, for a cantilever beam, the natural frequency is inversely proportional to the square of the beam length. In general, if the structure is more flexible, it is

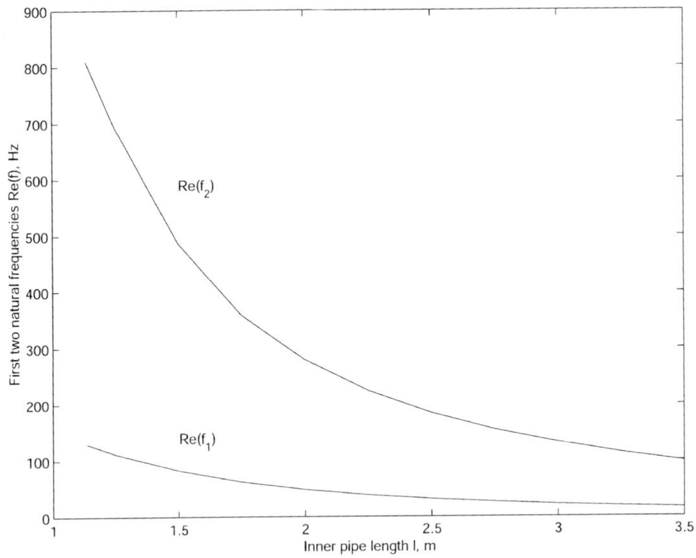


Figure 4.30. First two system frequencies vs. the inner pipe length $l$.

more susceptible to buckling or flutter. Figures 4.29 and 4.30 also show that the coupled system frequencies for the given design configuration with different pipe lengths are between the lower and upper bounds calculated for the two cases in which the inner and outer pipes have the same lengths of $l$ and $L$, respectively. In addition, we find in Figures 4.29 and 4.30 that the lengths of the pipes have a much larger effect on the second coupled system natural frequency than the first one. This result also matches physical understanding based on the dependence of mode shapes on pipe length. We also find that the inclination angle $\theta$ and the depth of the submerged pipe system $y_o$ do not significantly influence the characteristic behavior of the tubular beam. Concerning the inner pipe radius, results in Figure 4.31 with forty-two grid points show that for the constant volume flow rates that have been assumed, there exists in this case an optimal radius around 0.16 m, if we want to maximize the natural frequencies of the inner pipe. In practice we do want to avoid the low frequency consistency and pressure variations introduced by the low frequency vibration of the inner pipe, which may not be effectively attenuated. Figure 4.32 clearly indicates with forty-two grid points that the larger the outer pipe diameter for the constant volume flow rates, the larger the natural frequencies of the inner pipe. However, as the outer pipe diameter increases, its effect on the inner pipe diminishes and the natural frequencies of the inner pipe reach a plateau.

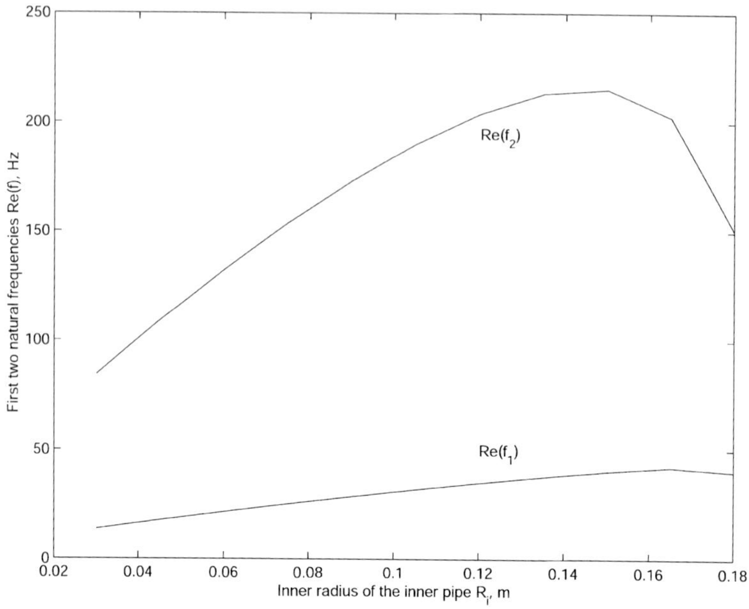


Figure 4.31. First two system frequencies vs. the inner radius of the inner pipe $R_i$.

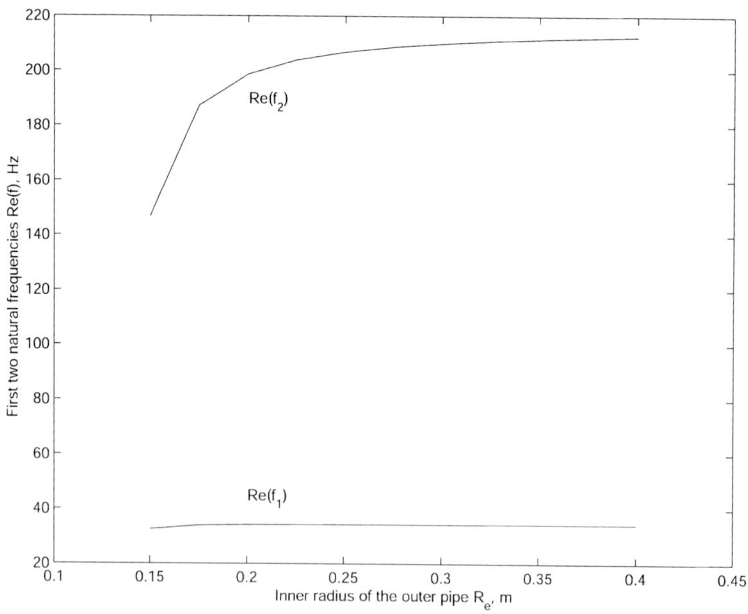


Figure 4.32. First two system frequencies vs. the inner radius of the outer pipe $R_e$.

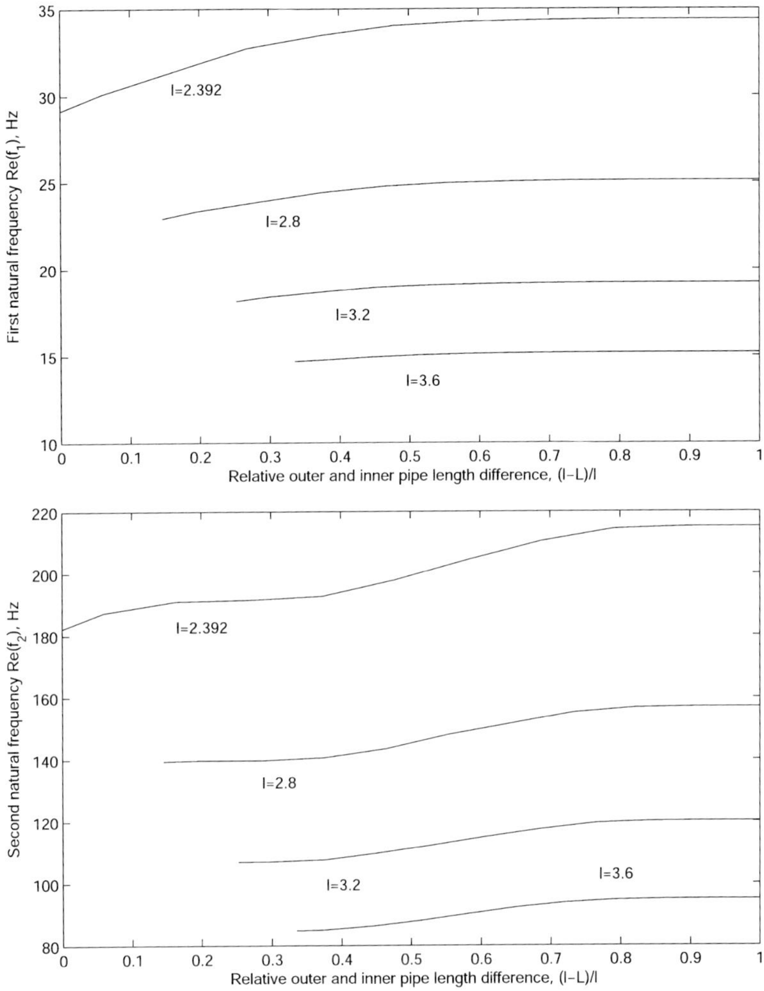


Figure 4.33. First two system frequencies vs. outer and inner pipe length ratio $(l - L)/l$.

In addition, to further assist in the design of the pipe system, with forty-two grid points, Figures 4.33, 4.34, 4.35, and 4.36 present the first two frequencies $\mathrm{Re}(f_1)$ and $\mathrm{Re}(f_2)$ and their corresponding damping ratios $\mathrm{Im}(f_1)/\mathrm{Re}(f_1)$ and $\mathrm{Im}(f_2)/\mathrm{Re}(f_2)$ as functions of the relative difference of outer and inner pipe length $(l - L)/l$ and radius $(R_e - R_o)/R_e$ at various outer pipe lengths $l$ and radii $R_e$, respectively. As shown in Figures 4.33 and 4.34, the longer the inner pipe, the lower the natural frequencies of the inner pipe and the larger the

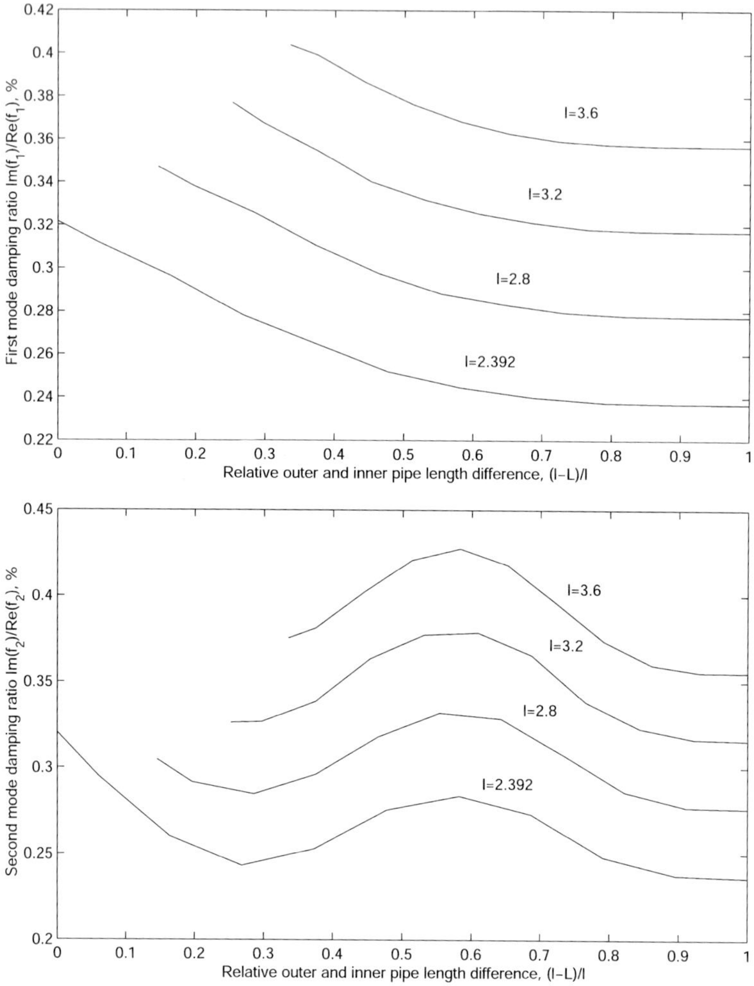


Figure 4.34. First two system damping ratios vs. outer and inner pipe length ratio $(l - L)/l$.

damping ratio needed for the natural attenuation of the inner pipe vibration. In Figures 4.35 and 4.36, the design information most worth mentioning is that with the same outer pipe inner diameter $R_e$, the smaller the inner pipe diameter represented by $R_o$, the smaller the natural frequencies of the inner pipe. However, the change of damping ratio is not monotonic and there exists a region around $(R_e - R_o)/R_e = 0.94$ such that the damping ratio is at its lowest level.

For the steady flow case, results in Figure 4.37 with twenty-one grid points show the loci of the fourth mode in the complex $\omega$ plane as functions of the di-

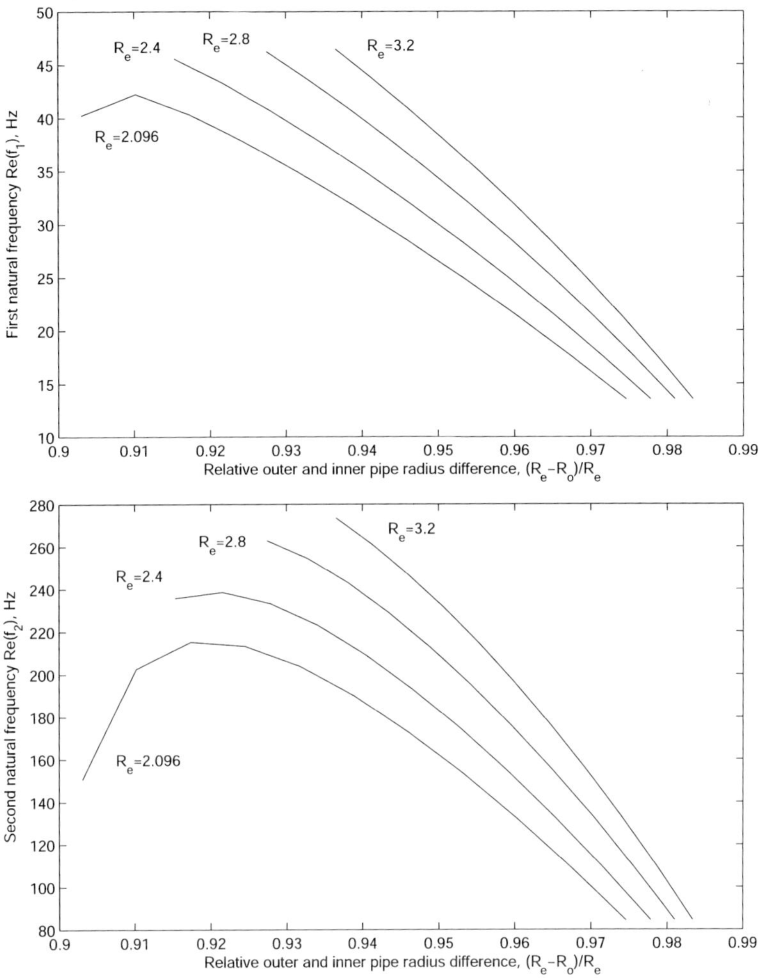


Figure 4.35. First two system frequencies vs. outer and inner pipe radius ratio $(R_e - R_o)/R_e$.

mensionless velocity $u = \sqrt{\rho A_i / EI} U_i l$, i.e., $U_i = 13.28u$. We find that the fourth mode is the mode in which, for a sufficiently high value of $u$, flutter instability can happen. We note that friction forces have positive effects on avoiding flutter instability, and the longer the outer pipe length, the less likely that flutter will occur. Similarly, as depicted in Figure 4.38 with twenty-one grid points, friction forces have positive effects on delaying buckling instability. However, the longer the outer pipe length, the more susceptible the inner pipe is to buckling. A more elaborate study on the dependence of the critical buckling velocity $u_c$ on

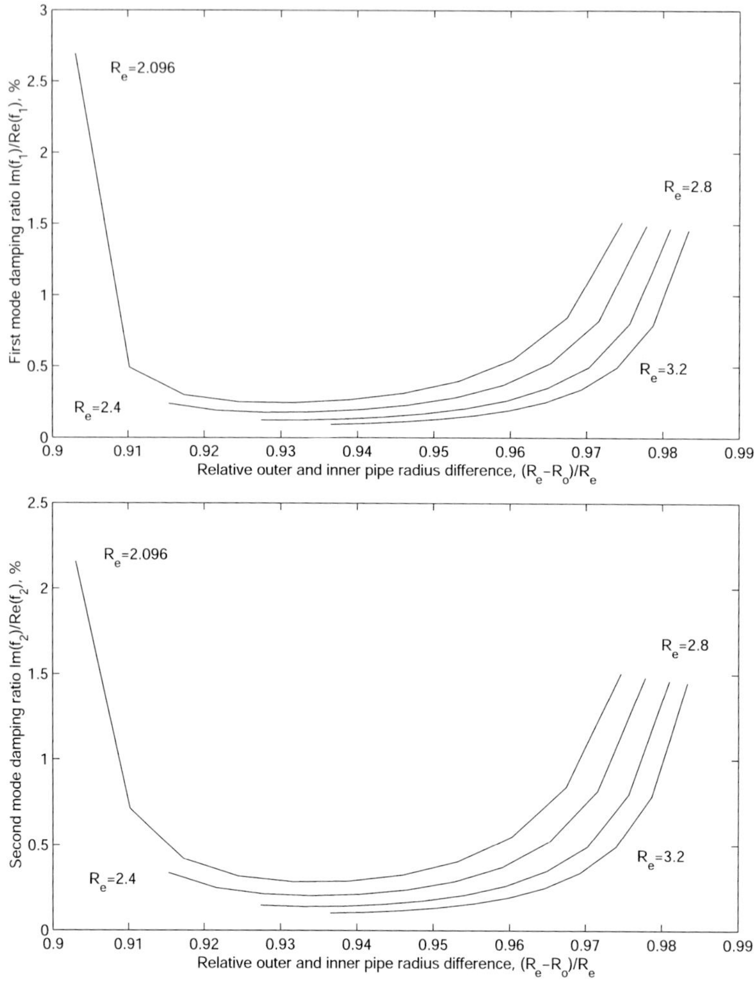


Figure 4.36. First two system damping ratios vs. outer and inner pipe length ratio $(R_e - R_o)/R_e$.

the outer pipe length, friction, and gravitational forces indicates, as shown in Figure 4.39 with twenty-one grid points, that there exists a transition region around $L/l = 0.6$. In general, the effects of gravitational forces are not as significant as those of frictional forces.

In practice, due to the existence of a pressure pulsation attenuator and other low pass filters, high frequency variations can be effectively reduced. However, we still need to separate other vibration frequencies from the predicted natural frequencies of the inner tubular pipe coupled with the surrounding fluids in order

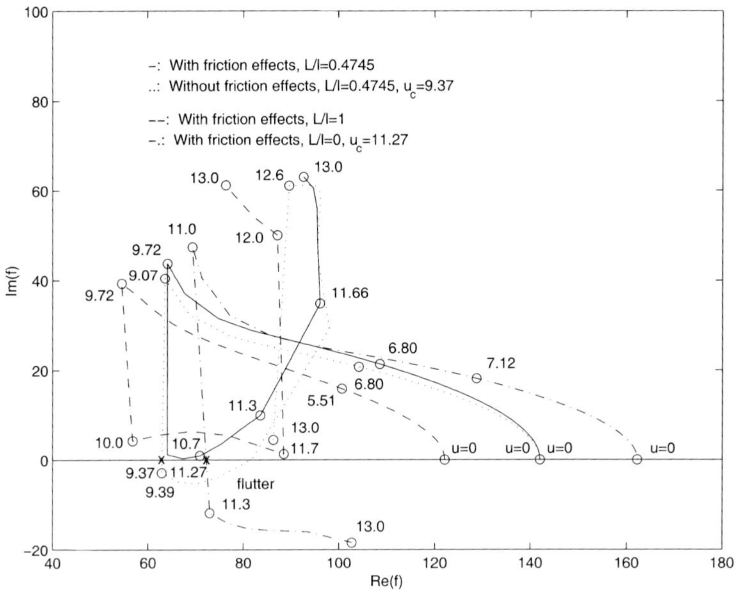


Figure 4.37. Flutter instability of the fourth mode with $U_i/U_e = 1.7$.

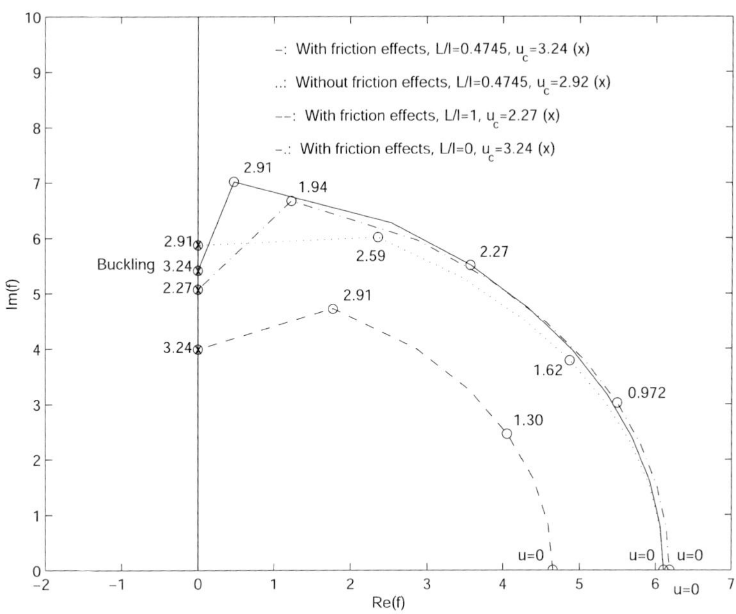


Figure 4.38. Buckling instability of the first mode with $U_i/U_e = 1.7$.

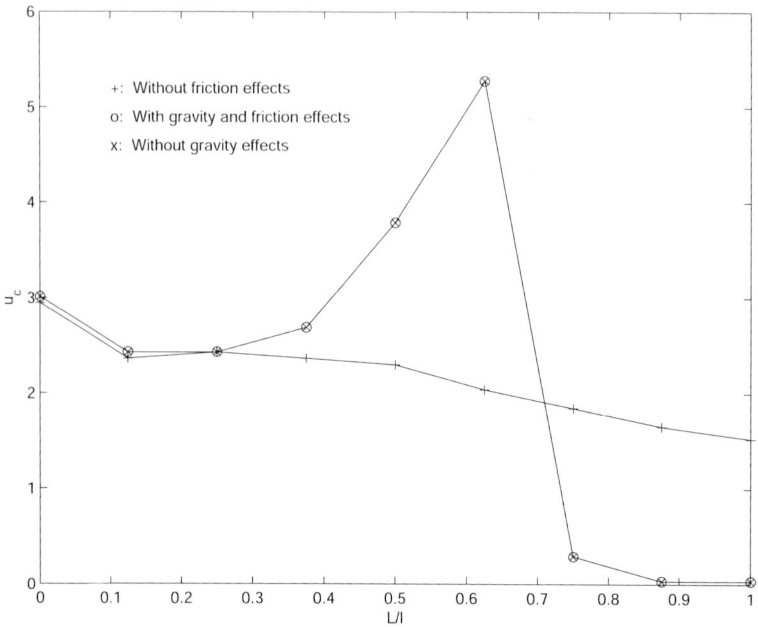


Figure 4.39. Critical buckling velocity vs. the outer pipe length with $U_i/U_e = 1.7$.

to avoid resonance. In addition, by varying different design parameters, we reach the following conclusions:

(1) By decreasing the inner or outer pipe lengths, we can increase the natural frequencies. However, changing the inner pipe length is more effective.
(2) The inclination angle, depth of the submerged pipe system, and gravity are not important design parameters as far as the pipe's dynamical behavior is concerned.
(3) For a piping system with fixed volume flow rates, there exists an optimal inner pipe radius.
(4) In general, the larger the outer pipe radius, the more stable the suspended pipe system will be. However, as the outer pipe radius increases, the inner pipe vibration model approaches that of the case of a singular flexible pipe immersed in an unconfined fluid.
(5) For the same relative outer and inner pipe length difference, the damping ratios increase with the inner pipe length, while for the same relative outer and inner pipe radius difference, the damping ratios decrease with the external pipe inner radius.

*Parametric instability and Bolotin method*

For the pulsatile internal flow cases, we denote $\omega_{oi}$ as the $i$th natural frequency of the corresponding steady flow case. Here we study parametric resonance using the Bolotin method [285]. As a consequence of Eq. (2.33), we employ

$$T - p_i A_i = (l - x)\rho A_i \bar{U}_i \omega_o \epsilon \sin \omega_o t, \tag{2.69}$$

while the coefficients $c_1$ to $c_6$ in Eq. (2.32) take on the following form:

$$\begin{aligned} c_1 &= EI, \\ c_2 &= \rho A_i \bar{U}_i^2 (1 + \epsilon \cos \omega_o t)^2 + (x - l)\rho A_i \bar{U}_i \omega_o \epsilon \sin \omega_o t, \\ c_3 &= 2\rho A_i \bar{U}_i (1 + \epsilon \cos \omega_o t), \quad \text{and} \quad c_4 = m + \rho A_i. \end{aligned} \tag{2.70}$$

In addition, we may rewrite $c_2$ and $c_3$ in the form

$$\begin{aligned} c_2 &= \bar{c}_2 + \tilde{c}_2^1 \epsilon \cos \omega_o t + (x - l)\tilde{c}_2^2 \epsilon \sin \omega_o t + \tilde{c}_2^3 \epsilon^2 \cos 2\omega_o t \quad \text{and} \\ c_3 &= \bar{c}_3 + \tilde{c}_3 \epsilon \cos \omega_o t, \end{aligned} \tag{2.71}$$

with

$$\begin{aligned} &\bar{c}_2 = \rho A_i \bar{U}_i^2 (1 + \epsilon^2/2), \qquad \tilde{c}_2^1 = 2\rho A_i \bar{U}_i^2, \qquad \tilde{c}_2^2 = \rho A_i \bar{U}_i \omega_o, \\ &\tilde{c}_2^3 = \rho A_i \bar{U}_i^2/2, \qquad \bar{c}_3 = 2\rho A_i \bar{U}_i, \qquad \tilde{c}_3 = 2\rho A_i \bar{U}_i, \end{aligned}$$

and obtain the following homogeneous governing partial differential equation:

$$\begin{aligned} &c_1 \frac{\partial^4 y}{\partial x^4} + \left(\bar{c}_2 + \tilde{c}_2^1 \epsilon \cos \omega_o t + (x - l)\tilde{c}_2^2 \epsilon \sin \omega_o t + \tilde{c}_2^3 \epsilon^2 \cos 2\omega_o t\right) \frac{\partial^2 y}{\partial x^2} \\ &\quad + (\bar{c}_3 + \tilde{c}_3 \epsilon \cos \omega_o t) \frac{\partial^2 y}{\partial x \partial t} + c_4 \frac{\partial^2 y}{\partial t^2} = 0. \end{aligned} \tag{2.72}$$

We express the solution of Eq. (2.72) in the form of a series $y(x,t) = \sum_{j=1}^{\infty} a_j(t)\phi_j(x)$, where $\phi_j$ is the $j$th eigenfunction corresponding to eigenvalue $\lambda_j$ for the system

$$\frac{d^4 \phi_j}{dx^4} = \lambda_j^4 \phi_j, \quad 0 < x < l, \tag{2.73}$$

with the boundary conditions $\phi_j(0) = \phi_j'(0) = 0$ and $\phi_j(l) = \phi_j'(l) = 0$.

Moreover, the eigenvalue $\lambda_j$ satisfies the transcendental algebraic equation

$$1 - \cos \lambda_j l \cos h\lambda_j l = 0, \quad j = 1, 2, \ldots, \tag{2.74}$$

and the corresponding $\phi_j$ is given by

$$\begin{aligned} \phi_j(x) &= (\cos \lambda_j x - \cos h\lambda_j x) + \sigma_j (\sin \lambda_j x - \sin h\lambda_j x), \\ \sigma_j &= (\sin \lambda_j l + \sin h\lambda_j l)/(\cos \lambda_j l - \cos h\lambda_j l). \end{aligned} \tag{2.75}$$

Of course, we also have the following orthogonal relations

$$\int_0^l \phi_j(x)\phi_k(x)\,dx = 0 \quad \text{and} \quad \int_0^l \phi_j''(x)\phi_k''(x)\,dx = 0, \qquad j \neq k. \tag{2.76}$$

If we substitute into Eq. (2.32), multiply the resulting equation by $\phi_k(x)$, and integrate from $x = 0$ to $x = l$, we obtain the following equation:

$$\begin{aligned} c_4\ddot{a}_k + \sum_{j=1}^{\infty}\Big\{(\bar{c}_3 + \tilde{c}_3\epsilon\cos\omega_o t)o_{kj}\dot{a}_j + \big(\bar{c}_2 + \tilde{c}_2^1\epsilon\cos\omega_o t \\ + \tilde{c}_2^3\epsilon^2\cos 2\omega_o t\big)p_{kj}a_j + \tilde{c}_2^2\epsilon q_{kj}a_j\sin\omega_o t\Big\} + c_1\lambda_k^4 a_k = 0, \end{aligned} \tag{2.77}$$

with

$$\begin{aligned} \mu_k &= \int_0^l \phi_k^2\,dx, \qquad o_{kj} = \int_0^l \phi_k\phi_j'\,dx/\mu_k, \\ p_{kj} &= \int_0^l \phi_k\phi_j''\,dx/\mu_k, \qquad q_{kj} = \int_0^l (x-l)\phi_k\phi_j''\,dx/\mu_k. \end{aligned} \tag{2.78}$$

Notice that some analytical expressions exist for Eq. (2.78). In this work, we choose to use numerical integrations. We select a two-term series expansion, i.e., $j = 1, 2$. In this manner, we obtain the following as the coupled equations for $a_1(t)$ and $a_2(t)$ resulting from Eq. (2.77): for $i = 1$ to 2

$$\begin{aligned} c_4\ddot{a}_1 + (\bar{c}_3 + \tilde{c}_3\epsilon\cos\omega_o t)o_{1i}\dot{a}_i + \big(\bar{c}_2 + \tilde{c}_2^1\epsilon\cos\omega_o t + \tilde{c}_2^3\epsilon^2\cos 2\omega_o t\big)p_{1i}a_i \\ + \tilde{c}_2^2\epsilon\sin\omega_o t\, q_{1i}a_i + c_1\lambda_1^4 a_1 = 0, \end{aligned} \tag{2.79}$$

$$\begin{aligned} c_4\ddot{a}_2 + (\bar{c}_3 + \tilde{c}_3\epsilon\cos\omega_o t)o_{2i}\dot{a}_i + \big(\bar{c}_2 + \tilde{c}_2^1\epsilon\cos\omega_o t + \tilde{c}_2^3\epsilon^2\cos 2\omega_o t\big)p_{2i}a_i \\ + \tilde{c}_2^2\epsilon\sin\omega_o t\, q_{2i}a_i + c_1\lambda_2^4 a_2 = 0. \end{aligned} \tag{2.80}$$

The coefficients in Eqs. (2.79) and (2.80) are periodic with the period $T_o$ and the transition from stability to instability is marked by the existence of a solution with the period $T_o$ or $2T_o$. The solutions associated with the primary instability at $\omega_o/\omega_{oi} = 2/k$, $k = 1, 3, \ldots,$ correspond to solutions with the period $2T_o$ and are constructed from the Fourier series:

$$a_i(t) = \sum_{j=1,3,\ldots}^{\infty}\{r_{ij}\cos j\omega_o t/2 + s_{ij}\sin j\omega_o t/2\}, \qquad i = 1, 2. \tag{2.81}$$

Solutions associated with the secondary instability at $\omega_o/\omega_{oi} = 2/k$, $k = 2, 4, \ldots,$ correspond to solutions with the period $T_o$ and are also constructed using the Fourier series:

$$a_i(t) = e_{i0}/2 + \sum_{j=2,4,\ldots}^{\infty} \{e_{ij} \cos j\omega_o t/2 + f_{ij} \sin j\omega_o t/2\}, \quad i = 1, 2. \tag{2.82}$$

The primary principal instability boundary is generated by solutions of the form (2.81) with $j = 1$. Thus, truncating the series in Eq. (2.81), we have

$$a_i(t) = r_i \cos \omega_o t/2 + s_i \sin \omega_o t/2, \quad i = 1, 2, \tag{2.83}$$

with $r_i \equiv r_{i1}$ and $s_i \equiv s_{i1}$.

If we substitute into Eqs. (2.79) and (2.80), use various trigonometric relations, and collect the coefficients of the linearly independent terms $\sin \omega_o t/2$ and $\cos \omega_o t/2$, we obtain the system of algebraic equations

$$\begin{bmatrix} M_{11} & M_{12} & M_{13} & M_{14} \\ M_{21} & M_{22} & M_{23} & M_{24} \\ M_{31} & M_{32} & M_{33} & M_{34} \\ M_{41} & M_{42} & M_{43} & M_{44} \end{bmatrix} \begin{bmatrix} r_1 \\ s_1 \\ r_2 \\ s_2 \end{bmatrix} = \begin{bmatrix} 0 \\ 0 \\ 0 \\ 0 \end{bmatrix}, \tag{2.84}$$

with

$$\begin{aligned}
M_{11} &= -c_4\omega_o^2/4 + \bar{c}_2 p_{11} + \epsilon\tilde{c}_2^1 p_{11}/2 + c_1\lambda_1^4, \\
M_{12} &= \bar{c}_3 o_{11}\omega_o/2 + \epsilon\tilde{c}_3 o_{11}\omega_o/4 + \epsilon\tilde{c}_2^2 q_{11}/2, \\
M_{13} &= \bar{c}_2 p_{12} + \epsilon\tilde{c}_2^1 p_{12}/2, \\
M_{14} &= \bar{c}_3 o_{12}\omega_o/2 + \epsilon\tilde{c}_3 o_{12}\omega_o/4 + \epsilon\tilde{c}_2^2 q_{12}/2; \\
M_{21} &= -\bar{c}_3 o_{11}\omega_o/2 + \epsilon\tilde{c}_3 o_{11}\omega_o/4 + \epsilon\tilde{c}_2^2 q_{11}/2, \\
M_{22} &= -c_4\omega_o^2/4 + \bar{c}_2 p_{11} - \tilde{c}_2^1 p_{11}/2 + c_1\lambda_1^4, \\
M_{23} &= -\bar{c}_3 o_{12}\omega_o/2 + \epsilon\tilde{c}_3 o_{12}\omega_o/4 + \epsilon\tilde{c}_2^2 q_{12}/2, \\
M_{24} &= \bar{c}_2 p_{12} - \tilde{c}_2^1 p_{12}/2; \\
M_{31} &= \bar{c}_2 p_{21} + \epsilon\tilde{c}_2^1 p_{21}, \\
M_{32} &= \bar{c}_3 o_{21}\omega_o/2 + \epsilon\tilde{c}_3 o_{21}\omega_o/4 + \epsilon\tilde{c}_2^2 q_{21}/2, \\
M_{33} &= -c_4\omega_o^2/4 + \bar{c}_2 p_{22} + \epsilon\tilde{c}_2^1 p_{22}/2 + c_1\lambda_2^4, \\
M_{34} &= \bar{c}_3 o_{22}\omega_o/2 + \epsilon\tilde{c}_3 o_{22}\omega_o/4 + \epsilon\tilde{c}_2^2 q_{22}/2; \\
M_{41} &= -\bar{c}_3 o_{21}\omega_o/2 + \epsilon\tilde{c}_3 o_{21}\omega_o/4 + \epsilon\tilde{c}_2^2 q_{21}/2, \\
M_{42} &= \bar{c}_2 p_{21} - \epsilon\tilde{c}_2^1 p_{21}/2, \\
M_{43} &= -\bar{c}_3 o_{22}\omega_o/2 + \epsilon\tilde{c}_3 o_{22}\omega_o/4 + \epsilon\tilde{c}_2^2 q_{22}/2, \\
M_{44} &= -c_4\omega_o^2/4 + \bar{c}_2 p_{22} - \epsilon\tilde{c}_2^1 p_{22}/2 + c_1\lambda_2^4.
\end{aligned}$$

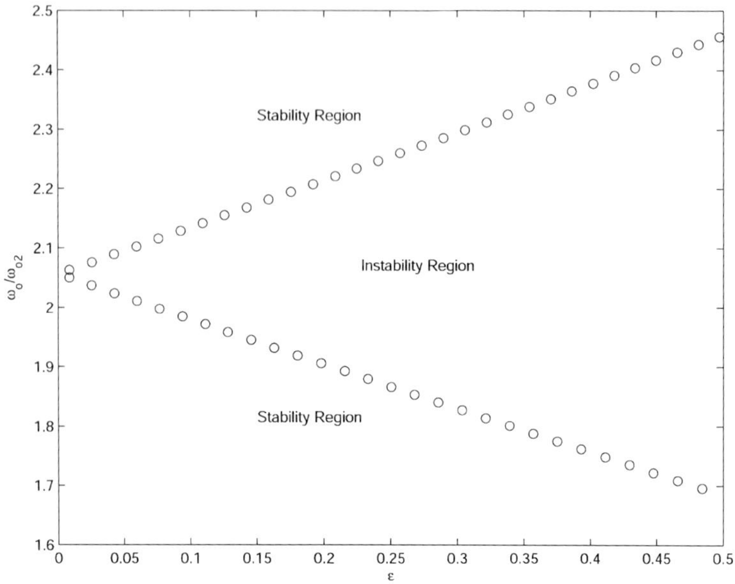


Figure 4.40. Results from the Bolotin method.

The existence of the solution of the form (2.83) corresponds to $\det M_{ij} = 0$. Using this criterion, we may obtain the dynamic stability information associated with various design parameters, such as $\epsilon$ and $\omega_o$. Figure 4.40 shows that the results derived from the Bolotin method with $l = 0.2375$ m, $R_i = 4.763$ mm, $R_o = 6.350$ mm, $\rho_s/\rho_f = 3$, and $\mu = 0.01$ Pa correctly predicted the instability around $\omega_o/\omega_{o2}$, even with the limited number of series expansion, which in this case is 2.

The simple mathematical model can be used to derive the on-set of instabilities for both steady and pulsatile flows. In particular, the Bolotin method with couple terms is sufficient to obtain the dynamic stability regions corresponding to the first coupled natural frequencies. For the pulsatile concentric flows, we introduce the Floquet method, named after Gaston Floquet (1917–2001), in lieu of the Bolotin method. Starting with the set of second-order ordinary differential equations in Eq. (2.39), for the case of pulsatile flow, we introduce a new solution vector $\mathbf{q} = \langle \mathbf{Y}, \dot{\mathbf{Y}} \rangle$, and replace Eq. (2.39) with the following system of $2N$ first-order differential equations with periodic coefficients:

$$\dot{\mathbf{q}} = \mathbf{A}(t)\mathbf{q}, \tag{2.85}$$

with

$$\mathbf{A}(t) = \begin{bmatrix} \mathbf{0} & \mathbf{I} \\ -\mathbf{M}^{-1}\mathbf{K} & -\mathbf{M}^{-1}\mathbf{C} \end{bmatrix}.$$

Of course, the stability analysis of the trivial solution of Eq. (2.85) also applies to the stability of solutions near the fixed point of the nonlinear dynamical system with the Jacobian matrix $\mathbf{A}(t)$. For this nonautonomous system, the matrix $\mathbf{A}(t)$ has a period $T_o$. According to the Floquet theory, the fundamental matrix satisfying Eq. (2.85) can be expressed as the function of a periodic nonsingular matrix $\mathbf{Q}(t)$, with the period $T_o$, and a constant matrix $\mathbf{D}$,

$$\mathbf{q}(t) = \mathbf{Q}(t)e^{t\mathbf{D}}. \tag{2.86}$$

Therefore, the eigenvalues of the matrix $\mathbf{D}$, denoted by $\beta$, determine the stability of Eq. (2.85), and we obtain

$$\mathbf{q}(t + T_o) = \mathbf{q}(t)\mathbf{N}, \tag{2.87}$$

with the monodromy matrix $\mathbf{N} = e^{T_o\mathbf{D}}$.

The eigenvalues of the matrix $\mathbf{N}$, denoted by $\lambda$, are related to the eigenvalues $\beta$ of the matrix $\mathbf{D}$ by

$$\beta = \frac{1}{T_o}\left(\log|\lambda| + i \arg\lambda\right), \quad i = \sqrt{-1}. \tag{2.88}$$

It is clear that only the real part of $\beta$ is uniquely defined. If we assign

$$\mathbf{q}(0) = \mathbf{I}, \tag{2.89}$$

where $\mathbf{I}$ is the identity matrix, we obtain,

$$\mathbf{N} = \mathbf{q}(T_o). \tag{2.90}$$

In order to derive the monodromy matrix $\mathbf{N}$, we numerically integrate Eq. (2.85), together with the initial condition (2.89), using the second-order Runge–Kutta method:

$$\mathbf{q}(t + \Delta t) = \mathbf{q}(t) + \Delta t(\mathbf{k}_1 + \mathbf{k}_2)/2, \tag{2.91}$$

with $\mathbf{k}_1 = \mathbf{A}(t)\mathbf{q}(t)$ and $\mathbf{k}_2 = \mathbf{A}(t + \Delta t)(\mathbf{q}(t) + \Delta t\mathbf{A}(t)\mathbf{q}(t))$.

The $i$th column of the matrix $\mathbf{N}$ corresponds to the numerical solution of Eq. (2.85) with the $i$th column of the identity matrix $\mathbf{I}$ as the initial condition. The construction of the monodromy matrix can be very expensive due to the explicit nature of the Runge–Kutta scheme. The detailed discussion on the selection of a proper time step is available in Ref. [299]. We study the stability issues for a particular pipe system design with the following physical parameters: $\rho = 1000$ kg/m$^3$; $m = 2.12$ kg/m$^3$; $l = 2.392$ m; $L = 1.135$ m; $x_o = 2.4$ m; $y_o = 6.155$ m; $R_i = 0.02$ m; $R_o = 0.025$ m; $R_e = 0.035$ m; $g = 9.8$ m/s$^2$; $E = 70$ GPa; $C_f^1 = 0.004\pi$; $C_f^2 = 0.5\pi R_i/l$; and $C_b = 0.0125\pi$.

For the pulsatile flow case, we denote $\omega_{oi}$ as the $i$th natural frequency of the corresponding steady flow case with both gravity and friction effects. As discussed in

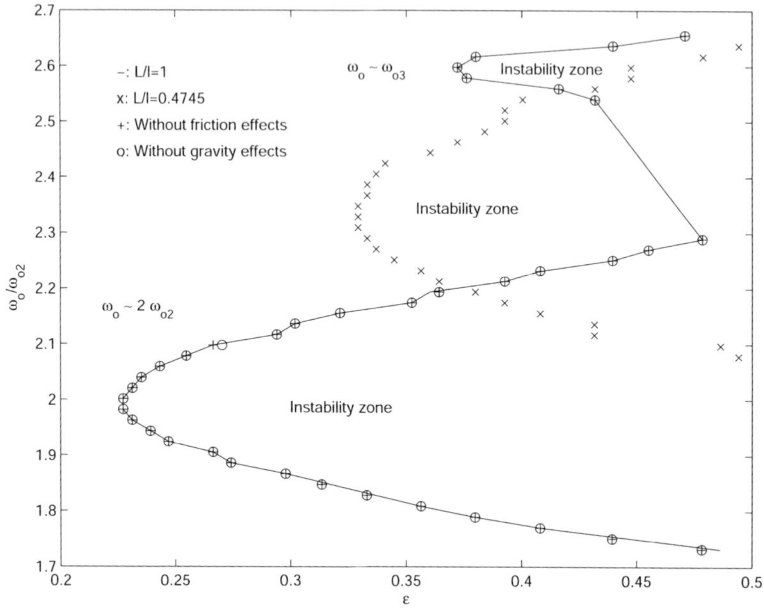


Figure 4.41. Dynamic instability regions.

Refs. [196,253,254], although the numerical Floquet analysis is lengthier than the Bolotin method, it includes both the parametric ($\omega_o/\omega_{oi} = 2/k, k = 1, 2, 3, \ldots$) and combination instabilities ($\omega_o/(\omega_{oi} - \omega_{oj}) = 1/k, k = 1, 2, 3, \ldots$). Results with five grid points and $\omega_{o1} = 26.13$ and $\omega_{o2} = 117.3$ rad/s in Figure 4.41 show the dynamic instability regions for different cases. As can be seen, gravity and friction effects are less significant than the outer pipe length $l$ with respect to the dynamic stability issues. In this section we consider the case of a pipe system with the same inner and outer pipe length. To simplify the expression for the eigenfunctions satisfying the beam equation and its associated boundary conditions, we ignore frictional forces, the axial load at the tip of the inner tubular beam, and, in general, the gravitational forces. As a consequence of Eq. (2.60), the forms of $\mathcal{G}_0$ and $\mathcal{G}_1$, and the assumed structure of the pulsatile inner pipe flow, we derive the same equation (2.62) along with the following coefficients $c_1$ to $c_4$:

$$c_1 = EI,$$
$$c_2 = \rho A_i \bar{U}_i^2 (1 + \epsilon \cos \omega_o t)^2 + (x - l)\rho A_i \bar{U}_i \omega_o \epsilon \sin \omega_o t + \chi \rho A_o U_e^2,$$
$$c_3 = 2\rho A_i \bar{U}_i (1 + \epsilon \cos \omega_o t) + 2\chi \rho A_o U_e,$$
$$c_4 = m + \rho A_i + \chi \rho A_o.$$

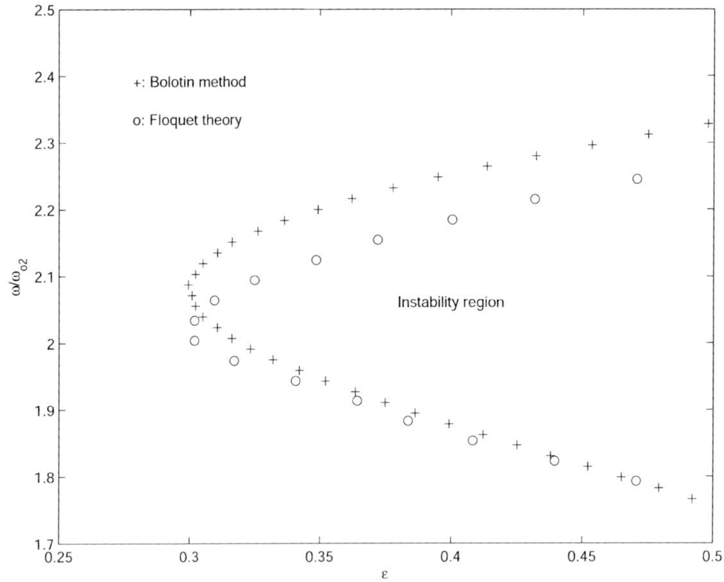


Figure 4.42. Results derived from the Floquet theory and the Bolotin method.

Likewise, we can rewrite $c_2$ and $c_3$ in the form (2.71) with

$$\bar{c}_2 = \rho A_i \bar{U}_i^2(1+\epsilon^2/2) + \chi\rho A_o U_e^2, \qquad \tilde{c}_2^1 = 2\rho A_i \bar{U}_i^2,$$
$$\tilde{c}_2^2 = \rho A_i \bar{U}_i \omega_o, \qquad \tilde{c}_2^3 = \rho A_i \bar{U}_i^2/2, \qquad \bar{c}_3 = 2\rho A_i \bar{U}_i + 2\chi\rho A_o U_e,$$
$$\tilde{c}_3 = 2\rho A_i \bar{U}_i,$$

and obtain the same governing partial differential equation (2.72), along with the same form of (2.73), with the boundary conditions $\phi_j(0) = \phi_j'(0) = 0$ and $\phi_j''(l) = \phi_j'''(l) = 0$.

Moreover, the eigenvalue $\lambda_j$ satisfies the transcendental algebraic equation

$$1 + \cos\lambda_j l \cos h\lambda_j l = 0, \quad j = 1, 2, \ldots, \tag{2.92}$$

and the corresponding $\phi_j$ is given by

$$\begin{aligned}\phi_j(x) &= (\cos\lambda_j x - \cos h\lambda_j x) + \sigma_j(\sin\lambda_j x - \sin h\lambda_j x),\\ \sigma_j &= (\sin\lambda_j l - \sin h\lambda_j l)/(\cos\lambda_j l + \cos h\lambda_j l).\end{aligned} \tag{2.93}$$

Notice the subtle differences in Eqs. (2.74) and (2.92), as well as Eqs. (2.75) and (2.93), due to different boundary conditions of the pipe systems. Results in Figure 4.42 with eleven grid points show that the results derived from the Bolotin method and the Floquet theory match. The advantage of using the Bolotin method is, of course, its simplicity and efficiency. However, when considering more terms in the series expansions, a derivation based on the Bolotin method can be extensive, and the method is limited to the parametric instability analysis [196].

### *4.2.3. Collapsible tube; tube law; transverse and axial oscillations*

The subject of flow through a collapsible (deformable) tube is ubiquitous in cardiovascular systems. During lung expiration phase, the rib cage and diaphragm contract to generate a pressure gradient from alveoli to mouth and to elevate the pressure in the tissues surrounding airways. In some cases, such as asthmatic attacks, airways can collapse, leading to difficulties in expiration. Forced expiration can also generate oscillations of the deformable airway walls (wheezing). In sleep, collapsible upper airways often cause snoring (obstructive sleep apnoea) and crackles (popping of liquid plugs in obstructed airways). Moreover, veins often function as collapsible tubes, which has an important effect on veins returning to right and left atriums and consequently on the overall dynamics of blood circulation. Of course, the standard technique for measuring blood pressure converts the brachial artery into a collapsible tube. In fact, Korotkoff sound, named after Nikolai Sergeievich Korotkov (1874–1920), often heard when blood pressure is measured, may be a consequence of the spontaneous oscillation. In some cases, external pressures also known as cuff pressure are imposed. For example, during a normal cardiac cycle, coronary arteries often collapse resulting in the regulation of blood flow within the heart muscle. One important aspect of coronary circulation is that blood flow to the left ventricle occurs primarily during the cardiac cycle in which the heart is relaxed. As in the modeling of other aspects of cardiovascular systems, analytical and computational methods are often used to complement experimental studies. In practice, hypotheses and assumptions are often used to formulate models with different levels of complexities. Important insights and key parameters can be derived and later employed in experimental validation and confirmation.

#### *Collapsible tube*

The collapse and self-excited oscillation of flexible tubes, as shown in Figure 4.43, have received considerable attention due to some interesting flow phenomena including flow separation, vortex shedding, and flow choking. Extensive theoretical and experimental studies have been published over the past two decades on this subject [57,178,268,275]. Although many key phenomena can be explained and predicted with one-dimensional flow (plug flow) models, efficient and effective three-dimensional computational models are still very much needed. In general, the lumped parameters include blood volume, blood flow rate, and pressure,

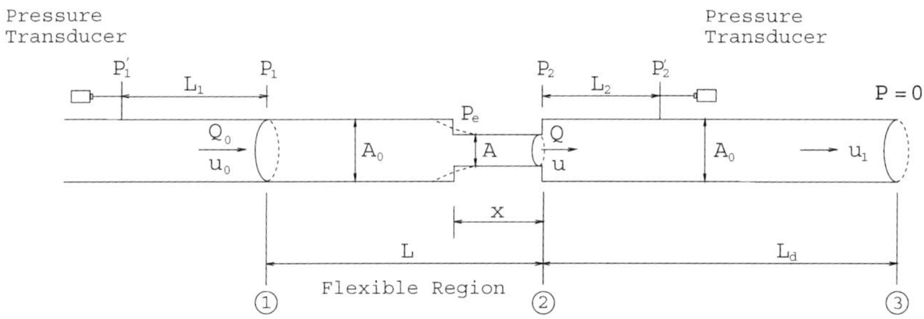


Figure 4.43. Schematic diagram of flow through a collapsible tube.

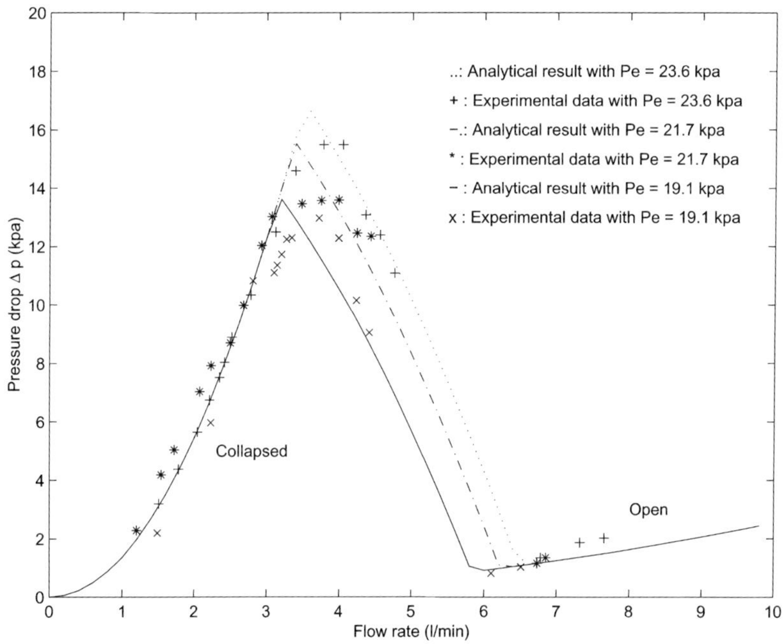


Figure 4.44. The relationship between pressure drop and flow rate.

which are analogous to charge, current, and voltage in a typical electrical circuit. Conrad and McQueen [286,289] were among the first to introduce systematic experimental pressure drop and flow rate curves, as shown in Figure 4.44, for both steady and unsteady conditions of a short Penrose tube. They also proposed that dynamic negative resistance was related to self-excited oscillations and presented an electrical circuit analogy. More recently, Bertram [37] demonstrated a complex unstable equilibrium behavior in a collapsible tube with thick walls by varying

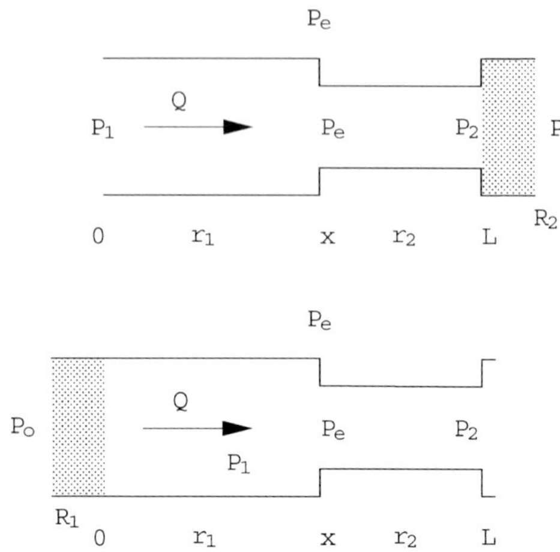


Figure 4.45. Two different approaches in collapsible tube experiments.

downstream resistance while keeping constant pressure inside the chamber of the collapsible tube system. Based on some simple resistive models for steady flow through collapsible tubes, Bertram confirms that it is possible to predict pressure drop and flow rate characteristics under all three conditions, namely, constant external pressure, constant upstream transmural pressure, and constant downstream transmural pressure [39]. To characterize the relationship between pressure drop and flow rate through such a collapsible tube, Starling resistor model, named after Ernest Henry Starling (1866–1927), is often employed, in which the external pressure, both upstream and downstream pressures are adjustable. The collapsible tube experimental setup can take on two distinct approaches as illustrated in Figure 4.45. In general, the flow rate controlled nonlinear resistance is defined as QNLR; while the pressure controlled nonlinear resistance is named as PNLR. Increasing progressively the upstream resistance $R_1$, while keeping the downstream resistance $R_2$ constant, results in a reduction of both flow rate and pressure inside the tube.

Complex static and dynamic behaviors of flow through collapsible tubes depend on many adjustable parameters. As shown in Figures 4.43 and 4.45, some of the key parameters include inlet and outlet pressures $P_1$ and $P_2$, external (cuff) pressure $P_e$, tube inner radius $r_i$, tube thickness $h$, tube length $L$, collapsed tube length $x$, tube wall Young's modulus $E$, tube axial stress $\sigma$, tube density $\rho_t$, fluid density $\rho$, fluid viscosity $\mu$, and flow rate $Q$. Of course, some of the other key physical parameters can be derived from these relatively independent parameters. The obvious dependent parameters are upstream and downstream flow resistances $R_1$ and $R_2$, which are related to fluid viscosity, inner tube radius and tube lengths;

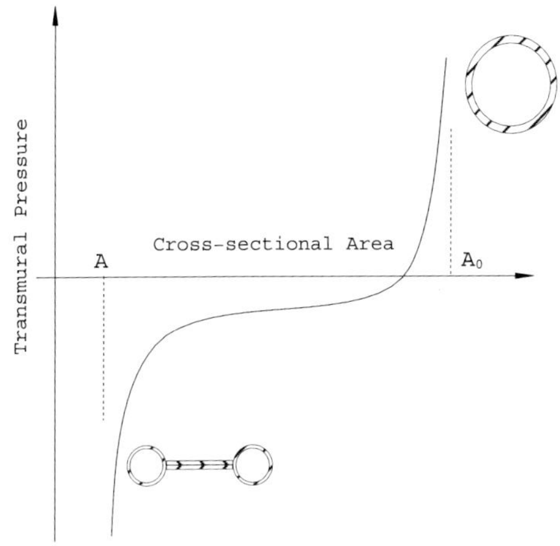


Figure 4.46. Transmural (internal minus external) pressure and cross-sectional area.

upstream and downstream flow inertias $I_1$ and $I_2$, which are dependent on fluid density, inner tube radius and tube lengths.

*Tube law*

To understand the complex FSI system, we establish simpler systems which may provide overall characteristics of the dynamical behaviors. The simplest one-dimensional models omit axial tension and bending stiffness of the tube wall. Instead, elastic properties are described by a tube law which links transmural pressure $P_t$ and cross-sectional area $A$ as shown in Figure 4.46. Flow through the collapsed region is similar to that through a venturi or flow nozzle. In the contraction section, flow can accelerate to high speed, which causes the external pressure to exceed the internal pressure and hence the further collapse of the flexible vessel. The system equation is then hyperbolic, controlled with another key parameter, the speed index $S$, defined as the ratio of averaged cross-sectional velocity $u$ to the axial velocity of radial wave propagation $c$:

$$c^2 = \frac{A}{\rho}\frac{dP_t}{dA}. \tag{2.94}$$

Flow limitation is due to a choke at $u = c$ followed by a downstream hydroelastic jump which eventually violates the one-dimensional assumption. Modified tube laws have also been proposed to include effects of axial tension and bending stiffness. In particular, for a two-dimensional membrane with a thickness $h$, we have the bending rigidity $D = Eh^3/12(1 - \nu^2)$, with the Poisson ratio $\nu$; for a three-dimensional tube with a thickness $h$, the bending rigidity is expressed as

$D = E\pi r_i^3 h$. Note that we assume the thickness to be small in comparison with the channel height or tube radius. Buckling begins theoretically when the negative transmural pressure exceeds a particular level. Collapsible tubes are easily deformed by changes in transmural pressure. Vessels subjected to positive transmural pressure are simply inflated and retain approximately cylindrical shape. In general, a typical relationship between transmural pressure and cross-sectional area is depicted in Figure 4.46. From right to left, it is clear that the tube is initially very stiff and remains circular with slight deformation as transmural pressure decreases. However, when a critical transmural pressure is passed, the tube buckles, becomes highly deformable, and changes very quickly from an elliptical cross-section to two lobes connected with completed collapsed opposite walls in contact with each other. From there on the tube becomes stiff again. From slight distention to fully collapsed states, the transmural pressure can be approximated as the following function of the cross-sectional area of the collapsible tube:

$$P_t = P - P_e = k\left(\alpha^{\beta_1} - \alpha^{\beta_2}\right), \tag{2.95}$$

with $\alpha = A/A_o$, and $k$ is experimental parameter related to the bending rigidity of the tube.

A typical set of parameters $\beta_1$ and $\beta_2$ are 20 and $-1.5$ [72]. We note that from previous experimental measurements that $\beta_1$ is always one magnitude higher than $\beta_2$. Hence, in some cases, the other tube law is approximated as

$$P_1 - P_e = \begin{cases} k_1(1 - \alpha^{-1.5}), & \text{if } \alpha \leqslant 1, \\ k_2(\alpha - 1), & \text{if } \alpha > 1, \end{cases} \tag{2.96}$$

where $k_1$, $k_2$, and $k$ are experimental parameters related to the bending rigidity of the tube at open and close phases.

Of course, all these tube laws are based on the assumption that the pressure varies sufficiently slowly, especially around the buckling (collapsing) region. Viscoelasticity will be expected when subjected to higher cyclic rates of pressure change, the effect of which will be manifested as hysteresis due to the difference between loading and unloading curves.

A typical pressure drop and flow rate curve as shown in Figure 4.44 is separated into different regions depending on the geometrical state of the conduit. When the tube is fully collapsed, i.e., $P_e > P_1 > P_2$, the resistance is a larger constant than when the tube is fully inflated. As the flow rate increases, $P_1$ will eventually become comparable to $P_e$ and the tube gradually opens. When $P_1$ and $P_2$ are both greater than $P_e$, the tube is fully open. When the tube is fully inflated (open), $\Delta P$ is linearly proportional to the steady flow rate $Q$ for fully developed laminar viscous flow and $P_1 > P_2 > P_e$. As $Q$ decreases, there is a sudden collapse of the tube starting from the outlet side (downstream) and propagating to the inlet side (upstream). The partially collapsed state corresponds to $P_1 > P_e > P_2$, and the collapsed region, with $x \in [0, L]$. The transition from these two states constitutes the most important, interesting, and complex phenomena in the study of flow through collapsible tubes. The dynamic aspects of the collapsible tube are related to the partially collapsed tube shifting from the other two extreme conditions. Experimental measurements have already revealed a rich collections of nonlinear oscillations. To measure the pressure drop and flow rate relationship, the tube is initially fully collapsed and the external pressure is kept constant. In the transition from a fully collapsed to partially collapsed stage, the pressure drop initially increases with the flow rate, but subsequently reaches an asymptotic value and then decreases while flow rate continues to increase. Naturally, when the tube becomes fully open, the pressure drop will again increase with the flow rate. Starting from the distended state, when the transmural pressure becomes low enough, the tube is set to self-induced oscillations. Before the onset of the oscillations, the tube takes the shape of a convergent-divergent nozzle. The location of this neck is the result of balance of various forces exerted upon the membrane of the tubes as well as its material properties, such as the forces due to the transmural pressure, fluid wall-shear stress, tube axial tension, axial, and circumferential tube wall bending stiffness.

Experiments also demonstrate that self-excited oscillations only occur when the steady-flow operating point is in the dynamic negative resistance region [288]. The downstream resistance is set to one middle value since the downstream valve is adjusted in the middle position. These curves can be fitted by following the work by Bertram [37], it is possible to fit the power-law relation in the form $\Delta p = kQ^b$. For the fully inflated case, the exponent $b$ is 1.936 and $k$ is 0.2107, whereas for the fully collapsed case, $k$ is 1.139 and $b$ is 1.956. Note that for the fully collapsed case $k$ is larger than the one in fully open condition, on the contrary, the exponent $b$ is quite close to the one in fully open state.

*Transverse and axial oscillations*

In particular, for the upstream domain, represented with a collapsible region of length $L$, which includes the collapsed region of length $x$, the input flow rate $Q_o$ is considered constant. For the downstream domain, represented with an open region of length $L_d$, the input flow rate $Q$ can be time-dependent along with the position of the collapsed region $x$ because of its axial shift. Moreover, since the cross-sectional area stays the same, the contained fluid volume is also constant, and consequently, the flow rate for the entire downstream region can be simply denoted as $Q$. Naturally, for steady flow cases, we have $Q = Q_o$ based on the mass conservation. For the collapsed region, according to Figure 4.46, we denote the cross-sectional area as $A$, whereas for both downstream and upstream uncollapsed tube, the cross-sectional area is denoted as $A_o$. We further denote $u_o$ and $u$ as the uniform flow velocities within open and collapsed sections of the upstream region, and $u_1$ as the equivalent uniform flow velocity within the downstream region.

In a one-dimensional model, the resistance of the collapsible tube is often defined as the ratio of the pressure drop between inlet and outlet of the tube and the steady flow rate. The dynamic resistance is defined as the rate of the change of the pressure drop with respect to the instantaneous flow rate. Because the flow through the partially collapsed tube is no longer steady, a dynamic resistance is helpful in the study of complex dynamical behaviors. Negative dynamic pressure resistance has also been reported in Conrad et al. [288] and attributed to Korotkoff sound used clinically to measure systolic and diastolic arterial blood pressure via sphygmomanometry.

Self-excited oscillation and pulse waves are generated by the pulsation of the heart. A typical one-dimensional approximation of flow through a collapsible tube is the Windkessel model, a model to relate pressure drop with flow rate, in which the compliant vessel acts as a capacitor in series with a resistor [312]. A dimensional analysis of the unsteady Navier–Stokes equations leads to a nondimensional number commonly referred to as the Womersley number, named after John R. Womersley (1907–1958), $Wo = R\sqrt{\rho\omega/\mu}$, where $\omega$ is the angular frequency. This number is introduced for flow induced by an oscillating plate, which may be interpreted as the ratio of characteristic length and the Stokes viscous layer thickness $\sqrt{\mu/\rho\omega}$ or oscillatory inertia to viscous forces. In addition, the Womersley number can be also viewed as a combination of the Reynolds number and Strouhal number $Wo = \sqrt{2\pi Re\, St}$. An interesting extension will be the Womersley solution of the velocity profile. Unsteady flow within the vessel depends on the Womersley number as well as the Reynolds number. For low Womersley numbers, the viscosity effects dominate and flow profile is primarily decided by the Reynolds number. For high Womersley numbers, the flow within the vessel essentially behaves as a plug flow [11].

Over a wide parameter range, self-excited oscillations are reported and many hypothesis requiring explanation are presented. However, further mathematical study of dissipative oscillation or limit cycle is still needed. In our simplified model, we adopt two possible states: open and closed, represented by cross-sectional area $A_o$ and $A$, respectively. Note that the closed state could be more accurately called the almost closed state. To account for the horizontal oscillation of collapsed region, we introduce an idealized abrupt change from $A_o$ to $A$. The actual buckled region can be viewed as an inclined line or surface, the center of which crosses this ideal transition line or surface. Hence, the axial shift of the collapsed region within the upstream domain is quantified with the axial location $x$ of the transition area. For brevity, considering the pressure drop $P_1 - P_2$ and averaged flow rate $Q$ through the collapsed region, the viscous fluid effect can be expressed as

$$P_1 - P_2 = -\gamma Q + \beta \frac{Q^3}{3}, \tag{2.97}$$

where $\gamma$ stands for the negative dynamic resistance of the collapsible tube and $\beta$ is a measure of the nonlinear pressure drop and flow rate relationship.

For fully developed laminar flow within a tube of diameter $R$, the linear resistance $\gamma$ can be easily derived as $8\mu/(\pi R^4)$. Furthermore, the inertial effect of the collapsible tube is represented as $L\,dQ/dt$ and the stiffness effect is quantified as $\int Q\,dt/C$, where $L$ and $C$ stand for the so-called inertance and compliance of the collapsible tube. Therefore, if we introduce $r$ as the resistance of the rest of the flow loop, the governing equation, analogous to a circuit, can be expressed as

$$L\ddot{Q} + r\dot{Q} + \frac{1}{C}Q + \dot{P}_1 - \dot{P}_2 = 0. \tag{2.98}$$

Employing Eq. (2.97) and assigning the nondimensional quantities $\omega_o = 1/\sqrt{LC}$, $\epsilon = (\gamma - r)\sqrt{C/L}$, $\tau = \omega_o t$, and $a = \sqrt{\beta/(\gamma - r)}Q$, Eq. (2.98) can be rewritten as the following van der Pol's equation:

$$\frac{d^2a}{d\tau^2} - \epsilon(1 - a^2)\frac{da}{d\tau} + a = 0. \tag{2.99}$$

As illustrated in Chapter 3, from the study of van der Pol's equation, the oscillation occurs when $\gamma > r$. For $0 < \epsilon \ll 1$, the oscillations are sinusoidal with a limit cycle amplitude of $2\sqrt{(\gamma - r)/\beta}$. For $\epsilon \gg 1$, the oscillations are nonsinusoidal relaxation the period of which can be expressed as $T \simeq 1.6\epsilon/\omega_o = 1.6(\gamma - r)C$. Of course, this van der Pol oscillator model represents only the transmural or transverse direction tube instability and oscillation, which are presented by the collapsed and open states switching at a specific point of the collapsible tube. However, the dynamical instability behavior governed by Eq. (2.99) does not include the horizontal or axial shift of the buckled region, the study of which includes the conservation laws for upstream and downstream regions as illustrated in Figure 4.43.

Following the initial work of Peskin and McQueen, from the mass conservation, we have for both upstream and downstream regions,

$$\frac{dV_{12}}{dt} = A_o u_o - Au, \tag{2.100}$$

$$\frac{dV_{23}}{dt} = Au - A_o u_1, \tag{2.101}$$

where the mass volumes $V_{12}$ and $V_{23}$ within the upstream and downstream regions are expressed as

$$V_{12} = A_o(L - x) + Ax \quad \text{and} \quad V_{23} = A_o L_d.$$

Since the fluid volume within the downstream domain $V_{23}$ is constant, Eq. (2.101) can be simply replaced with $u_1 = Au/A_o = Q/A_o$. In addition, Eq. (2.100) can be expressed as

$$(A - A_o)\frac{dx}{dt} = Q_o - Q. \tag{2.102}$$

From the linear momentum conservation, we have for both upstream and downstream domains,

$$\frac{dL_{12}}{dt} = P_1 A_o + \rho u_o u_o A_o - P_e(A_o - A) - P_2 A - \rho u u A - \gamma_1(L - x)u_o A_o^2 - \gamma_2 x u A^2, \tag{2.103}$$

$$\frac{dL_{23}}{dt} = P_2 A_o + \rho u u A - \rho u_1 u_1 A_o - \gamma_3 L_d u_1 A_o^2, \tag{2.104}$$

where $\gamma_1$, $\gamma_2$, and $\gamma_3$ stand for the friction coefficients with respect to the viscous effects in the upstream uncollapsed, collapsed, and downstream regions, respectively, and the linear momentums $L_{12}$ and $L_{23}$ within upstream and downstream regions can be expressed as

$$L_{12} = A_o(L - x)\rho u_o + Ax\rho u \quad \text{and} \quad L_{23} = A_o L_d \rho u_1.$$

Note that the inlet for the downstream region coincides with the transition line or surface and has a cross-sectional area $A$. Therefore, although the uniform velocity of the inlet is $u$ within the collapsed cross-sectional area, the normal velocity within the rest of the interface is zero. Furthermore, without the detriment to the physical model, we assume that the pressure has the same value $P_2$ across the interface. This assumption implies a positive loss of kinetic energy or cost which is denoted as $c(t)$ in the following energy conservation equations:

$$\frac{dE_{12}}{dt} = P_1 u_o A_o + \frac{1}{2}\rho u_o^2 u_o A_o + P_e(A_o - A)\frac{dx}{dt} - P_2 u A - \frac{1}{2}\rho u^2 u A - \gamma_1(L - x)(u_o A_o)^2 - \gamma_2 x (uA)^2, \tag{2.105}$$

$$\frac{dE_{23}}{dt} = P_2 u A + \frac{1}{2}\rho u^2 u A - \frac{1}{2}\rho u_1^2 u_1 A_o - \gamma_3 L_d (u_1 A_o)^2 - \delta(t), \tag{2.106}$$

where the energies within upstream and downstream regions can be expressed as

$$E_{12} = \rho u_o^2 A_o (L - x)/2 + \rho u^2 A x/2 \quad \text{and} \quad E_{23} = A_o L_d \rho u_1^2/2.$$

Similar to the treatment of mass conservation equations, from the linear momentum conservation equations (2.103) and (2.104), we have

$$\rho(Q - Q_o)\frac{dx}{dt} + \rho x \frac{dQ}{dt} = P_1 A_o - P_e(A_o - A) - P_2 A + \rho\frac{Q_o^2}{A_o} - \rho\frac{Q^2}{A} - \gamma_1(L - x)Q_o A_o - \gamma_2 x Q A, \tag{2.107}$$

$$\rho L_d \frac{dQ}{dt} = P_2 A_o + \rho \frac{Q^2}{A} - \rho \frac{Q^2}{A_o} - \gamma_3 L_d Q A_o, \tag{2.108}$$

and from the energy conservation equations (2.105) and (2.106), we have

$$\begin{aligned} &\frac{1}{2}\rho\left(\frac{Q^2}{A} - \frac{Q_o^2}{A_o}\right)\frac{dx}{dt} + \rho x \frac{Q}{A}\frac{dQ}{dt} \\ &\quad = P_1 Q_o + \frac{1}{2}\rho\frac{Q_o^3}{A_o^2} + P_e(A_o - A)\frac{dx}{dt} - P_2 Q \\ &\qquad - \frac{1}{2}\rho\frac{Q^3}{A^2} - \gamma_1(L - x)Q_o^2 - \gamma_2 x Q^2, \end{aligned} \tag{2.109}$$

$$\rho L_d \frac{Q}{A_o}\frac{dQ}{dt} = P_2 Q + \frac{1}{2}\rho\frac{Q^3}{A^2} - \frac{1}{2}\rho\frac{Q^3}{A_o^2} - \gamma_3 L_d Q^2 - c(t). \tag{2.110}$$

From Eqs. (2.108) and (2.110) we can easily derive the energy loss

$$c(t) = \frac{1}{2}\rho Q^3\left(\frac{1}{A} - \frac{1}{A_o}\right)^2 = \frac{1}{2}\rho Q(u - u_1)^2 > 0, \tag{2.111}$$

which confirms that there is always kinetic energy loss of relative motion.

Therefore, in general, given the upstream flow rate $Q_o$ and the external pressure $P_e$, four independent equations, namely, (2.102), (2.107), (2.108), and (2.109) are used for the solutions of the unsteady states, $P_1(t)$, $P_2(t)$, $Q(t)$, and $x(t)$. Of course, for steady solutions of $P_1$, $P_2$, and $x$, Eq. (2.102) simply yields $Q = Q_o$ and, directly obtained from Eqs. (2.107), (2.108), and (2.109), the following three equations are employed:

$$\begin{aligned} &P_1 A_o - P_e(A_o - A) - P_2 A + \rho Q_o^2\left(\frac{1}{A_o} - \frac{1}{A}\right) \\ &\quad - \gamma_1(L - x)Q_o A_o - \gamma_2 x Q_o A = 0, \\ &P_1 - P_2 + \frac{1}{2}\rho Q_o^2\left(\frac{1}{A_o^2} - \frac{1}{A^2}\right) - \gamma_1(L - x)Q_o - \gamma_2 x Q_o = 0, \\ &P_2 A_o + \rho Q_o^2\left(\frac{1}{A} - \frac{1}{A_o}\right) - \gamma_3 L_d Q_o A_o = 0. \end{aligned} \tag{2.112}$$

By eliminating $P_1$ and $P_2$, we obtain

$$P_e = (\gamma_2 x + \gamma_3 L_d)Q_o + \rho Q_o^2\left(\frac{A_o - A}{2A_o A^2} - \frac{A_o - A}{A_o^2 A}\right). \tag{2.113}$$

If we suppose that the collapsed tube has a much smaller cross-section area, i.e., $2A \ll A_o$, we have

$$\frac{A_o - A}{2A_o A^2} - \frac{A_o - A}{A_o^2 A} = \left(\frac{1}{A} - \frac{1}{A_o}\right)\left(\frac{1}{2A} - \frac{1}{A_o}\right) \simeq \frac{1}{2A^2}. \tag{2.114}$$

With these assumptions and simplifications, the results can be written as

$$\gamma_2 x Q_o = P_e - \gamma_3 L_d Q_o - \frac{\rho Q_o^2}{2A^2}. \tag{2.115}$$

Notice that for the flexible tube to collapse we must have $x > 0$ so that it imposes the following limitation for the flow rate:

$$P_e > \gamma_3 L_d Q_o + \frac{\rho Q_o^2}{2A^2}. \tag{2.116}$$

Now, by solving Eqs. (2.112), we can determine pressure-flow relationships under fully open, partially open, and fully collapsed states, respectively:

$$P_1 - P_2 = \begin{cases} \gamma_1 L Q_o, & \text{fully open, } x = 0, \\ \gamma_1 L Q_o - \gamma_3 L_d Q_o + P_e, & \text{partially open, } 0 < x < L, \\ \gamma_2 L Q_o, & \text{fully collapsed, } x = L. \end{cases} \tag{2.117}$$

In a typical experimental setup, because of the difficulty of setting pressure gauges at the end points of a flexible tube region, we employ the following modifications to transform the pressure drop along the tube to the locations where the pressures are actually measured. For the type of flow under consideration, pressure $P_1'$ and $P_2'$ can be expressed as $P_1' - P_1 = \gamma_1 L_1 Q_o$ and $P_2 - P_2' = \gamma_3 L_2 Q_o$. As a consequence, Eq. (2.117) is modified as

$$P_1' - P_2' = \begin{cases} \gamma_1 (L + L_1) Q_o + \gamma_3 L_2 Q_o, & \text{fully open, } x = 0, \\ \gamma_1 (L + L_1) Q_o & \\ \quad + \gamma_3 (L_2 - L_d) Q_o + P_e, & \text{partially open, } 0 < x < L, \\ \gamma_1 L_1 Q_o + \gamma_2 L Q_o + \gamma_3 L_2 Q_o, & \text{fully collapsed, } x = L. \end{cases} \tag{2.118}$$

Considering the oscillation state, similarly, we substitute $P_1$ and $P_2$ as functions of $dQ/dt$ and $dx/dt$, and obtain

$$\frac{dx}{dt} = \frac{Q - Q_o}{A_o - A}, \tag{2.119}$$

$$\begin{aligned} \frac{dQ}{dt} = \frac{1}{\alpha} \left\{ \left( \frac{Q}{A_o} - \frac{Q_o}{2A_o} - \frac{Q^2}{2Q_o A} \right) \frac{Q_o - Q}{A_o - A} + \frac{1}{2} \left( \frac{Q_o^2}{A_o^2} - \frac{2Q^2}{A_o A} + \frac{Q^3}{Q_o A^2} \right) \right\} \\ - \frac{1}{\alpha} \left\{ \left[ \frac{Q^2}{A_o} \left( \frac{1}{A_o} - \frac{1}{A} \right) + (\gamma_3 L_d + \gamma_2 x) \frac{Q}{\rho} - \frac{P_e}{\rho} \right] \left( \frac{A}{A_o} - \frac{Q}{Q_o} \right) \right\}, \end{aligned} \tag{2.120}$$

with $\alpha = (A/A_o - Q/Q_o)(x/A + L_d/A_o)$.

Eqs. (2.119) and (2.120) can be integrated numerically for the solutions of $x(t)$ and $Q(t)$. Now the key question is what happens if the collapsed section approaches zero. As $x$ becomes zero, the tube is forced to open. This of course involves the transverse instability motion governed by a van der Pol oscillator. For simplicity, if we focus on the axial oscillation of the buckled tube region, we can introduce a plug flow model with the consideration of collision during the opening of the tube between the upstream column of fluid with velocity $u_o$ and the downstream column of fluid with velocity $u_1$. During the collision, sudden changes in velocity occur, which introduced by impulsive pressure $P_1^*$ and $P_2^*$, that is $P_1(t) = P_1^* \Delta t$ and $P_2(t) = P_2^* \Delta t$, by applying the momentum conservation to two control volumes, we have

$$\begin{aligned} A_o P_1^* - A_o P_2^* &= \rho L \Delta Q_o, \\ A_o P_2^* &= \rho L_d \Delta Q. \end{aligned} \tag{2.121}$$

Note that the flow rate within the upstream is constant during the impact, i.e., $\Delta Q_o = 0$, and only the flow rate in downstream region is time-dependent. Hence, from Eq. (2.121), we have $P_1^* = P_2^* = \rho L_d \Delta Q / A_o$. Furthermore, if we express the energy of the entire fluid as

$$K = \frac{\rho}{2}\left(\frac{L Q_o^2}{A_o} + \frac{L_d Q^2}{A_o}\right), \tag{2.122}$$

the rate of change of energy can be expressed as

$$\Delta K = \frac{\rho L_d}{2A_o} \Delta Q^2. \tag{2.123}$$

According to energy conservation, the rate of work done by the pressure impulse on the plug fluid of the entire region is equivalent to the rate of change of the kinetic energy. We then have

$$P_1^* Q_o = \frac{\rho L_d}{2A_o} \Delta Q^2 = \frac{\rho L_d}{2A_o}\left[(Q^+)^2 - (Q^-)^2\right]. \tag{2.124}$$

Here, we assume that the flow rates before and after the collision, denoted as $Q^-$ and $Q^+$, are discontinuous, i.e., $Q^- \neq Q^+$. Employing $\Delta Q = Q^+ - Q^-$, we derive, from Eq. (2.124),

$$Q_o = \frac{1}{2}(Q^+ + Q^-). \tag{2.125}$$

Eq. (2.125) defines the mechanism for the oscillation of the axial shift of the buckled tube region, which is physically different from van der Pol like oscillation of close and open motions. However, if we consider the entire flexible tube area as a QNLR system, whether or not the instability involves the axial or transverse motions of the tube, the pressure drop $P_1 - P_2$ and flow rate $Q$ can be expressed as a certain nonlinear fashion which governs the state of oscillation. Furthermore, it is relatively straightforward to extend the laminar flow case to the turbulence flow case and include the effects of Womersley number. The parameters that need to be changed are the friction coefficients $\gamma_1$, $\gamma_2$, and $\gamma_3$, which is proportional to the square of flow rate according to the turbulence flow condition. We use $\alpha_1$, $\alpha_2$, and $\alpha_3$ to replace $\gamma_1$, $\gamma_2$, and $\gamma_3$ in the equations to delineate the turbulence from the laminar flow. The total dominated mass, linear momentum, and energy conversation equations can be rewritten as follows:

$$(A - A_o)\frac{dx}{dt} = Q_o - Q, \tag{2.126}$$

$$\begin{aligned} \rho(Q - Q_o)\frac{dx}{dt} + \rho x\frac{dQ}{dt} &= P_1 A_o - P_e(A_o - A) - P_2 A + \rho\frac{Q_o^2}{A_o} - \rho\frac{Q^2}{A} \\ &\quad - \alpha_1(L - x)Q_o^2 A_o - \alpha_2 x Q^2 A, \end{aligned} \tag{2.127}$$

$$\rho L_d\frac{dQ}{dt} = P_2 A_o + \rho\frac{Q^2}{A} - \rho\frac{Q^2}{A_o} - \alpha_3 L_d Q^2 A_o, \tag{2.128}$$

$$\begin{aligned} \frac{1}{2}\rho\left(\frac{Q^2}{A} - \frac{Q_o^2}{A_o}\right)\frac{dx}{dt} + \rho x\frac{Q}{A}\frac{dQ}{dt} &= P_1 Q_o + \frac{1}{2}\rho\frac{Q_o^3}{A_o^2} + P_e(A_o - A)\frac{dx}{dt} - P_2 Q \\ &\quad - \frac{1}{2}\rho\frac{Q^3}{A^2} - \alpha_1(L - x)Q_o^3 - \alpha_2 x Q^3, \end{aligned} \tag{2.129}$$

$$\rho L_d\frac{Q}{A_o}\frac{dQ}{dt} = P_2 Q + \frac{1}{2}\rho\frac{Q^3}{A^2} - \frac{1}{2}\rho\frac{Q^3}{A_o^2} - \alpha_3 L_d Q^3 - c(t). \tag{2.130}$$

Similarly, for the steady state, we can find the solutions of pressure drop in terms of flow rate and external pressure under fully open, partially open, and fully collapsed states, respectively

$$P_1 - P_2 = \begin{cases} \alpha_1 L Q_o^2, & \text{fully open, } x = 0, \\ \alpha_1 L Q_o^2 - \alpha_3 L_d Q_o^2 + P_e, & \text{partially open, } 0 < x < L, \\ \alpha_2 L Q_o^2, & \text{fully collapsed, } x = L. \end{cases} \tag{2.131}$$

For the oscillation state, we replace $P_1$ and $P_2$ as functions of $dQ/dt$ and $dx/dt$, and obtain

$$\frac{dx}{dt} = \frac{Q - Q_o}{A_o - A}, \tag{2.132}$$

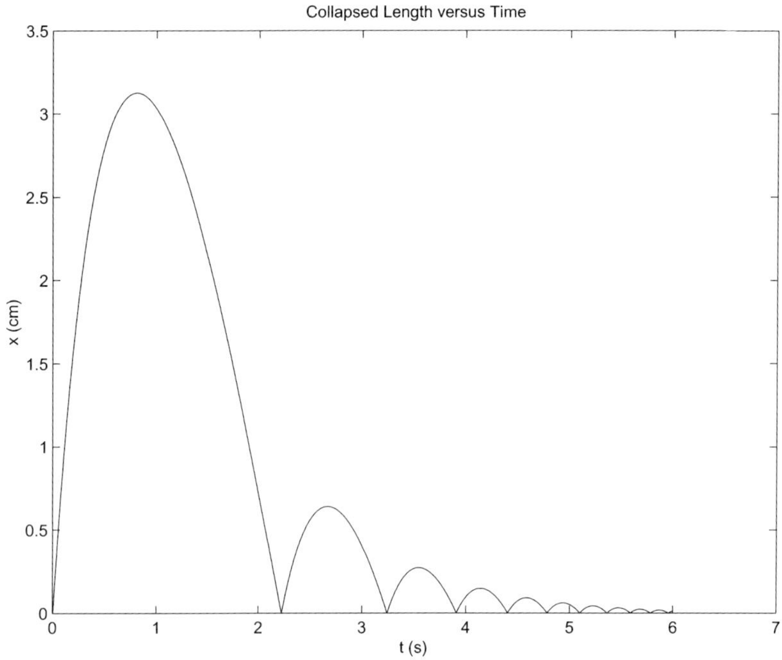


Figure 4.47. Time history of the length of the collapsed region.

$$\frac{dQ}{dt} = \frac{1}{\alpha}\left\{\left(\frac{Q}{A_o} - \frac{Q_o}{2A_o} - \frac{Q^2}{2Q_oA}\right)\frac{Q_o - Q}{A_o - A} + \frac{1}{2}\left(\frac{Q_o^2}{A_o^2} - \frac{2Q^2}{A_oA} + \frac{Q^3}{Q_oA^2}\right)\right\}$$
$$- \frac{1}{\alpha}\left\{\left[\frac{Q^2}{A_o}\left(\frac{1}{A_o} - \frac{1}{A}\right) + (\alpha_3 L_d + \alpha_2 x)\frac{Q^2}{\rho} - \frac{P_e}{\rho}\right]\left(\frac{A}{A_o} - \frac{Q}{Q_o}\right)\right\}, \tag{2.133}$$

with $\alpha = (A/A_o - Q/Q_o)(x/A + L_d/A_o)$.

Figure 4.47 shows the time history of the length of the collapsed region $x$ derived from Eqs. (2.132) and (2.133). In this test case with $Q_o = 12.33$ ml/s and $P_e = 1261.4$ Pa, the collapsible section will experience close and open shifts till the length of the collapsed region approaches zero indicating the tube is eventually fully open. In fact, it is possible to show that Eqs. (2.132) and (2.133) are always dissipative due to a positive energy loss $c(t)$. Moreover, Figure 4.48 shows that as the length of the collapsed region decreases the flow rate difference of the collapsed region and the outlet also decreases. Of course, when the tube is fully open, there will be no such flow rate difference. This is again confirmed in Figure 4.49 the flow rate at the collapsed region will eventually approach the flow rate of the fully open tube. Finally, in the phase diagram as shown in Figure 4.50,

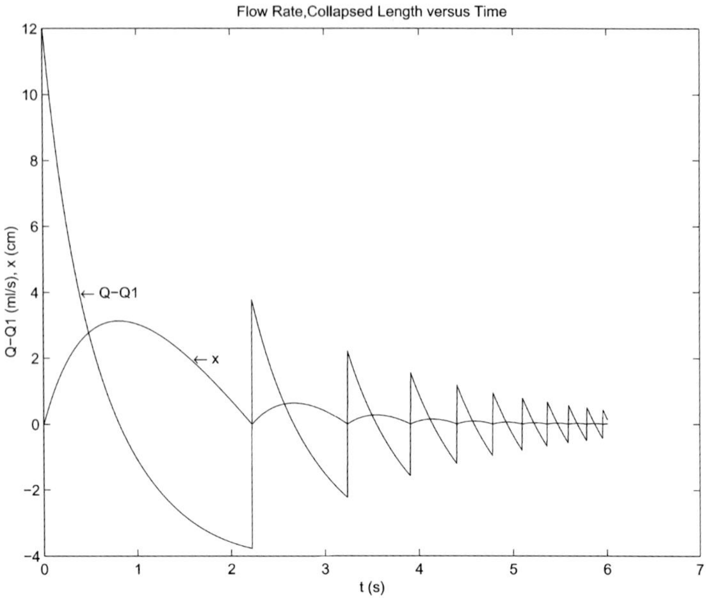


Figure 4.48. Time history of the flow rate difference.

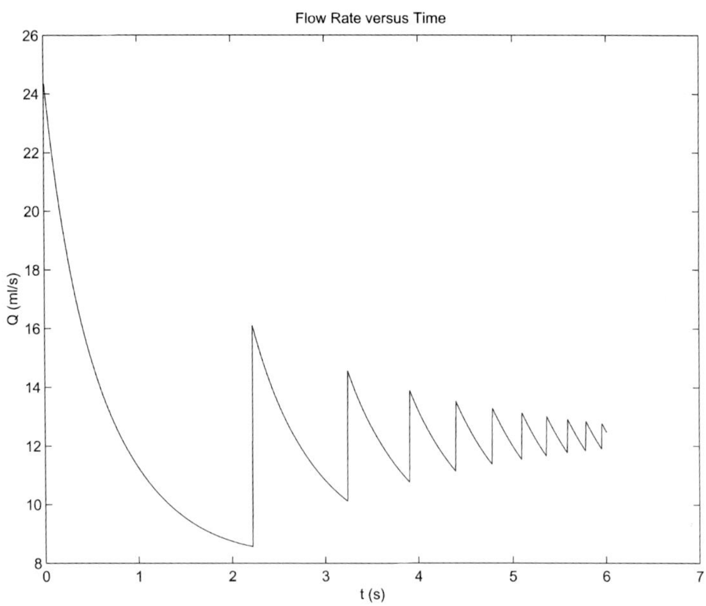


Figure 4.49. Time history of the flow rate.

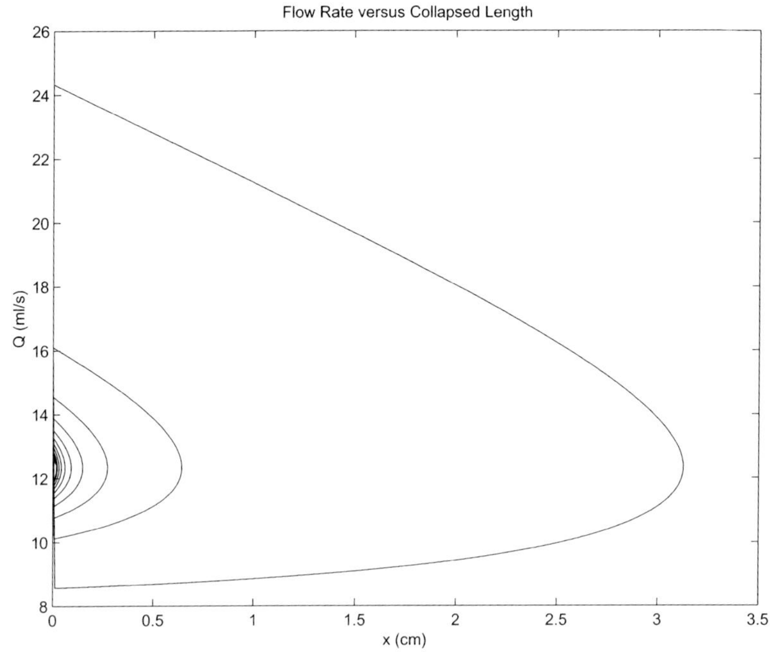


Figure 4.50. The flow rate vs. the length of the collapsed region.

the trajectory of the dynamical system is clearly dissipative and converges to the stationary point on the vertical axis.

## 4.3. Dynamics of thin structures

In the aerospace industry, structures vibrating in steady flows can introduce oscillatory aerodynamic forces which in turn create so-called galloping and flutter [316]. In practice, flutter corresponds to an instability induced by aerodynamic forces which are comparable to the weight and flexural or torsional rigidity of the cross section and are significant enough to alter the natural frequencies. Galloping is associated with an instability due to low structural damping combined with small oscillatory aerodynamic forces which do not shift the natural frequencies.

The flutter instability of flexible panels is of particular concern during high-speed flight. In fact, panel flutter is encountered in a number of industries involving the movement of thin materials, e.g., textile, plastic film, and chapter industries. One of two approaches can generally be adopted. The first is to consider the flexible plate stationary with airflow passing over the plate surface. This approach is often used to study the panel flutter of supersonic airplane wing surface plates, in which the airflow is over one side of the plate. The second approach, which is often used to study moving thin materials, is to consider the thin plate moving at a constant velocity.

### *4.3.1. Stationary and moving panel; MITC elements*

In the textile, paper, plastic film, and other industries involving moving thin materials, stress analysis is essential for the control of wrinkle, flutter, and sheet break. Similar vibration problems with moving plate and shell structures can be found in aeronautic and aerospace structures, computer hard disk drives, and rotating blades in turbomachinery. Although the mechanical behavior of axially moving materials has been studied for many years, little information is available on stress analysis including the transverse shear of moving orthotropic thin plates, such as paper sheets. Following the early work of Mote [188], significant research efforts have been directed toward studying the linear and nonlinear dynamical behavior of an axially moving string model [118–120]. In practice, in order to consider the dynamical behavior of moving materials coupled with surrounding fluid air, one often ignores boundary layer shear forces and introduces different added mass expressions for the Coriolis and centrifugal inertia terms [221]. For purposes of attenuation and guidance, coupling problems between moving material and fluid/air bearings have also been studied [33,201,291,315]. More recently, Heinrich and Connolly [121] carried out a three-dimensional analysis of an axially moving tape coupled with bearings by using the classical Kirchhoff thin plate assumptions. The importance and practical applications of stress analysis of moving materials were also discussed by Lee and Ng [112]. They applied a mode method with the Kirchhoff thin plate theory for isotropic materials and Lagrangian kinematic descriptions to avoid convective terms. It has been indicated in Ref. [302] that when the axial tension is not predominant, or the bending stiffness of the plate is significant, reliable numerical formulations are needed to accurately predict the transverse shear, as well as in-plane stress distributions of the moving thin plate and shell structures. So far, no work has been documented in predicting the stress distributions of moving orthotropic thin plates.

#### *Stationary and moving panel*

Since the inception of finite element methods, many plate bending formulations and elements have been investigated. Proper review of such topics is available in Refs. [12,259,264,281]. It has been widely recognized that mixed plate formulations based on the Mindlin/Reissner theory, named after Raymond D. Mindlin (1906–1987) and Eric Reissner (1913–1996), can effectively eliminate "shear locking" and predict stress levels of thin plates accurately and reliably. Recently, the MITC elements have been numerically proven, to satisfy the inf-sup condition [116]. The basic difficulty in choosing proper orders of interpolation for

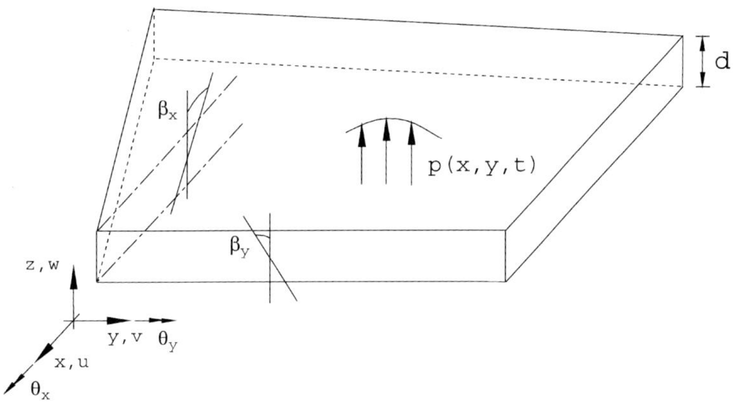


Figure 4.51. Plate deformation assumptions including transverse shear.

out-of-plane displacement, section rotations, and transverse shear strains which result in nonlocking behavior and optimal convergence of the element is summarized in Refs. [29,148,150].

In this section, we extend the mixed plate formulation, with MITC elements, to the analysis of axially moving orthotropic thin plates. In three-dimensional elasticity theory, there are six stress components that are expressed in terms of six strain components through material constitutive laws. For the general plate section illustrated in Figure 4.51, the top and bottom faces of the plate (at $z = \pm d/2$) are considered to be free from tangential traction, but under normal pressures $p^+$ and $p^-$, i.e.,

$$\begin{aligned} &\sigma_{xz}|_{z=\pm d/2} = 0, && \sigma_{yz}|_{z=\pm d/2} = 0; \\ &\sigma_{zz}|_{z=+d/2} = -p^+, && \sigma_{zz}|_{z=-d/2} = -p^-. \end{aligned} \tag{3.134}$$

For convenience, we set $p = p^- - p^+$. We recognize that in some cases we may have to modify the above assumptions concerning the surrounding air shear boundary layer. Corresponding to the five stress components, the bending and twisting moments, and the transverse shearing forces, all measured per unit of length, are defined as

$$(M_x, M_y, M_{xy}) = \int_{-d/2}^{+d/2} (\sigma_{xx}, \sigma_{yy}, \sigma_{xy}) z \, dz, \tag{3.135}$$

$$(Q_x, Q_y) = \int_{-d/2}^{+d/2} (\sigma_{xz}, \sigma_{yz}) \, dz. \tag{3.136}$$

Also shown in Figure 4.51, $\boldsymbol{\beta} = \langle \beta_x, \beta_y \rangle$ is the vector whose components are the rotations of the normal to the undeformed middle surface in the $x$, $z$ and $y$, $z$ planes, respectively; it is equivalent to $\boldsymbol{\theta}^{\perp} = \langle -\theta_y, \theta_x \rangle$, which is normal to the vector $\boldsymbol{\theta}$ of rotations about the Cartesian axes $x$ and $y$. In the classical Kirchhoff plate theory, the transverse strains and stresses are ignored and, consequently, $\nabla w = \boldsymbol{\beta}$. However, we assume in this book, according to the Mindlin/Reissner plate theory [184,230], that the $x$-direction and $y$-direction displacements $u$ and $v$ are proportional to $z$, and that the $z$-direction displacement $w$ is independent of $z$, i.e.,

$$u = -z\beta_x(x, y, t), \qquad v = -z\beta_y(x, y, t), \qquad w = w(x, y, t). \tag{3.137}$$

Moreover, the five strain components are introduced as follows:

$$\begin{bmatrix} \epsilon_{xx} \\ \epsilon_{yy} \\ \gamma_{xy} \end{bmatrix} = -z \begin{bmatrix} \frac{\partial \beta_x}{\partial x} \\ \frac{\partial \beta_y}{\partial y} \\ \frac{\partial \beta_x}{\partial y} + \frac{\partial \beta_y}{\partial x} \end{bmatrix} = -z\boldsymbol{\kappa}, \qquad \begin{bmatrix} \gamma_{xz} \\ \gamma_{yz} \end{bmatrix} = \begin{bmatrix} \frac{\partial w}{\partial x} - \beta_x \\ \frac{\partial w}{\partial y} - \beta_y \end{bmatrix} = \boldsymbol{\gamma}. \tag{3.138}$$

In general, for an orthotropic material, we have

$$\begin{bmatrix} \sigma_{xx} \\ \sigma_{yy} \\ \sigma_{xy} \end{bmatrix} = \begin{bmatrix} a_{11} & a_{12} & 0 \\ a_{21} & a_{22} & 0 \\ 0 & 0 & a_{33} \end{bmatrix} \begin{bmatrix} \epsilon_{xx} \\ \epsilon_{yy} \\ \gamma_{xy} \end{bmatrix} = -\mathbf{C}_{\epsilon} z \boldsymbol{\kappa}, \tag{3.139}$$

$$\begin{bmatrix} \sigma_{xz} \\ \sigma_{yz} \end{bmatrix} = \begin{bmatrix} a_{44} & 0 \\ 0 & a_{55} \end{bmatrix} \begin{bmatrix} \gamma_{xz} \\ \gamma_{yz} \end{bmatrix} = \mathbf{C}_{\gamma} \boldsymbol{\gamma}, \tag{3.140}$$

where the material constants are defined by

$$a_{11} = \frac{E_x}{1 - \nu_{xy}\nu_{yx}}, \qquad a_{12} = a_{21} = \frac{\nu_{xy} E_x}{1 - \nu_{xy}\nu_{yx}}, \qquad a_{22} = \frac{E_y}{1 - \nu_{xy}\nu_{yx}};$$

$$a_{33} = G_{xy}, \qquad a_{44} = G_{xz}, \qquad a_{55} = G_{yz}.$$

In addition to the usual necessary condition $E_x/\nu_{yx} = E_y/\nu_{xy}$, for paper materials, we also have the Campbell relationship $1/G_{xy} = (1 + \nu_{yx})/E_x + (1 + \nu_{xy})/E_y$ [125]. Of course, in the case of an isotropic material, we would have $E_x = E_y = E$, $\nu_{xy} = \nu_{yx} = \nu$, and $G_{xy} = G_{xz} = G_{yz} = E/2(1 + \nu)$.

Substituting Eqs. (3.139) and (3.140) into Eqs. (3.135) and (3.136), we obtain

$$M_x = -\left( a_{11} \frac{\partial \beta_x}{\partial x} + a_{12} \frac{\partial \beta_y}{\partial y} \right) \frac{d^3}{12}, \tag{3.141}$$

$$M_y = -\left( a_{21} \frac{\partial \beta_x}{\partial x} + a_{22} \frac{\partial \beta_y}{\partial y} \right) \frac{d^3}{12}, \tag{3.142}$$

$$M_{xy} = -a_{33} \left( \frac{\partial \beta_x}{\partial y} + \frac{\partial \beta_y}{\partial x} \right) \frac{d^3}{12}, \tag{3.143}$$

and

$$Q_x = kd\left(\frac{\partial w}{\partial x} - \beta_x\right)a_{44}, \tag{3.144}$$

$$Q_y = kd\left(\frac{\partial w}{\partial y} - \beta_y\right)a_{55}. \tag{3.145}$$

The constant $k$ plays a role in accounting for the nonuniform distribution of transverse shear stresses through the plate thickness, and in this book has the value $5/6$. A more elaborate discussion with regard to the selection of $k$ is available in Refs. [143,184]. Denoting $\rho$ as the mass density, the indices $i$ ($i = 1, 2, 3$) as the $x$-, $y$-, and $z$-directions, from the force equilibrium equations, we obtain

$$\frac{\partial M_x}{\partial x} + \frac{\partial M_{xy}}{\partial y} - Q_x = -\frac{\rho d^3}{12}\frac{d^2\beta_x}{dt^2}, \tag{3.146}$$

$$\frac{\partial M_{xy}}{\partial x} + \frac{\partial M_y}{\partial y} - Q_y = -\frac{\rho d^3}{12}\frac{d^2\beta_y}{dt^2}, \tag{3.147}$$

$$\frac{\partial Q_x}{\partial x} + \frac{\partial Q_y}{\partial y} + p - \rho g d = \rho d\frac{d^2 w}{dt^2}. \tag{3.148}$$

Of course, as indicated in Eq. (3.148), the only body force considered in this book is the gravitational force in the negative $z$-direction. Furthermore, in terms of out-of-plane displacement $w$, and rotations $\beta_x$ and $\beta_y$, we obtain from Eqs. (3.144) to (3.148)

$$a_{11}\frac{\partial^2\beta_x}{\partial x^2} + a_{12}\frac{\partial^2\beta_y}{\partial x\partial y} + a_{33}\frac{\partial^2\beta_x}{\partial y^2} + a_{33}\frac{\partial^2\beta_y}{\partial x\partial y} + a_{44}\left(\frac{\partial w}{\partial x} - \beta_x\right)\frac{12k}{d^2}$$
$$= \rho\frac{d^2\beta_x}{dt^2}, \tag{3.149}$$

$$a_{33}\frac{\partial^2\beta_x}{\partial x\partial y} + a_{33}\frac{\partial^2\beta_y}{\partial x^2} + a_{21}\frac{\partial^2\beta_x}{\partial x\partial y} + a_{22}\frac{\partial^2\beta_y}{\partial y^2} + a_{55}\left(\frac{\partial w}{\partial y} - \beta_y\right)\frac{12k}{d^2}$$
$$= \rho\frac{d^2\beta_y}{dt^2}, \tag{3.150}$$

$$a_{44}\left(\frac{\partial^2 w}{\partial x^2} - \frac{\partial\beta_x}{\partial x}\right)k + a_{55}\left(\frac{\partial^2 w}{\partial y^2} - \frac{\partial\beta_y}{\partial y}\right)k + \frac{p}{d} - \rho g = \rho\frac{d^2 w}{dt^2}. \tag{3.151}$$

We note here that when the plate is stationary, the material derivative $d/dt$ is the same as the spatial derivative $\partial/\partial t$; however, when the plate is moving with velocity $V$ in the $x$ direction, we have $d/dt = \partial/\partial t + V\partial/\partial x$. In general, for moving materials, we have $d/dt = \partial/\partial t + \mathbf{V}\cdot\nabla$.

For an isotropic and stationary plate, the following governing equations presented by Mindlin [184] are recovered:

$$\frac{D}{2}\left[(1-\nu)\nabla^2\beta_x + (1+\nu)\frac{\partial\phi}{\partial x}\right] + kGd\left(\frac{\partial w}{\partial x} - \beta_x\right) = \frac{\rho d^3}{12}\frac{\partial^2\beta_x}{\partial t^2}, \tag{3.152}$$

$$\frac{D}{2}\left[(1-\nu)\nabla^2\beta_y + (1+\nu)\frac{\partial\phi}{\partial y}\right] + kGd\left(\frac{\partial w}{\partial y} - \beta_y\right) = \frac{\rho d^3}{12}\frac{\partial^2\beta_y}{\partial t^2}, \tag{3.153}$$

$$kGd\left(\nabla^2 w - \phi\right) + p - \rho g d = \rho d\frac{\partial^2 w}{\partial t^2}, \tag{3.154}$$

where $\phi$ is defined as $\partial\beta_x/\partial x + \partial\beta_y/\partial y$ and the bending stiffness $D$ equals $Ed^3/12(1-\nu^2)$. Furthermore, the classical theory of stationary isotropic elastic thin plates based on the Kirchhoff assumptions (refer to [251]) yields

$$D\nabla^4 w + \rho d\frac{\partial^2 w}{\partial t^2} = p - \rho g d. \tag{3.155}$$

It was found by Timoshenko [249], Uflyand [319], Hencky [103], and Mindlin [184] that Eq. (3.155) cannot be used to predict sharp transients or frequencies of high-order vibrational modes. In order to predict such effects, transverse shear deformations have to be included.

Based on the classical Kirchhoff assumptions, for a moving isotropic thin plate with tension $T$, which often emerges in the analysis of moving tapes, saw bands, plastic films, and subsonic panel flutter problems [32,63,121,199], Eq. (3.155) is modified as

$$D\nabla^4 w + \rho d\left(\frac{\partial}{\partial t} + V\frac{\partial}{\partial x}\right)^2 w - T\frac{\partial^2 w}{\partial x^2} = p - \rho g d. \tag{3.156}$$

Nevertheless, it has recently been recognized that if the axial tension is not predominant or as the axial moving speed increases, a reliable mixed plate theory is needed to accurately predict the stress distributions within a moving thin plate [302].

*MITC elements*

In the mixed formulation based on the Mindlin/Reissner plate theory, we account for the effects of inertia, by applying Hamilton's principle with the following definition of kinetic energy:

$$\int_{\Omega}\left\{\frac{\rho d^3}{24}\left(\frac{d\beta_x}{dt}\right)^2 + \frac{\rho d^3}{24}\left(\frac{d\beta_y}{dt}\right)^2 + \frac{\rho d}{2}\left(\frac{dw}{dt}\right)^2\right\} d\Omega, \tag{3.157}$$

where $\Omega$ represents the midsurface area.

In actuality, we include in the body forces the following inertia terms:

$$\rho d\left(\frac{\partial}{\partial t}+V\frac{\partial}{\partial x}\right)^2 w \quad \text{and} \quad \frac{\rho d^3}{12}\left(\frac{\partial}{\partial t}+V\frac{\partial}{\partial x}\right)^2 \boldsymbol{\beta}.$$

Defining the spaces $\Theta = (\mathcal{H}_0^1(\Omega))^2$ and $\mathcal{W} = \mathcal{H}_0^1(\Omega)$, and assuming that a force function $f_w$ is given in $\mathcal{L}^2(\Omega)$, and a moment function $\mathbf{f}_\beta$ is given in $(\mathcal{L}^2(\Omega))^2$, our problem can be presented as follows:

$$\inf_{\boldsymbol{\beta}\in\times,\, w\in\mathcal{W}} \frac{d^3}{2}a(\boldsymbol{\beta},\boldsymbol{\beta})+\frac{kd}{2}\|\boldsymbol{\beta}-\nabla w\|_0^2-d^3(f_w,w)-d^3(\mathbf{f}_\beta,\boldsymbol{\beta}), \tag{3.158}$$

where the body forces (or moments) have the form

$$f_w=\left(p-\rho g d-\rho d\left(\frac{\partial}{\partial t}+V\frac{\partial}{\partial x}\right)^2 w\right)/d^3, \tag{3.159}$$

$$\mathbf{f}_\beta=-\frac{\rho}{12}\left(\frac{\partial}{\partial t}+V\frac{\partial}{\partial x}\right)^2\boldsymbol{\beta}, \tag{3.160}$$

and the bending strain energy $\frac{d^3}{2}a(\boldsymbol{\beta},\boldsymbol{\beta})$, which includes the contribution of the normal stress $\sigma_T(y)$ due to the axial tension, is expressed by

$$\frac{d^3}{2}a(\boldsymbol{\beta},\boldsymbol{\beta})=\frac{d^3}{24}\int_\Omega \boldsymbol{\kappa}^T\mathbf{C}_\epsilon\boldsymbol{\kappa}\,d\Omega+\int_\Omega\frac{d\sigma_T(y)}{2}\left(\frac{\partial w}{\partial x}\right)^2 d\Omega. \tag{3.161}$$

Given the finite element subspaces $\Theta_h \subset \Theta$ and $W_h \subset \mathcal{W}$, the discretized problem may be written as

$$\begin{aligned}&\inf_{\boldsymbol{\beta}_h\in\Theta_h,\, w_h\in W_h}\frac{d^3}{2}a(\boldsymbol{\beta}_h,\boldsymbol{\beta}_h)+\frac{kd}{2}\|R\boldsymbol{\beta}_h-\nabla w_h\|_0^2\\&\quad -d^3(f_w,w_h)-d^3(\mathbf{f}_\beta,\boldsymbol{\beta}_h),\end{aligned} \tag{3.162}$$

where the linear operator $R$ satisfies $R\nabla w_h=\nabla w_h, \forall w_h\in W_h$ and $\|R\boldsymbol{\beta}_h\|_1 \leqslant c\|\boldsymbol{\beta}_h\|_1$, $\forall\boldsymbol{\beta}_h \in \Theta_h$, and the transverse shear strains $\boldsymbol{\gamma}_h$ are expressed as $kd^{-2}(R\boldsymbol{\beta}_h-\nabla w_h)$.

Notice that the shear strain energy is given by $\frac{kd}{2}\|R(\boldsymbol{\beta}_h-\nabla w_h)\|_0^2$ instead of $\frac{kd}{2}\|\boldsymbol{\beta}_h-\nabla w_h\|_0^2$. In essence, the successful selection of $R$ ensures the reliability of the mixed formulation [74,148]. In fact, the need for such a mixed formulation is reflected in Eqs. (3.152) to (3.154), where the shear terms are predominant just as bulk modulus for almost incompressible materials [143]. When the plate is moving, as occurs during a transport process or the rotation of blades, computer hard disks, and space structures, additional difficulties arise when the Coriolis force breaks the self-adjoint property and along with the centrifugal force, introduces destabilizing effects.

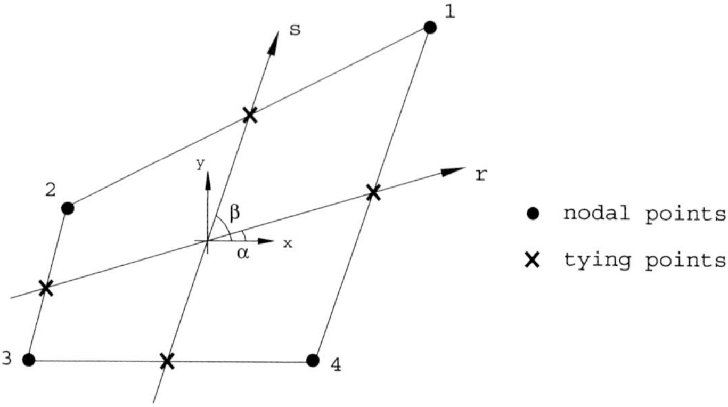


Figure 4.52. A general MITC4 plate element.

For the limiting case in which $d \to 0$, the limit $w$ corresponds to the Kirchhoff model and the variational problem assumes the form

$$a(\boldsymbol{\beta}_h, \bar{\boldsymbol{\beta}}_h) + (\boldsymbol{\gamma}_h, R\bar{\boldsymbol{\beta}}_h - \nabla \bar{w}_h) = (f_w, \bar{w}_h) + (\mathbf{f}_\beta, \bar{\boldsymbol{\beta}}_h),$$
$$\forall \bar{\boldsymbol{\beta}}_h \in \Theta_h, \ \forall \bar{w}_h \in W_h, \tag{3.163}$$
$$R\boldsymbol{\beta}_h = \nabla w_h. \tag{3.164}$$

In this formulation, mixed interpolated tensorial component (MITC) elements are employed. For a typical MITC4 element shown in Figure 4.52, the linear operator $R$ is selected in such a way that at the tying points (midpoints of each edge) we have $R\boldsymbol{\beta}_h = \boldsymbol{\beta}_h$ (Ref. [143]). Also as shown in Figure 4.52, the angles between the $r$ and $x$ axes and $s$ and $x$ axes are defined as $\alpha$ and $\beta$, and the shear strains are expressed as

$$\gamma_{xz} = \gamma_{rz} \sin\beta - \gamma_{sz} \sin\alpha, \tag{3.165}$$
$$\gamma_{yz} = -\gamma_{rz} \cos\beta + \gamma_{sz} \cos\alpha, \tag{3.166}$$

with

$$\gamma_{rz} = \lambda_1 \Bigg\{ (1+s)\left[\frac{w_1 - w_2}{2} - \frac{x_1 - x_2}{4}(\beta_x^1 + \beta_x^2) - \frac{y_1 - y_2}{4}(\beta_y^1 + \beta_y^2)\right]$$
$$+ (1-s)\left[\frac{w_4 - w_3}{2} - \frac{x_4 - x_3}{4}(\beta_x^4 + \beta_x^3) - \frac{y_4 - y_3}{4}(\beta_y^4 + \beta_y^3)\right]\Bigg\}$$
$$/8 \det \mathbf{J},$$

$$\gamma_{sz} = \lambda_2\Bigg\{(1+r)\left[\frac{w_1-w_4}{2} - \frac{x_1-x_4}{4}(\beta_x^1+\beta_x^4) - \frac{y_1-y_4}{4}(\beta_y^1+\beta_y^4)\right]$$
$$+ (1-r)\left[\frac{w_2-w_3}{2} - \frac{x_2-x_3}{4}(\beta_x^2+\beta_x^3) - \frac{y_2-y_3}{4}(\beta_y^2+\beta_y^3)\right]\Bigg\}$$
$$/8\det\mathbf{J},$$

$$\mathbf{J} = \begin{bmatrix} \partial x/\partial r & \partial y/\partial r \\ \partial x/\partial s & \partial y/\partial s \end{bmatrix},$$

$$\lambda_1 = \sqrt{(C_x + rB_x)^2 + (C_y + rB_y)^2},$$

$$\lambda_2 = \sqrt{(A_x + sB_x)^2 + (A_y + sB_y)^2};$$

$$A_x = x_1 - x_2 - x_3 + x_4, \qquad A_y = y_1 - y_2 - y_3 + y_4;$$
$$B_x = x_1 - x_2 + x_3 - x_4, \qquad B_y = y_1 - y_2 + y_3 - y_4;$$
$$C_x = x_1 + x_2 - x_3 - x_4, \qquad C_y = y_1 + y_2 - y_3 - y_4.$$

For the moving plates considered in this book, we use two built-in edge supports at $x = 0$ and $x = L$, and two free edges (zero shear forces, bending moments, and twisting moments) at $y = 0$ and $y = B$. For the mixed thin plate theory, general specific boundary considerations are discussed in Ref. [98]. Using the standard procedure of interpolating, and considering a typical finite element, we have

$$w = \mathbf{H}_w\hat{\mathbf{W}}, \qquad \frac{\partial w}{\partial x} = \mathbf{B}_w\hat{\mathbf{W}},$$
$$R\boldsymbol{\beta} - \nabla w = \mathbf{B}_{\gamma w}\hat{\mathbf{W}} + \mathbf{B}_{\gamma x}\hat{\boldsymbol{\beta}}_x + \mathbf{B}_{\gamma y}\hat{\boldsymbol{\beta}}_y,$$
$$\beta_x = \mathbf{H}_x\hat{\boldsymbol{\beta}}_x, \qquad \frac{\partial \beta_x}{\partial x} = \mathbf{B}_{x1}\hat{\boldsymbol{\beta}}_x, \qquad \frac{\partial \beta_x}{\partial y} = \mathbf{B}_{x2}\hat{\boldsymbol{\beta}}_x,$$
$$\beta_y = \mathbf{H}_y\hat{\boldsymbol{\beta}}_y, \qquad \frac{\partial \beta_y}{\partial x} = \mathbf{B}_{y1}\hat{\boldsymbol{\beta}}_y, \qquad \frac{\partial \beta_y}{\partial y} = \mathbf{B}_{y2}\hat{\boldsymbol{\beta}}_y,$$

where $\mathbf{H}_w$, $\mathbf{H}_x$, and $\mathbf{H}_y$ are the interpolation matrices for $w$, $\beta_x$, and $\beta_y$, respectively, and the solution vector is denoted as $\widehat{\mathbf{U}} = \langle \hat{\mathbf{W}}, \hat{\boldsymbol{\beta}}_x, \hat{\boldsymbol{\beta}}_y \rangle$.

The matrix equations derived from the standard Galerkin formulation yield

$$\mathbf{M}\ddot{\widehat{\mathbf{U}}} + \mathbf{C}\dot{\widehat{\mathbf{U}}} + \mathbf{K}\widehat{\mathbf{U}} = \mathbf{R}, \tag{3.167}$$

with

$$\mathbf{M} = \begin{bmatrix} \mathbf{M}_{ww} & \mathbf{0} & \mathbf{0} \\ \mathbf{0} & \mathbf{M}_{xx} & \mathbf{0} \\ \mathbf{0} & \mathbf{0} & \mathbf{M}_{yy} \end{bmatrix}, \qquad \mathbf{C} = \begin{bmatrix} \mathbf{C}_{ww} & \mathbf{0} & \mathbf{0} \\ \mathbf{0} & \mathbf{C}_{xx} & \mathbf{0} \\ \mathbf{0} & \mathbf{0} & \mathbf{C}_{yy} \end{bmatrix},$$

$$\mathbf{K} = \begin{bmatrix} \mathbf{K}_{ww} & \mathbf{K}_{wx} & \mathbf{K}_{wy} \\ \mathbf{K}_{xw} & \mathbf{K}_{xx} & \mathbf{K}_{xy} \\ \mathbf{K}_{yw} & \mathbf{K}_{yx} & \mathbf{K}_{yy} \end{bmatrix}, \qquad \mathbf{R} = \begin{bmatrix} \mathbf{R}_w \\ \mathbf{R}_x \\ \mathbf{R}_y \end{bmatrix},$$

and

$$\mathbf{K}_{xx} = \int_\Omega \frac{d^3}{12}\big(a_{11}\mathbf{B}_{x1}^T\mathbf{B}_{x1} + a_{33}\mathbf{B}_{x2}^T\mathbf{B}_{x2}\big)\,d\Omega + \int_\Omega kd\mathbf{B}_{\gamma x}^T\mathbf{C}_\gamma\mathbf{B}_{\gamma x}\,d\Omega - \int_\Omega \frac{\rho d^3 V^2}{12}\mathbf{B}_{x1}^T\mathbf{B}_{x1}\,d\Omega,$$

$$\mathbf{K}_{xy} = \int_\Omega \frac{d^3}{12}\big(a_{12}\mathbf{B}_{x1}^T\mathbf{B}_{y2} + a_{33}\mathbf{B}_{x2}^T\mathbf{B}_{y1}\big)\,d\Omega + \int_\Omega kd\mathbf{B}_{\gamma x}^T\mathbf{C}_\gamma\mathbf{B}_{\gamma y}\,d\Omega,$$

$$\mathbf{K}_{yx} = \int_\Omega \frac{d^3}{12}\big(a_{21}\mathbf{B}_{y2}^T\mathbf{B}_{x1} + a_{33}\mathbf{B}_{y1}^T\mathbf{B}_{x2}\big)\,d\Omega + \int_\Omega kd\mathbf{B}_{\gamma y}^T\mathbf{C}_\gamma\mathbf{B}_{\gamma x}\,d\Omega,$$

$$\mathbf{K}_{yy} = \int_\Omega \frac{d^3}{12}\big(a_{22}\mathbf{B}_{y2}^T\mathbf{B}_{y2} + a_{33}\mathbf{B}_{y1}^T\mathbf{B}_{y1}\big)\,d\Omega + \int_\Omega kd\mathbf{B}_{\gamma y}^T\mathbf{C}_\gamma\mathbf{B}_{\gamma y}\,d\Omega - \int_\Omega \frac{\rho d^3 V^2}{12}\mathbf{B}_{y1}^T\mathbf{B}_{y1}\,d\Omega,$$

$$\mathbf{K}_{ww} = \int_\Omega kd\mathbf{B}_{\gamma w}^T\mathbf{C}_\gamma\mathbf{B}_{\gamma w}\,d\Omega + \int_\Omega \big(\sigma_T(y)d - \rho d V^2\big)\mathbf{B}_w^T\mathbf{B}_w\,d\Omega,$$

$$\mathbf{R}_w = \int_\Omega (p - \rho g d)\mathbf{H}_w^T\,d\Omega,$$

$$\mathbf{K}_{xw} = \int_\Omega kd\mathbf{B}_{\gamma x}^T\mathbf{C}_\gamma\mathbf{B}_{\gamma w}\,d\Omega, \qquad \mathbf{K}_{yw} = \int_\Omega kd\mathbf{B}_{\gamma y}^T\mathbf{C}_\gamma\mathbf{B}_{\gamma w}\,d\Omega;$$

$$\mathbf{K}_{wx} = \int_\Omega kd\mathbf{B}_{\gamma w}^T\mathbf{C}_\gamma\mathbf{B}_{\gamma x}\,d\Omega, \qquad \mathbf{K}_{wy} = \int_\Omega kd\mathbf{B}_{\gamma w}^T\mathbf{C}_\gamma\mathbf{B}_{\gamma y}\,d\Omega;$$

$$\mathbf{C}_{xx} = \int_\Omega \frac{\rho d^3 V}{6}\mathbf{H}_x^T\mathbf{B}_{x1}\,d\Omega, \qquad \mathbf{C}_{yy} = \int_\Omega \frac{\rho d^3 V}{6}\mathbf{H}_y^T\mathbf{B}_{y1}\,d\Omega;$$

$$\mathbf{C}_{ww} = \int_\Omega 2\rho d V\mathbf{H}_w^T\mathbf{B}_w\,d\Omega, \qquad \mathbf{M}_{ww} = \int_\Omega \rho d\mathbf{H}_w^T\mathbf{H}_w\,d\Omega;$$

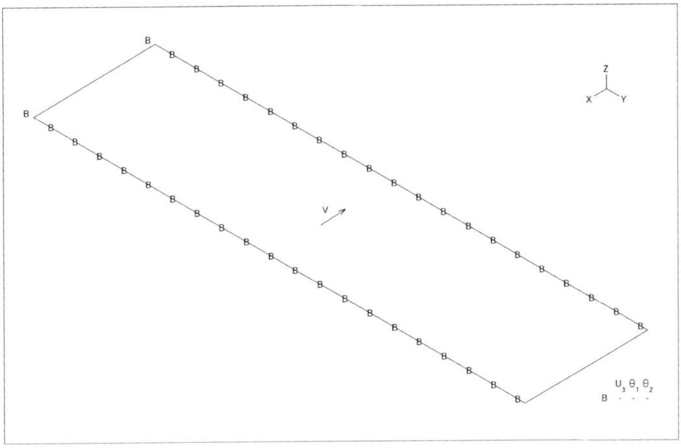


Figure 4.53. An orthotropic plate moving between two supports.

$$\mathbf{M}_{xx} = \int_{\Omega} \frac{\rho d^3}{12} \mathbf{H}_x^T \mathbf{H}_x \, d\Omega, \qquad \mathbf{M}_{yy} = \int_{\Omega} \frac{\rho d^3}{12} \mathbf{H}_y^T \mathbf{H}_y \, d\Omega;$$

and the contributions from distributed or concentrated moments in $x$- and $y$-directions are included in $\mathbf{R}_x$ and $\mathbf{R}_y$.

For illustration purposes, we analyze a typical orthotropic plate, namely a $2 \times 8$ m paper sheet as depicted in Figure 4.53 with the following physical parameters: $\rho = 700\ \text{kg/m}^3$; $d = 0.7$ mm; $L = 2$ m; $B = 8$ m; $E_x = 7.44$ GPa; $E_y = 3.47$ GPa; $\nu_{xy} = 0.149$; $G_{xy} = 2.04$ GPa; $G_{xz} = 0.099$ GPa; $G_{yz} = 0.137$ GPa. To evaluate the critical axial velocity of the moving plate, we have computed the first natural frequency $\omega$ with the proposed finite element formulation. Figure 4.54 shows that when the orthotropic plate represented with $5 \times 20$ MITC4 elements is not under pretension, the plate bending stiffness can sustain an axial velocity up to 2 m/s. In general, the first natural frequency (flexural mode) decreases continuously with increasing axial velocity, and the lower the first natural frequency, the more susceptible the plate is to flutter. Of course, as the tension increases, the critical velocity increases as well.

Table 4.3 lists the results for the first five natural frequencies for $5 \times 20$ MITC4 plate elements and zero traveling velocity. As can be seen, the proposed formulation with zero axial velocity agrees exactly with the reliable Mindlin/Reissner plate theory with MITC elements in ADINA. We note that the natural frequencies computed here include the effects of transverse shear. The natural frequency approximations for a stationary orthotropic thin plate which conforms to the Kirchhoff assumptions are available in Ref. [247].

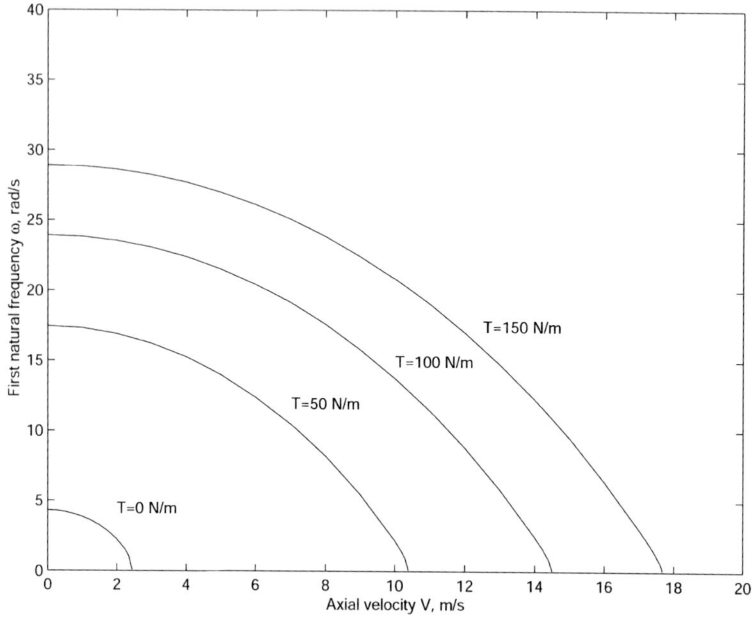


Figure 4.54. First natural frequency vs. axial velocity.

Table 4.3 Frequency comparison with ADINA for the zero traveling velocity case

| Mode # | Frequencies (rad/s) | |
|---|---|---|
| | ADINA | This work |
| 1 | 4.304 | 4.304 |
| 2 | 4.335 | 4.335 |
| 3 | 4.440 | 4.440 |
| 4 | 4.631 | 4.631 |
| 5 | 4.932 | 4.932 |

Figure 4.57 shows the first two natural frequency modes for the plate represented with $5 \times 20$ MITC4 elements moving at different velocities. In the transient dynamics analysis, we demonstrate the flexural wave propagation in the moving plate due to an impact load calculated with the proposed formulation. The plate is subjected to an impulse force $2f(t)$ applied at its center ($x = 1$ m, $y = 4$ m). The corresponding impulse time function $f(t)$ is shown in Figure 4.55. The analysis is carried out for 60 time steps with step size $\Delta t = 0.02$ s. The transient response of the plate center point is presented in Figure 4.56, and the MD (axial moving direction or machine direction) and CD (cross-sectional direction) dis-

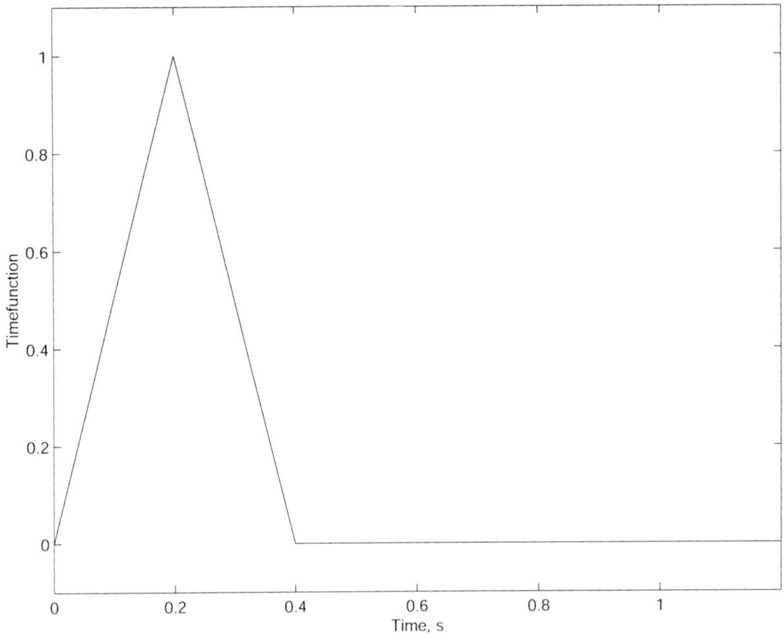


Figure 4.55. Impulse load time function $f(t)$.

placement profiles are shown in Figures 4.58 and 4.59, respectively. Once again, the transient response for a plate with zero velocity matches exactly with ADINA results.

Results at $t = 1.2$ s of the orthotropic plate represented with $10 \times 40$ MITC4 elements moving at different velocities in Figures 4.60 to 4.62 clearly show the Coriolis and centrifugal effects associated with axial motion on the out-of-plane displacement, $x$-rotation, and $y$-rotation. Moreover, numerical results for zero moving velocity cases are identical to ADINA results. No shear locking is observed in the five stress band plots in Figures 4.63 to 4.67. The stress bands in all cases are quite smooth, indicating accurate solutions. These sets of five stress band plots had not been obtained prior to the development of the mixed formulation presented here for the moving orthotropic thin plate.

The proposed mixed formulation based on the Mindlin/Reissner theory for a moving orthotropic plate is a natural extension of the reliable mixed formulation for stationary plates and shells. We consider the numerical tests used in this work to be comprehensive, and the natural frequencies and mode shapes, along with the five stress band plots, appear to be sufficient to demonstrate the accuracy and reliability of the proposed formulation. The employed MITC4 plate elements are demonstrated numerically to be effective for use in the analysis of a moving or-

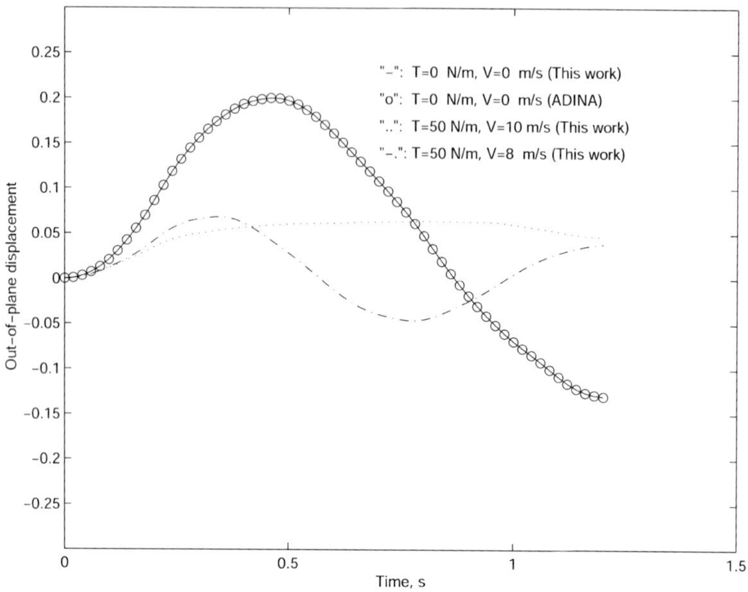


Figure 4.56. Out-of-plane displacement at the plate center.

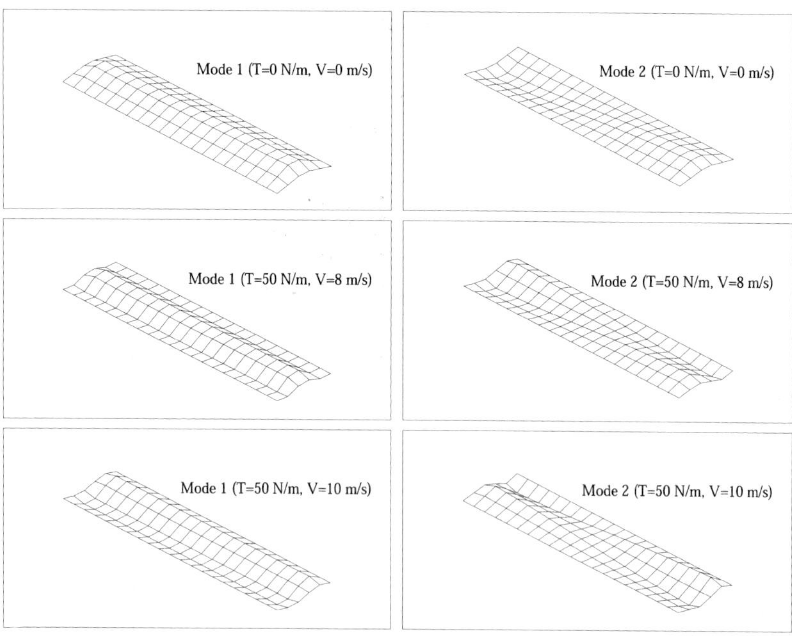


Figure 4.57. First two mode shapes of out-of-plane displacements.

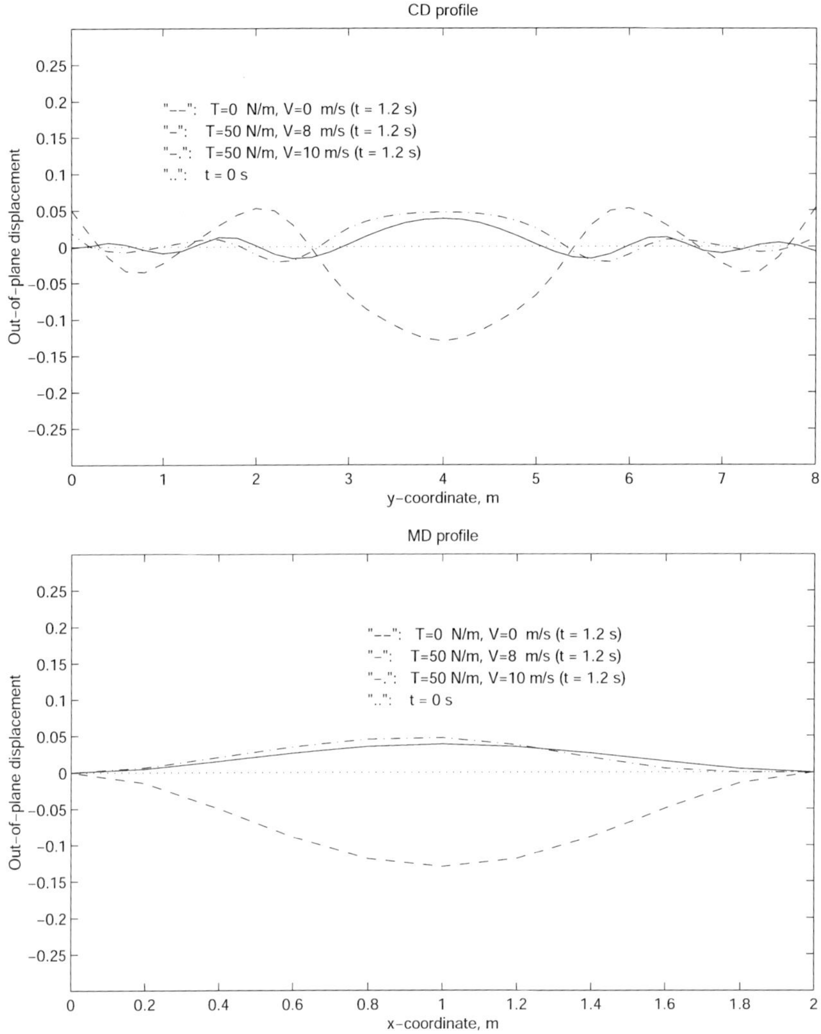


Figure 4.58. CD and MD out-of-plane displacement profiles for different axial velocities.

thotropic plate and other analogous problems. We expect other MITC elements to perform similarly. A main attribute of this work is the fact that critical axial velocity as well as normal and shear stress waves can be predicted for any orthotropic thin plate moving at various velocities.

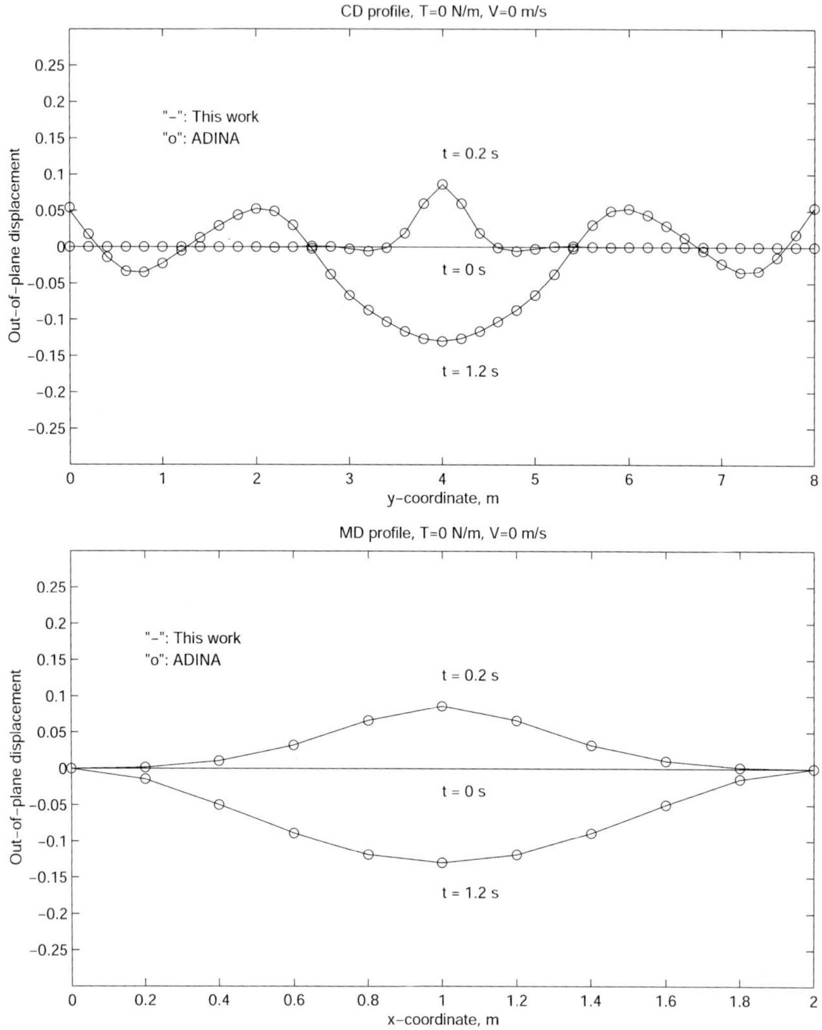


Figure 4.59. CD and MD out-of-plane displacement profiles for zero traveling velocity cases.

## *4.3.2. Bending and torsion; aeroelasticity models*

There is no beginning point more appropriate for the discussion of aeroelasticity than the study of instability modes of a cantilever wing. In practice, flutter of a cantilever wing often involves a combination of torsion and displacement modes; while galloping often happens to a single mode. Essentially, each cross

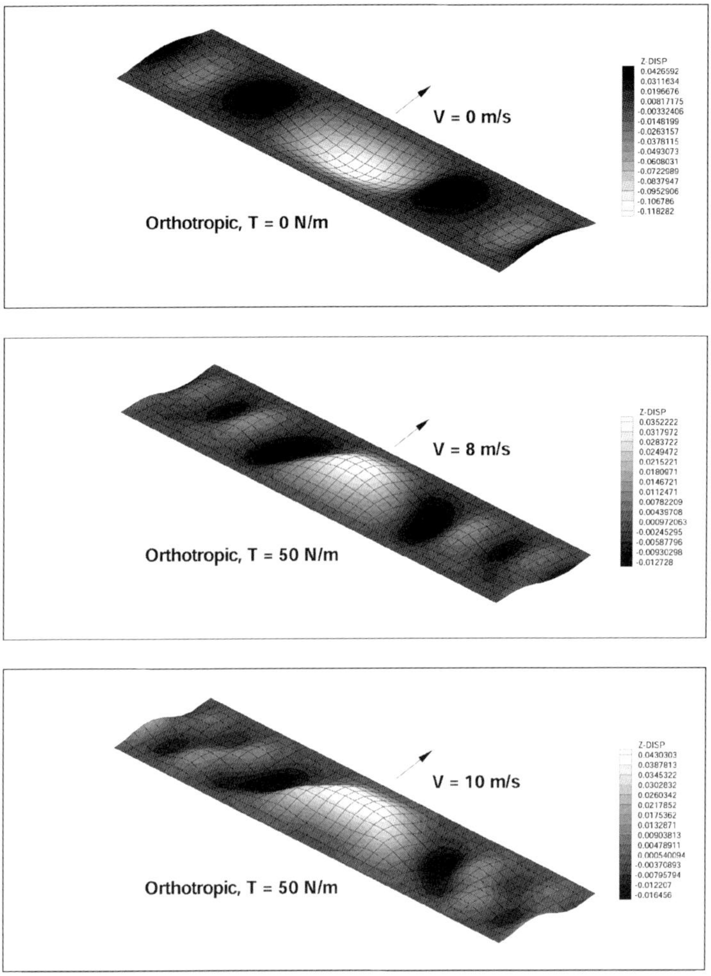


Figure 4.60. Out-of-plane displacement ($w$) band plots.

section of such a structure can be considered as a lifting surface, on which static or dynamic lifting force, drag force, and moment (torque) are exerted due to the surrounding air motion. Naturally, the force and moment introduced through the coupling between the flexible structure and the surrounding flow must also be dependent on the motion of the structure. The traditional way to simplify this fully coupled fluid–solid system is to separate the fluid and structure domains. In other words, the force and moment are computed based on the coupling of a lifting surface, which is discussed in detail in Chapter 5, whereas the static and

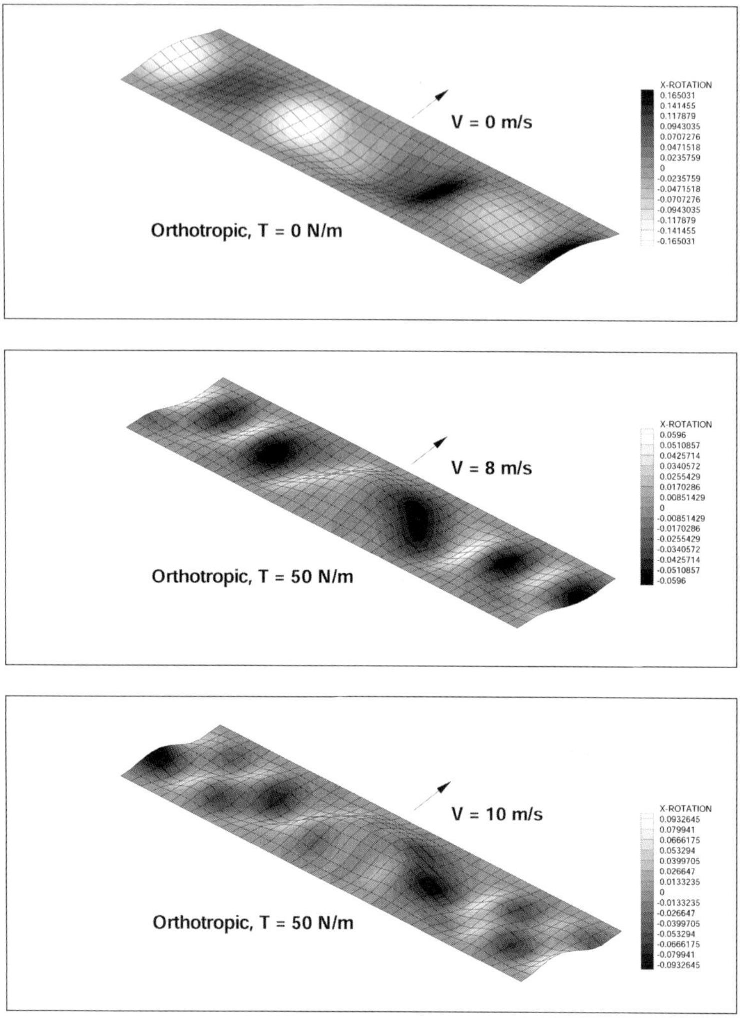


Figure 4.61. $X$-direction rotation ($\theta_x$) band plots.

dynamic behaviors of the structure will be considered based on the coupled fluid force and moment, acting as external loading. The study of the flexible structure is further simplified to elemental structural components, such as beams and shafts of thin-wall structures. Consequently, the governing equation of the flexible structure coupled with the surrounding fluid is either solved analytically or numerically. In most cases, light flexible wing-like structure involves thin-wall cross-sections. Before we simplify these types of FSI systems to one- or two-

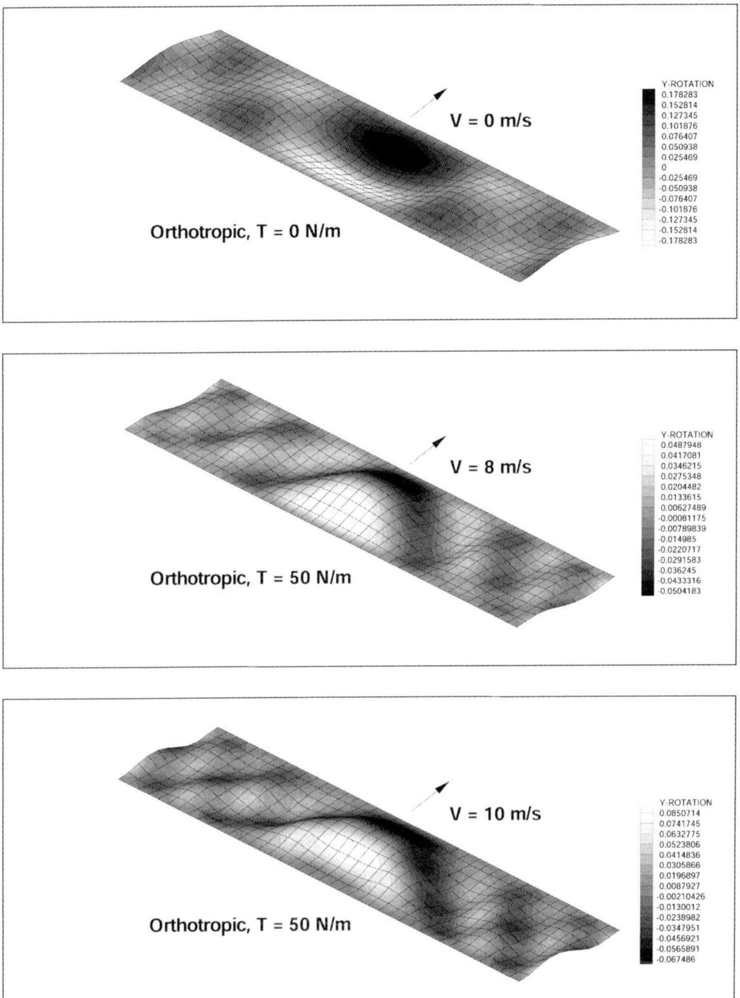


Figure 4.62. $Y$-direction rotation ($\theta_y$) band plots.

dimensional models, we would like to present some fundamental stress analysis of thin structures.

*Bending and torsion*

Consider first an open thin-wall section. The simplest form of such a structure is a beam with rectangular cross sections, as shown in Figure 4.68. In accor-

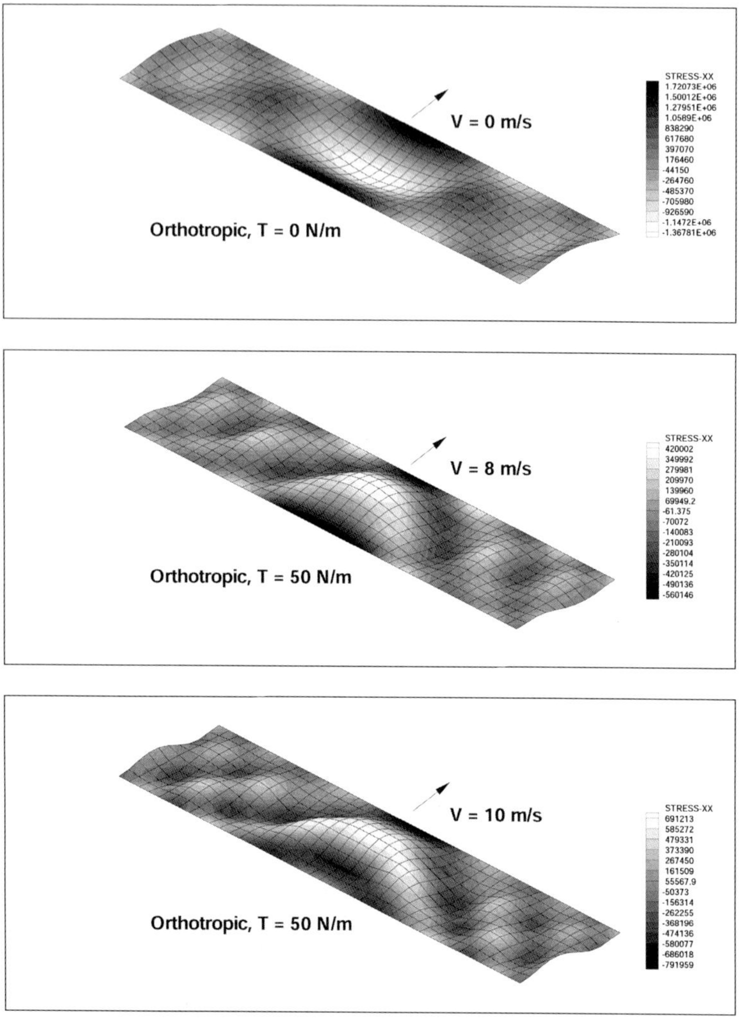


Figure 4.63. $X$-direction normal stress ($\sigma_{xx}$) band plots.

dance with convention in the study of lifting surfaces, we assign $z$-direction as the axial direction of the beam and $y$-direction as the transverse direction. For this thin structure, consider that the $y$-directional thickness $h$ is still small compared with $x$-directional width $b$. Of course, for the beam theory to hold, the length of the structure $l$ is dominant with respect to both $h$ and $b$, as shown in Figure 4.68. Thus, the bending is along $x$-direction, or the weaker bending rigidity direction. Assume we have the normal stress compo-

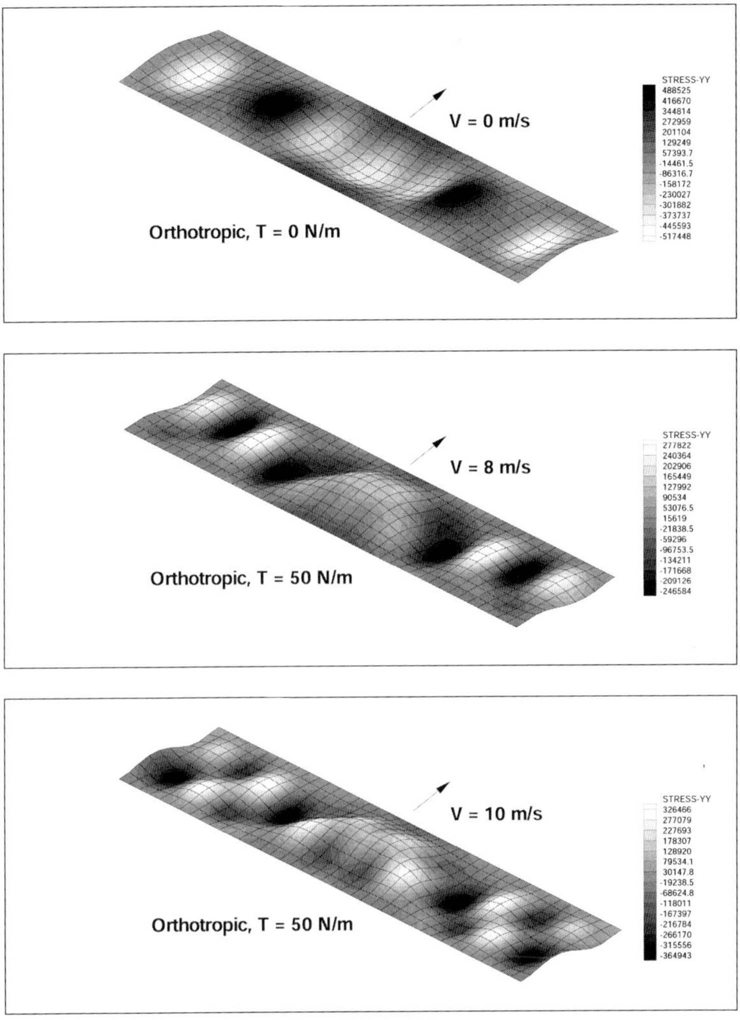


Figure 4.64. $Y$-direction normal stress ($\sigma_{yy}$) band plots.

nents $\sigma_{xx} = 0$ and $\sigma_{yy} = 0$, and assign the $z$-axis along the neutral axis or plane. Further denoting the positive transverse deflection or displacement $v$ as upwards, the positive bending moment on the right-hand side of the differential element $dz$ as depicted in Figure 4.68 is clockwise and the bending of the beam is convex, from which the curvature can be approximated in small displacement analysis as $\kappa = 1/\rho = -\partial^2 v/\partial z^2$. Consequently, we

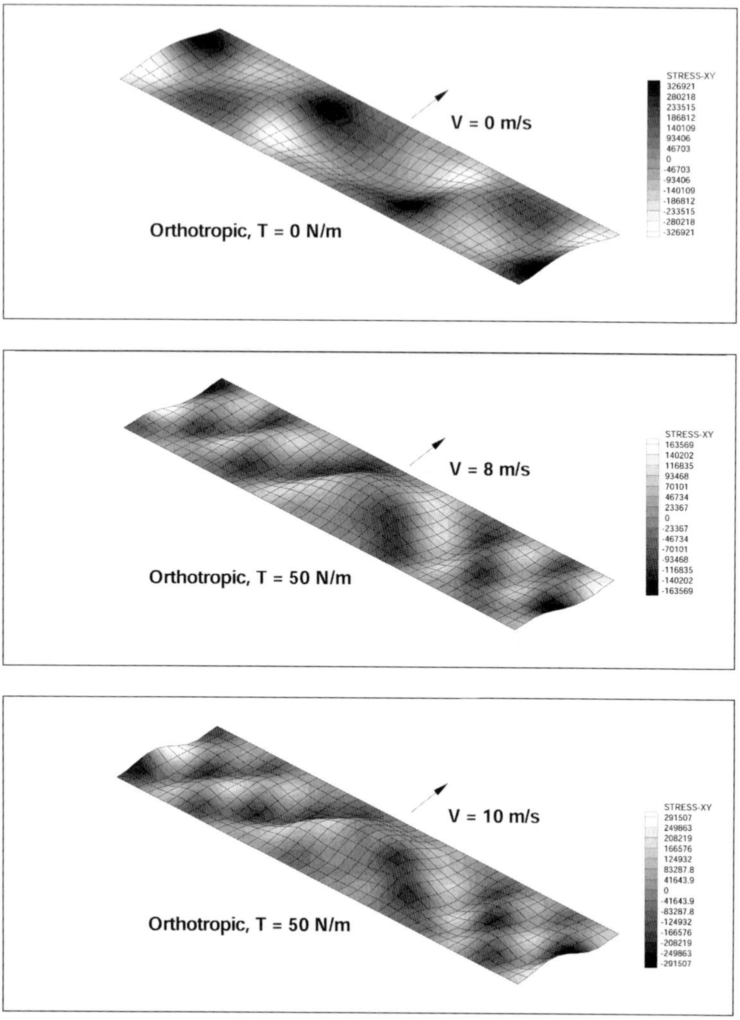


Figure 4.65. In-plane shear stress ($\sigma_{xy}$) band plots.

have

$$\sigma_{zz} = E\epsilon_{zz} = E\frac{(\rho + y)\,d\phi - \rho\,d\phi}{\rho\,d\phi} = \frac{Ey}{\rho} = -Ey\frac{\partial^2 y}{\partial z^2}, \tag{3.168}$$

where $\rho$ stands for the radius of curvature and $d\phi$ represents the differential rotation as shown in Figure 4.68.

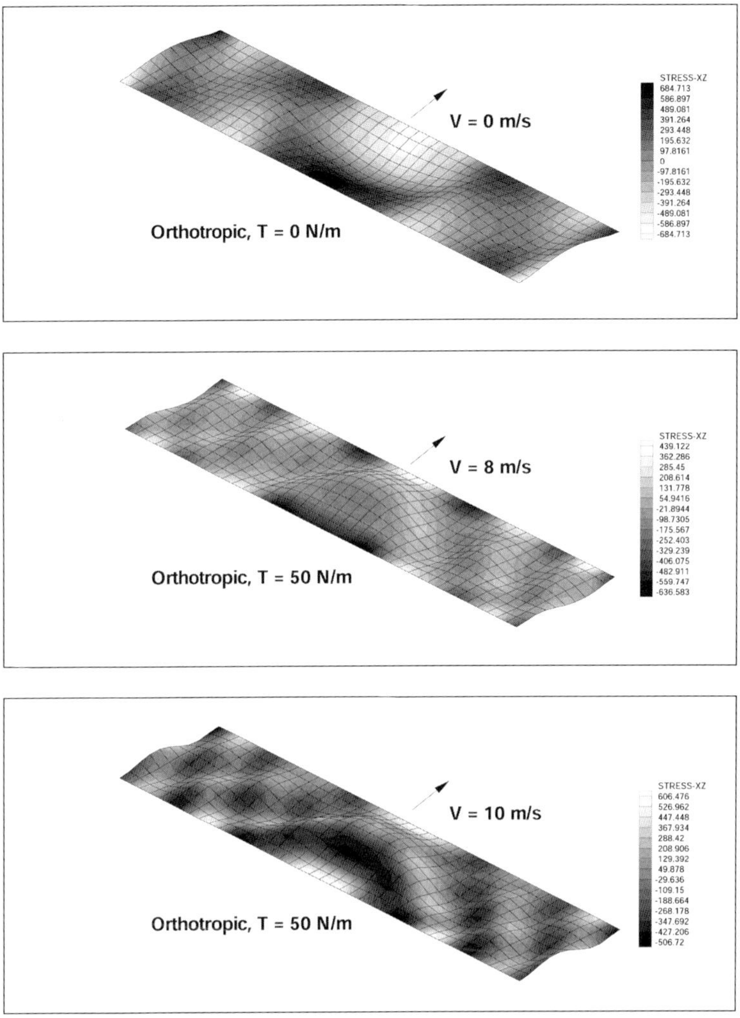


Figure 4.66. Transverse shear stress ($\sigma_{xz}$) band plots.

From Eq. (3.168), we have

$$M_x = \int_A y\sigma_{zz}\,dA = -E\frac{\partial^2 v}{\partial z^2}\int_A y^2\,dA = -EI_{xx}\frac{\partial^2 v}{\partial z^2}, \tag{3.169}$$

where $I_{xx}$ represents the area moment of inertia in $x$-direction.

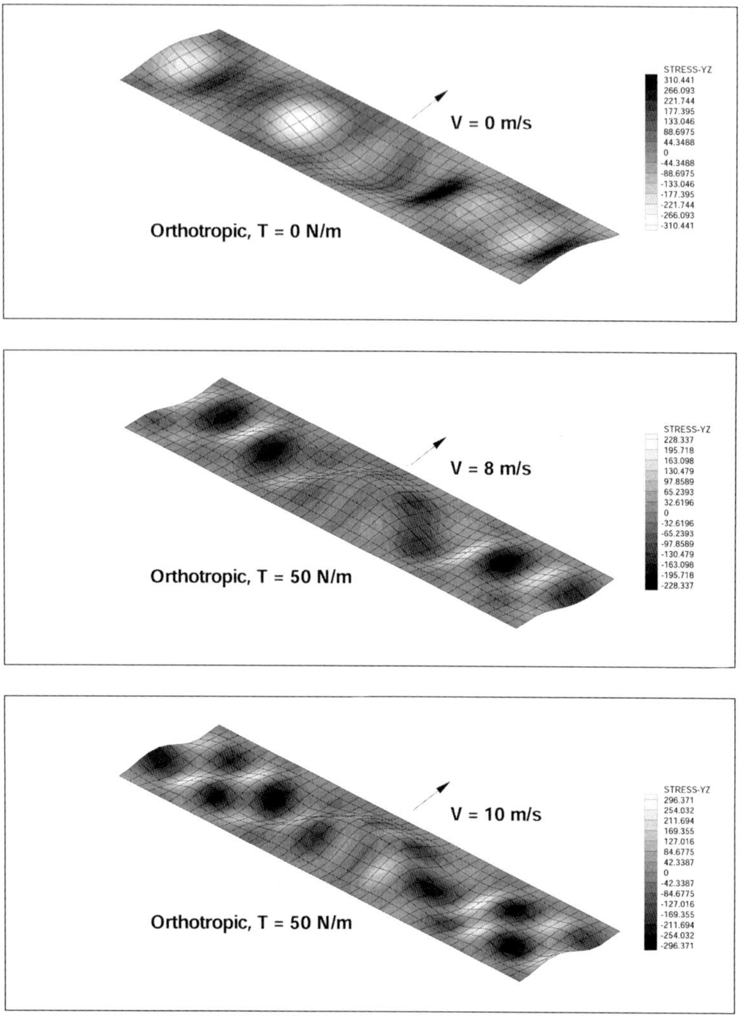


Figure 4.67. Transverse shear stress ($\sigma_{yz}$) band plots.

Substituting Eq. (3.169) into Eq. (3.168), we derive

$$\sigma_{xx} = M_x y / I_{xx}. \tag{3.170}$$

Finally, from the force and moment equilibria of the differential element $dz$, as depicted in Figure 4.68, we have

$$p = -\frac{\partial Q}{\partial z}, \tag{3.171}$$

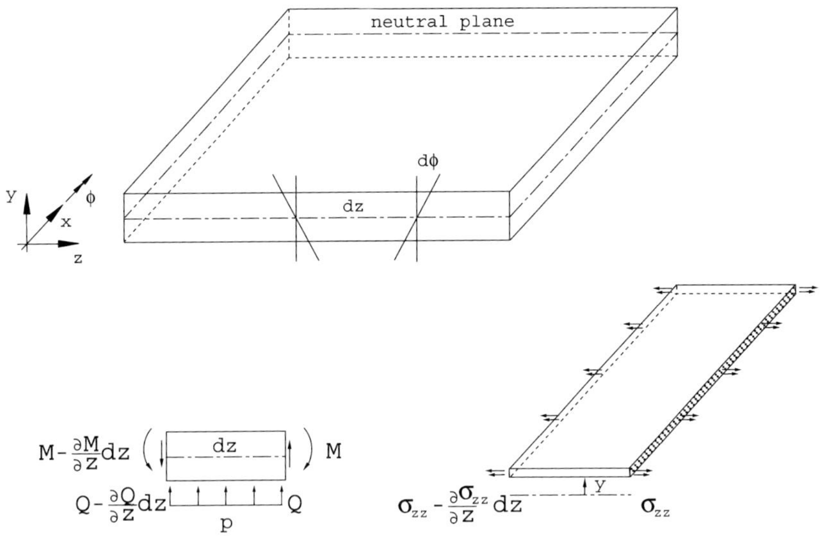


Figure 4.68. Geometrical definitions of a beam with thin-wall cross sections.

$$Q = \frac{\partial M}{\partial z}, \tag{3.172}$$

where $Q$ stands for the transverse shear force and $p$ represents the distributed force, which includes the inertia force.

Note that in accordance with convention, the shear force $Q$ is positive in the positive $y$-direction on the right-hand side of the differential element, and negative on the left-hand side. In addition, the distributed force $p$ is positive in the positive $y$-direction, the same as the transverse displacement. Therefore, combining Eqs. (3.171) and (3.172), we have

$$\frac{\partial^2}{\partial z^2}\left(EI_{xx}\frac{\partial^2 v}{\partial z^2}\right) = p. \tag{3.173}$$

For a cantilever wing as shown in Figure 4.69, we can easily integrate Eq. (3.173) and derive the effective stiffness for the flexible structure. Considering the force equilibrium of the upper part of the differential element also depicted in Figure 4.68, using Eqs. (3.170) and (3.172), the transverse shear stress $\sigma_{zy}$ can be expressed as

$$\sigma_{zy} = \frac{Q}{bI_{xx}}\int_{y}^{h/2} y\,dA. \tag{3.174}$$

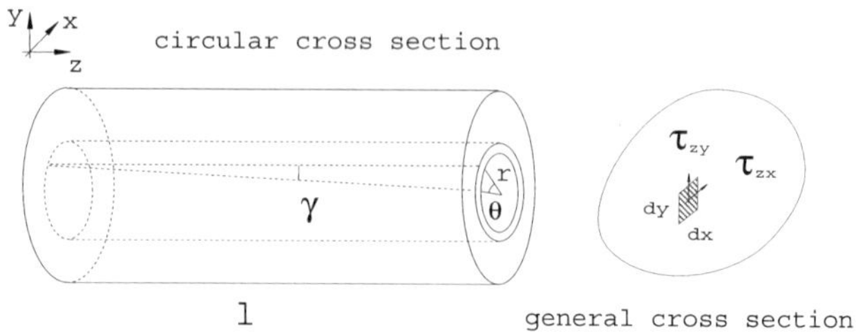


Figure 4.69. A prismatic shaft with circular and general cross sections.

In practice, the structure can bend with respect to both $y$- and $x$-axes, and such asymmetrical bending can be dealt with using the following three equations:

$$\int_A \sigma_{zz}\, dA = 0, \tag{3.175}$$

$$\int_A \sigma_{zz} y\, dA = M_x, \tag{3.176}$$

$$\int_A \sigma_{zz} x\, dA = -M_y, \tag{3.177}$$

which can be used to calculate the three constants of the axial normal stress expressed in the following $P^1$ form:

$$\sigma_{zz} = a + bx + cy. \tag{3.178}$$

From Eq. (3.175), it is clear that as long as we assign the $z$-axis to the center of geometry, the constant $a$ is equal to zero. Furthermore, if we rotate the $x$- and $y$-axes to a special position such that $I_{xy} = \int_A xy\, dA = 0$, namely the principal plane, the bending can be resolved into components lying in each of the two principal axes.

If the beam is also subjected to axial forces, namely, $\sigma_{zz} \neq 0$, within the linear range we can use the superposition principle to include the effect of normal stress. In particular, the right-hand side of Eq. (3.175) will no longer be zero. It will instead represent the total axial force. Likewise, the constant $a$ in Eq. (3.178) will be the average normal stress instead of zero.

For cantilever wing structures, the bending rigidity provides the stiffness for translational degrees of freedom. Similarly, the torsional rigidity provides the stiffness for rotational degrees of freedom. To start the discussion on torsion, we consider a circular shaft with radius $R$, with its axis oriented along the $z$-axis, as shown in Figure 4.70. It is obvious that the shear stress flow for the circular cross

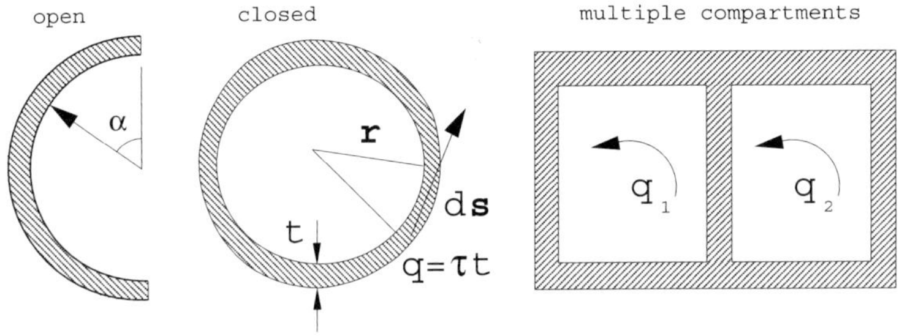


Figure 4.70. Various cross sections of thin-wall structures.

section should be tangential and the torque $M_{zz}$ can be expressed as

$$M_{zz} = \int_A r\tau \, dA = \int_A rG\gamma \, dA = \int_0^R 2\pi r^2 G\gamma \, dr, \tag{3.179}$$

where $\gamma$ and $\tau$ represent shear strain and tangential shear stress, respectively, and $G$ is the shear modulus.

As depicted in Figure 4.69, the tangential shear strain $\gamma$ is equivalent to the angle of twist per unit length of the shaft, multiplied by the radius expressed as

$$\gamma = \frac{r\,d\theta}{dl} = \frac{r\theta}{l}, \tag{3.180}$$

assuming there is no distributed torque within $z \in (0, l)$.

Therefore, from Eq. (3.179), the torque is calculated as

$$M_{zz} = G\frac{\theta}{l}J, \tag{3.181}$$

with $J = \pi R^4/2$, where $J$ stands for the polar moment of inertia and is expressed as $J = I_{xx} + I_{yy}$.

In general, for a primatic shaft with a general cross section, with the Saint-Venant assumptions we have for any cross section along the $z$-axis

$$u = -\frac{\theta}{l}zy, \qquad v = \frac{\theta}{l}zx, \qquad w = \frac{\theta}{l}\psi(x, y), \tag{3.182}$$

where $\psi(x, y)$ is the so-called warping function.

Eq. (3.182) yields the only two nonzero strain terms:

$$\begin{aligned} \gamma_{zx} &= \frac{\theta}{l}\left(\frac{\partial\psi}{\partial x} - y\right), \\ \gamma_{zy} &= \frac{\theta}{l}\left(\frac{\partial\psi}{\partial y} + x\right). \end{aligned} \tag{3.183}$$

Consequently, the shear stress components are expressed as $\sigma_{zx} = G\gamma_{zx}$ and $\sigma_{zy} = G\gamma_{zy}$. Furthermore, directly derived from the force equilibrium, we have

$$\frac{\partial \sigma_{zy}}{\partial y} + \frac{\partial \sigma_{zx}}{\partial x} = 0. \tag{3.184}$$

Therefore, we can introduce the so-called Prandtl stress function $\phi(x, y)$, named after Ludwig Prandtl (1875–1953), such that

$$\sigma_{zx} = \frac{\partial \phi}{\partial y}, \qquad \sigma_{zy} = -\frac{\partial \phi}{\partial x}. \tag{3.185}$$

Combining Eqs. (3.183) and (3.185), we have the following Prandtl membrane analogy:

$$\nabla^2 \phi = -2G\frac{\theta}{l}. \tag{3.186}$$

If we designate the boundary unit normal vector as $(dy/ds, -dx/ds)$, where $ds$ is the differential arc length, the no traction boundary condition yields

$$\frac{d\phi}{ds} = 0, \tag{3.187}$$

which means that the Prandtl stress function is constant and often set at zero, along the boundary of the cross section.

Thus, employing integrating by parts, the torque is calculated as,

$$M_{zz} = \int_A (x\sigma_{zy} - y\sigma_{zx})\, dA = 2\int_A \phi\, dA. \tag{3.188}$$

Therefore, using Eq. (3.186) and the boundary condition (3.187), we can solve for $\phi$ and relate $\phi$ to the twist angle per unit length $\theta/l$. For instance, consider the stress function $\phi$ expressed as

$$\phi = C\left(\frac{x^2}{a^2} + \frac{y^2}{b^2} - 1\right), \tag{3.189}$$

where $C$ is a constant to be determined.

The stress function in (3.189) obviously satisfies the boundary condition (3.187) of the elliptic cross section with a major $a$ and minor $b$ of a primatic bar. Therefore, substituting Eq. (3.189) into Eq. (3.186), we derive

$$C = -\frac{a^2 b^2 G}{a^2 + b^2}\frac{\theta}{l}, \tag{3.190}$$

and from Eq. (3.188), we obtain

$$M_{zz} = \frac{\pi a^3 b^3 G}{a^2 + b^2}\frac{\theta}{l}. \tag{3.191}$$

Table 4.4 Torsion of prismatic bar with rectangular cross section

| b/a | 1 | 2 | 4 | 6 | 8 | 10 | $\infty$ |
|---|---|---|---|---|---|---|---|
| $\beta_1$ | 0.208 | 0.246 | 0.282 | 0.299 | 0.307 | 0.312 | 0.333 |
| $\beta_2$ | 0.208 | 0.309 | 0.378 | 0.402 | 0.414 | 0.421 | N/A |
| $\beta_3$ | 0.141 | 0.229 | 0.281 | 0.299 | 0.307 | 0.312 | 0.333 |

Furthermore, we have the shear stress components expressed as

$$\begin{aligned}\sigma_{zx} &= \frac{\partial\phi}{\partial y} = -\frac{2a^2Gy}{a^2+b^2}\frac{\theta}{l},\\ \sigma_{zy} &= -\frac{\partial\phi}{\partial x} = \frac{2b^2Gx}{a^2+b^2}\frac{\theta}{l}.\end{aligned} \tag{3.192}$$

For a special case of a circular cross section, i.e., $a = b = R$, Eq. (3.191) is identical to (3.181). However, for rectangular cross sections, numerical procedures or analytical procedures involving Fourier series must be used to solve Poisson's equation (3.186). In engineering practice, assuming $b > a$, we rewrite Eqs. (3.191) and (3.192) as

$$\tau_A = \frac{M_{zz}}{\beta_1 ba^2}, \qquad \tau_B = \frac{M_{zz}}{\beta_2 ba^2}, \qquad \frac{\theta}{l} = \frac{M_{zz}}{\beta_3 Gba^3}, \tag{3.193}$$

where $A$ and $B$ stand for the middle points at the long and short sides, respectively, and $\beta_1$, $\beta_2$, and $\beta_3$ represent the constants based on the numerical solutions.

It is obvious from Table 4.4 that the maximum shear stress happens for a square cross section, i.e., $b/a = 1$, and the minimum value of the maximum shear stress within the rectangular cross section happens for an infinite long thin-wall cross section, i.e., $b/a = \infty$. Also as illustrated in Table 4.4, the maximum and minimum twist angle per unit length correspond to a square and an infinite long thin-wall cross section, respectively.

At this point, we are almost ready to simplify the thin-wall structure for flow-induced vibration analysis. We still have to discuss the concept of shear center, an important concept with respect to bending and torsion of cantilever wings. Eq. (3.174) will then be used to calculate shear centers of open thin-wall structures.

In essence, the shear center is the fictitious point within the cross section of the beam at which the applied shear force will not introduce a torque or an axial direction moment $M_{zz}$. In our discussion of beam bending, we mention that the neutral axis must pass the center of geometry, or centroid. Thus, it is particularly important to calculate the shear center if the cross section has only one axis of symmetry, and the shear center does not coincide with the centroid.

Note that in the calculation of the shear center of thin-wall structures, it is convenient to define shear flow $q$ as $\tau t$, where $t$ is the local wall thickness. Consider, for example, a semicircular open thin-wall cantilever wing as depicted in Figure 4.70. Employing Eq. (3.174) and assuming the shear force $Q$ is in the positive $y$-direction, the shear flow $q$ at the polar angle $\alpha$ can be calculated as follows:

$$q = \frac{Q}{I_{xx}} \int_0^\alpha t R^2 \cos\alpha \, d\alpha = \frac{Q R^2 t \sin\alpha}{I_{xx}}, \tag{3.194}$$

where $R$ is the radius of the semicircular cross section, and the area moment of inertia $I_{xx}$ is derived as

$$I_{xx} = \int_0^\pi (R\cos\alpha)^2 Rt \, d\alpha = \frac{\pi R^3 t}{2}. \tag{3.195}$$

According to the symmetry of the semicircular cross section, the shear center must be located along the $x$-axis. As depicted in Figure 4.70, we denote $e$ as the distance from the shear center to the center of the circle $o$, which can then be evaluated by the moment equilibrium about $o$,

$$e = \frac{1}{Q} \int_0^\pi q R^2 \, d\alpha = \frac{4R}{\pi} > R. \tag{3.196}$$

Eq. (3.196) shows that the shear center of the semi-circle is in fact outside the circle. If the thin-wall cantilever wing has a close cross-section as shown in Figure 4.70, the calculation of the shear flow and shear center will be quite different. The easiest way to derive the torque and shear flow relation is to use the following:

$$M_{zz}\mathbf{k} = \oint \mathbf{r} \times \boldsymbol{\tau} \, dA = q \oint \mathbf{r} \times d\mathbf{s} = 2qA\mathbf{k}, \tag{3.197}$$

where $\boldsymbol{\tau}$ is the shear stress or traction in the tangential direction, $\mathbf{k}$ represents the unit normal vector in the axial or $z$-direction, and $A$ stands for the total cross-sectional area enclosed by the thin-wall structure.

Of course, the relation (3.197) can be also derived from the Prandtl membrane analogy. In addition, to obtain the twist angle and torque relation, it is the best to use the Castigliano's theorem, named after Carlo Alberto Castigliano (1847–1884). Considering the cantilever wing with a uniform cross section and an axial length $l$ and employing Eq. (3.197), the total strain energy can be established as

$$U = l \oint \frac{\tau^2}{2G} t \, ds = \frac{M_{zz}^2 l}{8GA^2} \oint \frac{ds}{t}. \tag{3.198}$$

Therefore, using Castigliano's theorem, and employing Eq. (3.197), the angle of twist per unit length can be evaluated as

$$\frac{\theta}{l} = \frac{q}{2GA} \oint \frac{ds}{t}. \tag{3.199}$$

Furthermore, for thin-wall structures with $N$ compartments, we can solve for $q_i$, with $i = 1, 2, \ldots, N$ and $M_{zz}$ using the following $N + 1$ equations, which are derived from the fact that the total torque is contributed by all $N$ compartments and each compartment shares the same twist angle per unit length:

$$M_{zz} = 2 \sum_{i=1}^{N} A_i q_i, \tag{3.200}$$

$$\frac{\theta}{l} = \frac{1}{2GA_i} \left( \oint_i \frac{q_i}{t}\, ds - \sum \int_{l_k} \frac{q_k}{t}\, ds \right), \tag{3.201}$$

where $q_k$ represents the shear flow in the compartments adjacent to the $i$th compartment.

Here we must be particularly careful about the calculation of the shear center of the closed thin-wall cross section. Although the procedure still holds, we must first introduce a fictitious cut and release an additional unknown of the shear flow, as if the thin-wall cross section is under torsion. For clarity of presentation, we consider a thin-wall rectangular cross section as shown in Figure 4.71, with width $a$ and height $b$, and wall thickness $t$, except for the left side thickness $2t$. Since we can introduce any amount of total shear force which is ultimately irrelevant to the calculation of the shear center, an intrinsic geometrical property, we determine that the shear force for the left and right parts $Q_l$ and $Q_r$ are equivalent to their corresponding area moments of inertia $I_l$ and $I_r$. According to symmetry and the concept of shear flow, the axial moments or torques contributed by the upper and lower horizontal webs are identical. Assigning $q_w$, $q_{fl}$, and $q_{fr}$ as the shear flow within the upper web, left flange, and right flange, respectively, and $q_a$ as the additional shear flow released from the cut at the upper right corner, we have

$$q_w = \frac{b}{2} t x, \quad \text{with } 0 \leqslant x \leqslant a, \tag{3.202}$$

$$q_{fl} = \frac{b}{2} t a + \left( \frac{b^2}{4} - y^2 \right) t, \quad \text{with } -\frac{b}{2} \leqslant y \leqslant \frac{b}{2}, \tag{3.203}$$

$$q_{fr} = \left( \frac{b^2}{8} - \frac{y^2}{2} \right) t, \quad \text{with } -\frac{b}{2} \leqslant y \leqslant \frac{b}{2}. \tag{3.204}$$

Note here that we denote the introduced shear forces upwards, or in the positive $y$-direction, and the released shear flow counterclockwise. Therefore, the shear

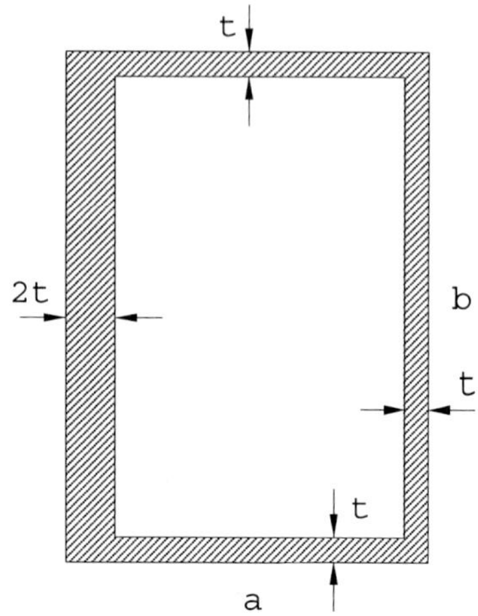


Figure 4.71. A thin-wall box beam.

flow introduced by the shear force in the left part is clockwise, and the shear flow in the right part is upwards. The whole idea here is to locate the shear force at the shear center such that there is no torsion or twist. Hence, we can use Eq. (3.199) to obtain the additional unknown $q_a$,

$$2\int_0^a \frac{q_w}{t}\,dx + \int_{-b/2}^{b/2} \frac{q_{fl}}{2t}\,dy - \int_{-b/2}^{b/2} \frac{q_{fr}}{t}\,dy - \frac{q_a}{t}(2a+b) - \frac{q_a}{2t}b = 0, \tag{3.205}$$

from which we have $q_a = abt(2a+b)/2(4a+3b)$.

Finally, employing the axis moment balance about the lower right corner, the shear center is located as

$$\begin{aligned} e &= \frac{1}{I_l + I_r}\left[2\int_0^a (q_w - q_a)b\,dx + \int_{-b/2}^{b/2} (q_{fl} - q_a)a\,dy\right] \\ &= \frac{a}{3}\frac{12a^2 + 23ab + 6b^2}{(4a+3b)(2a+b)}, \end{aligned} \tag{3.206}$$

with the area moments of inertia $I_l = b^3t/6 + ab^2t/2$ and $I_r = b^3t/12$.

*Aeroelasticity models*

Following the previous discussion on thin-wall structures, we are ready to consider a two-dimensional model consisting of one typical cross section of thin-wall cantilever wing, as shown in Figure 4.72. The pivot point is in fact the shear center

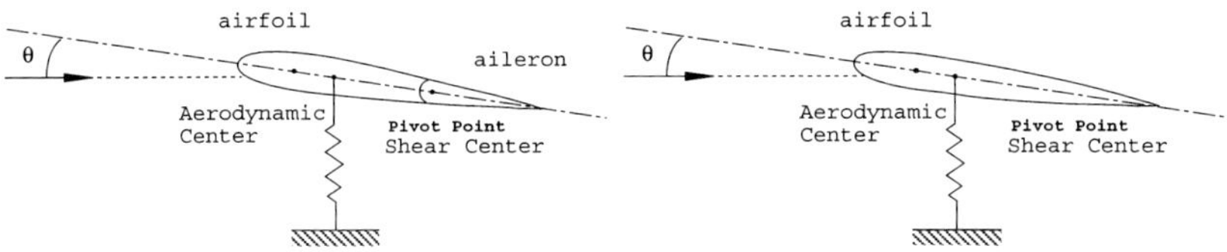


Figure 4.72. A typical two-dimensional model of a thin-wall cantilever wing.

where the shear force or transverse force does not introduce in-plane twist. The locus of shear centers of cross sections of a cantilever wing is called the elastic axis.

First, let us consider the torsionally buckling or divergence instability, in which dynamic forces and coupling are ignored. Assume the surrounding fluid introduces a lift and moment coefficients $C_L(\theta)$ and $C_M(\theta)$, and the center of lift force has a distance of $x_a$ away from the shear center, i.e., the pivot point.

A detailed discussion of lifting surfaces is available in Chapter 5. Suppose the wing section twist is an additional angle $\theta$ from its initial angle of attack $\theta_o$, and the torsional rigidity is $GJ$, namely, the rotational spring constant is $GJ/l$, where $l$ is the axial length of the cantilever wing at the location of the cross section. In addition, assign $J_\theta$ as the polar moment of inertia of the cross section, which can be different from $J$ depending on the support point. From the moment balance about the shear center, we have

$$\frac{\rho}{2}\bar{u}^2 b x_a\left[C_L(\theta_o) + \frac{\partial C_L}{\partial \theta}\theta\right] + \frac{\rho}{2}\bar{u}^2 b^2\left[C_M(\theta_o) + \frac{\partial C_M}{\partial \theta}\theta\right] = \frac{GJ}{l}(\theta_o + \theta) + J_\theta\ddot{\theta}, \tag{3.207}$$

where $\bar{u}$ is the uniform incoming flow velocity.

Note that in Eq. (3.207), we assume that at zero angle of attack, with respect to the $x$-axis, or zero-lift line, the surrounding fluid exerts no twist moment. Naturally, for the initial equilibrium, we have

$$\frac{\rho}{2}\bar{u}^2 b x_a C_L(\theta_o) + \frac{\rho}{2}\bar{u}^2 b^2 C_M(\theta_o) = \frac{GJ}{l}\theta_o. \tag{3.208}$$

In addition, it is clear that as the incoming velocity $\bar{u}$ increases, $\frac{\rho}{2}\bar{u}^2 b(x_a \partial C_L/\partial\theta + b\partial C_M/\partial\theta)$ will approach $GJ/l$ and as a consequence, the cross section will loose its torsional resistance. The natural frequency for the torsional degree of freedom is zero. Thus, the following critical velocity corresponds to the buckling or divergence instability:

$$\bar{u}_d^2 = \frac{2GJ}{\rho b l} \Big/ \left(x_a\frac{\partial C_L}{\partial \theta} + b\frac{\partial C_M}{\partial \theta}\right). \tag{3.209}$$

Consider an aileron control surface attached to the wing cross section via a hinge for which the rotational spring constant is $k_\beta$ and the polar moment of inertia is $J_\beta$. Introduce an additional angle of rotation $\beta$ of the aileron as shown in Figure 4.72. To include the lift and moment effects of the aileron with respect to rotation around the hinge, we define an aerodynamic hinge moment coefficient $C_H$. Thus, moment balance around the pivot and hinge can be written as

$$\frac{\rho}{2}\bar{u}^2 b x_a\left[C_L(\theta_o,\beta_o)+\frac{\partial C_L}{\partial\beta}\beta+\frac{\partial C_L}{\partial\theta}\theta\right] + \frac{\rho}{2}\bar{u}^2 b^2\left[C_M(\theta_o,\beta_o)+\frac{\partial C_M}{\partial\beta}\beta+\frac{\partial C_M}{\partial\theta}\theta\right] = \frac{GJ}{l}(\theta_o+\theta)+\rho J_\theta\ddot{\theta}, \tag{3.210}$$

$$\frac{\rho}{2}\bar{u}^2\bar{b}^2\left[C_H(\theta_o,\beta_o)+\frac{\partial C_H}{\partial\beta}\beta+\frac{\partial C_H}{\partial\theta}\theta\right] = k_\beta(\beta-\beta_o)+J_\beta\ddot{\beta}, \tag{3.211}$$

where $\bar{b}$ denotes the chord of the aileron.

Note that the rotational spring of the hinge involves relative rotation, whereas the additional lift generated by the aileron is measured by absolute rotation. Similarly, the buckling or divergence instability corresponds to the loss of stiffness or resistance to the rotations $\theta$ and $\beta$, and the critical velocity $\bar{u}$ is governed by

$$\det\begin{pmatrix} bx_a\frac{\partial C_L}{\partial\theta}+b^2\frac{\partial C_M}{\partial\theta}-\frac{2GJ}{l\rho\bar{u}^2} & bx_a\frac{\partial C_L}{\partial\beta}+b^2\frac{\partial C_M}{\partial\beta} \\ \bar{b}^2\bar{x}_a\frac{\partial C_H}{\partial\theta} & \bar{b}^2\frac{\partial C_H}{\partial\beta}-\frac{2k_\beta}{\rho\bar{u}^2} \end{pmatrix}=0. \tag{3.212}$$

Naturally, the critical velocity derived from the quadratic equation (3.212) is chosen as the lower positive value. To delineate the aeroelastic effects of the aileron, we assume a fictitious wing with $k_\beta\to\infty$. Therefore, from Eq. (3.211), we have $\beta=\beta_o$ and Eq. (3.210) is similar to Eq. (3.207) except for the additional lift force and moment. However, it is interesting to note that with $\beta=\beta_o$, Eq. (3.211) yields the same form of the critical velocity with respect to buckling or divergence instability, as from Eq. (3.207). Furthermore, defining another critical velocity corresponding with the aileron reversal, at which the change of lift due to an aileron rotational angle vanishes, by employing the initial equilibrium of (3.208), we have

$$\frac{\partial L}{\partial\beta}=\frac{\partial C_L}{\partial\theta}\frac{\partial\theta}{\partial\beta}+\frac{\partial C_L}{\partial\beta}=0. \tag{3.213}$$

Eq. (3.213) can be also considered as the measure of the aileron effectiveness. Moreover, differentiating both sides of Eq. (3.210) with respect to $\beta$, and employing (3.213), we have

$$\bar{u}_r^2=\frac{2GJ}{\rho b^2 l}\left[-\frac{\partial C_L}{\partial\beta}\left(\frac{\partial C_L}{\partial\theta}\right)^{-1}\right]\Bigg/\left[\frac{\partial C_M}{\partial\beta}-\frac{\partial C_M}{\partial\theta}\frac{\partial C_L}{\partial\beta}\left(\frac{\partial C_L}{\partial\theta}\right)^{-1}\right]. \tag{3.214}$$

Hence, we can define a so-called elastic efficiency of the aileron as $(1-\bar{u}^2/\bar{u}_r^2)/(1-\bar{u}^2/\bar{u}_d^2)$, which represents the ratio of the lift force produced by an elastic hinge and a fictitious rigid hinge. When $\bar{u}_r < \bar{u}_d$, the aileron efficiency decreases to zero as $\bar{u} \to \bar{u}_r$. On the other hand, when $\bar{u}_r > \bar{u}_d$, as $\bar{u} \to \bar{u}_d$, the effect of the aileron from Eq. (3.213) approaches to infinity, which can be easily confirmed by substituting the solution of $\theta$ of Eq. (3.207). However, once we pass the critical velocity $\bar{u}_d$, for $\bar{u} \in (\bar{u}_d, \bar{u}_r)$, the aileron efficiency becomes negative.

Note that if the cantilever wing structure represents the impeller of rotary machines, a centrifugal force must also be introduced. Consequently, the center of mass should also be considered along with the shear center and the center of aerodynamic force. In fact, the location of the center of mass will be used for the study of flutter instability.

To study flutter instability, we must first obtain the dynamical equations. Again, we consider a typical section of the cantilever wing, as shown in Figure 4.72. In general, the shear center does not coincide with the center of mass, where inertia force can be considered to be concentrated. Denote a distance $x_i$ between the shear center and the center of mass, and consider the typical cross section as a rigid body with mass $m$ and polar moment of inertia $J_\theta$ with respect to the center of mass. Furthermore, for convenience, we introduce the translational spring constant $k_u$ and the rotational spring constant $k_\theta$. Hence, the kinetic and potential energies can be expressed as

$$T = \frac{m}{2}(\dot{u}_c - x_i\dot{\theta})^2 + \frac{\rho J_\theta}{2}\dot{\theta}^2, \tag{3.215}$$

$$U = \frac{1}{2}k_u u_c^2 + \frac{1}{2}k_\theta \theta^2, \tag{3.216}$$

where $u_c$ and $\theta$ represent the vertical displacement of the shear center and the rotation, respectively.

Employing Lagrange's equation as discussed in Chapter 3, we can easily derive the governing equation

$$m\ddot{u}_c - mx_i\ddot{\theta} + k_u u_c = L, \tag{3.217}$$

$$-mx_i\ddot{u}_c + \left(\rho J_\theta + mx_i^2\right)\ddot{\theta} + k_\theta\theta = M, \tag{3.218}$$

where $L$ and $M$ are lift force and moment.

Since for stability analysis we only have to consider the increment of the lift and the moment, namely $\frac{\rho\bar{u}^2 b}{2}\frac{\partial C_L}{\partial\theta}\theta$ and $\frac{\rho\bar{u}^2 b}{2}(b\frac{\partial C_M}{\partial\theta} + x_a\frac{\partial C_L}{\partial\theta})\theta$, Eqs. (3.217) and (3.218) can be written as

$$\begin{bmatrix} m & -mx_i \\ -mx_i & mx_i^2 + \rho J_\theta \end{bmatrix}\begin{bmatrix} \ddot{u}_c \\ \ddot{\theta} \end{bmatrix} + \begin{bmatrix} k_u & k_{u\theta} \\ 0 & k_{\theta\theta} \end{bmatrix}\begin{bmatrix} u_c \\ \theta \end{bmatrix} = \begin{bmatrix} R_u \\ R_\theta \end{bmatrix}, \tag{3.219}$$

with $k_{\theta\theta} = k_\theta - \frac{\rho}{2}\bar{u}^2 b(b\frac{\partial C_M}{\partial\theta} + x_a\frac{\partial C_L}{\partial\theta})$ and $k_{u\theta} = -\frac{\rho}{2}\bar{u}^2 b\frac{\partial C_L}{\partial\theta}$.

It is clear from Eq. (3.219) that the divergence instability corresponds to the determinant of the stiffness matrix equal to zero, which means physically the cantilever wing loses its stiffness. Based on the discussion in Chapter 2, we have the following characteristic equation:

$$m\rho J_\theta \omega^4 - \left[mk_{\theta\theta} + k_u\left(mx_i^2 + \rho J_\theta\right) + k_{u\theta} m x_i\right]\omega^2 + k_u k_{\theta\theta} = 0. \tag{3.220}$$

According to the definition of flutter instability, $\mathrm{Im}(\omega) = 0$ and $\mathrm{Re}(\omega) > 0$, we define

$$\bar{u}_c^2 = \frac{2k_\theta}{\rho b} \Bigg/ \left(x_a \frac{\partial C_L}{\partial \theta} + b\frac{\partial C_M}{\partial \theta}\right), \tag{3.221}$$

if $\bar{u} < \bar{u}_c$,

$$k_u\left[k_\theta - \frac{1}{2}\rho\bar{u}^2 b\left(b\frac{\partial C_M}{\partial \theta} + x_a\frac{\partial C_L}{\partial \theta}\right)\right] > 0, \tag{3.222}$$

hence the critical point corresponds to

$$m\left[k_\theta - \frac{\rho}{2}\bar{u}^2 b\left(b\frac{\partial C_M}{\partial \theta} + x_a\frac{\partial C_L}{\partial \theta}\right)\right] + k_u\left(mx_i^2 + \rho J_\theta\right) - \frac{m}{2}\rho x_i \bar{u}^2 b\frac{\partial C_L}{\partial \theta} = 0. \tag{3.223}$$

In fact, if $\partial C_M/\partial\theta$ is ignored, flutter instability will not happen when the center of mass is ahead of the shear center, i.e., $x_i < 0$. Note that we ignore two important aspects. First, the axial direction variation of cross section is neglected. Second, the effects of the dynamical behavior of the cantilever wing on the lift and moment are not considered. In the study of the typical cross section, $L$ and $M$ represent the lift and moment per unit axial length. Hence, employing Lagrange's equation, we have the governing partial differential equations

$$\frac{\partial^2}{\partial z^2}\left(EI\frac{\partial^2 u_c}{\partial z^2}\right) + m(z)\frac{\partial^2 u_c}{\partial t^2} - m(z)x_i(z)\frac{\partial^2 \theta}{\partial t^2} - L = 0, \tag{3.224}$$

$$\frac{\partial}{\partial z}\left(GJ\frac{\partial \theta}{\partial z}\right) + m(z)x_i(z)\frac{\partial^2 u_c}{\partial t^2} - \left[\rho J_\theta + m(z)x_i^2(z)\right]\frac{\partial^2 \theta}{\partial t^2} + M = 0, \tag{3.225}$$

where the expressions of $L$ and $M$, including the oscillation of the wing are written as follows:

$$L = \frac{\rho}{2}\bar{u}^2 b\frac{\partial C_L}{\partial \theta}\bar{\theta}, \tag{3.226}$$

$$M = \frac{\rho}{2}\bar{u}^2 b^2\left(-\frac{b\pi}{8\bar{u}}\frac{\partial \theta}{\partial t}\right) + \left(\bar{x} - \frac{b}{x}\right)\frac{\partial C_L}{\partial \theta}L, \tag{3.227}$$

with

$$\bar{\theta} = \theta + \frac{1}{\bar{u}}\frac{\partial u_c}{\partial t} + \frac{1}{\bar{u}}\bar{x}\frac{\partial \theta}{\partial t}. \tag{3.228}$$

In general, the natural frequency of this system is a complex number, expressed as $a + bi$. The divergence instability corresponds to $a = 0$ and the flutter instability corresponds to $a > 0$ and $b = 0$. The divergence or buckling instability represents the loss of structural stiffness, and the flutter instability stands for a negative damping, from which additional energy is introduced to the system.

*Chapter 5*

# Boundary Integral Approaches

In this chapter, we deal with a particular set of FSI problems involving a rigid or linear elastic body interacting with potential flow, acoustic medium, or low Reynolds number flow [133,141,306]. The commonality of these FSI problems is the boundary integral formulations of the unknown function and its derivatives, which often warrant exact or analytical solutions. If, however, the shape of the geometrical domain physically occupied by the coupled media precludes analytical solutions, boundary element methods must be used to solve these integral formulations [219,255].

We start with a separate presentation of underlying physics and problem formulations for potential flow, acoustic medium, and low Reynolds number flow. In Section 5.2, we discuss the fundamental concepts of Green's functions, boundary integral formulations, and the corresponding boundary element methods. In Section 5.3, radiation and scattering of waves are presented with respect to potential flows.

## 5.1. Underlying physics and formulations

### *5.1.1. Potential flow*

Potential flow is often called ideal flow, where the bulk of the fluid is assumed to be incompressible and inviscid. The fundamental equations are the following continuity equation and Euler's equations:

$$\nabla \cdot \mathbf{v} = 0, \tag{1.1}$$

$$\rho\left(\frac{\partial \mathbf{v}}{\partial t} + \mathbf{v} \cdot \nabla \mathbf{v}\right) = -\nabla p + \mathbf{f}^b, \tag{1.2}$$

where $\mathbf{v}$, $p$, and $\mathbf{f}^b$ stand for the velocity vector, pressure, and body force vector, respectively.

An essential concept of potential flow is the circulation $\Gamma$, defined as the integrated tangential velocity around a counterclockwise material contour $C$:

$$\Gamma = \oint_C \mathbf{v} \cdot d\mathbf{s}. \tag{1.3}$$

By invoking Stokes' theorem in the form of Eq. (2.138), we derive

$$\int_S (\nabla \times \mathbf{v}) \cdot d\mathbf{S} = \oint_C \mathbf{v} \cdot d\mathbf{s}, \tag{1.4}$$

where $S$ represents any smooth surface bounded by the closed contour $C$.

It is clear that the circulation $\Gamma$ defined in Eq. (1.3) directly reflects the vorticity within the fluid. The material or substantial derivative of the circulation $\Gamma$ can be also expressed as

$$\frac{d\Gamma}{dt} = \oint_C \left(\frac{\partial}{\partial t} + \mathbf{v} \cdot \nabla\right)\mathbf{v} \cdot d\mathbf{s} = \oint_C \left(\frac{\partial \mathbf{v}}{\partial t} + \mathbf{v} \cdot \nabla \mathbf{v}\right) \cdot d\mathbf{s} + \oint_C \mathbf{v} \cdot d\mathbf{v}. \tag{1.5}$$

Suppose the body force is conservative, namely, $\mathbf{f} = -\nabla\Omega$, where $\Omega$ is the force potential. For example, if the gravity is in the negative direction of the $y$-axis, the gravitational potential can be expressed as $\Omega = \rho g y$. Hence, employing Eq. (1.2), we can rewrite Eq. (1.5) as

$$\begin{aligned}\frac{d\Gamma}{dt} &= -\frac{1}{\rho}\oint_C \nabla(p + \Omega) \cdot d\mathbf{s} + \frac{1}{2}\oint_C d(\mathbf{v} \cdot \mathbf{v}) \\ &= -\frac{1}{\rho}\oint_C d(p + \Omega) + \frac{1}{2}\oint_C d(\mathbf{v} \cdot \mathbf{v}) = 0.\end{aligned} \tag{1.6}$$

Kelvin's theorem, named after William Thomson Kelvin (1824–1907), states that for the prevalent ideal fluid acted upon by conservative forces, e.g., gravity, the circulation $\Gamma$ defined in (1.3) remains constant. The underlying physics for Eq. (1.6) is significant. Because there is no fluid viscosity, i.e., no shear stresses, the rotation rate of the fluid particles will not be altered. For instance, if the fluid motion starts from a state of rest, it will remain irrotational for all time and the circulation about any material contour will vanish.

Since Eqs. (1.4) and (1.6) hold for any material contour within the fluid domain at all times, the motion of the fluid is irrotational, i.e.,

$$\boldsymbol{\omega} = \nabla \times \mathbf{v} = \mathbf{0}, \tag{1.7}$$

where $\boldsymbol{\omega}$ stands for the vorticity vector.

If we consider an infinitesimally small material tube, the so-called vortex tube or filament, and employing Stokes' and Kelvin's theorems, we have

$$\Gamma = \omega A = C, \tag{1.8}$$

where $C$ is a time-independent constant, and $\omega$ and $A$ represent the vortex component perpendicular to the cross section of the tube and the cross-sectional area, respectively.

Denote the length of the vortex filament as $L$. Since the material tube encompasses incompressible fluid, the volume of the filament is invariably time-independent. Thus, we have

$$\frac{\omega A}{AL} = \frac{\omega}{L} = D, \tag{1.9}$$

where $D$ represents a time-independent constant.

REMARK 5.1.1. Eq. (1.9) is somehow similar to the conservation of angular momentum, with which, the increase of $L$ corresponds to the decrease of $A$ and the increase of $\omega$.

Applying the results of vector analysis in the form of (1.17) in Chapter 2 and introducing a velocity potential $\phi$, such that $\mathbf{v} = \nabla\phi$, we derive from the continuity equation (1.1) the following Laplace's equation, named after Pierre-Simon Laplace (1749–1827):

$$\nabla^2\phi = 0. \tag{1.10}$$

Similarly, for ideal fluid acted upon by conservative forces, Euler's equation (1.2) can then be rewritten in the following perfect differential form:

$$\nabla\left(\rho\frac{\partial\phi}{\partial t} + \frac{\rho}{2}|\nabla\phi|^2 + p + \Omega\right) = \mathbf{0}. \tag{1.11}$$

Thus, along the streamline, the tangent is parallel to the velocity vector $\mathbf{v}$, we obtain Bernoulli's equation, named after Daniel Bernoulli (1700–1782),

$$\rho\frac{\partial\phi}{\partial t} + \frac{\rho}{2}|\nabla\phi|^2 + p + \Omega = C(t), \tag{1.12}$$

where the time-dependent constant $C(t)$ can be absorbed into the velocity potential or set equal to zero.

Since the fluid velocity field is represented by the gradient of the velocity potential $\phi$, the original governing equations in the form of Eqs. (1.1) and (1.2) are then replaced with the scalar Laplace's and Bernoulli's equations (1.10) and (1.12). To better illustrate the concept of potential flow, we will discuss a few simple examples. We start with an elegant and effective representation of two-dimensional cases using complex variables. According to Chapter 2, for the harmonic function $\phi$, the velocity potential, we can introduce a conjugate harmonic function $\psi$, the so-called stream function, such that

$$u = \frac{\partial \phi}{\partial x} = \frac{\partial \psi}{\partial y}, \qquad v = \frac{\partial \phi}{\partial y} = -\frac{\partial \psi}{\partial x}, \tag{1.13}$$

where $u$ and $v$ stand for the horizontal and vertical velocity components, respectively.

Just as the velocity vector marks the contour gradient of the potential function, the stream function marks the stream lines. Hence, the wall boundary corresponds to the stream function contour, which we often set as $\psi = 0$. Furthermore, the contours of the potential and stream functions form an orthogonal curvilinear coordinate system.

To exploit a wide variety of mathematical tools in accordance with harmonic functions and complex analysis, we define a complex variable $z = x + iy$ and the complex potential $F(z)$:

$$F(z) = \phi + i\psi. \tag{1.14}$$

Based on Eq. (1.13), both real and imaginary parts of $F(z)$ satisfy the Cauchy–Riemann equations. Thus, the complex potential $F(z)$ is an analytic function with respect to the complex variable $z$. Employing Eq. (1.13), the derivative of the complex potential $F(z)$ can be simply expressed as

$$\frac{dF}{dz} = u - iv. \tag{1.15}$$

It is obvious that the case $F(z) = C$, where $C$ is a constant, represents a trivial case with no physical significance; whereas the case $F(z) = Cz$ corresponds to a uniform flow with $u = \mathrm{Re}(C)$ and $v = -\mathrm{Im}(C)$. As with the case $F(z) = Cz^n$, with $n \in \mathcal{Q}$, it is more convenient to express the complex potential function in a polar coordinate system. Employing de Moivre's equation (1.73) in Chapter 2, we have

$$F(r, \theta) = Cr^n(\cos n\theta + i \sin n\theta). \tag{1.16}$$

If we substitute Eq. (1.16) into the Cauchy–Riemann equation in polar coordinate system (3.230) in Chapter 2, it becomes clear that the potential flow represented by Eq. (1.16) corresponds to a corner flow within an angle $\theta, \theta = \pi/n$ with $n > 1/2$, as illustrated in Figure 5.1, where $n > 1$ and $n < 1$ represent the interior and exterior corner flows, respectively. In fact, for the interior corner flow case, namely, $n > 1$, the entire two-dimensional plane can be viewed as separated by walls at $\theta = m\pi/n$, with $0 < m < n$. Of course $n = 1$ corresponds to a degenerate case of uniform flow. Furthermore, unlike the interior flow, in which

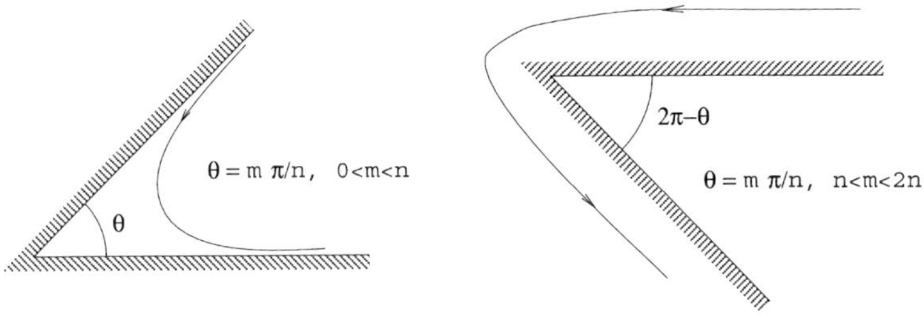


Figure 5.1. Interior and exterior corner flow representations.

the origin is a stagnation point, there exists an infinite velocity around the corner for the exterior flow with $n < m < 2n$.

Another important example of potential flow is $F(z) = (q/2\pi)\ln z$, which represents at the origin a source for $q > 0$, or a sink for $q < 0$. Again, expressing $F(z)$ in a polar coordinate system, we have

$$F(r, \theta) = \frac{q}{2\pi}\ln r + \frac{iq(\theta + 2k\pi)}{2\pi}, \quad \text{with } k \in \mathcal{Z}. \tag{1.17}$$

Note that in this case, there is a branch point at the origin. Again, applying Eq. (3.230) in Chapter 2, we can easily derive the tangential velocity $v_\theta = 0$ and the radial velocity $v_r = q/2\pi r$. Likewise, the case $F(z) = -(i\gamma/2\pi)\ln z$ represents a vorticity at the origin. Employing Eq. (3.230) in Chapter 2, we have $v_r = 0$ and $v_\theta = \gamma/2\pi r$. Suppose we enclose the origin with a circle of radius $r$, for the case with a source or sink, the net mass flow rate can be expressed as

$$\int_0^{2\pi} v_r r\, d\theta = q; \tag{1.18}$$

while for the singular vortex at the origin, the circulation around the origin is calculated as

$$\int_0^{2\pi} v_\theta r\, d\theta = \gamma. \tag{1.19}$$

In fact, Eqs. (1.18) and (1.19) illustrate the exact physical meaning of the source or sink $q$ and the vortex $\gamma$ situated at the origin. Furthermore, the case $F(z) = m/2\pi z$ represents a dipole with a dipole moment $m$, which can be viewed as a limit case of a sink and a source with the same strength $m/x_o$, located at the origin and $x = x_o$, respectively, with $x_o \to 0$, i.e.,

$$\lim_{x_o \to 0} \frac{m}{2\pi x_o}\ln(z + x_o) - \frac{m}{2\pi x_o}\ln z = \frac{m}{2\pi z}. \tag{1.20}$$

REMARK 5.1.2. Laplace's equation for potential flow is only satisfied for fluid domains excluding these singularity or branch points.

We are now ready to discuss the applications of these elementary examples of potential flow. For instance, the potential flow field of a uniform flow passing a rigid cylinder can be simply expressed as a combination of a uniform flow with a horizontal velocity $\bar{u}$ and a dipole moment flow with a moment $m$,

$$F(z) = \bar{u}z + \frac{m}{2\pi z}. \tag{1.21}$$

If we express Eq. (1.21) in a polar coordinate system, we then obtain

$$\begin{aligned} \phi &= \left(\bar{u}r + \frac{m}{2\pi r}\right)\cos\theta, \\ \psi &= \left(\bar{u}r - \frac{m}{2\pi r}\right)\sin\theta. \end{aligned} \tag{1.22}$$

Clearly that if we choose a particular dipole moment $m = 2\pi\bar{u}R^2$, where $R$ stands for the rigid cylinder radius, the interface between the potential flow and rigid impermeable cylinder is marked by $\psi = 0$ or $v_r = \partial\phi/\partial r = 0$. Furthermore, employing the concepts of conformal mapping discussed in Chapter 2, we can transform geometries in $z = x + iy$ complex plane, which is difficult to deal with, to $\zeta = \xi + i\eta$ complex plane, which can be expressed with simple analytical expressions of the complex potential function.

Again, we use the flow passing a rigid cylinder to illustrate the application of conformal mapping. According to the discussion in Chapter 2, the conformal mapping $\zeta = z + 1/z$ maps a unit circle to a finite line segment $[-1, 1]$. Therefore, the potential function for a uniform flow passing the cylinder with a unit radius in $z$ complex plane can be expressed as a uniform flow passing a line segment. Since the line segment has no thickness, it has no effect on the surrounding flow if the flow is oriented in the same direction as the line segment. Therefore, the complex potential can simply be written as $\bar{u}\zeta$ or $\bar{u}(z + 1/z)$, which is equivalent to Eq. (1.21) with $\bar{u} = 1$, $R = 1$, and $m = 2\pi$.

Although the complex potential form is not applicable to three-dimensional flows, some simple examples still bear certain resemblances. For example, the velocity potential of the uniform flow with a horizontal velocity $\bar{u}$ can be clearly expressed as $\phi = \bar{u}x$. However, derived from the mass conservation over a spherical surface, the velocity potential of a source, situated at the origin, is expressed as $\phi = -q/4\pi r$. Likewise, the potential for a dipole, situated at the origin and oriented in the $x$-direction, can be expressed as $\phi = mx/4\pi r^3$, where $m$ is the dipole moment. Analogously, the potential for a uniform flow passing a rigid sphere takes the similar form of Eq. (1.21),

$$\phi = \bar{u}x + \frac{mx}{4\pi r^3}, \tag{1.23}$$

or in a polar coordinate system

$$\phi = \bar{u} r \cos\theta + \frac{m \cos\theta}{4\pi r^2}. \tag{1.24}$$

Thus, a proper selection of $m$ must correspond $v_r = 0$ at the spherical surface, namely, $m = 2\pi \bar{u} R^3$, where $R$ is the radius of the sphere.

### *5.1.2. Acoustic medium*

Acoustics is the science of sound, including its production, transmission, and effects within various continuous media. It is therefore natural to include this subject in FSI systems [197]. Despite some similarities, however, acoustics are quite different from optics, since sound is mechanical, rather than electromagnetic.

Wave propagation can be categorized as the effect of a sharply applied localized disturbance in a medium soon to be transmitted or spread to other parts of the medium, in which energy is carried by partly kinematic and partly potential energies. Wave propagations, whether the transmission of sound in air or the spreading of ripples on a pond of water, all share common features and bear resemblances to each other in their treatments. Because the analytical approaches heavily discussed in this chapter involve special solutions, Green's functions, series expansions, and perturbation techniques, which are useful for very limited special geometries, we illustrate the boundary element methods as a natural extension to general integral equations. Other computational approaches such as the use of finite elements will be addressed in Chapter 6 and Ref. [114]. In aerospace and mechanical engineering, the interaction of acoustic wave (sound) with rigid or elastic body is the key aspect in understanding noise control and sonar techniques [3,62,217].

The key assumption for acoustic media (fluid and solid) is to ignore the shear effects. Consequently, the mass conservation is represented with the relationship between the pressure $p$ and the volumetric strain $\nabla \cdot \mathbf{u}$ as in Eq. (2.195) in Chapter 2, where $\mathbf{u}$ represents the displacement of the media. Thus, for both fluid and solid acoustic media, from Eq. (2.163) in Chapter 2, it is not difficult to identify the governing partial differential equations as

$$\rho \frac{d^2\mathbf{u}}{dt^2} = \nabla\sigma + \mathbf{f}, \tag{1.25}$$

where $\mathbf{f}$ stands for the body force and the stress tensor $\boldsymbol{\sigma}$ can be simply expressed as $-p\mathbf{I}$.

Employing Eq. (2.195), Eq. (1.25) can be rewritten as the following two key equations

$$\frac{\partial^2 p}{\partial t^2} = c^2 \nabla^2 p, \tag{1.26}$$

$$\frac{\partial^2 \mathbf{u}}{\partial t^2} = c^2 \nabla^2 \mathbf{u}, \tag{1.27}$$

where the wave speed $c$ is expressed as $\sqrt{\kappa/\rho}$.

Likewise, to match with the discussion in Chapter 6, for acoustic medium, we can also introduce a velocity potential $\phi$ and derive the following governing equation

$$\frac{\partial^2 \phi}{\partial t^2} = c^2 \nabla^2 \phi. \tag{1.28}$$

### 5.1.3. Stokes flow

The third group of problems are low Reynolds number hydrodynamics which mainly deals with the Stokes flow interacting with particles. More elaborate discussion can be found in Chapter 7 and Refs. [100,219]. It is important to note that the current boundary element methods share various names in different areas such as panel methods in ocean engineering.

The steady and unsteady Stokes flows, named after George Gabriel Stokes (1819–1903), are often written as follows:

$$\mathbf{0} = -\nabla p + \mu \nabla^2 \mathbf{v} + \mathbf{f}^b, \tag{1.29}$$

$$\rho \frac{\partial \mathbf{v}}{\partial t} = -\nabla p + \mu \nabla^2 \mathbf{v} + \mathbf{f}^b, \tag{1.30}$$

where the body force $\mathbf{f}^b$ is often denoted as the gravitational force $\rho \mathbf{g}$, and the mass conservation equation is written as the incompressible form $\nabla \cdot \mathbf{v} = 0$.

It is clear that Eq. (1.29) is directly derived from the incompressibility condition and Eq. (2.163) in Chapter 2 with $\sigma_{ij} = -p\delta_{ij} + \mu(\partial u_i/\partial x_j + \partial u_j/\partial x_i)$. Furthermore for the unsteady Stokes flow, in addition to $Re \ll 1$, we must also have $Re\, St \sim 1$. For clarity, we employ both indicial and vectorial notations. Suppose the specific body force can be expressed as $\mathbf{f}^b = -\nabla B$, where $B$ is a harmonic function, i.e., $\nabla^2 B = 0$. It is easy to derive from Eq. (1.29) the following three key equation forms:

$$\nabla^2 p = 0, \tag{1.31}$$

$$\nabla^4 \mathbf{v} = \mathbf{0}, \tag{1.32}$$

$$\nabla^2 \boldsymbol{\omega} = \mathbf{0}. \tag{1.33}$$

As a consequence, for a domain of Stokes fluid enclosed by a closed surface, we have

$$\mathbf{F} = \int_S \boldsymbol{\sigma} \mathbf{n} \, dS = -\int_V \rho \mathbf{g} \, dV, \tag{1.34}$$

$$\mathbf{M} = \int_S \mathbf{r} \times \boldsymbol{\sigma}\mathbf{n}\, dS = -\int_V \rho\mathbf{r} \times \mathbf{g}\, dV. \tag{1.35}$$

## 5.2. Green's functions and boundary integrals

### 5.2.1. Reciprocal relations

A proper beginning of the discussion on Green's functions, named after George Green (1793–1841), is the consideration of solutions to Laplace's equation. Based on the results of vector analysis in Chapter 2, we can easily derive the following two Green's identities:

$$\phi\nabla^2\psi = \nabla \cdot (\phi\nabla\psi) - \nabla\phi \cdot \nabla\psi, \qquad \text{Green's First Identity,} \tag{2.36}$$

$$\psi\nabla^2\phi - \phi\nabla^2\psi = \nabla \cdot (\psi\nabla\phi - \phi\nabla\psi), \qquad \text{Green's Second Identity.} \tag{2.37}$$

Moreover, for harmonic functions $\phi$ and $\psi$, namely, $\nabla^2\phi = \nabla^2\psi = 0$, Green's second identity can be rewritten as the following reciprocal relation:

$$\nabla \cdot (\psi\nabla\phi - \phi\nabla\psi) = 0. \tag{2.38}$$

Applying the divergence theorem, we can rewrite Eq. (2.38) in a differential form which is often called Green's theorem, in the surface integral form,

$$\int_S \phi\nabla\psi \cdot \mathbf{n}\, dS = \int_S \psi\nabla\phi \cdot \mathbf{n}\, dS, \tag{2.39}$$

where $\mathbf{n}$ is the unit surface normal vector pointing outward from the volume.

Essentially, Green's function $G(\mathbf{x}; \boldsymbol{\xi})$ of Laplace's equation can be considered as a special class of harmonic functions, singular at a point (or pole) $\boldsymbol{\xi}$ and satisfying the following singularly forced Laplace's equation:

$$\nabla^2 G(\mathbf{x}; \boldsymbol{\xi}) = q\delta(\mathbf{x} - \boldsymbol{\xi}), \tag{2.40}$$

where $\mathbf{x}$ represents an arbitrary field point.

In engineering practice, Green's function may be identified with the steady temperature (or concentration) field due to a point sink ($q > 0$) and source ($q < 0$) at the point $\boldsymbol{\xi}$. In addition, the solution corresponding to $G(\mathbf{x}; \boldsymbol{\xi})|_S = 0$ is called Green's function of the first kind, whereas the solution corresponds to $\partial G(\mathbf{x}; \boldsymbol{\xi})/\partial n|_S = 0$ is called Green's function of the second kind. Moreover, a Green's function on an infinite domain corresponds to a flow with no interior boundaries and is also called the free-space Green's function and often has an analytical expression [87]. For example, the three-dimensional free-space Green's function of Eq. (2.40) can be expressed as

$$G(\mathbf{x}; \boldsymbol{\xi}) = -\frac{q}{4\pi r}, \tag{2.41}$$

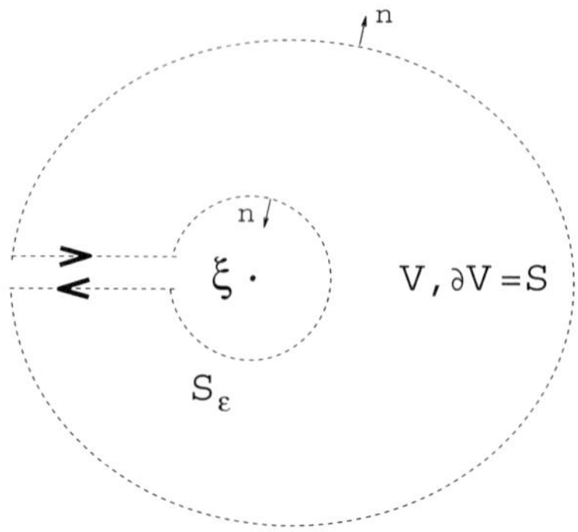


Figure 5.2. Green's function and singularities.

with $r = \sqrt{(x-\xi)^2 + (y-\eta)^2 + (z-\zeta)^2}$; whereas the corresponding two-dimensional solution takes the form of

$$G(\mathbf{x}; \boldsymbol{\xi}) = \frac{q}{2\pi} \ln r. \tag{2.42}$$

In general, to satisfy the boundary conditions, we often introduce in addition to the free-space Green's function a complementary regular function $H(\mathbf{x}; \boldsymbol{\xi})$, which is nonsingular in the entire domain $V$ including the boundary $\partial V = S$. Unfortunately, in general, there is only a limited set of special domains for which the regular function $H(\mathbf{x}; \boldsymbol{\xi})$ can be found. For example, suppose there is a plane wall at $x = \bar{x}$, at which $G(\mathbf{x}; \boldsymbol{\xi}) = 0$. The Green's function for this bounded domain is expressed as

$$G(\mathbf{x}; \boldsymbol{\xi}) = \frac{q}{4\pi} \left( \frac{1}{r} - \frac{1}{\bar{r}} \right), \tag{2.43}$$

with $\bar{r} = \sqrt{(x - 2\bar{x} + \xi)^2 + (y-\eta)^2 + (z-\zeta)^2}$.

Green's theorem in the form of Eq. (2.39) holds for all harmonic functions. Let $\psi$ be the Green's function with a unit source located at $\boldsymbol{\xi} = (\xi, \eta, \zeta)$, i.e., $\psi = 1/4\pi r$, with $r = \sqrt{(x-\xi)^2 + (y-\eta)^2 + (z-\zeta)^2}$. Supposing $\boldsymbol{\xi}$ is located within the fluid domain $V$, as shown in Figure 5.2, Eq. (2.39) will not hold for the selected $\phi$ because of its singularity at $\boldsymbol{\xi}$. However, if we surround the source point $\boldsymbol{\xi}$ with a spherical surface $S_\epsilon$ of a radius $\epsilon$, also shown in Figure 5.2, the singularity problem can be circumvented. Since the connection or tube between two separate surfaces $S$ and $S_\epsilon$ does not contribute to the surface integral, Eq. (2.39) holds for the fluid domain excluding the vicinity of the singularity point. For specially chosen $\psi$, we have

$$\frac{1}{4\pi} \int_{S+S_\epsilon} \left( \phi \frac{\partial}{\partial n} \frac{1}{r} - \frac{1}{r} \frac{\partial \phi}{\partial n} \right) dS = 0, \tag{2.44}$$

and consequently,

$$\frac{1}{4\pi}\int\limits_{S}\left(\phi\frac{\partial}{\partial n}\frac{1}{r}-\frac{1}{r}\frac{\partial\phi}{\partial n}\right)dS=-\frac{1}{4\pi}\int\limits_{S_\epsilon}\left(\phi\frac{\partial}{\partial n}\frac{1}{r}-\frac{1}{r}\frac{\partial\phi}{\partial n}\right)dS. \tag{2.45}$$

The area of $S_\epsilon$ is $4\pi\epsilon^2$ and the normal derivative is equivalent to $-\partial/\partial r$, in accordance with the unit normal vector definition. In the limit $\epsilon \to 0$, the contribution from the right-hand side integral over $S_\epsilon$ in Eq. (2.45) approaches finite limit $-\phi(\boldsymbol{\xi})$. Therefore,

$$\int\limits_{S}\left(\phi\frac{\partial}{\partial n}\frac{1}{r}-\frac{1}{r}\frac{\partial\phi}{\partial n}\right)dS=-4\pi\phi(\boldsymbol{\xi}), \tag{2.46}$$

if $\boldsymbol{\xi}$ is located inside the boundary surface $S$.

Likewise, suppose $\boldsymbol{\xi}$ is situated on the boundary surface $S$. A hemi-sphere can be used as the surrounding surface $S_\epsilon$, and Eq. (2.46) is then modified as

$$\int\limits_{S}\left(\phi\frac{\partial}{\partial n}\frac{1}{r}-\frac{1}{r}\frac{\partial\phi}{\partial n}\right)dS=-2\pi\phi(\boldsymbol{\xi}). \tag{2.47}$$

Of course, if the singularity point $\boldsymbol{\xi}$ moves outside the boundary surface $S$, there is no need for a surrounding surface $S_\epsilon$, because the entire domain becomes regular with no singularity point. Consequently,

$$\int\limits_{S}\left(\phi\frac{\partial}{\partial n}\frac{1}{r}-\frac{1}{r}\frac{\partial\phi}{\partial n}\right)dS=0. \tag{2.48}$$

Note that the integral in Eq. (2.47) is analogous to the principal value integral discussed in Chapter 2. In fact, Eqs. (2.46), (2.47), and (2.48) provide us with a base for boundary element methods. Given the effects of sources, sinks, dipoles, and multipoles on surrounding fluid, it seems plausible to have a finite collection of them in the representation of general body geometries. Thus, Fredholm integral equations, named after Erik Ivar Fredholm (1866–1927), can be introduced for the solution of FSI problems involving potential flow, acoustic medium, and low Reynolds number flow [139,218].

### *5.2.2. Boundary element approaches*

As illustrated in detail in Ref. [158], two sets of linear integral equations will be encountered:

$$\int\limits_{S} K(\mathbf{x};\mathbf{y})\psi(\mathbf{y})\,dS(\mathbf{y})=f(\mathbf{x}), \tag{2.49}$$

which is called the Fredholm Integral Equation of the first kind; whereas the Fredholm Integral Equation of the second kind is denoted as

$$\psi(\mathbf{x}) - \int\limits_S K(\mathbf{x}; \mathbf{y})\psi(\mathbf{y})\, dS(\mathbf{y}) = f(\mathbf{x}). \tag{2.50}$$

If we introduce the mapping $L$, the Fredholm Integral Equations can be written as the operator equations

$$L\psi = f \quad \text{and} \quad \psi - L\psi = f. \tag{2.51}$$

If the surface $S$ is a function of $\mathbf{x}$, we have the following Volterra Integral Equation of the first kind, named after Vito Volterra (1860–1940):

$$\int\limits_{S(\mathbf{x})} K(\mathbf{x}; \mathbf{y})\psi(\mathbf{y})\, dS(\mathbf{y}) = f(\mathbf{x}), \tag{2.52}$$

and the Volterra Integral Equation of the second kind

$$\psi(\mathbf{x}) - \int\limits_{S(\mathbf{x})} K(\mathbf{x}; \mathbf{y})\psi(\mathbf{y})\, dS(\mathbf{y}) = f(\mathbf{x}). \tag{2.53}$$

Introducing a mapping $\bar{L}$, the Fredholm Integral Equations can also be written as the operator equations

$$\bar{L}\psi = f \quad \text{and} \quad \psi - \bar{L}\psi = f. \tag{2.54}$$

Supposing that Green's function $G(\mathbf{x}; \boldsymbol{\xi})$ and the body velocity $v_n$ are given, the boundary integral method, or panel method, can be easily constructed. If we consider that $\boldsymbol{\xi}$ is located on the body surface, from Eq. (2.47), the discretized surface potential can be evaluated as follows:

$$2\pi\phi^{(l)} + \sum_{k=1,k\neq l}^{N} \phi^{(k)} A_{kl} = \sum_{k=1,k\neq l}^{N} v_n^{(k)} B_{kl}, \tag{2.55}$$

with

$$A_{kl} = \int\limits_{S_k} \frac{\partial G}{\partial n}\, dS \bigg|_{at\,\boldsymbol{\xi}^{(l)}} \quad \text{and} \quad B_{kl} = \int\limits_{S_k} G\, dS \bigg|_{at\,\boldsymbol{\xi}^{(l)}},$$

where $S_k$ and $\boldsymbol{\xi}^{(k)}$ denote the surface area and the coordinate of the geometrical center of the $k$th discretized surface or panel, and $u_n$ stands for the given normal velocity defined by $\partial\phi/\partial n$ on the surface.

Note that the principal value integral requires the exclusion of $\boldsymbol{\xi}$ in the surface integral, thus rendering $k \neq l$. Moreover, to evaluate the potential within the fluid domain, we employ Eq. (2.46) and obtain

$$\phi(\boldsymbol{\xi}) = -\frac{1}{4\pi}\sum_{k=1}^{N}\left[\phi^{(k)}\int_{S_k}\frac{\partial G}{\partial n}\,dS - v_n^{(k)}\int_{S_k} G\,dS\right]. \tag{2.56}$$

## 5.3. FSI systems with potential flows

In civil and ocean engineering, the surface gravity waves often introduce significant hydrodynamic loads on off-shore structures and ships [133,211]. In this section, we use FSI systems with potential flow as examples to illustrate concepts describing wave–object interaction, such as incident wave, reflected wave, transmitted wave, radiation, scattering, diffraction, and scattering.

### 5.3.1. *Interaction with rigid body; added mass; lift surface*

*Interaction with rigid body*

In the study of potential flow interacting with rigid body, the shear forces, which are induced by fluid viscosity, are completely neglected. However, a manifestation of the viscous effect through a circulation condition will be discussed separately in the study of lifting surfaces.

As a result of neglecting shear stresses, the surrounding ideal flow, as depicted in Figure 5.3, occupying the volume $V$ with $\partial V = S_b \cup S_c$, will exert force $\mathbf{F}_b$ and moment $\mathbf{M}_b$ on the submerged body,

$$\mathbf{F}_b = \int_{S_b} p\mathbf{n}\,dS, \tag{3.57}$$

$$\mathbf{M}_b = \int_{S_b} p(\mathbf{r}\times\mathbf{n})\,dS, \tag{3.58}$$

where the unit normal vector $\mathbf{n}$ points towards the body enclosed by the surface $S_b$.

If we introduce a velocity potential $\phi$ and ignore the hydrostatic pressure in Bernoulli's equation, we can rewrite Eqs. (3.57) and (3.58) as

$$\mathbf{F}_b = -\rho\int_{S_b}\left(\frac{\partial\phi}{\partial t} + \frac{1}{2}|\nabla\phi|^2\right)\mathbf{n}\,dS, \tag{3.59}$$

$$\mathbf{M}_b = -\rho\int_{S_b}\left(\frac{\partial\phi}{\partial t} + \frac{1}{2}|\nabla\phi|^2\right)(\mathbf{r}\times\mathbf{n})\,dS. \tag{3.60}$$

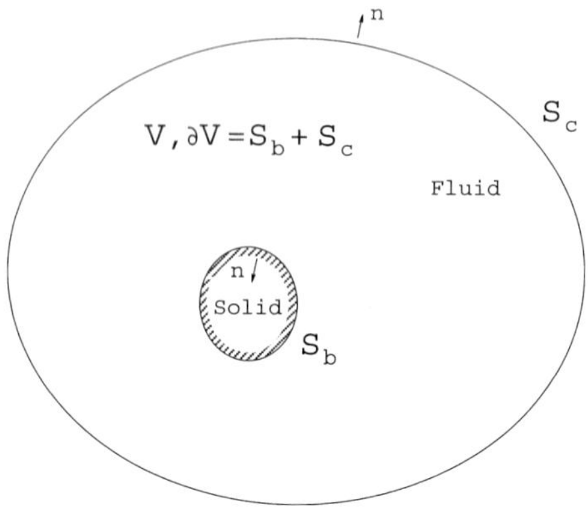


Figure 5.3. A typical control volume selection encircling a submerged body.

First, we consider Reynolds Transport theorems in Eqs. (2.149) and (2.151) in Chapter 2 and arrive at

$$\mathbf{F}_b + \mathbf{F}_c = -\frac{d}{dt}\int\limits_V \rho \nabla\phi \, dV - \rho \int\limits_{S_c} \frac{\partial \phi}{\partial n} \nabla\phi \, dS, \tag{3.61}$$

with

$$\frac{d}{dt}\int\limits_V \nabla\phi \, dV = \int\limits_V \nabla \frac{\partial \phi}{\partial t} \, dV + \int\limits_{S_b+S_c} \nabla\phi \mathbf{v} \cdot \mathbf{n} \, dS. \tag{3.62}$$

For convenience, we choose a fixed control surface with $\mathbf{v} \cdot \mathbf{n} = 0$, on $S_c$ and $\mathbf{v} \cdot \mathbf{n} = \partial\phi/\partial n$, on $S_b$, and $\mathbf{F}_c$ and $\mathbf{v}$ represent the force and the velocity vectors on the control surface $S_c$. Of course, in the discussion of potential flow the fluid density is a constant. By employing the divergence theorem in the form of Eqs. (2.140) and (2.141) in Chapter 2, we have from Eq. (3.62),

$$\frac{d}{dt}\int\limits_{S_b+S_c} \phi \mathbf{n} \, dS = \int\limits_{S_b+S_c} \left[ \frac{\partial \phi}{\partial t}\mathbf{n} + \nabla\phi(\mathbf{v} \cdot \mathbf{n}) \right] dS. \tag{3.63}$$

Obviously, for the chosen fixed control surface, we have

$$\frac{d}{dt}\int\limits_{S_c} \phi \mathbf{n} \, dS = \int\limits_{S_c} \frac{\partial \phi}{\partial t}\mathbf{n} \, dS, \tag{3.64}$$

and consequently, Eq. (3.63) provides

$$\frac{d}{dt}\int\limits_{S_b} \phi \mathbf{n} \, dS = \int\limits_{S_b} \left( \frac{\partial \phi}{\partial t}\mathbf{n} + \frac{\partial \phi}{\partial n}\nabla\phi \right) dS. \tag{3.65}$$

Hence, the expression for $\mathbf{F}_b$ in Eq. (3.61) can be rewritten as follows:

$$\mathbf{F}_b = -\mathbf{F}_c - \rho \frac{d}{dt} \int_{S_b} \phi \mathbf{n}\, dS - \rho \int_{S_c} \frac{\partial \phi}{\partial t} \mathbf{n}\, dS - \rho \int_{S_c} \frac{\partial \phi}{\partial n} \nabla \phi\, dS. \tag{3.66}$$

As a different approach, we substitute Eq. (3.65) into Eq. (3.59) and obtain

$$\mathbf{F}_b = -\rho \frac{d}{dt} \int_{S_b} \phi \mathbf{n}\, dS + \rho \int_{S_b} \left( \frac{\partial \phi}{\partial n} \nabla \phi - \frac{1}{2} \nabla \phi \cdot \nabla \phi \mathbf{n} \right) dS. \tag{3.67}$$

Again, if we employ the divergence theorem in the forms of Eqs. (2.140) and (2.141), we have

$$\int_{S_b+S_c} \left[ \frac{\partial \phi}{\partial n} \frac{\partial \phi}{\partial x_i} - \frac{1}{2} \left( \frac{\partial \phi}{\partial x_i} \right)^2 \mathbf{n} \right] dS = \int_V \left[ \frac{\partial}{\partial x_i} \left( \frac{\partial \phi}{\partial x_i} \nabla \phi \right) - \frac{1}{2} \nabla \left( \frac{\partial \phi}{\partial x_i} \right)^2 \right] dV$$
$$= \int_V \nabla \phi \frac{\partial^2 \phi}{\partial x_i \partial x_i}\, dV = 0. \tag{3.68}$$

Thus, Eq. (3.67) can be rewritten as

$$\mathbf{F}_b = -\rho \frac{d}{dt} \int_{S_b} \phi \mathbf{n}\, dS - \rho \int_{S_c} \left( \frac{\partial \phi}{\partial n} \nabla \phi - \frac{1}{2} \nabla \phi \cdot \nabla \phi \mathbf{n} \right) dS. \tag{3.69}$$

Observing the following definition,

$$\mathbf{F}_c = -\rho \int_{S_c} \left( \frac{\partial \phi}{\partial t} + \frac{1}{2} \nabla \phi \cdot \nabla \phi \right) \mathbf{n}\, dS, \tag{3.70}$$

we can easily show that Eq. (3.69) is identical to Eq. (3.66).

Similarly, by employing the divergence theorem in the form of Eq. (2.142) in Chapter 2, we may derive the alternative expression for the moment:

$$\mathbf{M}_b = -\rho \frac{d}{dt} \int_{S_b} \phi (\mathbf{r} \times \mathbf{n})\, dS - \rho \int_{S_c} \mathbf{r} \times \left( \frac{\partial \phi}{\partial n} \nabla \phi - \frac{1}{2} \nabla \phi \cdot \nabla \phi \mathbf{n} \right) dS. \tag{3.71}$$

*Added mass*

For a rigid body moving in a potential flow field, interaction is manifested in the total force and moment acting on the rigid body. Assuming that the fluid domain is unbounded. The contribution from the external surface $S_c$ vanishes and Eqs. (3.59) and (3.60) are simplified as

$$\mathbf{F}_b = -\rho \frac{d}{dt} \int_{S_b} \phi \mathbf{n} \, dS, \tag{3.72}$$

$$\mathbf{M}_b = -\rho \frac{d}{dt} \int_{S_b} \phi (\mathbf{r} \times \mathbf{n}) \, dS. \tag{3.73}$$

Consider the most general case of a moving rigid body with a translational velocity $\bar{\mathbf{v}}(t)$ and an angular velocity $\bar{\boldsymbol{\omega}}(t)$. In ocean engineering, the three translational components $\bar{v}_1$, $\bar{v}_2$, and $\bar{v}_3$ denote the surge, heave, and sway motions. The three rotational components $\bar{\omega}_1$, $\bar{\omega}_2$, and $\bar{\omega}_3$ denote the roll, yaw, and pitch motions. To simplify further discussion, we often assign $\bar{v}_4 = \bar{\omega}_1$, $\bar{v}_5 = \bar{\omega}_2$, and $\bar{v}_6 = \bar{\omega}_3$.

As discussed in Chapter 3, for a moving rigid body it is often convenient to attach a moving coordinate frame $\hat{o}\hat{x}\hat{y}\hat{z}$ with the origin $\hat{o}$ on the geometrical center. Thus, the Cartesian coordinate system $oxyz$ represents the fixed coordinate frame. Utilizing Eq. (1.11) in Chapter 2, we have

$$\frac{\partial \phi}{\partial n} = \bar{\mathbf{v}}(t) \cdot \mathbf{n} + \bar{\boldsymbol{\omega}}(t) \cdot (\hat{\mathbf{r}} \times \mathbf{n}), \qquad \text{on } S_b, \tag{3.74}$$

where the unit normal vector $\mathbf{n}$ and the position vector $\hat{\mathbf{r}}$ are attached to the moving coordinate system.

From Eq. (3.74), it is suggested that the total potential can be expressed as the following sum:

$$\phi(t) = \sum_{i=1}^{6} \bar{v}_i(t) \phi_i, \qquad \text{with } i = 1, 2, \ldots, 6, \tag{3.75}$$

where $\phi_i$ represents the velocity potential due to a body motion with unit velocity in the $i$th mode.

Note that because the translational and angular velocities have different units, the potentials $\phi_i$ are not homogeneous with respect to their dimensions. Moreover, because Eq. (3.75) holds for arbitrary velocity components $\bar{v}_i$, the potentials $\phi_i$ must be governed by Laplace's equation:

$$\nabla^2 \phi_i = 0, \qquad \text{with } i = 1, 2, \ldots, 6, \tag{3.76}$$

and satisfy the following boundary conditions

$$\frac{\partial \phi_i}{\partial n} = n_i, \qquad \text{with } i = 1, 2, 3, \tag{3.77}$$

$$\frac{\partial \phi_i}{\partial n} = (\hat{\mathbf{r}} \times \mathbf{n})_{i-3}, \qquad \text{with } i = 4, 5, 6. \tag{3.78}$$

REMARK 5.3.1. In Eqs. (3.72) and (3.73), the integrals are evaluated in the local coordinate system, whereas the force and moment, as well as the position vectors, are expressed in the fixed coordinate system. In general, the unit normal vector **n** is time-dependent. However, in the special case of steady translation, the unit normal vector is independent of time.

Hence, the potentials $\phi_i$, when expressed in the moving or body-fixed coordinate system, will depend only on the body geometry via the boundary conditions (3.77) and (3.78) and are independent of the velocities $\bar{v}_i$. Therefore, the decomposition (3.75) isolates the time dependence of the velocity potential.

In a special case of steady translation of a rigid body in an infinite, inviscid, and irrotational fluid, based on Eq. (3.72), it is easy to verify the so-called d'Alembert's paradox: the hydrodynamic force acting on the body is zero. Furthermore, using Eq. (3.68) in an infinite domain because the integrals on $S_c$ are neglected, the second term of Eq. (3.65) corresponds directly to the second term of Eq. (3.59). Therefore, d'Alembert's paradox is identical to the omission of the second term of Eq. (3.65) or Eq. (3.59).

REMARK 5.3.2. At this point, we would like to clarify a point about the time dependence of the unit normal vector **n**. In the discussion of Eqs. (3.61) to (3.71), we introduce transport theorems and various forms of the divergence theorem, in which the unit normal vector **n** is dependent of time with respect to the fixed, or reference coordinate system. Consequently, on $S_b$, we have $\partial\phi/\partial n = \mathbf{v}\cdot\mathbf{n}$, where **v** represents a general expression of the surface velocity vector. However, this does not conflict with the more specific definition of the normal velocity component on $S_b$ in the form of Eq. (3.74).

REMARK 5.3.3. According to the linear decomposition of Eq. (3.75), the velocity potential is linearly proportional to $\bar{v}_i$. Combined with the discussion of d'Alembert's paradox, or the omission of the second term of Eq. (3.65), we can deduce that the hydrodynamic force and moment must be linearly proportional to $\bar{v}_i$ as well.

Recognizing the following time derivative of the local variable **n** and $\hat{\mathbf{r}} \times \mathbf{n}$, which can be directly obtained from Eq. (2.115) in Chapter 2:

$$\frac{d\mathbf{n}}{dt} = \bar{\omega} \times \mathbf{n}, \tag{3.79}$$

$$\frac{d(\hat{\mathbf{r}} \times \mathbf{n})}{dt} = \bar{\omega} \times (\hat{\mathbf{r}} \times \mathbf{n}). \tag{3.80}$$

According to the discussion in Chapter 3, by employing Eq. (1.46) in Chapter 3 and the chain rule, we can rewrite Eqs. (3.72) and (3.73) as

$$\mathbf{F}_b = -\rho\dot{\bar{v}}_i \int_{S_b} \phi_i \mathbf{n}\, dS - \rho\bar{v}_i \bar{\boldsymbol{\omega}} \times \int_{S_b} \phi_i \mathbf{n}\, dS, \tag{3.81}$$

$$\mathbf{M}_b = \mathbf{r}_o \times \mathbf{F}_b - \rho\bar{v}_i \bar{\mathbf{v}} \times \int_{S_b} \phi_i \mathbf{n}\, dS - \rho\dot{\bar{v}}_i \int_{S_b} \phi_i (\hat{\mathbf{r}} \times \mathbf{n})\, dS$$
$$- \rho\bar{v}_i \bar{\boldsymbol{\omega}} \times \int_{S_b} \phi_i (\hat{\mathbf{r}} \times \mathbf{n})\, dS, \tag{3.82}$$

where $\mathbf{r}_o$ represents the position vector of the geometrical center in the fixed coordinate system.

Moreover, from the boundary conditions (3.77) and (3.78), the integrals involved in Eqs. (3.81) and (3.82) can be also expressed as

$$\int_{S_b} \phi_i n_j\, dS = \int_{S_b} \phi_i \frac{\partial \phi_j}{\partial n}\, dS, \tag{3.83}$$

$$\int_{S_b} \phi_i (\hat{\mathbf{r}} \times \mathbf{n})_j\, dS = \int_{S_b} \phi_i \frac{\partial \phi_{j+3}}{\partial n}\, dS, \tag{3.84}$$

with $j = 1, 2, 3$.

At this stage, we can define the added mass tensor as

$$m_{ij} = \rho \int_{S_b} \phi_j \frac{\partial \phi_i}{\partial n}\, dS, \tag{3.85}$$

and Eqs. (3.72) and (3.73) can be expressed as

$$F_j = -\dot{\bar{v}}_i m_{ji} - \epsilon_{jkl} \bar{v}_i \bar{\omega}_k m_{li}, \tag{3.86}$$
$$M_j = -\dot{\bar{v}}_i m_{j+3,i} - \epsilon_{jkl} \bar{v}_i \bar{\omega}_k m_{l+3,i} - \epsilon_{jkl} \bar{v}_i \bar{v}_k m_{li}, \tag{3.87}$$

with $i = 1, 2, \ldots, 6$ and $j, k, l = 1, 2, 3$, where $\epsilon_{jkl}$ is the permutation symbol or alternating tensor.

Note that the term $\mathbf{r}_o \times \mathbf{F}_b$ is avoided by defining the moment about the geometrical center. It is also important to point out that the added mass coefficients depend only on the body geometry. In fact, if the fluid is ideal and irrotational, and viscous forces are neglected, the added mass coefficients are regarded as the most important characteristics of such fluid–solid interaction problems. Furthermore, if there are no rigid body rotations, the total hydrodynamic force does not depend on the steady velocity components. This is a reflection of d'Alembert's paradox, named after Jean le Rond d'Alembert (1717–1783). Nevertheless, under the same condition, the total hydrodynamic moment is not zero due to the third term in Eq. (3.87), which is often called the Munk moment, named after Michael Max Munk (1890–1917).

Employing the divergence theorem along with the governing equations and boundary conditions of the potential $\phi_i$, Eq. (3.85) can be written as

$$m_{ij} = \rho \int_V \nabla\phi_i \cdot \nabla\phi_j \, dV. \tag{3.88}$$

Therefore, the total kinematic energy $T$ within the fluid domain $V$ can be expressed as

$$T = \frac{\rho}{2} \int_V \nabla\phi \cdot \nabla\phi \, dV = \frac{1}{2} \bar{v}_i \bar{v}_j m_{ij}. \tag{3.89}$$

Eq. (3.89) clearly illustrates the physical significance of the added mass, namely, the equivalent mass of the surrounding fluid carrying the corresponding kinetic energy.

*Lifting surfaces*

In the previous discussion, the fluid is assumed to be ideal and irrotational, and viscous effects are neglected. However, in the study of airplane wings and hydrofoils, we need to introduce an interesting variation of fluid–solid interactions involving potential flow, namely, lifting surfaces to account for the lift force.

A lifting surface is a mathematical model of a thin streamlined body tilted at a small angle from its moving direction as shown in Figure 5.4. The small angle is often called the angle of attack, and is directly related to the lift force resulting from the interaction between the lift surface and the surrounding fluid. In addition, a drag force, and in some cases, a moment can also be generated in such fluid–solid interactions. In this section, we focus on the quantitative understanding of two-dimensional lift surfaces. In engineering applications, if the aspect ratio of the transverse dimension ($z$-coordinate) to the cross-sectional dimension ($x$- and $y$-coordinates) is small, it is valid to consider each cross section as a two-dimensional lift surface. Otherwise, three-dimensional effects must be included.

The key assumptions delineating a lift surface from a bluff body are a small angle of attack $\alpha$, a small curvature or camber of the lift surface, and a smooth tangential trailing edge. Hence, viscous effects are confined to a thin boundary layer at a reasonably high Reynolds number. Moreover, if the flow does not leave the trailing edge in a smooth tangential manner, there must be an infinite velocity around the trailing edge in accordance with the discussion of corner flows. Therefore, an important mathematical condition, the so-called Kutta condition, named

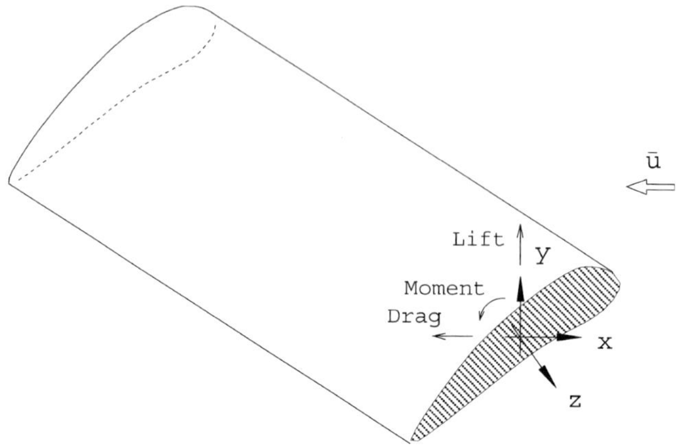


Figure 5.4. Geometrical definitions of a typical lift surface.

after Martin Wilhelm Kutta (1867–1944), must be imposed to render the velocity at the trailing edge finite. This mathematical condition requires a distribution of a positive, i.e., counterclockwise circulation around the lift surface, assuming the incoming uniform flow $\bar{u}$ is from right to left and depicted in Figure 5.4. The addition of such a circulation seems to contradict Kelvin's theorem. Nevertheless, if we consider an upstream material contour at time $t_1$ and $t_2$, as illustrated in Figure 5.6, with the lift surface piercing the material contour, the subsequent material contour surrounding the lift surface and a portion of its wake does indeed remain at zero. Furthermore, if the material contour is sufficiently large and surrounds the lift surface initially and at all subsequent times, the mathematically introduced circulation must then be balanced by an equal and opposite vortex, often considered as the starting vortex shed from the trailing edge into the wake during the initial acceleration. In actuality, unless we are interested in the transient behavior of the lift surface, the starting vortex can be disregarded. Although the vortex initially shed from the accelerating lift surface will eventually disappear due to the fluid viscosity, the circulation attached to the lift surface is essential to the generation of the lift force.

As shown in Figure 5.5, a typical lift surface is oriented in the $x$ direction, with upper and lower surfaces denoted as $y_u(x)$ and $y_l(x)$, respectively. For convenience, we assume the leading and trailing edges to be situated at $x = 1$ and $x = -1$. Moreover, we assign a perturbation velocity vector $\langle u, v \rangle$, representing the flow velocity due to the existence of the lift surface. Hence, the total fluid velocity vector is $\langle u - \bar{u}, v \rangle$. If we defining the perturbation velocity potential as $\phi(x, y)$ such that $\nabla \phi = \langle u, v \rangle$, the governing equation and boundary conditions

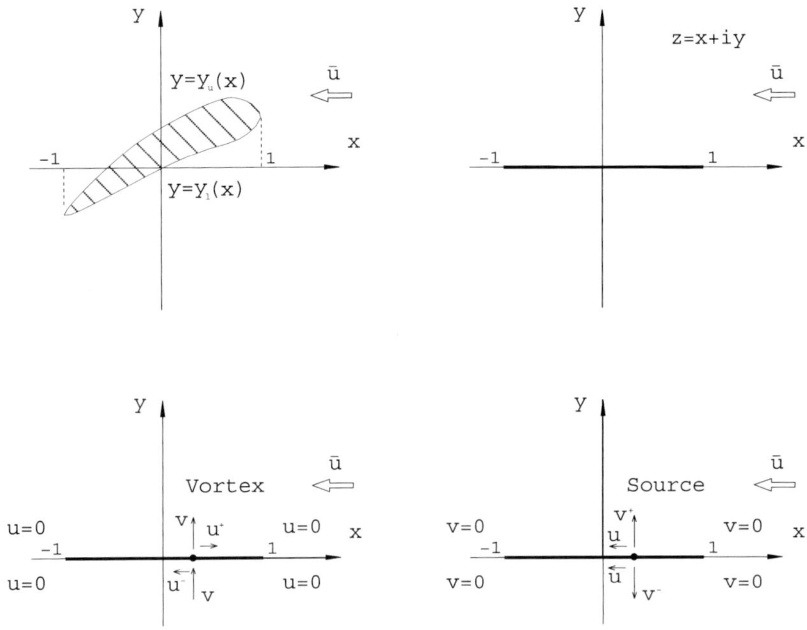


Figure 5.5. Perturbation velocity across the cut for odd and even problems.

can be written as

$$
\begin{aligned}
&\nabla^2\phi = 0, \\
&\frac{\partial \phi}{\partial n} = \bar{u} n_1, \quad \text{on } y = y_u(x) \text{ and } y_l(x), \\
&\boldsymbol{\nabla}\phi < \infty, \quad \text{at the trailing edge}, \\
&\boldsymbol{\nabla}\phi \to 0, \quad \text{at } \infty,
\end{aligned}
\tag{3.90}
$$

where $n_1$ stands for the $x$ component of the unit normal vector.

In this discussion, we limit ourselves to linearized lifting surfaces by assuming the following:

- $u \sim O(\epsilon)$ and $\bar{u} \sim O(1)$ with $\epsilon \ll 1$, or $u \ll \bar{u}$; and
- $y_u(x)$, $y_l(x) \sim O(\epsilon)$, and $\partial y_u/\partial x$, $\partial y_l/\partial x \sim O(\epsilon)$.

This, from the kinematic boundary condition, we can easily derive

$$
\frac{\partial \phi}{\partial y} = \left(\frac{\partial \phi}{\partial x} - \bar{u}\right)\frac{\partial y_u}{\partial x}, \quad \text{on } y = y_u(x), \tag{3.91}
$$

$$
\frac{\partial \phi}{\partial y} = \left(\frac{\partial \phi}{\partial x} - \bar{u}\right)\frac{\partial y_l}{\partial x}, \quad \text{on } y = y_l(x). \tag{3.92}
$$

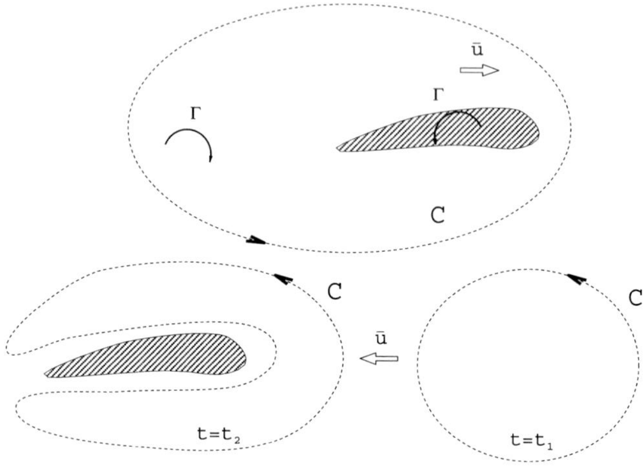


Figure 5.6. Kutta condition and Kelvin's theorem.

If we employ the linearization assumptions and Taylor's expansion, Eqs. (3.91) and (3.92) can be simplified as

$$\frac{\partial \phi}{\partial y} = -\bar{u}\frac{\partial y_u}{\partial x}, \quad \text{on } y = 0+, \tag{3.93}$$

$$\frac{\partial \phi}{\partial y} = -\bar{u}\frac{\partial y_l}{\partial x}, \quad \text{on } y = 0-. \tag{3.94}$$

REMARK 5.3.4. This linearization implies that in two-dimensional cases, the lift surface can be considered as a branch cut with upper and lower surfaces.

The simplification from Eqs. (3.91) and (3.92) to Eqs. (3.93) and (3.94) is crucial. Based on this we no longer have to deal with the actual geometry of the lifting surface; rather, a singular cut at $y = 0$ within a finite interval of $x$ coordinate, namely, $x \in (-1, 1)$ will suffice. Moreover, neglecting the nonlinear terms of $O(\epsilon^2)$, the dynamic pressure in the fluid is given by

$$p - p_\infty = -\frac{1}{2}\rho(\mathbf{v}\cdot\mathbf{v} - \bar{u}^2) \simeq \rho u \bar{u}. \tag{3.95}$$

Consequently, the lift force $L$, oriented in the vertical or $y$-direction, is derived with the same order of approximation, by integrating the pressure difference across the lift surface:

$$L = \oint (p - p_\infty)\, dx = \rho\bar{u} \oint u\, dx = \rho\bar{u}\Gamma, \tag{3.96}$$

where $\Gamma$ is the total circulation around the lift surface.

In fact, Eq. (3.96) can be generated as the Kutta–Joukowski theorem, which states that for any two-dimensional body, moving at a constant velocity in an unbounded inviscid fluid, the hydrodynamic pressure force, i.e., lift force, is normal to the velocity vector and is equal to the product of the fluid density, body velocity, and circulation about the body.

Here, to represent two distinct physical problems, we introduce an even potential $\phi_e$ and an odd potential $\phi_o$, such that

$$\begin{aligned}
\phi(x, y) &= \phi_e(x, y) + \phi_o(x, y), \\
\phi_e(x, y) &= \frac{1}{2}\bigl[\phi(x, y) + \phi(x, -y)\bigr], \\
\phi_o(x, y) &= \frac{1}{2}\bigl[\phi(x, y) - \phi(x, -y)\bigr].
\end{aligned} \tag{3.97}$$

Note that the operator $\partial/\partial y$ is odd with respect to $y$. Consequently, $\partial\phi_e/\partial y$ and $\partial\phi_o/\partial y$ are odd and even functions, respectively. Therefore, based on (3.97), boundary conditions (3.93) and (3.94) can be decomposed as

$$\begin{aligned}
\frac{\partial\phi_e}{\partial y} &= \mp\frac{\bar{u}}{2}\frac{\partial(y_u - y_l)}{\partial x}, \quad \text{on } y = 0\pm, \\
\frac{\partial\phi_o}{\partial y} &= -\frac{\bar{u}}{2}\frac{\partial(y_u + y_l)}{\partial x}, \quad \text{on } y = 0\pm.
\end{aligned} \tag{3.98}$$

As illustrated in Figure 5.5, the even potential $\phi_e$ corresponds to a source problem, presenting the effects of thickness, denoted as $h(x) = y_u - y_l$, at zero angle of attack, whereas the odd potential $\phi_o$ corresponds to a vorticity problem, representing the effects of camber and angle of attack, for which the so-called mean-camber line is defined by the center curve $c(x) = (y_u + y_l)/2$. It is clear that the pressure distribution is symmetric in the even potential problem, which has no direct contribution to the lift and moment. Nevertheless, thickness has indirect effects on pressure distribution through alteration at the separation or cavitation point. In particular, the thickness effect around the leading edge plays an important role in rendering the infinite velocity around the sharp edge of the mean-camber line which exists in the mathematical model as a more physical finite value. Note that the similar infinite velocity at the trailing edge is prevented by viscous effects in the real fluid via the Kutta condition.

To complete the mathematical model for both even and odd situations, we must set appropriate boundary conditions. As depicted in Figure 5.5, for the even problem, $u(x, y)$ and $v(x, y)$ are an even and odd function of $y$, respectively. Naturally, outside the cut, where the functions are continuous, $v(x, y)$ must vanish on the real axis. Also illustrated in Figure 5.5, for the odd problem, $u(x, y)$ and $v(x, y)$

are an odd and even function of $y$, respectively. Thus, $u(x, y)$ must vanish on the real axis outside the cut.

Consider first the odd problem with the potential $\phi_o$, and introduce a potential function

$$F(z) = -\frac{i}{2\pi}\int_{-1}^{1} \gamma(\xi)\ln(z-\xi)\,d\xi, \tag{3.99}$$

with the vortex distribution $\gamma(\xi)$, with $\xi \in (-1, 1)$.

To take the derivative of this odd potential function, we must consider the singularity point at $\xi = x$. Employing the semi-circular contours around $\xi = x$, for the upper and lower sides of the lift surface, we obtain

$$\begin{aligned} u - iv = \frac{dF(z)}{dz} &= \frac{i}{2\pi}\int_{-1}^{1} \frac{\gamma(\xi)}{\xi - z}\,d\xi \\ &= \frac{i}{2\pi}\int_{-1}^{1} \frac{\gamma(\xi)}{\xi - x}\,d\xi \mp \frac{\gamma(x)}{2}, \qquad \text{on } y = 0\pm. \end{aligned} \tag{3.100}$$

Therefore, directly from Eq. (3.100), we obtain

$$\begin{aligned} u^+ - u^- &= -\gamma(x), \\ v^+ = v^- &= -\bar{u}\frac{dc(x)}{dx} = -\frac{1}{2\pi}\int_{-1}^{1} \frac{\gamma(\xi)}{\xi - x}\,d\xi. \end{aligned} \tag{3.101}$$

Naturally, the total circulation required in the Kutta condition can be expressed as

$$\Gamma = \int_{-1}^{1} \gamma(\xi)\,d\xi. \tag{3.102}$$

Let us now consider the even problem with the potential $\phi_e$. We introduce the potential function

$$F(z) = \frac{1}{2\pi}\int_{-1}^{1} q(\xi)\ln(z-\xi)\,d\xi, \tag{3.103}$$

with the source distribution $q(\xi)$.

Likewise, to take a derivative of this even potential function, we also have to consider the singularity point at $\xi = x$. Again, employing the semi-circular contours for the upper and lower sides of the lift surface, we obtain

$$u - iv = \frac{dF(z)}{dz} = -\frac{1}{2\pi}\int_{-1}^{1}\frac{q(\xi)}{\xi - z}\,d\xi$$
$$= -\frac{1}{2\pi}\int_{-1}^{1}\frac{q(\xi)}{\xi - x}d\xi \mp \frac{iq(x)}{2}, \quad \text{on } y = 0\pm. \tag{3.104}$$

We then derive, directly from Eq. (3.104),

$$\begin{aligned} v^{+} - v^{-} &= -\bar{u}\frac{dh(x)}{dx} = q(x), \\ u^{+} = u^{-} &= -\frac{1}{2\pi}\int_{-1}^{1}\frac{q(\xi)}{\xi - x}\,d\xi. \end{aligned} \tag{3.105}$$

Combine the odd and even potential functions, we derive

$$F(z) = \phi + i\psi = \frac{1}{2\pi}\int_{-1}^{1}\big[q(\xi) - i\gamma(\xi)\big]\ln(z - \xi)\,d\xi, \tag{3.106}$$

and consequently,

$$u - iv = \frac{dF(z)}{dz} = -\frac{1}{2\pi}\int_{-1}^{1}\frac{q(\xi) - i\gamma(\xi)}{\xi - z}\,d\xi. \tag{3.107}$$

Hence, employing the semi-circular contours for the upper and lower sides of the lift surface, we obtain

$$u^{\pm} - iv^{\pm} = -\frac{1}{2\pi}\int_{-1}^{1}\frac{q(\xi) - i\gamma(\xi)}{\xi - x}\,d\xi$$
$$\mp \frac{iq(x) + \gamma(x)}{2}, \quad \text{on } y = 0\pm, \tag{3.108}$$

At this point, we must apply the boundary conditions depicted in Figure 5.5, and derive the following from Eqs. (3.98) and (3.108):

$$\begin{aligned} v^{\pm}(x) &= \pm\frac{\bar{u}}{2}\frac{dh(x)}{dx}, \quad \text{on } y = 0\pm, \\ u^{\pm}(x) &= \mp\frac{\gamma(x)}{2}, \quad \text{on } y = 0\pm. \end{aligned} \tag{3.109}$$

In accordance with convention, the positive vortex distribution $\gamma(x)$ corresponds to the counterclockwise rotation. To solve the lifting problem, it is necessary to determine the vortex and source strength. Apparently, given the thickness $h(x)$ and using Eq. (3.105), the source strength can be easily determined. On the other hand, given the camber line $c(x)$, the lifting problem is a mixed boundary-value problem governed by a singular integral equation, since the kernel or known part of the integrand is singular.

To determine the vortex strength $\gamma(x)$, we must first exploit the physical understanding of the conjugate nature of the source and vortex problems. Suppose the complex velocity $u - iv$ in the lifting problem is multiplied by a suitably chosen analytic function, such as $\sqrt{1-z^2}$, which is real on the cut and pure imaginary on the remainder of the real axis. The product will be a new analytic function whose imaginary part is known all along the real axis, just like the thickness problem. Clearly, $\sqrt{1-z^2}$ is multiple valued and the concept of branch point and branch cut discussed in Section 2.3 must be used. Denoting

$$\tilde{u} - i\tilde{v} = (u - iv)\sqrt{1-z^2}, \tag{3.110}$$

we have

$$\tilde{v}^{\pm} = \begin{cases} -\bar{u}\frac{dc(x)}{dx}\sqrt{1-x^2}, & \text{on } y = 0\pm,\ |x| < 1, \\ 0, & \text{on } y = 0\pm,\ |x| > 1. \end{cases} \tag{3.111}$$

Then, the converted problem is indistinguishable from the thickness problem, and the equivalent source strength can be expressed according to Eq. (3.105) as

$$\tilde{q}(x) = -2\bar{u}\frac{dc(x)}{dx}\sqrt{1-x^2}, \quad \text{for } |x| < 1. \tag{3.112}$$

Employing Eq. (3.98), we have

$$u - iv = \frac{1}{\pi}\frac{1}{\sqrt{1-z^2}}\int_{-1}^{1}\bar{u}\frac{dc(\xi)}{d\xi}\frac{\sqrt{1-\xi^2}}{\xi - z}\,d\xi. \tag{3.113}$$

Therefore, if we employ the semi-circular contour for the upper and lower sides of the lift surfaces, from Eq. (3.113), we can easily derive

$$\gamma(x) = -\frac{2}{\pi}\frac{1}{\sqrt{1-x^2}}\int_{-1}^{1}\bar{u}\frac{dc(\xi)}{d\xi}\frac{\sqrt{1-\xi^2}}{\xi - x}\,d\xi. \tag{3.114}$$

Note that here we have square-root singularities at both the leading and trailing edge with the multiplication factor $\sqrt{1-z^2}$. To anticipate the singularity at the leading edge, and to satisfy the Kutta condition, we can instead multiply the complex velocity $u - iv$ in the lifting problem by a suitably chosen analytic function, such as $\sqrt{(1-z)/(1+z)}$, and derive

$$\gamma(x) = -\frac{2}{\pi}\sqrt{\frac{1+x}{1-x}}\int_{-1}^{1}\bar{u}\frac{dc(\xi)}{d\xi}\sqrt{\frac{1-\xi}{1+\xi}}\frac{1}{\xi-x}\,d\xi. \tag{3.115}$$

It is important to recognize that Eqs. (3.114) and (3.115) represent two particular solutions of the singular integral equation (3.101) for $\gamma(x)$. Hence, to obtain the complete solution of Eq. (3.101), we need to find the solution for the following homogeneous equation:

$$-\frac{1}{2\pi}\int_{-1}^{1}\frac{\gamma(\xi)}{\xi-x}\,d\xi = 0. \tag{3.116}$$

It is not difficult to identify the solution of the homogeneous equation (3.116) as inversely proportional to the function $\sqrt{1-x^2}$ based on the discussions in Chapter 2 and Eq. (3.244). First of all, let us discuss the difference between the two particular solutions in Eqs. (3.114) and (3.115). From the following derivation, it is clear that the difference between the two particular solutions is inversely proportional to the function $\sqrt{1-x^2}$:

$$\begin{aligned}&\int_{-1}^{1}\sqrt{1-\xi^2}\left(1-\frac{1+x}{1+\xi}\right)\frac{dc(\xi)}{d\xi}\frac{\bar{u}}{\xi-x}\,d\xi\\ &\quad=\int_{-1}^{1}\sqrt{\frac{1-\xi}{1+\xi}}\bar{u}\frac{dc(\xi)}{d\xi}\,d\xi = D,\end{aligned} \tag{3.117}$$

where $D$ stands for the constant derived from the definite integral involving $\xi$.

Therefore, the complete solution of the singular integral equation (3.101) can be expressed as

$$\gamma(x) = \frac{2}{\pi}\frac{1}{\sqrt{1-x^2}}\left[\int_{-1}^{1}-\bar{u}\frac{\sqrt{1-\xi^2}}{\xi-x}\frac{dc(\xi)}{d\xi}\,d\xi + C\right], \tag{3.118}$$

where $C$ is an arbitrary constant.

Substituting Eq. (3.118) into the definition of the total circulation (3.102), we obtain

$$\Gamma = -2\frac{\bar{u}}{\pi}\int_{-1}^{1}\frac{dc(\xi)}{d\xi}\sqrt{1-\xi^2}\,d\xi\int_{-1}^{1}\frac{dx}{(\xi-x)\sqrt{1-x^2}} + 2C. \tag{3.119}$$

Since the homogeneous solution of Eq. (3.116) is inversely proportional to $\sqrt{1-x^2}$, implying that the last integral in Eq. (3.119) is zero, we have $C = \Gamma/2$. In addition, to obtain a finite value of $\gamma(x)$ for the trailing edge $x = -1$, we must have

$$C = \int_{-1}^{1} \bar{u} \frac{dc(\xi)}{d\xi} \frac{\sqrt{1-\xi^2}}{1+\xi} \, d\xi. \tag{3.120}$$

After imposing the Kutta condition at the trailing edge, we have the choice of $C$ or $\Gamma$ to complete the solution of the singular integral equation (3.101). Thus, the lift $L$ and the moment $M$ acting on the lifting surface follow directly from (3.119):

$$L = \rho\bar{u} \int_{-1}^{1} \gamma(x) \, dx,$$

$$M = \rho\bar{u} \int_{-1}^{1} \gamma(x) x \, dx. \tag{3.121}$$

In practice, we also define the nondimensional lift and moment coefficients as $C_L = 2L/\rho\bar{u}^2 l$ and $C_M = 2M/\rho\bar{u}^2 l^2$, with the chord length $l = 2$. Moreover, the center of pressure can be easily evaluated as $M/L$. It is clear that given the mean-camber line $c(x)$, we can directly derive $C_L$ and $C_M$ from Eq. (3.118). However, to evaluate the Cauchy principal value of the improper real integral in Eq. (3.118), we must employ the Hilbert transformation, named after David Hilbert (1862–1943). Let us first introduce the following definite integral

$$I_n = \int_{-1}^{1} \frac{x^n \, dx}{\sqrt{1-x^2}} = \int_{0}^{\pi} (\cos\theta)^n \, d\theta$$
$$= \begin{cases} 0, & \text{for odd } n, \\ \pi \frac{1 \cdot 3 \cdot 5 \ldots (n-1)}{2 \cdot 4 \cdot 6 \ldots (n)}, & \text{for even } n. \end{cases} \tag{3.122}$$

Hence, the Hilbert transformation is defined as

$$H_n(x) = \int_{-1}^{1} \frac{\xi^n \, d\xi}{(\xi - x)\sqrt{1-\xi^2}}. \tag{3.123}$$

Since the homogeneous solution for Eq. (3.116) is inversely proportional to $\sqrt{1-x^2}$, we have for $x \in (-1, 1)$, $H_o(x) = 0$. However, to derive $H_o(x)$ for

$x \in (-\infty, -1)$ or $(1, \infty)$, we must use the following complex integral:

$$H_o(z) = \int_{-1}^{1} \frac{d\xi}{(\xi - z)\sqrt{1-\xi^2}} = -\pi \frac{1}{\sqrt{z^2-1}}, \tag{3.124}$$

where $z$ is the complex variable and $H_o(x)$ corresponds to the real part of $H_o(z)$.

Naturally, according to the branch cut discussion in Chapter 2, we can see that, for $x \in (-1, 1)$ as $z \to x$ from upper and lower sides of the real axis, the real part of $H_o(z)$ vanishes, while the imaginary part yields $\pm\pi i/\sqrt{1-x^2}$ for $y = 0\pm$. In addition, for $x \in (-\infty, -1)$ or $(1, \infty)$, we have

$$H_o(x) = \begin{cases} -\pi/\sqrt{x^2-1}, & \text{for } x > 1, \\ \pi/\sqrt{x^2-1}, & \text{for } x < -1. \end{cases} \tag{3.125}$$

Furthermore, the recursion formula can be easily derived as

$$H_n(x) = \int_{-1}^{1} \frac{[(\xi - x)\xi^{n-1} + x\xi^{n-1}]\,d\xi}{(\xi - x)\sqrt{1-\xi^2}} = I_{n-1} + xH_{n-1}(x). \tag{3.126}$$

Therefore, following from the recursion formula (3.126), we have for $x \in (-1, 1)$, $H_1(x) = \pi$, $H_2(x) = \pi x$, and more generally, $H_n(x)$ is an $(n-1)$th polynomial, whereas for $x \notin (-1, 1)$, we have

$$H_1(x) = \pi + \begin{cases} -\pi x/\sqrt{x^2-1}, & x > 1, \\ \pi x/\sqrt{x^2-1}, & x < -1, \end{cases} \tag{3.127}$$

and

$$H_2(x) = \pi x + \begin{cases} -\pi x^2/\sqrt{x^2-1}, & x > 1, \\ \pi x^2/\sqrt{x^2-1}, & x < -1. \end{cases} \tag{3.128}$$

We are ready to consider some examples. Given an uncambered foil with an angle of attack $\alpha$, i.e., $dc(x)/dx = \alpha$ and employing Eq. (3.120), we have

$$\Gamma = 2\bar{u} \int_{-1}^{1} \alpha \frac{1-\xi}{\sqrt{1-\xi^2}}\, d\xi = 2\pi\bar{u}\alpha, \tag{3.129}$$

Consequently, $L = 2\rho\bar{u}^2\pi\alpha$ and $C_L = 2\pi\alpha$. Moreover, from Eqs. (3.118) and (3.121), we have

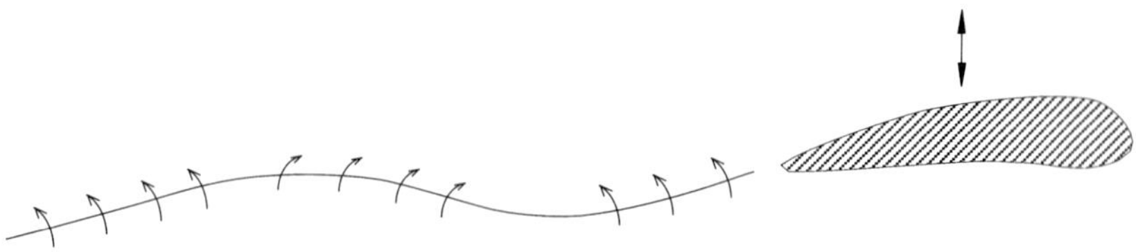

Figure 5.7. A typical oscillating lift surface.

$$M = \left[ -\frac{2\bar{u}}{\pi} \int_{-1}^{1} \frac{dc(\xi)}{d\xi} \sqrt{1-\xi^2}\, d\xi \int_{-1}^{1} \frac{x\, dx}{(\xi - x)\sqrt{1-x^2}} \right.$$

$$\left. + 4\bar{u}\alpha \int_{-1}^{1} \frac{x\, dx}{\sqrt{1-x^2}} \right] \rho \bar{u}. \tag{3.130}$$

Thus, using partial fractions and substituting the values of $I_n$ at $n = 0, 1, 2$, we have $M = \rho\pi\bar{u}^2\alpha$ and $C_M = \pi\alpha/2$. Likewise, for the mean-camber line $c(x) = \alpha x - \beta x^2/2$, we have $C_L = 2\pi\alpha + \pi\beta$ and $C_M = \pi\alpha/2$. At this stage, we can discuss the unsteady lifting surface theory, in which the downwash induced at the lifting surface is time-dependent, as illustrated in Figure 5.7. Here we assume that each individual vortex element induces a contribution to the downwash which diminishes with time as the vortex moves downstream. In this respect, a memory is associated with the motion of the lifting surface. Consequently, the lift force at any given time will depend on the integrated downwash of the entire wake as well as the previous time-history of the motion.

For the odd potential problem, employing the material derivative and linearization to account for the unsteady motion of the mean camber line, we modify Eq. (3.101) and obtain

$$v_o(x,t) = \frac{\partial c(x)}{\partial t} - \bar{u}\frac{\partial c(x)}{\partial x}, \quad \text{for } x \in (-1, 1), \tag{3.131}$$

where the vertical velocity components near the lifting surface are denoted as $v^+(x,t) = v^-(x,t) = v_o(x,t)$.

Based on previous discussion and Eq. (3.101), we have $u^{\pm}(x,t) = \mp\gamma(x,t)/2$. Furthermore, the dynamic pressure expression derived from Bernoulli's equation is also modified from (3.95) to the following:

$$p - p_\infty = -\rho\left(\frac{\partial \phi}{\partial t} - \bar{u}\frac{\partial \phi}{\partial x}\right). \tag{3.132}$$

Since in the vortex wake, there exists no pressure jump across the wake. If we take an $x$ derivative on both sides of Eq. (3.132), and use Eq. (3.101), we derive

$$\frac{\partial \gamma(x,t)}{\partial t} - \bar{u}\frac{\partial \gamma(x,t)}{\partial x} = 0, \quad \text{for } x \in (-\infty, -1). \tag{3.133}$$

Eq. (3.133) suggests that the solution of $\gamma(x,t)$ should have the following wave propagation form:

$$\gamma(x,t) = \gamma(x + \bar{u}t), \quad \text{for } x \in (-\infty, -1). \tag{3.134}$$

Nevertheless, unlike Eq. (3.101), the calculation of the vertical velocity $v_o(x,t)$ must take into consideration of the downwash vortex wake, and Eq. (3.109) is modified as the following singular integral equation:

$$v_o(x,t) = -\frac{1}{2\pi}\int_{-\infty}^{-1} \frac{\gamma(\xi + \bar{u}t)}{\xi - x}\, d\xi - \frac{1}{2\pi}\int_{-1}^{1} \frac{\gamma(\xi,t)}{\xi - x}\, d\xi,$$
$$\text{for } x \in (-1, 1). \tag{3.135}$$

Again, according to Kelvin's theorem, the total circulation of the lifting surface and its wake must amount to zero. Therefore, we have

$$\int_{-\infty}^{1} \gamma(x,t)\, dx = \int_{-\infty}^{-1} \gamma(x,t)\, dx + \Gamma(t) = 0. \tag{3.136}$$

In this case, the total circulation on the lifting surface $\Gamma(t)$ is time-dependent. Hence, if we take a partial derivative with respect to $t$ on both sides of Eq. (3.136), and utilize Eq. (3.133), we derive

$$\frac{d\Gamma(t)}{dt} + \int_{-\infty}^{-1} \bar{u}\frac{\partial \gamma(x,t)}{\partial x}\, dx = 0. \tag{3.137}$$

Assuming that the vortex at $x = -\infty$ is zero, Eq. (3.137) can then be rewritten as

$$\frac{d\Gamma(t)}{dt} = -\bar{u}\gamma(-1,t). \tag{3.138}$$

Eq. (3.138) relates the vorticity shed at the trailing edge to the rate of change of $\Gamma$, which is directly linked to the lift force. Moreover, employing the same technique in the solution of the singular integral equation (3.101), we derive the complete solution:

$$\gamma(x,t) = \frac{2}{\pi}\frac{1}{\sqrt{1-x^2}}\left\{\int_{-1}^{1}\frac{\sqrt{1-\xi^2}}{\xi-x}\left[v_o(\xi,t)\right.\right.$$
$$\left.\left. + \frac{1}{2\pi}\int_{-\infty}^{-1}\frac{\gamma(\zeta+\bar{u}t)}{\zeta-\xi}\,d\zeta\right]d\xi + \frac{\Gamma(t)}{2}\right\}. \tag{3.139}$$

Now, if we employ the Hilbert transformation, the double integral in Eq. (3.139) can be further simplified using partial fractions:

$$\int_{-1}^{1}\frac{\sqrt{1-\xi^2}}{\xi-x}\int_{-\infty}^{-1}\frac{\gamma(\zeta+\bar{u}t)}{\zeta-\xi}\,d\zeta\,d\xi$$
$$= \int_{-\infty}^{-1}\frac{\gamma(\zeta+\bar{u}t)}{\zeta-x}\int_{-1}^{1}\sqrt{1-\xi^2}\left(\frac{1}{\xi-x}-\frac{1}{\xi-\zeta}\right)d\xi\,d\zeta$$
$$= \pi\int_{-\infty}^{-1}\gamma(\zeta+\bar{u}t)\,d\zeta + \pi\int_{-\infty}^{-1}\sqrt{\zeta^2-1}\frac{\gamma(\zeta+\bar{u}t)}{\zeta-x}\,d\zeta, \tag{3.140}$$

where we use, for $\zeta \in (-\infty,-1)$, Eq. (3.128).

Employing Eq. (3.136), we can rewrite Eq. (3.139) as

$$\gamma(x,t) = \frac{2}{\pi\sqrt{1-x^2}}\left[\int_{-1}^{1}\frac{\sqrt{1-\xi^2}}{\xi-x}v_o(\xi,t)\,d\xi\right.$$
$$\left. + \frac{1}{2}\int_{-\infty}^{-1}\sqrt{\xi^2-1}\frac{\gamma(\xi+\bar{u}t)}{\xi-x}\,d\xi\right]. \tag{3.141}$$

Similar to the treatment of the steady solution, we invoke the Kutta condition. The terms in brackets must vanish at $x=-1$, thus,

$$-2\int_{-1}^{1}\sqrt{\frac{1-\xi}{1+\xi}}v_o(\xi,t)\,d\xi = \int_{-\infty}^{-1}\sqrt{\frac{\xi-1}{\xi+1}}\gamma(\xi+\bar{u}t)\,d\xi. \tag{3.142}$$

It is obvious that for the steady case, a degenerate case of Eq. (3.142), namely, $v_o(\xi,t) = -\bar{u}\,dc(\xi)/d\xi$ where a starting vortex $-\Gamma$, is assumed to be situated at $\xi=-\infty$. Eq. (3.142) is equivalent to Eq. (3.120). Again, based on Bernoulli's equation (3.132), the lift force is calculated as

$$L = \rho \int_{-1}^{1} \left( \frac{\partial[\phi]}{\partial t} - \bar{u}\frac{\partial[\phi]}{\partial x} \right) dx = \rho \int_{-1}^{1} \left( \frac{\partial[\phi]}{\partial t} + \bar{u}\gamma \right) dx, \tag{3.143}$$

where the jump [ ] represents the difference across the lifting surface and $[\phi] = \phi^+ - \phi^-$.

Notice that since there is a square-root infinity in the velocity component at $x = 1$, there must be a square-root zero of the velocity potential at $x = 1$. Therefore, integrating Eq. (3.143), we have

$$L = \rho \int_{-1}^{1} \left[ (1+x)\frac{\partial \gamma}{\partial t} + \bar{u}\gamma \right] dx. \tag{3.144}$$

Similarly, the moment is obtained as

$$M = \rho \int_{-1}^{1} \left[ -\frac{1}{2}(1-x^2)\frac{\partial \gamma}{\partial t} + \bar{u}\gamma x \right] dx. \tag{3.145}$$

### *5.3.2. Radiation and scattering*

One important aspect of potential flow interacting with rigid bodies is the need to account for surface wave effects. Of particular interest are dynamic loads on fixed or moving structures. Similar effects can also be found in FSI systems involving acoustic medium and Stokes flow [166]. Some of these problems, in particular, the nonlinear effects, are still open questions for research. In this section, we focus on the linearized wave theory.

In accordance with convention, we assign $y = \eta(x, z, t)$ as the free surface elevation. Hence, the kinematic boundary condition is expressed as

$$\frac{d(y-\eta)}{dt} = \frac{\partial \phi}{\partial y} - \frac{\partial \eta}{\partial t} - \frac{\partial \phi}{\partial x}\frac{\partial \eta}{\partial x} - \frac{\partial \phi}{\partial z}\frac{\partial \eta}{\partial z} = 0, \quad \text{on } y = \eta(x, z, t), \tag{3.146}$$

where $\phi$ represents the velocity potential.

Assuming that the pressure on the free surface is zero, with respect to the atmospheric pressure,

$$\frac{\partial \phi}{\partial t} + \frac{1}{2}\nabla\phi \cdot \nabla\phi + g\eta = 0, \quad \text{on } y = \eta(x, z, t). \tag{3.147}$$

Eq. (3.147) provides the dynamic boundary condition. Moreover, the material derivative of the pressure is zero, i.e.,

$$\left(\frac{\partial}{\partial t} + \nabla\phi \cdot \nabla\right)\left(\frac{\partial \phi}{\partial t} + \frac{1}{2}\nabla\phi \cdot \nabla\phi + g\eta\right) = 0,$$
$$\text{on } y = \eta(x, z, t). \tag{3.148}$$

It is clear that the boundary conditions (3.146) and (3.147) are nonlinear. To simplify the mathematical model, we often introduce the linearization procedure and consequently perturbation approaches. First, let us expand the velocity potential with respect to the plane $y = 0$:

$$\phi(x, y, z, t) = \phi(x, 0, z, t) + y\frac{\partial \phi}{\partial y}\bigg|_{y=0} + \frac{1}{2}y^2\frac{\partial^2 \phi}{\partial y^2}\bigg|_{y=0} + \cdots. \tag{3.149}$$

As with linearization in the study of lifting surfaces, we assume that the elevation of the free surface wave $\eta(x, z, t)$ is of $O(\epsilon)$, with $\epsilon \ll 1$. Hence, the $O(\epsilon^2)$ approximation of Eqs. (3.147) and (3.148) yield, on $y = 0$,

$$\frac{\partial^2 \phi}{\partial t^2} + g\frac{\partial \phi}{\partial y} = 0, \tag{3.150}$$

$$\frac{\partial \phi}{\partial t} + g\eta = 0. \tag{3.151}$$

Naturally, the governing equation for the fluid domain is still Laplace's equation, and the mathematical model can be summarized as follows:

$$\nabla^2\phi = 0, \quad \text{for } -\infty < y < 0, \tag{3.152}$$

$$\frac{\partial \phi}{\partial y} = 0, \quad \text{on } y = -\infty, \tag{3.153}$$

$$\frac{\partial^2 \phi}{\partial t^2} + g\frac{\partial \phi}{\partial y} = 0, \quad \text{on } y = 0, \tag{3.154}$$

where $h$ stands for the finite depth of the potential flow domain.

If we suppose the fluid depth is infinite and assign $k$ and $\omega$ as the wave number and frequency, the characteristic solution of the velocity potential can be expressed as

$$\phi = \text{Re}\left[Ce^{ky - ik(x\cos\theta + z\sin\theta) + i\omega t}\right], \tag{3.155}$$

where the polar angle $\theta$ in the horizontal $x$-$z$ plane represents the wave propagation direction, and the complex constant $C$ can be easily derived, based on (3.151), as $C = igA/\omega$, with the maximum wave elevation $A$.

Without detriment to the discussion of the key concepts, let us focus first on the two-dimensional problem. Deriving directly from the general solution (3.155) and ignoring the phase difference, we can express the potential as

$$\phi = \frac{gA}{\omega} e^{ky} \sin(kx - \omega t). \tag{3.156}$$

Hence, from Eq. (3.151), we derive the wave elevation as

$$\eta = A \cos(kx - \omega t). \tag{3.157}$$

Moreover, from the free surface boundary condition (3.154), we obtain the following dispersion relation:

$$\omega^2 = kg. \tag{3.158}$$

Hence, the velocity components are expressed as

$$\begin{aligned} u &= \frac{\partial \phi}{\partial x} = \omega A e^{ky} \cos(kx - \omega t), \\ v &= \frac{\partial \phi}{\partial y} = \omega A e^{ky} \sin(kx - \omega t). \end{aligned} \tag{3.159}$$

Note that the fluid velocity components decay exponentially with respect to their distance from the free surface. In addition, while the wave, or rather the wave phase, moves at the velocity $v_p$, the fluid particle moves with a much smaller velocity. In fact, it is not difficult to show that the fluid particles on the free surface orbit around their mean positions.

Assign $\lambda$ and $T$ as the wave length and period and the phase velocity $v_p$ can be written as

$$v_p = \frac{\lambda}{T} = \frac{\omega}{k} = \frac{g}{\omega} = \sqrt{\frac{g}{k}}. \tag{3.160}$$

To illustrate the concept of group velocity $v_g$, we consider a wave group of two waves $(\omega, k)$ and $(\omega + d\omega, k + dk)$ with the complex constants $C_1$ and $C_2$, respectively. Because of the linearization assumption, the total wave elevation can be expressed as

$$\eta = \mathrm{Re}\left[ C_1 e^{ky - ikx + i\omega t} \left( 1 + \frac{C_2}{C_1} e^{-idkx + id\omega t} \right) \right]. \tag{3.161}$$

Obviously, Eq. (3.161) represents two types of wave propagations. Here, the factor in parenthesis represents an amplitude modulation and has a slower variation in space and time compared with the first exponential term, the so-called carrier wave. Analogously, Eq. (3.161) represents the same phenomena, which is called beat, in mechanical vibrations and electromagnetic wave motion. In contrast to fast wave speed, namely, phase velocity $v_p$, which is evaluated in Eq. (3.160), slow wave speed, or group velocity $v_g$, is expressed as

$$v_g = \frac{d\omega}{dk} = \frac{g}{2\omega} = \frac{v_p}{2}. \tag{3.162}$$

In order to understand the physical significance of group velocity, we must first introduce wave energy. Physically, surface waves are the result of a balance between kinetic and gravitational potential energy in the fluid. For the same two-dimensional example, we denote a total energy $E$, or rather the energy density per unit area of a vertical column extending throughout the depth of the fluid and bounded by the mean free surface,

$$E = \rho \int_{-\infty}^{\eta} \left( \frac{u^2}{2} + \frac{v^2}{2} \right) dy + \rho \int_{0}^{\eta} gy\, dy = \frac{\rho \omega^2 A^2}{4k} e^{2k\eta} + \frac{\rho g \eta^2}{2}. \tag{3.163}$$

Note here that we ignore the constant gravitational potential energy below the mean free surface, which does not contribute to the wave motion. Moreover, using the linearization assumption and the dispersion relation, we can simplify Eq. (3.163) as

$$E = \frac{\rho g A^2}{4} + \frac{\rho g A^2}{2} \cos^2(kx - \omega t). \tag{3.164}$$

Since the time average of $\cos^2(kx - \omega t)$ is equal to $1/2$, in Eq. (3.164), the kinetic and potential energy are identical and the average total wave energy $\bar{E}$ is denoted as

$$\bar{E} = \frac{\rho g A^2}{2}. \tag{3.165}$$

Employing transport theorems discussed in Chapter 2, Bernoulli's equation, and the following:

$$\frac{\partial}{\partial t}\left( \frac{1}{2} \nabla \phi \cdot \nabla \phi \right) = \nabla \phi \cdot \nabla \frac{\partial \phi}{\partial t} = \nabla \cdot \left( \frac{\partial \phi}{\partial t} \nabla \phi \right), \tag{3.166}$$

we can easily derive the rate of wave energy flux across a typical vertical column, namely, one side of the control surface,

$$\frac{dE}{dt} = \rho \int_{-\infty}^{\eta} \frac{\partial \phi}{\partial t} \frac{\partial \phi}{\partial x} dy + \rho \int_{-\infty}^{\eta} \left( \frac{u^2}{2} + \frac{v^2}{2} \right) v^c\, dy + \rho \int_{0}^{\eta} g y v^c\, dy, \tag{3.167}$$

where the velocity $v^c$ is the velocity of the control surface.

Hence, we introduce the linearization assumption and Eq. (3.156). The time average of Eq. (3.179) can then be expressed as

$$\overline{\frac{dE}{dt}} = -\frac{\rho g^2 A^2}{4\omega} + \frac{\rho g A^2}{2} v^c. \tag{3.168}$$

Therefore, for a particularly selected velocity, which we can easily derive as $v_g$, the time average wave energy flux across a typical vertical column is zero.

This confirms that the physical meaning of the group velocity corresponds to the energy propagation velocity.

Considering the fluid domain with a finite depth $h$, we can easily modify the characteristic equation (3.155) as

$$\phi = \frac{gA}{\omega}\frac{\cosh k(y+h)}{\cosh kh}\sin(kx - \omega t), \tag{3.169}$$

and the velocity components can be expressed as

$$\begin{aligned} u &= \frac{\partial \phi}{\partial x} = \frac{gAk}{\omega}\frac{\cosh k(y+h)}{\cosh kh}\cos(kx - \omega t), \\ v &= \frac{\partial \phi}{\partial y} = \frac{gAk}{\omega}\frac{\sinh k(y+h)}{\cosh kh}\sin(kx - \omega t). \end{aligned} \tag{3.170}$$

Similarly, substituting Eq. (3.169) into the free surface interface condition (3.154), we derive the dispersion relationship as

$$\omega^2 = gk \tanh kh. \tag{3.171}$$

Hence, the group velocity can be expressed as

$$v_g = \frac{1}{2}\left(1 + \frac{2kh}{\sinh 2kh}\right)\frac{\omega}{k}. \tag{3.172}$$

For deep water or short wave, i.e., $kh \gg 1$, $\tanh kh \simeq 1$ and from Eqs. (3.171) and (3.172) we can easily recover Eq. (3.162). However, for shallow water, i.e., $kh \ll 1$, $\tanh kh \simeq kh$, Eqs. (3.171) and (3.172) yield

$$\omega^2 = k^2 gh, \tag{3.173}$$

$$v_g = \frac{\omega}{k}. \tag{3.174}$$

In this case, the group velocity is identical to the phase velocity. It is also interesting to note that for shallow water, horizontal velocity is uniform across the depth of the potential flow domain and vertical velocity vanishes linearly towards the bottom.

Let us now return to the discussion of three-dimensional wave motion. With the same linearization assumption, the two-dimensional results can be easily extended to three-dimensional motions by superimposing waves moving at the oblique angle $\theta$, and can be expressed as

$$\eta(x,z,t) = \mathrm{Re}\left[\int_0^\infty d\omega \int_0^{2\pi} d\theta\, A(\omega,\theta) e^{-ik(x\cos\theta + z\sin\theta) + i\omega t}\right]. \tag{3.175}$$

Consider the wave motions generated by an object, moving at a speed of $v^c$ in $x$ direction, we attach a reference system moving with the object. The wave elevation is then expressed as

$$\eta(x,z,t) = \mathrm{Re}\left[\int_0^{\infty} d\omega \int_0^{2\pi} d\theta\, A(\omega,\theta) e^{-ik(x\cos\theta + z\sin\theta) + i(\omega - kv^c\cos\theta)t}\right], \tag{3.176}$$

where $k(\omega)$ is the wavenumber corresponding to a given frequency $\omega$ in accordance with the dispersion relation for infinite water depth.

Hence, for a steady state with a constant velocity $v^c$, with respect to the moving reference frame, Eq. (3.176) should be time-independent. Thus, the phase velocity of each admissible wave component is given by

$$v_p = \frac{\omega}{k} = v^c \cos\theta. \tag{3.177}$$

Eq. (3.177) requires that $\cos\theta > 0$ for a given frequency and wave number. Moreover, from the dispersion relationship in deep water (3.158), Eq. (3.177) is modified as

$$k(\theta) = \frac{g}{(v^c)^2\cos^2\theta}. \tag{3.178}$$

Consequently, Eq. (3.176) is simplified as

$$\eta(x,z) = \mathrm{Re} \int_{-\pi/2}^{\pi/2} A(\theta) e^{-ik(\theta)(x\cos\theta + z\sin\theta)}\, d\theta. \tag{3.179}$$

It is important to recognize that in the moving reference coordinate, energy propagation velocity, i.e., group velocity, is zero. Therefore, we have

$$\frac{d}{dk}\left[k(\theta)(x\cos\theta + z\sin\theta)\right] = 0. \tag{3.180}$$

Now, substitute Eq. (3.178) into Eq. (3.180), and we have

$$x\sec^2\theta\sin\theta + z\sec^3\theta(1+\sin^2\theta) = 0, \tag{3.181}$$

or

$$\frac{z}{x} = -\frac{\cos\theta\sin\theta}{1+\sin^2\theta}. \tag{3.182}$$

It is obvious from Eq. (3.182) that there exist two wave systems: namely, a diverging wave system with $|\theta| > \sin^{-1}(1/\sqrt{3})$, and a transverse wave system with $|\theta| < \sin^{-1}(1/\sqrt{3})$. Both wave systems meet at the maximum or minimum values of $z/x = \pm 2^{-3/2}$. Such wave patterns are named Kelvin wave patterns and depicted in Figure 5.9.

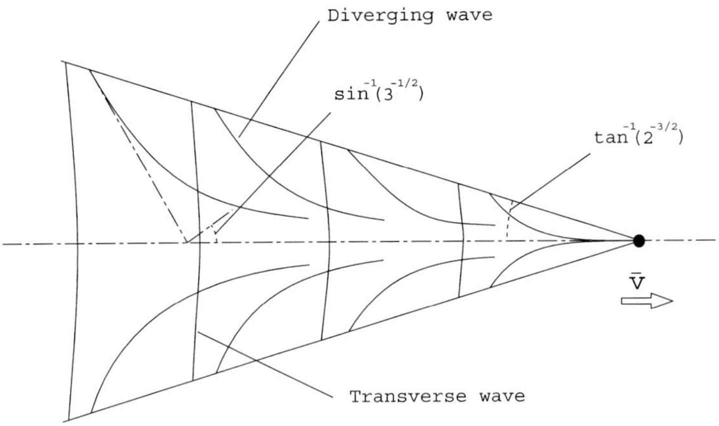


Figure 5.8. An illustration of the Kelvin wave pattern.

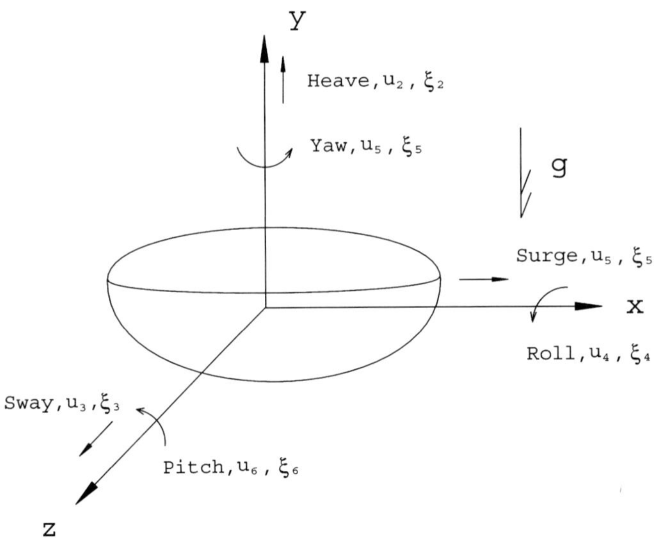


Figure 5.9. Kinematic definitions of a rigid body interacting with surrounding potential flow.

### 5.3.3. *Rigid body surface waves interactions*

The interaction between floating or submerged rigid body and regular surface waves is an important subject, bearing some resemblances to radiation and scattering for electromagnetic or acoustic waves. In accordance with convention, the six degrees of freedom are defined as surge, heave, and sway translational motions, and roll, yaw, and pitch rotational motions, as illustrated in Figure 5.9.

Again, we limit the discussion to plane progressive waves of small amplitude, with sinusoidal time dependence. Naturally, a linear superposition of such sinusoidal components can be used to represent the description of irregular waves. This superposition should also account for different facets of rigid body interacting with surface waves.

Similar to the discussion in Chapters 2 and 3, as depicted in Figure 5.8, we denote the complex amplitude of the displacement $u_j(t)$ as $\xi_j$ and the corresponding velocity as $v_j(t)$. If we employ the same principle used in the derivation of added mass coefficients and define the velocity potential $\phi_j$ corresponding to the degree of freedom $\xi_j$, we have

$$u_j(t) = \mathrm{Re}(\xi_j e^{i\omega t}), \tag{3.183}$$

$$v_j(t) = \mathrm{Re}(i\omega\xi_j e^{i\omega t}), \tag{3.184}$$

with $j = 1, 2, \ldots, 6$.

Here, $\phi_j$ represents the velocity potential of a rigid body motion with unit amplitude in the absence of incident waves. Moreover, the presence of the rigid body in the fluid results in diffraction of the incident wave and the addition of a disturbance to the velocity potential, the so-called scattering effect of the body. Therefore, the total velocity potential for the fluid-structure interaction system can be expressed as

$$\phi(\mathbf{x}, t) = \mathrm{Re}\left[\sum_{j=1}^{6} \xi_j \phi_j(\mathbf{x}) e^{i\omega t} + A\phi_a(\mathbf{x}) e^{i\omega t}\right], \tag{3.185}$$

where the velocity potential $\phi_a$ represents the incident wave and its interaction with the body.

In practice, the velocity potential $\phi_j$ can be considered as the forced fluid motion due to the six displacement components of the body. Therefore $\phi_j$ is also called the solution of the radiation problem. Just as in the discussion of added mass coefficients, the velocity potential components $\phi_j$ are governed by Laplace's equation, and on the body surface $S_b$ we have

$$\frac{\partial \phi_j}{\partial n} = i\omega n_j, \quad \text{with } j = 1, 2, 3; \tag{3.186}$$

$$\frac{\partial \phi_j}{\partial n} = i\omega(\mathbf{r} \times \mathbf{n})_{j-3}, \quad \text{with } j = 4, 5, 6, \tag{3.187}$$

where $\mathbf{n}$ is the unit normal vector on the body surface, directed into the body.

REMARK 5.3.5. The seemingly extra $i\omega$ is introduced because of the fact that $\phi_j$ represents the velocity potential, while $\xi_j$ stands for the displacement magnitude.

Likewise, the velocity potential $\phi_a$ also satisfies Laplace's equation with the boundary condition

$$\frac{\partial \phi_a}{\partial n} = 0, \quad \text{on } S_b. \tag{3.188}$$

The solution of the velocity potential $\phi_a$ represents the wave interacting with a fixed body, floating or submerged, and is often called the diffraction problem. If we denote the incident wave potential as $\phi_o$, in general the incident wave potential is given, either for infinite depth as in Eq. (3.179) or for finite depth as in Eq. (3.169). The diffraction velocity potential $\phi_a$ can be decomposed as

$$\phi_a = \phi_o + \phi_7, \tag{3.189}$$

where $\phi_7$ represents the scattering velocity potential.

Furthermore, in addition to the same boundary condition at the bottom of the fluid domain, on the free surface $y = 0$ we also have the same free surface boundary condition as in Eq. (3.154)

$$-\frac{\omega^2}{g}\phi_j + \frac{\partial \phi_j}{\partial y} = 0, \quad \text{for } j = 0, 1, \ldots, 6, 7, \tag{3.190}$$

different from the wave propagation in the infinite fluid domain, to properly set the boundary-value problem for FSI problems.

We must also prescribe the so-called radiation condition at infinity to account for the waves radiating away from the body. For instance, in the two-dimensional case, with the definition in Eq. (3.185), we must have

$$\phi_j \propto e^{\mp ikx}, \quad \text{as } x \to \pm\infty, \text{ for } j = 1, 2, \ldots, 7; \tag{3.191}$$

while the corresponding three-dimensional radiation boundary condition requires

$$\phi_j \propto \frac{1}{\sqrt{R}} e^{-ikR}, \quad \text{as } R \to \infty, \text{ for } j = 1, 2, \ldots, 7. \tag{3.192}$$

REMARK 5.3.6. The radiation boundary condition can be derived from the energy flux across a fixed control surface, namely, a column for two-dimensional cases, or a portion of a spherical surface for three-dimensional cases. The expression of the energy flux can be derived in the form of Eq. (3.167) or $\int_{S_c} p v_n \, d\Gamma$, where $v_n$ represents the normal velocity at the control surface $S_c$.

Substituting the velocity potential in (3.185) in Bernoulli's equation, the total pressure can be expressed as

$$p = -\rho \,\mathrm{Re}\left[\left(\sum_{j=1}^{6} \xi_j \phi_j + A\phi_a\right) i\omega e^{i\omega t}\right] - \rho g y. \tag{3.193}$$

REMARK 5.3.7. In order to include the hydrostatic pressure, we must introduce $y$ in the fixed coordinate system.

Hence, the force and moment acting on the body can be calculated, according to Eqs. (3.57) and (3.58), as

$$\mathbf{F} = -\rho \,\mathrm{Re}\left[ i\omega e^{i\omega t} \int_{S_b} \left( \sum_{j=1}^{6} \xi_j \phi_j + A\phi_a \right) \mathbf{n}\, dS \right] - \rho g \int_{S_b} \mathbf{n} y\, dS; \tag{3.194}$$

$$\begin{aligned} \mathbf{M} = & -\rho \,\mathrm{Re}\left[ i\omega e^{i\omega t} \int_{S_b} \left( \sum_{j=1}^{6} \xi_j \phi_j + A\phi_a \right) (\mathbf{r} \times \mathbf{n})\, dS \right] \\ & - \rho g \int_{S_b} (\mathbf{r} \times \mathbf{n}) y\, dS; \end{aligned} \tag{3.195}$$

where $S_b$ denotes the wetted surface around the body.

Note that the integration of the thin strip from the static equilibrium water line $y = 0$ into the free surface elevation $\eta$ is of order $O(\epsilon^2)$, and can be ignored in the linearization. Therefore, for the floating body, the surface of integration $S_b$ can be considered as the wetted body surface beneath $y = 0$.

Furthermore, the three integrals in Eqs. (3.194) and (3.195) represent three distinct physical significances. The first integral is reminiscent of the added mass coefficient calculation, except that in Eqs. (3.194) and (3.195) the potential $\phi_j$ is complex due to the free surface wave. The second integral represents the exciting force and moment, proportional to the incident wave amplitude. The third term is the hydrostatic component which is both elementary and important.

With the same understanding reached in Chapter 3, the resulting force and moment are defined generally in terms of the damping and added mass coefficients, corresponding to the real and imaginary parts of these integrals. Just as we absorb the rotational degrees of freedom into the translational degrees of freedom, we denote $F_4 = M_1$, $F_5 = M_2$, and $F_6 = M_3$. At this point, we can introduce the boundary conditions (3.186) and (3.187), and the first term of Eqs. (3.194) and (3.195) can be expressed as

$$F_k^1 = \mathrm{Re}\left( \sum_{j=1}^{6} \xi_j e^{i\omega t} \hat{F}_{kj} \right), \tag{3.196}$$

with $k = 1, 2, \ldots, 6$ and

$$\hat{F}_{kj} = -\rho \int_{S_b} \frac{\partial \phi_k}{\partial n} \phi_j\, dS. \tag{3.197}$$

The coefficient $\hat{F}_{kj}$ obviously satisfies the reciprocal relationship and represents the complex force or moment components in the direction $k$ or $k-3$, respectively, due to an $j$th sinusoidal motion component of unit amplitude. As a result of free surface waves, the coefficient $\hat{F}_{kj}$ is complex and can be expressed as

$$\hat{F}_{kj} = \omega^2 a_{kj} - i\omega b_{kj}. \tag{3.198}$$

Consequently, the first term is expressed as

$$F_k^1 = -\sum_{j=1}^{6} \left[ a_{kj}\ddot{u}(t)_j + b_{kj}\dot{u}(t)_j \right]. \tag{3.199}$$

The physical significance of the coefficients $a_{kj}$ is in fact identical to the concept of added mass. Of course, the added mass coefficients $m_{kj}$ as derived in Eq. (3.85) do not include the free surface effects. Moreover, the coefficients $b_{kj}$ are proportional to the body velocity and can be considered damping coefficients. However, unlike mechanical damping, which dissipates kinetic or potential energy into the surrounding material, the coefficients $b_{kj}$ represent the work done to radiate waves away from the oscillating body.

Likewise, the second term, which represents excitation force and moment, is expressed as

$$F_k^2 = \mathrm{Re}\left(Ae^{i\omega t}X_k\right), \quad \text{with } k = 1, 2, \ldots, 6, \tag{3.200}$$

$$X_k = -\rho \int_{S_b} \frac{\partial \phi_k}{\partial n}(\phi_o + \phi_7)\, dS. \tag{3.201}$$

In some cases, to simplify the computation, the Froude–Krylov hypothesis is introduced: namely, the diffraction potential $\phi_7$ is completely neglected in Eq. (3.201). Moreover, if we use Green's theorem and the diffraction boundary condition (3.188), Eq. (3.201) can be expressed as

$$X_k = -\rho \int_{S_b} \left( \phi_o \frac{\partial \phi_k}{\partial n} - \phi_k \frac{\partial \phi_o}{\partial n} \right) dS. \tag{3.202}$$

The most important feature of Eq. (3.202), which is often called the Haskind relations, is the absence of diffraction potential. Finally, we must discuss the hydrostatic term, or the third integral, written as

$$\mathbf{F}^3 = -\rho g \int_{S_b} \mathbf{n} y\, dS, \tag{3.203}$$

$$\mathbf{M}^3 = -\rho g \int_{S_b} (\mathbf{r} \times \mathbf{n}) y\, dS. \tag{3.204}$$

Based on physical intuition, it is not difficult to determine that the hydrostatic term will result in restoration forces and moments, and contribute to the stiffness term of the final governing equations. In particular, if the stiffness is nonpositive, we have a special problem called hydroinstability.

We first introduce a moving coordinate system attached to the gravity center of the body and the following transformation

$$\mathbf{r} = \hat{\mathbf{r}} + \mathbf{u}_t + \mathbf{u}_r \times \hat{\mathbf{r}}, \tag{3.205}$$

with translational displacement $\mathbf{u}_t = \langle u_1, u_2, u_3 \rangle$ and rotational displacement $\mathbf{u}_r = \langle u_4, u_5, u_6 \rangle$.

Naturally, without the effects of free surface waves, if the body is in static equilibrium, Eqs. (3.203) and (3.204) represent the static buoyancy force and moment, balanced by the body's weight. As previously discussed, the integration surface $S_b$ starts from the position $y = 0$ after the linearization. Thus, if we enclose the submerged body with a surface $y = 0$ and use the divergence theorem in the forms of (2.141) and (2.143) in Chapter 2 as well as the transformation in Eq. (3.205), we obtain

$$\mathbf{F}^3 = \mathbf{j}\rho g \left( \bar{V} - \int\limits_{S_o} (u_2 + u_6 \hat{x} - u_4 \hat{z})\, d\hat{x}\, d\hat{z} \right), \tag{3.206}$$

$$\hat{\mathbf{M}}^3 = \rho g \left\{ \int\limits_{V_o} \left[ \mathbf{k}(\hat{x} + u_5 \hat{z} - u_6 \hat{y}) - \mathbf{i}(\hat{z} + u_4 \hat{y} - u_5 \hat{x}) \right] dV - \int\limits_{S_o} (u_2 + u_6 \hat{x} - u_4 \hat{z})(\mathbf{k}\hat{x} - \mathbf{i}\hat{z})\, d\hat{x}\, d\hat{z} \right\}, \tag{3.207}$$

where $\bar{V}$ denotes the displaced water volume, and $\hat{\mathbf{M}}^3$ represents the moment with respect to the gravity center of the body.

Notice that the horizontal components of the hydrostatic force and the vertical component of the moment are zero. If we denote the translational and rotational spring with coefficients $c_{kj}$, the final expression for this fluid-structure coupling system can be expressed as

$$(M_{kj} + a_{kj})\ddot{\xi}_j(t) + b_{kj}\dot{\xi}_j(t) + c_{kj}\xi_j(t) = \mathrm{Re}\big(A X_k e^{i\omega t}\big). \tag{3.208}$$

The solution of Eq. (3.208) is of course followed by the discussion in Chapter 3.

EXAMPLE 5.1. A spherical bubble of radius $R(t)$ expands in an infinite, inviscid, incompressible fluid of density $\rho$. Find the total kinetic energy in the fluid as a function of $\rho$, $R$, and derivatives of $R$; find the pressure $p$ on the bubble's surface as a function of the same quantities and for the case of small oscillations about a mean position $R(t) = R_o + A\cos(\omega t)$, with $A \ll R_o$. Assuming the bubble is filled with a gas that satisfies the isothermal gas law, namely, $pV = C$, where $C$ is a constant, find the natural frequency of oscillation.

Due to symmetries, we have at $r = R$, $v_r = \dot{R}(t)$. Using Green's function and the source $Q = 4\pi R^2 v_r$, we have the velocity potential $\phi = -Q/4\pi r = -R^2\dot{R}(t)/r$. The total kinetic energy within the fluid is

$$E = \frac{\rho}{2}\int \mathbf{v}\cdot\mathbf{v}\,d\Omega = \frac{\rho}{2}\int_r^\infty \frac{R^4\dot{R}^2}{r^4}4\pi r^2\,dr = 2\pi\rho R^3\dot{R}^2.$$

Moreover, on the surface of the bubble, or $r = R$, from the Bernoulli's equation, we get

$$p - p_\infty = -\rho\left(\frac{\partial\phi}{\partial t} + \frac{v_r^2}{2}\right) = \rho\left(R\ddot{R} + \frac{3\dot{R}^2}{2}\right).$$

As given, $\dot{R} = -A\omega\sin(\omega t)$ and $\ddot{R} = -A\omega^2\cos(\omega t)$, we have

$$\begin{aligned} p - p_\infty &= \rho\left[-RA\omega^2\cos(\omega t) + \frac{3}{2}A^2\omega^2\sin^2(\omega t)\right] \\ &= -\rho R_o A\omega^2\cos(\omega t) + O(A^2). \end{aligned}$$

Therefore, we have

$$\dot{p} = \rho R_o A\omega^3\sin(\omega t) + O(A^2).$$

For isothermal gas, we have $V\dot{p} + \dot{V}p = 0$. Moreover, $\dot{V} = 4\pi R^2\dot{R} = -4\pi R_o^2 A\omega\sin(\omega t) + O(A^2)$. Now, if we take $O(A)$ term, we have

$$\omega = \sqrt{\frac{3p_\infty}{\rho R_o^2}}.$$

*Chapter 6*

# Computational Linear Models

In the preceding chapters, we have considered some computational and analytical approaches to FSI systems. However, the computational approaches discussed so far are based on simplified models of fluids and solids. These fundamentals lead to many engineering applications. Although more sophisticated computational approaches are becoming main stream in the study of fluid and solid mechanics, these simplified models are crucial in revealing the physics and establishing characteristic dimensionless parameters. As computational procedures for fluids and solids become more reliable and affordable, engineering competition and economics tend to push designers to optimize original designs, and in certain cases, to simulate various options [143,277]. Of course, the effective use of computational approaches depends largely on two factors: first, the possibility of a reliable procedure; and second, the cost of solutions.

In this chapter, we focus on linear acoustoelastic/slosh FSI systems and ignore material nonlinearities as well as geometric nonlinearities due to large displacements, rotations, and deformations. FSI interfacial motions are insignificant by definition in these linear models. Nevertheless different coupling conditions must be considered for various mathematical models or meshing conditions [136]. In particular, the requirements of both mass and momentum conservations must be satisfied on discretized interfacial configurations.

In the first section, we present both the velocity potential and mixed displacement-based formulations for acoustoelastic/slosh FSI systems. In the second section, direct integration and mode superposition methods are introduced. In the final section, we elaborate upon the inf-sup conditions with respect to mixed formulations.

## 6.1. Potential and displacement-based formulations

A number of finite element formulations have been proposed to model an acoustic fluid for the analysis of FSI problems, among them the displacement formulation (see Bathe and Hahn [154], Hamdi et al. [175], Olson and Bathe [170], Bathe et al. [151], Wang and Bathe [300,301]), the displacement potential and pressure formulation, and the velocity potential and pressure formulation (Morand and Ohayon [106], Everstine [85], Olson and Bathe [171], Felippa and Ohayon [31], MacNeal et al. [176]). A recent review of various approaches to FSI problems is available in Ref. [105]. In this section, we present two types of formulations: a velocity potential formulation and a displacement-based formulation.

### 6.1.1. $\phi$-$\mathbf{u}$ and $p$-$\phi$-$\mathbf{u}$ formulations

We assume an inviscid, irrotational compressible fluid with small motions in contact with elastic solids. The gravity effects are included in the body force. In the $\phi$-$\mathbf{u}$ formulation, we use the velocity potential as the state variable for fluids and the displacement for solids. In the $p$-$\phi$-$\mathbf{u}$ formulation, we replace one velocity potential unknown with a pressure unknown for each isolated fluid region in order to eliminate the zero frequency mode for constant pressure within each enclosed cavity. In fact, both potential-based formulations will yield the same nonzero natural frequencies of the fluid-structure system.

*$\phi$-$\mathbf{u}$ formulation*

In essence, if we introduce a velocity potential $\phi$, the same as discussed in Chapter 5, with $\mathbf{v} = \nabla\phi$, irrotational constraints are automatically satisfied. For the structure domain $\Omega_s$ with its natural boundary $\Gamma^{s,f}$ and the FSI interface $\Gamma^{fsi}$, the variational indicator $(V.I.)_s$ is defined as

$$\int_{t_1}^{t_2} \left\{ \frac{1}{2} \int_{\Omega_s} \boldsymbol{\epsilon}^T \mathbf{C}_s \boldsymbol{\epsilon}\, d\Omega - \frac{1}{2} \int_{\Omega_s} \rho_s \dot{\mathbf{u}}^T \dot{\mathbf{u}}\, d\Omega - \int_{\Gamma^{fsi}} \left(\mathbf{u}^{fsi}\right)^T \mathbf{f}^{sf}\, d\Gamma - \int_{\Gamma^{s,f}} \left(\mathbf{u}^{s,f}\right)^T \mathbf{f}^{s,p}\, d\Gamma - \int_{\Omega_s} \mathbf{u}^T \mathbf{f}^s\, d\Omega \right\} dt, \tag{1.1}$$

where $\mathbf{C}_s$, $\boldsymbol{\epsilon}$, $\rho_s$, $\mathbf{u}$, $\mathbf{f}^{s,p}$, $\mathbf{f}^{sf}$, and $\mathbf{f}^s$ stand for the stress-strain material matrix, strain tensor, solid density, displacement vector, surface traction vector, FSI interface traction vector, and body force vector, respectively.

For the fluid domain $\Omega_f$ with the FSI interface $\Gamma^{fsi}$, assuming the surface traction at its natural boundary $\Gamma^{f,f}$ is zero, the variational indicator $(V.I.)_f$ has the form

$$\int_{t_1}^{t_2} \left\{ \frac{1}{2} \int_{\Omega_f} \frac{1}{\kappa} (\rho_f \dot{\phi})^2\, d\Omega - \frac{1}{2} \int_{\Omega_f} \rho_f (\nabla\phi)^2\, d\Omega + \int_{\Gamma^{fsi}} \rho_f \dot{\phi}^{fsi} \left(\mathbf{u}^{fsi}\right)^T \mathbf{n}\, d\Gamma \right\} dt, \tag{1.2}$$

where $\phi$, $\kappa$, $\rho_f$, and $\mathbf{n}$ are the velocity potential, fluid bulk modulus, fluid density, and unit outward normal vector.

The key to linking fluid and solid domains is to recognize the following kinematic and dynamic matching conditions along the FSI interface:

$$\mathbf{f}^{sf} = \rho_f \dot{\phi}^{fsi}\mathbf{n}, \tag{1.3}$$

$$\dot{\mathbf{u}}^{fsi} \cdot \mathbf{n} = \frac{\partial \phi}{\partial n}. \tag{1.4}$$

Eqs. (1.3) and (1.4) combine the variational indicators for fluid and solid domains. The general kinematic and dynamic matching conditions along the FSI interface will be discussed further in Chapter 7. According to Hamiltonian principle, named after William Rowan Hamilton (1805–1865), all variations vanish at $t = t_1$ and $t = t_2$ and we obtain

$$\begin{aligned}
&\int_{\Omega_s} \rho_s \delta\mathbf{u}^T \ddot{\mathbf{u}}\, d\Omega + \int_{\Omega_s} \delta\boldsymbol{\epsilon}^T \mathbf{C}_s \boldsymbol{\epsilon}\, d\Omega - \int_{\Gamma^{s,f}} \left(\delta\mathbf{u}^{s,f}\right)^T \mathbf{f}^{s,p}\, d\Gamma \\
&\quad - \int_{\Omega_s} \delta\mathbf{u}^T \mathbf{f}^s\, d\Omega - \int_{\Gamma^{fsi}} \left(\delta\mathbf{u}^{fsi}\right)^T \rho_f \dot{\phi}^{fsi}\mathbf{n}\, d\Gamma = 0,
\end{aligned} \tag{1.5}$$

$$- \int_{\Omega_f} \rho_f (\nabla\delta\phi)\cdot(\nabla\phi)\, d\Omega - \int_{\Omega_f} \frac{{\rho_f}^2}{\kappa} \ddot{\phi}\delta\phi\, d\Omega - \int_{\Gamma^{fsi}} \rho_f \dot{\mathbf{u}}^{fsi}\cdot\mathbf{n}\delta\phi^{fsi}\, d\Gamma = 0. \tag{1.6}$$

The discretized forms of Eqs. (1.5) and (1.6) are

$$\begin{aligned}
&\delta\mathbf{U}: \quad \mathbf{M}_{ss}\ddot{\mathbf{U}} + \mathbf{C}_{fs}^T\dot{\boldsymbol{\Phi}} + \mathbf{K}_{ss}\mathbf{U} = \mathbf{R}_s, \\
&\delta\phi: \quad -\mathbf{M}_{ff}\ddot{\boldsymbol{\Phi}} + \mathbf{C}_{fs}\dot{\mathbf{U}} - \mathbf{K}_{ff}\boldsymbol{\Phi} = \mathbf{0}.
\end{aligned} \tag{1.7}$$

Based on Eqs. (1.5), (1.6), and (1.7), it is not difficult to identify the governing partial differential equations for the fluid domain (1.30) and for the solid domain (1.27) as illustrated in Chapter 5. We use standard isoparametric elements for both fluid and solid domains [143,171]. A typical $N$-node solid element has the following discretization relations:

$$\begin{aligned}
&x = \mathbf{hX}, \qquad y = \mathbf{hY}, \qquad z = \mathbf{hZ}, \\
&\mathbf{u} = \mathbf{HU}, \qquad \mathbf{u}^{fsi}\cdot\mathbf{n} = \mathbf{bU},
\end{aligned} \tag{1.8}$$

where the matrix $\mathbf{b}$ is used to select the boundary nodal displacement unknowns and the other matrices and vectors are defined as

$$\begin{aligned}
&\mathbf{h} = [h_1, h_2, \ldots, h_N], \qquad \mathbf{X} = \langle x_1, x_2, \ldots, x_N\rangle, \\
&\mathbf{Y} = \langle y_1, y_2, \ldots, y_N\rangle, \qquad \mathbf{Z} = \langle z_1, z_2, \ldots, z_N\rangle, \\
&\mathbf{H} = \begin{bmatrix} h_1 & 0 & 0 & h_2 & 0 & 0 & h_3 & 0 & 0 & \ldots & h_N & 0 & 0 \\ 0 & h_1 & 0 & 0 & h_2 & 0 & 0 & h_3 & 0 & 0 & \ldots & h_N & 0 \\ 0 & 0 & h_1 & 0 & 0 & h_2 & 0 & 0 & h_3 & 0 & 0 & \ldots & h_N \end{bmatrix}, \\
&\mathbf{U} = \langle u_1, v_1, w_1, u_2, v_2, w_2, \ldots, u_N, v_N, w_N\rangle.
\end{aligned}$$

Then, on the solid element level, we have

$$\mathbf{K}_{ss} = \int_{\Omega^s} \mathbf{B}^T \mathbf{C}_s \mathbf{B}\, d\Omega \quad \text{and} \quad \mathbf{M}_{ss} = \int_{\Omega^s} \rho_s \mathbf{H}^T \mathbf{H}\, d\Omega, \tag{1.9}$$

and

$$\mathbf{R}_s = \int_{\Omega^s} \mathbf{H}^T \mathbf{f}^s\, d\Omega + \int_{\Gamma^{s,f}} \left(\mathbf{H}^{s,f}\right)^T \mathbf{f}^{s,p}\, d\Gamma.$$

For a typical fluid element (without replacement of one velocity potential unknown by the pressure unknown), we have the following discretization relations:

$$\phi = \mathbf{h}\boldsymbol{\Phi}, \qquad \nabla\phi = \bar{\mathbf{B}}\boldsymbol{\Phi}, \qquad \phi^{fsi} = \mathbf{a}\boldsymbol{\Phi}, \tag{1.10}$$

where the matrix $\mathbf{a}$ is used to select the boundary nodal velocity potential unknowns and the other matrices and vectors are defined as

$$\boldsymbol{\Phi} = \langle \phi_1, \phi_2, \ldots, \phi_N \rangle,$$

$$\bar{\mathbf{B}} = \begin{bmatrix} \partial h_1/\partial x & \partial h_2/\partial x & \partial h_3/\partial x & \ldots & \partial h_N/\partial x \\ \partial h_1/\partial y & \partial h_2/\partial y & \partial h_3/\partial y & \ldots & \partial h_N/\partial y \\ \partial h_1/\partial z & \partial h_2/\partial z & \partial h_3/\partial z & \ldots & \partial h_N/\partial z \end{bmatrix}.$$

Therefore, on the fluid element level, we have

$$\mathbf{M}_{ff} = \int_{\Omega^f} \frac{\rho_f^2}{\kappa} \mathbf{h}^T \mathbf{h}\, d\Omega, \qquad \mathbf{K}_{ff} = \int_{\Omega^f} \rho_f \bar{\mathbf{B}}^T \bar{\mathbf{B}}\, d\Omega. \tag{1.11}$$

Finally, for both fluid and solid domains, the interaction matrix is defined as

$$\mathbf{C}_{fs}^T = -\int_{\Gamma^{fsi}} \rho_f \mathbf{b}^T \mathbf{a}\, d\Gamma. \tag{1.12}$$

*$p$-$\phi$-$\mathbf{u}$ formulation*

If we have $k$ separate fluid domains, we need to use $k$ independent hydrostatic pressure unknowns. For illustrative purposes, we consider one isolated fluid region $\Omega^f$ in contact with elastic solids. Thus, one hydrostatic pressure unknown $p$ is needed in the $p$-$\phi$-$\mathbf{u}$ formulation to replace one nodal velocity potential in the $\phi$-$\mathbf{u}$ formulation and the variational indicator has the form

$$\int_{t_1}^{t_2} \left\{ \frac{1}{2} \int_{\Omega^f} \frac{1}{\kappa} (p - \rho_f \dot{\phi})^2\, d\Omega - \frac{1}{2} \int_{\Omega^f} \rho_f (\nabla\phi)^2\, d\Omega \right.$$
$$\left. - \int_{\Gamma^{fsi}} \left(p - \rho_f \dot{\phi}^{fsi}\right) \left(\mathbf{u}^{fsi}\right)^T \mathbf{n}\, d\Gamma \right\} dt. \tag{1.13}$$

According to Hamiltonian principle, we obtain the following for fluid and solid domains:

$$\int_{\Omega^s} \rho_s \delta \mathbf{u}^T \ddot{\mathbf{u}}\, d\Omega + \int_{\Omega^s} \delta \boldsymbol{\epsilon}^T \mathbf{C}_s \boldsymbol{\epsilon}\, d\Omega - \int_{\Gamma^{fsi}} \left(\delta \mathbf{u}^{fsi}\right)^T \mathbf{f}^{sf}\, d\Gamma$$
$$- \int_{\Gamma^{s,f}} \left(\delta \mathbf{u}^{s,f}\right)^T \mathbf{f}^{s,p}\, d\Gamma - \int_{\Omega^s} \delta \mathbf{u}^T \mathbf{f}^s\, d\Omega = 0, \tag{1.14}$$

$$\int_{\Omega^f} \frac{p}{\kappa} \delta p\, d\Omega - \int_{\Omega^f} \frac{\rho_f}{\kappa} \dot{\phi} \delta p\, d\Omega + \int_{\Omega^f} \frac{\dot{p}}{\kappa} \rho_f \delta\phi\, d\Omega - \int_{\Omega^f} \rho_f (\nabla \delta\phi) \cdot (\nabla\phi)\, d\Omega$$
$$- \int_{\Omega^f} \frac{{\rho_f}^2}{\kappa} \ddot{\phi} \delta\phi\, d\Omega - \int_{\Gamma^{fsi}} \mathbf{u}^{fsi} \cdot \mathbf{n} \delta p\, d\Gamma - \int_{\Gamma^{fsi}} \rho_f \dot{\mathbf{u}}^{fsi} \cdot \mathbf{n} \delta\phi^{fsi}\, d\Gamma = 0. \tag{1.15}$$

Note that Eq. (1.3) is changed to

$$\mathbf{f}^{sf} = -\left(p - \rho_f \dot{\phi}^{fsi}\right)\mathbf{n}. \tag{1.16}$$

The discretized forms of Eqs. (1.14) and (1.15) are

$$\begin{aligned} \delta\mathbf{U}:&\quad \mathbf{M}_{ss}\ddot{\mathbf{U}} + \mathbf{C}_{fs}^T \dot{\boldsymbol{\Phi}} + \mathbf{K}_{ss}\mathbf{U} + \mathbf{K}_{ps}^T \mathbf{P} = \mathbf{R}_s, \\ \delta p:&\quad \mathbf{C}_{pf}\dot{\boldsymbol{\Phi}} + \mathbf{K}_{pp}\mathbf{P} + \mathbf{K}_{ps}\mathbf{U} = \mathbf{0}, \\ \delta\phi:&\quad -\mathbf{M}_{ff}\ddot{\boldsymbol{\Phi}} + \mathbf{C}_{fs}\dot{\mathbf{U}} + \mathbf{C}_{pf}^T \dot{\mathbf{P}} - \mathbf{K}_{ff}\boldsymbol{\Phi} = \mathbf{0}. \end{aligned} \tag{1.17}$$

Without the loss of generality, for the fluid element in which one velocity potential unknown is replaced by the pressure unknown, we assume the $N$th nodal velocity potential unknown is replaced by a pressure unknown and the discretization relations are expressed as follows:

$$\phi = \mathbf{h}'\boldsymbol{\Phi}, \qquad \nabla\phi = \bar{\mathbf{B}}'\boldsymbol{\Phi}, \qquad \phi^{fsi} = \mathbf{a}'\boldsymbol{\Phi}, \tag{1.18}$$

with $\boldsymbol{\Phi} = \langle \phi_1, \phi_2, \ldots, \phi_{N-1} \rangle$ and

$$\bar{\mathbf{B}}' = \begin{bmatrix} \partial h_1/\partial x & \partial h_2/\partial x & \partial h_3/\partial x & \ldots & \partial h_{N-1}/\partial x \\ \partial h_1/\partial y & \partial h_2/\partial y & \partial h_3/\partial y & \ldots & \partial h_{N-1}/\partial y \\ \partial h_1/\partial z & \partial h_2/\partial z & \partial h_3/\partial z & \ldots & \partial h_{N-1}/\partial z \end{bmatrix}.$$

For these types of elements, we have

$$\begin{aligned} \mathbf{C}_{pf} &= \int_{\Omega^f} \frac{\rho_f}{\kappa} \mathbf{h}'\, d\Omega, & \mathbf{C}_{fs}^T &= -\int_{\Gamma^{fsi}} \rho_f \mathbf{b}^T \mathbf{a}'\, d\Gamma; \\ \mathbf{M}_{ff} &= \int_{\Omega^f} \frac{\rho_f^2}{\kappa} \mathbf{h}'^T \mathbf{h}'\, d\Omega, & \mathbf{K}_{ff} &= \int_{\Omega^f} \rho_f \bar{\mathbf{B}}'^T \bar{\mathbf{B}}'\, d\Omega; \\ \mathbf{K}_{pp} &= -\int_{\Omega^f} \frac{1}{\kappa}\, d\Omega, & \mathbf{K}_{ps}^T &= \int_{\Gamma^{fsi}} \mathbf{b}^T\, d\Gamma. \end{aligned} \tag{1.19}$$

Note that, similar to the two equations in (1.7), the three equations in (1.17) have units of force, volume, and mass flow rate, respectively. Combining the units from their corresponding variations, they all have the unit of energy.

The matrix form of (1.17) is:

$$\begin{bmatrix} \mathbf{M}_{ss} & \mathbf{0} & \mathbf{0} \\ \mathbf{0} & \mathbf{0} & \mathbf{0} \\ \mathbf{0} & \mathbf{0} & -\mathbf{M}_{ff} \end{bmatrix} \begin{bmatrix} \ddot{\mathbf{U}} \\ \ddot{\mathbf{P}} \\ \ddot{\boldsymbol{\Phi}} \end{bmatrix} + \begin{bmatrix} \mathbf{0} & \mathbf{0} & \mathbf{C}_{fs}^T \\ \mathbf{0} & \mathbf{0} & \mathbf{C}_{pf} \\ \mathbf{C}_{fs} & \mathbf{C}_{pf}^T & \mathbf{0} \end{bmatrix} \begin{bmatrix} \dot{\mathbf{U}} \\ \dot{\mathbf{P}} \\ \dot{\boldsymbol{\Phi}} \end{bmatrix}$$
$$+ \begin{bmatrix} \mathbf{K}_{ss} & \mathbf{K}_{ps}^T & \mathbf{0} \\ \mathbf{K}_{ps} & \mathbf{K}_{pp} & \mathbf{0} \\ \mathbf{0} & \mathbf{0} & -\mathbf{K}_{ff} \end{bmatrix} \begin{bmatrix} \mathbf{U} \\ \mathbf{P} \\ \boldsymbol{\Phi} \end{bmatrix} = \begin{bmatrix} \mathbf{R}_s \\ \mathbf{0} \\ \mathbf{0} \end{bmatrix}. \tag{1.20}$$

If we change the sign of the second equation in (1.17), we get the new matrix form as:

$$\begin{bmatrix} \mathbf{M}_{ss} & \mathbf{0} & \mathbf{0} \\ \mathbf{0} & \mathbf{0} & \mathbf{0} \\ \mathbf{0} & \mathbf{0} & -\mathbf{M}_{ff} \end{bmatrix} \begin{bmatrix} \ddot{\mathbf{U}} \\ \ddot{\mathbf{P}} \\ \ddot{\boldsymbol{\Phi}} \end{bmatrix} + \begin{bmatrix} \mathbf{0} & \mathbf{0} & \mathbf{C}_{fs}^T \\ \mathbf{0} & \mathbf{0} & -\mathbf{C}_{pf} \\ \mathbf{C}_{fs} & \mathbf{C}_{pf}^T & \mathbf{0} \end{bmatrix} \begin{bmatrix} \dot{\mathbf{U}} \\ \dot{\mathbf{P}} \\ \dot{\boldsymbol{\Phi}} \end{bmatrix}$$
$$+ \begin{bmatrix} \mathbf{K}_{ss} & \mathbf{K}_{ps}^T & \mathbf{0} \\ -\mathbf{K}_{ps} & -\mathbf{K}_{pp} & \mathbf{0} \\ \mathbf{0} & \mathbf{0} & -\mathbf{K}_{ff} \end{bmatrix} \begin{bmatrix} \mathbf{U} \\ \mathbf{P} \\ \boldsymbol{\Phi} \end{bmatrix} = \begin{bmatrix} \mathbf{R}_s \\ \mathbf{0} \\ \mathbf{0} \end{bmatrix}. \tag{1.21}$$

Eqs. (1.20) and (1.21) represent the same FSI problem and should have the same eigenvalues and eigenvectors. From the kinematic boundary condition along the FSI interface (1.4), we find that, physically, $\boldsymbol{\Phi}$ has a $\pi/2$ phase shift from $\mathbf{U}$. Therefore we assign the $m$th eigenvector as $\mathbf{X}_m = \langle \mathbf{U}_m, \mathbf{P}_m, i\boldsymbol{\Phi}_m \rangle$, where $\mathbf{U}_m$, $\mathbf{P}_m$, and $\boldsymbol{\Phi}_m$ are real vectors. Let the $m$th eigensolution be $\mathbf{X} = \mathbf{X}_m e^{\lambda_m t}$, from Eqs. (1.20) and (1.21), we get

$$a_m \lambda_m^2 + b_m \lambda_m + c_m = 0, \tag{1.22}$$
$$a'_m \lambda_m^2 + b'_m \lambda_m + c'_m = 0, \tag{1.23}$$

with

$$\begin{aligned} a_m &= \mathbf{U}_m^T \mathbf{M}_{ss} \mathbf{U}_m + \boldsymbol{\Phi}_m^T \mathbf{M}_{ff} \boldsymbol{\Phi}_m, \\ b_m &= 2i\big(\mathbf{U}_m^T \mathbf{C}_{fs}^T \boldsymbol{\Phi}_m + \mathbf{P}_m^T \mathbf{C}_{pf} \boldsymbol{\Phi}_m\big), \\ c_m &= \mathbf{U}_m^T \mathbf{K}_{ss} \mathbf{U}_m + \boldsymbol{\Phi}_m^T \mathbf{K}_{ff} \boldsymbol{\Phi}_m + \mathbf{P}_m^T \mathbf{K}_{pp} \mathbf{P}_m + 2\mathbf{P}_m^T \mathbf{K}_{ps} \mathbf{U}_m, \end{aligned}$$

and

$$\begin{aligned} a'_m &= \mathbf{U}_m^T \mathbf{M}_{ss} \mathbf{U}_m + \boldsymbol{\Phi}_m^T \mathbf{M}_{ff} \boldsymbol{\Phi}_m, \\ b'_m &= 2i\mathbf{U}_m^T \mathbf{C}_{fs}^T \boldsymbol{\Phi}_m, \\ c'_m &= \mathbf{U}_m^T \mathbf{K}_{ss} \mathbf{U}_m + \boldsymbol{\Phi}_m^T \mathbf{K}_{ff} \boldsymbol{\Phi}_m + \mathbf{P}_m^T (-\mathbf{K}_{pp}) \mathbf{P}_m. \end{aligned}$$

Since $\mathbf{M}_{ss}$, $\mathbf{M}_{ff}$, $-\mathbf{K}_{pp}$, and $\mathbf{K}_{ss}$ are positive definite matrices, $a'_m$ and $c'_m$ should be positive. Furthermore, Eq. (1.23) gives pure imaginary eigenvalues of the FSI system. If we assign $\lambda_m = i\omega_m$, $m = 1, 2, \ldots, n$, where $n$ is the total number of degrees of freedom, substitute eigensolution $\mathbf{X}_m e^{i\omega_m t}$ into the left side of Eq. (1.20), we get

$$\begin{aligned}
&-\omega_m^2 \mathbf{M}_{ss}\mathbf{U}_m - \omega_m \mathbf{C}_{fs}^T \boldsymbol{\Phi}_m + \mathbf{K}_{ss}\mathbf{U}_m + \mathbf{K}_{ps}^T \mathbf{P}_m = \mathbf{0},\\
&-\omega_m \mathbf{C}_{pf}\boldsymbol{\Phi}_m + \mathbf{K}_{ps}\mathbf{U}_m + \mathbf{K}_{pp}\mathbf{P}_m = \mathbf{0},\\
&\omega_m^2 \mathbf{M}_{ff}\boldsymbol{\Phi}_m + \omega_m \mathbf{C}_{fs}\mathbf{U}_m + \omega_m \mathbf{C}_{pf}^T \mathbf{P}_m - \mathbf{K}_{ff}\boldsymbol{\Phi}_m = \mathbf{0},
\end{aligned} \tag{1.24}$$

with the matrix form

$$\begin{aligned}
-\omega_m^2 &\begin{bmatrix} \mathbf{M}_{ss} & \mathbf{0} & \mathbf{0} \\ \mathbf{0} & \mathbf{0} & \mathbf{0} \\ \mathbf{0} & \mathbf{0} & \mathbf{M}_{ff} \end{bmatrix} \begin{bmatrix} \mathbf{U}_m \\ \mathbf{P}_m \\ \boldsymbol{\Phi}_m \end{bmatrix} - \omega_m \begin{bmatrix} \mathbf{0} & \mathbf{0} & \mathbf{C}_{fs}^T \\ \mathbf{0} & \mathbf{0} & \mathbf{C}_{pf} \\ \mathbf{C}_{fs} & \mathbf{C}_{pf}^T & \mathbf{0} \end{bmatrix} \begin{bmatrix} \mathbf{U}_m \\ \mathbf{P}_m \\ \boldsymbol{\Phi}_m \end{bmatrix} \\
&+ \begin{bmatrix} \mathbf{K}_{ss} & \mathbf{K}_{ps}^T & \mathbf{0} \\ \mathbf{K}_{ps} & \mathbf{K}_{pp} & \mathbf{0} \\ \mathbf{0} & \mathbf{0} & \mathbf{K}_{ff} \end{bmatrix} \begin{bmatrix} \mathbf{U}_m \\ \mathbf{P}_m \\ \boldsymbol{\Phi}_m \end{bmatrix} = \begin{bmatrix} \mathbf{0} \\ \mathbf{0} \\ \mathbf{0} \end{bmatrix}.
\end{aligned} \tag{1.25}$$

In the $\phi$-$\mathbf{u}$ formulation, we get an equation similar to Eq. (1.20),

$$\begin{aligned}
&\begin{bmatrix} \mathbf{M}_{ss} & \mathbf{0} & \mathbf{0} \\ \mathbf{0} & -\mathbf{M}_{ff} & -\mathbf{M}_{af}^T \\ \mathbf{0} & -\mathbf{M}_{af} & -\mathbf{M}_{aa} \end{bmatrix} \begin{bmatrix} \ddot{\mathbf{U}} \\ \ddot{\boldsymbol{\Phi}} \\ \ddot{\boldsymbol{\Phi}}_a \end{bmatrix} + \begin{bmatrix} \mathbf{0} & \mathbf{C}_{fs}^T & \mathbf{C}_{as}^T \\ \mathbf{C}_{fs} & \mathbf{0} & \mathbf{0} \\ \mathbf{C}_{as} & \mathbf{0} & \mathbf{0} \end{bmatrix} \begin{bmatrix} \dot{\mathbf{U}} \\ \dot{\boldsymbol{\Phi}} \\ \dot{\boldsymbol{\Phi}}_a \end{bmatrix} \\
&\quad + \begin{bmatrix} \mathbf{K}_{ss} & \mathbf{0} & \mathbf{0} \\ \mathbf{0} & -\mathbf{K}_{ff} & -\mathbf{K}_{af}^T \\ \mathbf{0} & -\mathbf{K}_{af} & -\mathbf{K}_{ff} \end{bmatrix} \begin{bmatrix} \mathbf{U} \\ \boldsymbol{\Phi} \\ \boldsymbol{\Phi}_a \end{bmatrix} = \begin{bmatrix} \mathbf{R}_s \\ \mathbf{0} \\ \mathbf{0} \end{bmatrix},
\end{aligned} \tag{1.26}$$

where $\boldsymbol{\Phi}_a$ denotes the velocity potentials which will be replaced by pressure unknowns.

Eq. (1.26) has the same eigenvalues as Eq. (1.20) except for the zero energy mode. Using the same reasoning, the $m$th eigensolution of Eq. (1.26) is in the form $\mathbf{X} = \mathbf{X}_m e^{\lambda_m t}$, with $\mathbf{X}_m = \langle \mathbf{U}_m, i\boldsymbol{\Phi}'_m, i(\boldsymbol{\Phi}_a)_m \rangle$. Assigning $\lambda_m = i\omega_m$, we have

$$\begin{aligned}
-\omega_m^2 &\begin{bmatrix} \mathbf{M}_{ss} & \mathbf{0} & \mathbf{0} \\ \mathbf{0} & \mathbf{M}_{ff} & \mathbf{M}_{af}^T \\ \mathbf{0} & \mathbf{M}_{af} & \mathbf{M}_{aa} \end{bmatrix} \begin{bmatrix} \mathbf{U}_m \\ \boldsymbol{\Phi}'_m \\ (\boldsymbol{\Phi}_a)_m \end{bmatrix} - \omega_m \begin{bmatrix} \mathbf{0} & \mathbf{C}_{fs}^T & \mathbf{C}_{as}^T \\ \mathbf{C}_{fs} & \mathbf{0} & \mathbf{0} \\ \mathbf{C}_{as} & \mathbf{0} & \mathbf{0} \end{bmatrix} \begin{bmatrix} \mathbf{U}_m \\ \boldsymbol{\Phi}'_m \\ (\boldsymbol{\Phi}_a)_m \end{bmatrix} \\
&+ \begin{bmatrix} \mathbf{K}_{ss} & \mathbf{0} & \mathbf{0} \\ \mathbf{0} & \mathbf{K}_{ff} & \mathbf{K}_{af}^T \\ \mathbf{0} & \mathbf{K}_{af} & \mathbf{K}_{aa} \end{bmatrix} \begin{bmatrix} \mathbf{U}_m \\ \boldsymbol{\Phi}'_m \\ (\boldsymbol{\Phi}_a)_m \end{bmatrix} = \begin{bmatrix} \mathbf{0} \\ \mathbf{0} \\ \mathbf{0} \end{bmatrix}.
\end{aligned} \tag{1.27}$$

Illustrating that by replacing one nodal velocity potential unknown in the $\phi$-$\mathbf{u}$ formulation with one pressure unknown in the $p$-$\phi$-$\mathbf{u}$ formulation for each separate fluid domain, we get the same nonzero eigenvalues. We express the $\phi$-$\mathbf{u}$ formulation in Eq. (1.26) and the $p$-$\phi$-$\mathbf{u}$ formulation in the following form:

$$-\omega_m^2\begin{bmatrix}\mathbf{M}_{ss} & \mathbf{0} & \mathbf{0}\\ \mathbf{0} & \mathbf{M}_{ff} & \mathbf{0}\\ \mathbf{0} & \mathbf{0} & \mathbf{0}\end{bmatrix}\begin{bmatrix}\mathbf{U}_m\\ \boldsymbol{\Phi}_m\\ \mathbf{P}_m\end{bmatrix} - \omega_m\begin{bmatrix}\mathbf{0} & \mathbf{C}_{fs}^T & \mathbf{0}\\ \mathbf{C}_{fs} & \mathbf{0} & \mathbf{B}_p^T\\ \mathbf{0} & \mathbf{B}_p & \mathbf{0}\end{bmatrix}\begin{bmatrix}\mathbf{U}_m\\ \boldsymbol{\Phi}_m\\ \mathbf{P}_m\end{bmatrix}$$
$$+\begin{bmatrix}\mathbf{K}_{ss} & \mathbf{0} & \mathbf{A}_p^T\\ \mathbf{0} & \mathbf{K}_{ff} & \mathbf{0}\\ \mathbf{A}_p & \mathbf{0} & \mathbf{D}_p\end{bmatrix}\begin{bmatrix}\mathbf{U}_m\\ \boldsymbol{\Phi}_m\\ \mathbf{P}_m\end{bmatrix} = \begin{bmatrix}\mathbf{0}\\ \mathbf{0}\\ \mathbf{0}\end{bmatrix}. \tag{1.28}$$

For simplicity, we consider only one fluid domain, i.e., $\mathbf{P}_m = [p]$. The analogy to the case involving several fluid domains is straightforward. We notice that there exists one zero frequency mode in the $\phi$-$\mathbf{u}$ formulation, i.e. constant velocity potential and zero displacements, denoted as $\langle \mathbf{0}, \alpha\mathbf{i}, \alpha\rangle$ with $\mathbf{i} = \langle 1, 1, \ldots, 1\rangle$, and the dimension of the vector $\mathbf{i}$ represents the number of the remaining velocity potential unknowns. Substituting the zero mode to Eq. (1.26) yields

$$\begin{aligned}&\mathbf{K}_{ff}\mathbf{i} + \mathbf{K}_{af}^T = \mathbf{0},\\ &\mathbf{K}_{af}\mathbf{i} + \mathbf{K}_{aa} = \mathbf{0}.\end{aligned} \tag{1.29}$$

By assigning the $m$th eigensolution of Eq. (1.26) as $\langle \mathbf{U}_m, (\boldsymbol{\Phi}_m + \alpha\mathbf{i}), \alpha\rangle$ and using Eq. (1.29), we obtain the following equations:

$$-\omega_m^2\mathbf{M}_{ss}\mathbf{U}_m - \left(\mathbf{C}_{fs}^T\mathbf{i} + \mathbf{C}_{as}^T\right)\alpha\omega_m + \mathbf{K}_{ss}\mathbf{U}_m - \omega_m\mathbf{C}_{fs}^T\boldsymbol{\Phi}_m = \mathbf{0}, \tag{1.30}$$

$$-\omega_m^2\mathbf{M}_{ff}\boldsymbol{\Phi}_m - \left(\mathbf{M}_{ff}\mathbf{i} + \mathbf{M}_{af}^T\right)\alpha - \omega_m\mathbf{C}_{fs}\mathbf{U}_m + \mathbf{K}_{ff}\boldsymbol{\Phi}_m = \mathbf{0}, \tag{1.31}$$

$$\begin{aligned}&\left(\mathbf{i}^T\mathbf{C}_{fs} + \mathbf{C}_{as}\right)\mathbf{U}_m/\omega_m + \left(\mathbf{M}_{af} + \mathbf{i}^T\mathbf{M}_{ff}\right)\boldsymbol{\Phi}_m\\ &\quad + \alpha\left(\mathbf{i}^T\mathbf{M}_{ff}\mathbf{i} + 2\mathbf{M}_{af}\mathbf{i} + \mathbf{M}_{aa}\right) = \mathbf{0}.\end{aligned} \tag{1.32}$$

Replacing $\alpha$ with $p/\rho_f\omega_m$ and compare with Eq. (1.28), we have

$$\begin{aligned}\mathbf{A}_p &= -\left(\mathbf{i}^T\mathbf{C}_{fs} + \mathbf{C}_{as}\right)/\rho_f,\\ \mathbf{B}_p &= \left(\mathbf{i}^T\mathbf{M}_{ff} + \mathbf{M}_{af}\right)/\rho_f,\\ \mathbf{D}_p &= -\left(\mathbf{i}^T\mathbf{M}_{ff}\mathbf{i} + 2\mathbf{M}_{af}\mathbf{i} + \mathbf{M}_{aa}\right)/\rho_f^2.\end{aligned}$$

Note that from the consistency of the interpolation functions we have $\sum_{i=1}^N h_i = 1$, $\sum_{i=1}^N a_i = 1$, $\sum_{i=1}^{N-1} a_i' = 1$, and $\sum_{i=1}^{N-1} h_i' = 1 - h_N$. Therefore it is obvious that we have $\mathbf{K}_{pp} = \mathbf{D}_p$, $\mathbf{K}_{ps} = \mathbf{A}_p$, and $\mathbf{C}_{pf} = \mathbf{B}_p$. By using the $p$-$\phi$-$\mathbf{u}$ formulation, we in fact eliminate the zero frequency in the $\phi$-$\mathbf{u}$ formulation, and both formulations have the same nonzero eigenvalues.

One of the main application areas is the design of liquid storage tanks. It is therefore important to incorporate ground motion effects in these FSI systems. To include the relative fluid motion in the system, we modify the variational indicator for fluids $(V.I.)_f$ as follows:

$$\int_{t_1}^{t_2} \left\{ \frac{1}{2} \int_{\Omega^f} \frac{1}{\kappa}(p - \rho_f \dot{\phi})^2 \, d\Omega - \frac{1}{2} \int_{\Omega^f} \rho_f (\nabla \phi + \dot{\mathbf{u}}_g)^2 \, d\Omega - \int_{\Gamma^{fsi}} (p - \rho_f \dot{\phi}^{fsi}) (\mathbf{u}^{fsi})^T \mathbf{n} \, d\Gamma \right\} dt, \tag{1.33}$$

where the ground velocity $\dot{\mathbf{u}}_g$ is assumed to be a known quantity.

Similarly, we derive the dynamic equations of the FSI system, with $\mathbf{G} = \int_{\Omega^f} \rho_f \bar{\mathbf{B}}^T \, d\Omega$,

$$\begin{bmatrix} \mathbf{M}_{ss} & \mathbf{0} & \mathbf{0} \\ \mathbf{0} & \mathbf{0} & \mathbf{0} \\ \mathbf{0} & \mathbf{0} & -\mathbf{M}_{ff} \end{bmatrix} \begin{bmatrix} \ddot{\mathbf{U}} \\ \ddot{\mathbf{P}} \\ \ddot{\boldsymbol{\Phi}} \end{bmatrix} + \begin{bmatrix} \mathbf{0} & \mathbf{0} & \mathbf{C}_{fs}^T \\ \mathbf{0} & \mathbf{0} & \mathbf{C}_{pf} \\ \mathbf{C}_{fs} & \mathbf{C}_{pf}^T & \mathbf{0} \end{bmatrix} \begin{bmatrix} \dot{\mathbf{U}} \\ \dot{\mathbf{P}} \\ \dot{\boldsymbol{\Phi}} \end{bmatrix} + \begin{bmatrix} \mathbf{K}_{ss} & \mathbf{K}_{ps}^T & \mathbf{0} \\ \mathbf{K}_{ps} & \mathbf{K}_{pp} & \mathbf{0} \\ \mathbf{0} & \mathbf{0} & -\mathbf{K}_{ff} \end{bmatrix} \begin{bmatrix} \mathbf{U} \\ \mathbf{P} \\ \boldsymbol{\Phi} \end{bmatrix} = \begin{bmatrix} \mathbf{R}_s - \mathbf{M}_{ss} \ddot{\mathbf{u}}_g \\ \mathbf{0} \\ \mathbf{G} \ddot{\mathbf{u}}_g \end{bmatrix}. \tag{1.34}$$

In both $\phi$-$\mathbf{u}$ and $p$-$\phi$-$\mathbf{u}$ formulations, their scalar unknowns are used for acoustic fluids. The procedures to make the matrices symmetric enlarge bandwidth and consequently require more computation effort [22,85,209]. In addition, special elements are needed along the FSI interface to couple fluid and solid domains. Furthermore, since the stiffness and mass matrices for acoustic fluids are not physical ones (as for solids), it is difficult to apply dynamic loading and the spectrum method.

### 6.1.2. $\mathbf{u}/p$ and $\mathbf{u}$-$p$-$\boldsymbol{\Lambda}$ formulations

Historically, two interrelated constraints, i.e., the incompressibility and irrotationality constraints created some confusion for acoustic fluid models and formulations. Primitive variable formulations have received considerable attention because they do not require any special FSI interface conditions or new solution strategies in frequency calculations and response spectrum analysis. With the ever-increasing availability of high speed and large capacity computers, this approach shows great promise in general applications for the solution of a broad range of problems, in particular nonlinear problems. Unfortunately, some difficulties have remained unsolved using primitive variable finite element formulations in the analysis of fluid flows and fluid-structure interactions.

In linear analysis, it has been widely reported that the displacement-based fluid elements employed in frequency or dynamic analysis exhibit spurious nonzero frequency circulation modes [140,170,175]. Various approaches have been introduced to obtain improved formulations. The penalty method has been applied by Hamdi et al. [175] and has been shown to give good solutions for the cases considered in that reference. Subsequently, Olson and Bathe [170] demonstrated that the method *locks up* in the frequency analysis of a solid vibrating in a fluid cavity. They also showed that reduced integration performed on penalty formulation yields some improvement in results, but does not generally assure solution convergence. Wilson and Khalvati [71] developed a formulation with rotational constraints and a reduced integration technique based on the pure displacement formulation, and Chen and Taylor [102] proposed a 4-node element with a reduced integration technique and an element mass matrix projection. Recently, Bermúdez and Rodríguez used simple 3-node triangular *edge* elements to model the fluid [19]. This formulation is promising, but the degrees of freedom of fluid elements are not those of structure and coupling along the FSI interface again requires special considerations.

It has been proposed that spurious nonzero frequencies are caused by the irrotationality constraint [71,102,175]. However, it has actually been shown that the origins of the spurious nonzero frequencies are in the use of the pure displacement-based formulation, which includes penalty formulations and in the mishandling of the FSI interface [13,300]. We propose displacement/pressure based formulations with proper elements, namely, mixed elements which satisfy the inf-sup condition. In addition, we address some subtle points regarding boundary conditions at the FSI interface and free surfaces. In the solutions of some selected generic test problems, we demonstrate that by using displacement/pressure formulations with proper elements and boundary conditions, we no longer encounter any spurious nonzero frequency pressure and rotational modes.

In both **u**/p and **u**-p-$\boldsymbol{\Lambda}$ mixed formulations, displacements are the state variables for both fluid and solid domains. Therefore **M** and **K** matrices are the true physical mass and stiffness matrices. Consequently it is possible to adopt the standard spectrum method widely used in structural dynamics to general acoustoelastic/slosh FSI systems.

### **u**/*p formulation*

In the displacement-based mixed formulation, we define a variational indicator for acoustic fluid:

$$\Pi = \int_{\Omega^f} \left\{ \frac{p^2}{2\kappa} - \mathbf{u}\cdot\mathbf{f}^f - \lambda_p\left(\frac{p}{\kappa} + \nabla\cdot\mathbf{u}\right)\right\} d\Omega + \int_{\Gamma^{f,f}} \bar{p} u_n^{f,f}\, d\Gamma, \qquad (1.35)$$

where the variables are $p$, $\mathbf{u}$, and the Lagrange multiplier $\lambda_p$.

We note that the first two terms correspond to the usual strain energy given in terms of pressure and potential of externally applied body forces. The third term implies the mass conservation equation. The last term is the potential due to any applied pressure on the Neumann boundary $\Gamma^{f,f}$. Moreover, the inertial effect $-\rho\ddot{\mathbf{u}}$ are also included in the body force $\mathbf{f}^f$.

Invoking the stationarity of $\Pi$, we identify the Lagrange multiplier $\lambda_p$ as pressure $p$ and obtain the governing equations

$$\nabla p + \rho\ddot{\mathbf{u}} = \mathbf{0}, \tag{1.36}$$

$$\nabla \cdot \mathbf{u} + \frac{p}{\kappa} = 0, \tag{1.37}$$

and the boundary conditions

$$\begin{aligned} \mathbf{u}\cdot\mathbf{n} &= \bar{u}_n, && \text{on } \Gamma^{f,v}, \\ p &= \bar{p}, && \text{on } \Gamma^{f,f}. \end{aligned} \tag{1.38}$$

The pressure $\bar{p}$ is commonly assigned to be zero on the free surface, provided the surface gravity waves are ignored. Applying the standard Galerkin discretization procedure, we have for a typical finite element

$$\mathbf{u} = \mathbf{H}\mathbf{U}, \qquad p = \mathbf{H}_p\mathbf{P}, \quad \text{and} \quad \nabla\cdot\mathbf{u} = (\nabla\cdot\mathbf{H})\mathbf{U} = \mathbf{B}\mathbf{U},$$

where $\mathbf{H}$ and $\mathbf{H}_p$ are the interpolation matrices, and $\mathbf{U}$ and $\mathbf{P}$ are the vectors of solution variables.

The matrix equations of the $\mathbf{u}$/p formulation are given as

$$\begin{bmatrix} \mathbf{M} & \mathbf{0} \\ \mathbf{0} & \mathbf{0} \end{bmatrix}\begin{bmatrix} \ddot{\mathbf{U}} \\ \ddot{\mathbf{P}} \end{bmatrix} + \begin{bmatrix} \mathbf{0} & \mathbf{K}_{up} \\ \mathbf{K}_{up}^T & \mathbf{K}_{pp} \end{bmatrix}\begin{bmatrix} \mathbf{U} \\ \mathbf{P} \end{bmatrix} = \begin{bmatrix} \mathbf{R} \\ \mathbf{0} \end{bmatrix}, \tag{1.39}$$

with

$$\mathbf{M} = \int_{\Omega^f} \rho\mathbf{H}^T\mathbf{H}\,d\Omega, \qquad \mathbf{K}_{up} = -\int_{\Omega^f} \mathbf{B}^T\mathbf{H}_p\,d\Omega;$$

$$\mathbf{K}_{pp} = -\int_{\Omega^f} \frac{1}{\kappa}\mathbf{H}_p^T\mathbf{H}_p\,d\Omega, \qquad \mathbf{R} = -\int_{\Gamma^{f,f}} (\mathbf{H}_n^{f,f})^T\,\bar{p}\,d\Gamma.$$

In two-dimensional analysis, two effective elements are the 9/3 and 9/4-$c$ elements schematically depicted in Figure 6.1. For these elements, full numerical integration corresponds to $3\times 3$ Gauss integration. For the 9/3 element, we interpolate the pressure linearly as

$$p = p_1 + p_2 r + p_3 s, \tag{1.40}$$

and for the 9/4-$c$ element we use the bilinear interpolation

$$p = p_1 + p_2 r + p_3 s + p_4 rs. \tag{1.41}$$

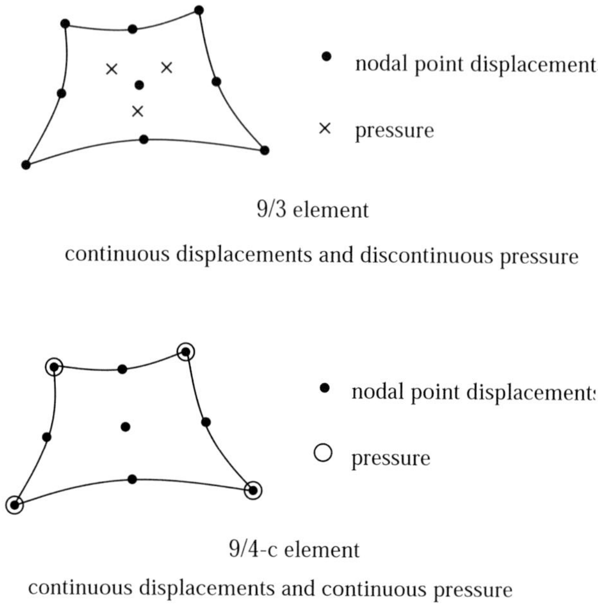


Figure 6.1. Two elements for **u**/p formulation.

If we are to use 4-node quadrilateral mixed elements in general meshes, the element proposed in Ref. [213] is the only one known to satisfy the inf-sup condition. The well-known 4/1 (Q1/P0) element is not recommended for use in general meshes because it does not satisfy the inf-sup condition. Of course, there are a few other elements available. It is interesting that, so far, there is no available simple 4-node element ($Q1/P0$ or $Q1/Q1$ element) satisfying the inf-sup condition with general meshes [213]. However, researchers have found that some special macroelements constructed with 4/1 elements pass the inf-sup condition [27,212]. Figure 6.2 shows the two types of macroelements reliable for mixed formulations. For these elements, full numerical integration corresponds to $2 \times 2$ Gauss integration. Using the **u**/p formulation, we no longer need to impose the irrotational constraint and can literally choose all the elements satisfying the inf-sup condition. Another benefit is that this **u**/p formulation can be easily extended to nonlinear analysis with the same primary unknowns. Since the irrotational constraint is not imposed in the **u**/p formulation, the solution gives many exact zero frequencies. It is conceivable that if we impose the irrotational constraint, we will derive the **u**-p-$\boldsymbol{\Lambda}$ formulation and the number of the zero frequency modes will be greatly reduced [151,300,301].

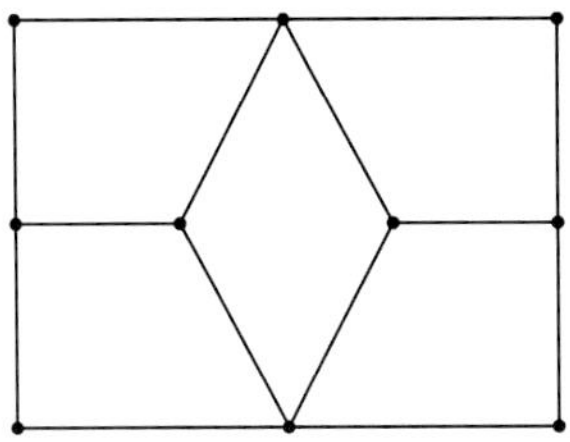

Type I macroelement with 4/1 and 4-1-1 elements

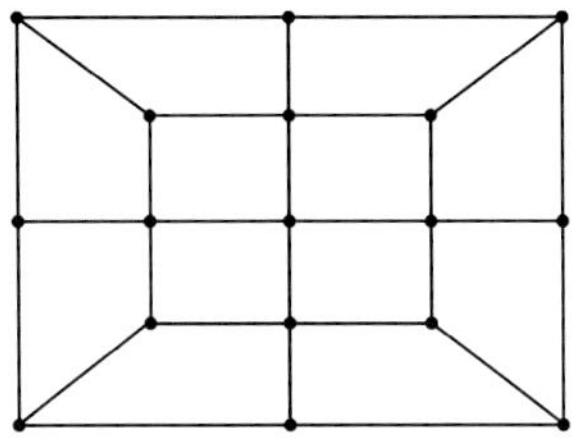

Type II macroelement with 4/1 and 4-1-1 elements

Figure 6.2. Two macroelements for $\mathbf{u}/p$ and $\mathbf{u}$-p-$\boldsymbol{\Lambda}$ formulations.

### $\mathbf{u}$-*p*-$\boldsymbol{\Lambda}$ *formulation*

We know that if we use the $\mathbf{u}/p$ formulation with proper elements, for example, 9/3 and 9/4-*c* elements, we will not have spurious nonzero frequency modes. To impose the irrotational constraint, penalty formulations, along with reduced integration techniques, are widely used historically. Consequently many researchers have encountered the so-called nonzero frequency spurious rotational modes and various remedy procedures have been proposed [71,102,175]. In the $\mathbf{u}$-p-$\boldsymbol{\Lambda}$ formulation, we introduce

$$\nabla \times \mathbf{u} = \frac{\boldsymbol{\Lambda}}{\alpha}, \tag{1.42}$$

where $\boldsymbol{\Lambda}$ is a vorticity moment.

The magnitude of $\boldsymbol{\Lambda}$ shall be small while $\alpha$ is a constant of large value. Based on our experience with the $\mathbf{u}/p$ formulation, we propose the following variational indicator for the finite element formulation of the fluid model considered here:

$$\begin{aligned}\Pi = \int_{\Omega^f} \bigg\{ & \frac{p^2}{2\kappa} - \mathbf{u} \cdot \mathbf{f}^f - \lambda_p \left( \frac{p}{\kappa} + \nabla \cdot \mathbf{u} \right) + \frac{\boldsymbol{\Lambda} \cdot \boldsymbol{\Lambda}}{2\alpha} \\ & - \boldsymbol{\lambda}_\Lambda \cdot \left( \frac{\boldsymbol{\Lambda}}{\alpha} - \nabla \times \mathbf{u} \right) \bigg\} \, d\Omega + \int_{\Gamma^{f,f}} \bar{p} u_n^{f,f} \, d\Gamma,\end{aligned} \tag{1.43}$$

where the variables are $p$, $\mathbf{u}$, $\boldsymbol{\Lambda}$, and the Lagrange multipliers $\lambda_p$ and $\boldsymbol{\lambda}_\Lambda$.

We note that the fourth term is included so that we may statically condense out the degrees of freedom of the vorticity moment, and the fifth term represents the constraint from Eq. (1.42). For the fourth and fifth terms we require that the constant $\alpha$ is large, and we use $\alpha = 1000\kappa$. However, from our numerical tests, we find that $\alpha$ can be any numerically reasonable value larger than $\kappa$, say $100\kappa \leqslant \alpha \leqslant 10^6\kappa$. The last term is the potential due to any applied boundary pressure on the Neumann boundary $\Gamma^{f,f}$.

Invoking the stationarity of $\Pi$, we identify the Lagrange multipliers $\lambda_p$ and $\boldsymbol{\lambda}_\Lambda$ as the pressure $p$ and vorticity moment $\boldsymbol{\Lambda}$, respectively, and we obtain the following field equations:

$$\nabla p - \mathbf{f}^f + \nabla \times \boldsymbol{\Lambda} = \mathbf{0}, \tag{1.44}$$

$$\nabla \cdot \mathbf{u} + \frac{p}{\kappa} = 0, \tag{1.45}$$

$$\nabla \times \mathbf{u} - \frac{\boldsymbol{\Lambda}}{\alpha} = \mathbf{0}, \tag{1.46}$$

and the boundary conditions

$$\begin{aligned} \mathbf{u} \cdot \mathbf{n} &= \bar{u}_n, && \text{on } \Gamma^{f,v}, \\ p &= \bar{p}, && \text{on } \Gamma^{f,f}, \\ \boldsymbol{\Lambda} &= \mathbf{0}, && \text{on } \Gamma. \end{aligned} \tag{1.47}$$

We use the standard Galerkin finite element discretization; and hence, for a typical element, we have the following additional interpolations in comparison with the $\mathbf{u}$/p formulation:

$$\boldsymbol{\Lambda} = \mathbf{H}_\Lambda \boldsymbol{\Lambda} \quad \text{and} \quad \nabla \times \mathbf{u} = (\nabla \times \mathbf{H})\mathbf{U} = \mathbf{D}\mathbf{U},$$

where $\mathbf{H}_\Lambda$ is the interpolation matrix and $\boldsymbol{\Lambda}$ is the vector of vorticity moment unknowns.

The matrix equations of our formulation are

$$\begin{bmatrix} \mathbf{M} & \mathbf{0} & \mathbf{0} \\ \mathbf{0} & \mathbf{0} & \mathbf{0} \\ \mathbf{0} & \mathbf{0} & \mathbf{0} \end{bmatrix} \begin{bmatrix} \ddot{\mathbf{U}} \\ \ddot{\mathbf{P}} \\ \ddot{\boldsymbol{\Lambda}} \end{bmatrix} + \begin{bmatrix} \mathbf{0} & \mathbf{K}_{up} & \mathbf{K}_{u\lambda} \\ \mathbf{K}_{up}^T & \mathbf{K}_{pp} & \mathbf{0} \\ \mathbf{K}_{u\lambda}^T & \mathbf{0} & \mathbf{K}_{\lambda\lambda} \end{bmatrix} \begin{bmatrix} \mathbf{U} \\ \mathbf{P} \\ \boldsymbol{\Lambda} \end{bmatrix} = \begin{bmatrix} \mathbf{R} \\ \mathbf{0} \\ \mathbf{0} \end{bmatrix}, \tag{1.48}$$

with

$$\mathbf{M} = \int_{\Omega^f} \rho \mathbf{H}^T \mathbf{H} \, d\Omega, \qquad \mathbf{K}_{up} = -\int_{\Omega^f} \mathbf{B}^T \mathbf{H}_p \, d\Omega;$$

$$\mathbf{K}_{u\lambda} = \int_{\Omega^f} \mathbf{D}^T \mathbf{H}_\Lambda \, d\Omega, \qquad \mathbf{K}_{pp} = -\int_{\Omega^f} \frac{1}{\kappa} \mathbf{H}_p^T \mathbf{H}_p \, d\Omega;$$

$$\mathbf{K}_{\lambda\lambda} = -\int_{\Omega^f} \frac{1}{\alpha} \mathbf{H}_\Lambda^T \mathbf{H}_\Lambda \, d\Omega, \qquad \mathbf{R} = -\int_{\Gamma^{f,f}} (\mathbf{H}_n^{f,f})^T \bar{p} \, d\Gamma.$$

Again, the key to success with finite element discretization is choosing appropriate interpolations for the displacements, pressure, and vorticity moment. Based on our experience with the $\mathbf{u}$/p formulation, we use displacement/pressure

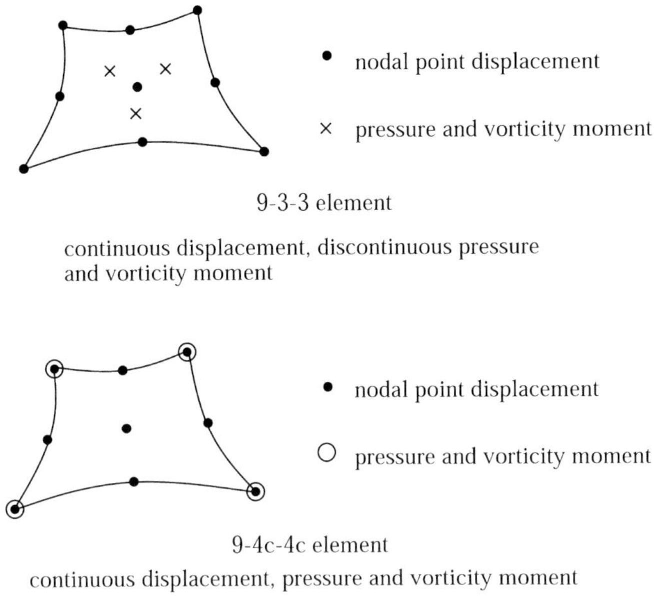


Figure 6.3. Two elements for **u**-p-$\boldsymbol{\Lambda}$ formulation.

interpolations that satisfy the inf-sup condition in the analysis of almost incompressible solids or acoustic fluids. For the vorticity moment we use the same or a lower order interpolation as we do for the pressure. Thus, some proposed elements for two-dimensional analysis are the 9-3-3, 9-3-1, and 9/4-$c$-4$c$ elements [151]. These elements are schematically depicted in Figure 6.3. For these elements, full numerical integration corresponds to $3 \times 3$ Gauss integration. These elements have displayed good predictive capabilities. However, while we can easily show that the 9-3-1 element satisfies the analytical inf-sup condition, in general we can only show that these elements pass the numerical inf-sup test. Limited number of elements can be analytically proven to satisfy the inf-sup condition [13,40,301].

Hence, for the 9-3-3 element, we interpolate the pressure and vorticity moment linearly as

$$p = p_1 + p_2 r + p_3 s, \tag{1.49}$$

$$\boldsymbol{\Lambda} = \boldsymbol{\Lambda}_1 + \boldsymbol{\Lambda}_2 r + \boldsymbol{\Lambda}_3 s, \tag{1.50}$$

and for the 9/4-$c$-4$c$ element we use the bilinear interpolations:

$$p = p_1 + p_2 r + p_3 s + p_4 rs, \tag{1.51}$$

$$\boldsymbol{\Lambda} = \boldsymbol{\Lambda}_1 + \boldsymbol{\Lambda}_2 r + \boldsymbol{\Lambda}_3 s + \boldsymbol{\Lambda}_4 rs. \tag{1.52}$$

Of course, additional elements could be proposed using this approach. In each case, the vorticity moment would be interpolated with functions that, when used for the pressure interpolation, give a stable element for almost incompressible conditions. For instance, the 9-3-1 element, which does not impose zero vorticity as strongly as the 9-3-3 element, is also reasonable. On the other hand, the 4-1-1 element with regular meshes is not recommended; however, some macroelements constructed with 4-1-1 elements can be applicable to certain cases. Full numerical integration must be used for all elements.

In this book, we ignore the surface tension and assign $\bar{p} = 0$ to the free surfaces. To apply both $\mathbf{u}/\mathrm{p}$ and $\mathbf{u}$-p-$\boldsymbol{\Lambda}$ formulations to the study of the free surface, we can simply introduce a free surface gravity wave potential $W_f$, denoted as

$$W_f = \int_{\Gamma^{f,f}} \frac{1}{2} u_s^2 \rho g \, d\Gamma, \tag{1.53}$$

where $u_s$ is the free surface elevation referring to its static equilibrium position $\Gamma^{f,f}$.

A similar approach has already been discussed in Ref. [71]. We often discretize $u_s$ with isoparametric elements of the same order on the free surface as in the fluid domain. For instance, in two-dimensional analysis, if we use 9-node elements for the fluid domain, we can use 3-node elements on the free surface. Therefore, we have $u_s = \mathbf{H}_s \mathbf{U}_s = \sum_{i=1}^{3} h_i U_s^i$ with $h_1 = (s^2 - s)/2$, $h_2 = (s^2 + s)/2$, and $h_3 = 1 - s^2$. From the variation of $W_f$, i.e. $\delta W_f = \int_{\Gamma^{f,f}} \delta u_s \rho g u_s \, d\Gamma$, we can easily derive the additional stiffness term for the nodal displacements $u_s$ on the free surface:

$$\mathbf{K}_{ss} = \int_{\Gamma^{f,f}} \rho g \mathbf{H}_s^T \mathbf{H}_s \, d\Gamma. \tag{1.54}$$

### *6.1.3. Solvability and stability; FSI interfaces; zero modes*

#### *Solvability and stability*

To analyze solvability and stability, we define the following finite element spaces:

$$\begin{aligned}
\mathcal{V}_h &= \left\{ \mathbf{v}_h \mid \mathbf{v}_h, \ \frac{\partial (v_h)_i}{\partial x_j} \in \mathcal{L}^2(\Omega), \ \mathbf{v}_h = \mathbf{0}, \ \text{on } \Gamma^v \right\}, \\
\mathcal{D}_h &= \{ q_h \mid q_h = \nabla \cdot \mathbf{v}_h, \ \text{for some } \mathbf{v}_h \in \mathcal{V}_h \}, \\
\mathcal{K}_h(q_h) &= \{ \mathbf{v}_h \mid \mathbf{v}_h \in \mathcal{V}_h, \ \nabla \cdot \mathbf{v}_h = q_h \},
\end{aligned} \tag{1.55}$$

where the mesh is represented by mesh size $h$.

For purposes of discussion, let us now recall the governing algebraic finite element equations derived from the displacement/pressure formulation for almost incompressible solids [13,143]:

$$\begin{bmatrix} (\mathbf{K}_{uu})_h & (\mathbf{K}_{up})_h \\ (\mathbf{K}_{up})_h^T & (\mathbf{K}_{pp})_h \end{bmatrix} \begin{bmatrix} \mathbf{U}_h \\ \mathbf{P}_h \end{bmatrix} = \begin{bmatrix} \mathbf{R}_h \\ \mathbf{0} \end{bmatrix}, \tag{1.56}$$

where $\mathbf{U}_h$ lists all unknown nodal point displacements and $\mathbf{P}_h$ lists pressure variables.

The mathematical analysis of the mixed formulation consists of a study of the solvability and stability of Eqs. (1.56), where the stability of the equations implies their solvability. For the corresponding dynamic problem, the governing equations are

$$\begin{bmatrix} (\mathbf{M}_{uu})_h & \mathbf{0} \\ \mathbf{0} & \mathbf{0} \end{bmatrix} \begin{bmatrix} \ddot{\mathbf{U}}_h \\ \ddot{\mathbf{P}}_h \end{bmatrix} + \begin{bmatrix} (\mathbf{K}_{uu})_h & (\mathbf{K}_{up})_h \\ (\mathbf{K}_{up})_h^T & (\mathbf{K}_{pp})_h \end{bmatrix} \begin{bmatrix} \mathbf{U}_h \\ \mathbf{P}_h \end{bmatrix} = \begin{bmatrix} \mathbf{R}_h \\ \mathbf{0} \end{bmatrix}. \tag{1.57}$$

As discussed elsewhere [28,143], the key to the stability of the formulation is the satisfaction of the following inf-sup condition:

$$\inf_{q_h \in D_h} \sup_{\mathbf{v}_h \in \mathcal{V}_h} \frac{\int_\Omega q_h \nabla \cdot \mathbf{v}_h \, d\Omega}{\|q_h\| \|\mathbf{v}_h\|} \geqslant \beta_o > 0, \tag{1.58}$$

where $\beta_o$ is a constant independent of $h$ and the bulk modulus.

If this condition is satisfied, the stability of the formulation is guaranteed and optimal error bounds are obtained for the selected displacement and pressure interpolations. If the bulk modulus is finite, we can statically condense out (at least in theory) the pressure unknowns, obtaining

$$(\mathbf{M}_{uu})_h \ddot{\mathbf{U}}_h + \left(\mathbf{K}_{uu}^*\right)_h \mathbf{U}_h = \mathbf{R}_h, \tag{1.59}$$

with

$$\left(\mathbf{K}_{uu}^*\right)_h = (\mathbf{K}_{uu})_h - (\mathbf{K}_{up})_h \left(\mathbf{K}_{pp}^{-1}\right)_h (\mathbf{K}_{up})_h^T. \tag{1.60}$$

Of course, in Eq. (1.60) the first part of the stiffness matrix is due to deviatoric strain energy and the second part is due to volumetric strain energy. Considering Eq. (1.60), for the shear modulus $G > 0$, with $q_h = \nabla \cdot \mathbf{v}_h \ll 1$, we have

$$(\mathbf{K}_{uu})_h \to \text{positive definite in } K_h(q_h), \tag{1.61}$$

$$(\mathbf{K}_{up})_h \left(\mathbf{K}_{pp}^{-1}\right)_h (\mathbf{K}_{up})_h^T \to \text{rank } m, \tag{1.62}$$

where $m$ is the number of pressure degrees of freedom.

In dynamic analysis, for each time step, the coefficient matrix is given as

$$\left(\mathbf{K}_{uu}^{**}\right)_h = \left(\mathbf{K}_{uu}^*\right)_h + c(\mathbf{M}_{uu})_h, \tag{1.63}$$

where the constant $c$ is a positive number related to the direct time integration scheme, for instance, $c = 4/\Delta t^2$ for the trapezoidal rule.

Since mass matrix $(\mathbf{M}_{uu})_h$ is a positive definite matrix, as a result of Eq. (1.63), $(\mathbf{K}_{uu}^{**})_h$ is always a positive definite matrix. Overall, we can now consider the following three categories of problems:

(1) the solid bulk modulus $\kappa$ and solid shear modulus $G$ are of the same order;
(2) $\kappa \gg G$ and $\kappa, G > 0$;
(3) $\kappa > 0$ and $G = 0$.

In category (1), the standard displacement formulation ensures solvability and stability. In category (2), i.e. almost incompressible material analysis, the displacement/pressure mixed formulation with mixed elements that satisfy the inf-sup condition is well established [28,143]. Category (3) includes the analysis of the inviscid acoustic fluid model discussed in this chapter. The loss of ellipticity introduces zero frequency modes corresponding to the zero deviatoric strain energy (note that for this case, $(\mathbf{K}_{uu})_h = \mathbf{0}$). A mathematical prediction of the *exact* number of zero frequencies is necessary to identify whether or not we have nonzero frequency spurious modes. For $n$ displacement unknowns, the exact number of zero frequencies is $n-m$, provided that the physical constant pressure mode arising with the enclosed boundary condition $\mathbf{u} \cdot \mathbf{n} = 0$ on $\Gamma$ has been eliminated.

Considering now an assemblage of elements, using Eqs. (1.39) and (1.48), we have the general governing equation

$$\begin{bmatrix} (\mathbf{M}_{uu})_h & \mathbf{0} \\ \mathbf{0} & \mathbf{0} \end{bmatrix} \begin{bmatrix} \ddot{\mathbf{U}}_h \\ \ddot{\mathbf{S}}_h \end{bmatrix} + \begin{bmatrix} (\mathbf{K}_{uu})_h & (\mathbf{K}_{us})_h \\ (\mathbf{K}_{us})_h^T & (\mathbf{K}_{ss})_h \end{bmatrix} \begin{bmatrix} \mathbf{U}_h \\ \mathbf{S}_h \end{bmatrix} = \begin{bmatrix} \mathbf{R}_h \\ \mathbf{0} \end{bmatrix}, \tag{1.64}$$

where now $\mathbf{U}_h$ lists all the unknown nodal point displacements and $\mathbf{S}_h$ lists all the pressure and, if applicable, vorticity moment variables. In our formulations for the acoustic fluid model, we have

*for* $\mathbf{u}/p$ *formulation*:

$$(\mathbf{K}_{us})_h = (\mathbf{K}_{up})_h \quad \text{and} \quad (\mathbf{K}_{ss})_h = (\mathbf{K}_{pp})_h,$$

*for* $\mathbf{u}$-$p$-$\boldsymbol{\Lambda}$ *formulation*:

$$(\mathbf{K}_{us})_h = \begin{bmatrix} (\mathbf{K}_{up})_h & (\mathbf{K}_{u\lambda})_h \end{bmatrix} \quad \text{and} \quad (\mathbf{K}_{ss})_h = \begin{bmatrix} (\mathbf{K}_{pp})_h & \mathbf{0} \\ \mathbf{0} & (\mathbf{K}_{\lambda\lambda})_h \end{bmatrix}.$$

Since $(\mathbf{M}_{uu})_h$ is positive definite and $(\mathbf{K}_{ss})_h$ is invertible, based on our inviscid acoustic fluid model, in frequency analysis, we will encounter exact zero frequencies corresponding to $(\mathbf{K}_{uu})_h = \mathbf{0}$. In a transient direct step-by-step solution, at each time step, we have

$$\begin{bmatrix} (\mathbf{K}_{uu}^{*})_h & (\mathbf{K}_{us})_h \\ (\mathbf{K}_{us})_h^T & (\mathbf{K}_{ss})_h \end{bmatrix} \begin{bmatrix} \mathbf{U}_h \\ \mathbf{S}_h \end{bmatrix} = \begin{bmatrix} \mathbf{R}_h \\ \mathbf{0} \end{bmatrix}, \tag{1.65}$$

where $\mathbf{R}_h$ is an effective load vector.

Notice that similar to Eq. (1.63), we also have

$$\left(\mathbf{K}_{uu}^{*}\right)_{h} = c(\mathbf{M}_{uu})_{h} \tag{1.66}$$

and the constant $c$ is a positive number related to the direct time integration scheme used.

Therefore $(\mathbf{K}_{uu}^{*})_h$ is always a positive definite matrix. For stability of our formulations, we need to choose spaces of displacements and pressure (and vorticity moment) such that we satisfy the following ellipticity and inf-sup conditions corresponding to Eq. (1.65)

*ellipticity condition*:

$$\exists c_1 > 0 \quad \text{such that} \quad \mathbf{V}_h^T \left(\mathbf{K}_{uu}^{*}\right)_h \mathbf{V}_h \geqslant c_1 \|\mathbf{V}_h\|^2 \quad \forall \mathbf{V}_h \in \mathcal{K}, \tag{1.67}$$

where $\mathcal{K} = Ker((\mathbf{K}_{us})_h^T)$

$$\mathcal{K} = \left\{\mathbf{V}_h \mid \mathbf{V}_h \in \mathcal{R}^n,\ (\mathbf{K}_{us})_h^T \mathbf{V}_h = \mathbf{0}\right\}; \tag{1.68}$$

inf-sup *condition*:

$$\inf_{\mathbf{S}_h} \sup_{\mathbf{U}_h} \frac{\mathbf{U}_h^T (\mathbf{K}_{us})_h \mathbf{S}_h}{\|\mathbf{U}_h\| \|\mathbf{S}_h\|} \geqslant c_2 > 0, \tag{1.69}$$

where the constant $c_2$ is independent of the mesh size $h$, the material property $\kappa$ and the constant $\alpha$.

The usual inf-sup condition is for the $\mathbf{u}/p$ formulation and a number of reliable finite elements have been proposed. However, little knowledge is available for the inf-sup condition given in Eq. (1.69) when the vorticity moment variables are included.

Since $(\mathbf{K}_{pp})_h$ and $(\mathbf{K}_{\lambda\lambda})_h$ are negative definite matrices, we need only to consider the stiffness matrix $(\mathbf{K}_{us})_h(\mathbf{K}_{ss}^{-1})_h(\mathbf{K}_{us})_h^T$ in order to understand the zero frequencies in our formulations. The key to the success of our mixed formulations is to choose appropriate interpolations for the displacements, pressure and vorticity moment. The results in Figure 6.5, with $N$ equal to the square root of the number of elements used, show that both 9-4$c$-4$c$ and 9-3-3 elements pass the numerical inf-sup test. In addition, this figure shows the results of the numerical inf-sup test performed for the 4-1-1 element, and as expected, the test is not passed. Figures 6.6 and 6.7 show the results of the numerical inf-sup test performed for the macroelements of Figure 6.2. We note that, as expected, the macroelements of the 4/1 element pass the test, but the macroelements of the 4-1-1 element do not pass the test.

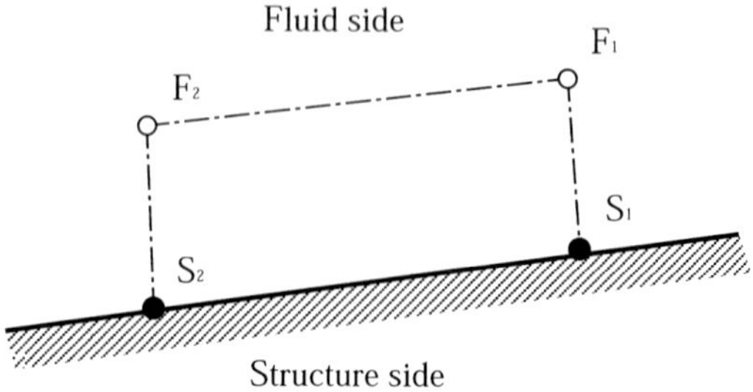


Figure 6.4. A typical FSI interface.

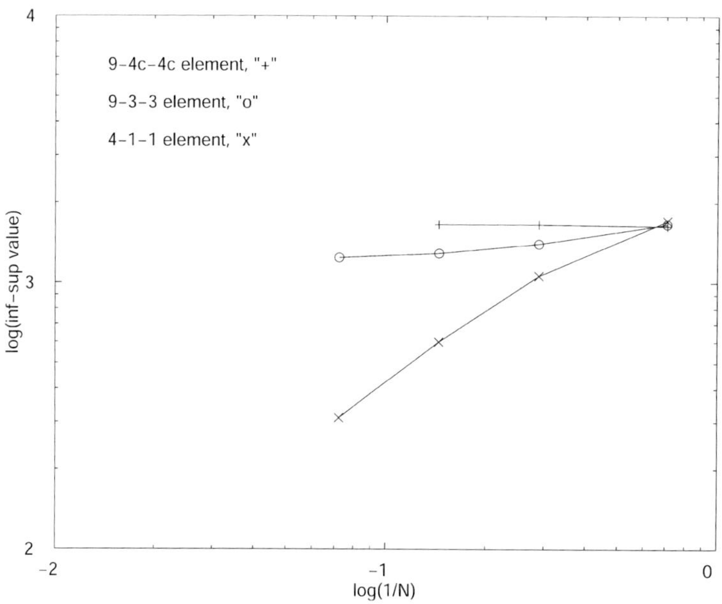


Figure 6.5. Numerical inf-sup test of 9-3-3, 9-4*c*-4*c*, and 4-1-1 elements of **u**-p-$\boldsymbol{\Lambda}$ formulation.

### *FSI interfaces*

In FSI problems, if there is no slip between different continuous media, the particles on both sides will share the same displacements and satisfy the force balance. For the interface between fluids and structures shown in Figure 6.4, if no additional assumptions are made for the fluids, while the force balance holds, the particles on both sides of the mesh points $S_1$ and $S_2$ will have the same displacements and velocities. We then have the following no slip conditions:

$$u_n^f|_{S_1,S_2} = u_n^s|_{S_1,S_2} \quad \text{and} \quad u_t^f|_{S_1,S_2} = u_t^s|_{S_1,S_2}. \tag{1.70}$$

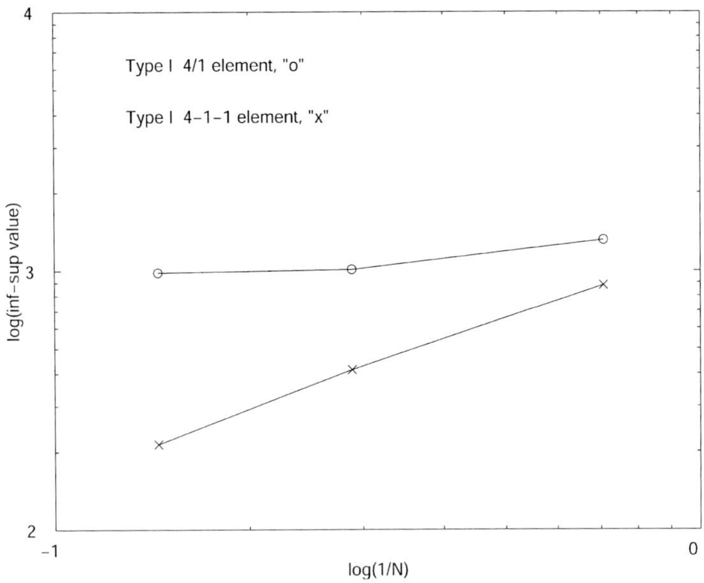


Figure 6.6. Numerical inf-sup test of Type I macroelements of $\mathbf{u}/p$ and $\mathbf{u}$-p-$\boldsymbol{\Lambda}$ formulations.

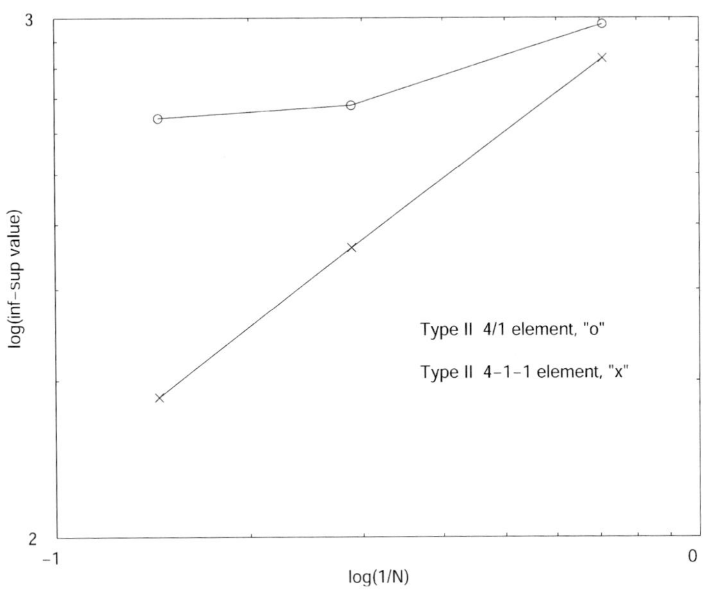


Figure 6.7. Numerical inf-sup test of Type II macroelements of $\mathbf{u}/p$ and $\mathbf{u}$-p-$\boldsymbol{\Lambda}$ formulations.

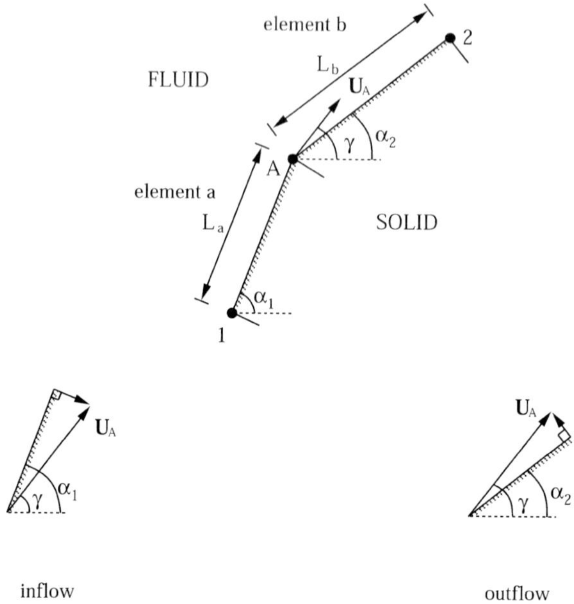


Figure 6.8. Tangential direction at node A for 3 or 4-node elements.

As fluid viscosity decreases, the layer $F_1F_2S_2S_1$ will get thinner and the normal direction gradient of the tangential velocity around the fluid-structure interface will increase. Therefore, in the limit case, namely inviscid fluid models, we have the following slip conditions:

$$u_n|_{F_1,F_2} = u_n^s|_{S_1,S_2} \quad \text{and} \quad u_t|_{F_1,F_2} \neq u_t^s|_{S_1,S_2}. \tag{1.71}$$

To satisfy mass conservation over the layer $F_1F_2S_2S_1$, we present the following discussion of slip boundary conditions for inviscid fluid models. Let us consider typical boundary lines composed of two adjacent 3 or 4-node elements (see Figure 6.8) and two 9-node elements (see Figure 6.9). Based on our assumption, the displacement component tangential to the solid is not restrained, while the displacement component normal to the solid is restrained to be equal to the displacement of the solid. The choice of directions of nodal tangential displacements, from which follow normal displacements, is critical.

Considering the solution of actual fluid flows, and fluid flows with structural interactions in which the fluid is modeled using the Navier–Stokes equations including wall turbulence effects or the Euler equations, we are accustomed to choosing the directions of nodal tangential velocities such that in the finite element discretization there is no transport of fluid across the fluid-structure interfaces. It is important that we employ the same concept in the definition of the

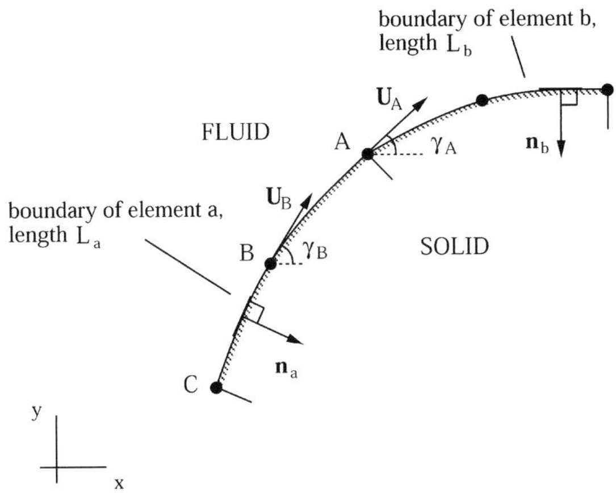


Figure 6.9. Tangential directions at nodes A and B for 9-node elements.

directions of nodal tangential displacements on the boundaries of the acoustic fluid considered here.

While we do not recommend using the $Q1/P0$ or $Q1/Q1$ element of the mixed formulation, lower-order elements, for example, 3-node triangular two-dimensional and 4-node tetrahedral three-dimensional elements, are effectively employed in fluid flow analysis and could be used in both **u**/p and **u**-p-$\boldsymbol{\Lambda}$ formulations [153,155]. Hence, consider first the boundary representation in Figure 6.8. This case has been discussed by Donea [59]. For simplicity, we use displacements to discuss mass conservation. The same concepts hold for velocities. For the displacement $\mathbf{U}_A$, we have the spurious fluxes $\Delta V_a$ and $\Delta V_b$ across the element lengths $L_a$ and $L_b$. Mass conservation requires that $\Delta V_a + \Delta V_b = 0$. Hence we must choose the effective tangential direction $\gamma$, given by

$$\tan \gamma = \frac{L_a \sin \alpha_1 + L_b \sin \alpha_2}{L_a \cos \alpha_1 + L_b \cos \alpha_2}. \tag{1.72}$$

We note that $\gamma$ is actually given by the direction of the line from node 1 to node 2. In general, the direction of $\mathbf{U}_A$ is not given by the mean of angles $\alpha_1$ and $\alpha_2$, but only in the specific case when $L_a = L_b$ do we have the familiar average

$$\gamma = (\alpha_1 + \alpha_2)/2. \tag{1.73}$$

We employ the same concept to establish the appropriate tangential directions at the typical nodes A and B of our 9-node elements in Figure 6.9. For node A, we need to have

$$\int_{L_a} \mathbf{u}_a^A \cdot \mathbf{n}_a \, dl + \int_{L_b} \mathbf{u}_b^A \cdot \mathbf{n}_b \, dl = 0, \tag{1.74}$$

with

$$\mathbf{n}_a\,dl = \left\langle -\frac{\partial y_a}{\partial s}\,ds,\ \frac{\partial x_a}{\partial s}\,ds \right\rangle \quad \text{and} \quad \mathbf{n}_b\,dl = \left\langle -\frac{\partial y_b}{\partial s}\,ds,\ \frac{\partial x_b}{\partial s}\,ds \right\rangle. \tag{1.75}$$

In the above equations, $x_a$, $y_a$, $x_b$ and $y_b$ are the interpolated coordinates on the boundaries of elements a and b, while $\mathbf{u}_a^A$ and $\mathbf{u}_b^A$ are the interpolated displacements corresponding to the displacement $\mathbf{U}_A$ at node $A$,

$$\mathbf{u}_a^A = \frac{s^2+s}{2}\mathbf{U}_A \quad \text{and} \quad \mathbf{u}_b^A = \frac{s^2-s}{2}\mathbf{U}_A. \tag{1.76}$$

For element $a$, mass conservation requires

$$\int_{L_a} \mathbf{u}_a^B \cdot \mathbf{n}_a\,dl = 0, \tag{1.77}$$

where $\mathbf{u}_a^B$ is the interpolated displacement on the boundary of element $a$ based on the nodal displacement $\mathbf{U}_B$ and is expressed as

$$\mathbf{u}_a^B = (1-s^2)\mathbf{U}_B. \tag{1.78}$$

The relation (1.77) shows that the appropriate tangential direction $\gamma_B$ at node B is given by

$$\tan\gamma_B = \frac{\partial y}{\partial s}\Big/\frac{\partial x}{\partial s}. \tag{1.79}$$

Our numerical experiments have shown that it is important to allocate appropriate tangential directions at all boundary nodes. Otherwise, spurious nonzero energy modes could emerge in the finite element solution. In nonlinear analysis, within each time increment, we still have to satisfy the mass conservation on the discretized slip boundary. However, after each incremental step, the material configuration will be updated as well as the mesh configuration. Therefore, the conditions of Eqs. (1.74) and (1.77) should be satisfied at every time step. This means we have to modify the flow tangential directions on FSI interfaces at every time step. It is somewhat involved, especially for high order elements, to impose the perfect inviscid mathematical model for practical problems. Nevertheless, the physical world always allows us to use viscous fluid models, and in practice, different meshing techniques could be used along the fluid boundaries and the FSI interface.

To clearly illustrate the coupling procedures, for both $\mathbf{u}/\mathrm{p}$ and $\mathbf{u}$-p-$\boldsymbol{A}$ formulations, we introduce the following generic governing equations for linear elastic solids:

$$\begin{bmatrix} \mathbf{M}_{ss} & \mathbf{M}_{sc} \\ \mathbf{M}_{cs} & \mathbf{M}_{cc} \end{bmatrix} \begin{bmatrix} \ddot{\mathbf{U}}_s \\ \ddot{\mathbf{U}}_c \end{bmatrix} + \begin{bmatrix} \mathbf{K}_{ss} & \mathbf{K}_{sc} \\ \mathbf{K}_{cs} & \mathbf{K}_{cc} \end{bmatrix} \begin{bmatrix} \mathbf{U}_s \\ \mathbf{U}_c \end{bmatrix} = \begin{bmatrix} \mathbf{R}_s \\ \mathbf{R}_c^s \end{bmatrix}, \tag{1.80}$$

where $\mathbf{U}_s$ and $\mathbf{U}_c$ refer to the nodal displacement unknowns in the solid interior and on the FSI interface, respectively.

Since the primary unknowns are used for both fluid and solid domains, the coupling is essentially established in the element assemblage process. To elaborate, we split the displacement unknowns for the acoustic fluid into $\mathbf{U}_f$ and $\mathbf{U}_c$, representing the nodal displacement unknowns in the fluid interior and on the FSI interface. It is obvious, based on the previous discussion, that $\mathbf{U}_c$ consist of normal displacements on the FSI interface, which is the same as $\mathbf{U}_c$ at all times from the solid side. Again, it is crucial to have the right tangential directions on the discretized FSI interface. For the $\mathbf{u}$/p formulation, we have

$$\begin{bmatrix} \mathbf{M}_{ff} & \mathbf{M}_{fc} & \mathbf{0} \\ \mathbf{M}_{cf} & \mathbf{M}_{cc}^{s} & \mathbf{0} \\ \mathbf{0} & \mathbf{0} & \mathbf{0} \end{bmatrix} \begin{bmatrix} \ddot{\mathbf{U}}_f \\ \ddot{\mathbf{U}}_c \\ \ddot{\mathbf{P}} \end{bmatrix} + \begin{bmatrix} \mathbf{K}_{ff} & \mathbf{K}_{fc} & \mathbf{K}_{fp} \\ \mathbf{K}_{cf} & \mathbf{K}_{cc}^{s} & \mathbf{K}_{cp} \\ \mathbf{K}_{pf} & \mathbf{K}_{pc} & \mathbf{K}_{pp} \end{bmatrix} \begin{bmatrix} \mathbf{U}_f \\ \mathbf{U}_c \\ \mathbf{P} \end{bmatrix} = \begin{bmatrix} \mathbf{R}_f \\ \mathbf{R}_c^{f} \\ \mathbf{0} \end{bmatrix}. \tag{1.81}$$

In the assemblage process, Eqs. (1.80) and (1.81) are combined and yield the final governing equation for the FSI system:

$$\begin{bmatrix} \mathbf{M}_{ss} & \mathbf{0} & \mathbf{M}_{sc} & \mathbf{0} \\ \mathbf{0} & \mathbf{M}_{ff} & \mathbf{M}_{fc} & \mathbf{0} \\ \mathbf{M}_{cs} & \mathbf{M}_{cf} & \mathbf{M}_{cc} & \mathbf{0} \\ \mathbf{0} & \mathbf{0} & \mathbf{0} & \mathbf{0} \end{bmatrix} \begin{bmatrix} \ddot{\mathbf{U}}_s \\ \ddot{\mathbf{U}}_f \\ \ddot{\mathbf{U}}_c \\ \ddot{\mathbf{P}} \end{bmatrix} + \begin{bmatrix} \mathbf{K}_{ss} & \mathbf{0} & \mathbf{K}_{sc} & \mathbf{0} \\ \mathbf{0} & \mathbf{K}_{ff} & \mathbf{K}_{fc} & \mathbf{K}_{fp} \\ \mathbf{K}_{cs} & \mathbf{K}_{cf} & \mathbf{K}_{cc} & \mathbf{K}_{cp} \\ \mathbf{0} & \mathbf{K}_{pf} & \mathbf{K}_{pc} & \mathbf{K}_{pp} \end{bmatrix} \begin{bmatrix} \mathbf{U}_s \\ \mathbf{U}_f \\ \mathbf{U}_c \\ \mathbf{P} \end{bmatrix} = \begin{bmatrix} \mathbf{R}_s \\ \mathbf{R}_f \\ \mathbf{0} \\ \mathbf{0} \end{bmatrix}, \tag{1.82}$$

with $\mathbf{M}_{cc} = \mathbf{M}_{cc}^{s} + \mathbf{M}_{cc}^{f}$ and $\mathbf{K}_{cc} = \mathbf{K}_{cc}^{s} + \mathbf{K}_{cc}^{f}$.

Likewise, for the $\mathbf{u}$-p-$\boldsymbol{\Lambda}$ formulation, we have

$$\begin{bmatrix} \mathbf{M}_{ss} & \mathbf{0} & \mathbf{M}_{sc} & \mathbf{0} & \mathbf{0} \\ \mathbf{0} & \mathbf{M}_{ff} & \mathbf{M}_{fc} & \mathbf{0} & \mathbf{0} \\ \mathbf{M}_{cs} & \mathbf{M}_{cf} & \mathbf{M}_{cc} & \mathbf{0} & \mathbf{0} \\ \mathbf{0} & \mathbf{0} & \mathbf{0} & \mathbf{0} & \mathbf{0} \\ \mathbf{0} & \mathbf{0} & \mathbf{0} & \mathbf{0} & \mathbf{0} \end{bmatrix} \begin{bmatrix} \ddot{\mathbf{U}}_s \\ \ddot{\mathbf{U}}_f \\ \ddot{\mathbf{U}}_c \\ \ddot{\mathbf{P}} \\ \ddot{\boldsymbol{\Lambda}} \end{bmatrix} + \begin{bmatrix} \mathbf{K}_{ss} & \mathbf{0} & \mathbf{K}_{sc} & \mathbf{0} & \mathbf{0} \\ \mathbf{0} & \mathbf{K}_{ff} & \mathbf{K}_{fc} & \mathbf{K}_{fp} & \mathbf{K}_{f\lambda} \\ \mathbf{K}_{cs} & \mathbf{K}_{cf} & \mathbf{K}_{cc} & \mathbf{K}_{cp} & \mathbf{K}_{c\lambda} \\ \mathbf{0} & \mathbf{K}_{pf} & \mathbf{K}_{pc} & \mathbf{K}_{pp} & \mathbf{0} \\ \mathbf{0} & \mathbf{K}_{\lambda f} & \mathbf{K}_{\lambda c} & \mathbf{0} & \mathbf{K}_{\lambda\lambda} \end{bmatrix} \begin{bmatrix} \mathbf{U}_s \\ \mathbf{U}_f \\ \mathbf{U}_c \\ \mathbf{P} \\ \boldsymbol{\Lambda} \end{bmatrix} = \begin{bmatrix} \mathbf{R}_s \\ \mathbf{R}_f \\ \mathbf{0} \\ \mathbf{0} \\ \mathbf{0} \end{bmatrix}. \tag{1.83}$$

In this coupled formulation for linear acoustoelastic FSI system, Eq. (1.82) is symmetric. $\mathbf{M}_{ab} = \mathbf{M}_{ba}^{T}$ and $\mathbf{K}_{ab} = \mathbf{K}_{ba}^{T}$, with $a$ and $b$, represent the index for solid ($s$), fluid ($f$), FSI interface ($c$), pressure ($p$), and vorticity moment ($\lambda$). Due to the kinematic and dynamic matching conditions at the FSI interface, in general we have $\mathbf{R}_c^f + \mathbf{R}_c^s = \mathbf{0}$. The validity and the expansion to nonlinear cases are discussed in Chapter 7. In Eq. (1.80), we use the standard displacement-based formulation for linear elastic solids. The same coupling procedure will also be valid for mixed formulations for almost incompressible solids.

*Zero modes*

Let us consider the case of Eq (1.56), where the number of displacement and pressure unknowns are $n$ and $m$. The left side matrix of Eq. (1.56) is denoted as $\mathbf{K}_h$. The mathematical analysis of the formulation resulting in Eq. (1.56) consists of a study of the solvability and stability of the equations or the structure of $\mathbf{K}_h$, where the stability of the equations implies their solvability [13].

From the discretized formulation, we have

$$\delta \mathbf{U}_h^T (\mathbf{K}_{up})_h \mathbf{P}_h = \int_{\Omega} p \nabla \cdot \mathbf{v}_h \, d\Omega, \tag{1.84}$$

where $\mathbf{v}_h$ can be any arbitrary admissible displacements and $\delta \mathbf{U}_h$ contains $n$ independent admissible vectors.

Assuming that $(\mathbf{K}_{uu})_h$ is a positive definite matrix, i.e., the ellipticity condition is satisfied, if we have

$$\int_{\Omega} p_o \nabla \cdot \mathbf{v}_h \, d\Omega = p_o \int_{\Gamma} \mathbf{v}_h \cdot \mathbf{n} \, d\Gamma = 0, \quad \forall \mathbf{v}_h \in \mathcal{V}_h, \tag{1.85}$$

where $p_o$ is a constant pressure, and $(\mathbf{K}_{pp})_h = \mathbf{0}$ (incompressible case) we can add an arbitrary constant pressure $p_o$ to any proposed solution of Eq. (1.56) [143].

In fact, Eq. (1.85) is equivalent to the following condition:

$$\sum_{j=1}^{m} (K_{up})_h^{ij} = 0, \quad i = 1, \ldots, n, \tag{1.86}$$

where $(K_{up})_h^{ij}$ is the $ij$th element of $(\mathbf{K}_{up})_h$.

Therefore, for the incompressible case, Eq. (1.85) implies that the rank of $\mathbf{K}_h$ is at most $n + m - 1$, and the mode corresponding to the zero eigenvalue contains constant pressure mode $\mathbf{P}_h = p_o \langle 1, 1, \ldots, 1 \rangle$. For almost incompressible fluid models, we will have the following stiffness matrix for the $\mathbf{u}$/p formulation:

$$\mathbf{K}_h = \begin{bmatrix} \mathbf{0} & (\mathbf{K}_{up})_h \\ (\mathbf{K}_{up}^T)_h & (\mathbf{K}_{pp})_h \end{bmatrix}, \tag{1.87}$$

where the artificial shear modulus $G$ is taken as zero in the limit case.

Since we are using the Galerkin formulation for almost incompressible fluids, $(\mathbf{K}_{pp})_h$ is a negative definite matrix. Therefore, we have

$$\mathbf{K}_h^* = -(\mathbf{K}_{up})_h(\mathbf{K}_{pp})_h^{-1}\left(\mathbf{K}_{up}^T\right)_h. \tag{1.88}$$

If Eq. (1.85) holds, it is clear from Eq. (1.86) that the rank of $(\mathbf{K}_{up})_h$ is at most $m-1$. From the discussion in Chapter 2, we conclude that the rank of $\mathbf{K}_h^*$ is also at most $m-1$. In actuality, to obtain a nonzero pressure $p_o$, Eq. (1.85) requires that the *discretized* boundary tangential directions satisfy Eqs. (1.74) and (1.77). This is the fundamental reason why spurious modes can be produced when Eqs. (1.74) and (1.77) are violated for inviscid acoustic fluids.

Likewise, in the $\mathbf{u}$-p-$\boldsymbol{\Lambda}$ formulation, we have

$$\mathbf{K}_h = \begin{bmatrix} \mathbf{0} & (\mathbf{K}_{up})_h & (\mathbf{K}_{u\lambda})_h \\ \left(\mathbf{K}_{up}^T\right)_h & (\mathbf{K}_{pp})_h & \mathbf{0} \\ \left(\mathbf{K}_{u\lambda}^T\right)_h & \mathbf{0} & (\mathbf{K}_{\lambda\lambda})_h \end{bmatrix}, \tag{1.89}$$

for which the comparable analysis holds. The corresponding mode contains constant pressure $\mathbf{P}_h = p_o\langle 1, 1, \ldots, 1\rangle$ along with zero vorticity moment mode $\boldsymbol{\Lambda}_h = \langle 0, 0, \ldots, 0\rangle$.

The mathematical models above appear to be rather natural when considering free surfaces and the interactions between different media, but we have to look more deeply into the boundary conditions. Different mathematical models will have different boundary conditions, and some subtle points arise in imposing the boundary conditions. Using the numerical examples in Chapter 6, we find the existence of nonzero frequency spurious mode when wrong boundary conditions are imposed.

For isentropic acoustic fluids, we have

$$\frac{dp}{d\rho} = c^2, \tag{1.90}$$

where $c$ is the compressible wave velocity or sound speed.

Substituting Eq. (1.90) into the mass conservation equation derived using Reynolds Transport theorems discussed in Chapter 2, we get

$$\nabla \cdot \mathbf{v} + \frac{\dot{p}}{\kappa} = 0, \tag{1.91}$$

with the bulk modulus $\kappa$ as $\rho c^2$.

In linear analysis, for any quantity $a$, the material time derivative $da/dt$ and the mesh referential derivative $a^*$ are the same as the spatial time derivative $\partial/\partial t$ and the equilibrium position stays the same as the original configuration. Therefore, we can use displacements instead of velocities as primitive variables in momentum and continuity equations. If we assume zero shear modulus for acoustic fluids, we have

$$\rho\ddot{\mathbf{u}} + \nabla p - \mathbf{f}^b = \mathbf{0}, \tag{1.92}$$

$$\nabla \cdot \mathbf{u} + \frac{p}{\kappa} = 0. \tag{1.93}$$

Since $\nabla \times \nabla\phi = \mathbf{0}$ for any smooth scalar valued function $\phi$, if the nonconservative forces are ignored, Eq. (1.92) implies

$$\frac{\partial}{\partial t}(\nabla \times \mathbf{v}) = \mathbf{0}. \tag{1.94}$$

Hence the motion is always circulation preserving, i.e., it is a motion in which the vorticity does not change with time.

If the fluid starts from rest, we have the irrotationality constraint

$$\nabla \times \mathbf{u} = \mathbf{0}. \tag{1.95}$$

For the isentropic and inviscid fluid models, in terms of displacements only, we have by substituting Eq. (1.92) into Eq. (1.93),

$$\kappa\nabla(\nabla \cdot \mathbf{u}) + \mathbf{f}^f = \mathbf{0}, \tag{1.96}$$

with $\mathbf{f}^f = -\rho\ddot{\mathbf{u}}$, if the other body forces are ignored.

The variational form of Eq. (1.96) can be written as

$$\int_{\Omega^f} \left\{\kappa(\nabla \cdot \mathbf{u})(\nabla \cdot \delta\mathbf{u}) - \mathbf{f}^f \cdot \delta\mathbf{u}\right\} d\Omega + \int_{\Gamma^{f,f}} \bar{p}\delta u_n^{f,f}\, d\Gamma = 0, \tag{1.97}$$

where $u_n^{f,f}$ is the displacement normal to the Neumann boundary $\Gamma^{f,f}$.

Eq. (1.97) is in fact the often used pure displacement-based formulation. It was widely reported that this formulation produces spurious nonzero frequency modes [140,170,175,263]. From the rigid cavity test problem in Ref. [175], Hamdi et al. concluded that the observed spurious nonzero frequency modes are rotational modes. Historically, many researchers also believed that nonzero frequency rotational modes were due to the constraint of Eq. (1.95). As pointed out in Refs. [169, 175], the irrotationality constraint is "lost" in the finite element formulation. Therefore, the following penalty formulation was proposed:

$$\int_{\Omega^f} \left\{\kappa(\nabla \cdot \mathbf{u})(\nabla \cdot \delta\mathbf{u}) + \alpha(\nabla \times \mathbf{u})(\nabla \times \delta\mathbf{u}) - \mathbf{f}^f \cdot \delta\mathbf{u}\right\} d\Omega$$
$$+ \int_{\Gamma^{f,f}} \bar{p}\delta u_n^{f,f}\, d\Gamma = 0, \tag{1.98}$$

where $\alpha$ is the large penalty parameter.

Interestingly, for some problems, the so-called spurious nonzero frequency rotational modes disappeared with the above formulation. Considering the fact that penalty formulations are too "stiff", researchers often use reduced integration methods for the terms

$$\int_{\Omega^f} \left\{\kappa(\nabla\cdot\mathbf{u})(\nabla\cdot\delta\mathbf{u}) + \alpha(\nabla\times\mathbf{u})(\nabla\times\delta\mathbf{u})\right\} d\Omega. \tag{1.99}$$

With reduced integration, the spurious nonzero frequency modes can be eliminated in some cases. However, a proper procedure is the replacement of the pure displacement-based formulation with a displacement/pressure formulation and the use of mixed elements that satisfy the inf-sup condition. Introducing a penalty term on the irrotationality constraint and applying the reduced integration method to the pure displacement-based formulation, the spurious nonzero frequency modes can be eliminated in some cases; however, while this actually somehow confirmed many researchers' misunderstanding of the causes of the reported spurious modes, we believe that it was coincidental. Since acoustic media are nearly incompressible with a high bulk modulus, it should be obvious that the pure displacement-based formulation and its penalty formulation introduce nonzero frequency spurious modes such as the familiar checkerboard pressure mode. In actuality, if the acoustic fluid models allow the existence of a small amount of shear modulus, we will no longer have the constraint of Eq. (1.95) and will also expect the rotational modes to be in the lowest end of the frequency spectrum. For acoustic fluid models with no shear modulus or inviscid acoustic fluid models, the rotational modes will have zero frequency. In conclusion, irrotationality constraint and incompressibility constraint have different physical significance.

In the frequency analysis without the effects of gravitational forces and other body forces, Eq. (1.96) can be written as

$$\kappa\nabla(\nabla\cdot\tilde{\mathbf{u}}) + \rho\omega^2\tilde{\mathbf{u}} = \mathbf{0}, \tag{1.100}$$

where $\tilde{\mathbf{u}}$ represents the eigenfunction.

Taking a curl on both sides of Eq. (1.100), we have $\kappa\nabla\times(\nabla(\nabla\cdot\tilde{\mathbf{u}})) = \mathbf{0}$, based on the discussion in Chapter 2. Consequently

$$\omega^2(\nabla\times\tilde{\mathbf{u}}) = \mathbf{0}. \tag{1.101}$$

It is apparent that in frequency analysis, for solutions corresponding to rotational motions ($\nabla\times\tilde{\mathbf{u}} \neq \mathbf{0}$), the frequencies are zero, while for solutions corresponding to irrotational motions ($\nabla\times\tilde{\mathbf{u}} = \mathbf{0}$), the frequencies can be nonzero. One may view the former as the way the system responds to rotational initial conditions. The analogy of this analysis to almost incompressible solid and FSI systems is straightforward. Based on the previous discussions, we need only consider the stiffness matrix $\mathbf{K}^* = \mathbf{K}_{up}\mathbf{K}_{pp}^{-1}\mathbf{K}_{up}^T$ to understand the zero modes. For the matrices we consider here, as illustrated in Chapter 2, SVD for any $n \times m$ matrix $\mathbf{K}_{up}$ can be expressed in the form

$$\mathbf{K}_{up} = \mathbf{PEQ}, \tag{1.102}$$

where $\mathbf{P}$ and $\mathbf{Q}$ are invertible matrices with dimensions $n \times n$ and $m \times m$, while the $n \times m$ $\mathbf{E}$ is in the form

$$\begin{bmatrix} \mathbf{I}_r & \mathbf{0} \\ \mathbf{0} & \mathbf{0} \end{bmatrix},$$

where $\mathbf{I}_r$ is a $r \times r$ dimension unit matrix and $r$ is the rank of $\mathbf{K}_{up}$.

From the consistency of our finite element discretizations, and due to the existence of a constant pressure distribution, we have $1 \leqslant r \leqslant m - 1$. Using Eq. (1.102), we get

$$\mathbf{K}^* = \mathbf{PEQK}_{pp}^{-1}\mathbf{Q}^T\mathbf{E}^T\mathbf{P}^T = \mathbf{PK}'_{pp}\mathbf{P}^T, \tag{1.103}$$

where $\mathbf{K}'_{pp}$ has the form

$$\begin{bmatrix} \mathbf{K}_r & \mathbf{0} \\ \mathbf{0} & \mathbf{0} \end{bmatrix},$$

and $\mathbf{K}_r$ is a $r \times r$ dimension invertible matrix.

It is then obvious that the number of zero modes is $k = n - r$, i.e.,

$$k \geqslant n - m + 1. \tag{1.104}$$

Note that we consider $n$ as the number of displacement degrees of freedom and $m$ as the number of pressure degrees of freedom. With the presence of free surfaces, the constant pressure mode no long exists, and we have the number of zero modes:

$$k \geqslant n - m. \tag{1.105}$$

Since the free surface is the material surface, we will have one implicit constraint imposed from the mass conservation. This implies that for a free surface with $i$ degrees of displacement freedom, the actual independent degree of freedom is $i - 1$. If we consider the gravitational effects, i.e., including the slosh modes, the number of zero modes will be

$$k \geqslant n - m - i + 1. \tag{1.106}$$

It is clear that the constraints in Eqs. (1.45) and (1.46) are very similar in nature and are not coupled in the formulation. Hence, we can reasonably assume that the two constraints should be imposed with the same interpolations in the element discretization. The treatment to include the surface waves in the

$\mathbf{u}$-p-$\boldsymbol{\Lambda}$ formulation is exactly the same as the one used for the $\mathbf{u}$/p formulation. The mathematical prediction of the number of zero frequency modes is also the same, however, the number of vorticity moment unknowns will be included in $m$.

If we compare Eq. (1.45) with Eq. (1.46), it is clear that the frequencies of the "vorticity modes" depend on $\alpha$ in the same way the frequencies of the "pressure modes" depend on $\kappa$. Since $\alpha$ is a numerically large value ($100\kappa \leqslant \alpha \leqslant 10^6\kappa$) in the $\mathbf{u}$-p-$\boldsymbol{\Lambda}$ formulation, we in fact shift the frequencies of the rotational modes to very high values. Correspondingly, the number of zero frequency modes is reduced by the number of discrete vorticity moment unknowns.

If we substitute Eq. (1.46) into Eq. (1.44), we have the following set of equations:

$$\nabla p - \mathbf{f}^f + \alpha\nabla \times (\nabla \times \mathbf{u}) = \mathbf{0}, \tag{1.107}$$

$$\nabla \cdot \mathbf{u} + \frac{p}{\kappa} = 0. \tag{1.108}$$

Since we have, based on the discussion in Chapter 2,

$$\nabla(\nabla \cdot \mathbf{u}) \equiv \nabla^2\mathbf{u} + \nabla \times (\nabla \times \mathbf{u}), \tag{1.109}$$

by substituting Eq. (1.108), we can rewrite Eq. (1.107) as

$$\left(1 - \frac{\alpha}{\kappa}\right)\nabla p - \mathbf{f}^f = \alpha\nabla^2\mathbf{u}. \tag{1.110}$$

Therefore, as we assign $\alpha$ a very small value *and* much smaller than $1/\kappa$, Eq. (1.110) approaches Eq. (1.36). Hence, while in general we would use $\alpha$ to be a numerically large value in the $\mathbf{u}$-p-$\boldsymbol{\Lambda}$ formulation, we recognize that by assigning a numerically small value to $\alpha$, the $\mathbf{u}$/p formulation is obtained. Moreover, we can see immediately that the term $\alpha\nabla^2\mathbf{u}$ is the stiffness term for $\mathbf{u}$. In the $\mathbf{u}$-p-$\boldsymbol{\Lambda}$ formulation, the large value of $\alpha$ means high stiffness for $\mathbf{u}$, and the modes introduced by this term shall be in the highest range of the frequency spectrum. Similarly, if we extend the same philosophy to the nonlinear problems, we will obtain

$$\nabla p - \frac{\alpha}{\kappa}\nabla\dot{p} - \mathbf{f}^f = \alpha\nabla^2\mathbf{v}, \tag{1.111}$$

$$\nabla \cdot \mathbf{v} + \frac{\dot{p}}{\kappa} = 0, \tag{1.112}$$

where $\mathbf{f}^f$ is the body force, including the inertia force, for the fluid domain.

However, the term $\alpha\nabla^2\mathbf{v}$ acts as the viscosity term instead of the stiffness term! If we apply the large value used for $\alpha$ in the $\mathbf{u}$-p-$\boldsymbol{\Lambda}$ formulation to the Navier–Stokes equations, we actually bring into the formulation a very large amount of viscosity. Now, if we change our philosophy to shifting the frequencies of rotational modes to the very lowest end of the spectrum, i.e., numerically sufficiently

Table 6.1 Frequency analysis of typical mixed elements

| Element | No. of zero modes | | First (rad/s) | Second (rad/s) | Third (rad/s) |
|---|---|---|---|---|---|
| | Theory | Result | | | |
| 9/3 | 15 | 15 | 3552.85 | 6153.71 | 6153.71 |
| 9-3-3 | 12 | 12 | 3552.85 | 6067.56 | 6067.56 |

small, around $10^{-5}$ to $10^{-3}$, instead of the highest end of the spectrum, we will have the equivalent of the $\mathbf{u}$/p formulation. To achieve this, we simply introduce a small variable $\alpha$ $(1/10^4\kappa \leqslant \alpha \leqslant 1/10\kappa)$. Since $1/\kappa$ is the small value of $O(\epsilon)$, $\alpha$ is $O(\epsilon)$ and $\alpha/\kappa$ is $O(\epsilon^2)$. Ignoring $O(\epsilon^2)$ terms, we have for the linear problems

$$\nabla p - \mathbf{f}^f = \alpha\nabla^2\mathbf{u}, \tag{1.113}$$

$$\nabla \cdot \mathbf{u} + \frac{p}{\kappa} = 0, \tag{1.114}$$

where $\alpha$ is identified as $G$ and the inertia force $-\rho\ddot{\mathbf{u}}$ is included in $\mathbf{f}^f$.

For one element (size $2 \times 2$) with $\kappa = 2.1 \times 10^9$ Pa, $\rho = 998.2$ kg/m$^3$ and $\alpha = 2.1 \times 10^{12}$ Pa, the first three nonzero frequencies as listed in Table 6.1 and all element eigenvectors as shown in Figures 6.10 and 6.11 are calculated with the $\mathbf{u}$/p and $\mathbf{u}$-p-$\boldsymbol{\Lambda}$ formulations. In Figures 6.10 and 6.11, * frequencies are physical and $^{\mathrm{o}}$ frequencies are due to the large parameter $\alpha$. It is clearly indicated that the highest three frequencies calculated with the $\mathbf{u}$-p-$\boldsymbol{\Lambda}$ formulation are introduced by the assigned $\alpha = 1000\kappa$. For nonlinear problems, which we will focus on in Chapter 7, we get the familiar pseudo-compressible Navier–Stokes equations (1.111) and (1.112), where $\alpha$ is identified as the dynamic viscosity $\mu$ and the inertia force $-\rho\, d\mathbf{v}/dt$ is included in $\mathbf{f}^f$.

## 6.2. Solution procedures and convergence issues

It has been amply recognized that many problems in fluid and solid mechanics cannot be solved efficiently using finite element discretization with only one unknown field variable. The common fundamental difficulties in such problems frequently arise because the solution variables are subject to some constraints. For example, in the case of the almost incompressible elasticity problem, the volumetric strains must be very small and approach zero, as the condition of total incompressibility is approached while the pressure is of the order of the boundary traction. Another example is the beam, plate and shell problems [147]. The key to whether or not a mixed formulation is actually valuable lies, of course, in the convergence properties of the formulation. These properties are governed by stability

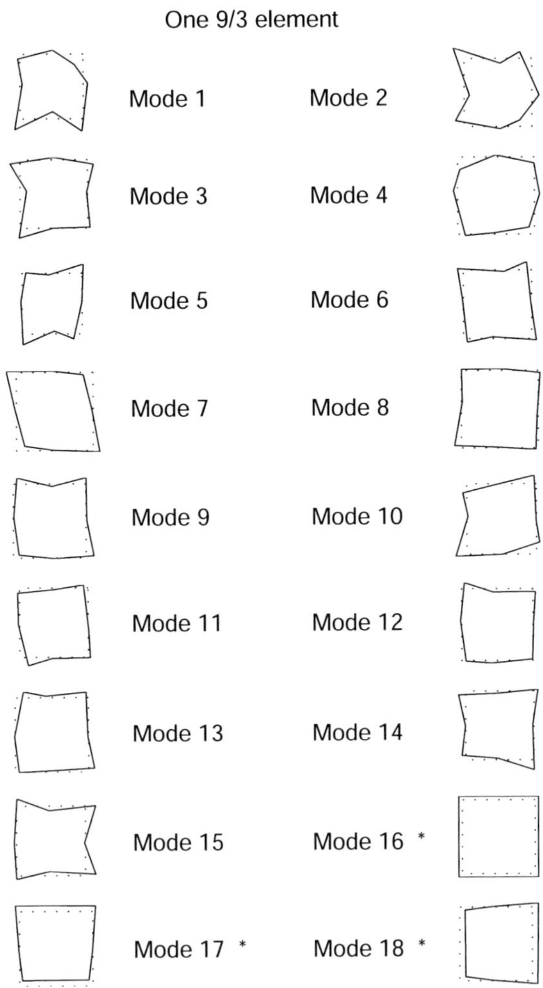


Figure 6.10. Mode shapes of one 9/3 element.

considerations as expressed in the ellipticity condition and the inf-sup condition [27]. Early engineering practice was to use simple constraint counts and comparisons of degrees of freedom. Although these can be effective for many engineering problems, a more rigorous numerical test has been proposed by Bathe et al. We must point out that for some problems both tests are equivalent [40]. However, the numerical inf-sup test is tied more to the inf-sup condition. Note that, mathematically speaking, the inf-sup condition is a sufficient condition which guarantees

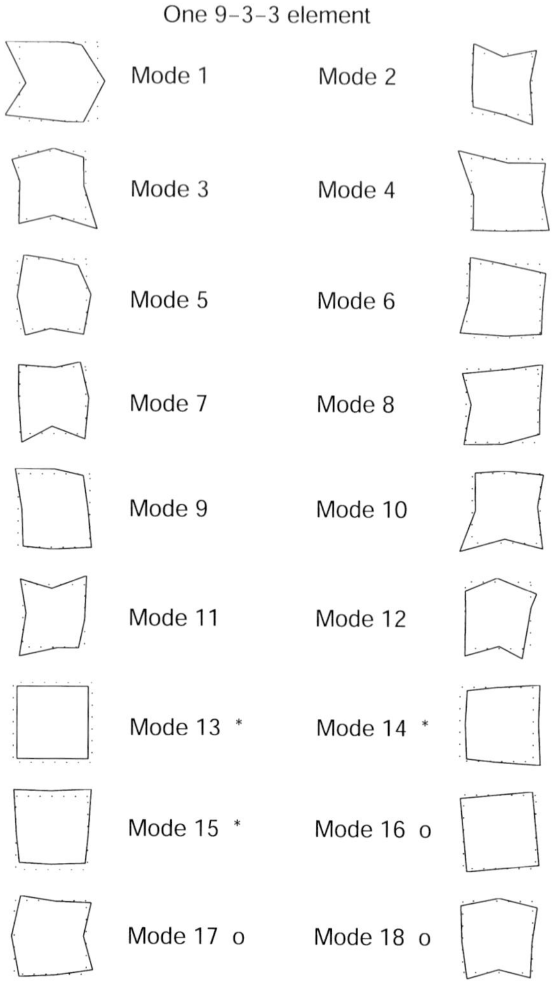


Figure 6.11. Mode shapes of one 9-3-3 element.

that for both displacements and pressures, optimal error estimates can be derived with coefficients independent of $\kappa$.

In this book we focus on effective displacement/pressure finite element formulations for the analysis of acoustoelastic/slosh FSI problems. In the $\mathbf{u}/p$ formulation we interpolate displacements and pressure as independent variables, and we employ elements that satisfy the inf-sup condition. If a discontinuous pressure approximation is used, such as for the 9/3 element, the pressure degrees of freedom are eliminated with static condensation on the element level, so that only the nodal displacement degrees of freedom are present in the assemblage process. It is also important that the slip boundary conditions are introduced, such that the requirements of mass and momentum conservation around the fluid boundaries and fluid-structure interfaces are satisfied.

A very important application area is linear analysis. In general, nonlinear analysis is to piecewise linearize the system and each incremental iteration resembles the corresponding linear system. For all numerical procedures, convergence study is needed to understand the effectiveness as well as the reliability of solution procedures. The most important aspect of the mixed displacement/pressure ($\mathbf{u}$/p) and displacement-pressure-vorticity moment ($\mathbf{u}$-p-$\boldsymbol{\Lambda}$) formulations is the inf-sup condition. Although the inf-sup condition for mixed formulations was proposed some time ago, an analytical proof of whether the inf-sup condition is satisfied by a specific element or discretization can be very difficult [27,143]. In practice, the numerical inf-sup test proposed in Ref. [40] is valuable.

### 6.2.1. Mode superposition; direct integration methods

In general, spatial discretization procedures yield symmetric and nonsymmetric systems of equations. For linear systems, the governing equation can be expressed as

$$\mathbf{M}\ddot{\mathbf{U}} + \mathbf{C}\dot{\mathbf{U}} + \mathbf{K}\mathbf{U} = \mathbf{R}(t), \tag{2.115}$$

where $\mathbf{K}$, $\mathbf{C}$, and $\mathbf{M}$ are matrices with respect to the unknown vector $\mathbf{U}$ and its first and second derivatives.

In most cases, if primary unknowns are used, $\mathbf{K}$, $\mathbf{C}$, and $\mathbf{M}$ represent the stiffness, damping, and mass matrices. However, for velocity potential formulations, these matrices can have different physical significance.

#### Mode superposition method

Denote the mode shape matrix and the generalized coordinate vector as $\boldsymbol{\Phi} = \langle \boldsymbol{\phi}_1, \boldsymbol{\phi}_2, \ldots, \boldsymbol{\phi}_n \rangle$ and $\mathbf{Y} = \langle Y_1, Y_2, \ldots, Y_n \rangle$, where $\boldsymbol{\phi}_j$ and $Y_j$ are $j$th mode shape vector and the corresponding modal amplitude, respectively. According to discussions in Chapter 2, we can then express the $n$-dimensional governing linear system of Eqs. (2.115) in a generalized coordinate system:

$$\mathbf{U}(t) = \sum_{j=1}^{n} \boldsymbol{\phi}_j Y_j(t) = \boldsymbol{\Phi}\mathbf{Y}(t). \tag{2.116}$$

We begin with a simple homogeneous case with $\mathbf{C} = \mathbf{0}$ and $\mathbf{R} = \mathbf{0}$. It is clear that the $j$th modal solution can be expressed as $\boldsymbol{\phi}_j[c_1 \cos(\omega_j t) + c_2 \sin(\omega_j t)]$ with real coefficients $c_1$ and $c_2$ or $\mathrm{Re}(ce^{i\omega_j t})$ with complex coefficient $c$, and from the governing equation (2.115), we have

$$-\omega_j^2 \mathbf{M}\boldsymbol{\phi}_j + \mathbf{K}\boldsymbol{\phi}_j = \mathbf{0}. \tag{2.117}$$

In addition, utilizing the orthogonal relations of different modes, if we scale the $j$th mode with a common factor $(\boldsymbol{\phi}_j^T \mathbf{M}\boldsymbol{\phi}_j)^{-1}$, we also have the following orthonormal relations:

$$\boldsymbol{\phi}_j^T \mathbf{M}\boldsymbol{\phi}_k = \begin{cases} 1, & \text{with } j = k, \\ 0, & \text{with } j \neq k, \end{cases} \tag{2.118}$$

and

$$\boldsymbol{\phi}_j^T \mathbf{K}\boldsymbol{\phi}_k = \begin{cases} \omega_j^2, & \text{with } j = k, \\ 0, & \text{with } j \neq k. \end{cases} \tag{2.119}$$

Therefore, with the following assumption for the damping matrix,

$$\boldsymbol{\Phi}^T \mathbf{C}\boldsymbol{\Phi} = \operatorname{diag}(2\xi_1\omega_1, 2\xi_2\omega_2, \ldots, 2\xi_n\omega_n), \tag{2.120}$$

where $\xi_j \in [0, 1)$ stands for the damping coefficient of the $j$th mode.

If we multiply by the transpose of the mode shape matrix on both sides of the governing equation system (2.115), we derive

$$\ddot{Y}_j + 2\xi_j\omega_j\dot{Y}_j + \omega_j^2 Y_j = r_j, \quad \text{with } j = 1, 2, \ldots, n, \tag{2.121}$$

where the $j$th generalized force is expressed as

$$r_j(t) = \boldsymbol{\phi}_j^T \mathbf{R}(t). \tag{2.122}$$

In fact, Eq. (2.122) can also be written as the conjugate form

$$\mathbf{R}(t) = \sum_{j=1}^{n} r_j(t)\mathbf{M}\boldsymbol{\phi}_j. \tag{2.123}$$

Naturally, if we solve for all the modal solutions using the single degree of freedom solution, the solution expressed in Eq. (2.116) will be identical to that from Eq. (2.115). However, in practice, for most types of loadings the contributions of the modes are greatest for the lowest modes and tend to decrease for higher frequencies. Thus, it is not usually necessary to include all the higher modes of vibration in the superposition process and Eq. (2.116) can be simplified as

$$\mathbf{U}^c(t) = \sum_{j=1}^{c} \boldsymbol{\phi}_j Y_j(t), \tag{2.124}$$

where $c$ represents the cut-off number of frequencies and is often much smaller than $n$.

Using Eq. (2.123), the mode superposition method yields the following error of the exciting force vector:

$$\Delta \mathbf{R} = \mathbf{R} - \sum_{j=1}^{c} r_j \mathbf{M} \boldsymbol{\phi}_j . \tag{2.125}$$

Here again, if we were to solve the following linear equation systems:

$$\mathbf{M}\Delta\ddot{\mathbf{U}} + \mathbf{C}\Delta\dot{\mathbf{U}} + \mathbf{K}\Delta\mathbf{U} = \Delta\mathbf{R}(t), \tag{2.126}$$

the original solution of (2.115) can be recovered as

$$\mathbf{U} = \mathbf{U}^c + \Delta\mathbf{U}. \tag{2.127}$$

However, if we consider the fact that the ignored modes have relatively high frequencies, a good approximation of the correction can be simply calculated from the following static correction equation:

$$\mathbf{K}\Delta\mathbf{U}(t) = \Delta\mathbf{R}(t). \tag{2.128}$$

Of course, if the assumption (2.120) does not hold, the so-called Rayleigh damping can be introduced:

$$\mathbf{C} = \alpha\mathbf{M} + \beta\mathbf{K}, \tag{2.129}$$

where coefficients $\alpha$ and $\beta$ can be determined with two given damping ratio with respect to two different frequencies, and the following:

$$\boldsymbol{\phi}_j^T(\alpha\mathbf{M} + \beta\mathbf{K})\boldsymbol{\phi}_j = 2\omega_j\xi_j . \tag{2.130}$$

For example, if we are given the damping ratio $\xi_1$ and $\xi_2$ with respect to $\omega_1$ and $\omega_2$, we have

$$\alpha + \beta\omega_1^2 = 2\omega_1\xi_1 \quad \text{and} \quad \alpha + \beta\omega_2^2 = 2\omega_2\xi_2, \tag{2.131}$$

which yields

$$\alpha = 2\omega_1\omega_2 \frac{\omega_2\xi_1 - \omega_1\xi_2}{\omega_2^2 - \omega_1^2} \quad \text{and} \quad \beta = \frac{2\omega_2\xi_2 - 2\omega_1\xi_1}{\omega_2^2 - \omega_1^2}. \tag{2.132}$$

At this point, the presented mode superposition method is often referred to as the spectrum method, which is applicable to systems with symmetric matrices, i.e., real eigenvalues and eigenvectors. With symmetric **M**, **K**, and **C** matrices (for conservative systems, the coupling matrix **C** has zero diagonal entries) in $\phi$-**u** and $p$-$\phi$-**u** formulations, a mode superposition method can be applied. However, we will also show that such a procedure is somewhat different from the common spectrum analysis in structural dynamics [182]. A determinant search method can be used to find the needed real eigenvalues $\omega_m$ from Eq. (1.25), where $m = 1, \ldots, c$

and $c$ stands for the number of modes below the cut-off frequency [171,210]. Note that the number of negative elements in the matrix $\mathbf{D}$ of the $\mathbf{LDL}^T$ factorization of $\mathbf{K} - \omega_m \mathbf{C} - \omega_m^2 \mathbf{M}$ is equal to the number of eigenvalues below $\omega_m^2$ plus the number of pressure unknowns.

Rewrite Eq. (1.20) in the following form:

$$\mathbf{AY} + \mathbf{B\dot{Y}} = \mathbf{F}, \tag{2.133}$$

with $\mathbf{X} = \langle \mathbf{U}, \mathbf{P}, \boldsymbol{\Phi} \rangle$, $\bar{\mathbf{F}} = \langle \mathbf{R}_s, \mathbf{0}, \mathbf{0} \rangle$, and

$$\mathbf{A} = \begin{bmatrix} \mathbf{K} & \mathbf{0} \\ \mathbf{0} & -\mathbf{M} \end{bmatrix}, \qquad \mathbf{Y} = \begin{bmatrix} \mathbf{X} \\ \dot{\mathbf{X}} \end{bmatrix},$$

$$\mathbf{B} = \begin{bmatrix} \mathbf{C} & \mathbf{M} \\ \mathbf{M} & \mathbf{0} \end{bmatrix}, \quad \text{and} \quad \mathbf{F} = \begin{bmatrix} \bar{\mathbf{F}} \\ \mathbf{0} \end{bmatrix}.$$

Without loss of generality, we assume the system has distinct eigenvalues. The following orthogonal relationships hold:

$$\mathbf{Y}_m^T \mathbf{B} \mathbf{Y}_k = \begin{cases} 0 & \text{if } m \neq k, \\ b_m + 2a_m \lambda_m & \text{if } m = k, \end{cases} \tag{2.134}$$

$$\mathbf{Y}_m^T \mathbf{A} \mathbf{Y}_k = \begin{cases} 0 & \text{if } m \neq k, \\ c_m - a_m \lambda_m^2 & \text{if } m = k. \end{cases} \tag{2.135}$$

If we have multiple eigenvalues, we can construct independent eigenvector pairs which still have orthogonality properties. Assign $\mathbf{Y}(t) = \mathbf{YQ}(t)$, where $\mathbf{Y} = (\mathbf{Y}_1, \mathbf{Y}_2, \dots, \mathbf{Y}_{2n})$ are the mode shapes and $\mathbf{Q}(t) = \langle q_1, q_2, \dots, q_{2n} \rangle$ is the generalized coordinate vector. Using the orthogonality relations, we have

$$\dot{q}_i + p_i q_i = h_i, \tag{2.136}$$

with $i = 1, 2, \dots, 2n$ and

$$p_i = \left(c_i - a_i \lambda_i^2\right) / (b_i + 2a_i \lambda_i) \quad \text{and} \quad h_i = \mathbf{Y}_i^T \mathbf{F} / (b_i + 2a_i \lambda_i). \tag{2.137}$$

The initial conditions for Eq. (2.133) are

$$q_i(0) = \mathbf{Y}_i^T \mathbf{B} \mathbf{Y}_o / (b_i + 2a_i \lambda_i). \tag{2.138}$$

Note that $\mathbf{Y}_o = \langle \mathbf{U}_o, \mathbf{P}_o, \boldsymbol{\Phi}_o, \dot{\mathbf{U}}_o, \dot{\mathbf{P}}_o, \dot{\boldsymbol{\Phi}}_o \rangle$ has to satisfy the second equation of (1.7). Define

$$\alpha_m = \left(\mathbf{U}_m^T \mathbf{V}_o^{(u)} + \mathbf{P}_m^T \mathbf{V}_o^{(p)} - \omega_m \boldsymbol{\Phi}_m^T \mathbf{V}_o^{(\dot{\Phi})}\right),$$

$$\beta_m = \left(\boldsymbol{\Phi}_m^T \mathbf{V}_o^{(\phi)} + \omega_m \mathbf{U}_m^T \mathbf{V}_o^{(\dot{u})} + \omega_m \mathbf{P}_m^T \mathbf{V}_o^{(\dot{p})}\right),$$

$$\gamma_m = 2\left(\boldsymbol{\Phi}_m^T \mathbf{C}_{fs} \mathbf{U}_m + \boldsymbol{\Phi}_m^T \mathbf{C}_{pf}^T \mathbf{P}_m\right),$$

$$\xi_m = 2\omega_m\left(\mathbf{U}_m^T \mathbf{M}_{ss} \mathbf{U}_m + \boldsymbol{\Phi}_m^T \mathbf{M}_{ff} \boldsymbol{\Phi}_m\right),$$

with $\mathbf{V}_o = \mathbf{B}\mathbf{Y}_o = \langle \mathbf{V}_o^{(u)}, \mathbf{V}_o^{(p)}, \mathbf{V}_o^{(\phi)}, \mathbf{V}_o^{(\dot{u})}, \mathbf{V}_o^{(\dot{p})}, \mathbf{V}_o^{(\dot{\phi})} \rangle$.

We then have

$$\begin{aligned} q_{(2m-1)}(0) &= \mathbf{Y}_{(2m-1)}^T \mathbf{V}_o/(b_m + 2\lambda_m a_m) \\ &= (-i\alpha_m + \beta_m)/(\gamma_m + \xi_m) = A_m + B_m i, \\ q_{(2m)}(0) &= -\mathbf{Y}_{2m}^T \mathbf{V}_o/(b_m + 2\lambda_m a_m) \\ &= (i\alpha_m + \beta_m)/(\gamma_m + \xi_m) = A_m - B_m i, \end{aligned}$$

with

$$A_m = \beta_m/(\gamma_m + \xi_m) \quad \text{and} \quad B_m = -\alpha_m/(\gamma_m + \xi_m).$$

Eq. (2.137) gives us

$$h_{(2m-1)} = C_m + D_m i \quad \text{and} \quad h_{2m} = C_m - D_m i,$$

with

$$D_m = -\left(\mathbf{U}_m^T \mathbf{F}^{(u)} + \mathbf{P}_m^T \mathbf{F}^{(p)}\right)/(\gamma_m + \xi_m) \quad \text{and} \quad C_m = \boldsymbol{\Phi}_m^T \mathbf{F}^{(\phi)}/(\gamma_m + \xi_m).$$

The solution to Eq. (2.136) becomes

$$q_i(t) = \int_0^t e^{p_i(\tau - t)} h_i(\tau)\, d\tau + q_i(0) e^{-p_i t}. \tag{2.139}$$

Eq. (2.133) has the following solution:

$$\mathbf{Y}(t) = \mathbf{Y}\mathbf{Q}(t) = \sum_{m=1}^{2n} \mathbf{Y}_m q_m(t). \tag{2.140}$$

Based on the previous discussion, we reconstruct the complex eigenvector and eigenvalue couples as follows:

$$\begin{aligned} \mathbf{Y}_{2m-1} &= \langle \mathbf{U}_m, \mathbf{P}_m, i\boldsymbol{\Phi}_m, i\omega_m \mathbf{U}_m, i\omega_m \mathbf{P}_m, -\omega_m \boldsymbol{\Phi}_m \rangle, \\ \mathbf{Y}_{2m} &= \langle \mathbf{U}_m, \mathbf{P}_m, -i\boldsymbol{\Phi}_m, -i\omega_m \mathbf{U}_m, -i\omega_m \mathbf{P}_m, -\omega_m \boldsymbol{\Phi}_m \rangle. \end{aligned} \tag{2.141}$$

The final form of the mode superposition solution is

$$\begin{bmatrix} \mathbf{U} \\ \mathbf{P} \\ \boldsymbol{\Phi} \\ \dot{\mathbf{U}} \\ \dot{\mathbf{P}} \\ \dot{\boldsymbol{\Phi}} \end{bmatrix} = \sum_{m=1}^{c} \begin{bmatrix} \mathbf{U}_m & \{F_m^u(t) + 2A_m \cos\omega_m t - 2B_m \sin\omega_m t\} \\ \mathbf{P}_m & \{F_m^p(t) + 2A_m \cos\omega_m t - 2B_m \sin\omega_m t\} \\ \boldsymbol{\Phi}_m & \{F_m^\phi(t) - 2A_m \sin\omega_m t - 2B_m \cos\omega_m t\} \\ \omega_m \mathbf{U}_m & \{F_m^{\dot{u}}(t) - 2A_m \sin\omega_m t - 2B_m \cos\omega_m t\} \\ \omega_m \mathbf{P}_m & \{F_m^{\dot{p}}(t) - 2A_m \sin\omega_m t - 2B_m \cos\omega_m t\} \\ \omega_m \boldsymbol{\Phi}_m & \{F_m^{\dot{\phi}}(t) - 2A_m \cos\omega_m t + 2B_m \sin\omega_m t\} \end{bmatrix}, \quad (2.142)$$

where $c$ denotes the number of real eigenvalues of interest to us and

$$\begin{aligned} F_m^u(t) &= F_m^p(t) = -F_m^{\dot{\phi}}(t) \\ &= \int_0^t \big[2C_m(\tau)\cos\omega_m(\tau - t) + 2D_m(\tau)\sin\omega_m(\tau - t)\big]\, d\tau, \\ F_m^\phi(t) &= F_m^{\dot{u}}(t) = F_m^{\dot{p}}(t) \\ &= \int_0^t \big[2C_m(\tau)\sin\omega_m(\tau - t) - 2D_m(\tau)\cos\omega_m(\tau - t)\big]\, d\tau. \end{aligned} \quad (2.143)$$

The mode superposition method outlined here provides a useful tool in the corresponding spectrum analysis of FSI problems. With the mode superposition method for generalized loads containing only certain frequencies, if the initial displacements and velocities are the linear combinations of a few modes, we could select a fairly small $c \ll n$ and drastically reduce the computation efforts compared with using the direct integration method. To improve the solution accuracy, similar static correction can also be applied:

$$\mathbf{A}\Delta\mathbf{Y} = \Delta\mathbf{F} = \mathbf{F} - \sum_{m=1}^{c} \mathbf{Y}_m^T \mathbf{F}(\mathbf{B}\mathbf{Y}_m)/(b_m + 2a_m\lambda_m). \quad (2.144)$$

In seismic analysis, the earthquake loading in some cases consists only of the lowest few modes, although the order of the system $n$ may be very large. However, in addition to ground accelerations, ground velocities are needed in Eq. (1.34). When we extend the proposed method to the study of dissipative systems, it is worth mentioning that the fluid-structure coupling discussed here contributes only to the imaginary part of the natural frequencies. Therefore, the damping treatment of the solids will be the same as discussed in Refs. [143,182].

*Direct integration method*

In the $\mathbf{u}/p$ and $\mathbf{u}$-p-$\boldsymbol{\Lambda}$ formulations, after we condense out the pressure and vorticity moment unknowns from Eqs. (1.82) and (1.83), we obtain:

$$\mathbf{M}\ddot{\mathbf{U}} + \mathbf{K}^*\mathbf{U} = \mathbf{R}(t), \tag{2.145}$$

where $\mathbf{M}$ is the positive definite mass (true physical mass) matrix, $\mathbf{K}^*$ is the stiffness (true physical stiffness) matrix, $\mathbf{U}$ is the nodal displacement unknown and $\mathbf{R}(t)$ is the discretized oscillatory force term. $\mathbf{K}^*$ is not necessarily a positive definite matrix, and

$$\mathbf{K}^* = \mathbf{K} - \mathbf{K}_{up}\mathbf{K}_{pp}^{-1}\mathbf{K}_{up}^T, \qquad \text{for the } \mathbf{u}/\text{p formulation,}$$

$$\mathbf{K}^* = \mathbf{K} - \mathbf{K}_{up}\mathbf{K}_{pp}^{-1}\mathbf{K}_{up}^T - \mathbf{K}_{u\lambda}\mathbf{K}_{\lambda\lambda}^{-1}\mathbf{K}_{u\lambda}^T, \qquad \text{for the } \mathbf{u}\text{-p-}\boldsymbol{\Lambda} \text{ formulation,}$$

where we assume the total degree of freedom is $n$.

The mode superposition method for Eq. (2.145) is the standard one for structural dynamics. In direct integration, the general governing equation (2.115) is integrated using a numerical step-by-step procedure. In essence, direct numerical integration aims to satisfy the governing equation (2.115) at discrete time stations with an interval $\Delta t$ apart. Depending on the discretizations of $\dot{\mathbf{U}}$ and $\ddot{\mathbf{U}}$, and the time marching schemes, various direct integration procedures can be introduced. The direct integration is called explicit if the next time step solution $\mathbf{U}^{n+1}$ is derived from the discretized governing equation at time step $n$, and implicit if at time step $n+1$. In this section, we discuss the central difference method, a typical explicit method, and the Newmark method, a typical implicit method. In the central difference method, we use the following second order discretization procedures:

$$\ddot{\mathbf{U}}^n = \frac{1}{\Delta t^2}\left(\mathbf{U}^{n+1} - 2\mathbf{U}^n + \mathbf{U}^{n-1}\right), \tag{2.146}$$

$$\dot{\mathbf{U}}^n = \frac{1}{2\Delta t}\left(\mathbf{U}^{n+1} - \mathbf{U}^{n-1}\right). \tag{2.147}$$

In addition, if we substitute Eqs. (2.146) and (2.147) into the governing equation systems (2.115) at time $t$, we obtain

$$\left(\frac{\mathbf{M}}{\Delta t^2} + \frac{\mathbf{C}}{2\Delta t}\right)\mathbf{U}^{n+1} = \mathbf{R}^n - \left(\mathbf{K} - \frac{2\mathbf{M}}{\Delta t^2}\right)\mathbf{U}^n - \left(\frac{\mathbf{M}}{\Delta t^2} - \frac{\mathbf{C}}{2\Delta t}\right)\mathbf{U}^{n-1}. \tag{2.148}$$

Obviously, from Eq. (2.148), we can solve for $\mathbf{U}^{n+1}$. Consequently, $\ddot{\mathbf{U}}^n$ and $\dot{\mathbf{U}}^n$ can be derived from Eqs. (2.146) and (2.147). Although the implementation of the central difference method is straightforward, especially for diagonal $\mathbf{M}$ and $\mathbf{C}$ matrices, this procedure has a restriction on the time step $\Delta t$.

Since Eq. (2.115) is identical to Eq. (2.121), the study of the stability of various direct integration schemes can be based on Eq. (2.121), and for the typical $j$th mode, the discretizations in Eqs. (2.146) and (2.147) can be expressed as

$$\ddot{Y}_j^n = \frac{1}{\Delta t^2}\big(Y_j^{n+1} - 2Y_j^n + Y_j^{n-1}\big), \tag{2.149}$$

$$\dot{Y}_j^n = \frac{1}{2\Delta t}\big(Y_j^{n+1} - Y_j^{n-1}\big). \tag{2.150}$$

Hence, the incremental equation for the $j$th modal solution can be expressed as

$$Y_j^{n+1} = \frac{\Delta t^2}{1+\xi_j\omega_j\Delta t} r_j^n + \frac{2-\omega_j^2\Delta t^2}{1+\xi_j\omega_j\Delta t} Y_j^n - \frac{1-\xi_j\omega_j\Delta t}{1+\xi_j\omega_j\Delta t} Y_j^{n-1}. \tag{2.151}$$

To study the numerical stability of the scheme in Eq. (2.151), we introduce a complex incremental factor $G_j$ with $Y_j^{n+1} = G_j Y_j^n$ to the homogeneous form of Eq. (2.151) with $r_j^n = 0$. The numerical stability requirement corresponds to

$$|G_j| \leqslant 1, \quad \text{with } j = 1, 2, \ldots, n. \tag{2.152}$$

Therefore, the governing equation for $G_j$, or, rather, the characteristic equation, is written as

$$(1 + \xi_j\omega_j\Delta t)G_j^2 - \big(2 - \omega_j^2\Delta t^2\big)G_j + (1 - \xi_j\omega_j\Delta t) = 0, \tag{2.153}$$

from which we obtain

$$G_j = \frac{(2-\omega_j^2\Delta t^2) \pm \sqrt{(2-\omega_j^2\Delta t^2)^2 - 4(1-\xi_j^2\omega_j^2\Delta t^2)}}{2(1+\xi_j\omega_j\Delta t)}. \tag{2.154}$$

For the case with no damping, i.e., $\xi_j = 0$, we use the solution for $G_j$ in the form of (2.154). We can then easily verify that $G_j$ must be complex in order to satisfy the stability requirement (2.152). Hence, the requirement (2.152) yields

$$\omega_j\Delta t \leqslant 2 \quad \text{or} \quad \Delta t \leqslant \frac{2}{\omega_j}, \quad \text{with } j = 1, 2, \ldots, n. \tag{2.155}$$

For the case with damping, using the same solution for $G_j$ and assuming $G_j$ is complex, we derive from the stability requirement (2.152)

$$\omega_j\Delta t \leqslant 2\sqrt{1-\xi_j^2} \quad \text{or} \quad \Delta t \leqslant \frac{2}{\omega_j}\sqrt{1-\xi_j^2}, \quad \text{with } j = 1, 2, \ldots, n. \tag{2.156}$$

The stability requirement (2.156) for the damped system is often considered as the counter part of the undamped system (2.155). Nevertheless, for high damping ratio $\xi_j \sim 1$, the stability requirement (2.156) becomes impossible to satisfy.

In the Newmark method, often considered as an extension of the linear acceleration method, we use the following discretization procedures:

$$\mathbf{U}^{n+1} = \mathbf{U}^n + \dot{\mathbf{U}}^n \Delta t + \left[\left(\frac{1}{2} - \alpha\right)\ddot{\mathbf{U}}^n + \alpha\ddot{\mathbf{U}}^{n+1}\right]\Delta t^2, \tag{2.157}$$

$$\dot{\mathbf{U}}^{n+1} = \dot{\mathbf{U}}^n + \left[(1-\beta)\ddot{\mathbf{U}}^n + \beta\ddot{\mathbf{U}}^{n+1}\right]\Delta t, \tag{2.158}$$

where $\alpha$ and $\beta$ are parameters governing integration accuracy and stability.

In general, we select $\alpha = 1/4$ and $\beta = 1/2$, which is often called constant-average-acceleration method or trapezoidal rule. In addition, substituting Eqs. (2.157) and (2.158) into the governing equation systems (2.115) at time $t + \Delta t$, we obtain

$$\left(\mathbf{K} + \frac{\mathbf{M}}{\alpha\Delta t^2} + \frac{\beta\mathbf{C}}{\alpha\Delta t}\right)\mathbf{U}^{n+1} = \hat{\mathbf{R}}^{n+1}, \tag{2.159}$$

where the effective load at time $t + \Delta t$ can be expressed as

$$\begin{aligned}\hat{\mathbf{R}}^{n+1} &= \mathbf{R}^{n+1} + \mathbf{M}\left(\frac{1}{\alpha\Delta t^2}\mathbf{U}^n + \frac{1}{\alpha\Delta t}\dot{\mathbf{U}}^n + \frac{1-2\alpha}{2\alpha}\ddot{\mathbf{U}}^n\right)\\ &\quad + \mathbf{C}\left(\frac{\beta}{\alpha\Delta t}\mathbf{U}^n + \frac{\beta-\alpha}{\alpha}\dot{\mathbf{U}}^n + \frac{\beta-2\alpha}{2\alpha}\Delta t\ddot{\mathbf{U}}^n\right).\end{aligned} \tag{2.160}$$

Obviously, from Eq. (2.159), we can solve for $\mathbf{U}^{n+1}$. Consequently, $\ddot{\mathbf{U}}^{n+1}$ can be derived from Eqs. (2.157) and (2.158) as

$$\ddot{\mathbf{U}}^{n+1} = \frac{1}{\alpha\Delta t^2}\left(\mathbf{U}^{n+1} - \mathbf{U}^n\right) - \frac{1}{\alpha\Delta t}\dot{\mathbf{U}}^n - \frac{1-2\alpha}{2\alpha}\ddot{\mathbf{U}}^n. \tag{2.161}$$

Although the implementation of the Newmark method is more complicated than the central difference method, the procedure is unconditionally stable. This can be easily confirmed using the governing equation in the form of Eq. (2.121). However, in practice, to achieve better accuracy for higher frequencies, we often have $\Delta t \leqslant T_n/10$, where $T_n$ stands for the smallest period.

If the governing equation systems are nonlinear, it is obvious that we can no longer use the traditional mode superposition method presented in this section. To use the direct integration, we must also combine the time incremental procedure with the Newton–Raphson iteration which will be elaborated in Chapter 7. Unlike the Newton–Raphson iteration for static nonlinear problems, the increment of the $k$th iteration yields

$$\mathbf{M}^{k-1}\Delta\ddot{\mathbf{U}}^k + \mathbf{C}^{k-1}\Delta\dot{\mathbf{U}}^k + \mathbf{K}^{k-1}\Delta\mathbf{U}^k = \Delta\mathbf{R}^k, \tag{2.162}$$

sharing the same form as Eq. (2.126).

REMARK 6.2.1. In Eq. (2.162), the matrices $\mathbf{M}^{k-1}$, $\mathbf{C}^{k-1}$, and $\mathbf{K}^{k-1}$ should be interpreted as incremental matrices of $k$th iteration for nonlinear problems. However, in the linearized dynamical system, those matrices are identical to the corresponding mass, damping, and stiffness matrices in linear systems.

We first introduce the starting values of the nonlinear iteration, often called the predictor phase:

$$
\begin{aligned}
\mathbf{U}^{n+1,0} &= \mathbf{U}^n + \Delta t \dot{\mathbf{U}}^n + \Delta t^2 \left( \frac{1}{2} - \alpha \right) \ddot{\mathbf{U}}^n, \\
\dot{\mathbf{U}}^{n+1,0} &= \dot{\mathbf{U}}^n + \Delta t (1 - \beta) \ddot{\mathbf{U}}^n, \\
\ddot{\mathbf{U}}^{n+1,0} &= \mathbf{0},
\end{aligned}
\tag{2.163}
$$

where $\alpha$ and $\beta$ are the same constants used for the Newmark method.

Then, we assign the incremental procedure of the nonlinear iteration, often called the corrector phase:

$$
\begin{aligned}
\ddot{\mathbf{U}}^{n+1,k} &= \ddot{\mathbf{U}}^{n+1,k-1} + \Delta \ddot{\mathbf{U}}^k, \\
\dot{\mathbf{U}}^{n+1,k} &= \dot{\mathbf{U}}^{n+1,0} + \beta \Delta t \ddot{\mathbf{U}}^{n+1,k}, \\
\mathbf{U}^{n+1,k} &= \mathbf{U}^{n+1,0} + \alpha \Delta t^2 \ddot{\mathbf{U}}^{n+1,k}.
\end{aligned}
\tag{2.164}
$$

Finally, the incremental equation (2.162) is replaced with

$$
\widetilde{\mathbf{M}}^{k-1} \Delta \ddot{\mathbf{U}}^k = \Delta \mathbf{R}^k, \tag{2.165}
$$

where the effective mass matrix is expressed as

$$
\widetilde{\mathbf{M}}^{k-1} = \mathbf{M}^{k-1} + \beta \Delta t \mathbf{C}^{k-1} + \alpha \Delta t^2 \mathbf{K}^{k-1}. \tag{2.166}
$$

### *6.2.2. Acoustoelastic/slosh FSI systems*

The formulations presented in this chapter have been implemented [151,300,301]. Two-dimensional linear elastic models are employed to couple with acoustic fluids in various acoustoelastic/slosh FSI systems. A series of examples are chosen to demonstrate the capability of different formulations and solution procedures.

We set the tall water column and rigid cavity problems to be the same as those used in Refs. [170] and [175] in order to compare our results with those reported using the pure displacement-based formulation and mixed formulations.

In all test problems, we want to evaluate the lowest frequencies of the complete FSI system. Of course, we do not calculate any zero frequencies and their mode shapes, but simply shift to the nonzero frequencies sought. The results are also found to be in good agreement with those calculated with the $\phi$-$\mathbf{u}$ formulation or analytical solutions [171], and the number of zero frequency modes is always exactly equal to the mathematical prediction. Further, in no case does a spurious nonzero frequency emerge with $\mathbf{u}$/p and $\mathbf{u}$-p-$\boldsymbol{\Lambda}$ mixed formulations. The pressure bands as an error measure are quite smooth, indicating accurate solutions. Figure 6.12 shows the tall water column considered in Ref. [170]. Using the acoustic wave speed $c = \sqrt{\kappa/\rho}$, similar to the displacement form (1.1) in Chapter 5 and

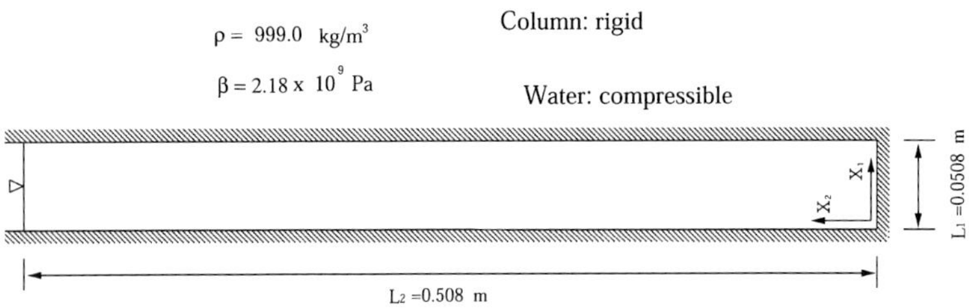


Figure 6.12. Tall water column problem.

the velocity potential form (1.28) in Chapter 5, the governing equations for the acoustic fluid in the pressure form are expressed as

$$\nabla^2 p = \frac{1}{c^2}\frac{\partial^2 p}{\partial t^2}, \tag{2.167}$$

$$\frac{\partial^2 \mathbf{u}}{\partial t^2} = -\frac{1}{\rho}\nabla p, \tag{2.168}$$

with the initial condition

$$\mathbf{u} = \mathbf{0}, \quad \text{at } t = 0,$$

and the boundary conditions

$$\begin{aligned} u_1 &= 0, \quad \text{at } x_1 = 0, L_1, \\ u_2 &= 0, \quad \text{at } x_2 = 0, \\ p &= 0, \quad \text{at } x_2 = L_2. \end{aligned}$$

The analytical solution of the above equations (refer to [96]) is

$$\begin{aligned} u_1 &= A\lambda_1 \sin(\lambda ct)\cos(\lambda_2 x_2)\sin(\lambda_1 x_1), \\ u_2 &= A\lambda_2 \sin(\lambda ct)\sin(\lambda_2 x_2)\cos(\lambda_1 x_1), \\ p &= -\rho\lambda^2 c^2 A \sin(\lambda ct)\cos(\lambda_2 x_2)\cos(\lambda_1 x_1), \end{aligned} \tag{2.169}$$

with

$$\begin{aligned} \lambda_1 &= \frac{n\pi}{L_1}, \quad n = 0, 1, 2, \ldots, \\ \lambda_2 &= \frac{m\pi}{2L_2}, \quad m = 1, 3, 5, \ldots, \\ \lambda^2 &= \lambda_1^2 + \lambda_2^2, \\ \omega &= c\lambda. \end{aligned} \tag{2.170}$$

Notice that the free surface condition is $p = 0$ with no gravity effects. If no further mathematical assumptions are made on the displacements of the free surface,

Figure 6.13. First four modes for the tall water column problem.

we shall expect surface elevations for some modes, indicated by the analytical solutions in Eqs. (2.169).

Here we use the same mesh as is used in Ref. [170]. However, we apply 9/4-$c$ and 9/3 elements for the $\mathbf{u}$/p formulation and 9/4-$c$-4$c$ and 9-3-3 elements for

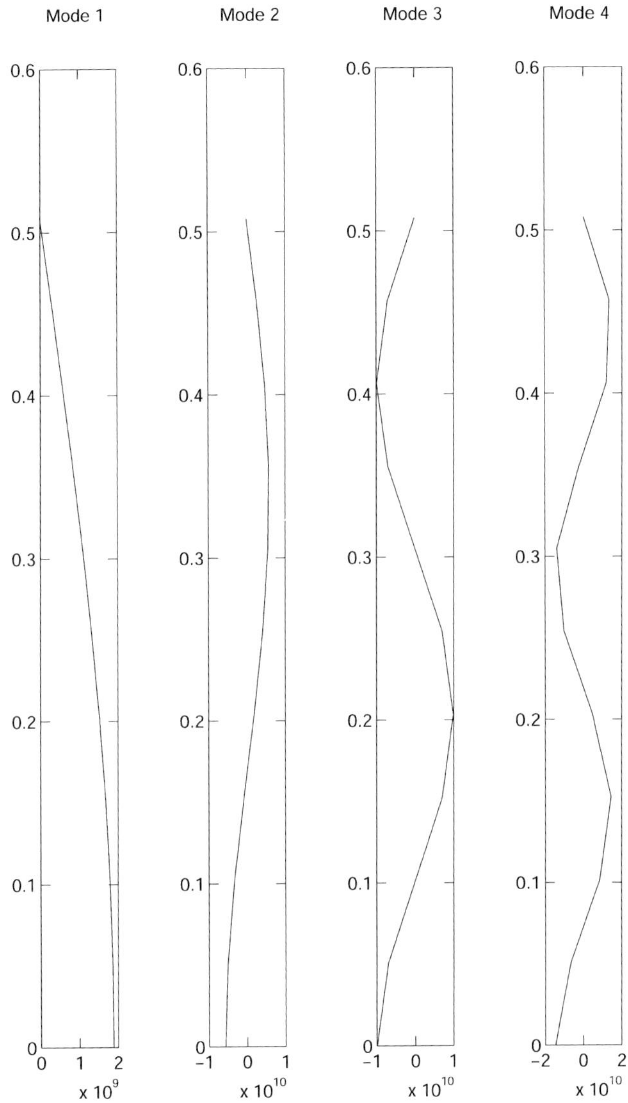


Figure 6.14. Pressure distributions of the first four modes for the tall water column problem.

the **u**-p-$\boldsymbol{\Lambda}$ formulation. The lowest four displacement modes with $2 \times 10$ 9/4-$c$ elements are shown in Figure 6.13. Figure 6.14 gives the corresponding pressure distributions. With mixed formulations, there is no spurious nonzero frequency mode observed. It is clear that reported nonzero frequency spurious modes were

Table 6.2 Analytic solution of acoustic frequencies of the tall water column problem

| Mode # | Frequencies (rad/s) | $n, m$ |
|---|---|---|
| 1 | 4567.74 | $n = 0, m = 1$ |
| 2 | 13703.2 | $n = 0, m = 3$ |
| 3 | 22838.7 | $n = 0, m = 5$ |
| 4 | 31974.2 | $n = 0, m = 7$ |

Table 6.3 Frequency analysis of the tall water column problem

| | Mixed finite element choice | | | | | | | |
|---|---|---|---|---|---|---|---|---|
| | 9/4-$c$ | 9/4-$c$-4$c$ | 9/3 | 9-3-3 | 4/1 | 4-1-1 | Type I 4/1 | Type I 4-1-1 |
| Zero theory | 130 | 97 | 103 | 43 | 83 | 3 | 103 | 3 |
| Zero result | 130 | 97 | 103 | 43 | 83 | 3 | 103 | 3 |
| First four acoustic modes (rad/s) | 4567.77 | 4567.77 | 4567.74 | 4567.74 | 4568.92 | 4568.91 | 4584.36 | 4568.49 |
| | 13709.8 | 13709.8 | 13703.7 | 13703.7 | 13734.9 | 13734.9 | 13769.6 | 13723.6 |
| | 22915.4 | 22915.4 | 22844.6 | 22844.6 | 22985.7 | 22985.7 | 23007.7 | 22933.4 |
| | 32332.9 | 32332.9 | 32004.8 | 32004.8 | 32378.3 | 32378.3 | 32342.2 | 32236.3 |

induced by the pure displacement-based formulation and its penalty formulation, which should not be used for incompressible or almost incompressible materials. Interestingly, the difference with the $\mathbf{u}/\mathrm{p}$ and $\mathbf{u}$-p-$\boldsymbol{\Lambda}$ formulations is mainly the number of zero frequency modes. In general, the number of zero frequency modes from the $\mathbf{u}/\mathrm{p}$ formulation is greater than the number of zero frequency modes from the $\mathbf{u}$-p-$\boldsymbol{\Lambda}$ formulation, and the difference between the two is the number of $\boldsymbol{\Lambda}$ unknowns. The analytical solution of the frequencies are listed in Table 6.2. Table 6.3 gives the frequency solutions using both the $\mathbf{u}/\mathrm{p}$ and $\mathbf{u}$-p-$\boldsymbol{\Lambda}$ formulations with different mixed finite elements. The numerical solutions of the frequencies are compared with the analytical values in Table 6.4.

In Ref. [175], Hamdi et al. tested the rigid cavity problem with $3 \times 4$ 9/4-$c$ elements. The analytical frequency solution of this rigid cavity problem as shown in Figure 6.15 is

$$\omega = c\pi\sqrt{\left(\frac{n}{a}\right)^2 + \left(\frac{m}{b}\right)^2}, \tag{2.171}$$

where $n$ and $m$ are integers and $c$ is the acoustic speed.

Figures 6.16 and 6.17 show the mode shapes and the pressure distributions of the first four frequencies calculated with a $3 \times 4$ mesh of 9/4-$c$ elements. Table 6.5

Table 6.4 Results in tall water column problem using the **u**/p formulation

| Mixed elements (mesh of 2 × 10 elements) | Frequencies (rad/s) | | | |
|---|---|---|---|---|
| | First | Second | Third | Fourth |
| 9/4-*c* | 4567.77 | 13709.8 | 22915.4 | 32332.9 |
| 9/3 | 4567.74 | 13703.7 | 22844.6 | 32004.8 |
| Analytical solution | 4567.74 | 13703.2 | 22838.7 | 31974.2 |

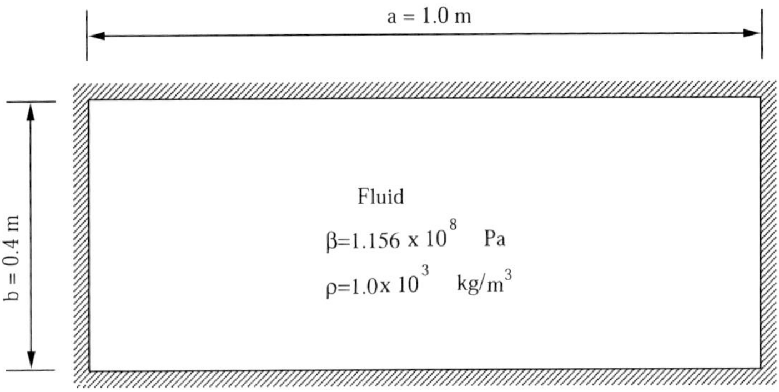


Figure 6.15. Rigid cavity problem.

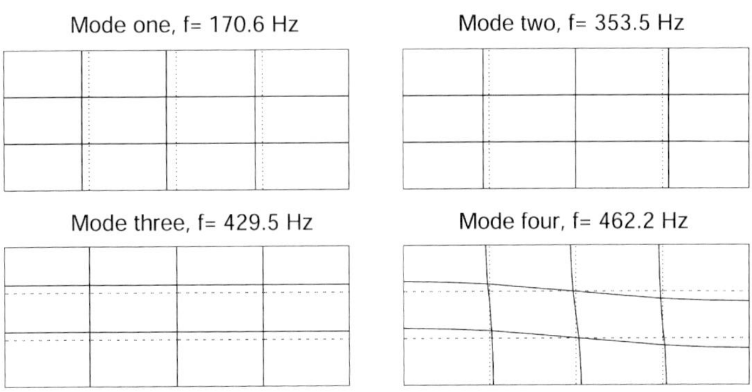


Figure 6.16. First four modes for the rigid cavity problem.

lists the frequencies calculated with both the **u**/p and **u**-p-$\boldsymbol{\Lambda}$ formulations with different types of mixed elements. The first four analytical frequencies calculated from Eq. (2.171) are listed in Table 6.6. Table 6.7 compares the numerical results with the analytical values.

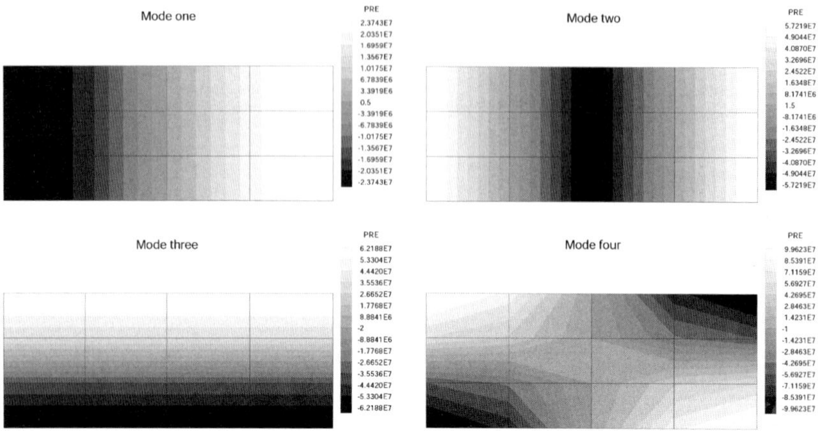


Figure 6.17. Pressure distributions of the first four modes for the rigid cavity problem.

Table 6.5 Frequency analysis of the rigid cavity problem

| | Mixed finite element choice | | | | | | | |
|---|---|---|---|---|---|---|---|---|
| | 9/4-*c* | 9/4-*c*-4*c* | 9/3 | 9-3-3 | 4/1 | 4-1-1 | Type I 4/1 (macro) | Type I 4-1-1 (macro) |
| Zero theory | 75 | 55 | 59 | 23 | 47 | −1 | 59 | −1 |
| Zero result | 75 | 55 | 59 | 23 | 47 | 0 | 59 | 0 |
| First four acoustic modes (Hz) | 170.6 | 170.6 | 170.0 | 170.0 | 171.1 | 171.1 | 172.6 | 170.4 |
| | 353.5 | 353.5 | 341.3 | 341.3 | 348.8 | 348.8 | 347.8 | 342.7 |
| | 429.5 | 429.5 | 425.3 | 425.3 | 429.9 | 429.9 | 429.5 | 428.1 |
| | 462.2 | 462.1 | 468.1 | 468.3 | 459.3 | 459.3 | 464.0 | 461.8 |

It has been reported that the displacement-based formulation introduced nonzero frequency spurious modes (which were identified as rotational modes), and these spurious modes could be eliminated by applying a penalty term of the irrotationality constraint of Eq. (1.95). From the calculated results, we conclude that if we use the $\mathbf{u}/p$ or $\mathbf{u}$-p-$\boldsymbol{\Lambda}$ formulation with elements satisfying the inf-sup condition, we will not encounter spurious nonzero frequency rotational modes. In actuality, spurious rotational modes are related to the familiar spurious pressure modes (through Eq. (1.109)).

For didactic reasons, a simple acoustic FSI model as depicted in Figure 6.18 is used to compare the mode superposition method with the direct time integration method. In this case, no physical damping is considered. The cross section area $A$,

Table 6.6 Analytic solution of acoustic frequencies of the rigid cavity problem

| Mode # | Frequencies (Hz) | $n, m$ |
|---|---|---|
| 1 | 170.0 | $n = 1, m = 0$ |
| 2 | 340.0 | $n = 2, m = 0$ |
| 3 | 425.0 | $n = 0, m = 1$ |
| 4 | 457.7 | $n = 1, m = 1$ |

Table 6.7 Results in the rigid cavity problem using the **u**/p formulation

| Mixed elements (3 × 4 mesh) | Frequencies (Hz) | | | |
|---|---|---|---|---|
| | First | Second | Third | Fourth |
| 9/4-*c* | 170.6 | 353.5 | 429.5 | 462.2 |
| 9/3 | 170.0 | 341.3 | 425.3 | 468.1 |
| Analytical solution | 170.0 | 340.0 | 425.0 | 457.7 |

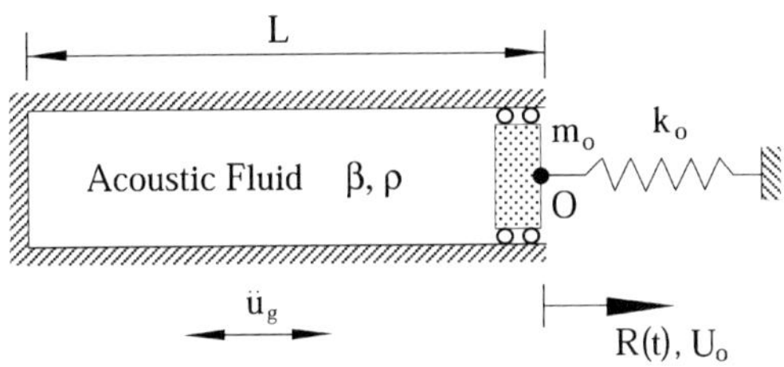


Figure 6.18. The acoustic fluid in a cavity interacting with a piston/spring system.

the spring constant $k_o$, and the rigid plate mass $m_o$ are assigned to be 1.0 $m^2$, $1.0 \times 10^7$ N/m, and 1000.0 kg, respectively. The other physical parameters are given as follows: $L = 10.0$ m, $\kappa = 2.1 \times 10^9$ Pa, and $\rho = 1000.0\ \text{kg/m}^3$.

From the solution of the coupled system eigenproblems, we obtain the first three nonzero natural frequencies: $\omega_1 = 211.8$, $\omega_2 = 744.0$, and $\omega_3 = 1677.4$ rad/s. In the $\phi$-**u** formulation, for the coupled system with one closed fluid domain, we have one zero frequency corresponding to $U_o = 0$ and constant $\phi_i$ over the fluid region. We notice that in this fluid-structure system, due to the fluid mass and compressibility, the lowest nonzero frequency of the coupled fluid-structure system is about 112 percent higher that the natural frequency of the piston/spring system without fluid coupling.

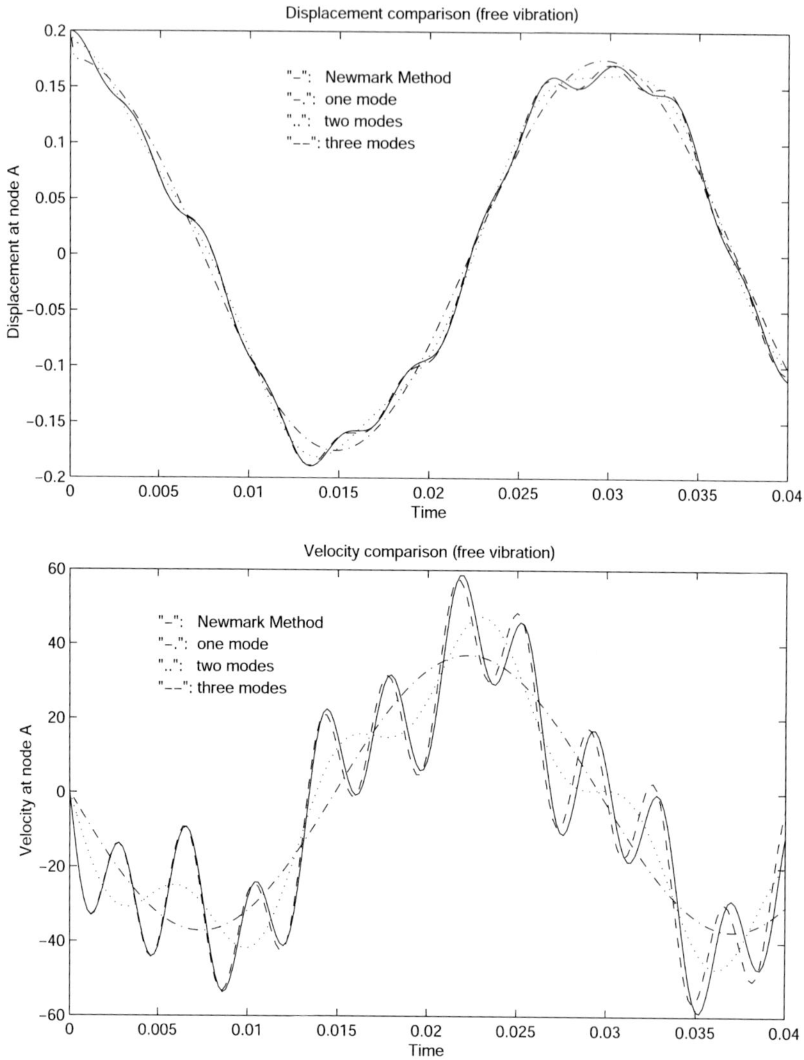


Figure 6.19. Test case one: displacement and velocity comparisons.

In test case one, an initial positive displacement of the plate ($U_o = 0.2$ cm) and an initial fluid pressure ($p = -4.2 \times 10^7$ Pa) are applied to the system. In test case two, in addition to initial conditions in case one, an excitation force $R(t) = 1.0 \times 10^8 \sin(200\pi t)$ N is applied to the plate. In test case three, a ground motion, $\ddot{u}_g = 20.0^2\pi \cos(100\pi t) + 8.0^2\pi \cos(200\pi t)$ m/s$^2$, is applied to the FSI system.

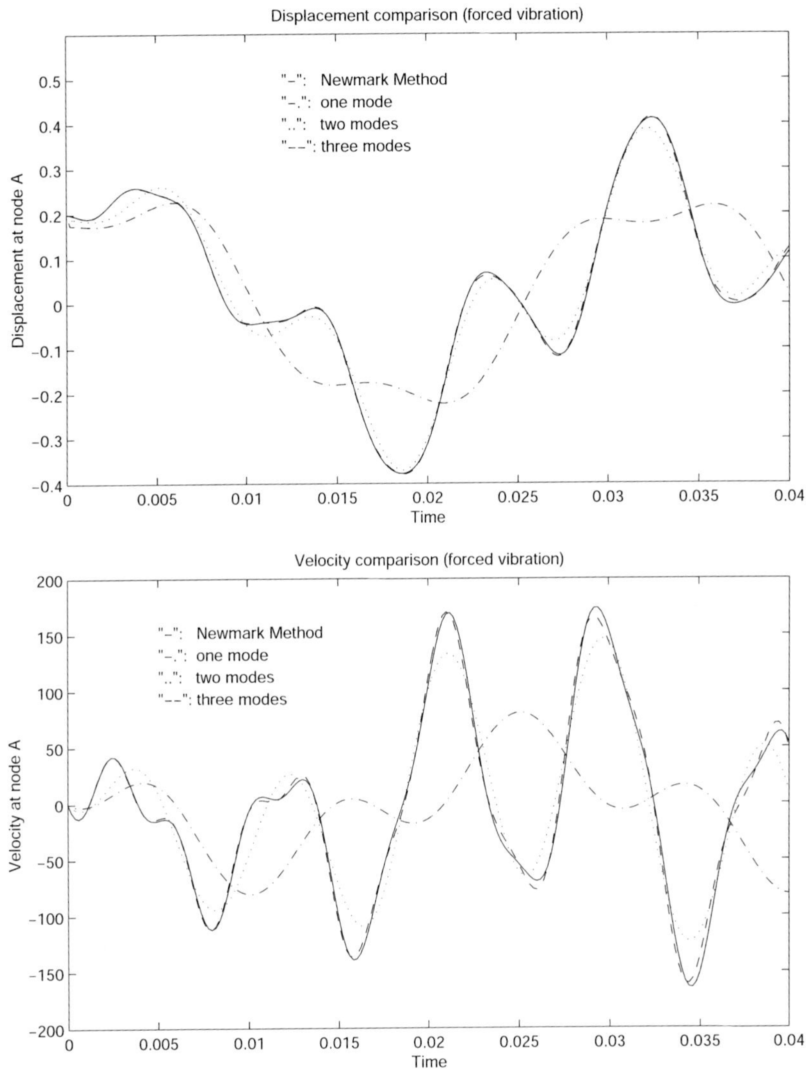


Figure 6.20. Test case two: displacement and velocity comparisons.

For the three test cases, the comparisons of the displacement and velocity at position A are shown in Figures 6.19, 6.20, and 6.21. As can be seen, the more modes we use, the closer the mode superposition solutions are to the direct time integration results. In addition, as illustrated in Figures 6.20 and 6.21, if the excitation forces or ground motions have certain frequencies closer to the coupled

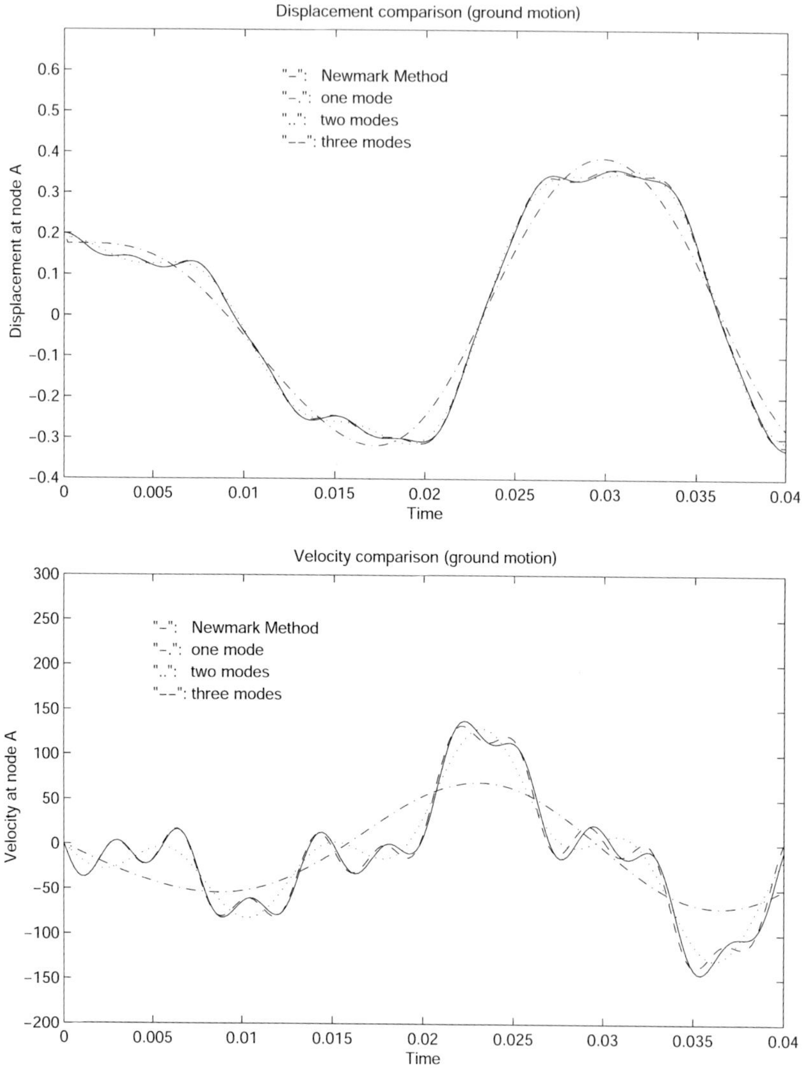


Figure 6.21. Test case three: displacement and velocity comparisons.

natural modes, as in test case two, and the excitation force frequency (200.0 Hz) is close to the second mode (211.8 Hz), these modes will have the dominating contribution in the mode superposition method.

To demonstrate the capability of predicting slosh modes, we analyze a rigid rectangular tank as shown in Figure 6.22. The analytical solution of the slosh

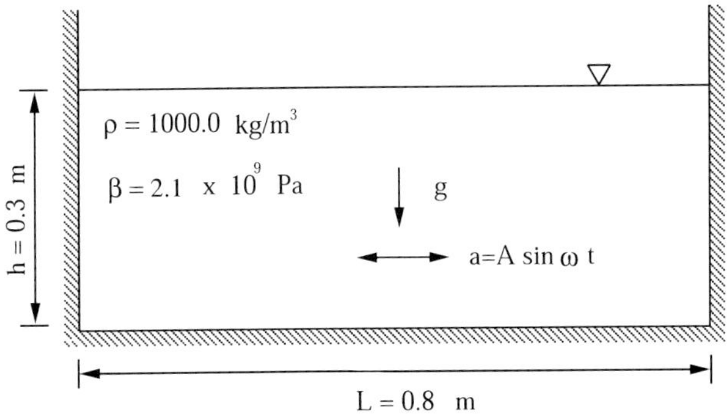


Figure 6.22. Slosh water tank problem.

Table 6.8 Frequency analysis of the slosh water tank problem

| | Mixed finite element choice | | | | | | | |
|---|---|---|---|---|---|---|---|---|
| | 9/4-*c* | 9/4-*c*-4*c* | 9/3 | 9-3-3 | 4/1 | 4-1-1 | Type I 4/1 | Type I 4-1-1 |
| Zero theory | 48 | 33 | 39 | 15 | 31 | −1 | 47 | −1 |
| Zero result | 48 | 33 | 39 | 15 | 31 | 0 | 47 | 0 |
| First four slosh modes (rad/s) | 5.657 | 5.656 | 5.713 | 5.712 | 5.684 | 5.683 | 5.711 | 5.639 |
| | 8.891 | 8.882 | 9.308 | 9.295 | 8.838 | 8.8170 | 8.833 | 8.680 |
| | 11.31 | 11.25 | 12.02 | 12.00 | 11.09 | 10.99 | 11.01 | 8.935 |
| | 12.52 | 12.45 | 17.08 | 16.65 | 13.17 | 12.85 | 13.20 | 10.69 |
| First four acoustic modes (rad/s) | 7600.3 | 7600.4 | 7589.6 | 7589.6 | 7636.5 | 7636.5 | 7643.8 | 7619.1 |
| | 9507.6 | 9506.5 | 9834.0 | 9817.6 | 9486.6 | 9483.7 | 9602.2 | 9541.4 |
| | 14063 | 14058 | 14861 | 14775 | 13783 | 13775 | 13993 | 13853 |
| | 20245 | 20234 | 20732 | 20607 | 19327 | 19317 | 19390 | 19127 |

frequencies is given as

$$\omega = \sqrt{gk \tanh(kh)}, \tag{2.172}$$

with $kL = n\pi, n = 1, 2, \ldots$.

The acoustic frequency can be calculated from Eq. (2.170). The computed frequencies are listed in Table 6.8, while the analytical frequency solutions for both the slosh and acoustic modes are listed in Table 6.9. The first four slosh and acoustic modes and their pressure bands are shown in Figures 6.23, 6.24, 6.25 and 6.26, respectively.

Figure 6.27 describes the tilted piston-container problem. The massless elastic piston moves horizontally. Figure 6.28 describes the problem of a rigid cylinder

Table 6.9 Analytic solution of acoustic frequencies of the slosh water tank problem

| Mode # | Slosh frequencies (rad/s) | Acoustic frequencies (rad/s) | $n, m$ |
|---|---|---|---|
| 1 | 5.641 | 7587.67 | $n = 0, m = 1$ |
| 2 | 8.695 | 9484.58 | $n = 1, m = 1$ |
| 3 | 10.74 | 13,678.9 | $n = 2, m = 1$ |
| 4 | 12.41 | 18,682.5 | $n = 3, m = 1$ |

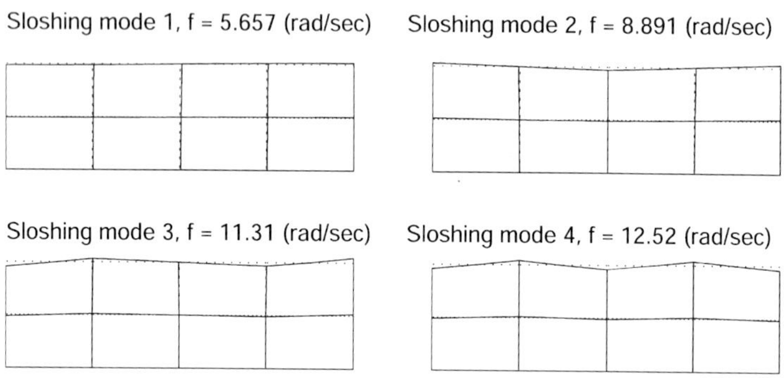


Figure 6.23. First four slosh modes for the slosh water tank problem.

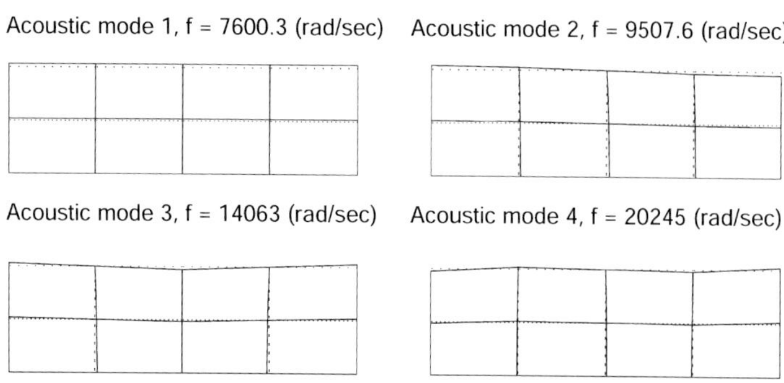


Figure 6.24. First four acoustic modes for the slosh water tank problem.

vibrating in an acoustic cavity. The cylinder is suspended from a spring and vibrates vertically in the fluid. Figure 6.29 shows a rigid ellipse on a spring in the same acoustic cavity. Figures 6.30, 6.31 and 6.32 show the typical meshes used in these problems. Tables 6.10 and 6.11 list the results obtained using the 9/3 element and 9-3-3 element, and Table 6.12 gives the results obtained with the $\phi$-$\mathbf{u}$

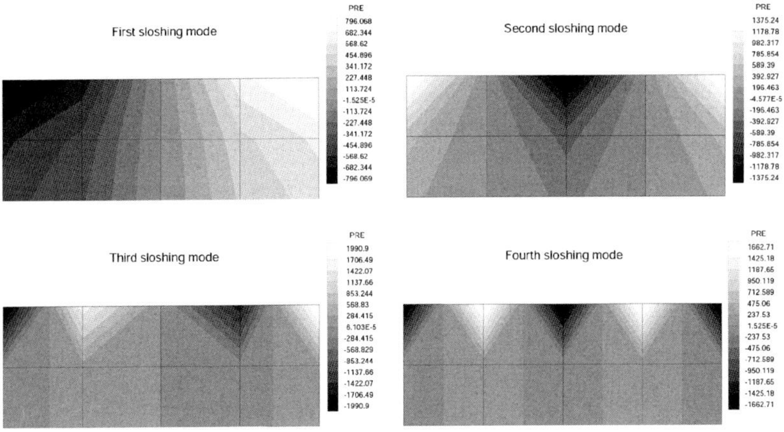


Figure 6.25. Pressure distributions of slosh modes for the slosh water tank problem.

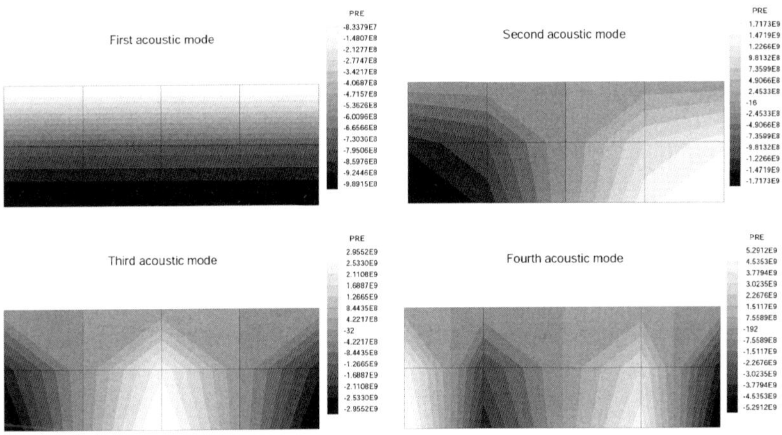


Figure 6.26. Pressure distributions of acoustic modes for the slosh water tank problem.

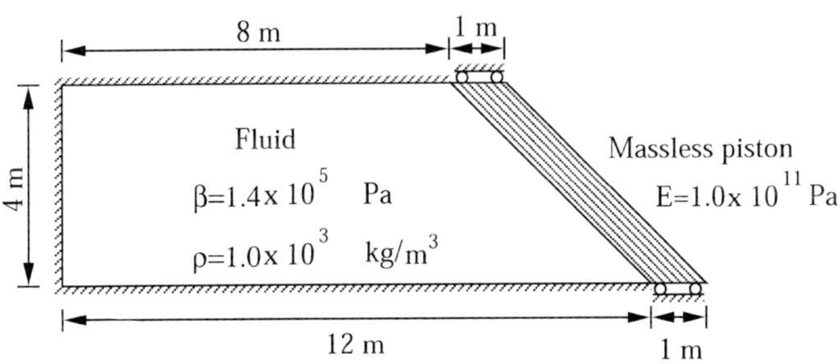


Figure 6.27. Tilted piston container system.

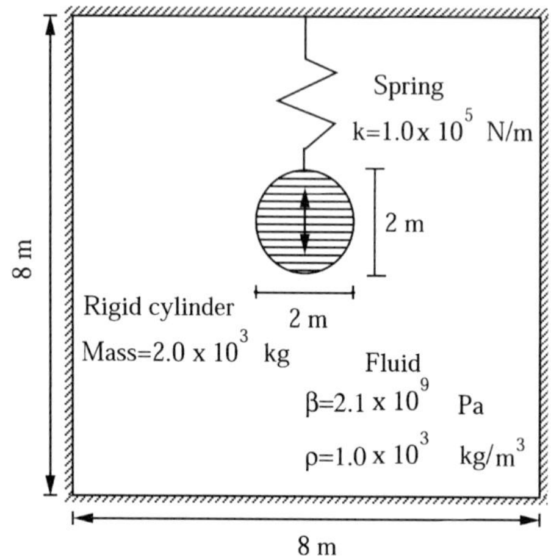


Figure 6.28. A rigid cylinder vibrating in an acoustic cavity.

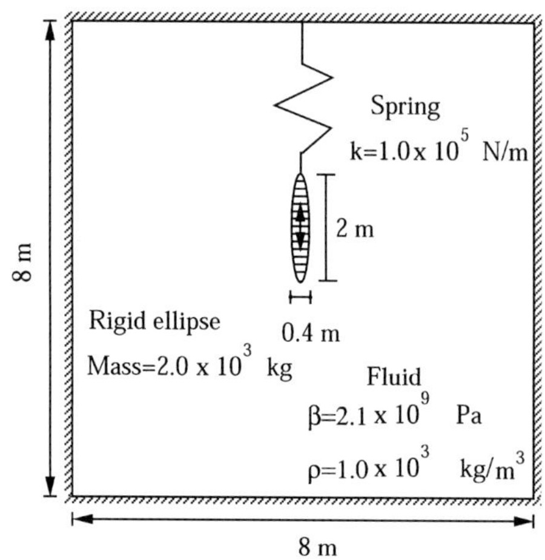


Figure 6.29. A rigid ellipse vibrating in an acoustic cavity.

formulation. The meshes used in this analysis have been derived by starting with coarse meshes and subdividing each element into two or four elements in each refinement.

Results in Figures 6.33, 6.34, and 6.35 with thirty-two 9-3-3 elements show the pressure band plots of the modes considered in Table 6.10. The frequencies from the incompressible fluid model with added mass effects are the upper bound of the calculated frequencies in Tables 6.10 and 6.11. This matches the mathematical prediction in Ref. [200]. To illustrate the importance of assigning appropriate

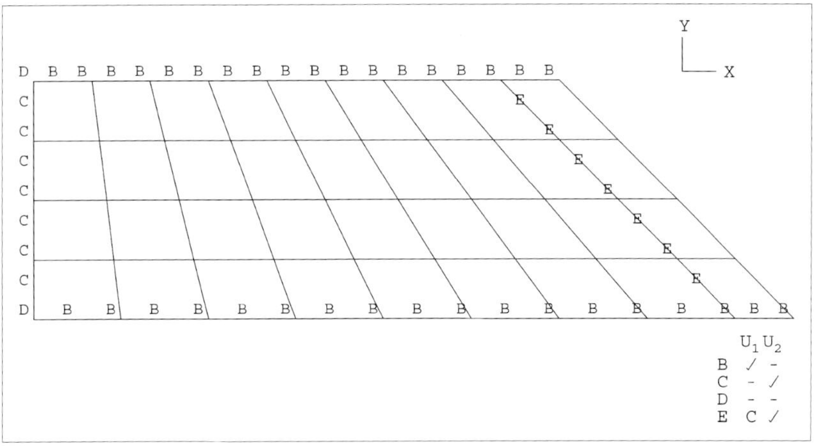


Figure 6.30. Typical mesh for the tilted piston container system.

Table 6.10 Analysis of test problems using the **u**/p formulation with 9/3 elements

| Test case | Mesh, no. of elements | Number of zero frequencies | | Frequencies (rad/s) | | | |
|---|---|---|---|---|---|---|---|
| | | Theory | Result | First | Second | Third | Fourth |
| Tilted piston-container | 4 | $\geqslant 19$ | 19 | 1.898 | 6.063 | 9.275 | 10.45 |
| | 16 | $\geqslant 79$ | 79 | 1.867 | 5.702 | 9.239 | 9.808 |
| | 32 | $\geqslant 159$ | 159 | 1.862 | 5.605 | 9.192 | 9.397 |
| Rigid cylinder problem | 2 | $\geqslant 7$ | 7 | 3.899 | 718.2 | 1178 | 1326 |
| | 8 | $\geqslant 37$ | 37 | 4.259 | 611.5 | 1193 | 1330 |
| | 32 | $\geqslant 157$ | 157 | 4.285 | 589.6 | 1138 | 1254 |
| Rigid ellipse problem | 2 | $\geqslant 7$ | 7 | 6.192 | 697.8 | 1216 | 1269 |
| | 8 | $\geqslant 37$ | 37 | 6.755 | 591.7 | 1229 | 1235 |
| | 32 | $\geqslant 157$ | 157 | 6.848 | 572.6 | 1157 | 1178 |

tangential displacement directions at the fluid nodal points on the FSI interface, we present the results of two test cases. In the first, the very coarse finite element model with two 9-3-1 elements, as shown in Figure 6.36 for the response analysis of the rigid cylinder, is considered. The assignment of correct tangential directions at nodes 10 and 11 is critical. If we use the actual cylinder geometry, the tangential directions are 60 and 120 degrees from the $x$-axis. However, using Eq. (1.77) we obtain 45 and 135 degrees, respectively. Table 6.13 lists some solution results and shows that a spurious nonzero frequency appears when incorrect tangential directions are assigned. In the rigid ellipse test case, the finite element model with

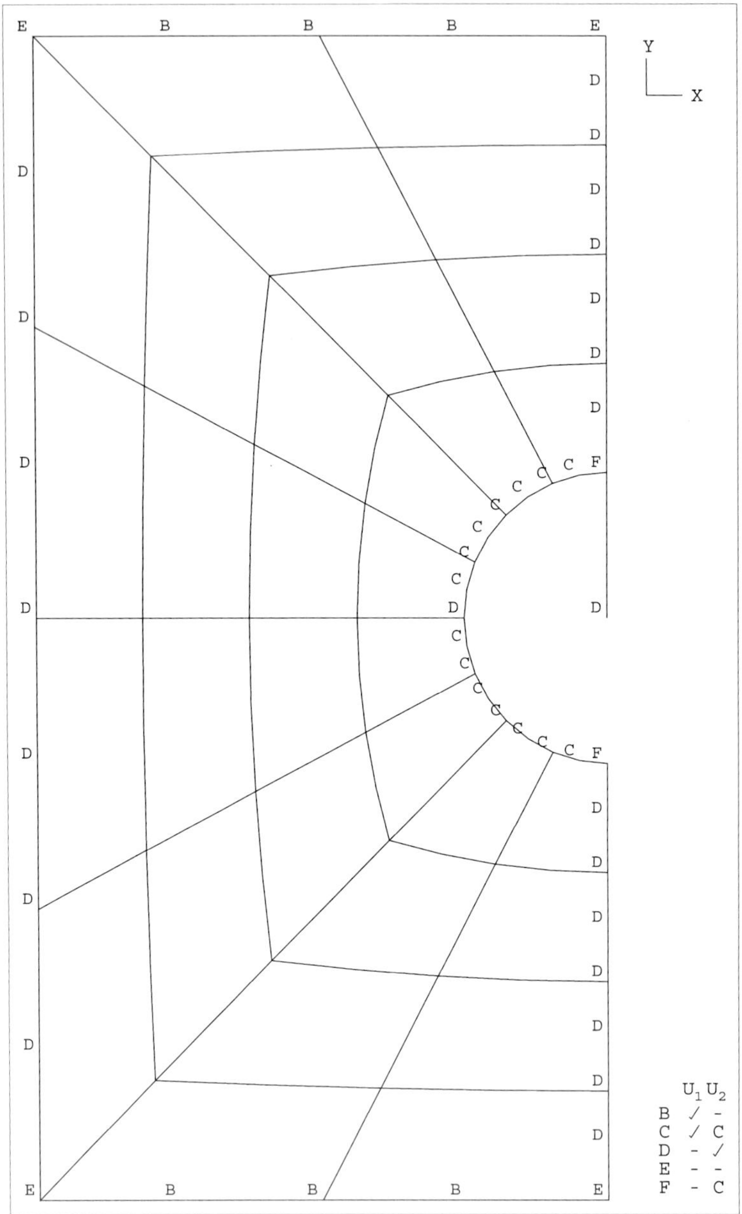


Figure 6.31. Typical mesh for the rigid cylinder problem.

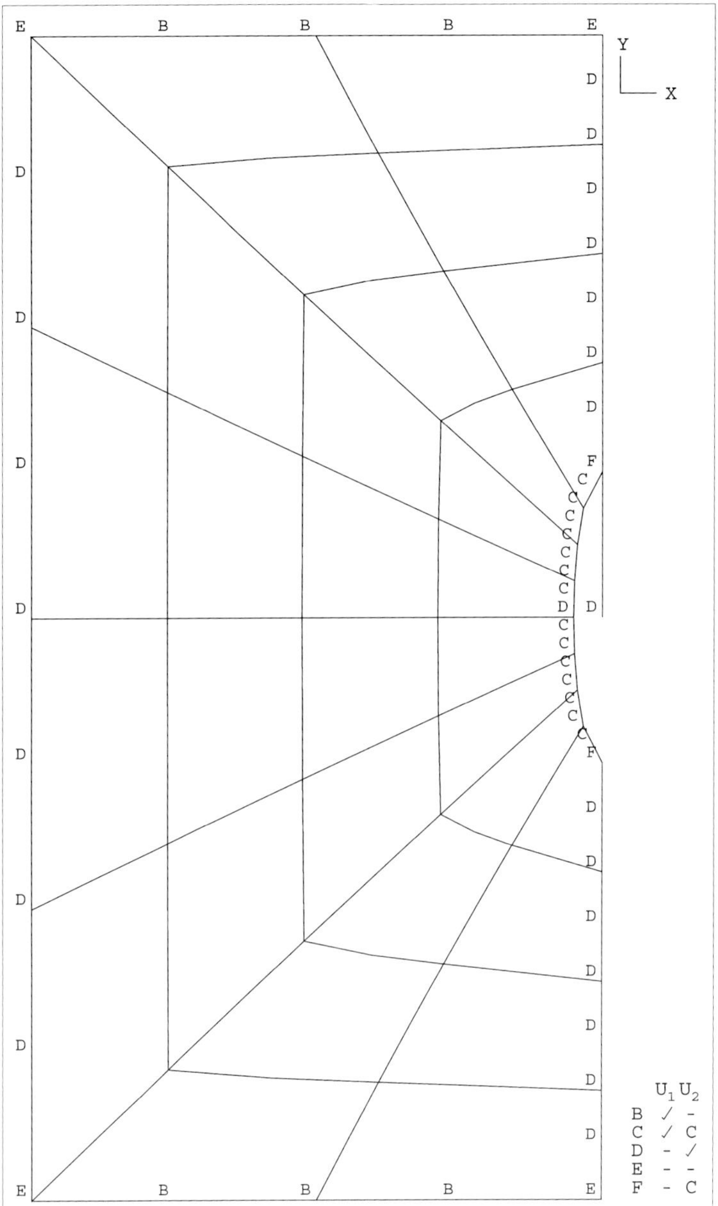


Figure 6.32. Typical mesh for the rigid ellipse problem.

Table 6.11 Analysis of test problems using the **u**-p-$\boldsymbol{\Lambda}$ formulation with 9-3-3 elements

| Test case | Mesh, no. of elements | Number of zero frequencies | | Frequencies (rad/s) | | | |
|---|---|---|---|---|---|---|---|
| | | Theory | Result | First | Second | Third | Fourth |
| Tilted piston-container | 4 | $\geqslant 7$ | 7 | 1.895 | 6.054 | 9.256 | 10.32 |
| | 16 | $\geqslant 31$ | 31 | 1.867 | 5.696 | 9.236 | 9.792 |
| | 32 | $\geqslant 63$ | 63 | 1.862 | 5.603 | 9.189 | 9.391 |
| Rigid cylinder problem | 2 | $\geqslant 1$ | 1 | 3.846 | 628.2 | 928.6 | 1341 |
| | 8 | $\geqslant 13$ | 13 | 4.228 | 609.5 | 1191 | 1326 |
| | 32 | $\geqslant 61$ | 61 | 4.281 | 589.4 | 1138 | 1252 |
| Rigid ellipse problem | 2 | $\geqslant 1$ | 1 | 5.511 | 629.9 | 932.6 | 1246 |
| | 8 | $\geqslant 13$ | 13 | 6.612 | 589.1 | 1220 | 1232 |
| | 32 | $\geqslant 61$ | 61 | 6.827 | 572.1 | 1155 | 1176 |

Table 6.12 Results obtained using the $\phi$-**u** formulation for analysis of three test problems

| Test case | Mesh, no. of elements | Frequencies (rad/s) | | | |
|---|---|---|---|---|---|
| | | First | Second | Third | Fourth |
| Tilted piston problem | 32 | 1.858 | 5.569 | 9.116 | 9.299 |
| Rigid cylinder problem | 32 | 4.269 | 581.8 | 1124 | 1224 |
| Rigid ellipse problem | 32 | 7.071 | 563.2 | 1138 | 1158 |

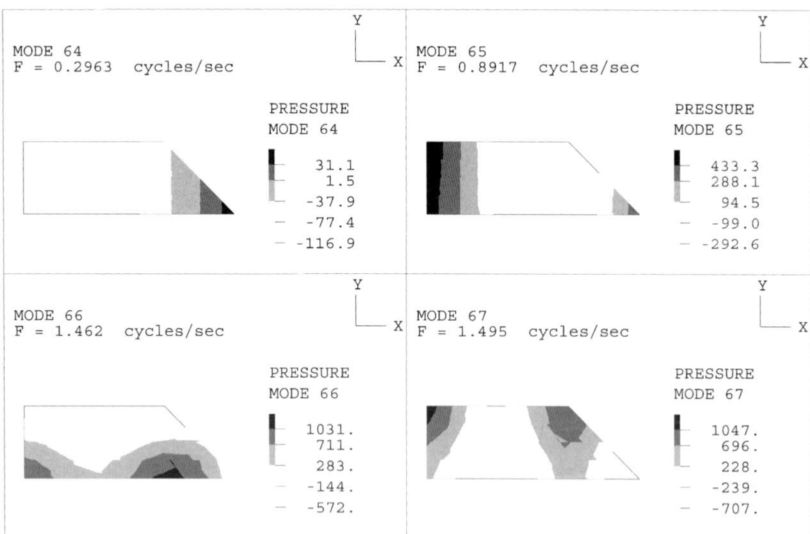


Figure 6.33. Pressure distributions of the first four modes for the tilted piston container problem.

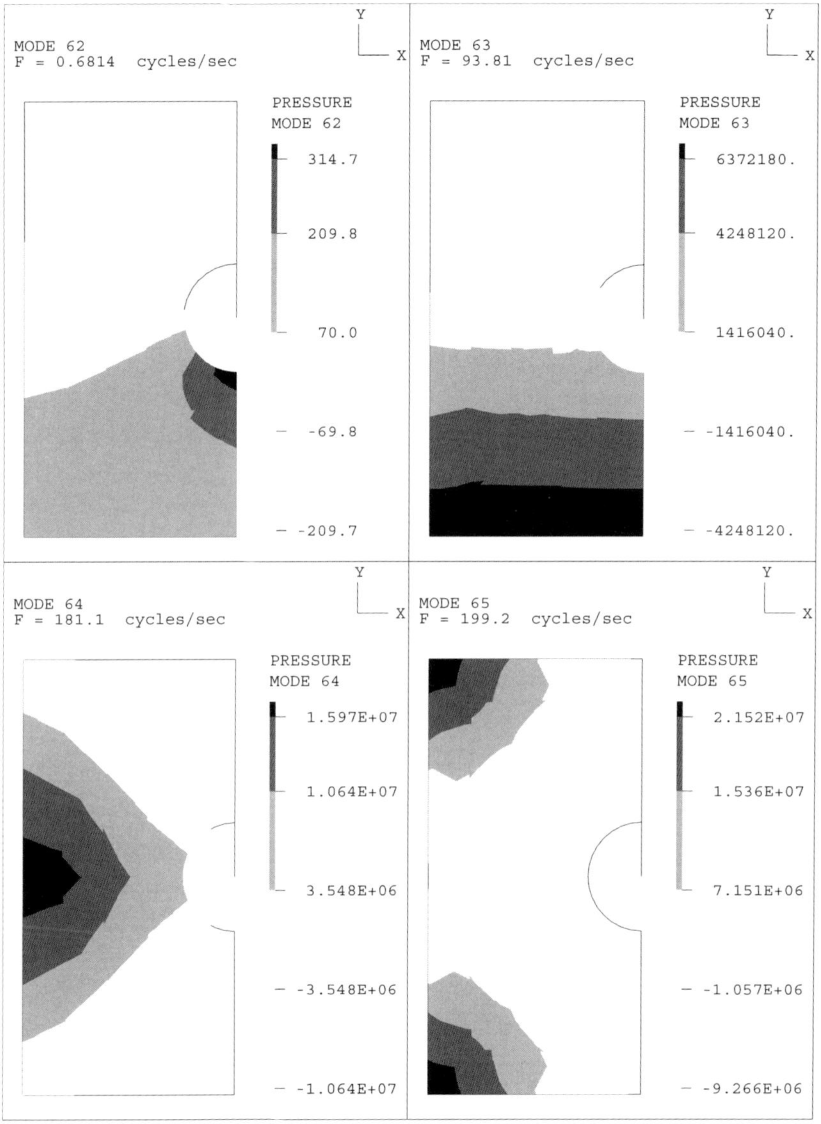


Figure 6.34. Pressure distributions of the first four modes for the rigid cylinder problem.

eight 9-3-3 elements, as shown in Figure 6.37, is used for the frequency solution of the rigid ellipse vibrating in the fluid-filled cavity. If we use the tangential directions given by the actual geometry of the ellipse, a nonzero spurious frequency

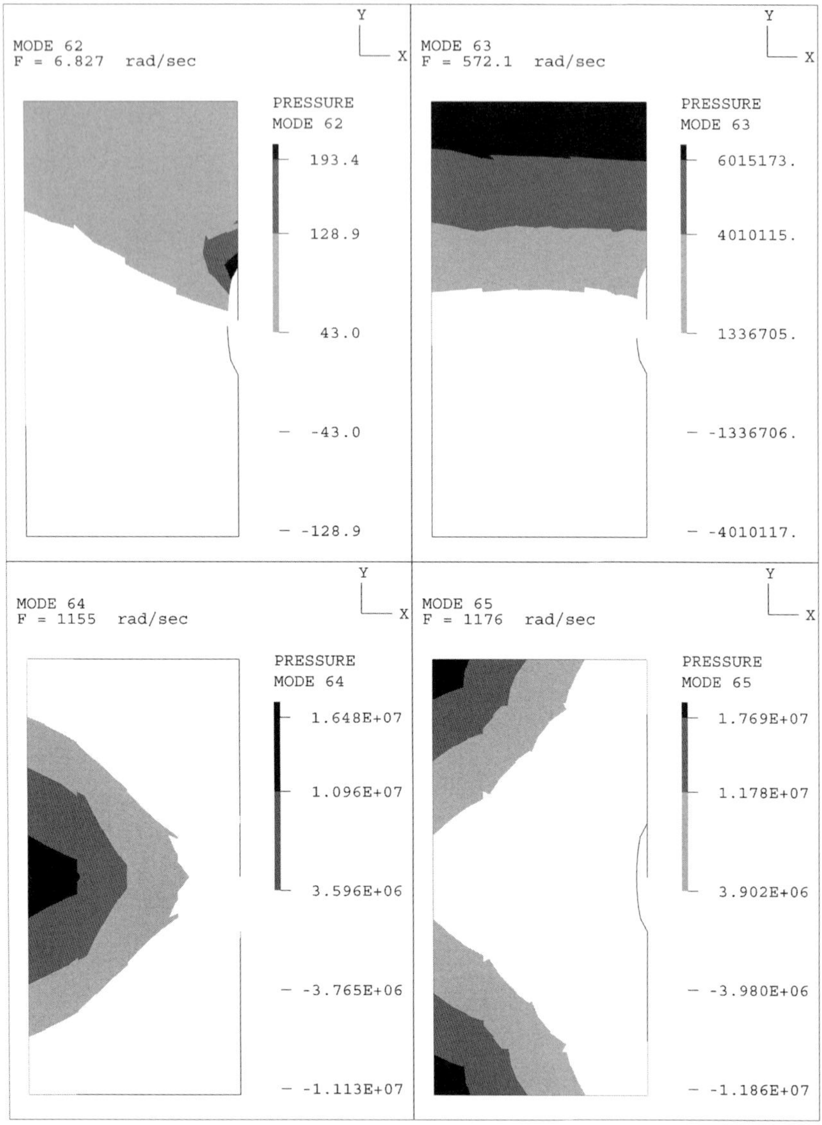


Figure 6.35. Pressure distributions of the first four modes for the rigid ellipse problem.

is calculated, whereas if the tangential directions are assigned using Eqs. (1.74) and (1.77), good solution results are obtained, as shown in Table 6.13. The comparison of the angles ($\gamma$) from the $x$-axis obtained using Eqs. (1.74) and (1.77) with the angles using the actual geometry is presented in Table 6.14.

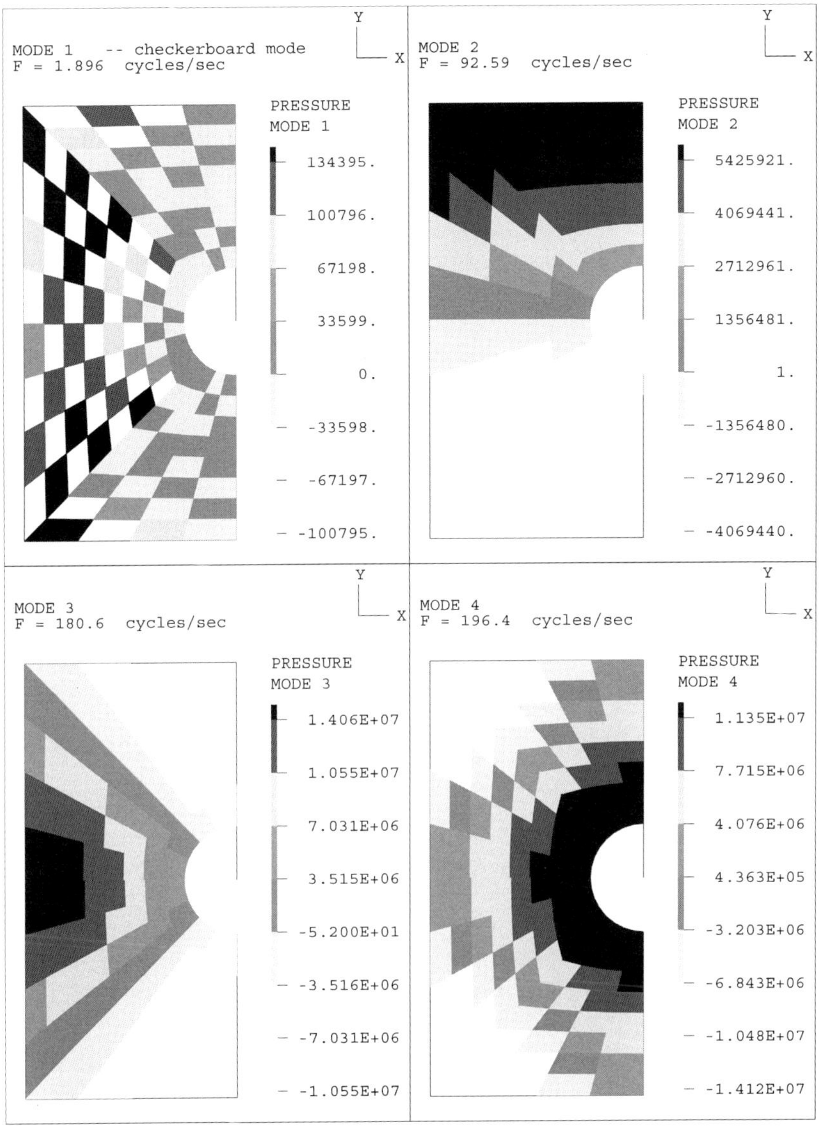


Figure 6.36. Response analysis of the rigid cylinder.

It is interesting to note that based on the mass transport requirement, angle $\gamma$ at node 7 is larger than at node 10, an unexpected result based on the actual geometry of the ellipse. We also show results obtained with the 4-1-1 element.

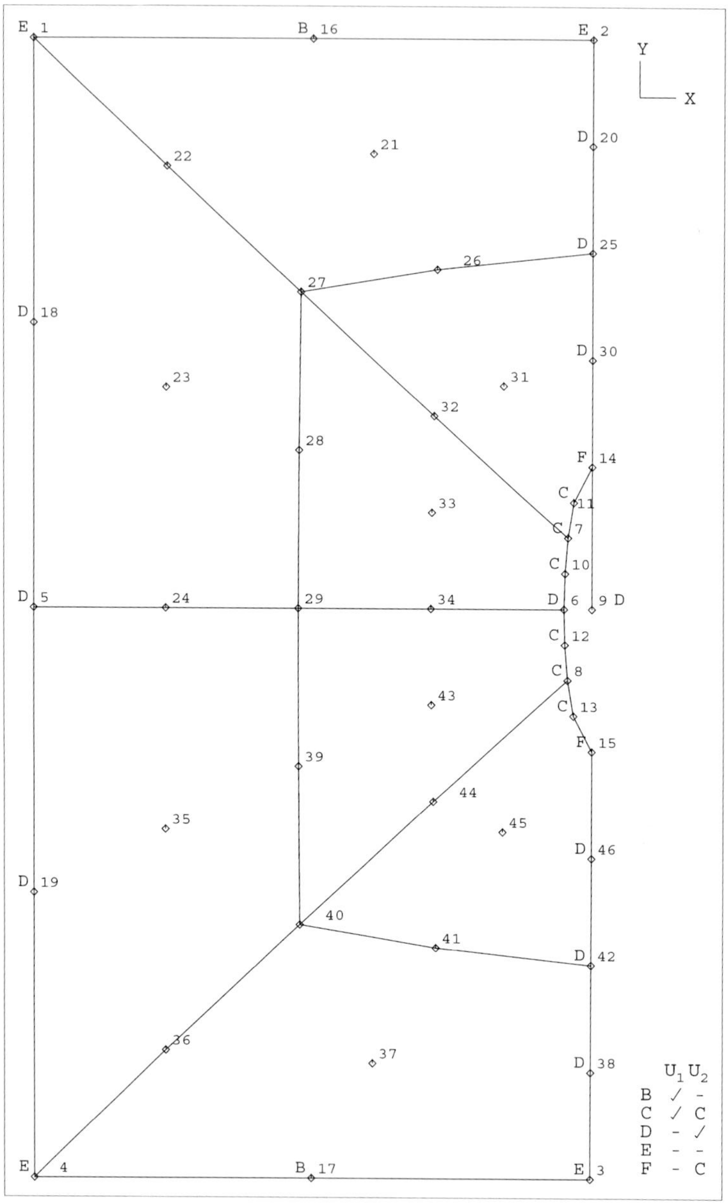


Figure 6.37. Response analysis of the rigid ellipse.

Table 6.13 Spurious modes

| Test case | Tangent direction | Number of zero frequencies | | Frequencies (rad/s) | | | |
|---|---|---|---|---|---|---|---|
| | | Theory | Result | First | Second | Third | Fourth |
| Rigid cylinder | incorrect | $\geqslant 5$ | 4 | 4.649 | 86.17* | 693.7 | 1154 |
| model in Figure 6.36 | correct | $\geqslant 5$ | 5 | 3.861 | 688.2 | 1169 | 1324 |
| Rigid ellipse | incorrect | $\geqslant 13$ | 12 | 6.719 | 34.65* | 591.4 | 1213 |
| model in Figure 6.37 | correct | $\geqslant 13$ | 13 | 6.612 | 589.1 | 1220 | 1232 |

* Due to incorrect tangential directions at the FSI interface.

Table 6.14 Comparison of model and geometry angles

| Node | Angle $\gamma\|_{\text{model}}$ | Angle $\gamma\|_{\text{geometry}}$ |
|---|---|---|
| 10 | 86.93° | 87.04° |
| 7 | 87.40° | 83.41° |
| 11 | 70.89° | 77.22° |

Table 6.15 Solution results using 4-1-1 elements

| Test case | Mesh, no. of elements | Frequencies (rad/s) | | | |
|---|---|---|---|---|---|
| | | First | Second | Third | Fourth |
| Rigid cylinder problem | 128 | 11.91* | 581.8 | 1135 | 1234 |
| Rigid ellipse problem | 128 | 557.5 | 988.2* | 1157 | 1180 |

* Frequencies are spurious.

Since the 4/1 displacement/pressure element, that is, the element with four corner nodes for the displacement interpolation and one constant pressure assumption for incompressible analysis does not satisfy the inf-sup condition, we expect that the 4-1-1 element will not provide a stable discretization. Table 6.15, Figures 6.38, and 6.39 show some solutions. These clearly indicate that the 4-1-1 element is not a reliable element, the predicted lowest frequencies are not accurate, and the checkerboard pressure bands obtained here are typical of those observed with the 4/1 element in almost incompressible solids.

However, some types of macroelements shown in Figure 6.2 with 4/1 elements have been proven mathematically to satisfy the inf-sup condition for **u**/p formulation. As a natural extension, we expected that the same macroelements with 4-1-1 elements would give good results. Tables 6.16 and 6.17 list the first four coupled frequencies. Surprisingly, we found that for both the rigid ellipse and rigid cylin-

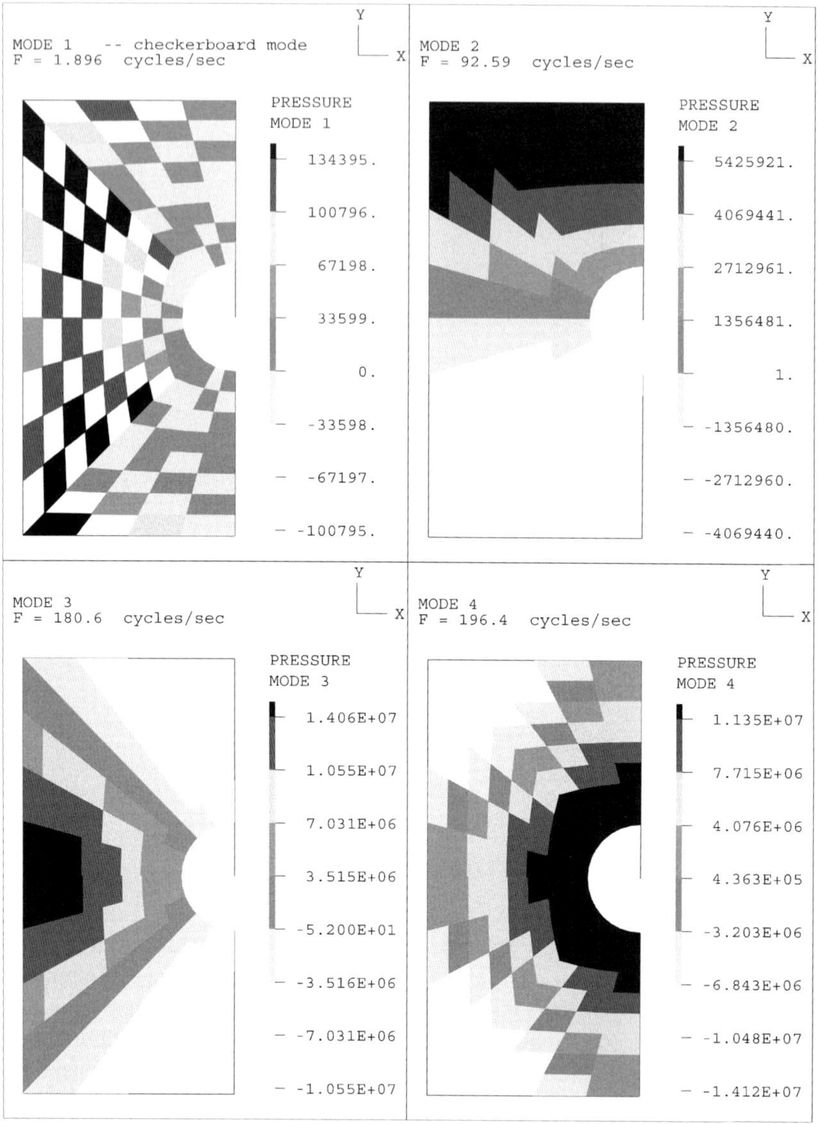


Figure 6.38. Checkerboard pressure of the rigid cylinder problem with 4-1-1 elements.

der problems, the structure modes predicted with type I and II macroelements as shown in Figure 6.2 in the **u**-p-$\boldsymbol{\Lambda}$ formulation are spurious. This indicates that the inf-sup condition for **u**-p-$\boldsymbol{\Lambda}$ formulation is not exactly the same as for **u**/p formulation, regardless of the similarities.

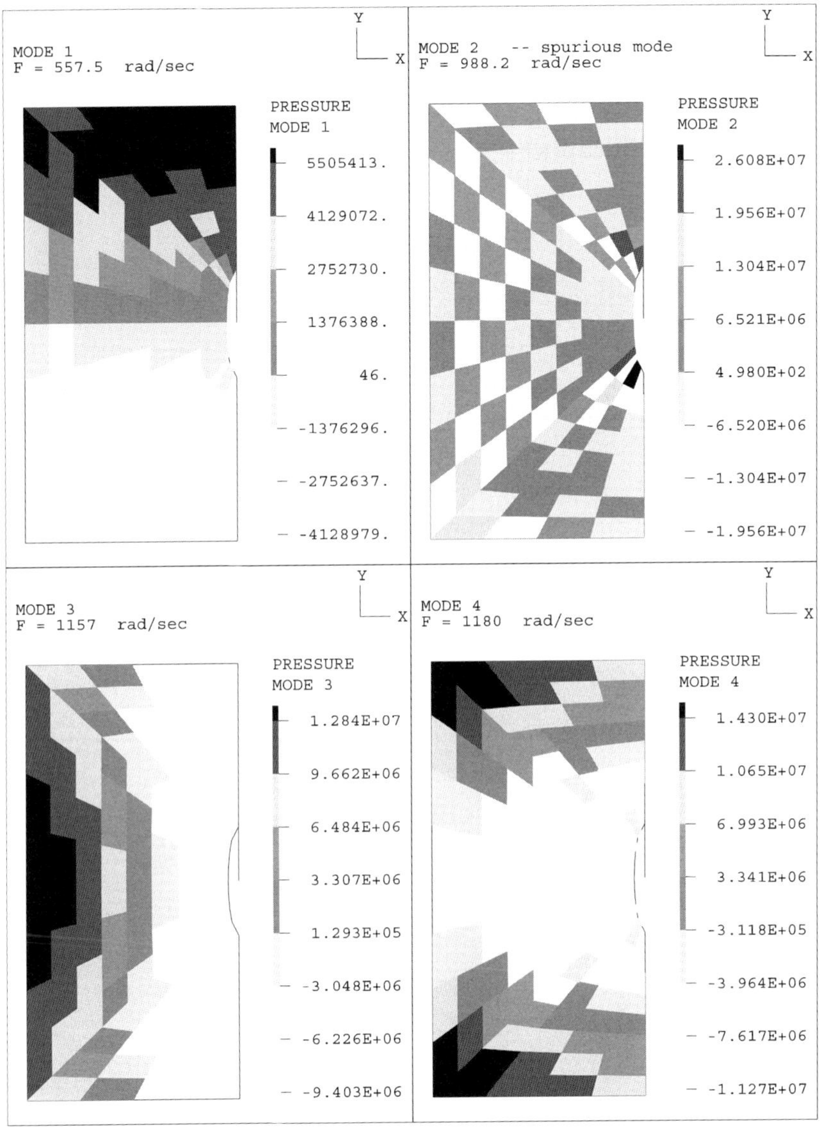


Figure 6.39. Checkerboard pressure of the rigid ellipse problem with 4-1-1 elements.

The explanation for this is that 4-1-1 elements give extra constraints indicated from the mathematical prediction of zero frequency modes, however, no mathematical proof is yet available. As expected, using the 4/1 macroelement in **u**/p

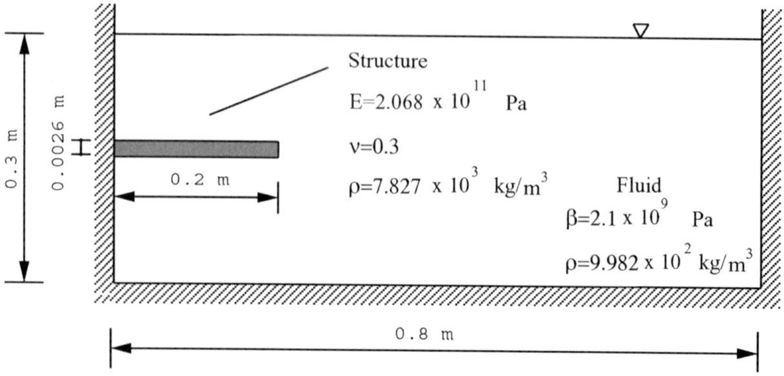


Figure 6.40. Acoustoelastic/slosh problem.

Table 6.16 Macroelements with 4-1-1 and 4/1 elements for the rigid ellipse problem

| Mode # | Type I | | Type II | |
|---|---|---|---|---|
| | (4/1) | (4-1-1) | (4/1) | (4-1-1) |
| 1 | 6.802 | 560.8 | 6.781 | 559.2 |
| 2 | 575.6 | 1035 | 579.0 | 1155 |
| 3 | 1167 | 1167 | 1165 | 1163 |
| 4 | 1210 | 1194 | 1195 | 1191 |

Table 6.17 Macroelements with 4-1-1 and 4/1 elements for the rigid cylinder problem

| Mode # | Type I | | Type II | |
|---|---|---|---|---|
| | (4/1) | (4-1-1) | (4/1) | (4-1-1) |
| 1 | 4.413 | 254.0 | 4.466 | 203.1 |
| 2 | 590.0 | 583.1 | 593.2 | 581.6 |
| 3 | 1157 | 1149 | 1156 | 1142 |
| 4 | 1280 | 1249 | 1262 | 1242 |

formulation, we can calculate the structure mode accurately. We come to the conclusion that the inf-sup condition can help to select good elements for both $\mathbf{u}/\mathrm{p}$ and $\mathbf{u}$-p-$\boldsymbol{\Lambda}$ formulations, but for $\mathbf{u}$-p-$\boldsymbol{\Lambda}$ formulation, 4-1-1 element and its macroelements are not recommended.

Figure 6.40 shows a system involving a submerged structure and the free surface. The submerged 2D plane strain structure has the dimension $0.2 \times 0.0026$ m.

Table 6.18 Acoustoelastic/slosh problem using the $\mathbf{u}/p$ formulation with 9/4-$c$ elements

| Mode | No. of elements | | Frequencies (rad/s) | |
|---|---|---|---|---|
| | Structure | Fluid | First | Second |
| Slosh | 1 | 11 | 5.4717 | 8.8003 |
| | 2 | 38 | 5.3901 | 8.4892 |
| | 4 | 140 | 5.3696 | 8.4533 |
| Structural | 1 | 11 | 190.85 | 6964.3 |
| | 2 | 38 | 150.27 | 1808.6 |
| | 4 | 140 | 129.07 | 1105.5 |
| Acoustic | 1 | 11 | 8867.4 | 12009 |
| | 2 | 38 | 7386.5 | 9214.8 |
| | 4 | 140 | 7509.3 | 9366.3 |

Table 6.19 Acoustoelastic/slosh problem using the $\mathbf{u}/p$ formulation with 9/3 elements

| Mode | No. of elements | | Frequencies (rad/s) | |
|---|---|---|---|---|
| | Structure | Fluid | First | Second |
| Slosh | 1 | 11 | 5.4717 | 8.8003 |
| | 2 | 38 | 5.3360 | 8.6016 |
| | 4 | 140 | 5.3435 | 8.4673 |
| Structural | 1 | 11 | 190.85 | 6964.3 |
| | 2 | 38 | 111.07 | 1800.7 |
| | 4 | 140 | 113.93 | 1013.3 |
| Acoustic | 1 | 11 | 8867.4 | 12009 |
| | 2 | 38 | 7381.3 | 9257.5 |
| | 4 | 140 | 7513.2 | 9382.8 |

Table 6.20 First four dry modes of the submerged structure

| No. of elements | Frequencies (rad/s) | | | |
|---|---|---|---|---|
| | First | Second | Third | Fourth |
| 1 | 457.432 | 36851.0 | 43851.5 | 155273 |
| 2 | 378.365 | 3802.02 | 42256.1 | 42882.7 |
| 4 | 363.097 | 2446.82 | 7852.46 | 20493.2 |

Table 6.20 lists the first four dry modes for the plane strain 2D structure with 9-node elements. Some preliminary analysis on similar systems are available in Refs. [1,185,198,202,248]. In certain acoustoelastic FSI problems, gravity effects are ignored. However, we will take the gravity effects into account in this exam-

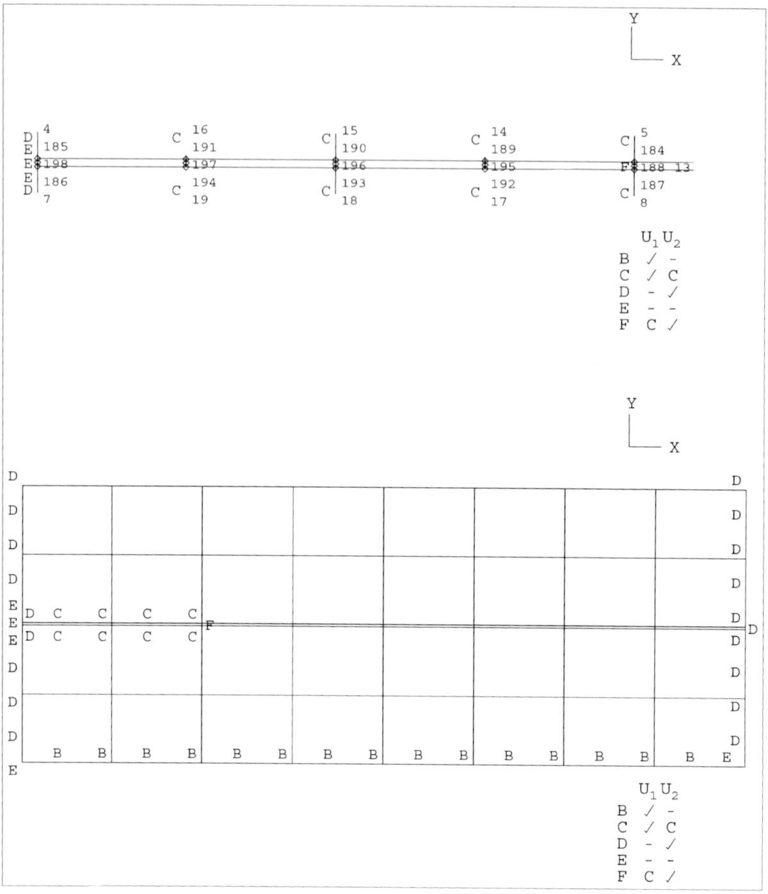


Figure 6.41. Typical mesh of the acoustoelastic/slosh problem.

ple and demonstrate that the **u**/p formulation can in one finite element analysis obtain slosh, structure and acoustic frequencies accurately. Results from convergence studies are listed in Tables 6.19 and 6.18. Figure 6.41 shows the typical mesh of this problem. Using the **u**/p formulation, the computational results of the number of zero frequency modes match the mathematical prediction. Figures 6.42 and 6.43 show the mode shapes and pressure bands of the first two slosh, structure and acoustic frequencies calculated with forty 9-node elements. Our numerical experiments have shown that it is important to allocate the appropriate tangential directions at all boundary nodes [151]. Otherwise, spurious nonzero energy modes are obtained in the finite element solution.

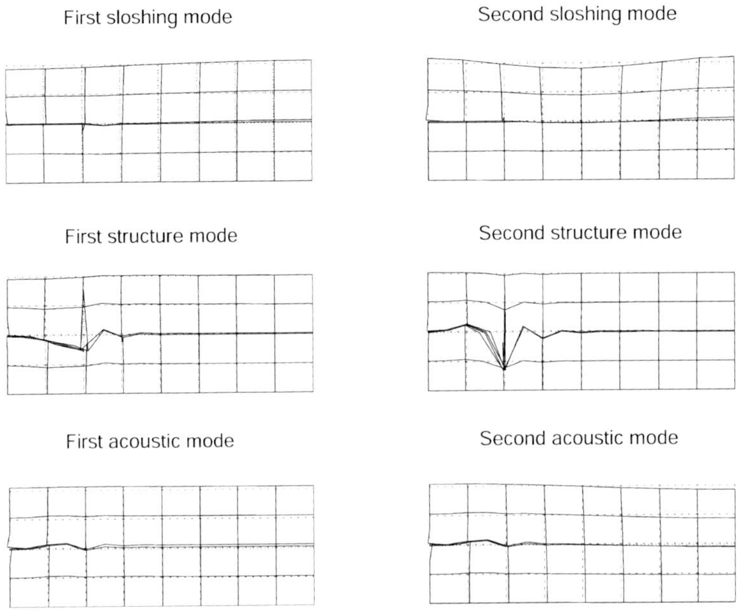


Figure 6.42. Mode shapes for the acoustoelastic/slosh problem.

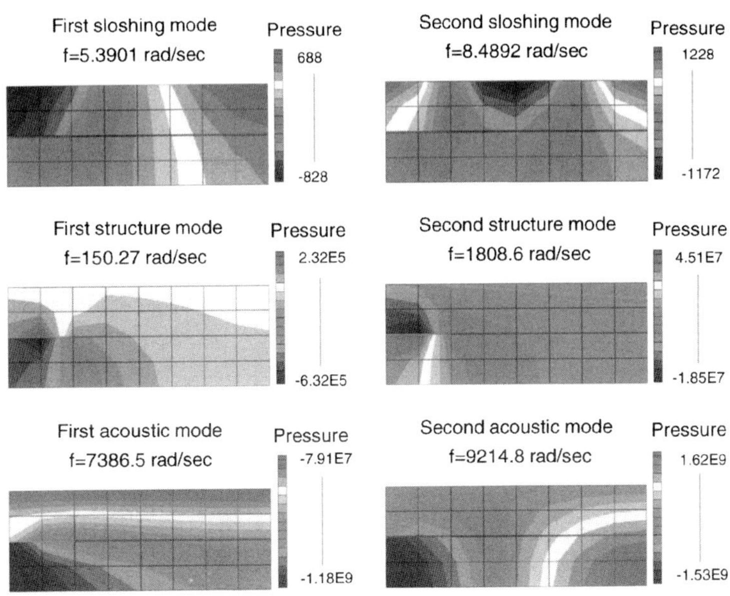


Figure 6.43. Pressure distributions for the acoustoelastic/slosh problem.

### 6.2.3. Inf-sup conditions of mixed formulations

To study the convergence of various numerical procedure, we first introduce a few concepts of Sobolev spaces. Consider a domain $\Omega \subset \mathcal{R}^n$ ($n = 2$ and 3 stand for two- and three-dimensional problems, respectively), with a sufficiently smooth boundary $\Gamma = \partial\Omega$, e.g., a Lipschitz continuous boundary. The space of square integrable scalar functions over $\Omega$ is given as

$$\mathcal{L}^2(\Omega) = \left\{ f \mid f \in \mathcal{R} \text{ and } \|f\|_o^2 = \int\limits_{\Omega} |f|^2 \, d\Omega < \infty \right\}.$$

Consequently, the space of square integrable vector functions over $\Omega$ is given as

$$\left(\mathcal{L}^2(\Omega)\right)^n = \left\{ \mathbf{v} \mid \mathbf{v} \in \mathcal{R}^n \text{ and } \|\mathbf{v}\|_o^2 = \int\limits_{\Omega} |\mathbf{v}|^2 \, d\Omega < \infty \right\}.$$

Thus, in general, we have the Sobolev space:

$$\mathcal{H}^m(\Omega) = \left\{ f \mid f \in \mathcal{R} \text{ and } D^k f \in \mathcal{L}^2(\Omega), \ \forall |k| \leqslant m \right\},$$
$$\left(\mathcal{H}^m(\Omega)\right)^n = \left\{ \mathbf{v} \mid \mathbf{v} \in \mathcal{R}^n \text{ and } D^k \mathbf{v} \in \left(\mathcal{L}^2(\Omega)\right)^n, \ \forall |k| \leqslant m \right\},$$

with $m \geqslant 0$ and $D^k = \partial^{|k|}/\partial^{k_1} x_1 \ldots \partial^{k_n} x_n$, $|k| = k_1 + \cdots + k_n$.

Naturally, we have $\mathcal{H}^0(\Omega) = \mathcal{L}^2(\Omega)$ and $(\mathcal{H}^0(\Omega))^n = (\mathcal{L}^2(\Omega))^n$. In some cases, the 0-norm $\|f\|_o$ or $\|\mathbf{v}\|_o$ is also written as $\|\mathbf{v}\|_{\mathcal{L}^2(\Omega)}$. Furthermore, the semi-norm and Sobolev norm of order $m$ are defined as

$$|f|_m^2 = |f|_{m,\Omega}^2 = \sum_{|k|=m} \left\| D^k f \right\|_o^2,$$

$$|\mathbf{v}|_m^2 = |\mathbf{v}|_{m,\Omega}^2 = \sum_{|k|=m} \left\| D^k \mathbf{v} \right\|_o^2,$$

$$\|\mathbf{v}\|_m^2 = \|\mathbf{v}\|_{m,\Omega}^2 = \sum_{|k| \leqslant m} |\mathbf{v}|_{k,\Omega}^2.$$

The other commonly used Sobolev spaces include

$$\mathcal{H}_o^1(\Omega) = \left\{ f \mid f \in \mathcal{H}^1(\Omega), \ f|_\Gamma = 0 \right\},$$
$$\left(\mathcal{H}_o^1(\Omega)\right)^n = \left\{ \mathbf{v} \mid \mathbf{v} \in \left(\mathcal{H}^1(\Omega)\right)^n, \ \mathbf{v}|_\Gamma = \mathbf{0} \right\},$$
$$\mathcal{H}_o^2(\Omega) = \left\{ f \mid f \in \mathcal{H}^2(\Omega), \ f|_\Gamma = 0, \ \left.\frac{\partial f}{\partial n}\right|_\Gamma = 0 \right\},$$
$$\left(\mathcal{H}_o^2(\Omega)\right)^n = \left\{ \mathbf{v} \mid \mathbf{v} \in \left(\mathcal{H}^2(\mathbf{v})\right)^n, \ \mathbf{v}|_\Gamma = \mathbf{0}, \ \left.\frac{\partial \mathbf{v}}{\partial n}\right|_\Gamma = \mathbf{0} \right\}.$$

In particular, consider a continuum media $\Omega$, with the body force $\mathbf{f}^b$. The displacement boundary condition (often called the Dirichlet boundary condition or the essential boundary condition) corresponds to $\mathbf{u}^{f,p} = \mathbf{u}_p$ on $\Gamma^{f,v}$ and the force boundary condition (often called the Neumann boundary condition or the natural boundary condition) corresponds to $\boldsymbol{\sigma}\mathbf{n} = \mathbf{f}^{f,p}$ on $\Gamma^{f,f}$, where the stress tensor is expressed in the tensor form and $\mathbf{n}$ represents the surface normal vector. The mathematical representation of problems in elasticity can be expressed as the following minimization problem:

$$\mathbf{u} \in \mathcal{U} \quad \text{and} \quad J(\mathbf{u}) = \inf_{\mathbf{v}\in\mathcal{U}} J(\mathbf{v}), \tag{2.173}$$

where the set $\mathcal{U}$ of admissible displacements is a closed convex subset of a Hilbert space $\mathcal{V}$ and $J$ stands for the energy of the system.

Comparing with Chapter 2, we can clearly identify that admissible displacements correspond to the displacements satisfying the displacement boundary conditions, and $J$ can be expressed as

$$J(\mathbf{v}) = \frac{1}{2}a(\mathbf{v}, \mathbf{v}) - f(\mathbf{v}), \tag{2.174}$$

where the symmetric bilinear form $a(\mathbf{v}, \mathbf{v})$ is $\int_\Omega \frac{1}{2}\boldsymbol{\epsilon}^T \mathbf{C}\boldsymbol{\epsilon}\, d\Omega$ and the linear form $f(\mathbf{v})$ is $\int_\Omega \mathbf{u}^T \mathbf{f}^{f,p}\, d\Gamma$.

THEOREM 6.2.1 *(Existence and uniqueness). If the space $\mathcal{V}$ is complete, $\mathcal{U}$ is a closed convex subset of $\mathcal{V}$, and the bilinear form $a(\mathbf{v}, \mathbf{v})$ is symmetric and $\mathcal{V}$-elliptic, i.e.,*

$$\exists\alpha > 0, \text{ such that } \forall\mathbf{v} \in \mathcal{V}, \quad a(\mathbf{v}, \mathbf{v}) \geqslant \alpha\|\mathbf{v}\|^2, \tag{2.175}$$

*then the minimization problem* (2.173) *has one and only one solution.*

Condition (2.175) is often called the ellipticity condition. Similarly, another important property of the bilinear form $a(.,.)$ is introduced as the continuity condition, i.e.,

$$\exists M > 0, \text{ such that } \forall\mathbf{v}_1, \mathbf{v}_2 \in \mathcal{V}, \quad a(\mathbf{v}_1, \mathbf{v}_2) \leqslant M\|\mathbf{v}_1\|\|\mathbf{v}_2\|. \tag{2.176}$$

In general, the norm $\|\mathbf{v}\|$ represents the 1-norm $\|\mathbf{v}\|_1$ and the constants $\alpha$ and $M$ depend on the material properties as well as geometry, but are independent of $\mathbf{v}$. Without loss of generality, to study the convergence of various formulations we assume the prescribed displacements $\mathbf{u}^{f,p} = \mathbf{0}$ and prescribed traction $\mathbf{f}^{f,p} = \mathbf{0}$. Thus, we can also introduce

$$\left(\mathcal{H}^1_{o,\Gamma^{f,v}}(\Omega)\right)^n = \left\{\mathbf{v} \mid \mathbf{v} \in \left(\mathcal{H}^1(\Omega)\right)^n,\ \mathbf{v}|_{\Gamma^{f,v}} = \mathbf{0}\right\}.$$

THEOREM 6.2.2 *(Lax–Milgram). Let $\mathcal{V}$ be a Hilbert space, let $a(.,.) : \mathcal{V} \times \mathcal{V} \rightarrow \mathcal{R}$ be a continuous $\mathcal{V}$-elliptic bilinear form, and let $f : \mathcal{V} \rightarrow \mathcal{R}$ be a continuous linear form. Then the abstract variational problem: Find an element $\mathbf{u} \in \mathcal{V}$ such that*

$$\forall \mathbf{v} \in \mathcal{V}, \quad a(\mathbf{u}, \mathbf{v}) = f(\mathbf{v}), \tag{2.177}$$

*has one and only one solution.*

Without restricting the essence of our exposition, in this section we only consider two-dimensional cases. The finite element formulations for three-dimensional cases can be directly constructed. Consider a bounded continuous domain $\Omega \subset \mathcal{R}^2$ with a sufficiently smooth boundary $\partial\Omega = \Gamma$, e.g., a Lipschitz continuous boundary. The components of the strain tensor $\boldsymbol{\epsilon}$ and the deviatoric strain tensor $\boldsymbol{\epsilon}'$ are defined as $\epsilon_{ij} = \frac{1}{2}(u_{i,j} + u_{j,i})$ and $\epsilon'_{ij} = \epsilon_{ij} - \frac{1}{3}\epsilon_{kk}\delta_{ij}$, where $\mathbf{u}$ stands for the displacement vector. Define the Sobolev space $(\mathcal{H}^1_{0,\Gamma^{f,v}}(\Omega))^2 = \{\mathbf{v} \mid \mathbf{v} \in (\mathcal{H}^1(\Omega))^2, \mathbf{v}|_{\Gamma^{f,v}} = \mathbf{0}\}$, where $\Gamma^{f,v}$ and $\Gamma^{f,f}$ stand for the Dirichlet and Neumann boundaries with $\Gamma = \Gamma^{f,f} \cup \Gamma^{f,v}$ and $\Gamma^{f,f} \cap \Gamma^{f,v} = \emptyset$. The variational discrete problem of the $\mathbf{u}/p$ formulation for nearly incompressible media with the bulk modulus $\kappa$ and the shear modulus $G$ can be expressed as

$$\min_{\mathbf{v}^h \in \mathcal{V}^h} \left\{ a\left(\mathbf{v}^h, \mathbf{v}^h\right) + \frac{\kappa}{2} \int_{\Omega} \left[P^h(\operatorname{div} \mathbf{v}^h)\right]^2 d\Omega - \int_{\Omega} \mathbf{f}^b \cdot \mathbf{v}^h \, d\Omega - \int_{\Gamma^{f,f}} \mathbf{f}^{f,p} \cdot \mathbf{v}^h \, d\Gamma \right\}, \tag{2.178}$$

where $a(\mathbf{v}^h, \mathbf{v}^h) = G \int_{\Omega} \boldsymbol{\epsilon}'(\mathbf{v}^h) : \boldsymbol{\epsilon}'(\mathbf{v}^h) \, d\Omega$, and the projection operator $P^h$ is defined by

$$\int_{\Omega} \left[P^h\left(\operatorname{div} \mathbf{v}^h\right) - \operatorname{div} \mathbf{v}^h\right] q^h \, d\Omega = 0, \quad \forall q^h \in \mathcal{O}^h, \tag{2.179}$$

with $\mathcal{V}^h \subset (\mathcal{H}^1_{0,\Gamma^{f,v}}(\Omega))^2$ and $\mathcal{O}^h \subset \mathcal{L}^2(\Omega)$.

The variational forms (2.178) and (2.179) can also be rewritten as

$$2G \int_{\Omega} \boldsymbol{\epsilon}'\left(\mathbf{u}^h\right) : \boldsymbol{\epsilon}'\left(\mathbf{v}^h\right) d\Omega - \int_{\Omega} p^h \operatorname{div} \mathbf{v}^h \, d\Omega = \int_{\Omega} \mathbf{f}^b \cdot \mathbf{v}^h \, d\Omega + \int_{\Gamma^{f,f}} \mathbf{f}^{f,p} \cdot \mathbf{v}^h \, d\Gamma, \quad \forall \mathbf{v}^h \in \mathcal{V}^h, \tag{2.180}$$

$$\int_{\Omega} \left( \frac{p^h}{\kappa} + \operatorname{div} \mathbf{u}^h \right) q^h \, d\Omega = 0, \quad \forall q^h \in \mathcal{O}^h. \tag{2.181}$$

Using the standard finite element interpolation procedure with $\mathbf{u}^h = \mathbf{HU}$, $p^h = \mathbf{H}_p\mathbf{P}$, $\epsilon'_{ij}(\mathbf{u}^h) = \mathbf{BU}$, and $\nabla \cdot \mathbf{u}^h = \bar{\mathbf{B}}\mathbf{U}$, where $\mathbf{H}$ and $\mathbf{H}_p$ are the interpolation matrices, and $\mathbf{U}$ and $\mathbf{P}$ are the solution vectors, respectively, Eqs. (2.180) and (2.181) can be written in the algebraic form as follows:

$$\begin{bmatrix} \mathbf{K}_{uu} & \mathbf{K}_{up} \\ \mathbf{K}_{pu} & \mathbf{K}_{pp} \end{bmatrix} \begin{bmatrix} \mathbf{U} \\ \mathbf{P} \end{bmatrix} = \begin{bmatrix} \mathbf{R} \\ \mathbf{0} \end{bmatrix}, \tag{2.182}$$

where

$$\mathbf{K}_{uu} = \int_{\Omega} 2G\mathbf{B}^T\mathbf{B}\, d\Omega, \qquad \mathbf{K}_{up} = -\int_{\Omega} \bar{\mathbf{B}}^T \mathbf{H}_p \, d\Omega;$$

$$\mathbf{K}_{pp} = -\int_{\Omega} \frac{1}{\kappa} \mathbf{H}_p^T \mathbf{H}_p \, d\Omega, \qquad \mathbf{R} = \int_{\Omega} \mathbf{H}^T \mathbf{f}^b \, d\Omega + \int_{\Gamma^{f,f}} \mathbf{H}^{f,f^T} \mathbf{f}^{f,p} \, d\Gamma;$$

with $\mathbf{K}_{pu} = \mathbf{K}_{up}^T$, and $\mathbf{H}^{f,f}$ obtained from $\mathbf{H}$.

According to Refs. [27,143], two solvability conditions must be satisfied, i.e., (a) $(\mathbf{v}^h)^T \mathbf{K}_{uu} \mathbf{v}^h > 0$, $\forall \mathbf{v}^h \in \ker(\mathbf{K}_{pu})$ and (b) $\ker(\mathbf{K}_{up}) = \{\mathbf{0}\}$. Obviously, a difficulty arises from the essential assumption for acoustic fluids that $G = 0$. However, as discussed in Ref. [301], when we consider the inertia force $-\rho\ddot{\mathbf{u}}$ in $\mathbf{f}^b$, the corresponding equation of motion can be expressed as

$$\begin{bmatrix} \mathbf{M}_{uu} & \mathbf{0} \\ \mathbf{0} & \mathbf{0} \end{bmatrix} \begin{bmatrix} \ddot{\mathbf{U}} \\ \ddot{\mathbf{P}} \end{bmatrix} + \begin{bmatrix} \mathbf{0} & \mathbf{K}_{up} \\ \mathbf{K}_{pu} & \mathbf{K}_{pp} \end{bmatrix} \begin{bmatrix} \mathbf{U} \\ \mathbf{P} \end{bmatrix} = \begin{bmatrix} \mathbf{R} \\ \mathbf{0} \end{bmatrix}, \tag{2.183}$$

where $\rho$ is the mass density, $\mathbf{n}$ is the unit normal vector pointing outwards, $\mathbf{M}_{uu} = \int_{\Omega} \rho \mathbf{H}^T \mathbf{H}\, d\Omega$, and $\mathbf{R} = -\int_{\Gamma^{f,f}} \mathbf{H}^{f,f^T}_n \bar{p}\, d\Gamma$, with $\bar{p} = -\mathbf{f}^{f,p} \cdot \mathbf{n}$.

As discussed in Ref. [300], assuming that the physical constant pressure mode arising from the boundary condition $\mathbf{u} \cdot \mathbf{n} = 0$ on $\Gamma$ has been eliminated and there is no spurious zero frequency [300], the solvability conditions are satisfied in a transient direct step-by-step solution, where at each time step, we have

$$\begin{bmatrix} \mathbf{K}_{uu}^* & \mathbf{K}_{up} \\ \mathbf{K}_{pu} & \mathbf{K}_{pp} \end{bmatrix} \begin{bmatrix} \mathbf{U} \\ \mathbf{P} \end{bmatrix} = \begin{bmatrix} \widehat{\mathbf{R}} \\ \mathbf{0} \end{bmatrix}, \tag{2.184}$$

with $\widehat{\mathbf{R}}$ the effective load vector and $\mathbf{K}_{uu}^* = \widehat{C}\mathbf{M}_{uu}$, with $\widehat{C}$ a positive constant associated with the direct time integration scheme, e.g., $\widehat{C} = 4/\Delta t^2$ for the trapezoidal rule.

In addition, for the frequency analysis, we have the eigenvalue problem $\mathbf{K}\boldsymbol{\phi} = \omega^2\mathbf{M}\boldsymbol{\phi}$, with $\mathbf{K} = -\mathbf{K}_{up}\mathbf{K}_{pp}^{-1}\mathbf{K}_{pu}$ and $\mathbf{M} = \mathbf{M}_{uu}$. Obviously, $\mathbf{M}$ is positive definite and the eigenvalue problem is well-posed. In fact, for $n$ displacement unknowns and $m$ pressure degrees of freedom, the number of zero frequencies is $n - m$.

Considering next the stability of a discretization scheme [27,143], the following two conditions must be satisfied:

$$\begin{aligned} &\text{Ellipticity:} \quad a(\mathbf{v}^h, \mathbf{v}^h) \geqslant C \|\mathbf{v}^h\|_1^2, \quad \forall \mathbf{v}^h \in \ker(\mathbf{K}_{pu}), \\ &\text{Inf-sup:} \quad \inf_{q^h \in \mathcal{O}^h} \sup_{\mathbf{v}^h \in \mathcal{V}^h} \frac{\int_\Omega q^h \nabla \cdot \mathbf{v}^h \, d\Omega}{\|\mathbf{v}^h\|_1 \|q^h\|_0} \geqslant \beta_o > 0, \end{aligned} \tag{2.185}$$

where $C$ is a positive constant independent of the mesh size $h$, and $\beta_o$ is a positive constant independent of both $h$ and $\kappa$.

Note that when $G \neq 0$, it is obvious that the ellipticity condition is satisfied [214]. Therefore, considering an acoustic fluid, the ellipticity condition can always be satisfied by some modifications to the variational formulation as discussed in Refs. [28,84,143]. Of course, in practice, a very small shear modulus compared with $\kappa$ to represent the acoustic fluid can be used.

If these modifications are not employed, the loss of ellipticity introduces zero frequency modes which correspond to zero deviatoric strain energy and can be effectively removed from the eigenvalue solutions in engineering computations. Therefore, the key stability requirement is the inf-sup condition for the selection of displacement and pressure interpolations, which governs the convergence of true physical nonzero frequency modes as confirmed in Refs. [300,301].

Furthermore, to reduce the number of zero eigenvalues, according to Refs. [84, 151,300], for acoustic fluids, we can use, the so-called displacement-pressure-vorticity moment ($\mathbf{u}$-$p$-$\boldsymbol{\Lambda}$) formulation. Assigning $\operatorname{rot}\lambda = \langle \partial\lambda/\partial x_2, -\partial\lambda/\partial x_1 \rangle$ and $\operatorname{rot}\mathbf{v} = \partial v_2/\partial x_1 - \partial v_1/\partial x_2$ in association with $\operatorname{grad} p = \langle \partial p/\partial x_1, \partial p/\partial x_2 \rangle$ and $\operatorname{div}\mathbf{v} = \partial v_1/\partial x_1 + \partial v_2/\partial x_2$, respectively, we can replace Eqs. (2.180) and (2.181) with

$$\begin{aligned} &\int_\Omega \lambda^h \operatorname{rot}\mathbf{v}^h \, d\Omega - \int_\Omega p^h \operatorname{div}\mathbf{v}^h \, d\Omega \\ &\quad = \int_\Omega \mathbf{f}^b \cdot \mathbf{v}^h \, d\Omega - \int_{\Gamma^{f,f}} \bar{p}^h (\mathbf{v}^h \cdot \mathbf{n}) \, d\Gamma, \quad \forall \mathbf{v}^h \in \mathcal{V}^h, \end{aligned} \tag{2.186}$$

$$\int_\Omega \left( \frac{p^h}{\kappa} + \operatorname{div}\mathbf{u}^h \right) q^h \, d\Omega = 0, \quad \forall q^h \in \mathcal{O}^h, \tag{2.187}$$

$$\int_\Omega \left( \frac{\lambda^h}{\alpha} - \operatorname{rot}\mathbf{u}^h \right) \mu^h \, d\Omega = 0, \quad \forall \mu^h \in \mathcal{P}^h, \tag{2.188}$$

where $\alpha$, the constant associated with the irrotational constraint, is a very large number, and $\mathcal{P}^h = \mathcal{O}^h$.

For **u**-$p$-$\boldsymbol{\Lambda}$ formulation, let $\lambda^h = \mathbf{H}_\lambda \boldsymbol{\Lambda}$ and $\operatorname{rot}\mathbf{u}^h = \widehat{\mathbf{B}}\mathbf{U}$, where $\mathbf{H}_\lambda$ is the interpolation matrix for the variable $\lambda^h$, and $\boldsymbol{\Lambda}$ is the solution vector for $\lambda^h$, we have the additional matrices

$$\mathbf{K}_{u\lambda} = \int_\Omega \widehat{\mathbf{B}}^T \mathbf{H}_\lambda \, d\Omega \quad \text{and} \quad \mathbf{K}_{\lambda\lambda} = -\int_\Omega \frac{1}{\alpha}\mathbf{H}_\lambda^T \mathbf{H}_\lambda \, d\Omega. \tag{2.189}$$

Of course, the key benefit of replacing $\mathbf{u}/p$ formulation with **u**-$p$-$\boldsymbol{\Lambda}$ formulation is the reduction of the number of zero frequencies to $n-m-k$, where $k$ is the number of vorticity moment degrees of freedom [151]. In addition, in the study of stability and optimal error bounds, it is necessary to use a modified **u**-$p$-$\boldsymbol{\Lambda}$ formulation in order to fulfill the ellipticity condition by considering both $\operatorname{div}\mathbf{v}^h$ and $\operatorname{rot}\mathbf{v}^h$, as discussed in Ref. [84].

The objective of this discussion is to present a comprehensive study of the solvability, stability, and optimal error bounds of mixed finite element formulations for acoustic fluids, including the interrelationship between the numerical inf-sup test and the eigenvalue problem pertaining to the natural frequencies of a coupled acoustoelastic system. In addition, we extend the technique proposed by Stenberg in Refs. [256–258] for the analytical proof of the inf-sup condition for Stokes flow to the study of the inf-sup condition for acoustic fluids and provide an analytical proof for the stability and optimal error bounds of a set of three-field mixed finite element discretizations.

Following the examples in Refs. [84,151,170], we consider the following eigenvalue problem: Find $\omega$, **u**, $p$, and $\lambda$ such that

$$\rho\omega^2\mathbf{u} - \operatorname{grad} p - \operatorname{rot}\lambda = 0 \quad \text{in } \Omega, \tag{2.190}$$

$$\frac{p}{\kappa} + \operatorname{div}\mathbf{u} = 0 \quad \text{in } d\Omega, \tag{2.191}$$

$$\frac{\lambda}{\alpha} - \operatorname{rot}\mathbf{u} = 0 \quad \text{in } \Omega, \tag{2.192}$$

$$\mathbf{u}\cdot\mathbf{n} = 0 \quad \text{on } d\Gamma^{f,v}, \tag{2.193}$$

$$\left(k - m\omega^2\right)\mathbf{u}\cdot\mathbf{n} = \int_{\Gamma^{f,f}} p\, d\Gamma \quad \text{on } d\Gamma^{f,f}, \tag{2.194}$$

$$\lambda = 0 \quad \text{on } d\Gamma, \tag{2.195}$$

where $k$ and $m$ are the mass and stiffness of the piston connected with the acoustic fluid at the boundary $\Gamma^{f,f}$ representing the elastic structure in the acoustoelastic problem, and the boundary condition (2.194) implies that $\mathbf{u}\cdot\mathbf{n}$ is constant along $\Gamma^{f,f}$.

According to Ref. [84], let

$$\mathcal{V} = \left\{\mathbf{v} \in \left(\mathcal{H}^1(\Omega)\right)^2 \mid \mathbf{v}\cdot\mathbf{n} = 0 \text{ on } \Gamma^{f,v},\ \mathbf{v}\cdot\mathbf{n} \text{ is constant on } \Gamma^{f,f}\right\},$$

$$\mathcal{V}_o = \{\mathbf{v} \in (\mathcal{H}^1(\Omega))^2 \mid \mathbf{v} \cdot \mathbf{n} = 0 \text{ on } \Gamma\},$$
$$\mathcal{O} = \mathcal{P} = \mathcal{L}^2(\Omega),$$
$$\mathcal{O}_o = \left\{q \in \mathcal{O} \,\middle|\, \int_\Omega q\, d\Omega = 0\right\}.$$

The corresponding variational form of (2.189)–(2.194) can be expressed as: Find $\omega \in \mathcal{R}$, $\mathbf{u} \in \mathcal{V}$, $p \in \mathcal{O}$, and $\lambda \in \mathcal{P}$ such that

$$\begin{aligned}
&a(\mathbf{u}, \mathbf{v}) + \left(\frac{\gamma_1}{\kappa} - 1\right) b(\mathbf{v}, p) - \left(\frac{\gamma_2}{\alpha} - 1\right) c(\mathbf{v}, \lambda) \\
&\quad = \rho\omega^2(\mathbf{u}, \mathbf{v}) + m\omega^2 n(\mathbf{u}, \mathbf{v}), \quad \forall \mathbf{v} \in \mathcal{V}, \\
&\frac{(p, q)}{\kappa} + b(\mathbf{u}, q) = 0, \quad \forall q \in \mathcal{O}, \\
&\frac{(\lambda, \mu)}{\alpha} - c(\mathbf{u}, \mu) = 0, \quad \forall \mu \in \mathcal{P},
\end{aligned} \tag{2.196}$$

where $\gamma_1$ and $\gamma_2$ are positive constants such that $0 \leqslant \gamma_1 \leqslant \kappa$ and $0 \leqslant \gamma_2 \leqslant \alpha$, $(\cdot,\cdot)$ is the usual inner product on $\mathcal{O} \times \mathcal{O}$, $\mathcal{P} \times \mathcal{P}$, or $\mathcal{V} \times \mathcal{V}$, and

$$n(\mathbf{u}, \mathbf{v}) = (\mathbf{u} \cdot \mathbf{n})|_{\Gamma f.f} (\mathbf{v} \cdot \mathbf{n})|_{\Gamma f.f}, \quad \forall \mathbf{u}, \mathbf{v} \in \mathcal{V}, \tag{2.197}$$

$$a(\mathbf{u}, \mathbf{v}) = \gamma_1 (\operatorname{div} \mathbf{u}, \operatorname{div} \mathbf{v}) + \gamma_2 (\operatorname{rot} \mathbf{u}, \operatorname{rot} \mathbf{v}) + kn(\mathbf{u}, \mathbf{v}), \quad \forall \mathbf{u}, \mathbf{v} \in \mathcal{V}, \tag{2.198}$$

$$b(\mathbf{v}, q) = (\operatorname{div} \mathbf{v}, q), \quad \forall \mathbf{v} \in \mathcal{V}, q \in \mathcal{O}, \tag{2.199}$$

$$c(\mathbf{v}, \mu) = (\operatorname{rot} \mathbf{v}, \mu), \quad \forall \mathbf{v} \in \mathcal{V}, \mu \in \mathcal{P}. \tag{2.200}$$

Note that in this formulation, the bilinear form $a(\mathbf{u}, \mathbf{v})$ is coercive on $\mathcal{V} \times \mathcal{V}$. While we give this formulation based on the constants $\gamma_1$ and $\gamma_2$ (to be selected), we do not recommend that it actually be used in practice. As noted above, the solution of the eigenvalues (or the transient response) can be obtained without use of these artificial constants. Whichever formulation is used, of key importance is the fact that for the bilinear forms $b(\mathbf{v}, q)$ and $c(\mathbf{v}, \mu)$, we have the following lemma:

LEMMA 6.2.1. *There exists a constant $C_o$ such that*

$$\sup_{\mathbf{v} \in \mathcal{V}_o \setminus \{\mathbf{0}\}} \frac{b(\mathbf{v}, q) + c(\mathbf{v}, \mu)}{\|\mathbf{v}\|_1} \geqslant C_o \big(\|q\|_0 + \|\mu\|_0\big), \quad \forall q \in \mathcal{O}_o, \ \mu \in \mathcal{P}. \tag{2.201}$$

PROOF. For any $q \in \mathcal{O}_o$, noting that the domain $\Omega$ is an open, bounded, convex domain with no re-entrant corners, there exists a unique $\phi \in \mathcal{H}^2(\Omega)$ satisfying (see [88] for details)

$$-\nabla^2\phi = q \quad \text{in } \Omega, \tag{2.202}$$

$$\frac{\partial\phi}{\partial n} = 0 \qquad \text{on } \Gamma, \tag{2.203}$$

with $\int_\Omega \phi \, d\Omega = 0$.

By elliptic regularity we have

$$\|\phi\|_2 \leqslant \bar{C}\|q\|_0, \tag{2.204}$$

where the positive constant $\bar{C}$ depends only on the area of $\Omega$. Analogously, for any $\mu \in \mathcal{P}$, there exists a unique $\psi \in \mathcal{H}^2(\Omega)$ satisfying (see [88] for details)

$$\nabla^2\psi = \mu \quad \text{in } \Omega, \tag{2.205}$$

$$\psi = 0 \qquad \text{on } \Gamma, \tag{2.206}$$

and by elliptic regularity we have

$$\|\psi\|_2 \leqslant \bar{C}\|\mu\|_0. \tag{2.207}$$

Let $\mathbf{z} = -\operatorname{grad}\phi - \operatorname{rot}\psi$ with $\mathbf{z} \in (\mathcal{H}^1(\Omega))^2$, and we can derive

$$\begin{aligned} \operatorname{div}\mathbf{z} &= -\nabla^2\phi = q, \\ \operatorname{rot}\mathbf{z} &= \nabla^2\psi = \mu, \end{aligned} \tag{2.208}$$

and

$$\mathbf{z}\cdot\mathbf{n}|_\Gamma = \left(-\frac{\partial\phi}{\partial n} - \frac{\partial\psi}{\partial\tau}\right)\bigg|_\Gamma = 0, \tag{2.209}$$

$$\|\mathbf{z}\|_1 \leqslant \|\phi\|_2 + \|\psi\|_2 \leqslant \bar{C}\big(\|q\|_0 + \|\mu\|_0\big), \tag{2.210}$$

where $\boldsymbol{\tau}$ is the unit tangent vector.

Hence, from (2.210), using the Schwarz inequality, we obtain

$$\sup_{\mathbf{v}\in\mathcal{V}_o\setminus\{\mathbf{0}\}} \frac{b(\mathbf{v},q) + c(\mathbf{v},\mu)}{\|\mathbf{v}\|_1} \geqslant \frac{\|q\|_0^2 + \|\mu\|_0^2}{\bar{C}(\|q\|_0 + \|\mu\|_0)} \geqslant C_o\big(\|q\|_0 + \|\mu\|_0\big),$$
$$\forall q \in \mathcal{O}_o,\ \mu \in \mathcal{P}, \tag{2.211}$$

with $C_o = 1/2\bar{C}$.

Let $\mathcal{V}^h$, $\mathcal{O}^h$, and $\mathcal{P}^h$ be finite element subspaces of $\mathcal{V}$, $\mathcal{O}$, and $\mathcal{P}$, respectively, the finite element approximation of the problem (2.196) becomes:

Find $\omega^h \in \mathcal{R}$, $\mathbf{u}^h \in \mathcal{V}^h$, $p^h \in \mathcal{O}^h$, and $\lambda^h \in \mathcal{P}^h$, such that

$$a\big(\mathbf{u}^h, \mathbf{v}^h\big) + \left(\frac{\gamma_1}{\kappa} - 1\right)b\big(\mathbf{v}^h, p^h\big) - \left(\frac{\gamma_2}{\alpha} - 1\right)c\big(\mathbf{v}^h, \lambda^h\big)$$
$$= \rho\big(\omega^h\big)^2\big(\mathbf{u}^h, \mathbf{v}^h\big) + m\big(\omega^h\big)^2 n\big(\mathbf{u}^h, \mathbf{v}^h\big), \quad \forall \mathbf{v}^h \in \mathcal{V}^h,$$

$$\frac{(p^h, q^h)}{\kappa} + b(\mathbf{u}^h, q^h) = 0, \quad \forall q^h \in \mathcal{O}^h,$$
$$\frac{(\lambda^h, \mu^h)}{\alpha} - c(\mathbf{u}^h, \mu^h) = 0, \quad \forall \mu^h \in \mathcal{P}^h.$$

Now, in order to have a good finite element method, the finite element spaces have to be chosen such that they inherit the property (2.201), i.e., they should satisfy

$$\sup_{\mathbf{v}^h \in \mathcal{V}_o^h \setminus \{\mathbf{0}\}} \frac{b(\mathbf{v}^h, q^h) + c(\mathbf{v}^h, \mu^h)}{\|\mathbf{v}^h\|_1} \geqslant C_o\big(\|q^h\|_0 + \|\mu^h\|_0\big),$$
$$\forall q^h \in \mathcal{O}_o^h,\ \mu^h \in \mathcal{P}^h, \tag{2.212}$$

and the theory of mixed methods provides the following optimal error estimate (see [183] for detail):

$$\left|\omega - \omega^h\right| \leqslant \tilde{C}\epsilon_h^2, \tag{2.213}$$

where $\tilde{C}$ is a positive constant independent of $h$ and material properties, $\omega$ and $\omega^h$ are solutions of the problems (2.196) and (2.212), respectively, and

$$\epsilon_h = \sup_{\|\mathbf{u}\|_1 + \|p\|_0 + \|\lambda\|_0 = 1} \ \inf_{\substack{(\mathbf{v}^h, q^h, \mu^h) \in \\ \mathcal{V}^h \times \mathcal{O}^h \times \mathcal{P}^h}} \left(\left\|\mathbf{u} - \mathbf{v}^h\right\|_1 + \left\|p - q^h\right\|_0 + \left\|\lambda - \mu^h\right\|_0\right).$$

In the following section, we will use the technique proposed by Stenberg [256–258] to check the discrete inf-sup condition (2.212). In order to provide a precise discussion, we have to define our concepts properly. Considering (2.212), we use $\mathcal{V}_o^h$ instead of $\mathcal{V}^h$, and if the inf-sup condition in the form of (2.212) is satisfied, since $\mathcal{V}_o^h \subset \mathcal{V}^h$, the inf-sup condition for the problem (2.212) is also satisfied, provided that the constant pressure mode is eliminated, i.e., we are using $\mathcal{O}_o^h$ instead of $\mathcal{O}^h$. Therefore, for simplicity and clarity, we adopt $\mathcal{V}_o^h$ and $\mathcal{O}_o^h$ in the rest of the section.

Let $\mathcal{T}^h$ be a partition of $\Omega$ which consists of either triangular or convex quadrilateral elements. Naturally, the partition is assumed to satisfy the usual compatibility and regularity conditions [143]. Let us further assume that the finite element polynomial spaces $\mathcal{V}_o^h$, $\mathcal{O}_o^h$, and $\mathcal{P}^h$ can be uniquely defined on $\mathcal{T}^h$ by using a reference element $\widehat{T}$, i.e., the unit triangle, or square element within the reference polynomial spaces, $\widehat{\mathcal{V}}$, $\widehat{\mathcal{O}}$, and $\widehat{\mathcal{P}}$. For $T \in \mathcal{T}^h$, let $\mathbf{F}_T : \widehat{T} \to T$ be an affine or bilinear mapping from $\widehat{T}$ onto $T$, i.e., $\mathbf{x} = \mathbf{F}_T(\hat{\mathbf{x}})$. We define

$$\mathcal{V}_o^h = \left\{\mathbf{v}^h \in \mathcal{V}_o \mid \mathbf{v}^h(\mathbf{x})|_T = \hat{\mathbf{v}}^h\big(\mathbf{F}_T^{-1}(\mathbf{x})\big),\ \hat{\mathbf{v}}^h \in \widehat{\mathcal{V}},\ \mathbf{x} \in T,\ \forall T \subset \Omega\right\},$$

and

$$\mathcal{O}_o^h = \left\{q^h \in \mathcal{O}_o \mid q^h(\mathbf{x})|_T = \hat{q}^h\big(\mathbf{F}_T^{-1}(\mathbf{x})\big),\ \hat{q}^h \in \widehat{\mathcal{O}},\ \mathbf{x} \in T,\ \forall T \subset \Omega\right\},$$

$$\mathcal{P}^h = \left\{\mu^h \in \mathcal{P} \mid \mu^h(\mathbf{x})|_T = \hat{\mu}^h\left(\mathbf{F}_T^{-1}(\mathbf{x})\right),\ \hat{\mu}^h \in \widehat{\mathcal{P}},\ \mathbf{x} \in T,\ \forall T \subset \Omega\right\};$$

or

$$\mathcal{O}_o^h = \left\{q^h \in \mathcal{O}_o \cap C(\Omega) \mid q^h(\mathbf{x})|_T = \hat{q}^h\left(\mathbf{F}_T^{-1}(\mathbf{x})\right),\ \hat{q}^h \in \widehat{\mathcal{O}},\ \mathbf{x} \in T,\ \forall T \subset \Omega\right\},$$

$$\mathcal{P}^h = \left\{\mu^h \in \mathcal{P} \cap C(\Omega) \mid \mu^h(\mathbf{x})|_T = \hat{\mu}^h\left(\mathbf{F}_T^{-1}(\mathbf{x})\right),\ \hat{\mu}^h \in \widehat{\mathcal{P}},\ \mathbf{x} \in T,\ \forall T \subset \Omega\right\};$$

where $C(\Omega)$ denotes the set of continuous functions on $\Omega$, and both $\mathbf{x}$ and $\hat{\mathbf{x}}$ represent the position vectors in the two-dimensional domain considered here and its reference domain, respectively.

The second choice for $\mathcal{O}_o^h$ and $\mathcal{P}^h$ is called the Taylor–Hood discretization. For $T \in \mathcal{T}^h$ we denote its characteristic diameter as $h_T$, its boundary as $\partial T$, its boundary unit normal vector as $\mathbf{n}_T$, its boundary unit tangent vector as $\boldsymbol{\tau}_T$, the set of its edges as $\ell(T)$, the characteristic length of its edges as $h_E$, and

$$\ell^h = \bigcup_{T \in \mathcal{T}^h} \ell(T) = \ell_\Omega^h \cup \ell_\Gamma^h,$$

with

$$\ell_\Omega^h = \left\{E \in \ell^h \mid E \cap \Omega \neq \emptyset\right\} \quad \text{and} \quad \ell_\Gamma^h = \left\{E \in \ell^h \mid E \subset \Gamma\right\}.$$

In general, by macroelement $M$ we mean the union of one or more neighboring elements in $\mathcal{T}^h$. Analogously, macroelement $M$ is said to be equivalent to a macroelement in the reference domain $\widehat{M}$ if there is a mapping $\mathbf{F}_M : \widehat{M} \rightarrow M$ satisfying the following conditions:

(i) $\mathbf{F}_M$ is continuous and bijective;
(ii) If $\widehat{M} = \bigcup_{j=1}^m \widehat{T}_j$, where $\widehat{T}_j$ and $T_j = \mathbf{F}_M(\widehat{T}_j)$ are elements in $\widehat{M}$ and $M$, respectively, with $j = 1, \ldots, m$; and
(iii) $\mathbf{F}_M|_{\widehat{T}_j} = \mathbf{F}_{T_j} \circ \mathbf{F}_{\widehat{T}_j}^{-1}$, where $\mathbf{F}_{\widehat{T}_j}$ and $\mathbf{F}_{T_j}$ are affine or bilinear mappings from the reference element $\widehat{T}$ onto $\widehat{T}_j$ and $T_j$, respectively, with $j = 1, \ldots, m$.

We denote the family of macroelements equivalent to $M$ and $\widehat{M}$ as $\mathcal{M}$ and $\widehat{\mathcal{M}}$, respectively. Hence, for a given macroelement $M$, analogous to the definition of finite element spaces, and $\ell^h$, $\ell_\Omega^h$, and $\ell_\Gamma^h$ on $\Omega$, we can define finite element spaces $V_M^h$, $Q_M^h$, and $\mathcal{P}_M^h$, along with $\ell_M^h$, $\ell_{M,\Omega}^h$, and $\ell_{M,\Gamma}^h$ on $M$. Notice here that $\mathcal{O}_M^h$ corresponds to $\mathcal{O}$ rather than $\mathcal{O}_o$ and $\mathcal{V}_M^h$ corresponds to $\mathcal{V}_o$ rather than $\mathcal{V}$ for the domain occupied by macroelement $M$.

We introduce the following norms in $\mathcal{O}_o^h$ and $\mathcal{P}^h$,

$$\|q^h\|_{\mathcal{V},\mathcal{O}}^2 = \sum_{T \in \mathcal{T}^h} h_T^2 \left\|\operatorname{grad} q^h\right\|_{0,T}^2 + \sum_{E \in \ell_\Omega^h} h_E \int_E \left|[q^h]\right|^2 d\Gamma, \quad \forall q^h \in \mathcal{O}_o^h, \tag{2.214}$$

$$\|\mu^h\|^2_{\mathcal{V},\mathcal{P}} = \sum_{T\in\mathcal{T}^h} h_T^2 \|\operatorname{rot}\mu^h\|^2_{0,T} + \sum_{E\in\ell^h_\Omega} h_E \int_E |[\mu^h]|^2\, d\Gamma + \sum_{E\in\ell^h_\Gamma} h_E \int_E |\mu^h|^2\, d\Gamma, \quad \forall\mu^h \in \mathcal{P}^h; \tag{2.215}$$

where $[a]$ denotes the jump of the variable $a$ across edges $E \in \ell^h_\Omega$, and analogously the following semi-norms in $\mathcal{O}^h_M$ and $\mathcal{P}^h_M$,

$$|q^h|^2_{M,\mathcal{O}} = \sum_{T\subset M} h_T^2 \|\operatorname{grad} q^h\|^2_{0,T} + \sum_{E\in\ell^h_M} h_E \int_E |[q^h]|^2\, d\Gamma, \tag{2.216}$$

$$|\mu^h|^2_{M,\mathcal{P}} = \sum_{T\subset M} h_T^2 \|\operatorname{rot}\mu^h\|^2_{0,T} + \sum_{E\in\ell^h_{M,\Omega}} h_E \int_E |[\mu^h]|^2\, d\Gamma + \sum_{E\in\ell^h_{M,\Gamma}} h_E \int_E |\mu^h|^2\, d\Gamma. \tag{2.217}$$

Furthermore, we define

$$V^h_{o,M} = \{\mathbf{v}^h \in \mathcal{V}^h_M \mid \mathbf{v}^h|_{\partial M\setminus\Gamma} = \mathbf{0},\ \mathbf{v}^h\cdot\mathbf{n}|_{\partial M\cap\Gamma} = 0\},$$
$$\mathcal{N}_M = \{(q^h,\mu^h) \in \mathcal{O}^h_M \times P^h_M \mid (\operatorname{div}\mathbf{v}^h, q^h)_M + (\operatorname{rot}\mathbf{v}^h, \mu^h)_M = 0,\ \forall\mathbf{v}^h \in \mathcal{V}^h_{o,M}\},$$

with

$$(\operatorname{div}\mathbf{v}^h, q^h)_M + (\operatorname{rot}\mathbf{v}^h, \mu^h)_M = \sum_{T\subset M}\int_T (q^h \operatorname{div}\mathbf{v}^h + \mu^h \operatorname{rot}\mathbf{v}^h)\, d\Omega,$$
$$\forall\mathbf{v}^h \in \mathcal{V}^h_{o,M},\ q^h \in \mathcal{O}^h_M,\ \mu^h \in \mathcal{P}^h_M. \tag{2.218}$$

Then we have the following lemma:

LEMMA 6.2.2. *There exist two positive constants $C_a$ and $C_b$ independent of $h$ and material properties such that*

$$\sup_{\mathbf{v}^h\in\mathcal{V}^h_o\setminus\{\mathbf{0}\}} \frac{b(\mathbf{v}^h,q^h)+c(\mathbf{v}^h,\mu^h)}{\|\mathbf{v}^h\|_1} \geqslant C_a(\|q^h\|_0 + \|\mu^h\|_0) - C_b(\|q^h\|_{\mathcal{V},\mathcal{O}} + \|\mu^h\|_{\mathcal{V},\mathcal{P}}),$$
$$\forall q^h \in \mathcal{O}^h_o,\ \mu^h \in \mathcal{P}^h. \tag{2.219}$$

PROOF. Let $(q^h, \mu^h) \in \mathcal{O}_o^h \times \mathcal{P}^h \subset \mathcal{O}_o \times \mathcal{P}$ be arbitrary, noting that the domain $\Omega$ is an open, bounded, convex domain with no re-entrant corners, then, from the inequality (2.201), there exists $\mathbf{v} \in \mathcal{V}_o$, such that

$$b(\mathbf{v}, q^h) + c(\mathbf{v}, \mu^h) \geqslant C_o(\|q^h\|_0 + \|\mu^h\|_0)^2, \tag{2.220}$$

$$\|\mathbf{v}\|_1 \leqslant \|q^h\|_0 + \|\mu^h\|_0. \tag{2.221}$$

We now interpolate $\mathbf{v}$ with $\mathbf{v}^h \in \mathcal{V}_o^h$, defined by the technique of Clemént [43], such that we have the following error estimates:

$$\sum_{T\in\mathcal{T}^h} h_T^{-2}\|\mathbf{v}-\mathbf{v}^h\|_{0,T}^2 + \sum_{E\in\ell^h} h_E^{-1}\int_E |\mathbf{v}-\mathbf{v}^h|^2\,d\Gamma \leqslant C_1|\mathbf{v}|_1^2, \tag{2.222}$$

and

$$\|\mathbf{v}^h\|_1 \leqslant C_2\|\mathbf{v}\|_1, \tag{2.223}$$

with $C_1$ and $C_2$ independent of $h$ and material properties.

Then, using integration by parts, we obtain

$$\begin{aligned}
& b(\mathbf{v}^h, q^h) + c(\mathbf{v}^h, \mu^h) \\
&\quad = b(\mathbf{v}^h - \mathbf{v}, q^h) + c(\mathbf{v}^h - \mathbf{v}, \mu^h) + b(\mathbf{v}, q^h) + c(\mathbf{v}, \mu^h) \\
&\quad \geqslant b(\mathbf{v}^h - \mathbf{v}, q^h) + c(\mathbf{v}^h - \mathbf{v}, \mu^h) + C_o(\|q^h\|_0 + \|\mu^h\|_0)^2 \\
&\quad = \sum_{T\in\mathcal{T}^h} (\mathbf{v}-\mathbf{v}^h, \operatorname{grad} q^h - \operatorname{rot} \mu^h) + C_o(\|q^h\|_0 + \|\mu^h\|_0)^2 \\
&\qquad + \sum_{E\in\ell_V^h} \int_E \left[(\mathbf{v}^h - \mathbf{v})\cdot \mathbf{n}_T([q^h]) + (\mathbf{v}^h - \mathbf{v})\cdot \boldsymbol{\tau}_T([\mu^h])\right] d\Gamma \\
&\qquad + \sum_{E\in\ell_S^h} \int_E (\mathbf{v}^h - \mathbf{v})\cdot \boldsymbol{\tau}_T \mu^h \, d\Gamma.
\end{aligned}$$

Furthermore, using the Hölder and Schwarz inequalities, (2.221), the fact that $|\mathbf{v}|_1^2 \leqslant \|\mathbf{v}\|_1^2$, and the following inequality based on (2.221) and (2.223),

$$\|\mathbf{v}^h\|_1 \leqslant C_2\|\mathbf{v}\|_1 \leqslant C_2(\|q^h\|_0 + \|\mu^h\|_0), \tag{2.224}$$

we have

$$\begin{aligned}
& b(\mathbf{v}^h, q^h) + c(\mathbf{v}^h, \mu^h) \\
&\quad \geqslant -\left(\sum_{T\in\mathcal{T}^h} h_T^{-2}\|\mathbf{v}-\mathbf{v}^h\|_{0,T}^2 + \sum_{E\in\ell^h} h_E^{-1}\int_E |\mathbf{v}-\mathbf{v}^h|^2\,d\Gamma\right)^{1/2},
\end{aligned}$$

$$
\begin{aligned}
&\big(\|q^h\|_{\mathcal{V},\mathcal{O}} + \|\mu^h\|_{\mathcal{V},\mathcal{P}}\big) + C_o\big(\|q^h\|_0 + \|\mu^h\|_0\big)^2 \\
&\quad \geqslant -\sqrt{C_1}|\mathbf{v}|_1\big(\|q^h\|_{\mathcal{V},\mathcal{O}} + \|\mu^h\|_{\mathcal{V},\mathcal{P}}\big) + C_o\big(\|q^h\|_0 + \|\mu^h\|_0\big)^2 \\
&\quad \geqslant -\sqrt{C_1}\big(\|q^h\|_0 + \|\mu^h\|_0\big)\big(\|q^h\|_{\mathcal{V},\mathcal{O}} + \|\mu^h\|_{\mathcal{V},\mathcal{P}}\big) \\
&\qquad + C_o\big(\|q^h\|_0 + \|\mu^h\|_0\big)^2 \\
&\quad = \big[C_o\big(\|q^h\|_0 + \|\mu^h\|_0\big) \\
&\qquad - \sqrt{C_1}\big(\|q^h\|_{\mathcal{V},\mathcal{O}} + \|\mu^h\|_{\mathcal{V},\mathcal{P}}\big)\big]\big(\|q^h\|_0 + \|\mu^h\|_0\big) \\
&\quad \geqslant \big[C_a\big(\|q^h\|_0 + \|\mu^h\|_0\big) - C_b\big(\|q^h\|_{\mathcal{V},\mathcal{O}} + \|\mu^h\|_{\mathcal{V},\mathcal{P}}\big)\big]\|\mathbf{v}^h\|_1,
\end{aligned}
$$

with $C_a = C_o/C_2$ and $C_b = \sqrt{C_1}/C_2$. □

LEMMA 6.2.3. *Suppose that for every $M \in \mathcal{M}$, the space $\mathcal{N}_M$, consisting of functions that are constant vectors on $M$, is two-dimensional if $\partial M \cap \Gamma = \emptyset$ or one-dimensional if $\partial M \cap \Gamma \neq \emptyset$. Then there exists a positive constant $\beta_{\mathcal{M}}$ such that the condition*

$$
\sup_{\mathbf{v}^h \in \mathcal{V}^h_{o,M}\setminus\{\mathbf{0}\}} \frac{b(\mathbf{v}^h, q^h) + c(\mathbf{v}^h, \mu^h)}{\|\mathbf{v}^h\|_1} \geqslant \beta_{\mathcal{M}}\big(|q^h|_{M,\mathcal{O}} + |\mu^h|_{M,\mathcal{P}}\big),
$$
$$
\forall q^h \in \mathcal{O}^h_M,\ \mu^h \in \mathcal{P}^h_M, \tag{2.225}
$$

*holds for every $M \in \mathcal{M}$.*

PROOF. Consider a fixed $M \in \mathcal{M}$. Define the constant $\beta_M$ as

$$
\begin{aligned}
\beta_M &= \inf_{\substack{(q^h,\mu^h)\in\mathcal{O}^h_M\times\mathcal{P}^h_M \\ |q^h|^2_{M,\mathcal{O}}+|\mu^h|^2_{M,\mathcal{P}}=1}} \ \sup_{\mathbf{v}^h\in\mathcal{V}^h_{o,M}\setminus\{\mathbf{0}\}} \frac{b(\mathbf{v}^h, q^h) + c(\mathbf{v}^h, \mu^h)}{\|\mathbf{v}^h\|_1} \\
&= \inf_{\substack{(q^h,\mu^h)\in\mathcal{O}^h_M\times\mathcal{P}^h_M \\ |q^h|^2_{M,\mathcal{O}}+|\mu^h|^2_{M,\mathcal{P}}=1}} \ \sup_{\substack{\mathbf{v}^h\in\mathcal{V}^h_{o,M}\setminus\{\mathbf{0}\} \\ \|\mathbf{v}^h\|_1=1}} \big[b\big(\mathbf{v}^h, q^h\big) + c\big(\mathbf{v}^h, \mu^h\big)\big].
\end{aligned}
$$

Since $\mathcal{N}_M$, consisting of functions that are constant vectors on $M$, is two-dimensional if $\partial M \cap \Gamma = \emptyset$ or one-dimensional if $\partial M \cap \Gamma \neq \emptyset$, and $\mathcal{O}^h_M$, $\mathcal{P}^h_M$, and $\mathcal{V}^h_{o,M}$ are finite-dimensional, it follows that $\beta_M > 0$.

Let us now prove that there is a constant $\beta_{\mathcal{M}}$ such that $\beta_M \geqslant \beta_{\mathcal{M}} > 0$ for every $M \in \mathcal{M}$. Let $\hat{\mathbf{x}}^1, \ldots, \hat{\mathbf{x}}^d$ be the vertices of the elements in $\widehat{M}$. Every $M \in \mathcal{M}$ is now uniquely defined by its vertices $\mathbf{x}^i = \mathbf{F}_M(\hat{\mathbf{x}}^i)$, with $i = 1, \ldots, d$, so we may write $\beta_M = \beta(\mathbf{x}^1, \ldots, \mathbf{x}^d)$. We will now consider the vertices as point $\mathbf{y} = (\mathbf{x}^1, \mathbf{x}^2, \ldots, \mathbf{x}^d)$ in $\mathcal{R}^{2d}$, and $\beta_M = \beta(\mathbf{y})$ as a function of $\mathbf{y}$. Let $h_M = \max_{T\subset M} h_T$. We may assume that $h_M = 1$ and that $\mathbf{x}^1$ coincides with the origin in $\mathcal{R}^2$. Since the general case can be handled by a scaling argument using the mapping $\mathbf{G}(\mathbf{x}) = h_M^{-1}(\mathbf{x} - \mathbf{x}^1)$, where $\mathbf{x}^1$ is chosen as the origin, all vertices $\mathbf{x}^2, \ldots, \mathbf{x}^d$ lie within a given distance from the origin. Further, every $T \subset M$ has a diameter less than or equal to unity and the triangulation $\mathcal{T}^h$ is regular. This means that point $\mathbf{y}$ belongs to a compact set, denoted by $\mathcal{D}$, in $\mathcal{R}^{2d}$. It can easily be proved that the function $\beta$ is continuous (see [256] for a similar argument), and since $\beta(\mathbf{y}) > 0$ for every $\mathbf{y} \in \mathcal{D}$, we conclude that there is a constant $\beta_{\mathcal{M}} > 0$, independent of $h$ and material properties, such that $\beta(\mathbf{y}) \geqslant \beta_{\mathcal{M}}$ for every $\mathbf{y} \in \mathcal{D}$. Thus,

$$\begin{aligned} &\inf_{\substack{(q^h,\mu^h)\in\mathcal{O}_M^h\times\mathcal{P}_M^h\\ |q^h|^2_{M,\mathcal{O}}+|\mu^h|^2_{M,\mathcal{P}}=1}} \sup_{\substack{\mathbf{v}^h\in\mathcal{V}^h_{o,M}\setminus\{\mathbf{0}\}\\ \|\mathbf{v}^h\|_1=1}} \left[b(\mathbf{v}^h, q^h) + c(\mathbf{v}^h, \mu^h)\right] \\ &\quad = \beta_M \geqslant \beta_{\mathcal{M}} > 0, \quad \forall M \in \mathcal{M}, \end{aligned} \tag{2.226}$$

and the desired inequality (2.225) follows from (2.226) directly. □

THEOREM 6.2.3. *Suppose that there is a fixed set of equivalent classes of macroelements $\mathcal{M}_i$, with $i = 1, \ldots, N$, a positive integer $L$, and a macroelement partition $\mathcal{M}^h$ for $V$, such that*

(M1) *For each $M \in \mathcal{M}_i$, with $i = 1, \ldots, N$, the space $\mathcal{N}_M$, two-dimensional if $\partial M \cap \Gamma = \emptyset$ or one-dimensional if $\partial M \cap \Gamma \neq \emptyset$, consists of functions that are constant on $M$.*
(M2) *Each $M \in \mathcal{M}^h$ belongs to one of the classes $\mathcal{M}_i$, with $i = 1, \ldots, N$.*
(M3) *Each $T \in \mathcal{T}^h$ is contained in at least one and not more than $L$ macroelements of $\mathcal{M}^h$.*
(M4) *Each $E \in \ell^h_\Omega$ is contained in the interior of at least one and not more than $L$ macroelements of $\mathcal{M}^h$.*
(M5) *Each $E \in \ell^h_\Gamma$ is contained on the boundary of at least one and not more than $L$ macroelements of $\mathcal{M}^h$. Then the inf-sup condition* (2.212) *holds.*

PROOF. Let $(q^h, \mu^h) \in \mathcal{O}^h_o \times \mathcal{P}^h$ be arbitrary. From Lemma 6.2.3 and (M1), we know that for every $M \in \mathcal{M}^h$ there exists $\mathbf{v}^h_M \in \mathcal{V}^h_{o,M}$, extending it with 0 outside of $M$, such that

$$\begin{aligned} b(\mathbf{v}^h_M, q^h) + c(\mathbf{v}^h_M, \mu^h) &= \sum_{T\subset M} \int_T \left(q^h \operatorname{div} \mathbf{v}^h_M + \mu^h \operatorname{rot} \mathbf{v}^h_M\right) d\Omega \\ &\geqslant C_3\left(|q^h|_{M,\mathcal{O}} + |\mu^h|_{M,\mathcal{P}}\right)^2, \end{aligned} \tag{2.227}$$

$$\|\mathbf{v}^h_M\|_1 = \|\mathbf{v}^h_M\|_{1,\Omega} = \|\mathbf{v}^h_M\|_{1,M} \leqslant |q^h|_{M,\mathcal{O}} + |\mu^h|_{M,\mathcal{P}}, \tag{2.228}$$

where $C_3$ is a constant independent of $M$, $h$, and material properties.

Let us define

$$\mathbf{v}^h = \sum_{M \in \mathcal{M}^h} \mathbf{v}_M^h, \tag{2.229}$$

and from (M3), (M4), and (M5), noting (2.227), we have

$$\begin{aligned} b(\mathbf{v}^h, q^h) + c(\mathbf{v}^h, \mu^h) &= \sum_{M \in \mathcal{M}^h} \big[b(\mathbf{v}_M^h, q^h) + c(\mathbf{v}_M^h, \mu^h)\big] \\ &\geqslant \sum_{M \in \mathcal{M}^h} C_3\big(|q^h|_{M,\mathcal{O}} + |\mu^h|_{M,\mathcal{P}}\big)^2 \\ &\geqslant C_3\big(\|q^h\|_{\mathcal{V},\mathcal{O}}^2 + \|\mu\|_{\mathcal{V},\mathcal{P}}^2\big). \end{aligned} \tag{2.230}$$

Furthermore, from (M4) and (M5), we know that each element $T \in \mathcal{T}^h$ is contained in at most $L$ macroelements. This, using Schwarz's inequality, named after Hermann Amandus Schwarz (1888–1951), and (2.228), gives

$$\begin{aligned} \|\mathbf{v}^h\|_1^2 &= \Big\| \sum_{M \in \mathcal{M}^h} \mathbf{v}_M^h \Big\|_1^2 \leqslant L \sum_{M \in \mathcal{M}^h} \|\mathbf{v}_M^h\|_1^2 \\ &\leqslant L \sum_{M \in \mathcal{M}^h} \big(|q^h|_{M,\mathcal{O}} + |\mu^h|_{M,\mathcal{P}}\big)^2 \\ &\leqslant 2\mathcal{L}^2\big(\|q^h\|_{\mathcal{V},\mathcal{O}}^2 + \|\mu^h\|_{\mathcal{V},\mathcal{P}}^2\big). \end{aligned} \tag{2.231}$$

Therefore we obtain, using Schwarz's inequality,

$$\begin{aligned} &\sup_{\mathbf{w}^h \in \mathcal{V}_o^h \setminus \{\mathbf{0}\}} \frac{b(\mathbf{w}^h, q^h) + c(\mathbf{w}^h, \mu^h)}{\|\mathbf{w}^h\|_1} \\ &\quad \geqslant \frac{b(\mathbf{v}^h, q^h) + c(\mathbf{v}^h, \mu^h)}{\|\mathbf{v}^h\|_1} \geqslant \frac{\sqrt{2}C_3}{2L}\big(\|q^h\|_{\mathcal{V},\mathcal{O}}^2 + \|\mu^h\|_{\mathcal{V},\mathcal{P}}^2\big)^{1/2} \\ &\quad \geqslant C_4\big(\|q^h\|_{\mathcal{V},\mathcal{O}} + \|\mu^h\|_{\mathcal{V},\mathcal{P}}\big), \quad \forall q^h \in \mathcal{O}_o^h,\ \mu^h \in \mathcal{P}^h, \end{aligned} \tag{2.232}$$

with $C_4 = C_3/2L$.

Combining Lemma 6.2.2 and the above inequality, we have that

$$\begin{aligned} &\sup_{\mathbf{v}^h \in \mathcal{V}_o^h \setminus \{\mathbf{0}\}} \frac{b(\mathbf{v}^h, q^h) + c(\mathbf{v}^h, \mu^h)}{\|\mathbf{v}^h\|_1} \\ &\quad = \xi \sup_{\mathbf{v}^h \in \mathcal{V}_o^h \setminus \{\mathbf{0}\}} \frac{b(\mathbf{v}^h, q^h) + c(\mathbf{v}^h, \mu^h)}{\|\mathbf{v}^h\|_1} \\ &\qquad + (1-\xi) \sup_{\mathbf{v}^h \in \mathcal{V}_o^h \setminus \{\mathbf{0}\}} \frac{b(\mathbf{v}^h, q^h) + c(\mathbf{v}^h, \mu^h)}{\|\mathbf{v}^h\|_1} \end{aligned}$$

$$\begin{aligned}
&\geqslant \xi\big[C_a\big(\|q^h\|_0 + \|\mu^h\|_0\big) - C_b\big(\|q^h\|_{\mathcal{V},\mathcal{O}} + \|\mu^h\|_{\mathcal{V},\mathcal{P}}\big)\big] \\
&\quad + C_4(1-\xi)\big(\|q^h\|_{\mathcal{V},\mathcal{O}} + \|\mu^h\|_{\mathcal{V},\mathcal{P}}\big) \\
&\geqslant \frac{C_a C_4}{C_b + C_4}\big(\|q^h\|_0 + \|\mu^h\|_0\big), \quad \forall q^h \in \mathcal{O}_o^h,\ \mu^h \in \mathcal{P}^h,
\end{aligned} \tag{2.233}$$

where we choose $\xi = C_4/(C_b + C_4) > 0$.

As defined earlier, the mass matrix $\mathbf{M}_{uu}$ corresponds to the $\mathcal{L}^2$-norm $\|\cdot\|_0$ on $V_o^h$, i.e., $\|\mathbf{u}^h\|_0^2 = \frac{1}{\rho}\mathbf{U}^T\mathbf{M}_{uu}\mathbf{U}$, and the matrices $\mathbf{K}_{pp}$ and $\mathbf{K}_{\lambda\lambda}$ correspond to the 0-norm $\|\cdot\|_0$ on $\mathcal{O}_o^h$ and $\mathcal{P}^h$, respectively. For clarity, we introduce two matrices $\tilde{\mathbf{K}}_{pp} = -\kappa\mathbf{K}_{pp}$ and $\tilde{\mathbf{K}}_{\lambda\lambda} = -\alpha\mathbf{K}_{\lambda\lambda}$, such that $\|p^h\|_0^2 = \mathbf{P}^T\tilde{\mathbf{K}}_{pp}\mathbf{P}$ and $\|\lambda^h\|_0^2 = \boldsymbol{\Lambda}^T\tilde{\mathbf{K}}_{\lambda\lambda}\boldsymbol{\Lambda}$. Moreover, we define a matrix $\mathbf{S}_{uu}$ to represent the 1-norm $\|\cdot\|_1$ on $\mathcal{V}_o^h$, i.e., $\|\mathbf{u}^h\|_1^2 = \mathbf{U}^T\mathbf{S}_{uu}\mathbf{U}$. In addition, we know that $\mathbf{K}_{up}$ and $\mathbf{K}_{u\lambda}$ are related to the bilinear forms $b(\mathbf{u}^h, p^h)$ and $c(\mathbf{u}^h, \lambda^h)$, i.e.,

$$b\big(\mathbf{u}^h, p^h\big) = -\mathbf{U}^T\mathbf{K}_{up}\mathbf{P} \quad \text{and} \quad c\big(\mathbf{u}^h, \lambda^h\big) = \mathbf{U}^T\mathbf{K}_{u\lambda}\boldsymbol{\Lambda}.$$

In the frequency analysis of acoustic fluids, as discussed in Refs. [151,300], for the $\mathbf{u}/p$ formulation, we need to solve the following eigenvalue problem for the nonzero eigenvalues:

$$\mathbf{K}_a\mathbf{U} = \omega_a^2\mathbf{M}_{uu}\mathbf{U}, \tag{2.234}$$

with $\mathbf{K}_a = -\mathbf{K}_{up}\mathbf{K}_{pp}^{-1}\mathbf{K}_{pu}$.

On the other hand, for the $\mathbf{u}$-$p$-$\boldsymbol{\Lambda}$ formulation, we need to solve the following eigenvalue problem for the nonzero eigenvalues:

$$\mathbf{K}_b\mathbf{U} = \omega_b^2\mathbf{M}_{uu}\mathbf{U}, \tag{2.235}$$

with $\mathbf{K}_b = -\mathbf{K}_{up}\mathbf{K}_{pp}^{-1}\mathbf{K}_{pu} - \mathbf{K}_{u\lambda}\mathbf{K}_{\lambda\lambda}^{-1}\mathbf{K}_{\lambda u}$ and $\mathbf{K}_{\lambda u} = \mathbf{K}_{u\lambda}^T$.

As discussed in detail in Ref. [300], some of the nonzero eigenvalues of (2.235), if $\alpha \gg \kappa$, located in the higher spectrum and representing the rotational modes, are in fact zero frequency modes in (2.234). Obviously, using (2.235) instead of (2.234) can significantly reduce the number of zero frequency modes.

Recall that in the numerical inf-sup tests for both the $\mathbf{u}/p$ and $\mathbf{u}$-$p$-$\boldsymbol{\Lambda}$ formulations, as presented in Ref. [301], we need to solve two similar eigenvalue problems by replacing the mass matrix $\mathbf{M}_{uu}$ with $\mathbf{S}_{uu}$, and modifying the stiffness matrices, i.e., we solve

$$\tilde{\mathbf{K}}_a\mathbf{U} = \lambda_a\mathbf{S}_{uu}\mathbf{U}, \tag{2.236}$$

$$\tilde{\mathbf{K}}_b\mathbf{U} = \lambda_b\mathbf{S}_{uu}\mathbf{U}, \tag{2.237}$$

with $\tilde{\mathbf{K}}_a = \frac{1}{\kappa}\mathbf{K}_a$ and $\tilde{\mathbf{K}}_b = -\frac{1}{\kappa}\mathbf{K}_{up}\mathbf{K}_{pp}^{-1}\mathbf{K}_{pu} - \frac{1}{\alpha}\mathbf{K}_{u\lambda}\mathbf{K}_{\lambda\lambda}^{-1}\mathbf{K}_{\lambda u}$.

Notice that the matrices $\mathbf{K}_a$, $\tilde{\mathbf{K}}_a$, $\mathbf{K}_b$, $\tilde{\mathbf{K}}_b$, $\mathbf{M}_{uu}$, and $\mathbf{S}_{uu}$ are all symmetric. In addition, $\mathbf{M}_{uu}$ and $\mathbf{S}_{uu}$ are positive definite. To have stable and reliable finite element discretizations for both mixed formulations, we must have

$$\inf_{q^h \in \mathcal{Q}_o^h \backslash \{0\}} \sup_{\mathbf{v}^h \in \mathcal{V}_0^h \backslash \{\mathbf{0}\}} \frac{b(\mathbf{v}^h, q^h)}{\|\mathbf{v}^h\|_1 \|q^h\|_0} = \sqrt{\hat{\lambda}_a(h)}, \tag{2.238}$$

with $\lim_{h \to 0} \sqrt{\hat{\lambda}_a(h)} = \sqrt{a_\lambda} > 0$; and

$$\inf_{(q^h, \mu^h) \in \mathcal{Q}_o^h \times \mathcal{P}^h \backslash \{(0,0)\}} \sup_{\mathbf{v}^h \in \mathcal{V}_0^h \backslash \{\mathbf{0}\}} \frac{b(\mathbf{v}^h, q^h) + c(\mathbf{v}^h, \mu^h)}{\|\mathbf{v}^h\|_1 (\|q^h\|_0 + \|\mu^h\|_0)} = \sqrt{\hat{\lambda}_b(h)}, \tag{2.239}$$

with $\lim_{h \to 0} \sqrt{\hat{\lambda}_b(h)} = \sqrt{b_\lambda} > 0$, where $\hat{\lambda}_a(h)$ and $\hat{\lambda}_b(h)$ are the smallest nonzero eigenvalues of the problems (2.236) and (2.237), respectively, and $a_\lambda$ and $b_\lambda$ are positive constants independent of $h$ (and, of course, material properties).

Comparing Eq. (2.234) with Eq. (2.236), it is easy to confirm that

$$\inf_{q^h \in \mathcal{Q}_o^h \backslash \{0\}} \sup_{\mathbf{v}^h \in \mathcal{V}_0^h \backslash \{\mathbf{0}\}} \frac{b(\mathbf{v}^h, q^h)}{\|\mathbf{v}^h\|_0 \|q^h\|_0} = \frac{\sqrt{\rho}}{\sqrt{\kappa}} \widehat{\omega}_a(h),$$

where $\widehat{\omega}_a$ is the smallest nonzero eigenvalue of the problem (2.234).

Since $\|\mathbf{v}^h\|_0 \leqslant \|\mathbf{v}^h\|_1$, $\forall \mathbf{v}^h \in \mathcal{V}_o^h$, we obtain

$$\frac{\sqrt{\rho}}{\sqrt{\kappa}} \widehat{\omega}_a(h) \geqslant \sqrt{\hat{\lambda}_a(h)}.$$

Thus, we have

$$\lim_{h \to 0} \widehat{\omega}_a(h) \geqslant \frac{\sqrt{\kappa a_\lambda}}{\sqrt{\rho}} > 0,$$

and because of the same matrix structure of $\mathbf{K}_a$ and $\tilde{\mathbf{K}}_a$, the number of zero frequencies of the eigenvalue problems (2.234) and (2.236) are the same.

Therefore, we can in this case simply calculate the lowest frequency of the problem in (2.234) for increasingly refined meshes. If this frequency approaches zero, the inf-sup test is not passed. On the other hand, if the smallest frequency does not approach zero, we cannot, strictly, say anything about the discretization scheme.

Considering the $\mathbf{u}$-$p$-$\boldsymbol{\Lambda}$ formulation, assume that $\alpha = K\kappa$, then we have $\mathbf{K}_b = \kappa(\mathbf{K}_{up}\tilde{\mathbf{K}}_{pp}^{-1}\mathbf{K}_{pu} + K\mathbf{K}_{u\lambda}\tilde{\mathbf{K}}_{\lambda\lambda}^{-1}\mathbf{K}_{\lambda u})$ and $\tilde{\mathbf{K}}_b = \mathbf{K}_{up}\tilde{\mathbf{K}}_{pp}^{-1}\mathbf{K}_{pu} + \mathbf{K}_{u\lambda}\tilde{\mathbf{K}}_{\lambda\lambda}^{-1}\mathbf{K}_{\lambda u}$. Therefore, it is obvious that we could only have a similar relationship when $K = 1$, based then on $\mathbf{K}_b = \kappa\tilde{\mathbf{K}}_b$, between the numerical inf-sup test value and the vibration frequency for the $\mathbf{u}$-$p$-$\boldsymbol{\Lambda}$ formulation. However, in practice, we want $K \gg 1$.

Let us look at two examples.

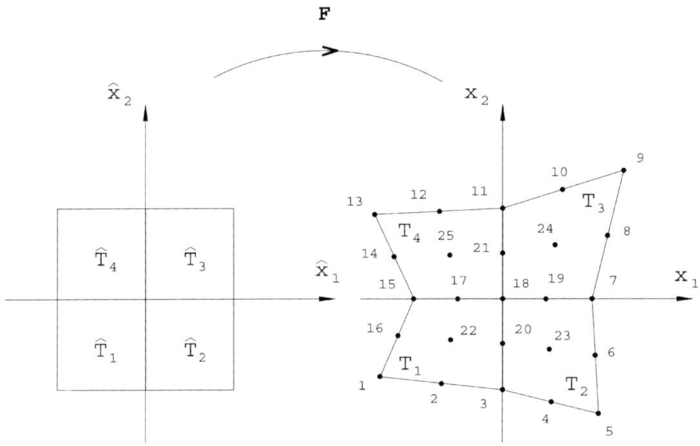


Figure 6.44. A typical macroelement.

EXAMPLE 6.1 *(The* 9-3-1 *element).* Let $\mathcal{T}^h$ be a partition of $V$ comprised of convex quadrilateral elements, and let the finite element spaces be defined as

$$\mathcal{V}_o^h = \{\mathbf{v}^h \in \mathcal{V}_o \mid \mathbf{v}^h|_T \in (\mathcal{O}_2(T))^2,\ T \in \mathcal{T}^h\},$$
$$\mathcal{O}_o^h = \{q^h \in \mathcal{O}_o \mid q^h|_T \in \mathcal{P}_1(T),\ T \in \mathcal{T}^h\},$$
$$\mathcal{P}^h = \{\mu^h \in \mathcal{P} \mid \mu^h|_T \in \mathcal{P}_o(T),\ T \in \mathcal{T}^h\}.$$

Suppose that the partition $\mathcal{T}^h$ satisfies the following: For each $T \in \mathcal{T}^h$, there exists at least one node of $T$ in $\Omega$. Then, for this method, the macroelement condition in Lemma 6.2.3 is valid for a macroelement consisting of four elements, as shown in Figure 6.44.

To prove this, we consider a macroelement $M = T_1 \cup T_2 \cup T_3 \cup T_4$ and the corresponding reference macroelement $\widehat{M} = \widehat{T}_1 \cup \widehat{T}_2 \cup \widehat{T}_3 \cup \widehat{T}_4$ as shown in Figure 6.44. Let $\mathbf{F} = (F_1, F_2)$ be the continuous piecewise bilinear mapping from $\widehat{M}$ onto $M$. Suppose $(q^h, \mu^h) \in \mathcal{N}_M$, i.e.,

$$\left(\operatorname{div}\mathbf{v}^h, q^h\right)_M + \left(\operatorname{rot}\mathbf{v}^h, \mu^h\right)_M = 0, \quad \forall \mathbf{v}^h \in \mathcal{V}_{o,M}^h.$$

Using integration by parts, we have

$$\begin{aligned}
&\left(\operatorname{div}\mathbf{v}^h, q^h\right)_M + \left(\operatorname{rot}\mathbf{v}^h, \mu^h\right)_M \\
&\quad = -\left(\mathbf{v}^h, \operatorname{grad} q^h\right)_M + \left(\mathbf{v}^h, \operatorname{rot}\mu^h\right)_M \\
&\qquad + \sum_{E \in \ell_{M,\Omega}^h} \int_E \left[\mathbf{v}^h \cdot \mathbf{n}_T\left([q^h]\right) + \mathbf{v}^h \cdot \boldsymbol{\tau}_T\left([\mu^h]\right)\right] d\Gamma \\
&\qquad + \sum_{E \in \ell_{M,\Gamma}^h} \int_E \mathbf{v}^h \cdot \boldsymbol{\tau}_T \mu^h \, d\Gamma = 0.
\end{aligned} \tag{2.240}$$

Given $\mathcal{Q}_2$, a biquadratic space for quadrilateral elements, since $\operatorname{grad} q^h|_{T_i}$ for $i = 1, \ldots, 4$ are constants, $\mathbf{v}^h$, $\operatorname{grad} q^h|_{T_i} \in \mathcal{Q}_2(T_i)$ for $i = 1, \ldots, 4$ and the composite integration (9-point integration) gives the exact values for the integrals $(\mathbf{v}^h, \operatorname{grad} q^h)_{T_i}$. Following the procedure of Ref. [258] and choosing $\mathbf{v}^h \in \mathcal{V}^h_{o,M}$ such that the only nonvanishing degrees of freedom are the values of both components at the center nodes $\mathbf{x}^{22}$, $\mathbf{x}^{23}$, $\mathbf{x}^{24}$, and $\mathbf{x}^{25}$ of $T_1$, $T_2$, $T_3$, and $T_4$, respectively, we obtain $\operatorname{grad} q^h|_{T_i} = \mathbf{0}$ for $i = 1, \ldots, 4$, which implies that both $q^h$ and $\mu^h$ are piecewise constant. Then we choose $\mathbf{v}^h \in \mathcal{V}^h_{o,M}$ such that the only nonvanishing degrees of freedom are $\mathbf{v}^h \cdot \mathbf{n}_1$, $\mathbf{v}^h \cdot \mathbf{n}_2$, $\mathbf{v}^h \cdot \mathbf{n}_3$, and $\mathbf{v}^h \cdot \mathbf{n}_4$, where $\mathbf{n}_1$, $\mathbf{n}_2$, $\mathbf{n}_3$, and $\mathbf{n}_4$ are the normals of the segments $\overline{\mathbf{x}^3\mathbf{x}^{18}}$, $\overline{\mathbf{x}^7\mathbf{x}^{18}}$, $\overline{\mathbf{x}^{11}\mathbf{x}^{18}}$, and $\overline{\mathbf{x}^{15}\mathbf{x}^{18}}$, evaluated at the points $\mathbf{x}^{20}$, $\mathbf{x}^{19}$, $\mathbf{x}^{21}$, and $\mathbf{x}^{17}$, respectively. It is not difficult to confirm that $q^h$ is constant on $M$. Therefore,

$$\begin{aligned}
&\left(\operatorname{div} \mathbf{v}^h, q^h\right)_M + \left(\operatorname{rot} \mathbf{v}^h, \mu^h\right)_M \\
&\quad = \sum_{E \in \ell^h_{M,\Omega}} \int_E \mathbf{v}^h \cdot \boldsymbol{\tau}_T\left(\left[\mu^h\right]\right) ds + \sum_{E \in \ell^h_{M,\Gamma}} \int_E \mathbf{v}^h \cdot \boldsymbol{\tau}_T \mu^h \, d\Gamma = 0. \qquad (2.241)
\end{aligned}$$

Now we choose $\mathbf{v}^h \in \mathcal{V}^h_{o,M}$ such that the only non-vanishing degrees of freedom are $\mathbf{v}^h \cdot \boldsymbol{\tau}_1$, $\mathbf{v}^h \cdot \boldsymbol{\tau}_2$, $\mathbf{v}^h \cdot \boldsymbol{\tau}_3$, and $\mathbf{v}^h \cdot \boldsymbol{\tau}_4$, where $\boldsymbol{\tau}_1$, $\boldsymbol{\tau}_2$, $\boldsymbol{\tau}_3$, and $\boldsymbol{\tau}_4$ are tangents of the segments $\overline{\mathbf{x}^3\mathbf{x}^{18}}$, $\overline{\mathbf{x}^7\mathbf{x}^{18}}$, $\overline{\mathbf{x}^{11}\mathbf{x}^{18}}$, and $\overline{\mathbf{x}^{15}\mathbf{x}^{18}}$, evaluated at the points $\mathbf{x}^{20}$, $\mathbf{x}^{19}$, $\mathbf{x}^{21}$, and $\mathbf{x}^{17}$, respectively. We confirm that $\mu^h$ is constant on $M$. If $\partial M \cap \Gamma = \emptyset$, we obtain the desired macroelement condition, including the fact that $\mathcal{N}_M$ is two-dimensional. If $\partial M \cap \Gamma \neq \emptyset$, without loss of generality, suppose $E_o = \overline{\mathbf{x}^1\mathbf{x}^3} \in \ell^h_{M,\Gamma}$. We need to choose $\mathbf{v}^h \in \mathcal{V}^h_{o,M}$ such that the only non-vanishing degree of freedom is $\mathbf{v}^h \cdot \boldsymbol{\tau}_{E_o}$ evaluated at the point $\mathbf{x}^2$. Then we obtain $\mu^h = 0$ on $M$, i.e., $\mathcal{N}_M$ is one-dimensional. Thus the desired macroelement condition is proved for a macroelement of four elements, and based on Lemma 6.2.3 and Theorem 6.2.3, the inf-sup condition of (2.212) is satisfied.

EXAMPLE 6.2 *(The* 9/4-*c*-1 *element).* Again, for the sake of simplicity, we consider a special case. For the general case, one has to check the macroelement condition numerically. We assume $\Omega$ to be a rectangle and $\mathcal{T}^h$ be a partition of $\Omega$ containing squares with the same size.

$$\begin{aligned}
\mathcal{V}^h_o &= \left\{\mathbf{v}^h \in \mathcal{V}_o \mid \mathbf{v}|_T \in \left(\mathcal{O}_2(T)\right)^2,\ T \in \mathcal{T}^h\right\}, \\
\mathcal{O}^h_o &= \left\{q^h \in \mathcal{O}_o \cap C(\Omega) \mid q^h|_T \in \mathcal{O}_1(T),\ T \in \mathcal{T}^h\right\}, \\
\mathcal{P}^h &= \left\{\mu^h \in \mathcal{P} \mid \mu^h|_T \in \mathcal{P}_o(T),\ T \in \mathcal{T}^h\right\}.
\end{aligned}$$

Suppose the partition $\mathcal{T}^h$ satisfies the following: For each $T \in \mathcal{T}^h$, there exists at least one node of $\mathcal{T}$ in $\Omega$. Then, we proceed to prove that the macroelement condition in Lemma 6.2.3 is valid for a macroelement consisting of four elements.

Following the procedure above, we choose $\mathbf{v}^h \in \mathcal{V}^h_{o,M}$ such that the only non-vanishing degrees of freedom are the values of both components at the center points of $T_1$, $T_2$, $T_3$, and $T_4$, respectively, as shown in Figure 6.44. Since here the elements $T_1$, $T_2$, $T_3$, and $T_4$ are squares, from $\operatorname{grad} q^h = 0$, we get

$$q^h(\mathbf{x}^1) = q^h(\mathbf{x}^{18}) = q^h(\mathbf{x}^5) = q^h(\mathbf{x}^9) = q^h(\mathbf{x}^{13}) = a,$$
$$q^h(\mathbf{x}^3) = q^h(\mathbf{x}^7) = q^h(\mathbf{x}^{11}) = q^h(\mathbf{x}^{15}) = b,$$

where $a$ and $b$ are constants.

Then we choose $\mathbf{v}^h \in \mathcal{V}^h_{o,M}$ such that the only non-vanishing degrees of freedom are $\mathbf{v}^h \cdot \boldsymbol{\tau}_1$, $\mathbf{v}^h \cdot \boldsymbol{\tau}_2$, $\mathbf{v}^h \cdot \boldsymbol{\tau}_3$, and $\mathbf{v}^h \cdot \boldsymbol{\tau}_4$, evaluated at the points $\mathbf{x}^{20}$, $\mathbf{x}^{19}$, $\mathbf{x}^{21}$, and $\mathbf{x}^{17}$, respectively. Without much difficulty, using $\mathbf{v}^h \cdot \operatorname{grad} q^h = \mathbf{v}^h \cdot \mathbf{n} \partial q^h / \partial n + \mathbf{v}^h \cdot \boldsymbol{\tau} \partial q^h / \partial \tau$, we obtain $a = b$ and $\mu^h|_{T_1} = \mu^h|_{T_2} = \mu^h|_{T_3} = \mu^h|_{T_4} =$ constant. Therefore, $q^h$ and $\mu^h$ are always constant functions on $M$. If $\partial M \cap \Gamma \neq \emptyset$, we suppose that one edge, say $E_o$, of $M$ is such that $E_o = \overline{\mathbf{x}^1\mathbf{x}^3} \in \ell^h_{M,\Gamma}$. Then we choose $\mathbf{v}^h \in \mathcal{V}^h_{o,M}$ such that the only non-vanishing degree of freedom is $\mathbf{v}^h \cdot \boldsymbol{\tau}_{E_o}$ evaluated at the point $\mathbf{x}^2$, and obtain $\mu^h = 0$ on $M$, i.e., $\mathcal{N}_M$ is one-dimensional. Hence, the desired macroelement condition is proved for a macroelement of four elements and the inf-sup condition (2.212) is satisfied based on Lemma 6.2.3 and Theorem 6.2.3.

*Chapter 7*

# Computational Nonlinear Models

As computational hardware and software become more reliable and affordable, engineering competition and economics tend to push designers to optimize original designs, and in certain cases, to simulate various nonlinear options. It is well known that nonlinearities which include both spatial discretization formulations and temporal integration algorithms need to be treated judiciously. In fact, one of the limitations of computational approaches is the effective and efficient handling of large dynamical systems.

To some extent, general procedures for combining FEM with CFD in the study of nonlinear FSI systems still rely on staggered iterations [145]. Effective solution strategies for stability analysis of extremely large nonlinear dynamical systems derived from semi-discretized FSI systems have not been well established. The discussion of recent computational procedures for nonlinear FSI systems will set the stage for general purpose instability analysis of large FSI systems.

In this chapter, we consider actual fluid flows with viscosity and slight compressibility. For high speed flows, full-fledged Euler equations coupled with the state equation must be considered. In addition, turbulent flows are often introduced by additional equations, for instance, two additional equations in $k$-$\epsilon$ or $k$-$\omega$ models and six additional equations in Reynolds stress model [53,165]. This chapter focuses on the fundamental issues related to coupled formulations for FSI systems, explicit and implicit time integration schemes, and direct and iterative solution strategies.

## 7.1. Upwind and stabilization

Convective terms present a major hurdle in the simulation of fluid flow interacting with structures. These hyperbolic terms introduce nonsymmetric discretized coefficient matrices and remain the source of nonphysical oscillatory solutions similar to Gibbs' phenomenon [311]. To circumvent the skew matrix derived with the standard Galerkin finite element, finite volume or central difference methods, researchers have proposed various alternative discretization procedures which are often called upwind or stabilization schemes. The basic idea of upwind, which was proposed by Courant et al. [46] in 1952, is to assign more weight to the nodal solution in the upstream direction than in the downstream direction. The initial upwind finite element schemes were outlined by Heinrich et al. [122,123] and Christie et al. [42]. Noble control volume finite element upwind formulations include the quadrilateral elements by Schneider and Raw [86] and the 4/3-$c$ triangular and 5/4-$c$ tetrahedral elements by Bathe et al. [155]. In addition, cubic-polynomial interpolation (CIP) and flow-condition-based interpolation (FCBI) techniques have also been introduced to eliminate nonphysical oscillation with the least numerical diffusivity [157,314]. Some other upwind approaches, such as the Lax–Wendroff/Taylor–Galerkin formulation [58], the Galerkin least squares method [279], and the Galerkin method with bubble functions were also developed recently [26]. Most importantly, the well-known Streamline Upwind Petrov/Galerkin (SUPG) method, originally developed by Brooks and Hughes [7], was studied and analyzed extensively [134,149,280].

It is possible to achieve exact nodal solutions for the one-dimensional model problem. This idea has been widely used in various upwind schemes such as the exponential schemes developed in control volume finite difference procedures (see Spalding [52], Patankar [266], and Minkowycz et al. [305]). Solutions for multi-dimensional cases generally exhibit either excessive diffusion or oscillatory behavior. In fact, all upwind schemes are essentially equivalent to the standard Galerkin or central difference method with a so-called artificial diffusivity.

### *7.1.1. One-dimensional model*

To elaborate upon the inner relationships among various upwind techniques, we begin with the analogous and simple forms of upwind schemes for one-dimensional convection–diffusion models governed with the following differential equation:

$$v\frac{d\theta}{dx} - \alpha\frac{d^2\theta}{dx^2} = 0, \tag{1.1}$$

with the boundary conditions

$$\theta = 0, \quad \text{at } x = 0,$$
$$\theta = 1, \quad \text{at } x = 1,$$

where $\alpha$ is the thermal diffusivity and $v$ is the prescribed velocity.

Although Eq. (1.1) is a simple constant coefficient ordinary differential equation with the exact solution $\theta = (e^{vx/\alpha} - 1)/(e^{v/\alpha} - 1)$, we recognize that the basic observations of discretization procedures are applicable to the solution of multi-dimensional cases and to the Navier–Stokes equations.

The typical $i$th finite difference equation, with central differencing for both the convective and diffusive terms, takes the form

$$-\left(\frac{v}{2}+\frac{\alpha}{h}\right)\theta_{i-1}+2\frac{\alpha}{h}\theta_i+\left(\frac{v}{2}-\frac{\alpha}{h}\right)\theta_{i+1}=0, \tag{1.2}$$

where $h$ denotes the mesh size.

The oscillatory nature of Eq. (1.2), when the element Peclet number $Pe^e = v\alpha/h > 2$, has been widely analyzed. To illustrate the remedy designed by Courant et al. [46], we assume without loss of generality that $v$ is positive. If we discretize the convective term with a backward Euler scheme, a so-called classical upwind scheme, we arrive at the following $i$th equation:

$$-\left(v+\frac{\alpha}{h}\right)\theta_{i-1}+\left(v+2\frac{\alpha}{h}\right)\theta_i-\frac{\alpha}{h}\theta_{i+1}=0. \tag{1.3}$$

Using Eq. (1.3), the oscillatory solution behavior is no longer present. It is not difficult to determine that in order to obtain Eq. (1.3), we must add an artificial diffusivity to Eq. (1.2):

$$\frac{v}{2}(\theta_{i-1}-2\theta_i+\theta_{i+1}), \tag{1.4}$$

and Eq. (1.3) corresponds to a modified problem:

$$v\frac{d\theta}{dx}=\left(\alpha+\frac{vh}{2}\right)\frac{d^2\theta}{dx^2}. \tag{1.5}$$

The control volume finite element method is a rather straightforward approach for this one-dimensional problem. The $i$th equation corresponds to satisfying the equilibrium of flux $v\theta - \alpha d\theta/dx$ for the $i$th control volume between the stations $i-1/2$ and $i+1/2$. In the standard control volume method without upwind, with

$$\theta_{i-1/2}=(\theta_i+\theta_{i-1})/2 \quad \text{and} \quad \theta_{i+1/2}=(\theta_i+\theta_{i+1})/2,$$

we find that the $i$th equation is exactly the same as Eq. (1.2).

In the control volume method with upwind, assuming $v$ is positive, we obtain Eq. (1.3) with

$$\theta_{i-1/2}=\theta_{i-1} \quad \text{and} \quad \theta_{i+1/2}=\theta_i.$$

In the standard Galerkin finite element formulation, the same trial functions are employed to express the weighting and the solution. However, in principle, different functions may be used in the variational formulation. The modified weight function in the SUPG method includes a first derivative term and can be written, in the one-dimensional case, as

$$\delta\bar{\theta}=\delta\theta+\frac{\xi h}{2}\frac{d\delta\theta}{dx}, \tag{1.6}$$

where $\xi$ is a parameter to be adjusted.

For a typical two-node element, we obtain the following stiffness matrix:

$$\mathbf{K}_e = \begin{bmatrix} -\frac{v}{2} + (\frac{\alpha}{h} + \frac{v\xi}{2}) & \frac{v}{2} - (\frac{\alpha}{h} + \frac{v\xi}{2}) \\ -\frac{v}{2} - (\frac{\alpha}{h} + \frac{v\xi}{2}) & \frac{v}{2} + (\frac{\alpha}{h} + \frac{v\xi}{2}) \end{bmatrix}, \tag{1.7}$$

from which, after the element assemblage, the $i$th equation becomes

$$-\left[\frac{v}{2} + \left(\frac{\alpha}{h} + \frac{v\xi}{2}\right)\right]\theta_{i-1} + \left(2\frac{\alpha}{h} + v\xi\right)\theta_i + \left[\frac{v}{2} - \left(\frac{\alpha}{h} + \frac{v\xi}{2}\right)\right]\theta_{i+1} = 0. \tag{1.8}$$

We note that with $\xi = 0$, the standard Galerkin finite element equation is recovered. When $\xi = 1$, the classic upwind scheme is obtained. In particular, for this one-dimensional model problem, $\xi$ can be evaluated so that nodal exact values are obtained for all values of $Pe^e$ or $vh/\alpha$,

$$\xi = \coth\left(\frac{Pe^e}{2}\right) - \frac{2}{Pe^e}. \tag{1.9}$$

An *ad hoc* generalization is applied to multi-dimensional elements based on directional Peclet numbers. In the numerical implementation of Eq. (1.9), doubly asymptotic approximations or critical approximations are often used:

$$\xi = \begin{cases} Pe^e/6, & -6 \leqslant Pe^e \leqslant 6, \\ \operatorname{sgn}(Pe^e), & |Pe^e| > 6, \end{cases}$$

or

$$\xi = \begin{cases} -1 - 2/Pe^e & Pe^e < -1, \\ 0 & -1 \leqslant Pe^e \leqslant 1, \\ 1 - 2/Pe^e & Pe^e > 1. \end{cases}$$

In fact, Eq. (1.8) represents a general form of upwind schemes for this one-dimensional model problem. For example, the Galerkin least squares method results in the same $\xi$ as in Eq. (1.9). With Eq. (1.9), Eq. (1.8) becomes equivalent to Eq. (1.3), in the hyperbolic limit as $\alpha \to 0$.

If we consider the type of elements shown in Figure 7.1 with $h_1 = 1/2 - x/h$, $h_2 = 1/2 + x/h$, and $\tilde{h}_3 = 1 - 4x^2/h^2$, it is not difficult to prove that the bubble function $\tilde{h}_3$, in essence, introduces an artificial diffusivity with $\xi = vh/6\alpha$. This magic number is in fact the same as in the doubly asymptotic approximation of the SUPG method. A comprehensive mathematical study of the general connection between the standard Galerkin method with bubble functions and the SUPG method is available in Ref. [26].

To study the stability of Eqs. (1.2), (1.3), and (1.8), we solve the corresponding constant-coefficient homogeneous difference equations. Assuming $\theta_i = CG^i$ with exponent $i$, we have for Eq. (1.8), the following quadratic characteristic equa-

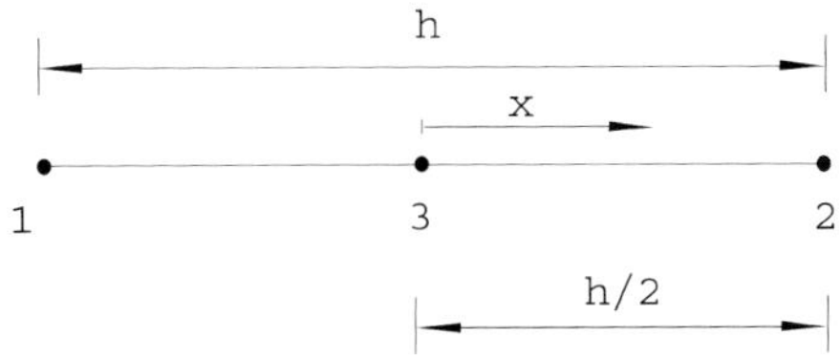


Figure 7.1. Discretization of convection–diffusion equation with a bubble function.

tion:

$$-\left[\frac{v}{2}+\left(\frac{\alpha}{h}+\frac{v\xi}{2}\right)\right]+\left(2\frac{\alpha}{h}+v\xi\right)G+\left[\frac{v}{2}-\left(\frac{\alpha}{h}+\frac{v\xi}{2}\right)\right]G^2=0, \tag{1.10}$$

which yields two roots

$$G_1=1 \quad \text{and} \quad G_2=\frac{2\alpha+(v+v\xi)h}{2\alpha-(v-v\xi)h}. \tag{1.11}$$

Considering the standard Galerkin finite element method, or the central difference method ($\xi = 0$). Since $G_2 < 0$, if $Pe^e > 2$, it is obvious that the solution of Eq. (1.2) contains oscillations. In particular, if $\alpha \to 0$, i.e., $G_2 \to -1$, we observe the sawtooth profile similar to the checkerboard pressure modes for almost incompressible material [27]. Furthermore, we notice that the coefficient matrix based on Eq. (1.2), though satisfying the consistency requirement is not diagonally dominant for $Pe^e > 2$. In the classical upwind method ($\xi = 1$), $G_2$ is always positive for all $Pe^e$, and the coefficient matrix based on Eq. (1.3) is diagonally dominant (though not strictly diagonally dominant). Therefore, the solution of Eq. (1.3) does not contain nonphysical oscillations. It is also interesting to note that if $0 < \xi < 1$, for $Pe^e > 2/(1-\xi)$, the solution retains oscillations and the corresponding coefficient matrix is not diagonally dominant.

### 7.1.2. Multi-dimensional model

To introduce finite element approximations in multi-dimensional cases, we consider the homogeneous convection–diffusion problem

$$\mathbf{v}\cdot\nabla\theta=\nabla\cdot(\alpha\nabla\theta)+f, \quad \text{in } \Omega, \tag{1.12}$$

$$\theta=0, \quad \text{on } \Gamma, \tag{1.13}$$

where for the bounded $n$-dimensional domain $\Omega \subset \mathcal{R}^n$ with a Lipschitz continuous boundary $\Gamma$, the given data are the source function $f$, the velocity field $\mathbf{v}$ with $\nabla\cdot\mathbf{v}=0$, and the diffusivity $\alpha$. With the following Sobolev space:

$$\mathcal{V}=\mathcal{H}^1_{0,\Gamma}(\Omega)=\left\{\theta \mid \theta\in\mathcal{H}^1(\Omega),\ \theta|_\Gamma=0\right\},$$

the standard variational form of Eq. (1.12) can be defined as:

Find $\theta \in \mathcal{V}$ such that

$$(\mathbf{v}\cdot\nabla\theta, \delta\theta) + (\alpha\nabla\theta, \nabla\delta\theta) = (f, \delta\theta), \quad \forall\delta\theta \in \mathcal{V}, \tag{1.14}$$

with

$$(\mathbf{v}\cdot\nabla\theta, \delta\theta) = \int_\Omega \delta\theta\mathbf{v}\cdot\nabla\theta\, d\Omega, \qquad (\alpha\nabla\theta, \nabla\delta\theta) = \int_\Omega \alpha\nabla\delta\theta\cdot\nabla\theta\, d\Omega,$$

$$\text{and} \quad (f, \delta\theta) = \int_\Omega \delta\theta f\, d\Omega.$$

Given a sequence of finite-dimensional subspaces $\mathcal{V}_h \subset \mathcal{V}$, we can obtain the following finite-dimensional approximation:

$$(\mathbf{v}\cdot\nabla\theta_h, \delta\theta_h) + (\alpha\nabla\theta_h, \nabla\delta\theta_h) = (f, \delta\theta_h), \quad \forall\delta\theta_h \in \mathcal{V}_h, \tag{1.15}$$

where $h$ is the mesh size parameter indicating the length of the side of a generic element, or the diameter of a circle encompassing that element.

Moreover, we can derive from Eqs. (1.14) and (1.15) the relation

$$\big(\mathbf{v}\cdot\nabla(\theta-\theta_h), \delta\theta_h\big) + \big(\alpha\nabla(\theta-\theta_h), \nabla\delta\theta_h\big) = 0, \quad \forall\delta\theta_h \in \mathcal{V}_h. \tag{1.16}$$

To establish an error estimate, we employ $|\delta\theta_h|^2_{1,\Omega} = (\nabla\delta\theta_h, \nabla\delta\theta_h)$. Since $\delta\theta_h$ is zero on $\Gamma$ and $\nabla\cdot\mathbf{v} = 0$ in $\Omega$, it follows that

$$\begin{aligned}(\mathbf{v}\cdot\nabla\delta\theta_h, \delta\theta_h) &= \int_\Omega \delta\theta_h\mathbf{v}\cdot\nabla\delta\theta_h\, d\Omega = \int_\Omega \mathbf{v}\cdot\nabla\left(\frac{\delta\theta_h^2}{2}\right) d\Omega \\ &= \int_\Omega \left\{\nabla\cdot\left(\mathbf{v}\frac{\delta\theta_h^2}{2}\right) - \frac{\delta\theta_h^2}{2}\nabla\cdot\mathbf{v}\right\} d\Omega \\ &= \int_\Gamma \frac{\delta\theta_h^2}{2}\mathbf{v}\cdot\mathbf{n}\, d\Gamma - \int_\Omega \frac{\delta\theta_h^2}{2}\nabla\cdot\mathbf{v}\, d\Omega = 0.\end{aligned} \tag{1.17}$$

Therefore, based on the semi-norm definition, we have

$$|\theta|_{m,\Omega} = \left(\sum_{|k|=m}\int_\Omega |\partial^k\theta|^2\, d\Omega\right)^{1/2},$$

and moreover,

$$\alpha|\delta\theta_h|^2_{1,\Omega} = (\alpha\nabla\delta\theta_h, \nabla\delta\theta_h) - (\mathbf{v}\cdot\nabla\delta\theta_h, \delta\theta_h). \tag{1.18}$$

Define $I_h\theta$ as the interpolation function of $\theta$, i.e., an element in $\mathcal{V}_h$ that, at the finite element nodes, has the exact value of the unknown solution $\theta$ and geometrically corresponds to a function close to $\theta$, we then obtain the estimates of the interpolation errors:

$$|\theta - I_h\theta|_{0,\Omega} \leqslant C_1 h^{k+1} \|\theta\|_{k+1}, \tag{1.19}$$

$$|\theta - I_h\theta|_{1,\Omega} \leqslant C_2 h^{k} \|\theta\|_{k+1}, \tag{1.20}$$

where $C_1$ and $C_2$ are constants independent of $h$, and $k$ is the order of interpolation functions [143,214].

Substituting $\delta\theta_h = \theta_h - I_h\theta$ into Eq. (1.16), we have

$$\big(\mathbf{v}\cdot\nabla(\theta-\theta_h), \theta_h - I_h\theta\big) + \big(\alpha\nabla(\theta-\theta_h), \nabla(\theta_h - I_h\theta)\big) = 0, \tag{1.21}$$

and

$$\begin{aligned} &\alpha|\theta_h - I_h\theta|_{1,\Omega}^2 \\ &\quad = \big(\alpha\nabla(\theta_h - I_h\theta), \nabla(\theta_h - I_h\theta)\big) - \big(\mathbf{v}\cdot\nabla(\theta_h - I_h\theta), \theta_h - I_h\theta\big). \end{aligned} \tag{1.22}$$

The first term in Eq. (1.21) can be rewritten as follows:

$$\begin{aligned} \big(\mathbf{v}\cdot\nabla(\theta-\theta_h), \theta_h - I_h\theta\big) &= \int_\Omega (\theta_h - I_h\theta)\mathbf{v}\cdot\nabla(\theta-\theta_h)\, d\Omega \\ &= \int_\Omega \nabla\cdot\big(\mathbf{v}(\theta-\theta_h)(\theta_h - I_h\theta)\big)\, d\Omega \\ &\quad - \int_\Omega (\theta_h - I_h\theta)(\theta-\theta_h)\nabla\cdot\mathbf{v}\, d\Omega \\ &\quad - \int_\Omega (\theta-\theta_h)\mathbf{v}\cdot\nabla(\theta_h - I_h\theta)\, d\Omega \\ &= -\big(\mathbf{v}\cdot\nabla(\theta_h - I_h\theta), \theta-\theta_h\big). \end{aligned} \tag{1.23}$$

Therefore, combining Eqs. (1.21), (1.22), and (1.23) yields

$$\begin{aligned} &\alpha|\theta_h - I_h\theta|_{1,\Omega}^2 \\ &\quad = \big(\alpha\nabla(\theta - I_h\theta), \nabla(\theta_h - I_h\theta)\big) - \big(\mathbf{v}\cdot\nabla(\theta_h - I_h\theta), \theta - I_h\theta\big). \end{aligned} \tag{1.24}$$

Then, the following inequality can be derived by using the Cauchy–Schwarz inequality:

$$\int_\Omega fg\, d\Omega \leqslant \left(\int_\Omega f^2\, d\Omega\right)^{1/2} \left(\int_\Omega g^2\, d\Omega\right)^{1/2}. \tag{1.25}$$

Let us choose $\delta\theta_h = \theta_h - I_h\theta$, and apply Eqs. (1.16) and (1.18). We obtain, based on the Cauchy–Schwarz inequality, the estimate

$$|\theta_h - I_h\theta|_{1,\Omega} \leqslant |\theta - I_h\theta|_{1,\Omega} + \frac{\|\mathbf{v}\|}{\alpha}|\theta - I_h\theta|_{0,\Omega}. \tag{1.26}$$

Therefore, an error estimate is established in the semi-norms of $\mathcal{H}^1_{0,\Gamma}$ of the form

$$
\begin{aligned}
|\theta - \theta_h|_{1,\Omega} &\leqslant |\theta - I_h\theta|_{1,\Omega} + |\theta_h - I_h\theta|_{1,\Omega} \\
&\leqslant 2|\theta - I_h\theta|_{1,\Omega} + \frac{\|\mathbf{v}\|}{\alpha}|\theta - I_h\theta|_{0,\Omega} \\
&\leqslant \left(2C_2 + \frac{C_1 h\|\mathbf{v}\|}{\alpha}\right)h^k\|\theta\|_{k+1}. \qquad (1.27)
\end{aligned}
$$

Note that, if $\alpha \to 0$, the inequality (1.27) does not yield convergence; however, for a finite $\alpha$, with a sufficiently refined mesh $h \simeq O(\alpha)$, the inequality (1.27) guarantees the convergence of $\theta_h$ to $\theta$. In fact, all upwind schemes introduce the notion of an artificial diffusivity $\alpha^*$. For instance, in the classic upwind scheme, the artificial diffusivity $\alpha^* = C_o\|\mathbf{v}\|h$, and we have, with $\alpha \to 0$,

$$
|\theta - \theta_h|_{1,\Omega} \leqslant \left(2C_2 + \frac{C_1}{C_o}\right)h^k\|\theta\|_{k+1}. \qquad (1.28)
$$

In general, the monotonic and often over diffusive classical upwind scheme with $O(h)$ can be written in the form

$$
\begin{aligned}
&(\mathbf{v}\cdot\nabla\theta_h, \delta\theta_h) + \big((\alpha + C_o\|\mathbf{v}\|h)\nabla\theta_h, \nabla\delta\theta_h\big) = (f, \delta\theta_h), \\
&\forall\delta\theta_h \in \mathcal{V}_h, \qquad (1.29)
\end{aligned}
$$

while the nonmonotonic and not too diffusive SUPG formulation can be written as

$$
\begin{aligned}
&(\mathbf{v}\cdot\nabla\theta_h, \delta\theta_h) + (\alpha\nabla\theta_h, \nabla\delta\theta_h) + \sum_{\Omega_e}\big(\mathbf{v}\cdot\nabla\theta_h - \nabla\cdot(\alpha\nabla\theta_h), \tau\mathbf{v}\cdot\nabla\delta\theta_h\big) \\
&\quad = (f, \delta\theta_h) + \sum_{\Omega_e}(f, \tau\mathbf{v}\cdot\nabla\delta\theta_h), \quad \forall\delta\theta_h \in \mathcal{V}_h, \qquad (1.30)
\end{aligned}
$$

where $\tau = h_e\xi_e/2\|\mathbf{v}\|$ and the element subdomain $\Omega_e$ satisfies $\bigcup_e \Omega_e = \Omega$ and $\bigcap_e \Omega_e = \emptyset$.

We recognize that to *completely* eliminate nonphysical oscillatory solutions, we need to use monotonic schemes. Nevertheless, in order to achieve high-order accuracy for all ranges of finite diffusivities, upwind schemes should approach the standard Galerkin formulation with sufficiently refined meshes. Although many of the upwind schemes work well for selected examples, it is often the case that in solving practical problems with distorted meshes, the stability and accuracy of such formulations need to be verified. The proposed numerical test in this section will be used to compare any given upwind scheme, for example, the SUPG formulation, with two limit cases, i.e., the classical upwind schemes, such as the Schneider and Raw scheme [86] for two-dimensional cases, and the Courant's scheme [46] for one-dimensional cases along with the standard Galerkin formulation. In order to measure the difference between a given upwind scheme and the standard Galerkin formulation with a finite diffusivity, we construct the following semi-norms:

$$|\theta_h^c - \theta_h^g|_{1,\Omega} = \left(\nabla(\theta_h^c - \theta_h^g), \nabla(\theta_h^c - \theta_h^g)\right), \tag{1.31}$$

$$|\theta_h^c - \theta_h^g|_{0,\Omega} = \left(\theta_h^c - \theta_h^g, \theta_h^c - \theta_h^g\right), \tag{1.32}$$

where the superscripts $c$ and $g$ stand for the solutions of a given upwind scheme and the standard Galerkin formulation, respectively.

It is worth noting that we can use the same approach to compare any given upwind scheme with the SUPG formulation. In addition, to estimate oscillatory behaviors, a generic eigenvalue test is designed to check whether the coefficient matrix is diagonally dominant for the hyperbolic limit. Although numerical tests are not as affirmative as analytical proofs, in practice a properly designed numerical evaluation is very likely to be effective. Similar ideas are used when analytical evaluations are not achieved when studying incompatible displacement formulations, the effects of element geometric distortions, and, most recently, the inf-sup condition of incompressible analysis. The following steps are designed in the proposed numerical test:

(1) Loop every row of the assembled coefficient matrix and select the rows satisfying $\sum_{j=1}^{N} K_{ij} = 0$, where $N$ is the number of nodal unknowns and $K_{ij}$ stands for the $ij$th element of the matrix.
(2) Check whether the selected rows (assume $N_d$ such rows) are diagonally dominant.
(3) If the selected rows are not diagonally dominant, check whether or not the ratio $F_d = \sqrt{\sum_{i=1}^{N_d} f_i^2}$ approaches 1, as $\alpha \to 0$, where $f_i = |K_{ii}| / \sum_{j=1, j\neq i}^{N} |K_{ij}|$.
(4) Normalize the coefficient matrix, for every $1 \leqslant i \leqslant N$, $K'_{ij} = K_{ij} / \max_{j=1}^{N}(|K_{ij}|)$, and calculate the maximum and minimum moduli of the eigenvalues of the normalized matrix.
(5) Select a proper finite value of diffusivity ($Pe^e \simeq O(1)$), and study the rate of convergence of the semi-norms defined in Eqs. (1.31) and (1.32) with a nested sequence of meshes. The finer meshes should contain the coarser meshes.

Step 1 is used to separate boundary effects from the upwind scheme, whereas Steps 2–4 are designed to evaluate whether or not the given upwind scheme is monotonic for the hyperbolic limit. It is known that with a finite diffusivity, at best, different upwind schemes can converge as fast as the corresponding standard Galerkin formulation, provided a sufficiently refined mesh is taken. Therefore, Step 5 is introduced to evaluate the accuracy of the given upwind scheme by comparing it with the standard Galerkin formulation.

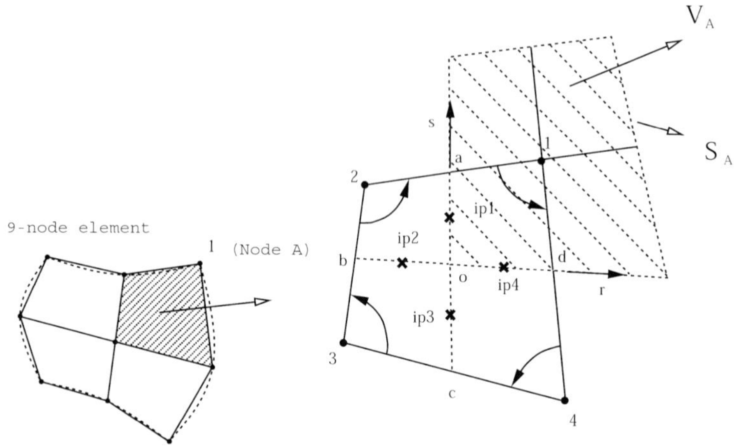


Figure 7.2. Typical 4-node element in the CVFEM upwind scheme.

In finite element formulation, many forms of upwind techniques can be used, though not completely satisfactory, to eliminate the spatial oscillations of the standard Galerkin formulation for convection dominated problems [104]. To develop a high-order element for general fluid flow conditions, we embed a control volume upwind scheme [86] in 9-node elements, including the 9/3 and 9/4-$c$ mixed elements, by treating one 9-node element as four 4-node elements for the convective terms only. We treat one 9-node element as four 4-node elements in dealing with the convective terms only. Figure 7.2 shows the typical 4-node element for the upwind control volume finite element method. The commonly used testing problem is the heat transfer problem governed by

$$\mathbf{v} \cdot \nabla \theta - \alpha \nabla^2 \theta = 0, \tag{1.33}$$

where $\mathbf{v}$, $\theta$, and $\alpha$ stand for the velocity vector, temperature, and thermal diffusivity.

Let $V$ and $S$ be the domain of interest and its boundary. The variational form of Eq. (1.33) is

$$\int_V \delta\theta \big(\mathbf{v} \cdot \nabla \theta - \alpha \nabla^2 \theta\big)\, dV = 0. \tag{1.34}$$

In the standard Galerkin formulation, $\delta\theta$ has the same interpolation function as $\theta$. For the SUPG formulation, $\delta\theta$ can be assigned as the combination of the same interpolation function for $\theta$ and an elementwise continuous interpolant, which is often the derivative of the interpolant for $\theta$, to suit its purpose of carrying the sign of the "wind" direction [7]. In the artificial viscosity approach, some

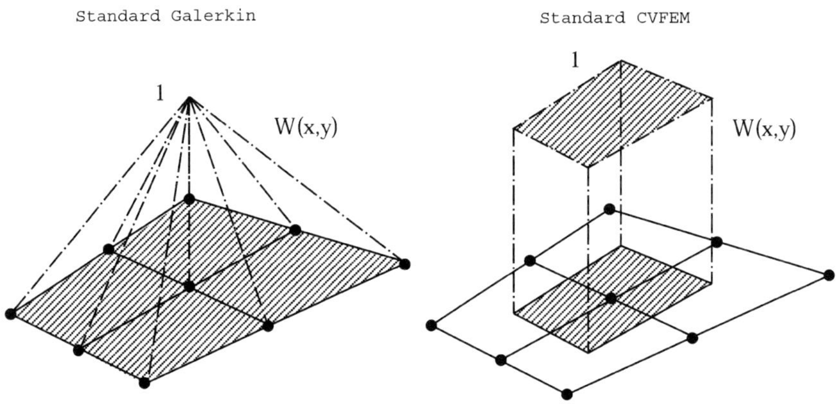


Figure 7.3. Typical interpolant functions.

constants, e.g., isotropic and anisotropic viscosity models, are introduced to the diffusion terms and the standard Galerkin formulation is used. Other approaches, such as the Galerkin least squares method, can also be applied. Interestingly, in the one-dimensional cases, many of the above methods will give the same set of discretized equations which provide the exact solutions on the nodal points, a so-called super convergence.

As shown in Figure 7.3, the control volume finite element method (CVFEM) is different from the Galerkin method. The approximation is directed to each node, upon which the conservation equations hold for the integral over the control volume,

$$\int_V \left(\mathbf{v} \cdot \nabla\theta - \alpha\nabla^2\theta\right) dV = 0. \tag{1.35}$$

In fact, we can obtain Eq. (1.35) by setting $\delta\theta = 1$ in the control volume around individual nodes in the variational form of Eq. (1.33). For the heat transfer problem with incompressible fluids, we can derive

$$\int_S (\mathbf{v} \cdot \mathbf{n}\theta - \alpha\nabla\theta \cdot \mathbf{n})\, dS = 0. \tag{1.36}$$

In the element assemblage process, the contributions from the element boundaries of adjacent elements cancel each other except on the domain boundary $S$. Hence, in Eq. (1.36), the important parts left are the evaluations on the inner element lines illustrated by dashed lines in Figure 7.2. Since as far as the convective terms are concerned, we assume that the unknown $\theta$ is interpolated using 4-node elements, we can use one-point Gauss numerical integration along these lines for Eq. (1.36). In the standard control volume finite element method [86], $\theta$ is interpolated with isoparametric interpolation functions. To achieve an optimal diagonally dominating discretized equation, i.e., the diagonal entity is positive and all the off-diagonal entities are negative due to the upwind scheme, the values of $\theta$ at the interpolation points, $ip1(r = 0,\ s = 0.5)$, $ip2(r = -0.5,\ s = 0)$, $ip3(r = 0,\ s = -0.5)$, and $ip4(r = 0.5,\ s = 0)$ are calculated according to the mass flux in and out of the control volumes. Taking the control volume for node 1 as an example, the contributions from the convective terms are

$$\dot{m}|_{ao}\theta_{ip1} + \dot{m}|_{od}\theta_{ip4}. \tag{1.37}$$

For the standard CVFEM, we have

$$\theta_{ip1} = \sum_{i=1}^{4} h_i|_{ip1}\theta_i, \tag{1.38}$$

with

$$h_1 = \frac{(1+r)(1+s)}{4}, \qquad h_2 = \frac{(1-r)(1+s)}{4},$$
$$h_3 = \frac{(1-r)(1-s)}{4}, \qquad h_4 = \frac{(1+r)(1-s)}{4}.$$

Notice here that we use the counterclockwise convention to define the normal directions on the inner surfaces of the control volume, and the mass flux is

$$\dot{m} = \mathbf{v} \cdot \mathbf{n}, \tag{1.39}$$

where the positive sign of $\dot{m}$ indicates the flux out of the control volume.

In the control volume finite element method, If $\dot{m}_{ao} < 0$, indicating that the flow direction is from left to right, we define

$$\theta_{ip1} = (1-f)\theta_2 + f\theta_{ip2}, \tag{1.40}$$

with the ratio of the mass fluxes $f = -\dot{m}_{bo}/\dot{m}_{oa}$ and $0 \leqslant f \leqslant 1$.

If $\dot{m}_{ao} > 0$, which indicates that the flow comes from right to left, we have

$$\theta_{ip1} = (1-f)\theta_1 + f\theta_{ip4}, \tag{1.41}$$

with $f = -\dot{m}_{od}/\dot{m}_{ao}$ and $0 \leqslant f \leqslant 1$.

The same approach is taken for $\theta_{ip2}$, $\theta_{ip3}$, and $\theta_{ip4}$. To extend this approach to solve nonlinear FSI problems with ALE formulations, we simply use the mass flux of

$$\dot{m} = \left(\mathbf{v} - \mathbf{v}^m\right) \cdot \mathbf{n}, \tag{1.42}$$

where $\mathbf{v}^m$ stands for the mesh velocity.

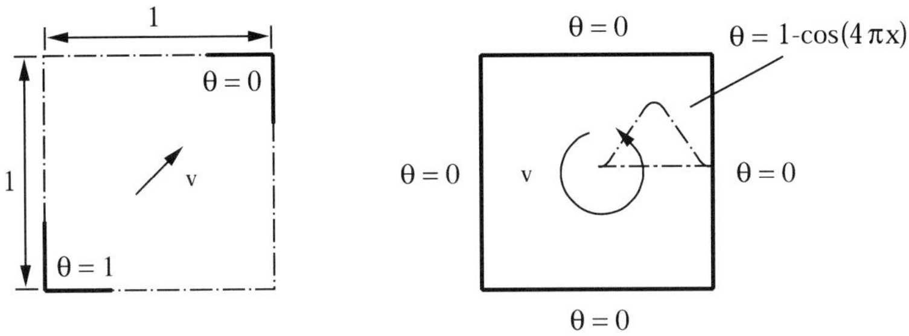


Figure 7.4. Two convection-dominated heat transfer problems.

Since the convective terms are nonlinear, iterative procedures must be used in the ALE formulation of the Navier–Stokes equations. In order to achieve a faster convergence rate of iterative procedures, we could use the true tangent stiffness matrix or include a line search option. In the successive iteration procedure, $\mathbf{v}-\mathbf{v}^m$ in Eq. (1.42) is taken from the last iteration step. In the control volume finite element method, for a typical node $A$ centered in the control volume $V_A$ with $\bigcup_A V_A = V$ and $\bigcap_A V_A = \emptyset$, we have Eqs. (1.34) and (1.35) for $V_A$ and $S_A$, the schematics of which are shown in Figure 7.2 for node 1 (equal to node $A$).

To test the proposed upwind formulation for 9-node elements, we analyze the two convection-dominated heat transfer examples shown in Figure 7.4. The Peclet number is $6.25 \times 10^9$ for both cases. In the diagonal flow problem, the flow is uniform ($|\mathbf{v}| = 1.0$) in the diagonal direction, while in the rotating cosine hill problem, the flow is rotational ($v_1 = -x_2$ and $v_2 = x_1$). Figures 7.5 and 7.6 give the results obtained with $\alpha = 10^{-8}$ using the standard Galerkin formulation and 9-node elements, the SUPG formulation and 4-node elements, and the proposed upwind formulation and 9-node elements. It is seen that the SUPG formulation retains spatial oscillations, while the proposed mixed upwind formulation for 9-node elements performs very well in the diagonal flow problem but introduces crosswind diffusion in the rotating cosine hill problem. This crosswind diffusion, of course, decreases as the mesh is refined.

The results in Figures 7.7 and 7.8 show the changes in the diagonal and off-diagonal ratio $F_d$. It is not surprising to find that in the Schneider and Raw scheme, $F_d = 1$ holds for all ranges of finite diffusivities, which means the solutions are smooth and monotonic. In addition, Figures 7.5 and 7.6 indicate that the standard Galerkin formulation gradually loses diagonal dominance as diffusivity approaches zero, and that the SUPG method, with $F_d \simeq 1$, should produce solutions between the two limit cases. Of course, the distribution of $1 - f_i$, corresponding to the nodal unknown $\theta_i$, implies possible nonphysical spatial oscillations along node $i$.

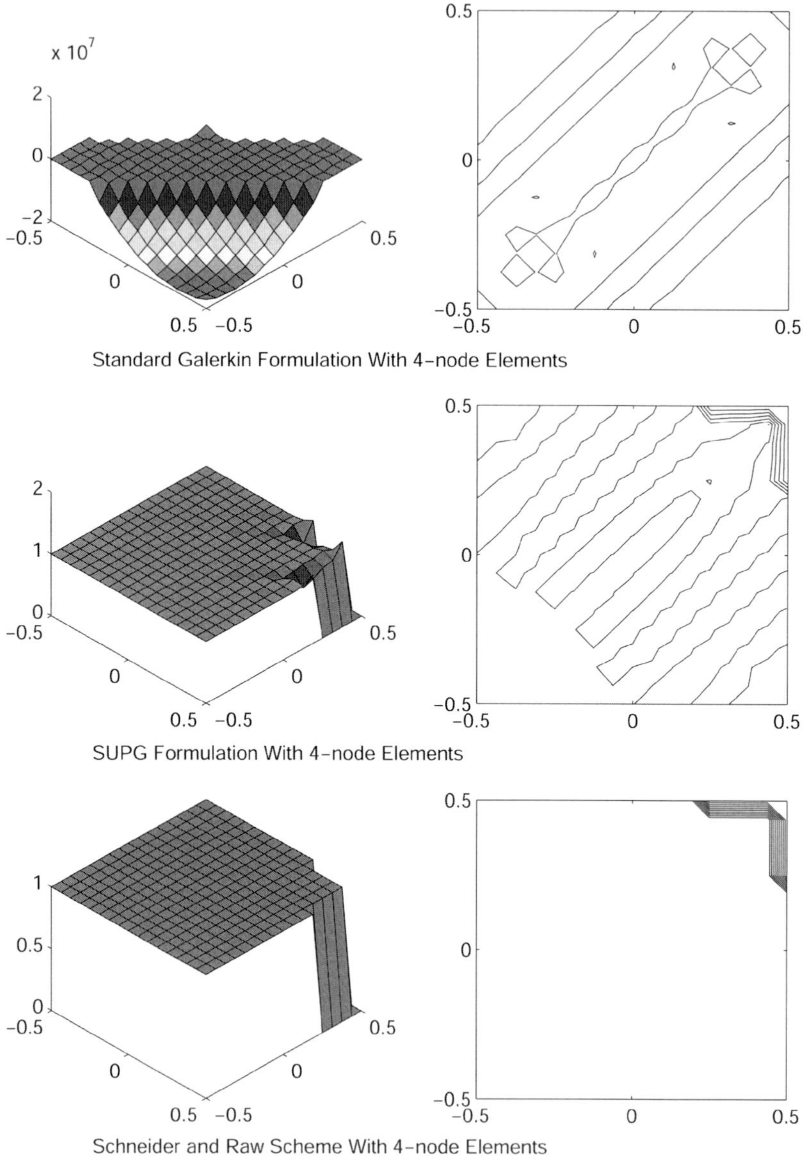


Figure 7.5. Solutions of the diagonal flow problem.

The rates of convergence of the 1-norm and 0-norm defined in Eqs. (1.31) and (1.32) with $\alpha = 0.005625$ are shown in Figures 7.1 and 7.9. It is not difficult to infer that the SUPG method has higher accuracy than the Schneider and Raw

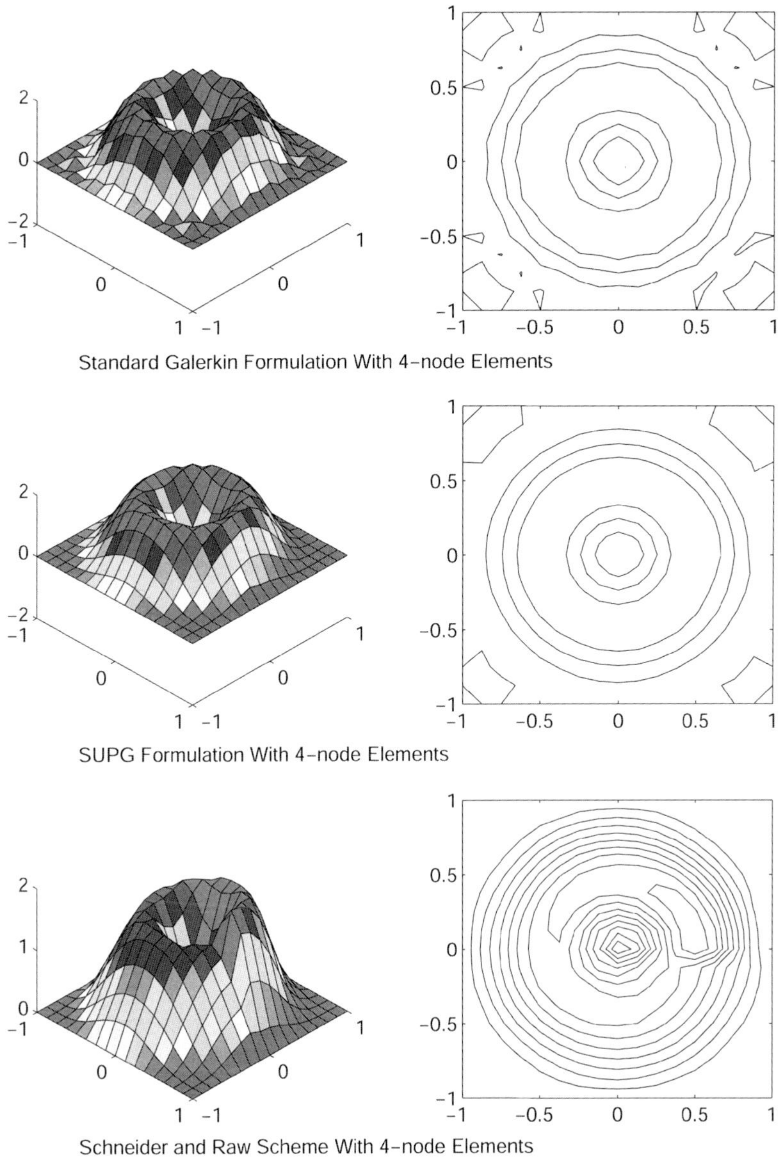


Figure 7.6. Solutions of the rotating cosine hill problem.

scheme, although the latter provides smoother solutions. Figures 7.10 and 7.11 give the results of the proposed normalized eigenvalue problem with different diffusivities, whereas Figures 7.12 and 7.13 show the standard eigenvalues of the

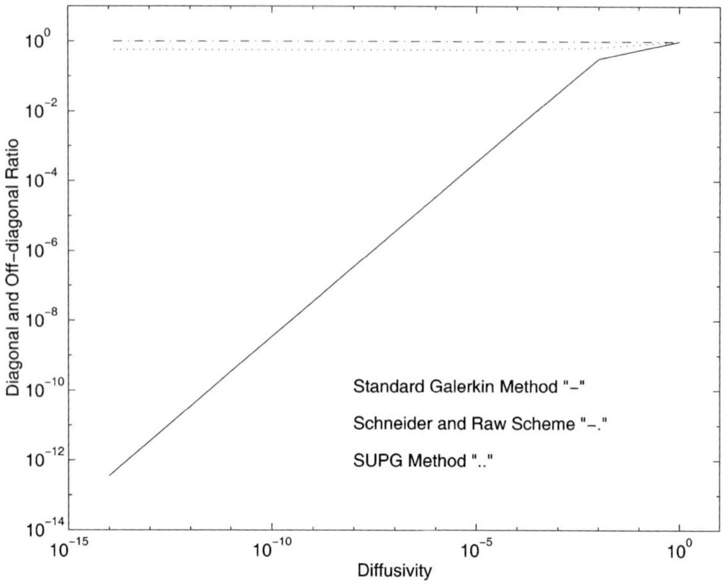


Figure 7.7. Diagonal and off-diagonal ratio of the diagonal flow problem.

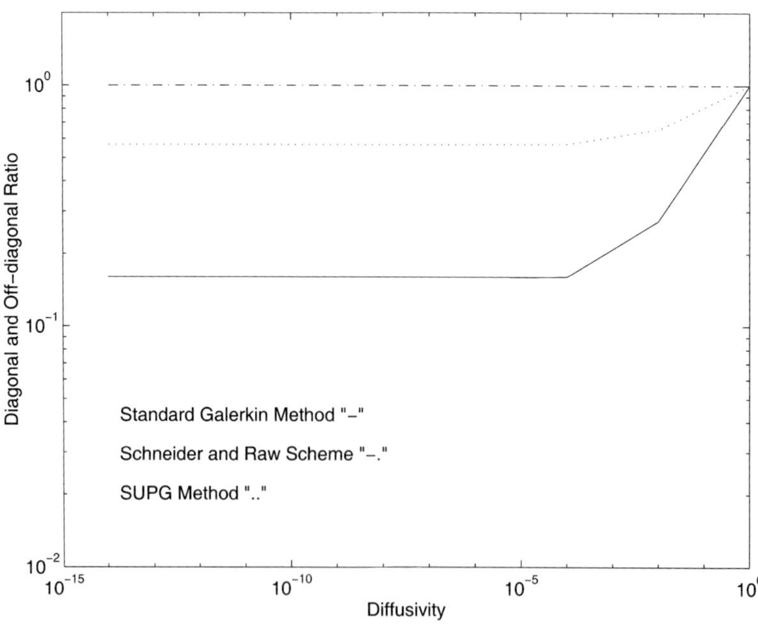


Figure 7.8. Diagonal and off-diagonal ratio of the rotating cosine hill problem.

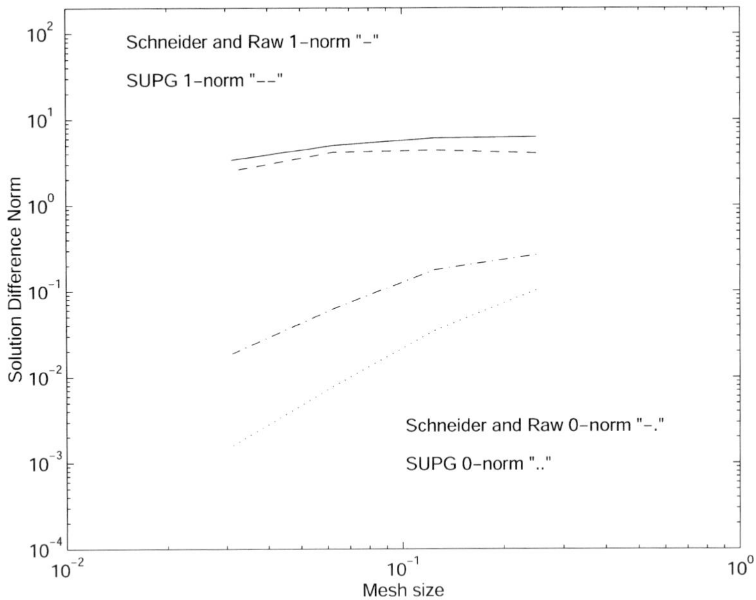


Figure 7.9. 1-norm and 0-norm convergence test of the rotating cosine hill problem.

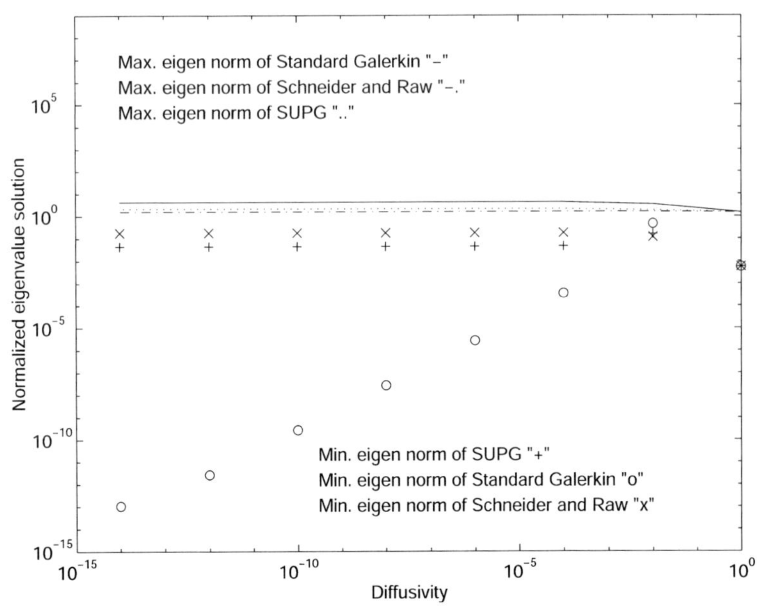


Figure 7.10. Normalized eigenvalue test of the diagonal flow problem.

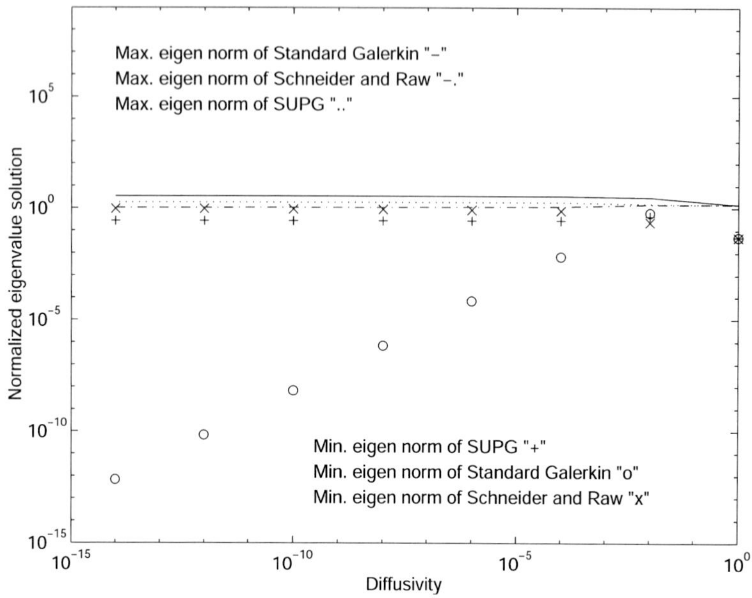


Figure 7.11. Normalized eigenvalue test of the rotating cosine hill problem.

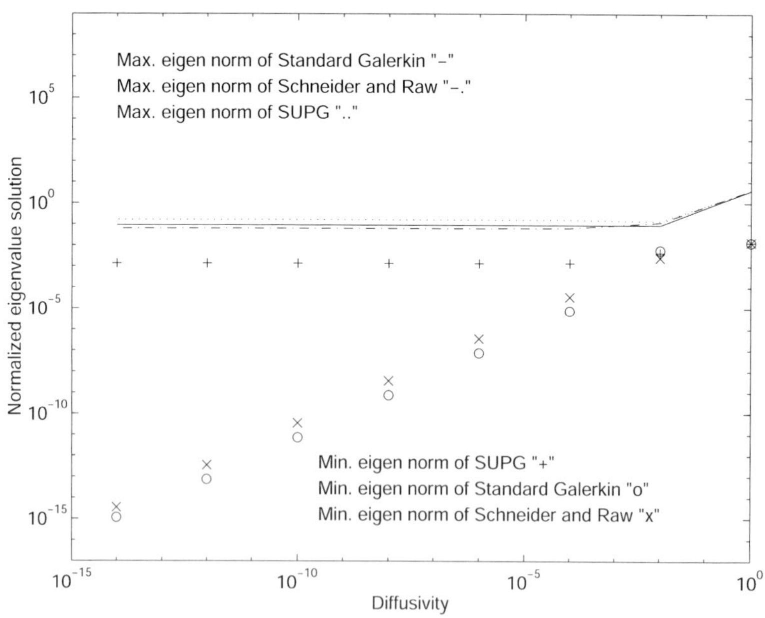


Figure 7.12. Standard eigenvalue test of the diagonal flow problem.

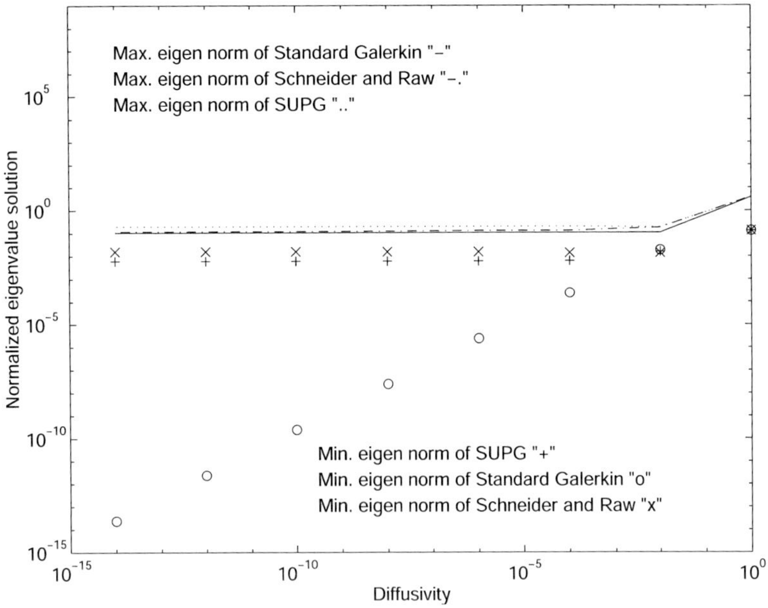


Figure 7.13. Standard eigenvalue test of the rotating cosine hill problem.

coefficient matrix. Note that with the normalization of the coefficient matrix, the lowest and highest moduli of the eigenvalues of the SUPG formulation are between the results of the standard Galerkin formulation and the Schneider and Raw scheme. However, the standard eigenvalues do not exhibit such relationships. It is obvious that Figures 7.10 and 7.11 match well with Figures 7.7 and 7.8, and that the normalized eigenvalue test again measures the likelihood of nonphysical spatial oscillations.

In summary, it is possible to improve upwind schemes, although for convection–diffusion problems, it appears that there is always a trade-off between the order of accuracy and stability. We conclude that the numerical evaluation proposed in this section may be of importance with respect to determining the advantages and disadvantages of any upwind discretizations and may eventually help in the development of optimal upwind schemes. The proposed numerical tests are cheaply computable and are very effective in comparison with two extreme cases with and without distorted elements: first, the monotonic classical upwind approach for the hyperbolic limit, and second, the standard Galerkin formulation with a finite diffusivity. Along with the diagonal and off-diagonal ratio convergence test, the proposed normalized eigenvalue problem will help to identify whether or not a given upwind scheme is monotonic. The convergence of two proposed semi-norms can be used to evaluate the accuracy of given upwind schemes. The

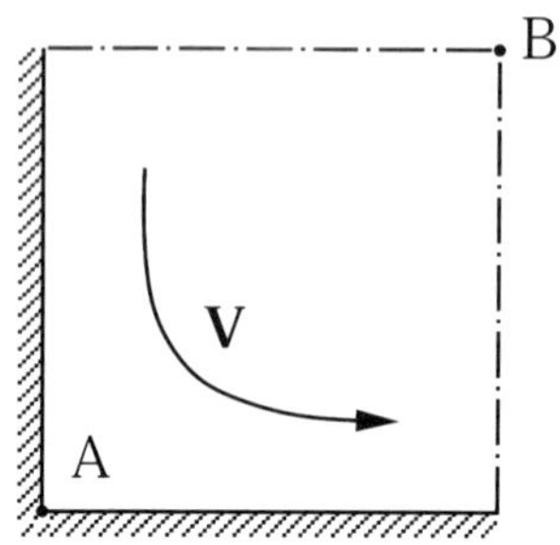


Figure 7.14. Corner flow problem.

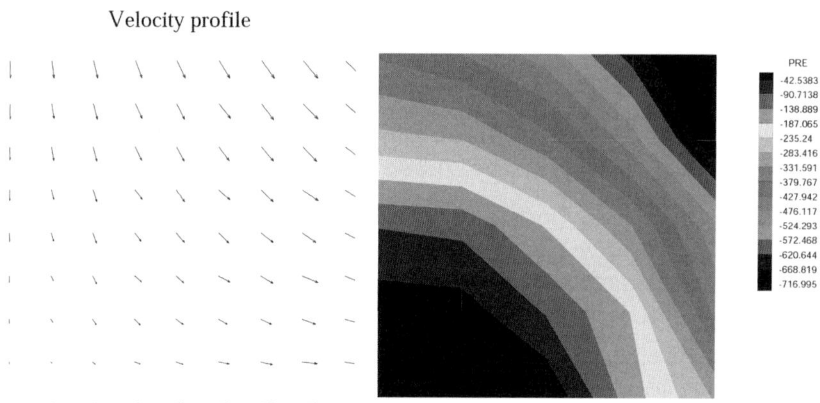


Figure 7.15. Velocity profile and pressure band of the corner flow problem.

numerical test is also applicable for the coefficient matrices derived from specific approaches such as wavelets and other interpolation functions.

Figure 7.14 illustrates the turning flow test problem for the control volume upwind scheme in convection dominated flow problems. In this test problem, we assign a known corner flow field:

$$u_1 = x_1 \quad \text{and} \quad u_2 = -x_2. \tag{1.43}$$

The corresponding pressure distribution is

$$p = -\frac{\rho}{2}\left(x_1^2 + x_2^2\right), \tag{1.44}$$

where the nodal pressure at node $A$ is designated zero.

From the analytic solution, node $B$ should have the pressure $-1000.0$. Using the proposed upwind formulation with sixteen 9-node elements, the calculated pressure for node $B$ is $-717.0$. Figure 7.15 shows that the pressure distribution is not exactly the circular shape as indicated in Eq. (1.44), due to crosswind diffusion. Nevertheless, with that coarse mesh, the flow field is smooth and fairly accurate.

## 7.2. Nonlinear finite element formulations for FSI systems

A typical FSI system is shown in Figure 7.16. The fluid domain of interest is denoted as $\Omega^f$, and its boundary $\partial\Omega^f$ or $\Gamma^f$ consisting of three types of mutually exclusive boundaries, namely, the Dirichlet boundary $\Gamma^{f,v}$, the Neumann boundary $\Gamma^{f,f}$, and the Cauchy boundary for the FSI interface $\Gamma^{fsi}$. Likewise, the solid domain of interest is $\Omega^s$, its boundary $\partial\Omega^s$ or $\Gamma^s$ consisting of three types of mutually exclusive boundaries, namely, the Dirichlet boundary $\Gamma^{s,v}$, the Neumann boundary $\Gamma^{s,f}$, and the Cauchy boundary for the FSI interface $\Gamma^{fsi}$.

### 7.2.1. *Mixed formulations for fluid and solid domains*

In this section, we adopt a finite volume upwind procedure along with an ALE description in the velocity/pressure ($\mathbf{v}/p$) mixed finite element formulation for viscous fluids. A similar displacement/pressure ($\mathbf{u}/p$) mixed finite element formulation is also introduced for flexible structures/solids in contact with viscous fluids. For both fluid and solid domains, high-order mixed elements satisfying the inf-sup conditions are used. With this approach, we are able to introduce a slight compressibility into viscous fluids. In addition, this approach accommodates the requirements for both the inf-sup condition and the upwind technique in a self-adaptive finite element analysis. Two distinct solution strategies are used: a direct

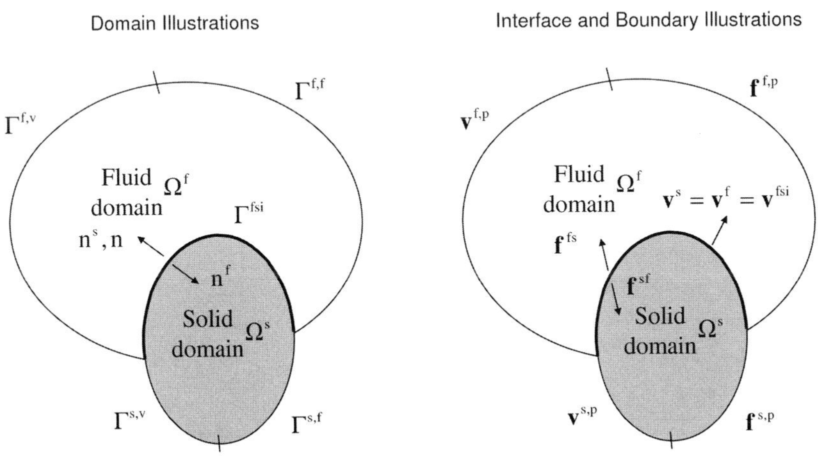


Figure 7.16. Illustration of a typical FSI system.

solver for fully coupled Newton–Raphson iterations, and an iterative matrix-free Newton–Krylov solver with a diagonal block preconditioner. Numerical results along with pros and cons for both approaches are discussed.

For slightly compressible viscous fluid flow, using indicial notations with the usual summation convention, the governing equations (momentum and continuity equations) are given by the following system of partial differential equations:

$$\rho^f \dot{v}_i^f - \sigma_{ij,j}^f - f_i^f = 0, \quad \text{in } \Omega^f, \tag{2.45}$$

$$v_{i,i}^f + \frac{\dot{p}^f}{\kappa^f} = 0, \quad \text{in } \Omega^f, \tag{2.46}$$

with the boundary conditions

$$v_i^f = v_i^{f,p}, \quad \text{on } \Gamma^{f,v}, \tag{2.47}$$

$$\sigma_{ij}^f n_j^f = f_i^{f,p}, \quad \text{on } \Gamma^{f,f}, \tag{2.48}$$

$$v_i^f = v_i^s, \quad \text{on } \Gamma^{fsi}, \tag{2.49}$$

$$\sigma_{ij}^f n_j^f = f_i^{sf}, \quad \text{on } \Gamma^{fsi}, \tag{2.50}$$

where $\rho^f$, $\mathbf{v}^f$, $\mathbf{v}^{f,p}$, $p^f$, $\mathbf{f}^f$, $\mu$, $\kappa^f$, $\mathbf{f}^{f,p}$, and $\mathbf{f}^{sf}$ are the fluid density, velocity, prescribed velocity at its Dirichlet boundary, pressure, body force, dynamic viscosity, bulk modulus, prescribed surface traction at its Neumann boundary, and surface traction attributed to the surrounding solid at the FSI interface, respectively, and the components of velocity strain and stress tensors are defined as $e_{ij}^f = (v_{i,j}^f + v_{j,i}^f)/2$ and $\sigma_{ij}^f = -p^f \delta_{ij} + 2\mu(e_{ij}^f - e_{kk}^f \delta_{ij}/3)$.

Note that the FSI force $\mathbf{f}^{fs}$ stands for the surface traction acting on the solid side of the interface from the surrounding fluid. Likewise, $\mathbf{f}^{sf}$ represents the surface traction acting on the fluid side of the interface from the surrounding solid. Naturally, according to Newton's first law, we have $\mathbf{f}^{fs} = -\mathbf{f}^{sf}$. As shown in Figure 7.16, on the fluid-structure interface $\Gamma^{fsi}$ where the boundary conditions can be considered as the Cauchy type for both fluid and solid domains, we have the kinematic matching condition $[v_i] = 0$, or (2.49). In addition, the dynamic matching condition $[\sigma_{ij} n_j] = 0$ or (2.50) can be expressed as the surface traction on the solid boundary:

$$\sigma_{ij}^f n_j^f = -\sigma_{ij}^s n_j^s, \quad \text{on } \Gamma^{fsi}, \tag{2.51}$$

where the surface normal vector $\mathbf{n}$ is aligned with that of the solid domain $\mathbf{n}^s$, and is opposite to that of the fluid domain $\mathbf{n}^f$.

Therefore, the condition in Eq. (2.51) implies that around the FSI interface, we have $\boldsymbol{\sigma}^s = \boldsymbol{\sigma}^f$, or $\mathbf{f}^{sf} = -\mathbf{f}^{fs}$. In the analysis of slightly compressible fluids, we assume that the compressibility measured by bulk modulus $\kappa^f$ is constant.

Therefore, we have the following relationship between the density and pressure of the fluid domain:

$$\frac{dp^f}{d\rho^f} = c^2 = \frac{\kappa^f}{\rho^f}. \tag{2.52}$$

From Eq. (2.52), it is straightforward to derive the following:

$$p^f(t) - p^f(0) = \kappa^f \ln \frac{\rho^f(t)}{\rho^f(0)}. \tag{2.53}$$

Nevertheless, in some formulations, $\kappa^f$ is sufficiently large in comparison with the other parameters, hence, the density $\rho^f$ is artificially kept constant. To account for the change of the fluid domain due to its interaction with the flexible structure/solid, we adopt an ALE kinematic description [59,293]. If we assign $\mathbf{v}^m$ as the arbitrary mesh velocity, we obtain

$$\dot{\mathbf{v}} = \check{\mathbf{v}} + \left(\mathbf{v} - \mathbf{v}^m\right) \cdot \nabla \mathbf{v}, \tag{2.54}$$

$$\dot{p} = \check{p} + \left(\mathbf{v} - \mathbf{v}^m\right) \cdot \nabla p, \tag{2.55}$$

where the mesh referential derivative of any quantity $a$ with respect to time can be expressed as $\check{a} = \partial a/\partial t + \mathbf{v}^m \cdot \nabla a$, and the differentiations are carried out with respect to the mesh positions.

If we denote $\mathbf{x}$, $\mathbf{y}$, and $\mathbf{z}$ as the material, spatial, and mesh point coordinates, the time rate of change of the spatial derivative $\partial/\partial y_i$ following the mesh point is defined as $-\partial/\partial y_k \partial v_k^m/\partial y_i$, with the mesh velocity $\mathbf{v}^m = \dot{\mathbf{u}}^m$, where $\mathbf{u}^m$ represents the mesh displacement vector. For the solid domain in contact with the fluid, we employ a Lagrangian kinematic description. The FSI interface will then be tracked automatically by the position of solid particles. In this work, we discuss a nonlinear solid mechanics model with both the geometrical and material nonlinearities [143,303,321]. We adopt the concepts of the deformation gradient $\mathbf{D}$, the Cauchy–Green deformation tensor $\mathbf{C}$ and their invariants $I_i$ and $J_i$ with $i = 1, 2, 3$, the Green–Lagrangian strain $\boldsymbol{\epsilon}$, and the second Piola–Kirchhoff stress $\mathbf{S}$ exclusively for the total Lagrangian representation of the solid domain. Denoting $\mathbf{x}^s(t)$ and $\mathbf{x}^s(0)$ as the generic material particle position at time $t$ and the corresponding initial position, we then express the displacement at time $t$ as $\mathbf{u}^s(t) = \mathbf{x}^s(t) - \mathbf{x}^s(0)$, and the deformation gradient as $D_{ij} = \partial x_i^s(t)/\partial x_j^s(0)$. Moreover, in the Lagrangian description of the solid domain, there is no need for convective terms. Hence, the material and referential derivatives are the same as the time derivative, and the solid velocity vector $\mathbf{v}^s$ and acceleration vector $\dot{\mathbf{v}}^s$ can be expressed as $\mathbf{v}^s = \dot{\mathbf{u}}^s$ and $\dot{\mathbf{v}}^s = \ddot{\mathbf{u}}^s$.

In the total Lagrangian description, the energy conjugate stress and strain pair with respect to the initial configuration $\Omega^s(0)$ is the Green–Lagrangian strain $\epsilon_{ij}$ and the second Piola–Kirchhoff stress $S_{ij}$. In the current configuration $\Omega^s(t)$, the energy conjugate stress and strain pair are the Cauchy stress $\sigma^s_{ij}$ and the small strain $e^s_{ij}$. Note that the definition of the small strain $e^s_{ij}$ with respect to the current solid configuration has the same form as the velocity strain defined for the fluid domain, namely, $e^s_{ij} = (u^s_{i,j} + u^s_{j,i})/2$. In the variational form or the weak form we have the following equivalence:

$$\int_{\Omega^s(0)} \delta\epsilon_{ij} S_{ij}\, d\Omega = \int_{\Omega^s(t)} \delta e^s_{ij} \sigma^s_{ij}\, d\Omega. \tag{2.56}$$

In order to match the expressions in Eqs. (2.45) and (2.46), we still prefer to use the Cauchy stress for the current solid configuration. However, the following expression must be used to convert the solid Cauchy stress from the second Piola–Kirchhoff stress:

$$\sigma^s_{ij} = \frac{1}{\det(\mathbf{D})} D_{im} S_{mn} D_{jn}. \tag{2.57}$$

Define the Green–Lagrangian strain as $\boldsymbol{\epsilon} = (\mathbf{C} - \mathbf{I})/2$, where $\mathbf{C}$ is the Cauchy–Green deformation tensor expressed as $\mathbf{D}^T\mathbf{D}$ and $\mathbf{I}$ is the identity matrix. For compressible solid media, using indicial notations, the governing equations (momentum and continuity equations) are given by the following system of partial differential equations:

$$\rho^s \ddot{u}^s_i - \sigma^s_{ij,j} - f^s_i = 0, \quad \text{in } \Omega^s, \tag{2.58}$$

$$J_3 - 1 + \frac{p^s}{\kappa^s} = 0, \quad \text{in } \Omega^s, \tag{2.59}$$

with the boundary conditions

$$\begin{aligned} u^s_i &= u^{s,p}_i, && \text{on } \Gamma^{s,v}, \\ \sigma^s_{ij} n^s_j &= f^{s,p}_i, && \text{on } \Gamma^{s,f}, \\ v^s_i &= v^f_i, && \text{on } \Gamma^{fsi}, \\ \sigma^s_{ij} n^s_j &= f^{fs}_i, && \text{on } \Gamma^{fsi}, \end{aligned} \tag{2.60}$$

where $\rho^s$, $\mathbf{u}^s$, $\mathbf{u}^{s,p}$, $\mathbf{f}^s$, $J_3$, $\kappa^s$, $\mathbf{f}^{s,p}$, and $\mathbf{f}^{fs}$ are the solid density, displacement vector, prescribed displacement at its Dirichlet boundary, body force, determinant of the deformation gradient, bulk modulus, prescribed surface traction at its Neumann boundary, and the surface traction contributed by the surrounding fluid at the FSI interface, respectively.

Note that solid density $\rho^s$ is with respect to the current solid configuration and should be denoted as $\rho^s(t)$. Thus, the solid density in the original configuration will be denoted as $\rho^s(0)$. We must also point out that the solid domain $\Omega^s$ and the material point position $\mathbf{x}^s$ all refer to the current solid configurations, and for clarity they could be denoted as $\Omega^s(t)$ and $\mathbf{x}^s(t)$, respectively. To properly introduce the material constitutive laws, we must first introduce elastic energy $\bar{W}$, which is often related to the invariants of the Cauchy–Green deformation tensor $\mathbf{C}$. A typical hyperelastic material can be described by the Mooney–Rivlin material model [265], which includes large displacements and large strains. The fundamental assumption of this material model is the strain energy potential $\bar{W}$ at time $t$, defined as

$$\bar{W} = C_1(J_1 - 3) + C_2(J_2 - 3) + \kappa^s(J_3 - 1)^2/2, \tag{2.61}$$

where the invariants $J_i$ are functions of the invariants $I_i$, namely, $J_1 = I_1 I_3^{-1/3}$, $J_2 = I_2 I_3^{-2/3}$, and $J_3 = I_3^{1/2}$, with $I_1 = C_{kk}$, $I_2 = [(I_1)^2 - C_{ij}C_{ij}]/2$, and $I_3 = \det(\mathbf{C})$.

Moreover, to account for the mass conservation, namely, the continuity equation, an additional elastic energy term $\bar{Q}$, or $-[p^s + \kappa^s(J_3 - 1)]^2/2\kappa^s$ is added to $\bar{W}$, along with solid unknown pressure $p^s$. In the mixed finite element formulation, $p^s$ is an independent pressure unknown. To delineate this pressure unknown from the pressure calculated based on the volumetric strain and the bulk modulus, we introduce

$$\bar{p} = -\kappa^s(J_3 - 1). \tag{2.62}$$

In comparison with Eq. (2.59), the continuity equation becomes a constraint between pressures $p^s$ and $\bar{p}$. In the displacement-based finite element formulation, such a constraint is strictly enforced, namely, $p^s = \bar{p}$. However, for almost incompressible materials with large bulk modulus, which covers almost all biological materials, the displacement-based finite element formulation will introduce so-called checkerboard pressure distributions and need to be replaced with mixed finite element formulations or combined with reduced integration techniques [40, 146]. Therefore, the second Piola–Kirchhoff stress can be expressed as

$$S_{kl} = \frac{\partial W}{\partial \epsilon_{kl}} = \frac{\partial \bar{W}}{\partial \epsilon_{kl}} - \left[p^s + \kappa^s(J_3 - 1)\right]\frac{\partial J_3}{\partial \epsilon_{kl}}, \tag{2.63}$$

with $W = \bar{W} + \bar{Q}$.

In the ALE formulation for the fluid domain, the state variables are attached to the mesh points and the mesh referential velocity is related to the material velocity. In the Lagrangian formulation for the solid domain, the state variables are also attached to the mesh points, nevertheless, the mesh referential velocity is the same as the material velocity. Here, since we solve the coupled governing equations for both fluid and solid domains, to formulate the incremental analysis we denote the time rate of change of a typical variable $a$ as $\dot{a}$. Of course, for the fluid domain $\dot{a}$ stands for the mesh referential derivative, whereas for the solid domain $\dot{a}$ stands for the material derivative. If we define the Sobolev spaces, the weak forms or variational forms are derived as: $\forall \delta p^f \in \mathcal{L}^2(\Omega^f)$, $\delta p^s \in \mathcal{L}^2(\Omega^s)$, $\delta \mathbf{v}^f \in (\mathcal{H}^1_{0,\Gamma^{f,v}}(\Omega^f))^d$, $\delta \mathbf{u}^s \in (\mathcal{H}^1_{0,\Gamma^{s,v}}(\Omega^s))^d$, find $p^s$ and $\mathbf{u}^s$ in $\Omega^s$, and $p^f$ and $\mathbf{v}^f$ in $\Omega^f$ such that for the fluid domain $\Omega^f$:

$$\delta W_v^f = \delta R_v^f \quad \text{and} \quad \delta W_p^f = 0, \tag{2.64}$$

with

$$\delta W_v^f = \int_{\Omega^f} \rho^f \delta \mathbf{v}^f \cdot \left[\dot{\mathbf{v}}^f + \left(\mathbf{v}^f - \mathbf{v}^m\right) \cdot \nabla \mathbf{v}^f\right] d\Omega + \int_{\Omega^f} \delta \mathbf{e}^f : \boldsymbol{\sigma}^f \, d\Omega, \tag{2.65}$$

$$\delta W_p^f = -\int_{\Omega^f} \delta p^f \left\{\nabla \cdot \mathbf{v}^f + \left[\dot{p}^f + \left(\mathbf{v}^f - \mathbf{v}^m\right) \cdot \nabla p^f\right]/\kappa^f\right\} d\Omega, \tag{2.66}$$

$$\delta R_v^f = \int_{\Omega^f} \delta \mathbf{v}^f \cdot \mathbf{f}^f \, d\Omega + \int_{\Gamma^{f,f}} \delta \mathbf{v}^{f,f} \cdot \mathbf{f}^{f,p} \, d\Gamma + \int_{\Gamma^{fsi}} \delta \mathbf{v}^{fsi} \cdot \mathbf{f}^{sf} \, d\Gamma, \tag{2.67}$$

and for the solid domain $\Omega^s$:

$$\delta W_u^s = \delta R_u^s \quad \text{and} \quad \delta W_p^s = 0, \tag{2.68}$$

with

$$\delta W_u^s = \int_{\Omega^s(0)} \rho^s(0) \delta \mathbf{u}^s \cdot \ddot{\mathbf{u}}^s \, d\Omega + \int_{\Omega^s(0)} \delta \boldsymbol{\epsilon} : \mathbf{S} \, d\Omega, \tag{2.69}$$

$$\delta W_p^s = -\int_{\Omega^s(0)} \delta p^s \left\{(J_3 - 1) + p^s/\kappa^s\right\} d\Omega, \tag{2.70}$$

$$\delta R_u^s = \int_{\Omega^s} \delta \mathbf{u}^s \cdot \mathbf{f}^s \, d\Omega + \int_{\Gamma^{s,f}} \delta \mathbf{u}^{s,f} \cdot \mathbf{f}^{s,p} \, d\Gamma + \int_{\Gamma^{fsi}} \delta \mathbf{u}^{fsi} \cdot \mathbf{f}^{fs} \, d\Gamma. \tag{2.71}$$

Note that $\delta \mathbf{v}^f$ denotes the arbitrary admissible velocity of the fluid domain with $\delta \mathbf{v}^f = \mathbf{0}$ on $\Gamma^{f,v}$, whereas $\delta \mathbf{u}^s$ denotes the arbitrary admissible displacement of the solid domain with $\delta \mathbf{u}^s = \mathbf{0}$ on $\Gamma^{s,v}$. Furthermore, the continuity equations are referred back to the original configuration for simplicity in the derivation of incremental matrices. An important aspect of the modeling of FSI systems is to simultaneously solve for the state variables of both fluid and solid domains. The coupling between these two domains is traced backed to the virtual power term $\int_{\Gamma^{fsi}} \delta \mathbf{v}^{fsi} \cdot \mathbf{f}^{sf} \, d\Gamma$ for the fluid domain and the virtual work term $\int_{\Gamma^{fsi}} \delta \mathbf{u}^{fsi} \cdot \mathbf{f}^{fs} \, d\Gamma$ for the solid domain. In fact, considering the nonslip boundary condition at the FSI interface $\Gamma^{fsi}$, namely, $\mathbf{v}^f = \mathbf{v}^s = \dot{\mathbf{u}}^s$ or $\mathbf{v}^{fsi} = \dot{\mathbf{u}}^{fsi}$, based on the dynamic matching $\mathbf{f}^{sf} = -\mathbf{f}^{fs}$, the virtual power/work input into the fluid domain exerted from the solid domain is exactly equal to the virtual power/work input into the solid domain exerted from the fluid domain. Since we are using the Lagrangian formulation for the solid domain, the variation of the displacement unknown is traditionally introduced.

In order to link with the virtual power representation for fluid domain, we have the following modifications:

$$\delta W_v^s = \int_{\Omega^s(0)} \rho^s(0)\delta \mathbf{v}^s \cdot \ddot{\mathbf{u}}^s \, d\Omega + \int_{\Omega^s(0)} \delta\dot{\boldsymbol{\epsilon}} : \mathbf{S} \, d\Omega, \tag{2.72}$$

$$\delta R_v^s = \int_{\Omega^s} \delta \mathbf{v}^s \cdot \mathbf{f}^s \, d\Omega + \int_{\Gamma^{s,f}} \delta \mathbf{v}^{s,f} \cdot \mathbf{f}^{s,p} \, d\Gamma + \int_{\Gamma^{fsi}} \delta \mathbf{v}^{fsi} \cdot \mathbf{f}^{fs} \, d\Gamma. \tag{2.73}$$

Using the kinematic matching condition along the FSI interface $\Gamma^{fsi}$, namely, $\delta \mathbf{v}^s = \delta \mathbf{v}^f = \delta \mathbf{v}^{fsi}$, and the dynamic matching condition, we have

$$\int_{\Gamma^{fsi}} \delta \mathbf{v}^{fsi} \cdot \mathbf{f}^{sf} \, d\Gamma + \int_{\Gamma^{fsi}} \delta \mathbf{v}^{fsi} \cdot \mathbf{f}^{fs} \, d\Gamma = 0. \tag{2.74}$$

In order to explicitly incorporate the matching of the velocity and pressure unknowns around the FSI interface, we denote

$$\delta W_v = \delta W_v^f + \delta W_v^s, \qquad \delta W_p = \delta W_p^f + \delta W_p^s, \quad \text{and} \quad \delta R_p = 0, \tag{2.75}$$

with $\delta R_v$ expressed as the addition of $\delta R_v^f$ and $\delta R_v^s$:

$$\delta R_v = \int_{\Omega^f} \delta \mathbf{v}^f \cdot \mathbf{f}^f \, d\Omega + \int_{\Gamma^{f,f}} \delta \mathbf{v}^{f,f} \cdot \mathbf{f}^{f,p} \, d\Gamma + \int_{\Omega^s} \delta \mathbf{v}^s \cdot \mathbf{f}^s \, d\Omega + \int_{\Gamma^{s,f}} \delta \mathbf{v}^{s,f} \cdot \mathbf{f}^{s,p} \, d\Gamma. \tag{2.76}$$

In mixed formulations, the high bulk moduli for both fluid and solid domains demand an implicit time integration for the coupled equations, in particular, the continuity equations. For both solution strategies, namely, direct method with Jacobian matrix, and matrix-free Newton–Krylov iterative method, we employ the Newton–Raphson iteration, and apply the Newmark time integration schemes as illustrated in Eqs. (2.138) and (2.139). The details of this time integration scheme for linear and nonlinear systems have been elaborated in Chapter 6 as well as Refs. [152,291,292].

As with linear systems, we separate the interior unknowns from those on the FSI interface in order to better accommodate the difference between displacement unknowns for solid domain and velocity unknowns for fluid domain. We denote $\mathbf{V}^c$ as the velocity unknown of the interface and $\mathbf{V}^f$ as the interior fluid velocity unknown. Likewise, we denote $\mathbf{U}^c$ as the displacement unknown of the interface, hence $\mathbf{U}^s$ is limited to the interior solid displacement unknown. In addition, in the ALE formulation, we introduce $\mathbf{U}^m$ to represent the mesh displacement in the interior region of the fluid domain. In order to maintain the regularity of the fluid mesh, instead of independently assigning the mesh displacement, we often relate the fluid mesh velocity with the corresponding FSI interface velocity $\mathbf{V}^c$. Of course, the FSI interface velocity $\mathbf{V}^c$ can be also expressed as $\dot{\mathbf{U}}^c$. Introducing the relation $L$ with which the mesh regularity is preserved during the time evolutions, we have

$$\Delta\mathbf{V}^m = L\Delta\mathbf{V}^c \quad \text{and} \quad \Delta\mathbf{U}^m = L\Delta\mathbf{U}^c. \tag{2.77}$$

Likewise, for the pressure unknowns along the FSI interface, if we use discontinuous pressure discretizations, we can condense out the solid pressure unknowns at the element level using the continuity equation in the context of the Lagrangian description [13,300]. In this chapter, we assume in the derivation continuous pressure discretizations for both fluid and solid domains. Hence, along the FSI interface, we define the nodal pressure unknown $\mathbf{P}^c$, whereas $\mathbf{P}^f$ and $\mathbf{P}^s$ represent the interior pressure unknowns for fluid and solid regions, respectively.

Around the FSI interface, we must accommodate the fact that $\mathbf{U}^c$ is the primary unknown for solids, whereas $\mathbf{V}^c$ is the primary unknown for fluids. Furthermore, using the Newmark scheme as illustrated in Eqs. (2.156) and (2.157), we also have, $\Delta\mathbf{V}^c = \beta\Delta\mathbf{U}^c/\alpha\Delta t$ and $\Delta\dot{\mathbf{V}}^c = \Delta\mathbf{V}^c/\beta\Delta t$. Of course, the coupling of the fluid and solid domains implies the unknowns $\mathbf{P}^c$ and $\mathbf{V}^c$ are used in both fluid and solid domains, namely, they are continuous through the FSI interface. With this primitive variable formulation for fluids, the coupling between fluids and structures is rather natural. If nonslip boundary conditions are used, the coupling between the fluids and structures can be directly achieved through the element assemblage process. For the problems with unknown material boundaries, such as large amplitude free surface and FSI interface motions, the mesh configurations cannot be assigned arbitrarily. Consequently, additional unknowns of mesh displacements on such material surfaces have to be introduced. For example, on the 2D free surface, where $v_1^m = 0$ and $v_2^m = v_2$, the kinematical relation $\breve{u}_s = v_2 - v_1 \partial u_s/\partial y_1$ provides the link between the mesh displacements over the fluid domain and free surface displacements $u_s$.

In the incremental analysis using the Newton–Raphson iteration, the equilibrium conditions of the spatially discretized or semi-discretized system representing the domain under consideration at any time $t$ can be expressed as

$$\delta\mathbf{W} - \delta\mathbf{R} = \mathbf{0}, \tag{2.78}$$

with $\delta\mathbf{W} = \langle \delta W_v, \delta W_p \rangle$ and $\delta\mathbf{R} = \langle \delta R_v, \delta R_p \rangle$.

In weak forms, the variations $\delta\mathbf{v}^f$, $\delta p^f$, $\delta\mathbf{v}^s$, $\delta p^s$, $\delta\mathbf{v}^c$, and $\delta p^c$ all belong to infinite-dimensional space $\mathcal{R}^\infty$. Nevertheless, after the finite element discretization, the variations of nodal or discretized variables $\delta\mathbf{V}^f$, $\delta\mathbf{P}^f$, $\delta\mathbf{U}^s$, $\delta\mathbf{P}^s$, $\delta\mathbf{U}^c$, and $\delta\mathbf{P}^c$ are within the finite-dimensional space $\mathcal{R}^{M_{vf}}$, $\mathcal{R}^{M_{pf}}$, $\mathcal{R}^{M_{us}}$, $\mathcal{R}^{M_{ps}}$, $\mathcal{R}^{M_{uc}}$, and $\mathcal{R}^{M_{pc}}$, respectively. Here, $M_{vf}$, $M_{pf}$, $M_{us}$, $M_{ps}$, $M_{ufsi}$, and $M_{pfsi}$ stand for the number of unknowns within $\mathbf{V}^f$, $\mathbf{P}^f$, $\mathbf{U}^s$, $\mathbf{P}^s$, $\mathbf{U}^c$, and $\mathbf{P}^c$, respectively.

Denote $\delta\mathbf{W}$ and $\delta\mathbf{R}$ as the left- and right-hand sides of equilibrium equations in variational forms, similar to the internal and external virtual works. In general, to accommodate the large boundary motions (particularly in FSI interaction problems), both $\delta\mathbf{W}$ and $\delta\mathbf{R}$ depend on the current configurations of fluid and solid domains. To solve this type of nonlinear system, we need to employ Newton–Raphson iterations. In the incremental analysis of Eq. (2.78), at any time $t$, we have

$$\begin{aligned}\Delta(\delta\mathbf{W} - \delta\mathbf{R}) &= (\delta\mathbf{W} - \delta\mathbf{R})(t + \Delta t) - (\delta\mathbf{W} - \delta\mathbf{R})(t)\\ &\simeq (\delta\check{\mathbf{W}} - \delta\check{\mathbf{R}})\Delta t.\end{aligned} \tag{2.79}$$

Naturally, based on the Newmark scheme illustrated in Eqs. (2.156) and (2.157), in the implicit scheme we can easily relate the increments of $\check{a}(t + \Delta t)$ and $\check{\check{a}}(t + \Delta t)$ to the increment of $a(t + \Delta t)$. In the implementation, the incremental analysis of the fluid and solid domains described by the ALE and Eulerian descriptions, respectively, are the same. Therefore, for the remaining incremental formulations, we simply replace $\check{a}$ with $\partial a/\partial t$ or $\dot{a}$. Furthermore, to use the chain differentiation and variational forms, we have

$$\begin{aligned}\Delta(\delta\mathbf{W} - \delta\mathbf{R}) \simeq{}& \frac{\partial(\delta\mathbf{W} - \delta\mathbf{R})}{\partial\boldsymbol{\Theta}}\Delta\boldsymbol{\Theta} + \frac{\partial(\delta\mathbf{W} - \delta\mathbf{R})}{\partial\dot{\boldsymbol{\Theta}}}\Delta\dot{\boldsymbol{\Theta}}\\ &+ \frac{\partial(\delta\mathbf{W} - \delta\mathbf{R})}{\partial\ddot{\boldsymbol{\Theta}}}\Delta\ddot{\boldsymbol{\Theta}},\end{aligned} \tag{2.80}$$

with $\boldsymbol{\Theta} = \langle\mathbf{V}^f, \mathbf{U}^c, \mathbf{U}^s, \mathbf{P}^f, \mathbf{P}^c, \mathbf{P}^s\rangle$.

More specifically, based on the discussions in Chapter 2, we can easily derive

$$\begin{aligned}\delta\dot{W}_v^f = \int_{\Omega^f} & \rho^f\delta v_i^f\left[\ddot{v}_i^f + \left(\dot{v}_k^f - \dot{v}_k^m\right)\frac{\partial v_i^f}{\partial y_k} + \left(v_k^f - v_k^m\right)\frac{\partial\dot{v}_i^f}{\partial y_k}\right.\\ &\left. - \left(v_k^f - v_k^m\right)\frac{\partial v_i^f}{\partial y_j}\frac{\partial v_j^m}{\partial y_k}\right]d\Omega\\ - \int_{\Omega^f} & \left(\dot{p}^f\frac{\partial\delta v_k^f}{\partial y_k} - p^f\frac{\partial\delta v_k^f}{\partial y_j}\frac{\partial v_j^m}{\partial y_k} + p^f\frac{\partial\delta v_n^f}{\partial y_n}\frac{\partial v_k^m}{\partial y_k}\right.\\ &\left. - \mu\frac{\partial\delta v_k^f}{\partial y_n}\frac{\partial\dot{v}_k^f}{\partial y_n}\right)d\Omega\end{aligned}$$

$$- \int_{\Omega^f} \mu \left( \frac{\partial \delta v_k^f}{\partial y_i} \frac{\partial v_i^m}{\partial y_n} \frac{\partial v_k^f}{\partial y_n} + \frac{\partial \delta v_k^f}{\partial y_n} \frac{\partial v_k^f}{\partial y_i} \frac{\partial v_i^m}{\partial y_n} - \frac{\partial \delta v_k^f}{\partial y_n} \frac{\partial v_k^f}{\partial y_n} \frac{\partial v_i^m}{\partial y_i} \right) d\Omega$$

$$+ \int_{\Omega^f} \delta v_i^f \left\{ \rho^f \left[ \dot{v}_i^f + \left( v_k^f - v_k^m \right) \frac{\partial v_i^f}{\partial y_k} \right] - f_i^f \right\} \frac{\partial v_n^m}{\partial y_n} \, d\Omega, \tag{2.81}$$

$$\delta \dot{W}_p = - \int_{\Omega^f} \delta p^f \left\{ \frac{\partial \dot{v}_k^f}{\partial y_k} - \frac{\partial v_k^f}{\partial y_n} \frac{\partial v_n^m}{\partial y_k} + \frac{1}{\kappa^f} \left[ \ddot{p}^f + \left( \dot{v}_k^f - \dot{v}_k^m \right) \frac{\partial p^f}{\partial y_k} \right] \right\} d\Omega$$

$$+ \int_{\Omega^f} \frac{\delta p^f}{\kappa^f} \left[ \left( v_k^f - v_k^m \right) \frac{\partial p^f}{\partial y_n} \frac{\partial v_n^m}{\partial y_k} - \left( v_k^f - v_k^m \right) \frac{\partial \dot{p}^f}{\partial y_k} \right] d\Omega$$

$$- \int_{\Omega^f} \delta p^f \left\{ \frac{\partial v_k^f}{\partial y_k} + \frac{1}{\kappa^f} \left[ \dot{p}^f + \left( v_k^f - v_k^m \right) \frac{\partial p^f}{\partial y_k} \right] \right\} \frac{\partial v_n^m}{\partial y_n} \, d\Omega, \tag{2.82}$$

$$\delta \dot{W}_v^s = \int_{\Omega^s(0)} \rho^s(0) \delta \mathbf{v}^s \cdot \ddot{\mathbf{v}}^s d\Omega + \int_{\Omega^s(0)} \frac{\partial \epsilon_{ij}}{\partial \mathbf{u}^s} \delta \mathbf{v}^s L_{kl}(S_{ij}) \frac{\partial \epsilon_{kl}}{\partial \mathbf{u}^s} \dot{\mathbf{u}}^s \, d\Omega, \tag{2.83}$$

$$\delta \dot{W}_p^s = - \int_{\Omega^s(0)} \delta p^s \left( \frac{\partial J_3}{\partial \mathbf{u}^s} \dot{\mathbf{u}}^s + \frac{\dot{p}^s}{\kappa^s} \right) d\Omega. \tag{2.84}$$

### 7.2.2. Direct method with Jacobian matrix

If we compare the discretized governing equation with the incremental form in (2.65), (2.67), (2.76), and (2.73), it is not difficult to identify the following:

$$\delta(\mathbf{V}^f)^T \mathbf{F}_{v^f} + \delta(\mathbf{V}^c)^T \mathbf{F}_{u^c} + \delta(\mathbf{V}^s)^T \mathbf{F}_{u^s} = \delta W_v - \delta R_v,$$
$$\delta(\mathbf{P}^f)^T \mathbf{F}_{p^f} + \delta(\mathbf{P}^c)^T \mathbf{F}_{p^c} + \delta(\mathbf{P}^s)^T \mathbf{F}_{p^s} = \delta W_p.$$

Notice that in Eqs. (2.82) and (2.84), for simplicity and conciseness, we do not separate the interior fluid velocity and the interior solid displacement from those on the fluid–solid interface. The benefit of the total Lagrangian formulation in the derivation of the Jacobian matrix is obvious. With the integration domain fixed as the original configuration, there is no complication as in the Leibniz derivative with moving boundaries for the fluid domain. However, for nonlinear solids, the complication comes with nonlinear strain and stress definitions.

In order to enforce a symmetric operation for the incremental analysis of the solid domain, we first introduce the following linear operator $L_{ij}$ to replace $\partial/\partial\epsilon_{ij}$ with $i, j = 1, 2, 3$:

$$L_{ij} = \frac{1}{2} \left( \frac{\partial}{\partial \epsilon_{ij}} + \frac{\partial}{\partial \epsilon_{ji}} \right) = \frac{\partial}{\partial C_{ij}} + \frac{\partial}{\partial C_{ji}}.$$

Hence, if we denote $\epsilon_{ijk}$ as the standard permutation symbol for cross product, the second Piola–Kirchhoff stress can be derived as

$$S_{ij} = L_{ij}(W) = C_1 L_{ij}(J_1) + C_2 L_{ij}(J_2) - p^s L_{ij}(J_3),$$

with

$$L_{ij}(J_1) = L_{ij}^1 I_3^{-1/3} - L_{ij}^3 I_1 I_3^{-4/3}/3,$$

$$L_{ij}(J_2) = L_{ij}^2 I_3^{-2/3} - 2L_{ij}^3 I_2 I_3^{-5/3}/3,$$

$$L_{ij}(J_3) = L_{ij}^3 I_3^{-1/2}/2,$$

and

$$L_{ij}^1 = L_{ij}(I_1) = 2\delta_{ij},$$

$$L_{ij}^2 = L_{ij}(I_2) = 2\delta_{ij} I_1 - (C_{ij} + C_{ji}),$$

$$L_{ij}^3 = L_{ij}(I_3) = (\epsilon_{ikl}\epsilon_{jmn} + \epsilon_{jkl}\epsilon_{imn}) C_{km} C_{ln}/2.$$

In the incremental analysis, in order to assemble the Jacobian matrix, the following operations will also be involved, based on Eq. (2.83):

$$C_{ijkl} = L_{kl}(S_{ij}) = C_1 L_{kl}\big(L_{ij}(J_1)\big) + C_2 L_{kl}\big(L_{ij}(J_2)\big) + \kappa^s L_{kl}(J_3) L_{ij}(J_3) + \kappa^s (J_3 - 1) L_{kl}\big(L_{ij}(J_3)\big),$$

with

$$L_{kl}\big(L_{ij}(J_1)\big) = -\big(L_{kl}^3 L_{ij}^1 + L_{kl}^1 L_{ij}^3 + I_1 L_{klij}^3\big) I_3^{-4/3}/3 + 4 I_1 I_3^{-7/3} L_{kl}^3 L_{ij}^3/9,$$

$$L_{kl}\big(L_{ij}(J_2)\big) = L_{klij}^2 I_3^{-2/3} - 2\big(L_{ij}^2 L_{kl}^3 + L_{kl}^2 L_{ij}^3 + I_2 L_{klij}^3\big) I_3^{-5/3}/3 + 10 I_2 I_3^{-8/3} L_{kl}^3 L_{ij}^3/9,$$

$$L_{kl}\big(L_{ij}(J_3)\big) = -L_{kl}^3 L_{ij}^3 I_3^{-3/2}/4 + L_{klij}^3 I_3^{-1/2}/2,$$

and

$$L_{klij}^1 = L_{kl}\big(L_{ij}^1\big) = 0,$$

$$L_{klij}^2 = L_{kl}\big(L_{ij}^2\big) = 4\delta_{ij}\delta_{kl} - 2(\delta_{ik}\delta_{jl} + \delta_{il}\delta_{jk}),$$

$$L_{klij}^3 = L_{kl}\big(L_{ij}^3\big) = (\epsilon_{ibc}\epsilon_{jdf} + \epsilon_{jbc}\epsilon_{idf}) \times (C_{bd}\delta_{ck}\delta_{fl} + C_{cf}\delta_{bk}\delta_{dl} + C_{bd}\delta_{cl}\delta_{fk} + C_{cf}\delta_{bl}\delta_{dk})/2.$$

Along the FSI interface, we have to satisfy both kinematic and dynamic boundary conditions. For example, along the nonslip interface, we use $\boldsymbol{\sigma}^s \mathbf{n}^s = -\boldsymbol{\sigma}^f \mathbf{n}^f$ and $\mathbf{u}_s^c = \mathbf{u}_f^c$. With primitive variables for both solids and fluids, the coupling is accomplished during the matrix assemblage processes. Using the mapping depicted in Eqs. (2.77), we can easily identify that the matrix $\mathbf{C}^m_{v^f}$ will be combined into the matrix $\mathbf{C}_{v^f u^c}$, the matrix $\mathbf{C}^m_{u^c}$ into the matrix $\mathbf{C}_{u^c u^c}$, the matrix $\mathbf{C}^m_{p^f}$ into the matrix $\mathbf{C}_{p^f u^c}$, the matrix $\mathbf{C}^m_{p^c}$ into the matrix $\mathbf{C}_{p^c u^c}$, the matrix $\mathbf{K}^m_{v^f}$ into the matrix $\mathbf{K}_{v^f u^c}$, the matrix $\mathbf{K}^m_{u^c}$ into the matrix $\mathbf{K}_{u^c u^c}$, the matrix $\mathbf{K}^m_{p^f}$ into the matrix $\mathbf{K}_{p^f u^c}$, and the matrix $\mathbf{K}^m_{p^c}$ into the matrix $\mathbf{K}_{p^c u^c}$. If we compare the discretized governing equation with the incremental form in (2.81), (2.82), (2.83), and (2.84), it is not difficult to identify the following:

$$\left[\delta(\mathbf{V}^f)^T \mathbf{C}^m_{v^f} + \delta(\mathbf{V}^c)^T \mathbf{C}^m_{u^c}\right]\Delta\mathbf{V}^m = -\int_{\Omega^f} \rho^f \delta v_i^f \frac{\partial v_i^f}{\partial y_k} \Delta v_k^m \, d\Omega,$$

$$\left[\delta(\mathbf{P}^f)^T \mathbf{C}^m_{p^f} + \delta(\mathbf{P}^c)^T \mathbf{C}^m_{p^c}\right]\Delta\mathbf{V}^m = \int_{\Omega^f} \delta p^f \frac{\partial p^f}{\partial y_k} \frac{\Delta v_k^m}{\kappa^f} \, d\Omega,$$

$$\delta(\mathbf{V}^f)^T \left(\mathbf{C}_{v^f v^f} \Delta\dot{\mathbf{V}}^f + \mathbf{M}_{v^f u^c} \Delta\ddot{\mathbf{U}}^c\right) + \delta(\mathbf{V}^c)^T \left(\mathbf{C}_{u^c v^f} \Delta\dot{\mathbf{V}}^f + \mathbf{M}_{u^c u^c} \Delta\ddot{\mathbf{U}}^c\right)$$
$$= \int_{\Omega^f} \rho^f \delta v_i^f \Delta\dot{v}_i^f \, d\Omega,$$

$$\delta(\mathbf{V}^f)^T \left(\mathbf{K}_{v^f p^f} \Delta\mathbf{P}^f + \mathbf{K}_{v^f p^c} \Delta\mathbf{P}^c\right) + \delta(\mathbf{V}^c)^T \left(\mathbf{K}_{u^c p^f} \Delta\mathbf{P}^f + \mathbf{K}_{u^c p^c} \Delta\mathbf{P}^c\right)$$
$$= -\int_{\Omega^f} \frac{\partial \delta v_k^f}{\partial y_k} \Delta p^f \, d\Omega,$$

$$\delta(\mathbf{P}^f)^T \left(\mathbf{K}_{p^f p^f} \Delta\mathbf{P}^f + \mathbf{K}_{p^f p^c} \Delta\mathbf{P}^c\right) + \delta(\mathbf{P}^c)^T \left(\mathbf{K}_{p^c p^f} \Delta\mathbf{P}^f + \mathbf{K}_{p^c p^c} \Delta\mathbf{P}^c\right)$$
$$= -\int_{\Omega^f} \left(v_k^f - v_k^m\right) \frac{\delta p^f}{\kappa^f} \frac{\partial \Delta p^f}{\partial y_k} \, d\Omega,$$

$$\delta(\mathbf{P}^f)^T \left(\mathbf{C}_{p^f p^f} \Delta\dot{\mathbf{P}}^f + \mathbf{C}_{p^f p^c} \Delta\dot{\mathbf{P}}^c\right) + \delta(\mathbf{P}^c)^T \left(\mathbf{C}_{p^c p^f} \Delta\dot{\mathbf{P}}^f + \mathbf{C}_{p^c p^c} \Delta\dot{\mathbf{P}}^c\right)$$
$$= -\int_{\Omega^f} \delta p^f \frac{\Delta\dot{p}^f}{\kappa^f} \, d\Omega,$$

$$\delta(\mathbf{V}^s)^T \left(\mathbf{M}_{u^s u^s} \Delta\ddot{\mathbf{U}}^s + \mathbf{M}_{u^s u^c} \Delta\ddot{\mathbf{U}}^c\right) + \delta(\mathbf{V}^c)^T \left(\mathbf{M}_{u^c u^s} \Delta\ddot{\mathbf{U}}^s + \mathbf{M}_{u^c u^c} \Delta\ddot{\mathbf{U}}^c\right)$$
$$= \int_{\Omega^s(0)} \rho^s(0) \delta\mathbf{v}^s \cdot \ddot{\mathbf{v}}^s \, d\Omega,$$

$$\delta(\mathbf{V}^s)^T(\mathbf{K}_{u^s u^s}\Delta\mathbf{U}^s + \mathbf{K}_{u^s u^c}\Delta\mathbf{U}^c) + \delta(\mathbf{V}^c)^T(\mathbf{K}_{u^c u^s}\Delta\mathbf{U}^s + \mathbf{K}_{u^c u^c}\Delta\mathbf{U}^c)$$
$$= \int_{\Omega^s(0)} \frac{\partial\epsilon_{ij}}{\partial\mathbf{u}^s}\delta\mathbf{v}^s L_{kl}(S_{ij})\frac{\partial\epsilon_{kl}}{\partial\mathbf{u}^s}\dot{\mathbf{u}}^s\,d\Omega,$$

$$\delta(\mathbf{V}^s)^T(\mathbf{K}_{u^s p^s}\Delta\mathbf{P}^s + \mathbf{K}_{u^s p^c}\Delta\mathbf{P}^c) + \delta(\mathbf{V}^c)^T(\mathbf{K}_{u^c p^s}\Delta\mathbf{P}^s + \mathbf{K}_{u^c p^c}\Delta\mathbf{P}^c)$$
$$= -\int_{\Omega^s(0)} \frac{\partial J_3}{\partial\mathbf{u}^s}\delta\mathbf{v}^s\dot{p}^s\,d\Omega,$$

$$\delta(\mathbf{P}^s)^T(\mathbf{K}_{p^s p^s}\Delta\dot{\mathbf{P}}^s + \mathbf{K}_{p^s p^c}\Delta\dot{\mathbf{P}}^c) + \delta(\mathbf{P}^c)^T(\mathbf{K}_{p^c p^s}\Delta\dot{\mathbf{P}}^s + \mathbf{K}_{p^c p^c}\Delta\dot{\mathbf{P}}^c)$$
$$= -\int_{\Omega^s(0)} \delta p^s\frac{\dot{p}^s}{\kappa^s}\,d\Omega,$$

$$\delta(\mathbf{P}^s)^T(\mathbf{K}_{p^s u^s}\Delta\mathbf{U}^s + \mathbf{K}_{p^s u^c}\Delta\mathbf{U}^c) + \delta(\mathbf{P}^c)^T(\mathbf{K}_{p^c u^s}\Delta\mathbf{U}^s + \mathbf{K}_{p^c u^c}\Delta\mathbf{U}^c)$$
$$= -\int_{\Omega^s(0)} \delta p^s\frac{\partial J_3}{\partial\mathbf{u}^s}\dot{\mathbf{u}}^s\,d\Omega,$$

$$\delta(\mathbf{P}^f)^T(\mathbf{K}_{p^f v^f}\Delta\mathbf{V}^f + \mathbf{C}_{p^f u^c}\Delta\dot{\mathbf{U}}^c) + \delta(\mathbf{P}^c)^T(\mathbf{K}_{p^c v^f}\Delta\mathbf{V}^f + \mathbf{C}_{p^c u^c}\Delta\dot{\mathbf{U}}^c)$$
$$= -\int_{\Omega^f} \delta p^f\left(\frac{\partial\Delta v_k^f}{\partial y_k} + \frac{\Delta v_k^f}{\kappa^f}\frac{\partial p^f}{\partial y_k}\right)d\Omega,$$

$$\delta(\mathbf{V}^f)^T(\mathbf{K}_{v^f v^f}\Delta\mathbf{V}^f + \mathbf{C}_{v^f u^c}\Delta\dot{\mathbf{U}}^c) + \delta(\mathbf{V}^c)^T(\mathbf{K}_{u^c v^f}\Delta\mathbf{V}^f + \mathbf{C}_{u^c u^c}\Delta\dot{\mathbf{U}}^c)$$
$$= \int_{\Omega^f}\left\{\rho^f\delta v_i^f\frac{\partial v_i^f}{\partial y_k}\Delta v_k^f + (v_k^f - v_k^m)\frac{\partial\Delta v_i^f}{\partial y_k} + \mu\frac{\partial\delta v_k^f}{\partial y_n}\frac{\partial\Delta v_k^f}{\partial y_n}\right\}d\Omega,$$

$$[\delta(\mathbf{V}^f)^T\mathbf{K}_{v^f}^m + \delta(\mathbf{V}^c)^T\mathbf{K}_{u^c}^m]\Delta\mathbf{U}^m$$
$$= \int_{\Omega^f}\left[p^f\frac{\partial\delta v_k^f}{\partial y_j}\frac{\partial\Delta u_j^m}{\partial y_k} - \rho^f\delta v_i^f(v_k^f - v_k^m)\frac{\partial v_i^f}{\partial y_j}\frac{\partial\Delta u_j^m}{\partial y_k}\right]d\Omega$$
$$- \int_{\Omega^f}\left(p^f\frac{\partial\delta v_n^f}{\partial y_n}\frac{\partial\Delta u_k^m}{\partial y_k} + \mu\frac{\partial\delta v_k^f}{\partial y_i}\frac{\partial\Delta u_i^m}{\partial y_n}\frac{\partial v_k^f}{\partial y_n}\right)d\Omega$$
$$- \int_{\Omega^f}\mu\left(\frac{\partial\delta v_k^f}{\partial y_n}\frac{\partial v_k^f}{\partial y_i}\frac{\partial\Delta u_i^m}{\partial y_n} - \frac{\partial\delta v_k^f}{\partial y_n}\frac{\partial v_k^f}{\partial y_n}\frac{\partial\Delta u_i^m}{\partial y_i}\right)d\Omega$$
$$+ \int_{\Omega^f}\delta v_i^f\left\{\rho^f\left[\dot{v}_i^f + (v_k^f - v_k^m)\frac{\partial v_i^f}{\partial y_k}\right] - f_i^f\right\}\frac{\partial\Delta u_n^m}{\partial y_n}\,d\Omega,$$

$$[\delta(\mathbf{P}^f)^T \mathbf{K}^m_{p^f} + \delta(\mathbf{P}^c)^T \mathbf{K}^m_{p^c}]\Delta \mathbf{U}^m$$
$$= \int_{\Omega^f} \left\{ \delta p^f \frac{\partial v_k^f}{\partial y_n} \frac{\partial \Delta u_n^m}{\partial y_k} + \frac{\delta p^f}{\kappa^f} (v_k^f - v_k^m) \frac{\partial p^f}{\partial y_n} \frac{\partial \Delta u_n^m}{\partial y_k} \right\} d\Omega$$
$$- \int_{\Omega^f} \delta p^f \left\{ \frac{\partial v_k^f}{\partial y_k} + \frac{1}{\kappa^f} \left[ \dot{p}^f + (v_k^f - v_k^m) \frac{\partial p^f}{\partial y_k} \right] \right\} \frac{\partial \Delta u_n^m}{\partial y_n} d\Omega.$$

Based on Eq. (2.80), apply the aforementioned discretizations, we get

$$\mathbf{M}\Delta\ddot{\boldsymbol{\Theta}} + \mathbf{C}\Delta\dot{\boldsymbol{\Theta}} + \mathbf{K}\Delta\boldsymbol{\Theta} = \Delta\mathbf{W} - \Delta\mathbf{R}, \tag{2.85}$$

with

$$\mathbf{M} = \begin{bmatrix} \mathbf{0} & \mathbf{M}_{v^f u^c} & \mathbf{0} & \mathbf{0} & \mathbf{0} & \mathbf{0} \\ \mathbf{0} & \mathbf{M}_{u^c u^c} & \mathbf{M}_{u^c u^s} & \mathbf{0} & \mathbf{0} & \mathbf{0} \\ \mathbf{0} & \mathbf{M}_{u^s u^c} & \mathbf{M}_{u^s u^s} & \mathbf{0} & \mathbf{0} & \mathbf{0} \\ \mathbf{0} & \mathbf{0} & \mathbf{0} & \mathbf{0} & \mathbf{0} & \mathbf{0} \\ \mathbf{0} & \mathbf{0} & \mathbf{0} & \mathbf{0} & \mathbf{0} & \mathbf{0} \\ \mathbf{0} & \mathbf{0} & \mathbf{0} & \mathbf{0} & \mathbf{0} & \mathbf{0} \end{bmatrix},$$

$$\mathbf{C} = \begin{bmatrix} \mathbf{C}_{v^f v^f} & \mathbf{C}_{v^f u^c} & \mathbf{0} & \mathbf{0} & \mathbf{0} & \mathbf{0} \\ \mathbf{C}_{u^c v^f} & \mathbf{C}_{u^c u^c} & \mathbf{0} & \mathbf{0} & \mathbf{0} & \mathbf{0} \\ \mathbf{0} & \mathbf{0} & \mathbf{0} & \mathbf{0} & \mathbf{0} & \mathbf{0} \\ \mathbf{C}_{p^f v^f} & \mathbf{C}_{p^f u^c} & \mathbf{0} & \mathbf{C}_{p^f p^f} & \mathbf{C}_{p^f p^c} & \mathbf{0} \\ \mathbf{C}_{p^c v^f} & \mathbf{C}_{p^c u^c} & \mathbf{0} & \mathbf{C}_{p^c p^f} & \mathbf{C}_{p^c p^c} & \mathbf{0} \\ \mathbf{0} & \mathbf{0} & \mathbf{0} & \mathbf{0} & \mathbf{0} & \mathbf{0} \end{bmatrix},$$

$$\mathbf{K} = \begin{bmatrix} \mathbf{K}_{v^f v^f} & \mathbf{K}_{v^f u^c} & \mathbf{0} & \mathbf{K}_{v^f p^f} & \mathbf{K}_{v^f p^c} & \mathbf{0} \\ \mathbf{K}_{u^c v^f} & \mathbf{K}_{u^c u^c} & \mathbf{K}_{u^c u^s} & \mathbf{K}_{u^c p^f} & \mathbf{K}_{u^c p^c} & \mathbf{K}_{u^c p^s} \\ \mathbf{0} & \mathbf{K}_{u^s u^c} & \mathbf{K}_{u^s u^s} & \mathbf{0} & \mathbf{K}_{u^s p^c} & \mathbf{K}_{u^s p^s} \\ \mathbf{K}_{p^f v^f} & \mathbf{K}_{p^f u^c} & \mathbf{0} & \mathbf{K}_{p^f p^f} & \mathbf{K}_{p^f p^c} & \mathbf{0} \\ \mathbf{K}_{p^c v^f} & \mathbf{K}_{p^c u^c} & \mathbf{K}_{p^c u^s} & \mathbf{K}_{p^c p^f} & \mathbf{K}_{p^c p^c} & \mathbf{K}_{p^c p^s} \\ \mathbf{0} & \mathbf{K}_{p^s u^c} & \mathbf{K}_{p^s u^s} & \mathbf{0} & \mathbf{K}_{p^s p^c} & \mathbf{K}_{p^s p^s} \end{bmatrix}.$$

The coupling between the fluid and solid domains is the velocity and pressure unknowns with the superscript *fsi*. For simplicity, we will henceforth denote the superscript *fsi* with a superscript $c$ in contrast to $f$ for the interior fluid unknowns and $s$ for the interior solid unknowns. This approach provides the true dynamic solution of the FSI interaction system and can be directly employed for the dissemination of the complex nonlinear dynamics of the coupled system. According to the kinematic and dynamic matching conditions around the FSI interface $\Gamma^{fsi}$, the variations of the displacement $\delta u^{fsi}$, the velocity $\delta v^{fsi}$, and the pressure $\delta p^{fsi}$ share the same infinite-dimensional space and, after the finite element discretizations, the same finite-dimensional space. As a consequence, we have the following equivalence. Finally, using the mapping relationships in Eq. (2.77) to relate the mesh velocity with the boundary velocity, we can derive the incremental equations for the couple system in the forms (2.156) and (2.157).

Of course, the key is to formulate matrices **M**, **C**, and **K** in the fully coupled incremental analysis, in which both the fluid and solid unknowns are solved simultaneously with either the direct or iterative methods. Without detriment to our presentation, we do not delve into the detailed numerical procedures for slip and nonslip boundaries or the selection of normal vectors [151]. Regarding the time integration procedure along with the Newton–Raphson iteration, we employ a Newark or predictor–corrector scheme. The tangent matrices as presented in Eq. (2.116) provide a link to the instability analysis.

### *7.2.3. Matrix-free Newton–Krylov; multigrid*

In general, for large FSI systems with one million degrees of freedoms, this Jacobian matrix requires a terabyte ($10^{12}$) memory, which is beyond the limit of computational facilitates available for most scientific researches. In this section, we discuss matrix-free Newton–Krylov and multigrid iterative methods. It has been confirmed that a geometrical multigrid solver is much more efficient than the conventional iterative methods. This speeds up the solution of linear systems of equations and is especially true for subsonic pressure solutions governed by elliptical partial differential equations. However, unlike the black box solver, geometrical multigrid solver requires a connection with the grid coefficient generator. Nevertheless, such a cascade of grid coarsening steps only doubles the computation effort for the original problem setup, which is negligible in comparison with the savings in solution time. Furthermore, for the rectangular box computational domain, a uniform mesh coarsening coupled with the final exact solution, using either a tridiagonal or nonsymmetric column solver, is found to be slightly more efficient than a sequential three-dimensional coarsening coupled with two-dimensional then, one-dimensional coarsening.

#### *Matrix-free Newton–Krylov*

In the matrix-free Newton–Krylov iteration, we do not form the Jacobian matrix. Based on this understanding, we would also like to design a preconditioning technique without the use of the Jacobian matrix [51,138,180].

In the $k$th Newton–Raphson iteration at time step $m+1$ of the nonlinear residual equation (3.296), from $\mathcal{R}^N$ to $\mathcal{R}^N$, with $N$ as the number of the total unknowns, we start with a first guess of the incremental unknowns $\Delta\boldsymbol{\Theta}^{k,0}$, namely, $\Delta\mathbf{V}^{f,0}$, $\Delta\mathbf{U}^{c,0}$, $\Delta\mathbf{U}^{s,0}$, $\Delta\mathbf{P}^{f,0}$, $\Delta\mathbf{P}^{c,0}$, and $\Delta\mathbf{P}^{s,0}$, which often are zero vectors. Then the residual of the linearized systems of equations at the $k$th Newton–Raphson iteration is evaluated as

$$\mathbf{p} = -\mathbf{r}^{m+1,k-1} - \mathbf{r}^{m+1,k-1}_{,v^f}\Delta\mathbf{V}^{f,0} - \mathbf{r}^{m+1,k-1}_{,u^c}\Delta\mathbf{U}^{c,0} - \mathbf{r}^{m+1,k-1}_{,u^s}\Delta\mathbf{U}^{s,0}$$
$$- \mathbf{r}^{m+1,k-1}_{,p^f}\Delta\mathbf{P}^{f,0} - \mathbf{r}^{m+1,k-1}_{,p^c}\Delta\mathbf{P}^{c,0} - \mathbf{r}^{m+1,k-1}_{,p^s}\Delta\mathbf{P}^{s,0}. \tag{2.86}$$

If we denote $\mathbf{J}$ as the $N \times N$ Jacobian matrix evaluated at time step $m + 1$, the error vector is used to construct the $n$-dimensional Krylov subspace $\mathcal{K}^n$ or $\mathrm{span}\{\mathbf{p}, \mathbf{Jp}, \ldots, \mathbf{J}^{n-1}\mathbf{p}\}$ and the $k$th Newton–Raphson iteration of the nonlinear residual equation can be rewritten as

$$\mathbf{J} = \left(\mathbf{r}^{m+1,k-1}_{,v^f}, \mathbf{r}^{m+1,k-1}_{,u^c}, \mathbf{r}^{m+1,k-1}_{,u^s}, \mathbf{r}^{m+1,k-1}_{,p^f}, \mathbf{r}^{m+1,k-1}_{,p^c}\mathbf{r}^{m+1,k-1}_{,p^s}\right). \tag{2.87}$$

The approximate solution $\Delta\boldsymbol{\Theta}$ is written as the combination of the initial guess $\Delta\boldsymbol{\Theta}^{k,0}$ and $\mathbf{z}^n$, with $\mathbf{z}^n \in \mathcal{K}^n$. Note that the dimension of the subspace $\mathcal{K}^n$ is $n$ which is much smaller than the dimension $N$ of the unknown vector $\Delta\boldsymbol{\Theta}$. The $N$-dimensional unknown vector $\Delta\boldsymbol{\Theta}$ or rather $\mathbf{z}^n$ is represented with $\mathbf{V}_n\mathbf{y}$, where $\mathbf{y}$ is a much smaller $n$-dimensional unknown vector. In the Generalized Minimum Residual (GMRES) method, modified Gram–Schmidt orthogonalization procedures are used to derive a set of orthonormal vectors $v_i$, with $1 \leqslant i \leqslant n$ in the Krylov space $\mathcal{K}^n$ and an $(n + 1) \times n$ upper-Hessenberg matrix $\bar{\mathbf{H}}_n$. If we define $\mathbf{V}_n = (\mathbf{v}_1, \mathbf{v}_2, \ldots, \mathbf{v}_n)$ and $\mathbf{V}_{n+1} = (\mathbf{v}_1, \mathbf{v}_2, \ldots, \mathbf{v}_{n+1})$, we have the following:

$$\mathbf{JV}_n = \mathbf{V}_{n+1}\bar{\mathbf{H}}_n, \tag{2.88}$$

and the remaining process in the GMRES method is to solve the least square problem:

$$\min_{\mathbf{z}\in\mathcal{K}^n} \|\mathbf{p} - \mathbf{Jz}\| \quad \text{or} \quad \min_{\mathbf{y}\in\mathcal{R}^n} \|\mathbf{p} - \mathbf{JV}_n\mathbf{y}\|. \tag{2.89}$$

Assuming $\gamma$ is the length of the initial residual vector $\mathbf{p}$ and $\mathbf{e}_1$ is the unit vector representing the first column of $(n + 1) \times (n + 1)$ identity matrix, substituting Eq. (2.88), we can show that Eq. (2.89) is equivalent to the following minimization within a much smaller space:

$$\min_{\mathbf{y}\in\mathcal{R}^n} \|\gamma\mathbf{e}_1 - \bar{\mathbf{H}}_n\mathbf{y}\|, \tag{2.90}$$

First of all, the initial residual vector $\mathbf{p}$ in the $k$th Newton–Raphson iteration at time step $m + 1$ of the nonlinear residual equation (3.296) is normalized as $\mathbf{v}_1$ with the length $\gamma = \|\mathbf{p}\|_2$. Using Eq. (2.90), we have the corresponding $(n + 1)$-dimensional residual vector $\mathbf{b} = \gamma\mathbf{e}_1$. We introduce a preconditioning matrix $\boldsymbol{\Lambda}$, for $i = 1$ to $n$, using the modified Gram–Schmidt orthogonalization process; we then have $\mathbf{q}_i = \boldsymbol{\Lambda}^{-1}\mathbf{v}_i$ and $\mathbf{w} = \mathbf{Jq}_i$, and for $j = 1$ to $i$, we have $h_{ji} = \mathbf{w}^T\mathbf{v}_j$ and $\mathbf{w}$ is updated with $\mathbf{w} - h_{ji}\mathbf{v}_j$. Consequently, we obtain $h_{(i+1)i} = \|\mathbf{w}\|_2$ and $\mathbf{v}_{i+1} = \mathbf{w}/h_{(i+1)i}$. An important procedure in the matrix-free Newton–Krylov is the replacement of $\mathbf{w} = \mathbf{J}\mathbf{q}_i$ with a finite difference based calculation,

$$\mathbf{J}\mathbf{q}_i \simeq \left[\mathbf{r}\left(\boldsymbol{\Theta}^{m+1,k-1} + e\mathbf{q}_i\right) - \mathbf{r}\left(\boldsymbol{\Theta}^{m+1,k-1}\right)\right]/e, \tag{2.91}$$

where $e$ is often set to be around the square root of the machine error [51].

After we establish the elements of an upper $n \times n$ Hessenberg matrix $\mathbf{H}_n$ as well as an upper $(n+1) \times n$ Hessenberg matrix $\bar{\mathbf{H}}_n$, for $j = 1$ to $n$, and $i = 1$ to $j - 1$, a factorization of $\mathbf{H}_n$ is carried out through the following rotation matrix operations:

$$h_{ij} = c_i h_{ij} + s_i h_{(i+1)j} \quad \text{and} \quad h_{(i+1)j} = -s_i h_{ij} + c_i h_{(i+1)j}, \tag{2.92}$$

where the entities of the rotation processes are calculated as

$$r = \sqrt{h_{jj}^2 + h_{(j+1)j}^2}, \qquad c_j = h_{jj}/r, \quad \text{and} \quad s_j = h_{(j+1)j}/r. \tag{2.93}$$

Through this rotation process, the upper Hessenberg matrix is converted to a diagonal matrix with the coefficients defined as: for $j = 1$ to $n$

$$h_{jj} = r, \quad p_j = c_j b_j, \quad \text{and} \quad p_{j+1} = -s_j b_j. \tag{2.94}$$

Finally, the termination criteria of the GMRES iteration will rest at the absolute value of $b_{n+1}$ in comparison with a given error $\epsilon$. If $|b_{n+1}| < \epsilon$, the solution vector $\Delta\boldsymbol{\Theta}$, or rather $\Delta\mathbf{V}^f$, $\Delta\mathbf{U}^c$, $\Delta\mathbf{U}^s$, $\Delta\mathbf{P}^f$, $\Delta\mathbf{P}^c$, and $\Delta\mathbf{P}^s$ is expressed as

$$\Delta\boldsymbol{\Theta}^{k,n} = \Delta\boldsymbol{\Theta}^{k,0} + \sum_{i=1}^{n} y_i \mathbf{q}^i, \tag{2.95}$$

or

$$\langle \Delta\mathbf{V}^f, \Delta\mathbf{U}^c, \Delta\mathbf{U}^s, \Delta\mathbf{P}^f, \Delta\mathbf{P}^c, \Delta\mathbf{P}^s \rangle = \Delta\boldsymbol{\Theta}^{k,0} + \sum_{i=1}^{n} y_i \mathbf{q}^i.$$

As a final remark, if the initial guess $\Delta\boldsymbol{\Theta}^{k,0}$ does not produce a good estimate within a sufficiently small Krylov subspace $\mathcal{K}^n$, $\Delta\boldsymbol{\Theta}^{k,n}$ will be introduced as an updated initial guess and the GMRES iteration procedure will continue until a solution with the desired accuracy is obtained.

### *Multigrid*

In this section, we explore the use of $V$-cycle multigrid solver for three-dimensional FSI problems. Based on comparison with a few generic one-, two-, and three-dimensional test cases, we confirm that the $V$-cycle multigrid solver is in general much more efficient than corresponding iterative solvers. The improvement in solution time with comparable accuracy is particularly significant ($\sim$ two magnitudes) for low spatial wave numbers.

There are several popular iterative methods for the solution of linear systems of equations, among which Jacobi, weighted or damped Jacobi, Gauss–Seidel, and Successive Overrelaxation (SOR) methods are prevalently used in engineering practice. An overview of various iterative solution techniques and their variations are available in Ref. [10]. In essence, all iterative solution methods for the solution of an elliptic pressure equation suffer from poor convergence rate for low spatial wave numbers. Because in engineering practice, subsonic pressure fields are almost always smooth, there is a need to introduce the multigrid solution technique which has been proven to be effective in eliminating those low wave number errors [242] and significantly speeding up the solution process.

Although in multigrid methods, the sweep or iteration is based on the conventional iterative methods, the coarsening of the fine grid, often by a factor of 2 in all directions, transforms the low wave number error into a high wave number error at all subsequent grid levels. Another benefit of the multigrid method is that the solution time is of order $N\log(N)$, where $N$ represents the number of grid unknowns. On the other hand, the Gauss–Seidel solution time is of order $N^2$. Since most iterative methods are effective in reducing high wave number errors, the coarsening of the fine grid results in tremendous speed for the solver.

There are a number of issues which have effects on the performance of multigrid accelerations. The key issue is the grid coarsening. In general, geometric coarsening is much more effective than algebraic coarsening and carries less than two times the work of a regular fine grid generation. Nevertheless, the geometric multigrid solver is not a strictly black box solver and special treatment of the mesh coarsening and the generation of the coefficients must be considered. This is of particular importance if the computational domain has different mesh densities in three directions.

In this section, we employ a $V$-cycle multigrid method with various mesh coarsening schemes. At each grid level, we perform $m$ (often 1 or 2) Gauss–Seidel or Successive Overrelaxation sweeps. The coarsening starts with a factor of 2 in all directions until the first dimension is exhausted. Subsequently, we adopt three different coarsening procedures for the cubic and rectangular boxes. We elaborate the $V$-cycle multigrid solver based on the coarsening of the geometrical grids. We also present some numerical examples based on the implemented multigrid solver.

In order to handle Dirichlet, Neumann, and periodic boundary conditions within the same program and the same array numbering in the actual implementation of multigrid solution procedures, we always solve the station values at the region

$$1 \leqslant i \leqslant N_1 - 1, \quad 1 \leqslant j \leqslant N_2 - 1, \quad 1 \leqslant k \leqslant N_3 - 1,$$

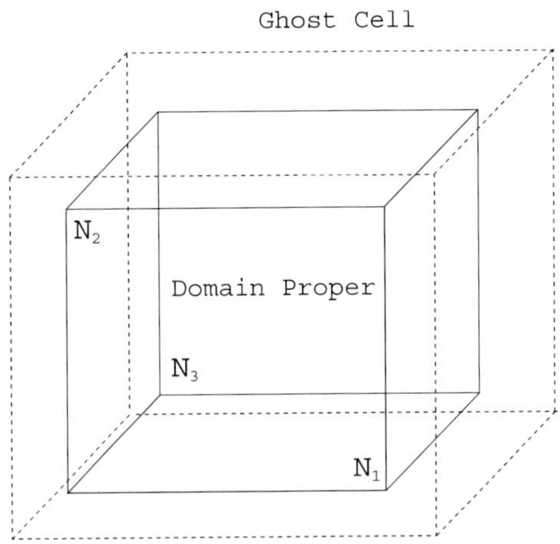


Figure 7.17. Domain proper and ghost region (cell).

where the stations $i = 0$ or $N_1$, $j = 0$ or $N_2$, $k = 0$ or $N_3$ represent the boundary stations for Dirichlet boundary conditions, or the ghost stations for Neumann and periodic boundary conditions, respectively.

In general, we call the stations from 1 to $N_1 - 1$, $N_2 - 1$, and $N_3 - 1$ in three directions, as the domain proper illustrated in Figure 7.17, and the stations at 0, $N_1$, $N_2$, and $N_3$ as the ghost region for Neumann or periodic boundary conditions and the boundary points for Dirichlet boundary conditions, respectively. The subtle change of the grid point definitions for different boundary conditions are illustrated in Figure 7.18. For example, if both ends of one dimension length $L_1$ have Dirichlet boundary conditions, with $N_1 = 2^{n_1}$ divisions, the grid size $h_1$ is defined as $L_1/N_1$. In the same computer program, in order to assign the special values to the points 0 and $N_1$ and retain the same numbering structure in the program, the known or given points are 0 and $N_1$ points and the station unknowns are from 1 to $N_1 - 1$. Therefore, if both ends have Neumann boundary conditions, the physical length in the $x$-direction starts at 1 and ends at $N_1$. Thus, we have the mesh size $h_1 = L_1/(N_1 - 2)$. Moreover, the station values at 0 and $N_1$ correspond to 2 and $N_1 - 2$, hence, named as the ghost points, whereas 1 and $N_1 - 1$ are defined as the boundary points. Naturally, as $N_1$ increases, such a discrepancy becomes insignificant. Likewise, for periodic boundary conditions, ghost points 0 and $N_1$ are assigned with the station values at $N_1 - 1$ and 1, respectively. For the case with mixed Dirichlet and Neumann boundary conditions, for instance, 0 is the Dirichlet boundary and $N_1 - 1$ is the Neumann boundary which provides a link between the station values at $N_1$ and $N_1 - 2$, and the grid size is defined as $h_1 = L_1/(N_1 - 1)$.

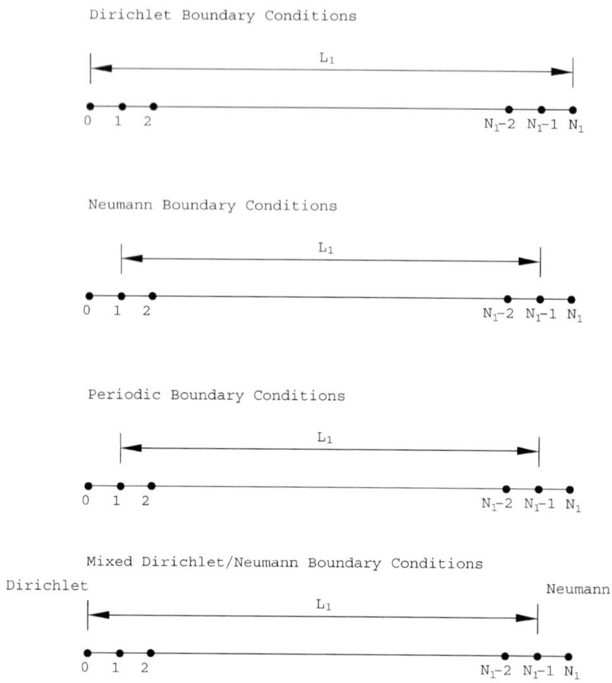


Figure 7.18. Dirichlet, Neumann, and periodic boundary treatments.

To implement a multigrid program, we need to choose a manageable data structure. The overall data structure of our three-dimensional $V$-cycle multigrid program is illustrated in Figure 7.19. In the implementation, we store the right-hand side source or boundary terms in the working array **f**. In order to use various iteration or relaxation schemes, we use the array **v** to store the iteration results **u** based on $\mathbf{Au} = \mathbf{f}$, and the buffer array **w** to calculate the error $\mathbf{r} = \mathbf{f} - \mathbf{Au}$ and to store the intermediate result during the iteration process.

Convention dictates that we store contiguously the solutions and the right-hand side vectors on the nested grids in single arrays. The data structures of **f**, **w**, and **v** are identical with the cascade of hierarchical meshes stacked on top of each other. The buffer array **w** is particularly needed for iterative procedures like SOR. We implement three different mesh coarsening strategies. For a cubic domain with an equal number of divisions in three directions, at the level $k$, with $1 \leqslant k \leqslant n$, the size of the working array is $(2^{n-k+1}+1)(2^{n-k+1}+1)(2^{n-k+1}+1)$ which is stored from the position $ntt$ onward, with

$$ntt = \left(2^{3n-3k+3} - 2^3\right)/7 + 2^{2n-2k+2} - 2^2 + 3\left(2^{n-k+1} - 2\right) + n - k. \tag{2.96}$$

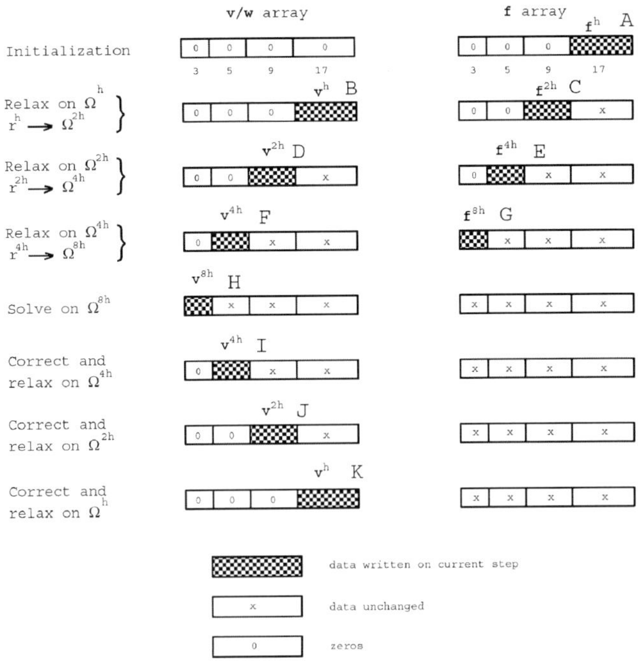


Figure 7.19. Data structures of one-dimensional multigrid procedures from A to K.

If the computational domain is a rectangular box, we have two different approaches. Suppose the size of the domain is $(2^{n_1} + 1) \times (2^{n_2} + 1) \times (2^{n_3} + 1)$, with $n_1 \leqslant n_2 \leqslant n_3$, in the first approach, we coarsen the mesh uniformly in three dimensions from the level 1 to $n_1$. Then, we solve exactly the $n_1$ level mesh $(2^1 + 1) \times (2^{n_2 - n_1 + 1} + 1) \times (2^{n_3 - n_1 + 1} + 1)$ using either the tridiagonal solver if $n_1 = n_2$, or the nonsymmetric column solver if $n_1 < n_2$. In general, if the division numbers in three directions are not extremely different, such a procedure will be very efficient. For example, if the second dimension has 8 times the division number and the third dimension has 128 times the division number in comparison with the first dimension, we only need to solve a linear system of equations with 3825 unknowns.

The other approach to dealing with a rectangular box domain is more involved in the construction of the data structures of the working arrays. In essence, such a coarsening procedure includes coarsening uniformly in all three dimensions from level 1 to $n_1$, coarsening uniformly in two larger dimensions from level $n_1 + 1$ to $n_2$, if $n_2 > n_1$, and finally coarsening in the largest dimension from level $n_2 + 1$ to $n_3$, if $n_3 > n_2$. Therefore, such a scheme includes one-dimensional, two-dimensional, and three-dimensional $V$-cycle multigrid procedures in the same program.

With respect to the data structures, the three-dimensional sweep at the level $k$, with $1 \leqslant k \leqslant n_1$, the size of the working array is $(2^{n_1-k+1}+1)(2^{n_2-k+1}+1) \times (2^{n_3-k+1}+1)$ which is stored from the position *ntt* onward, where

$$\begin{aligned} ntt = nttb &+ \left(2^{n_1+n_2+n_3-3k+3} - 2^{n_2+n_3-2n_1+3}\right)/7 \\ &+ \left(2^{n_2+n_1-2k+2} - 2^{n_2-n_1+2}\right)/3 + \left(2^{n_3+n_1-2k+2} - 2^{n_3-n_1+2}\right)/3 \\ &+ \left(2^{n_2+n_3-2k+2} - 2^{n_3+n_2-2n_1+2}\right)/3 + 2^{n_1-k+1} - 2 \\ &+ 2^{n_2-k+1} - 2^{n_2-n_1+1} + 2^{n_3-k+1} - 2^{n_3-n_1+1} + n_1 - k, \end{aligned} \tag{2.97}$$

with

$$\begin{aligned} nttb = ntta &+ 2^{n_3+n_2-2n_1+2} - 2^{n_3-n_2+2} + 3\left(2^{n_2-n_1+1} - 2^1\right) \\ &+ 3\left(2^{n_3-n_1+1} - 2^{n_3-n_2+1}\right) + 3(n_2 - n_1), \end{aligned} \tag{2.98}$$

$$ntta = 9\left(2^{n_3-n_2+1} - 2 + n_3 - n_2\right). \tag{2.99}$$

Likewise, the two-dimensional sweep at the level $n_1 + k$, if $n_2 > n_1$, with $1 \leqslant k \leqslant n_2 - n_1$, the size of the working array is $(2^1+1)(2^{n_2-n_1-k+1}+1) \cdot (2^{n_3-n_1-k+1}+1)$ which is stored from the position *ntt* onward, where

$$\begin{aligned} ntt = ntta &+ 2^{n_3+n_2-2n_1-2k+2} - 2^{n_3-n_2+2} + 3\left(2^{n_2-n_1-k+1} - 2\right) \\ &+ 3\left(2^{n_3-n_1-k+1} - 2^{n_3-n_2+1}\right) + 3(n_2 - n_1 - k). \end{aligned} \tag{2.100}$$

Finally, the one-dimensional sweep at the level $n_2 + k$, if $n_3 > n_2$, with $1 \leqslant k \leqslant n_3 - n_2$, the size of the working array is $(2^1+1)(2^1+1)(2^{n_3-n_2-k+1}+1)$ which is stored from the position *ntt* onward, where

$$ntt = 9\left(2^{n_3-n_2-k+1} - 2 + n_3 - n_2 - k\right). \tag{2.101}$$

Consider a typical $V$-cycle multigrid scheme and assign $\nu_1$ and $\nu_2$ as the iteration numbers of the down and up cycle sweeps. The computational flow can then be illustrated as follows:

$$\mathbf{v}^h \leftarrow MV^h\left(\mathbf{v}^h, \mathbf{f}^h\right)$$

(1) Relax $\nu_1$ times on $\mathbf{K}^h\mathbf{b}^h = \mathbf{f}^h$ with given initial guess $\mathbf{v}^h = \mathbf{0}$.
(2) If $\Omega^h$ coarsest grid, then go to (4), else

$$\begin{aligned} \mathbf{f}^{2h} &\leftarrow \mathbf{I}_h^{2h}\left(\mathbf{f}^h - \mathbf{K}^h\mathbf{v}^h\right), \\ \mathbf{v}^{2h} &\leftarrow \mathbf{0}, \\ \mathbf{v}^{2h} &\leftarrow MV^{2h}\left(\mathbf{v}^{2h}, \mathbf{f}^{2h}\right). \end{aligned}$$

(3) Convert $\mathbf{v}^h \leftarrow \mathbf{v}^h + \mathbf{I}_{2h}^h\mathbf{v}^{2h}$.
(4) Relax $\nu_2$ times on $\mathbf{K}^h\mathbf{b}^h = \mathbf{f}^h$ with initial guess $\mathbf{v}^h$.

Assuming $n_1 = n_2 = n_3 = n$, we assign the grid number $N$ as $2^n$ and consequently, as we move from level 1 to level $n$, the grid size evolves from $h$ to $2^{n-1}h$. The multigrid scheme can be illustrated as follows:

(1) Relax on $\mathbf{K}^{2^{l-1}h}\mathbf{u}^{2^{l-1}h} = \mathbf{f}^{2^{l-1}h}$ $\nu_1$ times with initial guess $\mathbf{v}^{2^{l-1}h} = \mathbf{0}$.
(2) Employ the full weighting restriction operator $\mathbf{I}_{2^{l-1}h}^{2^l h}$ and compute $\mathbf{f}^{2^l h} = \mathbf{I}_{2^{l-1}h}^{2^l h}\mathbf{r}^{2^{l-1}h}$.
(3) Repeat (1) and (2) from level 1 to $n-1$.
(4) Solve exactly $\mathbf{K}^{2^{n-1}h}\mathbf{u}^{2^{n-1}h} = \mathbf{f}^{2^{n-1}h}$.
(5) Employ the interpolation operator $\mathbf{I}_{2^l h}^{2^{l-1}h}$ and correct $\mathbf{v}^{2^{l-1}h}$ with $\mathbf{v}^{2^{l-1}h} + \mathbf{I}_{2^l h}^{2^{l-1}h}\mathbf{v}^{2^l h}$.
(6) Relax on $\mathbf{A}^{2^{l-1}h}\mathbf{u}^{2^{l-1}h} = \mathbf{f}^{2^{l-1}h}$ $\nu_2$ times with initial guess $\mathbf{v}^{2^{l-1}h}$.
(7) Repeat (5) and (6) from level $n-1$ to 1.

In the three-dimensional case, the full weighting operator $\mathbf{I}_h^{2h}$ is defined as

$$\begin{aligned}
P_{i,j,k}^{2h} = \frac{1}{64}\big(&P_{2i-1,2j-1,2k-1}^h + P_{2i+1,2j-1,2k-1}^h + P_{2i-1,2j+1,2k-1}^h \\
&+ P_{2i+1,2j+1,2k-1}^h + P_{2i-1,2j-1,2k+1}^h + P_{2i+1,2j-1,2k+1}^h \\
&+ P_{2i-1,2j+1,2k+1}^h + P_{2i+1,2j+1,2k+1}^h + 2P_{2i-1,2j-1,2k}^h \\
&+ 2P_{2i+1,2j-1,2k}^h + 2P_{2i-1,2j+1,2k}^h + 2P_{2i+1,2j+1,2k}^h \\
&+ 2P_{2i-1,2j,2k-1}^h + 2P_{2i+1,2j,2k-1}^h + 2P_{2i,2j-1,2k-1}^h \\
&+ 2P_{2i,2j+1,2k-1}^h + 2P_{2i-1,2j,2k+1}^h + 2P_{2i+1,2j,2k+1}^h \\
&+ 2P_{2i,2j-1,2k+1}^h + 2P_{2i,2j+1,2k+1}^h + 4P_{2i+1,2j,2k}^h \\
&+ 4P_{2i,2j-1,2k}^h + 4P_{2i-1,2j,2k}^h + 4P_{2i,2j+1,2k}^h \\
&+ 4P_{2i,2j,2k-1}^h + 4P_{2i,2j,2k+1}^h + 8P_{2i,2j,2k}^h\big),
\end{aligned}$$

whereas the interpolation operator $\mathbf{I}_{2h}^h$ is provided as follows:

$$\begin{aligned}
P_{2i,2j,2k}^h &= P_{i,j,k}^{2h}, \\
P_{2i+1,2j,2k}^h &= \frac{1}{2}\big(P_{i,j,k}^{2h} + P_{i+1,j,k}^{2h}\big), \\
P_{2i,2j+1,2k}^h &= \frac{1}{2}\big(P_{i,j,k}^{2h} + P_{i,j+1,k}^{2h}\big), \\
P_{2i,2j,2k+1}^h &= \frac{1}{2}\big(P_{i,j,k}^{2h} + P_{i,j,k+1}^{2h}\big), \\
P_{2i+1,2j+1,2k}^h &= \frac{1}{4}\big(P_{i,j,k}^{2h} + P_{i+1,j,k}^{2h} + P_{i,j+1,k}^{2h} + P_{i+1,j+1,k}^{2h}\big),
\end{aligned}$$

$$P^h_{2i+1,2j,2k+1} = \frac{1}{4}\left(P^{2h}_{i,j,k} + P^{2h}_{i+1,j,k} + P^{2h}_{i,j,k+1} + P^{2h}_{i+1,j,k+1}\right),$$

$$P^h_{2i,2j+1,2k+1} = \frac{1}{4}\left(P^{2h}_{i,j,k} + P^{2h}_{i,j+1,k} + P^{2h}_{i,j,k+1} + P^{2h}_{i,j+1,k+1}\right),$$

$$P^h_{2i+1,2j+1,2k+1} = \frac{1}{8}\left(P^{2h}_{i+1,j,k} + P^{2h}_{i,j+1,k} + P^{2h}_{i,j,k+1} + P^{2h}_{i,j,k} + P^{2h}_{i+1,j+1,k} + P^{2h}_{i+1,j,k+1} + P^{2h}_{i,j+1,k+1} + P^{2h}_{i+1,j+1,k+1}\right).$$

In the two-dimensional case, the full weighting operator $\mathbf{I}_h^{2h}$ is defined as

$$P^{2h}_{j,k} = \frac{1}{16}\left[P^h_{2j-1,2k-1} + P^h_{2j-1,2k+1} + P^h_{2j+1,2k-1} + P^h_{2j+1,2k+1} + 2\left(P^h_{2j,2k-1} + P^h_{2j,2k+1} + P^h_{2j-1,2k} + P^h_{2j+1,2k}\right) + 4P^h_{2j,2k}\right],$$

whereas the interpolation operator $\mathbf{I}_{2h}^h$ is provided as follows:

$$P^h_{2j,2k} = P^{2h}_{j,k},$$

$$P^h_{2j+1,2k} = \frac{1}{2}\left(P^{2h}_{j,k} + P^{2h}_{j+1,k}\right),$$

$$P^h_{2j,2k+1} = \frac{1}{2}\left(P^{2h}_{j,k} + P^{2h}_{j,k+1}\right),$$

$$P^h_{2j+1,2k+1} = \frac{1}{4}\left(P^{2h}_{j,k} + P^{2h}_{j+1,k} + P^{2h}_{j,k+1} + P^{2h}_{j+1,k+1}\right).$$

In the one-dimensional case, the full weighting operator $\mathbf{I}_h^{2h}$ is defined as

$$P^{2h}_k = \frac{1}{4}\left(P^h_{2k-1} + 2P^h_{2k} + P^h_{2k+1}\right),$$

whereas the interpolation operator $\mathbf{I}_{2h}^h$ is provided as follows:

$$P^h_{2k} = P^{2h}_k \quad \text{and} \quad P^h_{2k+1} = \frac{1}{2}\left(P^{2h}_k + P^{2h}_{k+1}\right).$$

To illustrate the basic procedures of multigrid solvers, we present first a simple one-dimensional model problem:

$$-\frac{\partial^2 u(x)}{\partial x^2} + \sigma u(x) = f(x), \tag{2.102}$$

with $0 \leqslant x \leqslant l$ and $\sigma > 0$.

If we have the Dirichlet boundary condition $u(0) = u(l) = 0$ and $f(x) = C\sin(k\pi x/l)$, the analytical solution for Eq. (2.102) can be expressed as

$$u(x) = \frac{C}{k^2\pi^2/l^2 + \sigma}\sin(k\pi x/l), \tag{2.103}$$

where $k$ stands for the wave number.

Supposing we discretize the domain into $N$ segments, namely, $N+1$ points, the boundary points $x=0$ and $x=l$ correspond to $j=0$ and $j=N+1$, respectively, the grid size $h$ can be expressed as $h=l/N$. Thus, the discretized $f(x)$ can be represented as $f_j^h = C\sin(jk\pi/N)$ and the exact solution is simply written as $u_j = Cl^2\sin(jk\pi/N)/(k^2\pi^2+l^2\sigma)$.

Replacing the derivative of Eq. (2.102) with a second-order finite difference leads to the following system of linear equations:

$$-\frac{u_{j-1}+u_{j+1}-2u_j}{h^2}+\sigma u_j = f_j, \quad \text{with } 1 \leqslant j \leqslant N-1. \tag{2.104}$$

Naturally, from the lexicographic ordering, such a system of equations can be written in a tridiagonal matrix form and the direct solver can be very effectively used.

We denote the linear system in a matrix form $\mathbf{Ku}=\mathbf{f}$, the matrix $\mathbf{K}$ can be decomposed as

$$\mathbf{K}=\mathbf{D}-\mathbf{L}-\mathbf{U}, \tag{2.105}$$

where $\mathbf{D}$, $-\mathbf{L}$, and $-\mathbf{U}$ are the diagonal of $\mathbf{K}$, strictly lower, and strictly upper triangular parts of $\mathbf{K}$, respectively.

Furthermore, Eq. (2.104) can be rewritten as

$$\mathbf{u}=\mathbf{P}_J\mathbf{u}+\mathbf{D}^{-1}\mathbf{f}, \tag{2.106}$$

with $\mathbf{P}_J=\mathbf{D}^{-1}(\mathbf{L}+\mathbf{U})$.

Eq. (2.106) is the starting point of various relaxation schemes. In this book, for demonstration purpose, we also illustrate a weighted Jacobi iteration method, one of the relaxation schemes, for the solution of the system equation (2.104):

$$\mathbf{v}^{(i+1)}=\left[(1-\omega)\mathbf{I}+\omega\mathbf{P}_J\right]\mathbf{v}^{(i)}+\omega\mathbf{D}^{-1}\mathbf{f}, \tag{2.107}$$

where $\omega$ is a weighting factor.

Note that with $\omega=1$, we have the original Jacobi iteration. Of course, as the iteration number $i$ increases, the approximation $\mathbf{v}^i$ approaches the solution $\mathbf{u}$ and the residual vector $\mathbf{r}^i=\mathbf{f}-\mathbf{K}\mathbf{v}^i$ approaches zero. The relaxation scheme, like Eq. (2.107), can be easily changed in the multigrid solution scheme.

Employing the discrete $L_2$ norm, we define the error as

$$\|\mathbf{u}^h-\mathbf{v}^h\|=\sqrt{\sum_{j=1}^{N-1}\left(u_j^h-v_j^h\right)^2/N}. \tag{2.108}$$

Note that even in the case of direct solvers, we can seldom solve the discrete problem exactly. Generally, the quantity or error norm measuring the difference between the direct solver solution and the exact solution is called the global error.

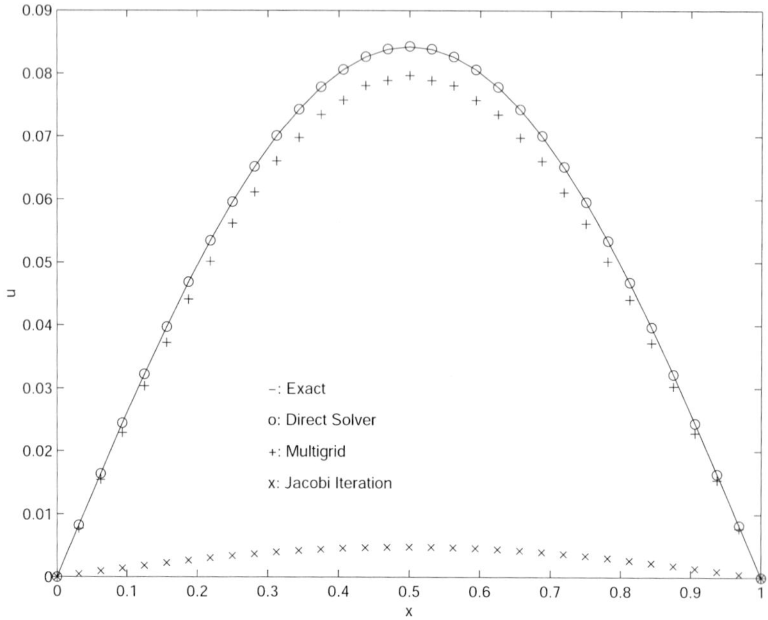


Figure 7.20. Comparison of results of mode 1.

The global error measures the truncation error of the discretization process as well as the round-off error associated with computation precision. Moreover, the error norm measuring the difference between the relaxation or multigrid solution and the exact solution is called the algebraic error. Naturally, the algebraic error measures how well our iterative approximations agree with the exact solution.

For the one-dimensional problem presented here, we compare the performance of the tridiagonal, Jacobi method, and multigrid method for modes 1, 2, and 4. Without detriment to the illustrative purpose, we choose, for convenience, $l = 1$, $C = 1$, $\sigma = 2$, $\nu_1 = \nu_2 = 2$, and $\omega = 1$. In addition, the grid size $h$ is assigned as $1/2^5$ and the number of iteration of the Jacobi method is set at 10.

It is clearly shown in Figures 7.20, 7.21, and 7.22 that for both low and high wave numbers the direct solver provides very accurate solutions, namely, the global errors are small. To be more specific, in this set of examples, the global errors are $3.979984 \times 10^{-5}$, $5.222148 \times 10^{-5}$, and $5.654029 \times 10^{-5}$ with respect to modes 1, 2, and 4, respectively. Notice that the global error slightly increases for higher modes. Furthermore, for 10 iterations, the algebraic errors of the Jacobi method are $5.621204 \times 10^{-2}$, $1.389512 \times 10^{-2}$, and $1.952700 \times 10^{-3}$ with respect to modes 1, 2, and 4. Consistent with theoretical prediction, the convergence rate of the iteration solution procedures for the lower modes is significantly lower

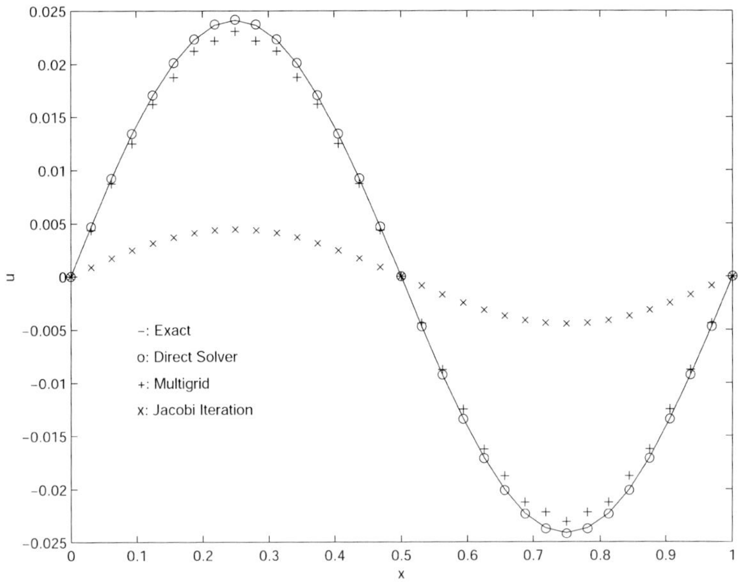


Figure 7.21. Comparison of results of mode 2.

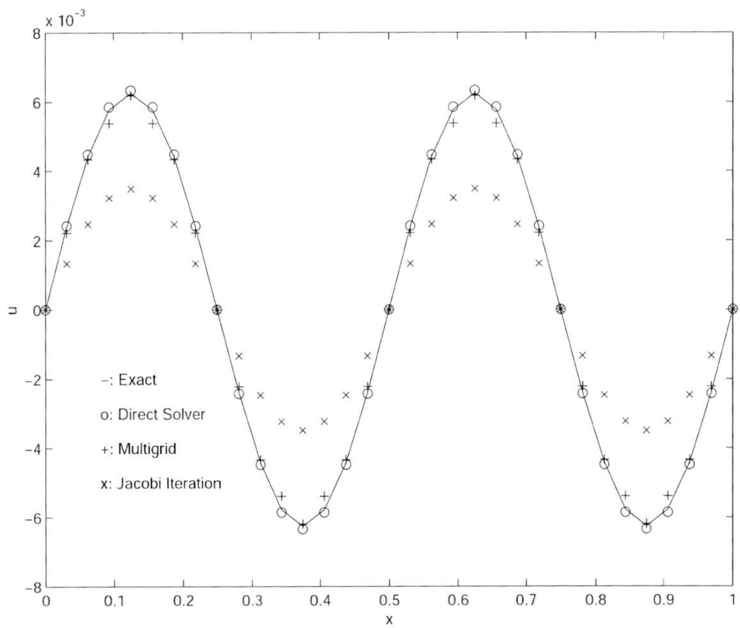


Figure 7.22. Comparison of results of mode 4.

than that of the higher modes. Although this trend still exists, since the multigrid relaxation sweep is the Jacobi method, the algebraic errors of the multigrid, $3.466863 \times 10^{-3}$, $9.529109 \times 10^{-4}$, and $2.1873990 \times 10^{-4}$ with respect to modes 1, 2, and 4, are much lower than those of the Jacobi method.

We now move on to the following three-dimensional model problem:

$$\frac{\partial^2 u}{\partial x^2} + \frac{\partial^2 u}{\partial y^2} + \frac{\partial^2 u}{\partial z^2} = f(x, y, z), \tag{2.109}$$

with $0 \leqslant x \leqslant l_1$, $0 \leqslant y \leqslant l_2$, and $0 \leqslant z \leqslant l_3$.

Suppose we have again the Dirichlet boundary condition $u = 0$ at 0, $l_1$, $l_2$, and $l_3$ and $f(x, y, z) = C \sin(k_1 \pi x/l_1) \sin(k_2 \pi y/l_2) \sin(k_3 \pi z/l_3)$. The analytical solution for Eq. (2.109) can be expressed as

$$\begin{aligned} u = &-\frac{C}{k_1^2\pi^2/l_1^2 + k_2^2\pi^2/l_2^2 + k_3^2\pi^2/l_3^2} \\ &\times \sin(k_1\pi x/l_1)\sin(k_2\pi y/l_2)\sin(k_3\pi z/l_3), \end{aligned} \tag{2.110}$$

where $k_1$, $k_2$, and $k_3$ stand for the wave numbers in $x$-, $y$-, and $z$-directions, respectively.

Supposing we discretize the domain into $N_1 \times N_2 \times N_3$ segments, namely, $(N_1 + 1) \times (N_2 + 1) \times (N_3 + 1)$ points, the boundary points 0, $l_1$, $l_2$, and $l_3$ correspond to 0, $N_1$, $N_2$, and $N_3$, respectively, the discretized $f(x, y, z)$ can be represented as $f^h_{i,j,k} = C \sin(ik_1\pi/N_1) \sin(jk_2\pi/N_2) \sin(kk_3\pi/N_3)$ and the exact solution is simply written as

$$\begin{aligned} u_{i,j,k} = &-\frac{C}{(k_1^2\pi^2/l_1^2 + k_2^2\pi^2/l_2^2 + k_3^2\pi^2/l_3^2)} \\ &\times \sin(ik_1\pi/N_1)\sin(jk_2\pi/N_2)\sin(kk_3\pi/N_3). \end{aligned} \tag{2.111}$$

If we replace the derivative of Eq. (2.109) with the 19-point finite difference scheme, the system of equations can be written in a tridiagonal block matrix form. However, in this case, the traditional tridiagonal solver will not be applicable, since the tridiagonal proper will break down in the course of the solution process. Here, we employ the successive over-relaxation (SOR) iteration method and the corresponding $V$-cycle multigrid iteration method. Due to the sequence of updating, the SOR scheme could be slightly different. For example, if we choose the data structure to loop from $j$, via $k$, to $i$, the SOR scheme can be simply illustrated as

$$\begin{aligned} u_{i,j,k}^{(m+1)} = \frac{\omega}{C_1}\big(f_{i,j,k} &- C_{10}u_{i-1,j+1,k}^{(m+1)} - C_{14}u_{i-1,j,k+1}^{(m+1)} - C_3u_{i-1,j,k}^{(m+1)} \\ &- C_{11}u_{i-1,j-1,k}^{(m+1)} - C_{15}u_{i-1,j,k-1}^{(m+1)} - C_7u_{i,j,k-1}^{(m+1)} - C_{13}u_{i+1,j,k-1}^{(m+1)} \\ &- C_{17}u_{i,j+1,k-1}^{(m+1)} - C_{19}u_{i,j-1,k-1}^{(m+1)} - C_{18}u_{i,j-1,k+1}^{(m+1)} - C_5u_{i,j-1,k}^{(m+1)} \end{aligned}$$

$$
\begin{aligned}
&- C_2 u^{(m)}_{i+1,j,k} - C_4 u^{(m)}_{i,j+1,k} - C_6 u^{(m)}_{i,j,k+1} - C_8 u^{(m)}_{i+1,j+1,k} \\
&- C_{12} u^{(m)}_{i+1,j,k+1} - C_{16} u^{(m)}_{i,j+1,k+1} - C_9 u^{(m)}_{i+1,j-1,k}\big) \\
&+ (1-\omega) u^{(m)}_{i,j,k},
\end{aligned} \tag{2.112}
$$

where $\omega$ is a weighting factor.

Note that with $\omega = 1$, we have the original Jacobi iteration. For the three-dimensional problem presented here, we compare the performance of the SOR method and multigrid method for modes 1 and 2. Without detriment to the illustrative purpose, we choose, for convenience, $l = 1$, $C = 1$, $\nu_1 = \nu_2 = 2$, and $\omega = 1.1$. In addition, the grid size $h_1$, $h_2$, and $h_3$ are assigned as $1/2^6$ and the number of iteration of the SOR method is set to be 100.

For a low wave number, mode 1, it takes 570 seconds to obtain the algebraic error $9.385221 \times 10^{-3}$ using the SOR iterative method, whereas it takes 37 seconds to reach $1.721074 \times 10^{-3}$ accuracy. For mode 2, with the same number of iterations, the SOR method yields an accuracy of $1.140422 \times 10^{-3}$, which is better than $2.342916 \times 10^{-3}$ produced with the multigrid solver. However, the multigrid solver takes 36 seconds with the same number of sweeps (2), whereas the SOR method takes 566 seconds. To be more specific, in this set of examples, the global errors are $3.979984 \times 10^{-5}$, $5.222148 \times 10^{-5}$, and $5.654029 \times 10^{-5}$ with respect to modes 1, 2, and 10, respectively. Furthermore, for 100 iterations, the algebraic errors of the Jacobi method are $5.621204 \times 10^{-2}$, $1.389512 \times 10^{-2}$, and $1.952700 \times 10^{-3}$ with respect to modes 1, 2, and 10, respectively. Consistent with the theoretical prediction, the convergence rate for the lower modes is significantly lower than that of the higher modes.

Finally, as shown in Figures 7.23, 7.24, 7.25, 7.26, and 7.27, the multigrid solution closely resembles the exact solution and the computation is more than 30 times more efficient than the Jacobi iterative solver, given a comparable accuracy. Of course, as predicted in theory, the Jacobi iteration provides better solution for higher modes than lower modes given the same number of iterations.

### *7.2.4. Laminar and turbulent flows; porous interface; benchmark tests*

In this section, we study two simple FSI models with a combination of analytical and computational approaches. Both models are derived from conventional paper machines in which wet paper sheets are conveyed through a series of dryer sections and the residual water is vaporized and carried out in the form of moisture. The thermal energy for drying is transferred to the paper by wrapping a series of large-diameter, rotating steam-filled cylinders. It is therefore not inconceivable that the most expensive part of a paper machine is its massive dryer sections. In fact, in addition to their enormous capital cost, drying sections relate directly to the paper machine energy consumption. Any improvement or even a better un-

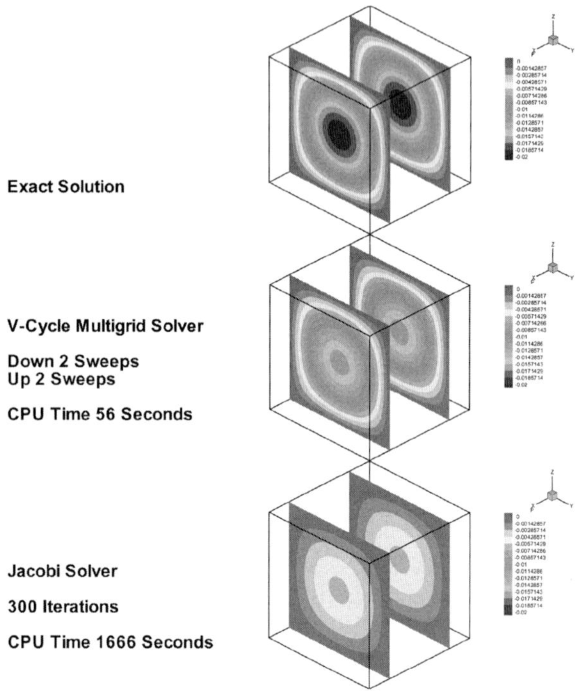


Figure 7.23. Cubic and first mode comparisons of multigrid and Jacobi solvers.

derstanding of dryer pocket ventilation can significantly reduce the papermaking operation cost.

The goal of increased evaporation rate is usually accomplished by various pneumatic devices, the so-called blow boxes, which are currently used in the industry. In general, in order to enhance the circulation in the pocket and to efficiently flush out the moisture, hot air is introduced from the blow box to the dryer pocket via moving permeable felt, where most of the evaporation occurs. In the past, the design of the dryer pocket and blow box has been largely based on trial and error. In fact, lack of understanding of coupled turbulent flow with mass and heat transfers within the dryer pocket has severely limited the application of various blow box designs and hampered the optimization of air movement, which is essential to moisture removal.

Recent works related to the design of the dryer pocket and blow box can be categorized into two subjects: sheet runnability and pocket ventilation. On the subject of sheet runnability, Roisum [60,61] and DeCrosta, Jr. [67] provided a comprehensive summary of paper runnability which includes air movement, vibration, and sheet properties, etc. Pramila [220] formulated an analytical model of the dynam-

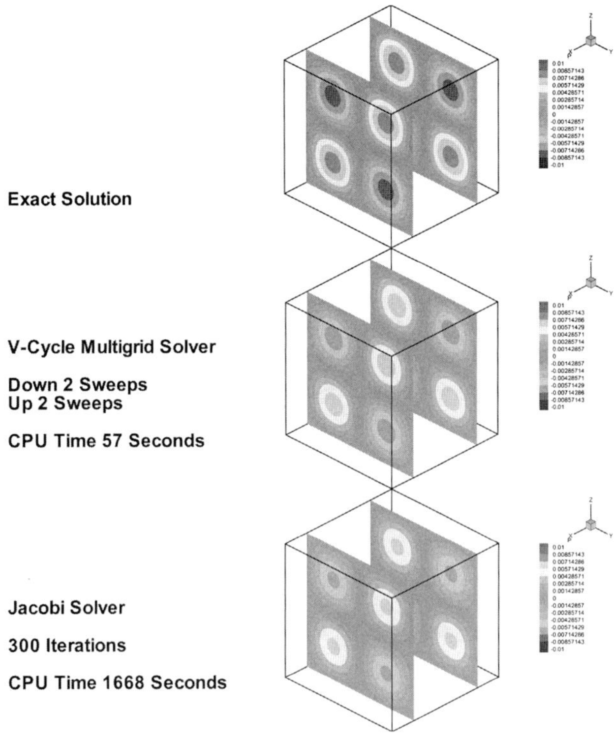


Figure 7.24. Cubic and second mode comparisons of multigrid and Jacobi solvers.

ics of sheet flutter coupled with the surrounding aerodynamic flow. Wang [291] proposed an idea of flutter suppression using air-bearings, and Wahren [290] discussed a simple model which relates wet web tensile data to basis weight, dryness, adhesion, machine speed, and draw. On the subject of ventilation, DeCrosta, Jr. [66] illustrated some fundamental theories pertaining to the evaporative drying process and Race et al. [223] raised some concerns on sheet flutter caused by spontaneous air flow through the felt fabric with high permeability. Although flow through porous media is a traditional subject in civil, mechanical, and chemical engineering [14], because of its complexity there has not yet been a reported simulation which includes the coupling of a highly turbulent flow with a permeable moving boundary as well as both mass and heat transfers. Furthermore, research attention has often been directed towards the flow within porous media and how their physical and chemical composition alter the macroscopic behaviors with respect to different conditions, and little information is available on the spontaneous flow through a moving porous medium, which is of significant importance in textile and papermaking industries.

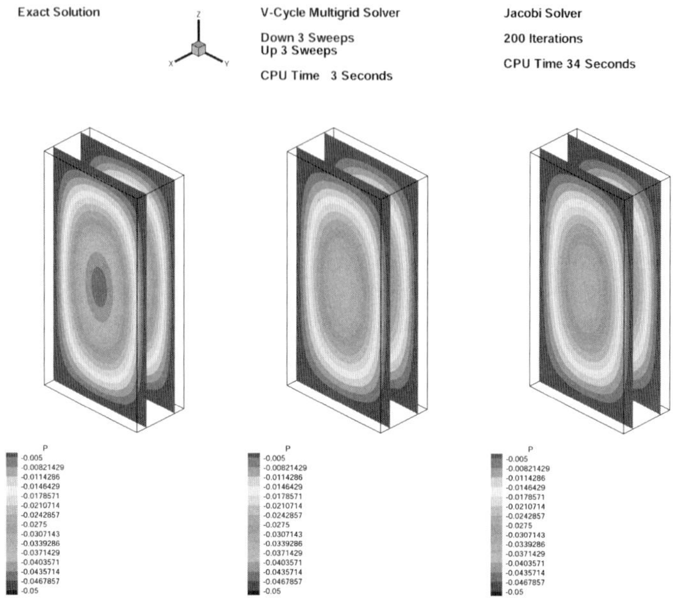


Figure 7.25. Rectangular box and first mode comparisons of multigrid and Jacobi solvers.

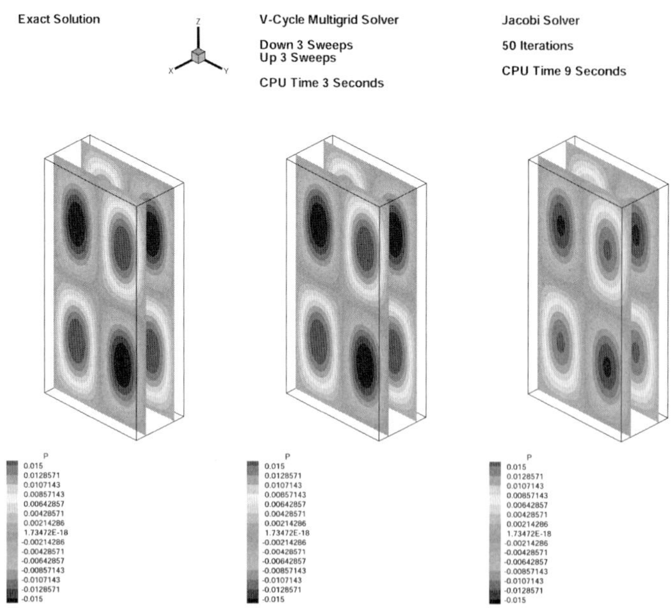


Figure 7.26. Rectangular box and second mode comparisons of multigrid and Jacobi solvers.

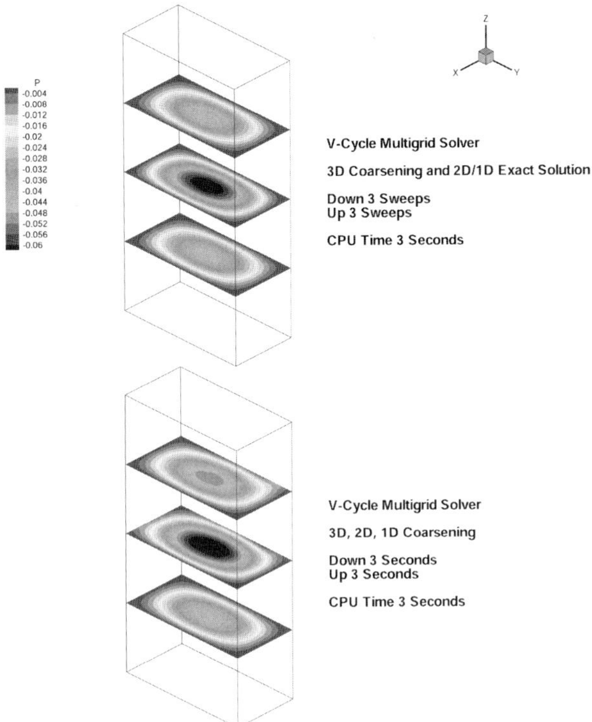


Figure 7.27. Two coarsening strategies in $V$-cycle multigrid solver.

These two FSI problems involve fluid flows passing over the surfaces of a structure which introduce both the normal (pressure) and the tangential (shear) traction. The deformation and motion of the structure alter the flow domains, which will in turn change the traction exerted on the structure. We start with a simple FSI model in which an immersed thin structure is under the influence of the surrounding, fully developed, incompressible Newtonian channel flows. By comparing analytical and computational procedures and the corresponding results, we intend to illustrate the advantages and disadvantages of analytical methods for such FSI problems. In another more sophisticated FSI model, we demonstrate the need to combine full-fledged FSI computational approaches with simple analytical models. The purpose of the second FSI example is to quantify the spontaneous air flow through the permeable felt, and to determine important factors which could significantly affect the quality of products as well as energy consumption. Here, we employ the Navier–Stokes equations for the calculation of the pressure and velocity distributions and Darcy's law for the derivation of the fluid flow through the moving porous medium.

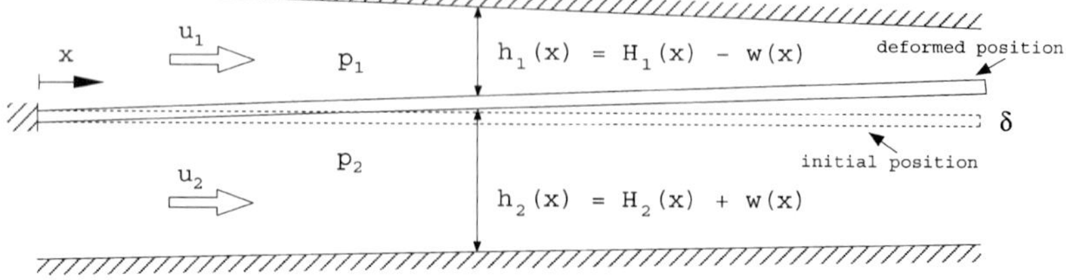


Figure 7.28. The mathematical model of the coupled system.

*Laminar and turbulent flows*

In order to focus on the coupling treatment of FSI systems using both analytical and computational approaches, we only consider the static response of the structure under the influence of steady fluid flows. The dynamical responses associated with FSI systems are discussed in other sections of this chapter as well as in Refs. [291,300,301] and the dynamical stability issues associated with the pulsatile flows can be found in Chapter 4 and Refs. [70,297,299].

We consider a thin structure, as illustrated in Figure 7.28, with one end ($x = 0$) fixed and the other end ($x = L$) free. The two fluid domains, separated by the thin structure with a uniform thickness $\delta$, are governed by widths $H_1(x)$ and $H_2(x)$, where $x$ stands for the axial coordinate starting from the build-in support of the thin structure. If we assume that the structure has a finite deformation $w(x)$, the widths of the flow regions can be expressed as

$$h_1(x) = H_1(x) - w(x) \quad \text{and} \quad h_2(x) = H_2(x) + w(x). \tag{2.113}$$

To incorporate the deformation of the structure, and consequently, the change of fluid domains, we extend the results of the fully developed laminar and turbulent flows in channels with uniform cross-sections to channels with variable cross-sections [5,272]. Because there exist pressure gradient and velocity component perpendicular to the axial flow direction, due to cross-sectional variations, to minimize these effects, and simplify the analytical approaches, we assume that the elastic structure experiences a small deformation $w(x)$, and the channel aspect ratios are small, i.e., $H_i/L \ll O(1)$, with $i = 1$ and 2 representing the upper and lower regions, respectively. Thus, we denote $u_i$ and $p_i$ as the axial velocities and pressures for the channel flows, $\rho$ as the fluid density, and $\mu$ as the fluid dynamic viscosity.

Although we use the two-dimensional plane strain solid model to represent the thin structure in the computational simulation, in the analytical approaches, we use the cantilever beam assumption for the thin elastic structure. The flexural rigidity $D$ is denoted as $EI/(1-\nu^2)$, where $E$ is the Young's modulus, $\nu$ is the Poisson's ratio, and $I = \delta^3/12$ is the moment of inertia of the beam cross-section.

Hence, the governing equation for the static equilibrium of the structure can be written as,

$$EI\frac{d^4w}{dx^4} - T(w,x)\frac{d^2w}{dx^2} = p_2(w,x) - p_1(w,x), \quad x \in [0, L], \tag{2.114}$$

with the boundary conditions

$$\begin{aligned} &w = 0, \qquad \frac{\partial w}{\partial x} = 0, && \text{at } x = 0, \\ &\frac{\partial^2 w}{\partial x^2} = 0, \qquad EI\frac{\partial^3 w}{\partial x^3} + \bar{p}\delta\frac{\partial w}{\partial x} = 0, && \text{at } x = L, \end{aligned} \tag{2.115}$$

where $T(w,x)$ represents the axial tension within the structure, $p_i(w,x)$ stand for the pressures acting on the surfaces of the structure, and $\bar{p}\delta$ denotes the axial load at the tip of the beam, often called the form drag [189].

It is clear that fluid forces are included in the expressions for axial tension $T(w,x)$ and pressure $p_i(w,x)$, and that Eq. (2.114) is nonlinear. Because the final equilibrium position of the elastic structure has to satisfy the balance of linear momentum for both fluid and solid domains, an iteration process is generally needed [291]. Using the small deformation assumption, without detriment to our main objective, we introduce the following Taylor's expansions:

$$\begin{aligned} &p_i(w,x) \simeq p_i(0,x) + w\left.\frac{\partial p_i(w,x)}{\partial w}\right|_{w=0} + O(w^2), \\ &T(w,x) \simeq T(0,x) + O(w). \end{aligned} \tag{2.116}$$

Therefore, we can easily derive the first-order approximation of Eq. (2.114) in the following form:

$$EI\frac{d^4w}{dx^4} - g_1(x)\frac{d^2w}{dx^2} + g_2(x)w = g_3(x), \quad x \in [0, L], \tag{2.117}$$

with

$$\begin{aligned} &g_1(x) = T(0,x), \\ &g_2(x) = \left.\frac{\partial p_1(w,x)}{\partial w}\right|_{w=0} - \left.\frac{\partial p_2(w,x)}{\partial w}\right|_{w=0}, \\ &g_3(x) = p_2(0,x) - p_1(0,x). \end{aligned} \tag{2.118}$$

Note that we retain the first-order approximation of $T(w,x)$, such that the approximation of $T(w,x)d^2w/dx^2$ is of $O(w^2)$, and the final solution is of $O(w^2)$. In addition, we should point out that in the dynamical analysis of FSI systems, both additional stiffness and mass terms can be introduced [105,296]. In the following two subsections, we will discuss different expressions for $g_j(x)$, $j = 1, 2$, and 3, in Eq. (2.117), for both laminar and turbulent flows.

For the fully developed laminar channel flows, we have the governing equation

$$\frac{\partial p_i}{\partial x} = \mu \frac{\partial^2 u_i}{\partial y^2}, \tag{2.119}$$

where $y$ represents local cross-sectional coordinates of the channel flow regions, and $y = 0$ and $h_i$ corresponds to the surfaces of the thin structure and the channel walls, respectively.

Substituting into Eq. (2.119) the following boundary conditions:

$$u_i = 0, \quad \text{at } y = 0 \text{ and } h_i, \tag{2.120}$$

we can easily derive the solutions of Eq. (2.119):

$$u_i = -\frac{1}{2\mu}\left(yh_i - y^2\right)\frac{\partial p_i}{\partial x}, \quad y \in [0, h_i]. \tag{2.121}$$

Furthermore, the flow rates $Q_i$ can be expressed as

$$Q_i = \int_0^{h_i} u_i \, dy = -\frac{h_i^3}{12\mu}\frac{\partial p_i}{\partial x}, \tag{2.122}$$

from which the pressure gradient can be derived as,

$$\frac{\partial p_i}{\partial x} = -\frac{12\mu Q_i}{h_i^3}. \tag{2.123}$$

For the incompressible flows considered here, the flow rates $Q_i$ are constant according to the mass conservation. Because the solutions $u_i$ and $p_i$ depend on the fluid domains defined by $h_i$ in Eq. (2.113), both $u_i$ and $p_i$ are analytical functions of the structure deformation $w$. Thus, from Eq. (2.123), we obtain the governing equation for the pressure difference:

$$\frac{\partial}{\partial x} p_1(w, x) - \frac{\partial}{\partial x} p_2(w, x) = -\frac{12\mu Q_1}{h_1^3} + \frac{12\mu Q_2}{h_2^3}. \tag{2.124}$$

We assume that the upper and lower fluid flows merge together at the right end of the structure, i.e.

$$p_1 = p_2 = 0, \quad \text{at } x = L, \tag{2.125}$$

and by integrating Eq. (2.124), we derive the following expression for the pressure difference $\Delta p(w, x)$:

$$\begin{aligned} \Delta p(w, x) &= p_1(w, x) - p_2(w, x) \\ &= 12\mu \int_x^L \left[ -\frac{Q_1}{h_1^3(s)} + \frac{Q_2}{h_2^3(s)} \right] ds. \end{aligned} \tag{2.126}$$

If we denote $\tau_i(x)$ as the shear stresses along the structure surfaces, from the $x$-direction force balance of the channel flows, we can easily obtain

$$\tau_i = -\frac{h_i}{2}\frac{\partial p_i}{\partial x}. \tag{2.127}$$

In addition, for the fully developed laminar flows, the shear stresses $\tau_i(x)$ can also be written as

$$\tau_i = \mu \frac{\partial u_i}{\partial y}\bigg|_{y=0}. \tag{2.128}$$

It is not difficult to prove that Eqs. (2.127) and (2.128) are equivalent, and that the shear stresses on the surfaces of the structure can be expressed as

$$\tau_i = \frac{6\mu Q_i}{h_i^2}. \tag{2.129}$$

Therefore, from the $x$-direction force equilibrium of the beam, we obtain

$$\frac{\partial T}{\partial x} = -(\tau_1 + \tau_2), \tag{2.130}$$

and as a consequence, with $T(w, L) = -\bar{p}\delta$, we obtain

$$\begin{aligned} T(w, x) &= T(w, L) + \int_x^L \left[\tau_1(s) + \tau_2(s)\right] ds \\ &= -\bar{p}\delta + 6\mu \int_x^L \left[\frac{Q_1}{h_1^2(s)} + \frac{Q_2}{h_2^2(s)}\right] ds. \end{aligned} \tag{2.131}$$

Finally, using the Taylor's expansions illustrated in Eq. (2.116), we obtain

$$\begin{aligned} g_1(x) &= -\bar{p}\delta + 6\mu \int_x^L \left[\frac{Q_1}{H_1^2(s)} + \frac{Q_2}{H_2^2(s)}\right] ds, \\ g_2(x) &= 36\mu \int_x^L \left[\frac{Q_1}{H_1^4(s)} + \frac{Q_2}{H_2^4(s)}\right] ds, \\ g_3(x) &= -12\mu \int_x^L \left[\frac{Q_1}{H_1^3(s)} - \frac{Q_2}{H_2^3(s)}\right] ds. \end{aligned} \tag{2.132}$$

For the fully developed turbulent channel flows, we define the shear stresses along the structural surfaces as $\tau_i = \rho(u_i^*)^2$, where $u_i^*$ stands for the corresponding friction velocity. Introducing $\Lambda_i = 8(u_i^*/U)^2$, with $U$ representing the inlet average velocity, we can easily obtain from the friction laws as discussed in Refs. [5,82]

$$\frac{1}{\sqrt{\Lambda_i}} = A \log_{10}(Re_i \sqrt{\Lambda_i}\,) + B, \tag{2.133}$$

where $A$ and $B$ are constants, and $Re_i$ is the Reynolds number defined as $Re_i = \rho u_i h_i/\mu = \rho Q_i/\mu$.

Because of mass conservation, the Reynolds number $Re_i$ is constant throughout the channel, and as a consequence based on Eq. (2.133), $\Lambda_i$ are also constant. Therefore, the wall shear stress can be expressed as

$$\tau_i(x) = \rho \Lambda_i U^2/8, \tag{2.134}$$

and moreover, from Eq. (2.127), we have,

$$\frac{\partial p_i}{\partial x} = -\frac{C_i}{h_i}, \tag{2.135}$$

with $C_i = \rho \Lambda_i U^2/4$.

If we compare Eq. (2.123) with Eq. (2.135), we can clearly identify the different physical attributes of the laminar and the turbulent flows. Eqs. (2.134) and (2.135) show that in turbulent channel flows, the pressure gradients and shear stresses depend mainly upon fluid inertia rather than viscous effects as in laminar channel flows. Again, from Eq. (2.135), we can derive the pressure difference

$$\Delta p(w,x) = p_1(w,x) - p_2(w,x) = -\int_x^L \left[\frac{C_1}{h_1(s)} - \frac{C_2}{h_2(s)}\right] ds. \tag{2.136}$$

Similarly, the axial tension within the beam can be expressed as

$$\begin{aligned}(Tw,x) &= T(w,L) + \int_x^L \big[\tau_1(s) + \tau_2(s)\big]\, ds \\ &= -\bar{p}\delta + \int_x^L \frac{C_1 + C_2}{2}\, ds.\end{aligned} \tag{2.137}$$

Finally, by virtue of the small deformation assumption for the structure, we obtain

$$\begin{aligned} g_1(x) &= -\bar{p}\delta + \frac{C_1 + C_2}{2}(L - x), \\ g_2(x) &= \int_x^L \left[\frac{C_1}{H_1(s)} + \frac{C_2}{H_2(s)}\right] ds, \\ g_3(x) &= -\int_x^L \left[\frac{C_1}{H_1(s)} - \frac{C_2}{H_2(s)}\right] ds. \end{aligned} \tag{2.138}$$

For the simple structural model discussed in this section, we employ the method of finite differences illustrated in (2.36) in Chapter 4 to convert the partial differential equation (2.117) into a set of linear algebraic equations. In addition, by employing the same finite difference procedure, we obtain the following boundary conditions corresponding to Eq. (2.115):

$$
\begin{aligned}
&W^0 = 0, \qquad W^{-1} = W^1, \qquad W^{N+1} = 2W^N - W^{N-1}, \\
&W^{N+2} = \left(4 - \frac{2\bar{p}\delta h^2}{EI}\right)\left(W^N - W^{N-1}\right) + W^{N-2}.
\end{aligned}
\tag{2.139}
$$

Therefore, the discretized characteristic equation based on the equilibrium equation (2.117) can be written as

for node $k$ $(1 \leqslant k \leqslant N)$:

$$
\begin{aligned}
&\frac{EI}{h^4}W^{k+2} - \left(\frac{g_1^k}{h^2} + \frac{4EI}{h^4}\right)W^{k+1} + \left(g_2^k + \frac{6EI}{h^4} + \frac{2g_1^k}{h^2}\right)W^k \\
&\quad - \left(\frac{g_1^k}{h^2} + \frac{4EI}{h^4}\right)W^{k-1} + \frac{EI}{h^4}W^{k-2} = g_3^k.
\end{aligned}
\tag{2.140}
$$

Note that the variable coefficients $g_j(x)$, $j = 1, 2$, and 3, in Eq. (2.117) are functions of $x$ and are denoted as $g_j^k$, $j = 1, 2$, and 3, at the $k$th nodal point. The final discretized approximation of $w(x)$ can be obtained by solving the algebraic equations (2.139) and (2.140). To validate the analytical approaches discussed in this section, we employ the general purpose commercial code ADINA. We briefly summarize the governing equations used for these examples. We represent both continuous media with a Cartesian coordinate system, where $i = 1, 2$, and 3 stand for $x$-, $y$-, and $z$-coordinates. For the homogeneous, viscous, incompressible laminar flow with constant properties, in computational approaches, we solve the following governing equations derived from the mass and momentum conservation laws with finite element and finite volume methods:

$$
\begin{aligned}
&\frac{\partial v_i}{\partial x_i} = 0, \\
&\rho\left(\frac{\partial v_i}{\partial t} + v_j\frac{\partial v_i}{\partial x_j}\right) = -\frac{\partial p}{\partial x_i} + \frac{\partial}{\partial x_j}\left(\mu\frac{\partial v_i}{\partial x_j}\right),
\end{aligned}
\tag{2.141}
$$

where $\mathbf{v}$ and $p$ stand for the fluid flow velocity vector and pressure, respectively.

By representing the fluctuating parts in the eddy dynamic viscosity $\mu_t$, turbulent kinetic energy $k$, and turbulent dissipation rate $\epsilon$, we can obtain for turbulent flows, the same governing equations as Eq. (2.141), except that $\mu$ is replaced with $\mu + \mu_t$, and $\mathbf{v}$ and $p$ stand for the corresponding time-average quantities. We use in this work the standard $k$-$\epsilon$ turbulence model governed by the following two additional equations:

$$\begin{aligned} \rho \frac{\partial k}{\partial t} + v_j \frac{\partial k}{\partial x_j} &= \frac{\partial}{\partial x_j}\left(\left(\mu + \frac{\mu_t}{\sigma_k}\right)\frac{\partial k}{\partial x_j}\right) + \mu_t \Phi - \frac{k}{T}, \\ \rho \frac{\partial \epsilon}{\partial t} + v_j \frac{\partial \epsilon}{\partial x_j} &= \frac{\partial}{\partial x_j}\left(\left(\mu + \frac{\mu_t}{\sigma_\epsilon}\right)\frac{\partial \epsilon}{\partial x_j}\right) + a_1 \frac{1}{T}\mu_t \Phi - a_2 \frac{\epsilon}{T}, \end{aligned} \tag{2.142}$$

where $a_1$, $a_2$, $\sigma_k$, and $\sigma_\epsilon$ are designated constants, $\Phi$ denotes the inner product of the velocity strain tensor $2e_{ij}e_{ij}$ with $e_{ij} = (\partial v_i/\partial x_j + \partial v_j/\partial x_i)/2$, and the turbulent time scale $T$ and viscosity $\mu_t$ are expressed as

$$T = \frac{k}{\epsilon} + \sqrt{\frac{\mu}{\rho\epsilon}} \quad \text{and} \quad \mu_t = \rho c_\mu k T, \tag{2.143}$$

with a constant $c_\mu$.

All the constants associated with turbulent flows which we have indicated above are closely connected with experimental validations [165,215,231]. Without loss of generality, in this book we select $a_1 = 1.44$, $a_2 = 1.92$, $\sigma_k = 1.0$, $\sigma_\epsilon = 1.3$, and $c_\mu = 0.09$. In addition, for the turbulent flow, we also introduce the following inlet conditions for the turbulent kinetic energy and dissipation:

$$k_i = C_i^1 U_i^2 \quad \text{and} \quad \epsilon_i = \frac{\sqrt{k_i^2}}{C_i^2 H_i(0)}, \tag{2.144}$$

where $C_i^1$ and $C_i^2$ are chosen to be $3/1000$ and $1/6$.

For the thin structure considered in this work, we also use the linear two-dimensional plane strain model derived from the general three-dimensional model in Chapter 2, i.e., $\epsilon_{33} = 0$, $\gamma_{31} = 0$, and $\gamma_{32} = 0$, and

$$\begin{bmatrix} \sigma_{11} \\ \sigma_{22} \\ \sigma_{12} \end{bmatrix} = \frac{E(1-\nu)}{(1+\nu)(1-2\nu)} \begin{bmatrix} 1 & \frac{\nu}{1-\nu} & 0 \\ \frac{\nu}{1-\nu} & 1 & 0 \\ 0 & 0 & \frac{1-2\nu}{2(1-\nu)} \end{bmatrix} \begin{bmatrix} \epsilon_{11} \\ \epsilon_{22} \\ \gamma_{12} \end{bmatrix}. \tag{2.145}$$

For this routine linear elasticity model, we use the displacement-based formulation and the linear strain-displacement relations in Chapter 2.

*Porous interface*

Our main interest with the second simple FSI model is to find out when the relative movement between the porous material and the surrounding air arises, why and how the air is drawn through the moving porous material, and what the significant influencing factors are. We establish two mathematical models as shown in Figure 7.29. In model (a), the flow is assumed to be laminar, whereas in model (b), the flow is considered turbulent. Furthermore, model (a) represents a long and narrow channel with a constant pressure drop across the entire length of the

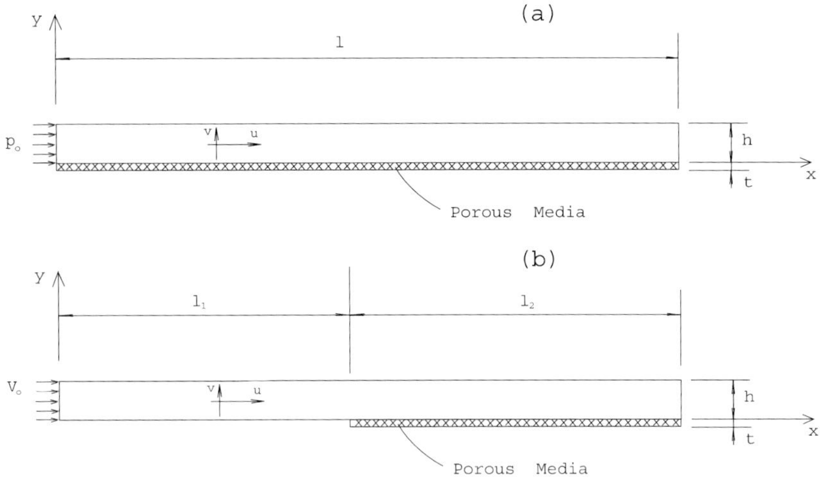


Figure 7.29. The mathematical models of flow passing a porous bottom.

model, whereas model (b) represents a long channel with a finite width and a constant inlet flow rate. Note that in model (a), the longitude size $l$ is much larger than the corresponding lateral size $h$. The rationale for the use of a narrow channel is two-fold. First, it represents an assumption that the effect of spontaneous flow is limited to the boundary layer near the porous material. Second, it greatly simplifies the calculation of the velocity field as well as the pressure distribution, and this assumption can be confirmed by comparing with the simulation results of model (a). In model (b), with a finite geometry and relatively high velocity, the fluid flow is turbulent. Moreover, to ensure a fully developed turbulent flow before the porous bottom, we attach an entrance zone with a length $l_1$.

For simplicity, the analytical approach is only introduced to model (a), in which the ambient gauge pressure is zero. According to Darcy's law, the fluid flow through the porous medium is dependent on the pressure gradient across the porous material. Here, we consider the thickness of the porous layer to be relatively small, similar to that of felt, and consequently, flow rate through the porous medium is proportional to the gauge pressure on the channel bottom. Thus, the spontaneous velocity distribution along the channel length is directly related to the corresponding pressure distribution along the porous layer.

At the left end of model (a), a definite pressure $p_o$ is applied. The right end of the channel keeps a zero pressure. Naturally, the applied pressure difference will drive the fluid from left to right, which is similar to an applied flow rate at the left end of model (b).

For model (b), the standard incompressible Navier–Stokes equations are introduced along the corresponding $k$-$\epsilon$ turbulent model. Needless to say, the full-fledged analytical solution is very complicated and difficult to obtain. In model (a), because of the large ratio of longitudinal and lateral dimensions, we can greatly simplify the governing equation. In model (a), based on the continuity equation, the velocity in the vertical direction $v$ is one order of magnitude smaller than the horizontal velocity $u$. Consequently, the term $\partial^2 u/\partial x^2$ is much smaller than the term $\partial^2 u/\partial y^2$. Moreover, $u\partial u/\partial x$ and $v\partial u/\partial y$ are higher order terms which can be neglected. Since the term $\partial p/\partial y$ is of a higher order term than the term $\partial p/\partial x$, the pressure $p$ is a function of $x$ and remains constant along $y$-direction. In addition, the term $\partial^2 u/\partial y^2$ is dominant and will balance the pressure gradient.

Based on the above discussion, the incompressible Navier–Stokes equations can be rewritten as

$$\frac{dp}{dx} = \mu \frac{\partial^2 u}{\partial y^2}, \tag{2.146}$$

$$\frac{\partial u}{\partial x} + \frac{\partial v}{\partial y} = 0, \tag{2.147}$$

where $\mu$ is the dynamic viscosity.

From the following integration of Eq. (2.147) with respect to $y$:

$$\frac{\partial}{\partial x}\int_0^h u\,dy + \int_0^h \frac{\partial v}{\partial y}\,dy = 0, \tag{2.148}$$

we derive,

$$\frac{dQ}{dx} - v^w = 0, \tag{2.149}$$

where $Q$ is the channel flow rate and $v^w$ is the vertical velocity along the border of the porous layer.

From Eq. (2.146) and the boundary condition $u = 0$ at $y = 0, h$, we obtain the expression of the velocity $u$:

$$u = \frac{1}{2\mu}\frac{dp}{dx}\left(y^2 - hy\right). \tag{2.150}$$

Hence, based on Eq. (2.150), the flow rate can be expressed as

$$Q = \int_0^h u\,dy = -\frac{h^3}{12\mu}\frac{dp}{dx}. \tag{2.151}$$

Now, if we substitute Eq. (2.151) into Eq. (2.149), we obtain

$$-\frac{h^3}{12\mu}\frac{d^2 p}{dx^2} - v^w = 0, \tag{2.152}$$

and by employing Darcy's law, we have

$$v^w = -K\frac{p(x)}{t}, \tag{2.153}$$

where $K$ is the permeability of the porous material, $t$ is the thickness of the layer, and the minus sign is introduced to account for the fact that when the pressure $p(x)$ is positive, the direction of the vertical velocity $v^w$ is downward.

Hence, from Eqs. (2.152) and (2.153), we derive

$$\frac{d^2 p}{dx^2} - \frac{12\mu K}{h^3 t} p = 0. \tag{2.154}$$

Eq. (2.154) is a second order linear homogeneous equation with constant coefficients, and its general solution can be expressed as

$$p(x) = A \sinh \omega x + B \cosh \omega x, \tag{2.155}$$

with $\omega = \sqrt{12\mu K/h^3 t}$, where $A$ and $B$ are constants dependent of the left and right pressure boundary conditions.

From the boundary conditions, $p = p_o$ at $x = 0$ and $p = 0$ at $x = l$, we can easily derive

$$A = -\frac{\cosh \omega l}{\sinh \omega l} p_o \quad \text{and} \quad B = p_o. \tag{2.156}$$

Consequently, the pressure distribution along the porous layer can be expressed as

$$p(x) = p_o\left(\cosh \omega x - \frac{\cosh \omega l}{\sinh \omega l} \sinh \omega x\right), \tag{2.157}$$

and the velocity $v^w(x)$ of the fluid flow through the porous layer can be written as

$$v^w = -\frac{K p_o}{t}\left(\cosh \omega x - \frac{\cosh \omega l}{\sinh \omega l} \sinh \omega x\right). \tag{2.158}$$

Depending on the mass conservation, the following relationship must hold:

$$Q_i = Q_o + Q_p, \tag{2.159}$$

where $Q_i$, $Q_o$, and $Q_p$ are the flow rates of the channel inlet, outlet, and porous bottom, and can be expressed as $Q_i = \int_0^h u(0, y)\, dy$, $Q_o = \int_0^h u(l, y)\, dy$, and $Q_p = \int_0^l v^w(x)\, dx$, respectively.

Based on the preceding results, we derive the following:

$$\frac{dp}{dx} = \left(\sinh \omega x - \frac{\cosh \omega l}{\sinh \omega l} \cosh \omega x\right)\omega p_o, \tag{2.160}$$

$$Q_i = -\frac{h^3}{12\mu}\frac{dp}{dx}\bigg|_{x=0} = \frac{\omega h^3 p_o \cosh \omega l}{12\mu \sinh \omega l}, \tag{2.161}$$

$$Q_o = -\frac{h^3}{12\mu}\frac{dp}{dx}\bigg|_{x=l} = \frac{\omega h^3 p_o}{12\mu \sinh \omega l} = \frac{Q_i}{\cosh \omega l}, \tag{2.162}$$

with

$$Q_p = \int_0^l K\frac{p(x)}{t}\,dx = K\frac{p_o}{\omega t}\left(\frac{\cosh \omega l - 1}{\sinh \omega l}\right). \tag{2.163}$$

In fact, from the mass conservation we obtain

$$Q_p = Q_i - Q_o = Q_i(1 - 1/\cosh \omega l), \tag{2.164}$$

which can be also derived from the definition of $\omega$ and Eq. (2.161).

Eqs. (2.163) and (2.164) explicitly show that the total flow rate out of the porous layer is linearly proportional to pressure $p_o$ or inlet flow rate $Q_i$. In addition, it is also shown in Eq. (2.164) that the correlation with channel length $l$, channel width $h$, porous layer thickness $t$, permeability $K$, and viscosity $\mu$ is implicit and complex. Finally, the distribution of spontaneous flow velocity $v^w$ can be derived from Eq. (2.158).

*Benchmark tests*

To compare analytical approaches with computational approaches, we consider two profiles of the system as shown in Figure 7.28: uniform cross-section system ($H_1 = 0.02$ m, $H_2 = 0.04$ m) and variable cross-section ($H_1(x) = 0.03 - x/75$ m and $H_2(x) = 0.04 - 2x/75$ m). Both systems have the following physical parameters: $\rho = 1000\,\text{kg/m}^3$; $L = 0.75$ m; $\delta = 0.01$ m; $E = 2.0 \times 10^{11}\,\text{N/m}^2$; and $\nu = 0.3$. Results in Figure 7.30 for uniform and variable cross-sections demonstrate that for both laminar and turbulent flow conditions, the first-order results of the structure deformation and the pressures exerted on the structure surfaces, derived from the analytical approaches, match ADINA FSI solutions. In particular, for the cases with uniform cross-sections, analytical and computational approaches produce nearly identical results, indicating that our first-order approximations are accurate and sufficient. To study the effects of the tip form drag $\bar{p}\delta$, we input the pressure difference calculated with ADINA into our analytical model. We find that in this case the effect on the lateral deformation of the structure is negligible.

We note that in ADINA FSI analysis, we are actually solving the two-dimensional Navier–Stokes equations coupled with a two-dimensional plane strain structure; moreover, we use the two-equation turbulence model ($k$-$\epsilon$ model). Because we did not include the velocity component and the pressure gradient perpendicular to the channel axes in our analytical approaches, there are as expected more discrepancies in the solutions obtained with different approaches for the

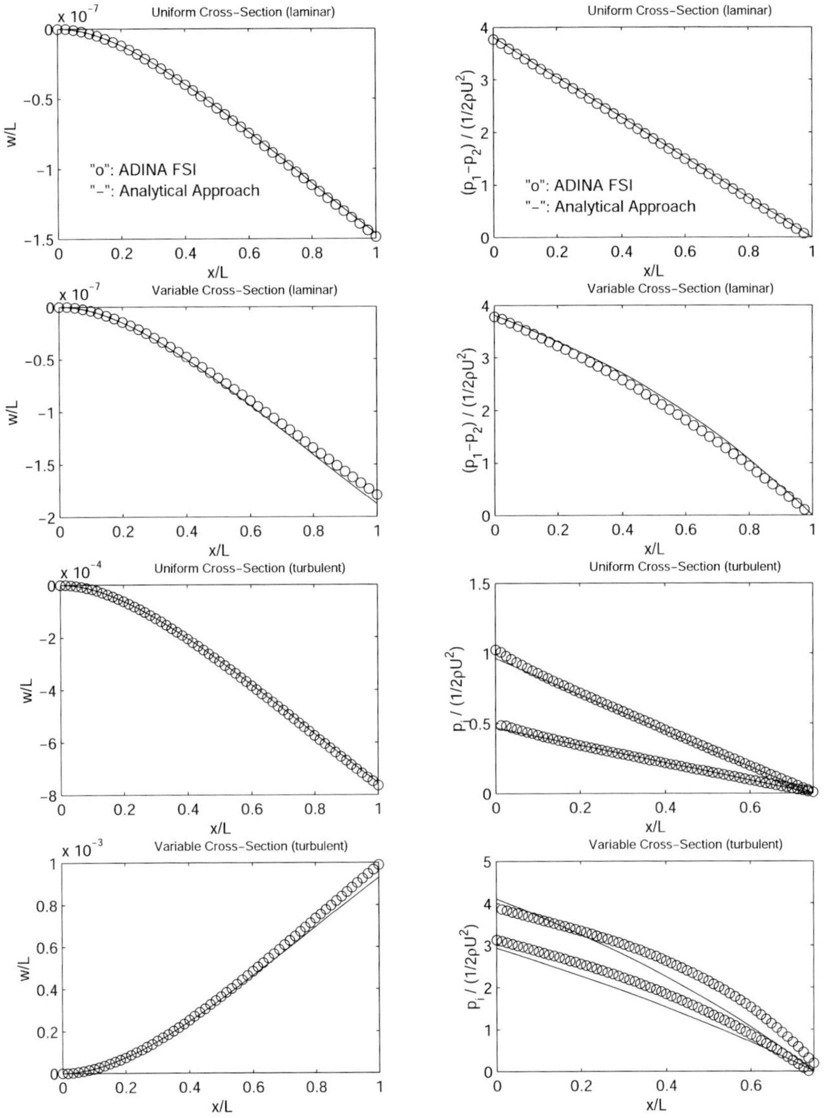


Figure 7.30. Comparison between analytical and computational approaches.

variable cross-section system than for uniform cross-section systems. Nonetheless, due to the small aspect ratios $h_i/L$, the results obtained with ADINA still closely match the analytical solutions. Figures 7.31 and 7.32 also confirm that two-dimensional effects for both laminar and turbulent cases are not strong. How-

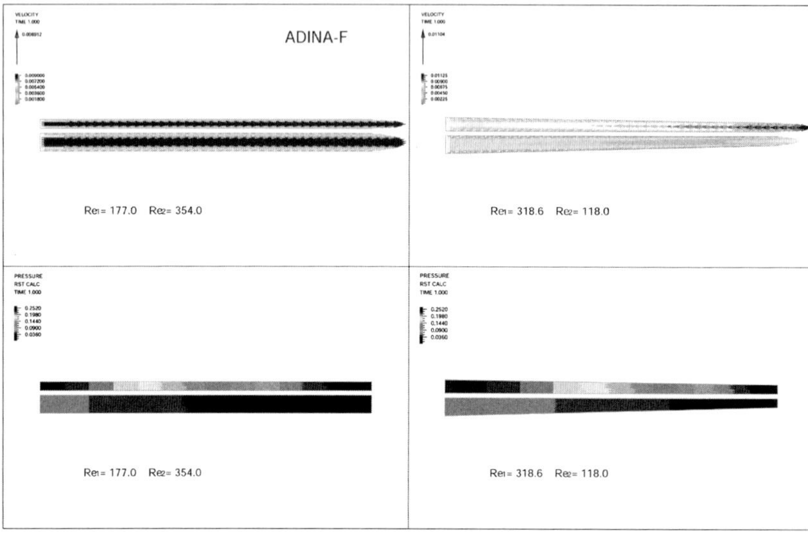


Figure 7.31. Computational results for laminar flow conditions.

ever, it is obvious that when small deformation and aspect ratios assumptions no longer hold, analytical approaches require the use of two-dimensional models and Newton–Raphson iterations. In addition, the assumption (2.125) is no longer valid, nor is the derivation of the pressure difference; therefore, analytical approaches as illustrated in this book will be both difficult to implement and inaccurate, making computational approaches necessary. To demonstrate this point further, we present ADINA simulation results in Figure 7.33. It is clear that the structure deformation is significant and Newton–Raphson iterations are required. In fact, to preserve the finite element mesh regularity, we also have to use the ALE description for viscous fluids along structures experiencing large displacements.

In general areas of turbulent mixing, many research efforts have been directed to experimental investigations. Some of the recent advances in applying CFD techniques to the chemical process industry are documented in Refs. [243,231]. Although numerical studies have been performed on two- or three-dimensional turbulent flow with mass or heat transfer, few studies are available in dealing with moving porous boundaries [161,298].

Here, we present a systematic study of a spontaneous viscous flow passing a moving porous layer. We consider both of the mathematical models illustrated in Figure 7.29. In model (a), laminar flow conditions are used along with a finite pressure drop between the inlet and outlet. In model (b), the turbulent $k$-$\epsilon$ model is employed with a prescribed flow rate at the inlet. The variation of the physical

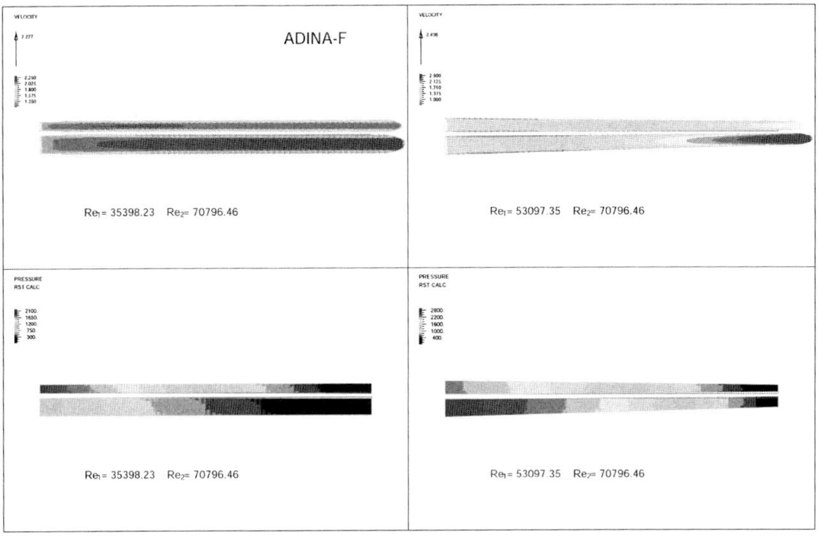


Figure 7.32. Computational results for turbulent flow conditions.

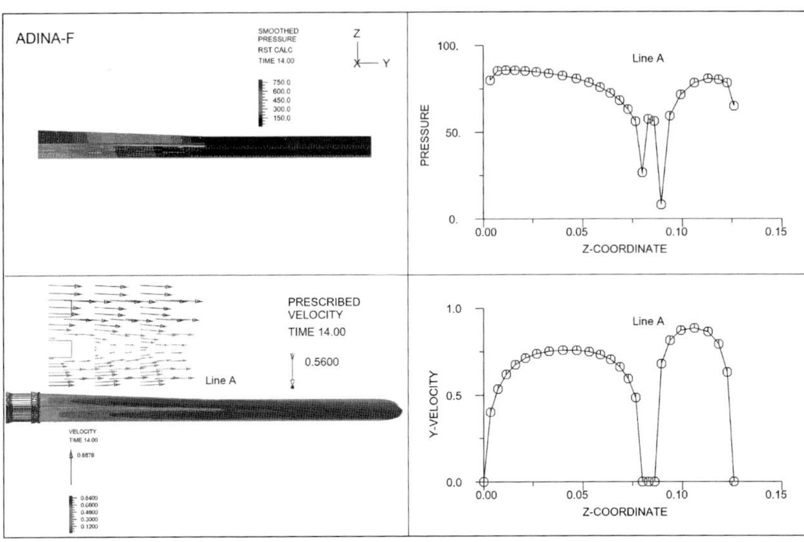


Figure 7.33. Computational results of a FSI model with large structure motions.

parameters includes flow velocity, which represents the moving velocity of the porous layer, and permeability. Note that for the computer simulation, there is no need to simplify the governing partial differential equations as we discussed in the

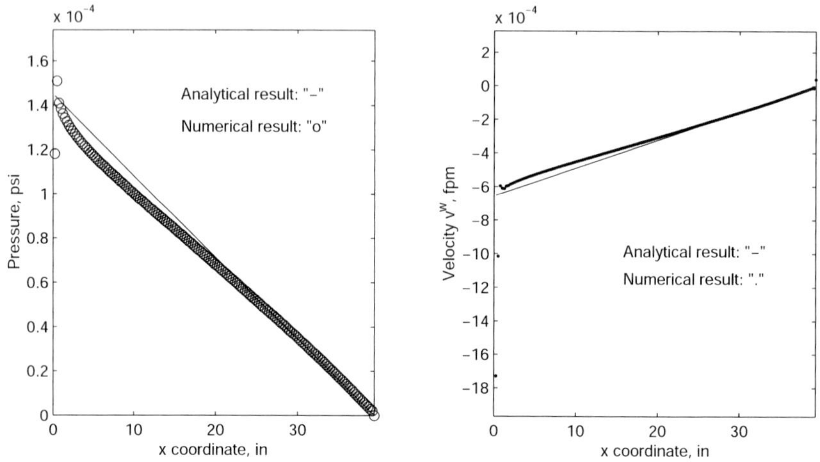


Figure 7.34. The pressure and the velocity along the porous bottom.

previous section for model (a). Furthermore, for the permeable layer, ignoring the gravitational and thermal effects, the governing equations for the two-dimensional isotropic porous flow can be written as

$$u = -K\frac{\partial p}{\partial x} \quad \text{and} \quad v = -K\frac{\partial p}{\partial y}, \tag{2.165}$$

where $K$ stands for the permeability constant.

Because of the simplicity of the computation geometry, we adopt a uniform mesh. In a reasonable refined mesh, the computation takes a few seconds. The simulation results of models (a) and (b) are compared with analytical results and existing experimental data. For model (a), we have $h = 1.968$ in, $l = 39.36$ in, $\rho = 0.0621$ lbm/ft$^3$, and $t = 0.3936$ in, whereas for model (b), we have $h = 3.936$ in, $t = 0.3149$ in, $l_1 = 11.808$ in, $l_2 = 19.7981$ in, and $\rho = 0.0621$ lbm/ft$^3$. We will first discuss the results of model (a). In Figure 7.34 with $p_o = 0.000145$ psi, $\mu = 1.207 \times 10^{-10}$ lb/ft$^2$ h, and $K = 0.245$ cfm/ft$^2$ per 0.5 in W.G., the comparison of analytical and computational results clearly validates the analytical approaches provided that the assumptions for laminar flow conditions and analytical approximation procedures are valid. Also indicated in Figure 7.34, the simulation results tend to have an edge effect around the inlet.

Of course, very low pressure at the channel inlet ensures low channel fluid flow velocity and, consequently, laminar flow assumptions. With the increase of pressure, channel flow velocity will increase as well as fluid flow out of the porous layer. Nevertheless, it is clearly shown in Figure 7.35 that as the pressure drop increases and the viscosity decreases, the laminar flow assumption will no longer

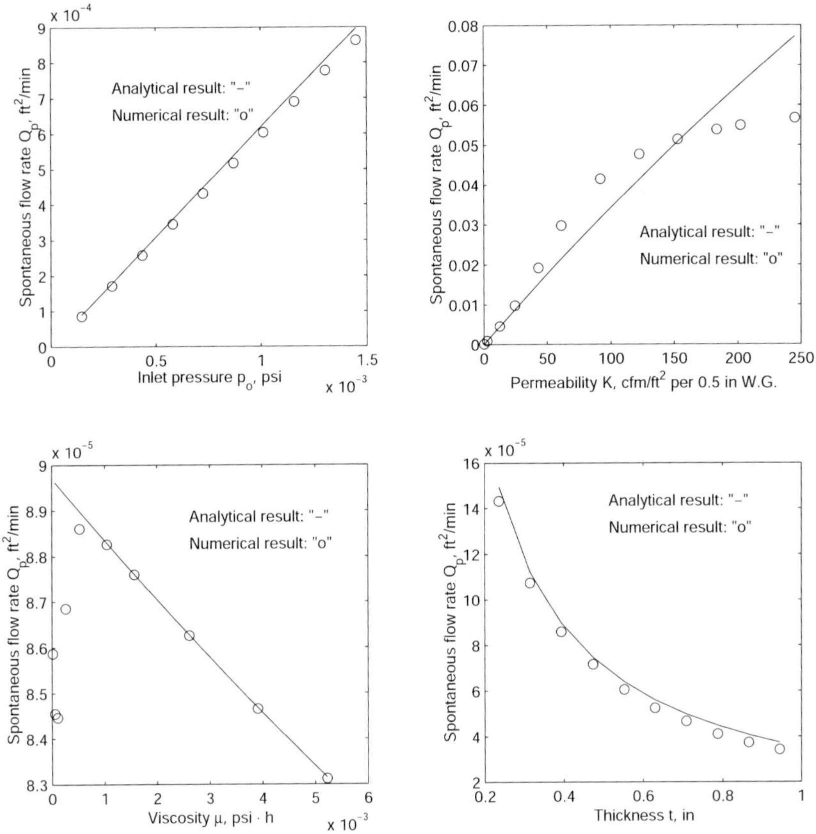


Figure 7.35. The spontaneous flow rates.

hold. Moreover, the one-dimensional channel flow assumption will also be invalid, which explains the difference between analytical and computational results for high pressure drop, low viscosity, and high permeability regions, respectively. Finally, Figure 7.35 also shows that as the porous layer thickness increases, the spontaneous flow rate decreases, suggesting that the change of the thickness will alter the distribution of pressure and, consequently, the flow field around the porous bottom.

Similar to model (a), we employ various physical parameters in model (b), such as the inlet flow rate, represented by the velocity input, and the permeability of the porous layer. As shown in Figure 7.36 with inlet velocity $V = 590.4$ fpm, at low permeability $K$, the spontaneous velocity profile is close to a linear monotonic distribution. Note that in order to eliminate the end effects, we exclude a few

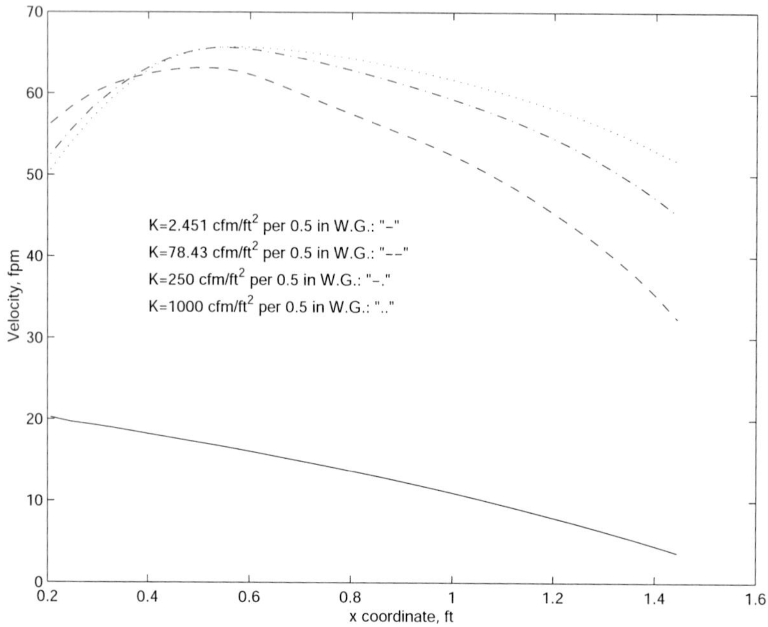


Figure 7.36. The spontaneous flow velocity profile as a function of the permeability.

points around the boundaries. Thus, we do not include in Figure 7.36 the outlet spontaneous velocity, which is zero due to the same ambient pressure at the outlet and the other side of the porous medium. Also shown in Figure 7.36, as the permeability increases, the velocity profile becomes increasingly flat. The physical understanding of this particular profile evolution can be attributed to the saturation phenomena of spontaneous flow. In fact, the saturation of the penetration flow is obvious based on the fact that the further increase of the permeability does not necessarily increase spontaneous flow as it does at the initial stage of low permeability. Because the penetration velocity is higher closer to the inlet, the saturation effect will be gradually experienced from left to right, as shown in Figure 7.36, which clearly explains the velocity plateau.

In Figure 7.37 with permeability $K = 2.451$ cfm/ft$^2$ per 0.5 in W.G., the spontaneous flow velocity profile is shown at four different inlet velocities. Since permeability $K$ is small, as discussed above, this set of velocity profiles share the same linear monotonic distribution. However, with the increase of the inlet velocity, which in fact represents the felt moving velocity or the inlet flow rate, spontaneous flow will increase correspondingly. To focus our attention on the total spontaneous flow rate instead of the velocity profile, we introduce Figure 7.38, which displays the flow rate as a function of the permeability at four different inlet

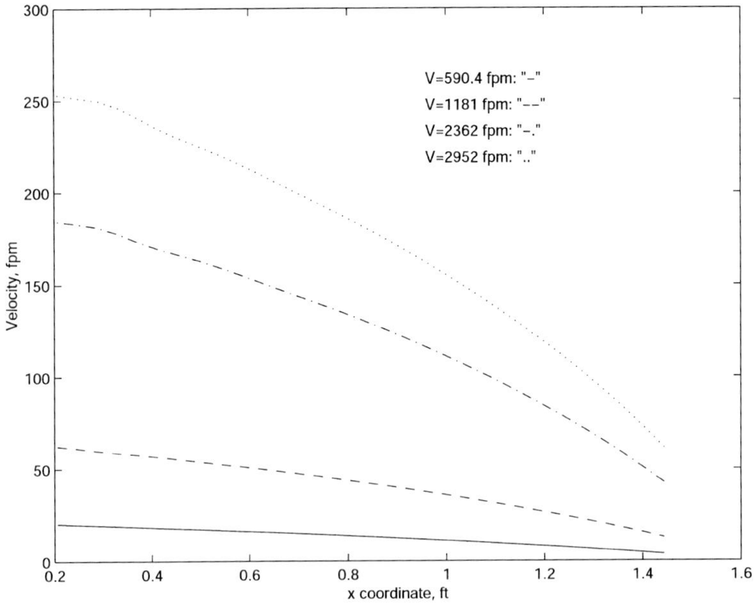


Figure 7.37. The spontaneous flow velocity profile at different inlet velocities.

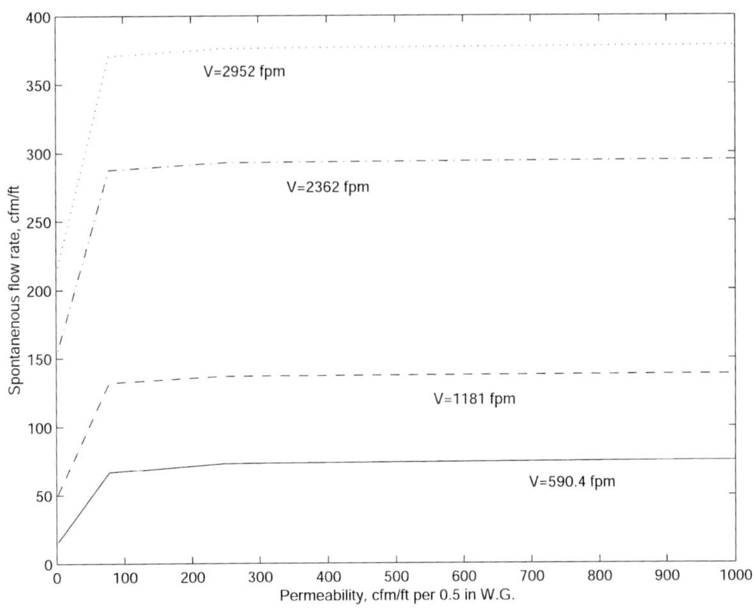


Figure 7.38. Spontaneous flow rate as a function of permeability at different felt velocities.

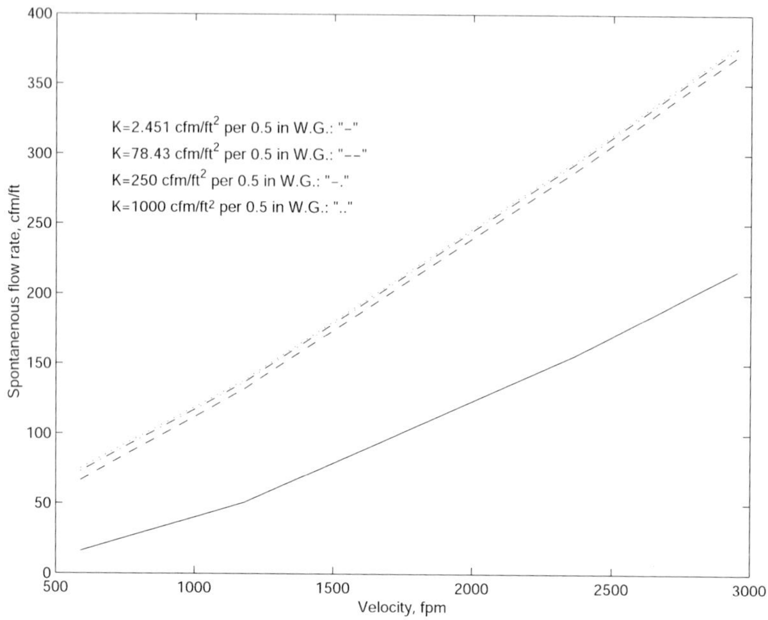


Figure 7.39. Spontaneous flow rate as a function of the felt velocity at different permeability.

velocities. First, we notice in Figure 7.38, that there exists an initial ramp before the permeability reaches 100 cfm/ft$^2$ per 0.5 in W.G. Once the saturation effects become dominant, the flow rates cease to increase according to the permeability. Second, it is clear that the saturated spontaneous flow rate will always increase with respect to the felt velocity given a constant permeability.

In Ref. [83], the existing experimental data are presented in a set of curves which reflect spontaneous flow rate as a function of the felt speed at different permeabilities. Here, we introduce Figures 7.39 and 7.40 in the same layout as used by Smook [83]. As shown in Figure 7.39, the spontaneous flow rate is almost linearly proportional to the felt velocity, which is consistent with the experimental data presented by Smook [83]. However, as we increase further the felt permeability, we can no longer achieve a significant increase of the spontaneous flow rate after we pass a certain permeability threshold, which we call the saturation point. Although the detailed convergent path to a saturated stage in our simulation results is not identical to the experimental data presented by Smook [83], the final saturated spontaneous flow rate as a function of the felt speed is, as confirmed in Figure 7.40 with $K = 1000$ cfm/ft$^2$ per 0.5 in W.G., very close to the experimental data in Ref. [83]. We should note that in our simulation we do not include the woven structure of the synthetic fabrics used to obtain the experimental data as well as the steam effects around the moving felt. The fact that the simulation

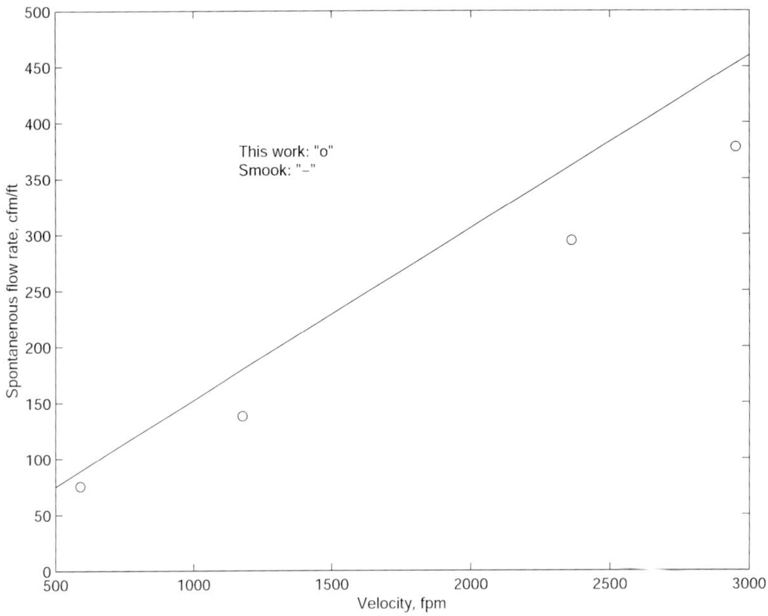


Figure 7.40. A comparison of the simulated spontaneous flow rate.

results are actually close to the experimental data confirms the validity of our computation models.

To demonstrate the capability of the proposed mixed $\mathbf{v}/p$ formulation with the ALE kinematic description and the developed upwind formulation for 9-node elements, we analyze a problem involving large free surface motions as well as a problem of the Navier–Stokes flow interacting with a hyperelastic structure under large displacements and large strains. In the region around the structure and free surfaces, with the help of the ALE description, we can relate the fluid mesh points to the structure nodes or assign additional mesh velocity unknowns in order to regularize the mesh inside the fluid domain. We use 9/3 or 9/4-$c$ elements and apply the developed upwind formulation.

In order to better illustrate the ALE descriptions for the viscous fluid, we present in Figure 7.41 a flow-through collapsible tube problem. The setting is similar to Ref. [240]. However, due to the relative length of the collapsible tube, the behaviors of the tube in coupling with the viscous flow are quite different. We present a set of preliminary results of a collapsible tube experiencing large displacements and large strains using the formulations illustrated in this section. Notice that there is no external fluid as introduced in Ref. [181] and the displacements are continuous around the nonslip FSI interface or the collapsible boundary.

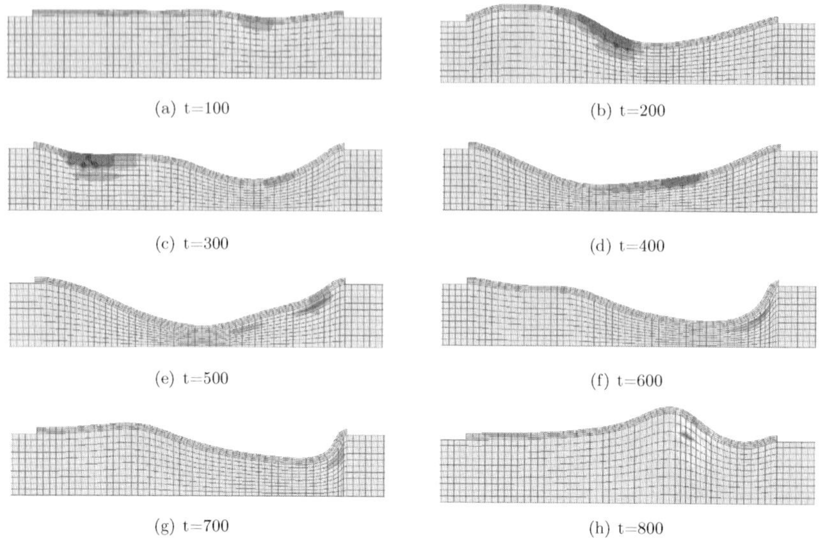


Figure 7.41. Vertical velocity snapshots of the collapsible tube problem.

The applications of the finite element method with velocity potential formulations for nonlinear free surface waves have been discussed in Refs. [137,167,224, 225]. Here we use the proposed $\mathbf{v}/p$ formulation with 9-node mixed elements. To avoid excessive mesh distortion, we apply the ALE description to the $\mathbf{v}/p$ formulation [205]. We subject the rigid tank with a free slosh surface shown in Figure 6.19 to the horizontal sinusoidal motion $a \sin \omega t$, where $w = 0.89$ rad/s and $a = 0.098$ g. The free surface profiles are shown in Figure 7.42 with eight 9-node elements and kinematic viscosity $\nu = 0.05$. Figures 7.44 and 7.45 show the velocity profiles and the pressure bands with eight 9-node elements. Figure 7.43 gives amplitudes with eight 9-node elements. In a linear analysis, both downward and upward motions of the slosh tank will have the same magnitude. However, as indicated by nonlinear analysis, the amplitudes of upward displacements and velocities are much higher than the downward ones. Note that with the influence of viscosities and crosswind diffusion due to the coarse mesh, free surface amplitudes are less than the results with the potential formulation.

The oscillations of node $A$ calculated with two relatively coarse 9-node element meshes are presented in Figure 7.48. With the coarse mesh used, the developed formulation still gives the convergence and provides reasonable results with no spatial oscillation and checkerboard pressure bands. In practice, before a large number of finite elements are used, it is beneficial to employ coarse meshes to obtain a first estimation for complicated problems. Figure 7.46 shows the first testing problem. Through this test example, we want to demonstrate that our $\mathbf{v}/p$

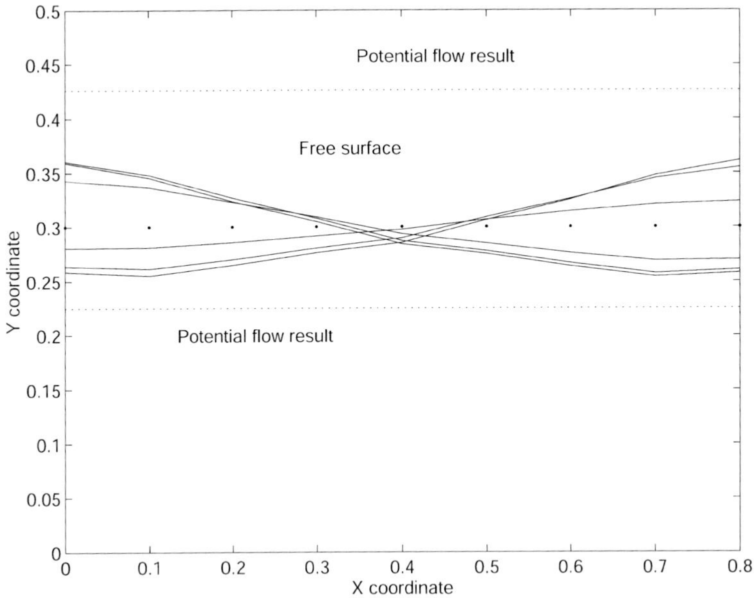


Figure 7.42. Free surface profiles.

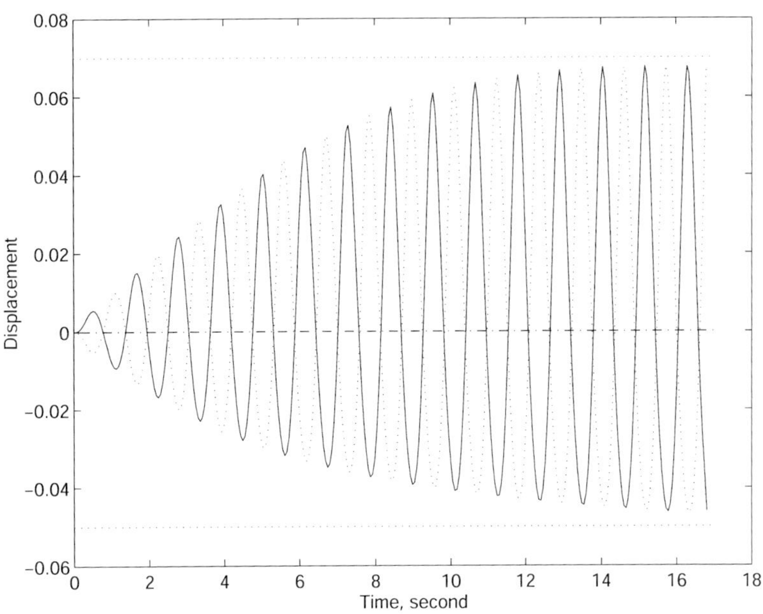


Figure 7.43. Amplitudes of the nonlinear slosh water tank problem.

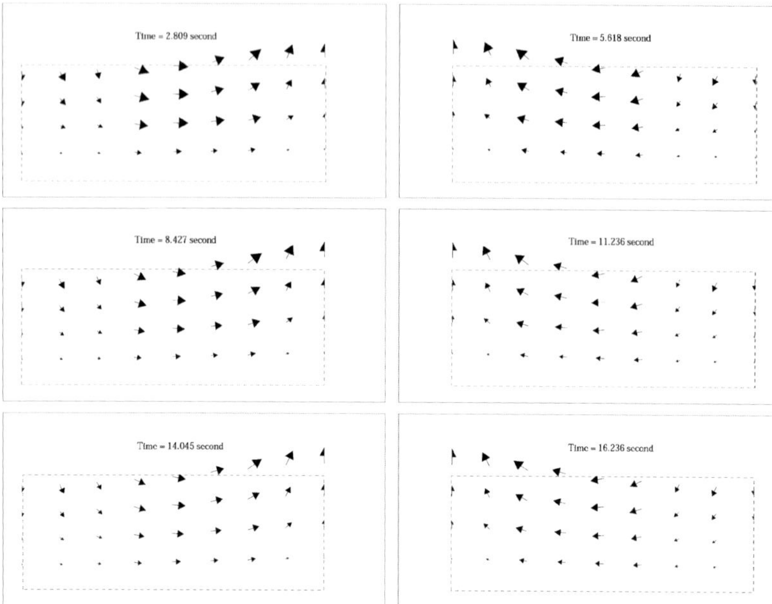


Figure 7.44. Velocity profiles of the nonlinear slosh water tank.

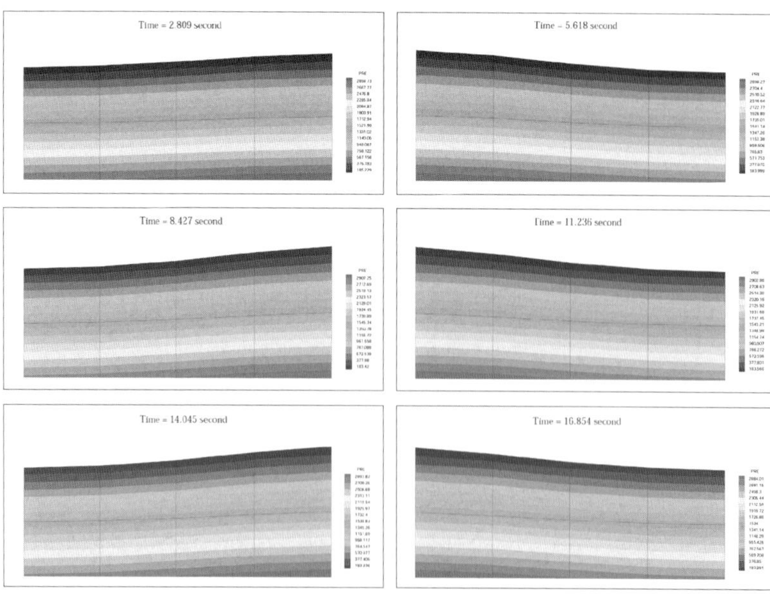


Figure 7.45. Pressure distributions of the nonlinear slosh water tank.

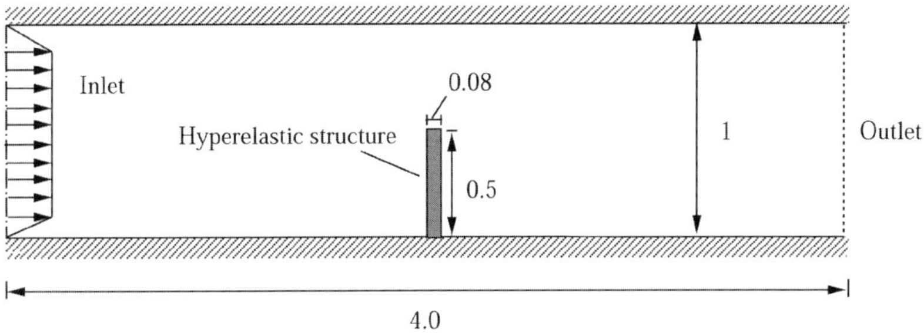


Figure 7.46. Navier–Stokes flow interacting with a hyperelastic structure problem.

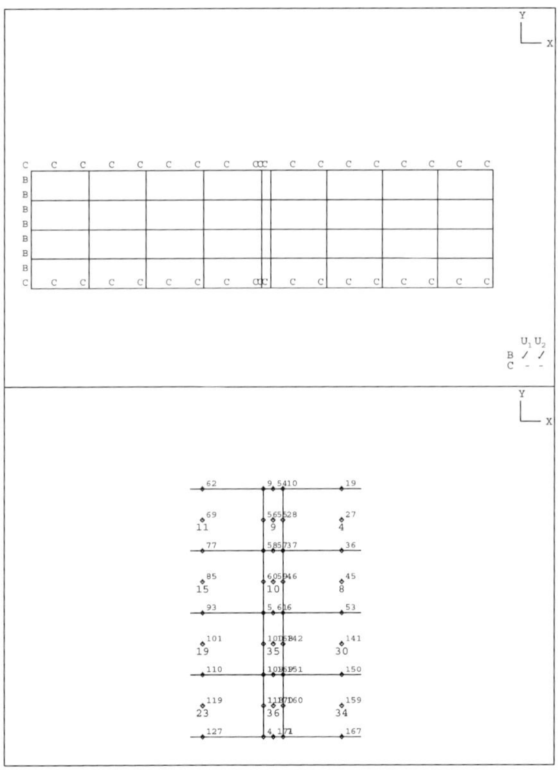

Figure 7.47. Typical mesh of the flow interacting with a hyperelastic structure problem.

formulation with the ALE description and newly developed upwind formulation can be successfully applied to convection dominated problems involving almost incompressible fluids and solids with large displacements and large strains. The typical mesh (a very coarse one) is given in Figure 7.47. Interestingly, with the

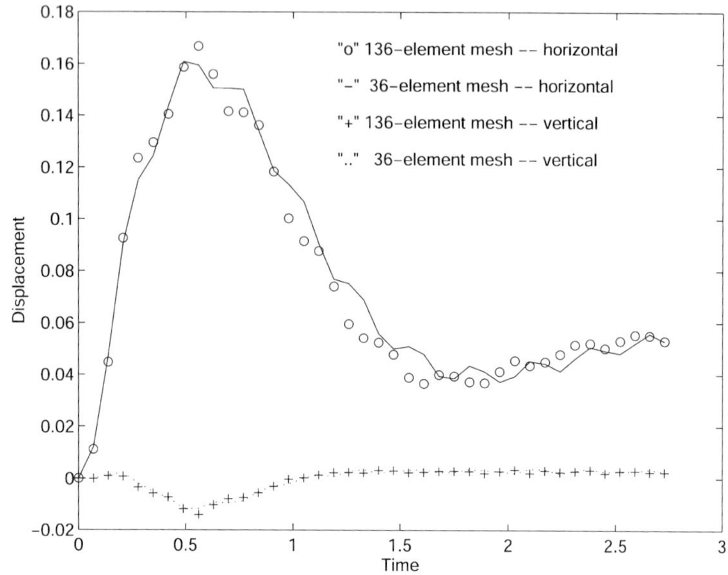


Figure 7.48. Displacement responses at node $A$ predicted with two relatively coarse meshes.

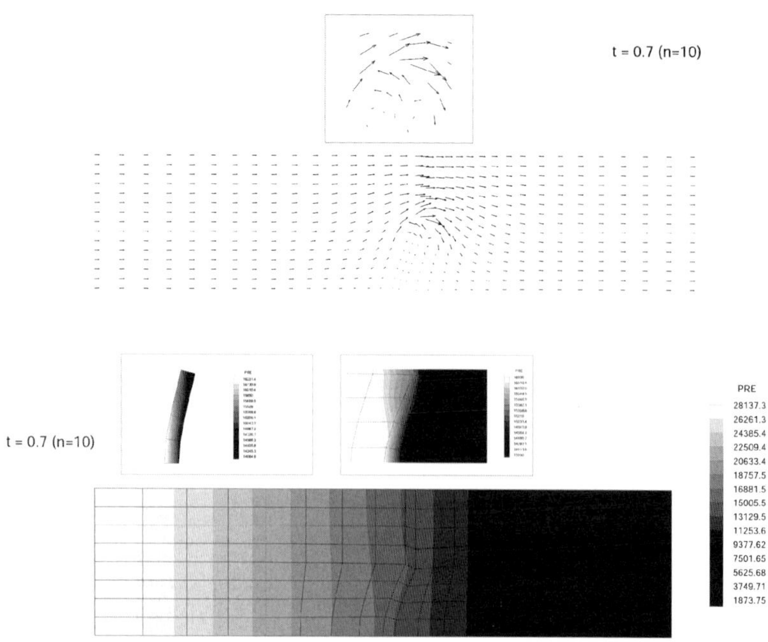


Figure 7.49. Velocity and pressure distributions at $t = 0.7$.

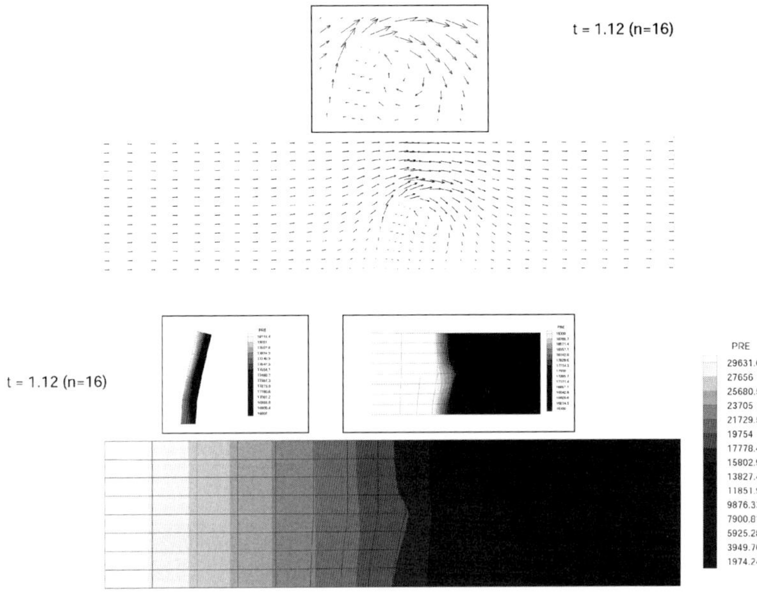


Figure 7.50. Velocity and pressure distributions at $t = 1.12$.

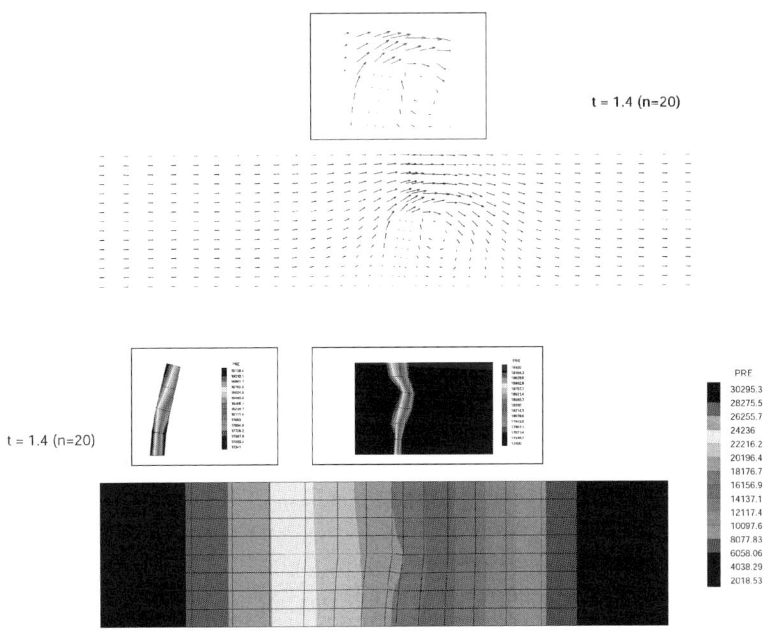


Figure 7.51. Velocity and pressure distributions at $t = 1.4$.

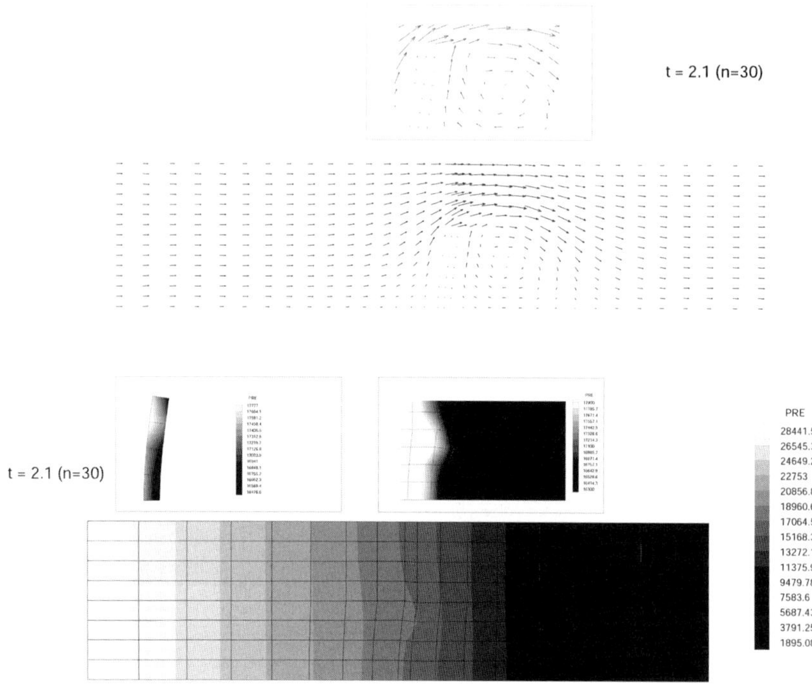


Figure 7.52. Velocity and pressure distributions at $t = 2.1$.

coarse mesh used, the developed ALE and upwind formulation gives the convergence and provides reasonable results with no spatial oscillation and checkerboard pressure modes. In practice, before a large number of finite elements is used, this type of coarse meshes can be applied to get a first estimation of a rather complicated problem. Figure 7.48 shows the displacement time history of node 5. The results at four different time snapshots with relatively fine meshes are shown in Figures 7.49, 7.50, 7.51 and 7.52. As indicated by the numerical results, the proposed upwind formulation with ALE descriptions provides reliable solutions, and neither checkerboard pressure bands nor spatial oscillations are observed. We conclude that the velocity/pressure formulation with mixed upwind treatment and ALE descriptions can be successfully adapted to the 9-node mixed elements satisfying the inf-sup condition. We recommend this approach as an alternative to accommodate the requirements for both the inf-sup condition and upwind schemes with high-order mixed elements. Nevertheless, more mathematical understanding of upwind issues is needed.

## 7.3. Immersed methods

The immersed boundary method was originally developed by Peskin [48] for the computation of blood flows interacting with the heart and heart valves. It has since then been successfully extended to three-dimensional heart-flow interactions and a variety of other biomechanics problems, which included the design of prosthetic cardiac valves [55], swimming motions of marine worms [78], wood pulp fiber dynamics [260], wave propagation in cochlea [238], and platelet aggregation and biofilm processes [6,54]. A recent review with comprehensive descriptions of the immersed boundary method and its applications has been presented by Peskin [49]. In this book, we only summarize the key aspects of the immersed boundary method.

The main advantages of the immersed boundary method are its simplicity and geometric flexibility. It is a mixed Eulerian–Lagrangian scheme that combines the computational efficiency inherent in using a fixed Eulerian grid for the fluid motion with the ease of tracking the immersed boundary represented by a set of moving Lagrangian points. In essence, the influence of the immersed elastic boundary is relegated to an inhomogeneous forcing term that can be distributed to the fluid surrounding the FSI interface. If the immersed structure is narrow and thin, the displaced volume is negligible. The FSI interface is modeled as a set of material points linked with springs or fibers. This moving interface treatment facilitates the handling, without the use of ALE descriptions or adaptive meshing, of immersed boundaries composed of nearly arbitrary shapes, sizes and configurations. However, Beyer and LeVeque [239] reported that Peskin's original method is limited to first order spatial accuracy by the Dirac delta function approximation.

Stockie and Wetton [260] analyzed the stiffness encountered in the immersed boundary method. It was found that very stiff "spring-like" links between boundary points require very restrictive time step selections. Thus many recent efforts have been made to improve the accuracy and computational efficiency of the immersed boundary method. Noticeably, Peskin and Printz [50] introduced a finite difference divergence operator to substitute the previous divergence derived from central difference schemes, which resulted a dramatic improvement in over-all volume conservation.

LeVeque and Li [232] considered the jump in variables that are discontinuous across boundaries and presented a second order accurate Cartesian grid method for solving elliptic equations whose solutions are not smooth across the interface due to discontinuous coefficients or singular source terms in the equation. The modified scheme incorporates the known jumps in the solution or its derivatives into finite difference schemes, and achieves second order accuracy at all points on the uniform grid even for arbitrary interfaces.

McQueen and Peskin [56] implemented a shared memory parallel vector computing scheme to compute the blood flow in the beating mammalian heart. Roma et al. [235] presented an adaptive version of the immersed boundary method employing a hierarchical, nested adaptive mesh refinement for the improvement of resolution and accuracy. Accuracy enhancement is accomplished by using a sequence of nested, progressively finer rectangular grid patches which locally cover the immersed boundary vicinity and dynamically follow the immersed boundary motion. Implicit formulations of the immersed boundary method are also used to free the method from its time step restriction [75].

Cortez and Minion [45] replaced the Dirac delta function with a cutoff function, or blob, to regularize the force field exerted by the membranes on the fluid. Velocity field induced is directly computed on a regular Cartesian grid via a smoothed dipole potential. The results demonstrate better volume conservation and higher order convergence. In addition, Goldstein et al. [95] used a feedback forcing method, which employed boundary body forces that allow the imposition of boundary conditions on a given surface not coinciding with the computational grid to compute the two-dimensional startup flow around a circular cylinder and three-dimensional plane.

Saiki and Biringen [73] used the forcing method of [95] to compute the flow around steady and rotating circular cylinders using fourth-order central difference approximations. Recently Mohd-Yusof [187] and Fadlun et al. [77] derived an alternative formulation of forcing that does not affect the stability of the discrete-time equations. In Ref. [187], the new forcing was combined with B-splines to compute the laminar flow over a three-dimensional ribbed channel, showing substantial improvements to the previous formulations. Moreover, with respect to high Reynolds number flow, Briscolini and Santangelo [30] used an immersed boundary approach referred to as the mask method to compute the unsteady two-dimensional flow around circular and square cylinders at Reynolds numbers up to 1000. Lai and Peskin [162] presented a formally second order accurate immersed boundary method with results showing less numerical viscosity and the potential to be used for high Reynolds number flows. Some approaches used in Refs. [20, 73,77,95,187], the so-called virtual boundary method, share the same spirit with the immersed boundary method but solve a rigid boundary problem rather than an elastic boundary problem.

In this section, in addition to an overview of the immersed boundary/continuum methods and their finite element formulations, explicit vs. implicit and incompressible vs. compressible issues are discussed. The recent finite element formulations retain the same strategies employed in the original immersed boundary method, namely, the independent Lagrangian solid mesh moves on top of a fixed or prescribed background Eulerian fluid mesh. The added features in recent finite element formulations mostly relate to the immersed solid which can occupy a finite volume in the fluid and be impermeable, compressible, and highly deformable. Furthermore, a matrix-free Newton–Krylov iterative solution technique also resolves the time step limitation issues related to stiff spring supports from the boundary and the high elasticity moduli of the immersed solid. This implicit iterative approach allows the application of immersed methods to many engineering problems, some of which are documented in this book for illustrative purposes.

Current computational methods rely on frequent remeshing, often at every time step, which can dramatically increase the computation cost and limit the accuracy. Especially in the study of biosystems, where complex three-dimensional structures are involved along with strong fluid-structure interactions, the excessive computational efforts involved in mesh adjustment often render the realistic simulation impossible. In the developed method, the mesh distortion of fluid surrounding immersed objects is completely eliminated, and the stress analysis of the structure/solid can be easily performed with finite element methods.

The key attribute of the immersed methods presented in this section is to eliminate FSI interfaces and, consequently, fluid mesh distortion problems by distributing nodal structural forces, calculated with finite element methods, through a delta function as equivalent body forces in the fluid domain. This approach enables the use of Fast Fourier Transform (FFT), parallel algorithm, and higher order schemes. The idea of combining the immersed boundary method with the finite element methods stems from proof that the nodal force calculation in the current immersed boundary method is equivalent to a traditional nonlinear finite element formulation for truss elements using the second Piola–Kirchhoff stress and the Green–Lagrangian strain [142]. Of course, the physical interpretation of finite element methods can be simplified as replacing a continuous medium with a collection of nodes or material points linked through a stiffness matrix. If the structure is best approximated with a fiber network, the stiffness matrix can be derived from truss elements or a finite difference scheme, as used in the current immersed boundary method. Otherwise, a traditional finite element formulation must be used to represent two- or three-dimensional elastic structures interacting with the surrounding fluids.

### *7.3.1. Immersed boundary method; nodal forces; incompressible continuum*

#### *Immersed boundary method*

Consider a neutrally buoyant immersed flexible structure contained in a viscous incompressible fluid. Let $\mathbf{F}(s,t)$ denote the elastic fiber point force (force per unit length, as in the surface tension definition) at the Lagrangian parametric coordinate $\mathbf{x}^s(s,t) = (x_1(s,t), x_2(s,t), x_3(s,t))$ as a function of the arc length $s$ and the time $t$:

$$\mathbf{F}(s,t) = \partial\big(T(s,t)\mathbf{t}(s,t)\big)/\partial s, \tag{3.166}$$

where $T$ and $\mathbf{t}$ represent the tension within the fiber and the unit tangent vector, respectively.

If we denote $\mathbf{f}(s, t)$ as the corresponding effective body force in the fluid domain at the spatial position $\mathbf{x}$, $\Omega$ as the entire fluid domain, the governing equations of FSI systems involving a single neutrally buoyant smooth submerged elastic fiber $\Gamma_s$ can be stated as

$$\rho\left(\frac{\partial \mathbf{v}}{\partial t} + \mathbf{v} \cdot \nabla \mathbf{v}\right) = -\nabla p + \mu \nabla^2 \mathbf{v} + \mathbf{f}, \tag{3.167}$$

$$\nabla \cdot \mathbf{v} = 0, \tag{3.168}$$

$$\mathbf{f}(\mathbf{x}, t) = \int\limits_{\Gamma_s} \mathbf{F}(s, t)\delta(\mathbf{x} - \mathbf{x}^s)\, d\Gamma, \tag{3.169}$$

$$\mathbf{v}^s = \int\limits_{\Omega} \mathbf{v}(\mathbf{x}, t)\delta(\mathbf{x} - \mathbf{x}^s)\, d\Omega, \tag{3.170}$$

where $\mathbf{F}(s, t)$ is given in Eq. (3.166), and $\mathbf{v}^s$, $\mathbf{v}(\mathbf{x}, t)$, $p(\mathbf{x}, t)$, $\rho$, and $\mu$ represent the velocity of the immersed structural/solid point $\mathbf{x}^s$, fluid velocity, pressure, density, and dynamic viscosity, respectively.

Since the elastic fiber does not occupy any volume, the net effects on the surrounding fluid can be represented by a set of forces associated with the immersed fiber points. Therefore, FSI systems can be simply depicted as a conventional fluid mechanics problem with an Eulerian kinematic description. Of course, all the difficulties are hidden in the representation of the inhomogeneous body force $\mathbf{f}(\mathbf{x}, t)$. Note that at a typical fiber point $k$, the resultant elastic or internal force $\mathbf{F}_k$ is determined by the fiber configuration and material property. Furthermore, if there are $I$ arcs passing through point $k$, assuming that each fiber has a tension $T_i$ and an increment of the arc length $\Delta s_i$, the $j$th component of the nodal force $\mathbf{F}_k(s, t)$ is calculated as

$$\sum_{i=1}^{I} F_{kj}^i(s, t) = \sum_{i=1}^{I} \frac{\partial (T_i t_{kj}^i)}{\partial s} \Delta s_i, \tag{3.171}$$

where $\mathbf{t}_k^i$ represents the unit tangent vector of the $i$th arc at the point $k$.

Let us consider a typical point $k$ as illustrated in Figure 7.53. The nodal force at the point $k$, denoted as $\mathbf{F}_k$, is determined by tensions within the two adjacent fibers. If the immersed boundary has the same mass density as the surrounding fluid, i.e., is neutrally buoyant, at the submerged point $k$, there is no attached additional mass, and the resultant fluid force vector $\mathbf{R}_k^f$ and the external solid force vector $\mathbf{R}_k^e$ must equal the resultant elastic or internal nodal force $\mathbf{F}_k$. However, if the mass density of the elastic structure $\rho_s$ is different from the surrounding fluid density $\rho_f$, $\mathbf{R}_k^f$ and $\mathbf{R}_k^e$ must achieve a dynamic equilibrium with the resultant elastic nodal force and the inertial or d'Alembert's force. Denoting $M$ and $\mathbf{a}_k$ as

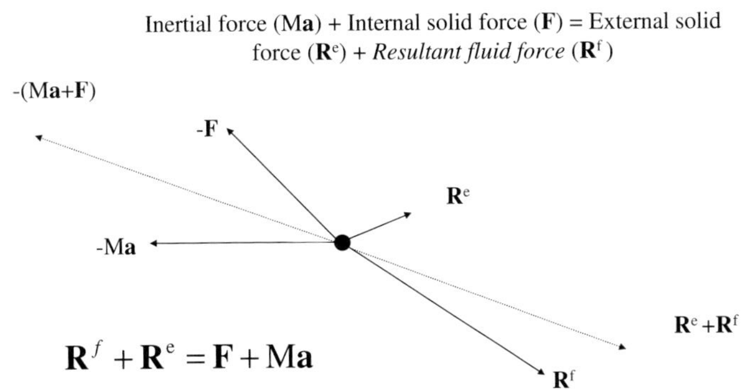


Figure 7.53. Force balance at a typical submerged point.

the lumped mass and the acceleration at the point $k$, we have

$$\mathbf{R}_k^f + \mathbf{R}_k^e = \begin{cases} \mathbf{F}_k, & \rho_s = \rho_f, \\ \mathbf{F}_k + \mathbf{M}\mathbf{a}_k, & \rho_s \neq \rho_f. \end{cases} \tag{3.172}$$

Again, this equilibrium depicted by Eq. (3.172) applies for *both* the boundary and the interior material points. Moreover, employing the discretized delta function, we can write the discretized form of Eqs. (3.169) and (3.170) as

$$\mathbf{f} = \sum_{k=1}^{K} \delta_h(\mathbf{x} - \mathbf{x}_k^s)(-\mathbf{R}_k^f), \tag{3.173}$$

$$\mathbf{v}_k^s = \sum_{\mathbf{x}_i \in \Omega_o^k} \delta_h(\mathbf{x}_i - \mathbf{x}_k^s)\mathbf{v}_i h^3, \tag{3.174}$$

where $K$ is the total number of submerged points, $\Omega_o^k$ stands for the finite support domain of the discretized delta function surrounding and centered at the $k$th submerged point, and $h$ is the grid size of the uniform background grid covering the entire fluid domain.

As shown in Figure 7.55, various flexible structures as well as immersed points can be modeled by the immersed boundary method. To better illustrate why the immersed boundary method works, we employ the familiar variational principles. Consider a fluid domain $\Omega$ enclosed with a sufficiently smooth boundary, $\partial\Omega = \Gamma_v \cup \Gamma_f$, as depicted in Figure 7.54, where $\Gamma_v$ and $\Gamma_f$ stand for the Dirichlet and Neumann boundaries, respectively. Suppose there exists an enclosed elastic boundary $\Gamma_s$ (a line for two-dimensional cases and a surface for three-dimensional cases) as the generic representation of the immersed boundary, and the fluid domain $\Omega$ is subdivided into two regions: the interior region $\Omega_i$ and the exterior region $\Omega_e$. The boundaries of interior and exterior regions can then be simply expressed as $\partial\Omega_i = \Gamma_s$ and $\partial\Omega_e = \Gamma_s \cup \Gamma_v \cup \Gamma_f$.

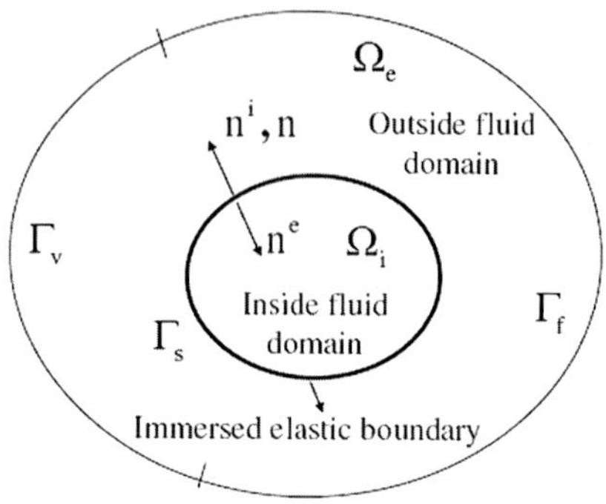


Figure 7.54. Immersed boundary illustration.

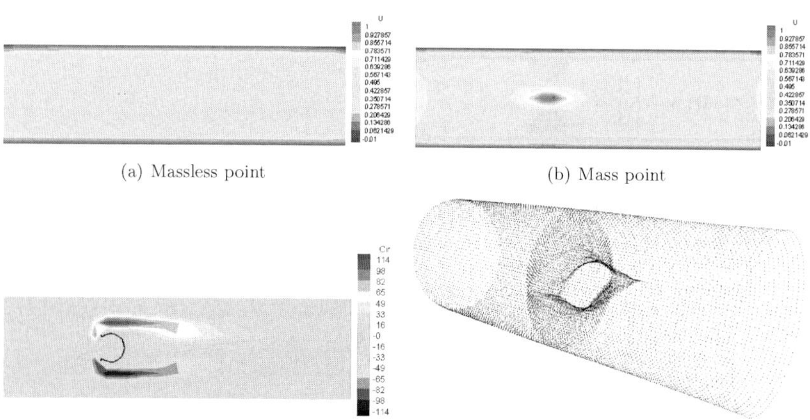

(a) Massless point (b) Mass point

(c) Fiber (d) Venous valve

Figure 7.55. Simple and complicated immersed structures constructed with immersed points.

If we denote $\boldsymbol{\sigma}$ as the stress tensor, $\mathbf{v}$ as the velocity vector, and $\rho$ as the density in the fluid domain, we establish the following governing set of equations (strong form):

$$\rho \dot{v}_i = \sigma_{ij,j} + f_i^e, \qquad \text{in } \Omega_i (\text{or } \Omega \setminus \Omega_e), \tag{3.175}$$

$$\rho \dot{v}_i = \sigma_{ij,j} + f_i^e, \qquad \text{in } \Omega_e, \tag{3.176}$$

$$[v_i] = 0, \qquad \text{on } \Gamma_s, \text{ kinematic matching}, \tag{3.177}$$

$$[\sigma_{ij} n_j] = f_i^s + m\ddot{u}_i^s, \qquad \text{on } \Gamma_s, \text{ dynamic matching}, \tag{3.178}$$

where the external body force $\mathbf{f}^e$ will be replaced by $\rho\mathbf{g}$, with $\mathbf{g}$ as the gravitational acceleration, $\mathbf{f}^s$ and $m$ stand for the elastic force and the mass density of the immersed boundary $\Gamma_s$ (per unit length for two-dimensional cases and per unit area for three-dimensional cases), $\mathbf{u}^s$ denotes the interface displacement, and the surface normal vector $\mathbf{n}$ is aligned with that of interior fluid domain $\mathbf{n}^i$ and opposite to that of exterior fluid domain $\mathbf{n}^e$.

At this point, we can derive a number of numerical methods to solve Eqs. (3.175) to (3.178). A straightforward approach is to represent exterior and interior fluid domains with different meshes and to match them accordingly at interface $\Gamma_s$. This approach represents the traditional treatment of FSI problems, in which the solid mesh is coupled with the fluid mesh around the FSI interface [155,274].

Defining the Sobolev space $(\mathcal{H}^1_{0,\Gamma_v}(\Omega))^d = \{\mathbf{w} \mid \mathbf{w} \in (\mathcal{H}^1(\Omega))^d,\ \mathbf{w}|_{\Gamma_v} = \mathbf{0}\}$, where $d$ represents the spatial dimensions, we express Eqs. (3.175) to (3.178) in the variational form (weak form):

$$\forall \mathbf{w} \in \left(\mathcal{H}^1_{0,\Gamma_v}(\Omega)\right)^d,$$

$$\int_{\Omega_i} w_i\left[\rho(\dot{v}_i - g_i) - \sigma_{ij,j}\right] d\Omega + \int_{\Omega_e} w_i\left[\rho(\dot{v}_i - g_i) - \sigma_{ij,j}\right] d\Omega = 0. \tag{3.179}$$

REMARK 7.3.1. In the variational form, $\mathbf{w} \in (\mathcal{H}^1_{0,\Gamma_v}(\Omega))^d$ implies that the kinematic matching at the interface $\Gamma_s$ written as Eq. (3.177) is satisfied for all $\mathbf{w}$.

Furthermore, using integration by parts and the divergence theorem, introducing dynamic matching at the FSI interface $\Gamma_s$, and combining interior and exterior fluid domains with $\Omega_e \cup \Omega_i = \Omega$, Eq. (3.179) can be rewritten as:

$$\forall \mathbf{w} \in \left(\mathcal{H}^1_{0,\Gamma_v}(\Omega)\right)^d$$

$$\int_{\Omega} \left[w_i \rho(\dot{v}_i - g_i) + w_{i,j}\sigma_{ij}\right] d\Omega + \int_{\Gamma_s} w_i^s\left(f_i^s + m\ddot{u}_i^s\right) d\Gamma - \int_{\Gamma_f} w_i f_i^{f,p}\, d\Gamma = 0. \tag{3.180}$$

REMARK 7.3.2. In Eq. (3.180), the term involving the given surface traction $\mathbf{f}^{f,p}$ will remain the same as if the variational form is carried out for the entire fluid domain instead of the interior and exterior parts. Thus the focus will be on the submerged interface $\Gamma_s$.

REMARK 7.3.3. In Eq. (3.180), the external work comes from the external body force $\mathbf{f}^e$ or $\rho\mathbf{g}$, surface traction $\mathbf{f}^{f,p}$ at the Neumann boundary $\Gamma_f$, and elastic and inertial forces around the FSI interface $\Gamma_s$. Moreover, in Eq. (3.180), we do not stipulate the material derivative $d\mathbf{v}/dt$ and the stress $\boldsymbol{\sigma}$. Hence the turbulent and the nonNewtonian fluid models can eventually be incorporated. Finally, the kinematic matching at the submerged interface $\Gamma_s$ also implies that the submerged interface will move at the same velocity as that of the fluid particles in the immediate vicinity.

In the immersed boundary method, we introduce the following two key equations:

$$f_i^{fsi} = -\int_{\Gamma_s} (f_i^s + m\ddot{u}_s)\delta(\mathbf{x} - \mathbf{x}^s)\, d\Gamma, \tag{3.181}$$

$$v_i^s = \int_{\Omega} v_i \delta(\mathbf{x} - \mathbf{x}^s)\, d\Omega, \tag{3.182}$$

where $\mathbf{f}^{fsi}$ is the so-called equivalent body force.

REMARK 7.3.4. In both Eqs. (3.181) and (3.182), the Dirac delta function is positioned at the current interface position $\mathbf{x}^s$. Before the discretization of the Dirac delta function, Eq. (3.182) can be simply interpreted as the evaluation of the fluid velocity at the submerged interface. In the discretized form, the Dirac delta function in Eq. (3.182) is equivalent to the shape function or kernel of the meshfree method.

Note that $\Gamma_s$ represents the current configuration of the submerged interface, and nonlinear mechanics are employed to relate the elastic force $\mathbf{f}^s$ with the interfacial position $\mathbf{x}^s$ or the displacement $\mathbf{u}^s$. It is also clear that as long as we use the same delta function for both Eqs. (3.181) and (3.182), the virtual power input from the submerged elastic boundary (or the immersed boundary) to the fluid domain can be expressed as

$$\begin{aligned}\int_{\Omega} w_i f_i^{fsi}\, d\Omega &= -\int_{\Gamma_s}\int_{\Omega} w_i \delta(\mathbf{x} - \mathbf{x}^s)(f_i^s + m\ddot{u}_i^s)\, d\Omega\, d\Gamma \\ &= -\int_{\Gamma_s} w_i^s (f_i^s + m\ddot{u}_i^s)\, d\Gamma. \end{aligned}\tag{3.183}$$

Because Eq. (3.183) holds for all $\mathbf{w} \in (\mathcal{H}^1_{0,\Gamma_v}(\Omega))^d$, the effect of the submerged elastic boundary can be simply replaced with the equivalent body force $\mathbf{f}^{fsi}$. Hence the governing equations (3.175) to (3.178) can be rewritten as

$$\rho \dot{v}_i = \sigma_{ij,j} + \rho g_i + f_i^{fsi}, \quad \text{in } \Omega; \tag{3.184}$$

and the variational equations (3.179) and (3.180) are modified as

$$\int_{\Omega} \left[w_i\left(\rho\dot{v}_i - \rho g_i - f_i^{fsi}\right) + w_{i,j}\sigma_{ij}\right] d\Omega - \int_{\Gamma_f} w_i f_i^{f,p}\, d\Gamma = 0. \tag{3.185}$$

REMARK 7.3.5. Eq. (3.185) provides us with the foundation of the key advantage of the immersed boundary method, namely, the independent Lagrangian solid mesh moves on top of a background Eulerian fluid mesh. We must also point out that the background fluid mesh could be a fixed Eulerian mesh or an ALE mesh with a prescribed mesh motion. In practice, such a mesh motion could follow moving structures or solids as well as conform to boundary deformation.

*Nodal forces*

Nodal force calculation is an important step in the immersed boundary method. The more accurate the nodal force distribution we calculate from the structure model, the better the simulation result we can expect. In previous structural models, elastic structures are normally discretized into certain patterns of networks that are composed of fibers. For example, in the simulation of hemodynamic interaction with heart muscles, the fiber network is generated in such a way that each fiber's direction is consistent with the muscle fiber direction. Nodal force at each fiber point is determined by tensions within fibers that are linked to this point. Therefore, in the immersed boundary method, nodal forces are calculated around a continuous arc and additional nodal points occupying the same location are introduced to represent the fiber crossing. Note that these fiber crossing points always move at the same velocity and will never separate. In the three-dimensional simulation, a large number of redundant degrees of freedom could significantly increase the computation cost. In general, finite element methods for immersed continuum lead to better stress predictions.

If the fibers are assumed to be massless, inertial forces are neglected. Therefore, the resultant force $\mathbf{F}_k$ is entirely dependent upon the tension $T_{k-1}$ and $T_k$ and should be balanced by the force applied by the fluid. In other words, the force applied to the fluid is equal to the resultant force $\mathbf{F}_k$. This resultant force $\mathbf{F}_k$, or nodal force, is determined by the fiber configuration and material property. In the immersed boundary method, the program loops all fibers to evaluate the tension forces, then calculates the nodal force at every fiber point. For a typical connection at fiber point $k$, denoting two adjacent fibers as $k-1$ and $k$, we can either have constant tension or specify a linear elastic modulus $K$ attribute to each fiber connection. For these two cases, the force calculation scheme is as follows:

(a) Constant tension $T$

$$\begin{aligned} \mathbf{F}_k &= \partial(T\mathbf{t}_k)/\partial s = (T_{k+1/2}\mathbf{t}_{k+1/2} - T_{k-1/2}\mathbf{t}_{k-1/2})/\Delta s \\ &= T(\mathbf{t}_{k+1/2} - \mathbf{t}_{k-1/2})/\Delta s, \end{aligned} \tag{3.186}$$

with

$$\mathbf{t}_{k-1/2} = \frac{\mathbf{x}_k^s - \mathbf{x}_{k-1}^s}{|\mathbf{x}_k^s - \mathbf{x}_{k-1}^s|} \quad \text{and} \quad \mathbf{t}_{k+1/2} = \frac{\mathbf{x}_{k+1}^s - \mathbf{x}_k^s}{|\mathbf{x}_{k+1}^s - \mathbf{x}_k^s|}, \tag{3.187}$$

where $\mathbf{t}_{k-1/2}$ and $\mathbf{t}_{k+1/2}$ represent the unit tangent vectors of fiber $k-1$ and $k$, respectively; likewise, the tension $T_{k-1/2}$ and $T_{k+1/2}$ within these fibers are constant and denoted as $T$.

(b) Linear elastic modulus $K$

$$\begin{aligned}\mathbf{F}_k &= \left[\left(K\frac{\partial \mathbf{x}}{\partial s}\right)_{k+1/2} - \left(K\frac{\partial \mathbf{x}}{\partial s}\right)_{k-1/2}\right]/\Delta s \\ &= K\left(\mathbf{x}_{k+1}^s + \mathbf{x}_{k-1}^s - 2\mathbf{x}_k^s\right)/(\Delta s)^2.\end{aligned} \tag{3.188}$$

Of course, as long as we represent the immersed objects with Lagrangian descriptions, the immersed boundary method will also be applicable for the modeling of one type of fluid immersed in another type of fluid. It also seems to be a natural extension to employ various finite element models for the nodal force calculation [303]. In a test example, one viscous fluid modeled with a collection of fluid particles is injected into another viscous fluid. Due to the surface tension effects, viscous fluid exiting the tube tends to form a droplet. Preliminary simulation results are illustrated in Figure 7.56.

*Incompressible continuum*

Instead of representing the immersed structures/solids with fibers as illustrated in Figure 7.55, the goal of the extended immersed boundary method or the immersed finite element method [303,321] is to represent the immersed incompressible solid with nonlinear finite element formulations. In addition, these incompressible solids have finite mass and volume. The initial attempt of connecting a traditional linear elasticity model with the immersed boundary method can be traced back to Sulsky and Brackbill [262], in which a stress function is transferred to the background fluid grid. In recent finite element extensions of the immersed boundary method, a more direct connection between fluid and solid domains is accomplished by employing the internal nodal forces calculated in the context of finite element methods. In these new attempts, submerged solids can experience large displacements and deformations. We must point out that, contrary to the traditional method of tracking the FSI interface, both the interior and the FSI interface of the submerged solid are modeled as submerged material points in contact with the background fluid, namely, the background fluid is everywhere.

Note that although the entire submerged solid domain is decomposed into a collection of submerged material points, due to the distributed nodal forces, with a sufficiently dense solid mesh, the surrounding fluid will not penetrate into the interior of the solid and the FSI interface will be automatically defined by the

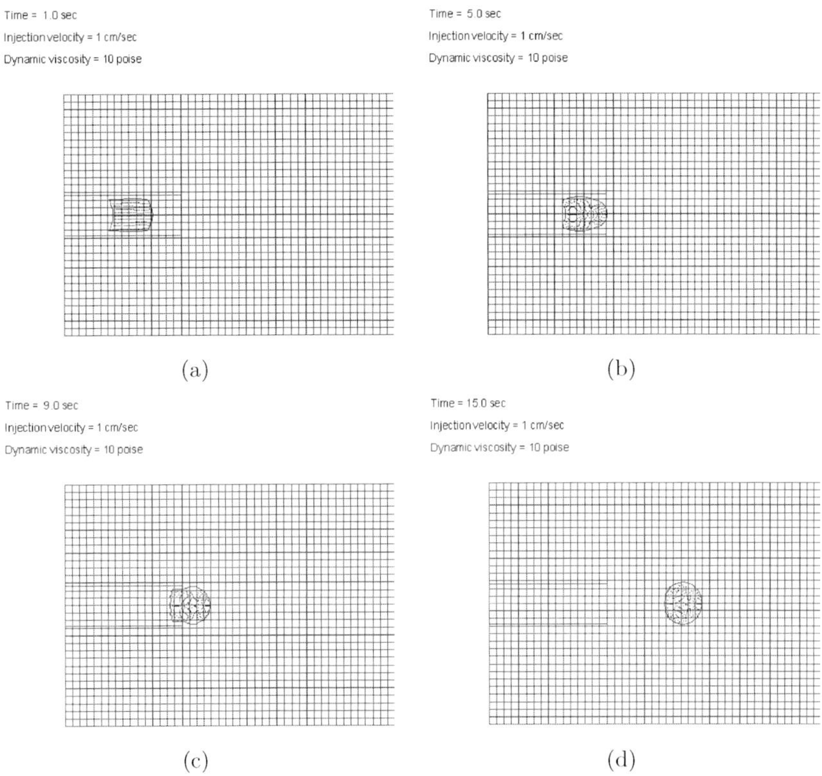


Figure 7.56. One viscous fluid injected into another viscous fluid.

submerged material points enclosing the solid domain. The physical interpretation of finite element methods can be simplified as replacing a continuous medium with a collection of nodes or material points linked through a stiffness matrix. Therefore, if the structure is approximated with an equivalent fiber network, the stiffness matrix can be derived from truss elements or a finite difference scheme as used in the immersed boundary method. In fact, it has been shown that the nodal force calculation in the immersed boundary is equivalent to a traditional nonlinear finite element formulation for truss elements using the second Piola–Kirchhoff stress and the Green–Lagrangian strain [294].

If the surrounding fluid is viscous and incompressible, the immersed solid must be incompressible in immersed methods. In addition, there are two views of immersed solids. The first view, matching the original understanding of the immersed boundary method, is that the immersed solid is wet and permeated with the same fluid as the surrounding. Therefore, the elasticity forces will be the additional force due to the solid portion of the immersed solid. This understanding is realistic for some biological system modeling, since tissues are mostly FSI systems and elastic parts are contributed by elastomer, collagen, or other solid constituents. The second view of the immersed solid is more in tune with traditional FSI systems in which the immersed solid is dry, impermeable, and completely separated from the surrounding fluid. In this context, if we were to use the immersed methods, the immersed solid must be incompressible to match with the surrounding incompressible fluid. Furthermore, a fictitious domain concept needs to be employed [94].

In the initial versions of the finite element formulations of immersed methods [23,24,303,321], the immersed solids are assumed to be incompressible. Furthermore, if the immersed solids are impermeable, the additional elastic forces in comparison with the viscous counterparts calculated with the material properties of the surrounding fluid are very large, hence both views for immersed solids yield the same forces. In the extended immersed boundary method, the fluid solver is still the same spectrum solver with periodic boundary conditions based on Chorin's algorithm and FFT solver [48]. Since the background fluid mesh is uniform, the same discretized delta function as derived in the immersed boundary method is employed. In the immersed finite element method, the fluid solver is replaced with the stabilized finite element formulation. In this case, the background fluid mesh is unstructured. Therefore, the discretized delta function is replaced with the kernel functions for meshless methods [172,310].

Consider a general three-dimensional incompressible hyperelastic material model with the following incompressible elastic energy potential:

$$W = C_1(J_1 - 3) + C_2(J_2 - 3), \tag{3.189}$$

with the solid deformation gradient $D_{ij} = \partial x_i^s(t)/\partial x_j^s(0)$, the invariants $J_1 = I_1$, $J_2 = I_2$, $I_1 = C_{kk}$, and $I_2 = (I_1^2 - C_{ij}C_{ij})/2$, where $\mathbf{C}$ is the Cauchy–Green deformation tensor.

Note that since the solid displacements are mapped from the background fluid, if the surrounding fluid is incompressible, the solid must also be incompressible, which corresponds to $J_3 = I_3^{1/2} = (\det(\mathbf{C}))^{1/2} = 1$. For solids with large displacements and deformations, the 2nd Piola–Kirchhoff stress $\mathbf{S}$ and the Green–Lagrangian strain $\boldsymbol{\epsilon}$ are used along with a total Lagrangian formulation. Hence, employing Eq. (3.189), we derive

$$S_{kl} = \frac{\partial W}{\partial \epsilon_{kl}} \quad \text{and} \quad \epsilon_{ij} = \frac{1}{2}(C_{ij} - \delta_{ij}). \tag{3.190}$$

Furthermore, the Cauchy stress can be calculated using Eq. (2.57). Thus, the equivalent internal force for the material points of the flexible structure/solid can be derived as

$$\mathbf{F}_k = \frac{\partial}{\partial \mathbf{x}_k^s}\left(\int\limits_{\Omega^s(t)} W\, d\Omega\right) = \int\limits_{\Omega^s(0)} S_{ml}\frac{\partial \epsilon_{ml}}{\partial \mathbf{x}_k^s}\, d\Omega = \int\limits_{\Omega^s(0)} S_{ml}\frac{\partial \epsilon_{ml}}{\partial \mathbf{u}_k^s}\, d\Omega, \tag{3.191}$$

where $\mathbf{x}_k^s$, $\mathbf{u}_k^s$, and $\mathbf{F}_k$ stand for the current position vector, displacement vector, and internal nodal force vector of the $k$th submerged node, respectively, and $\Omega^s(t)$ and $\Omega^s(0)$ represent the current and the original volume of the submerged solid.

Note that if the nonlinear structural material has a density $\rho_s$ different from the fluid density $\rho_f$, we should include the inertial force. Similar to Eq. (3.172), the resultant node force vector $\mathbf{R}_k$ can be expressed for a typical material point $k$:

$$\mathbf{R}_k = \mathbf{R}_k^f + \mathbf{R}_k^e = \begin{cases} \mathbf{F}_k, & \rho_s = \rho_f, \\ \mathbf{F}_k + M_{kj}\ddot{\mathbf{u}}_j, & \rho_s \neq \rho_f, \end{cases} \tag{3.192}$$

where $\mathbf{u}_j$ represents the nodal displacement vector at the $j$th node, and the consistent mass matrix $\mathbf{M}$ is defined as

$$\mathbf{M} = \int\limits_{\Omega^s(0)} (\rho_s - \rho_f)\mathbf{H}^T\mathbf{H}\det(\mathbf{D})\, d\Omega, \tag{3.193}$$

with $\mathbf{H}$ as the interpolation matrix.

Consequently, for the immersed solid, Eqs. (3.173) and (3.174) are still applicable for the modeling of the entire FSI system which are governed by Eqs. (3.167) and (3.168). Again, Eq. (3.173) is carried out for the current configuration of the submerged solid.

### *7.3.2. Particulate flow and blood modeling; mapping and kernel*

#### *Particulate flow and blood modeling*

To confirm these concepts, we present a set of numerical examples of a deformable cylinder or disk with a diameter of $2a$ free falling in a viscous fluid channel with a dimension of $2L \times 8L$. The physical parameters of this set of test cases are given as follows: acceleration due to gravitation $g = 9.81\ \mathrm{m/s^2}$, dynamic viscosity $\mu = 1\ \mathrm{dyne/cm^2\,s}$, fluid density $\rho_f = 1\ \mathrm{g/cm^3}$, and $L = 2$ cm. To implement the effect of gravity, an external body force is only applied to the cylinder. The buoyancy is captured by the definition of the mass matrix in Eq. (3.193). If the cylinder is rigid, the terminal velocity can be expressed as, according to Ref. [100],

$$U = \frac{(\rho_s - \rho_f)ga^2}{4\mu}\left[\ln\left(\frac{L}{a}\right) - 0.9157 + 1.7244\left(\frac{a}{L}\right)^2 - 1.7302\left(\frac{a}{L}\right)^4\right]. \tag{3.194}$$

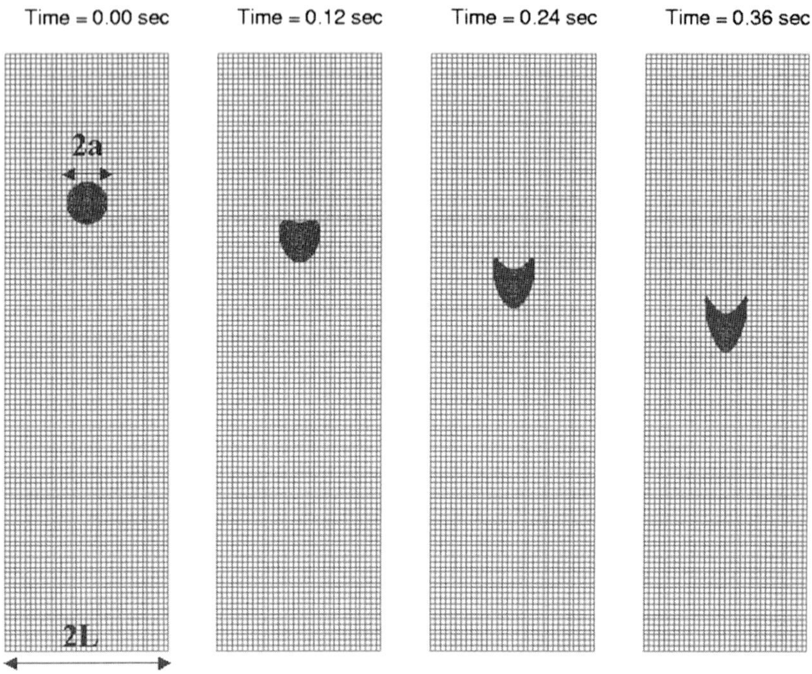


Figure 7.57. A uniform background fluid mesh and the submerged sphere.

Figure 7.57 shows the solid mesh as well as the background fluid mesh at different time snapshots. In Figure 7.58, with a diameter ratio $a/L = 0.25$ and a $64 \times 256$ fluid grid, it is clearly shown that the deformation of the submerged solid has a significant effect on the terminal velocity. In general, the flexibility of the cylinder decreases the surrounding fluid forces (viscous shear, form drag, etc.) and as a consequence increases terminal velocity. In this example, the submerged solid is made of an almost incompressible rubber material with the material constants $C_1 = 29300$, $C_2 = 17700$, and the density $\rho_s = 3\ \text{g/cm}^3$.

To increase accuracy in a subsequent three-dimensional falling sphere case, we employ a nonuniform fluid mesh as illustrated in Figure 7.59. To verify accuracy, we compare the solution of terminal velocity with the experimental data. It is explained in Ref. [44] that for a particle moving with steady terminal velocity $U_T$ in a gravitational field, the drag force $F_D$ balances the difference between the weight and buoyancy:

$$F_D = g(\rho^s - \rho^f)(\pi/6)D^3, \tag{3.195}$$

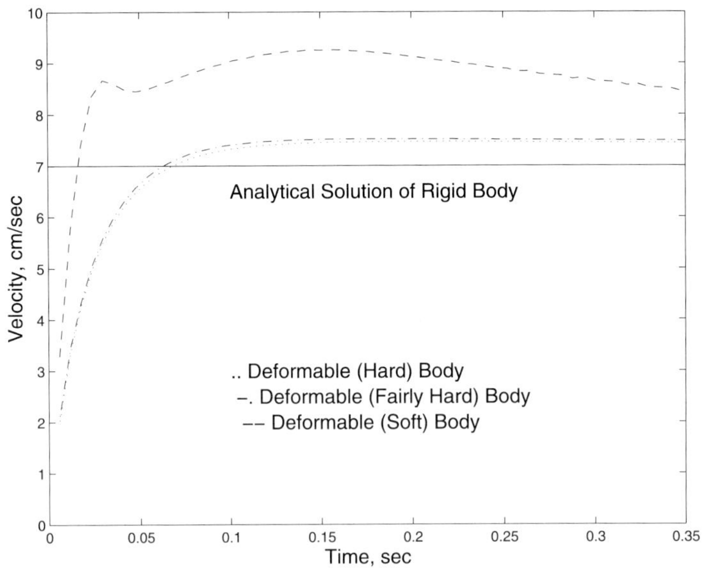


Figure 7.58. The velocity history of a moving object.

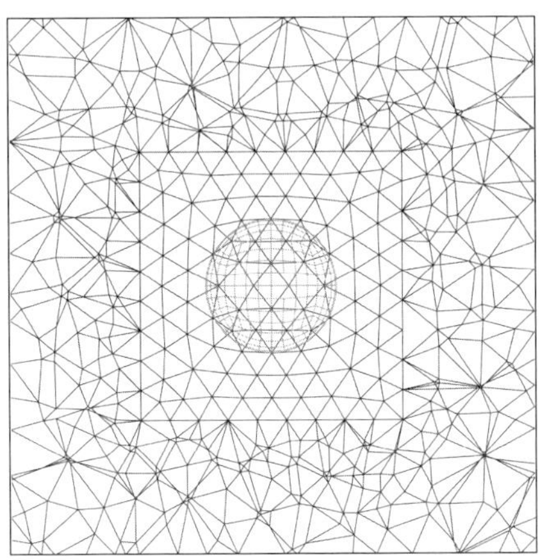

Figure 7.59. A gradient background mesh surrounding the submerged sphere.

and as a consequence the drag coefficient $C_D$ is calculated as

$$C_D = 4(\rho^s - \rho^f) g D / 3\rho^f U_T^2 = 4\rho^f (\rho^s - \rho^f) g D^3 / 3\mu^2 Re_T^2, \qquad (3.196)$$

where $Re_T$ is the Reynolds number at the terminal velocity.

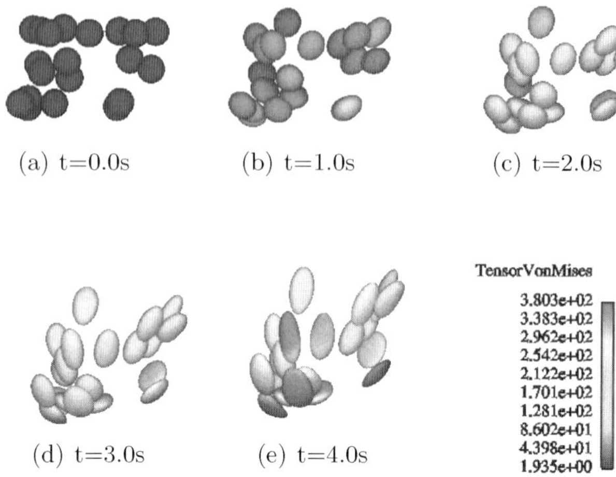


Figure 7.60. Stress distribution of 20 spheres dropping in a channel.

Since both the drag coefficient and Reynolds number are dependent of the terminal velocity $U_T$, an iterative procedure is needed. It is more convenient to express $Re$ as a function of the so-called best number $N_D$, which is defined as

$$N_D = C_D Re_T^2 = 4\rho^f(\rho^s - \rho^f) g D^3 / 3\mu^2. \tag{3.197}$$

Finally, the Reynolds number can be expressed as a function of $N_D$. In this case, for $580 \leqslant N_D \leqslant 1.55 \times 10^7$, the correlation for $Re$ as a function of $N_D$ is expressed as

$$\log_{10} Re = -1.81391 + 1.34671W - 0.12427W^2 + 0.006344W^3, \tag{3.198}$$

with $W = \log_{10} N_D$.

Using the parameters shown in the example, the experimental terminal velocity is found to be 2.203 cm/s with the terminal Reynolds number 88.11. The calculated terminal velocity from the IFEM solution is found to be 2.24412 cm/s, which deviates only 1.87% from the experimental value.

Now reset the material moduli for the soft deformable solid as $C_1 = 2.93 \times 10^2$ g/cm/s$^2$ and $C_2 = 1.77 \times 10^2$ g/cm/s$^2$. As expected from the simulation, the sphere experiences a higher pressure at the bottom and a lower pressure on the top as it tries to immerse itself deeper into the fluid. To demonstrate the advantage of immersed methods, we model a case with 20 soft spheres dropping in a channel. These same-sized spheres are placed randomly near the top of the channel. The spheres experience the gravitational and internal forces as well as the interaction forces with the surrounding fluid. To clearly illustrate the deformations of the spheres, we present in Figure 7.60 the enlarged images of the spheres and the pressure distributions at different time steps.

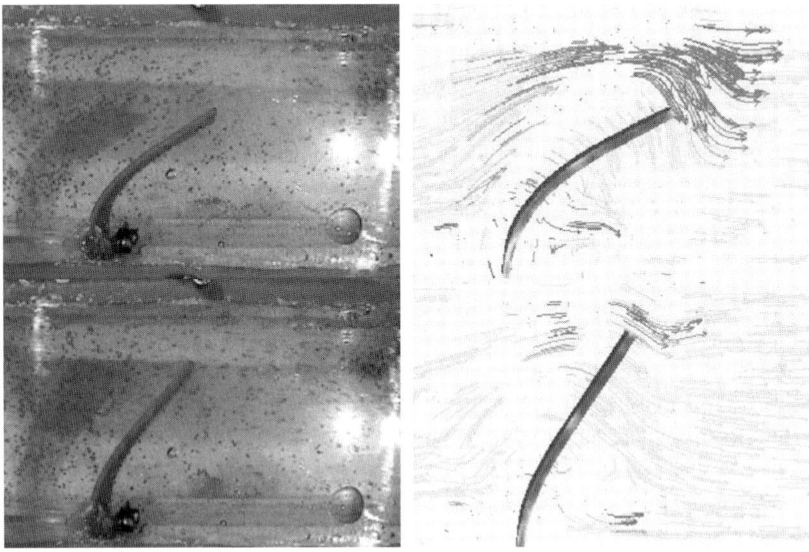

Figure 7.61. Experimental and computational results of a 3D FSI model.

In the simple setup of a pulsatile flow in a channel with a square cross section, a shell-like flexible structure similar to the heart valve is installed. The shell-like structure is made of incompressible rubber. As shown in Figure 7.61, the simulation results are very much the same as the experimental observation at individual time steps. In the experiment, water was pulsed through a column with a square cross section ($5 \times 5$ cm) at a frequency of 1 Hz. In the simulation, both velocity vectors of the fluid flow and stress tensors of the flexible structure are captured for the entire fluid and solid domains, whereas the preliminary experimental observation can only provide us with the dynamic behavior of the structure and overall flow patterns. This brings up an important point: for the actual valve design, simulation tools are very much required since many areas of the valve are difficult to visualize and only limited amounts of data can be collected experimentally.

One important application of immersed methods is the modeling of the human cardiovascular system. Human blood circulatory systems have evolved to supply nutrients and oxygen to, and carry the waste away from, the cells of multicellular organisms through the transport of blood, a complex fluid composed of deformable cells, proteins, platelets, and plasma. Currently, cardiovascular diseases represent the leading cause of death in developed countries. The lack of understanding of short and long-term development and evolution of many of the arterial and vascular diseases directly limits disease diagnosis and prevention as well as the planning of therapeutic approaches. It is therefore of significant clinical relevance to understand blood composition and its rheological behaviors in the context of multiscale and multiphysics hemodynamics.

Red blood cells (erythrocytes) typically comprise approximately 40% of blood by volume. Red blood cells (RBC) are normally 8–10 μm in diameter and 7.5–10 nm in thickness. RBC density is approximately 15% higher than that of water. Blood plasma can be accurately modeled with the Newtonian fluid model, yet blood flows do exhibit nonNewtonian or viscoelastic behaviors, especially under low Reynolds numbers. In fact, human blood is a typical thixotropic material of which the viscoelastic characteristics may vary significantly with stress and strain levels. Although in most arteries, blood behaves in a Newtonian fashion with a roughly constant viscosity, in microcirculatory system, nonNewtonian behavior is prevalent, in particular for low shear rates. A direct simulation of such non-Newtonian behaviors utilizing immersed methods is reported in Refs. [174,321].

The physiological functions of large blood vessels and micro- and capillary vessels are distinctly different. In general, blood is transported to critical locations within the body via large vessels. It has been recognized that flow conditions near vessel bifurcation points and low shear stress areas are among the important factors in the development of arterial and vascular diseases [275]. Furthermore, the modeling of macro-scale cardiovascular flows is directly beneficial to preoperative planning of surgical procedures for various cardiovascular diseases [34, 222]. On the other hand, the exchange of materials through the vessel walls with surrounding cells is accomplished within micro- and capillary vessels which account for, in adult cardiovascular and pulmonary circulations, a majority of the $10^{11}$ or so blood vessels [317,318]. As depicted in Fahraeus–Lindqvist effects, blood viscosity is lower in small vessels than in large vessels. In fact, the viscosity in capillaries is less than half of that in large vessels, due in part to red cells moving together in single files through small vessels. While theories of suspension rheology generally focus on homogeneous flows in infinite domains, the important phenomena of blood flows in microcirculation depend on the combined effects of vessel geometries, cell deformability, wall compliance, flow shear rates, and many micro-scale chemical, physiological, and biological factors [113,246]. There have been studies on shear flow effects on one or two cells [313], leukocytes adhesion to vascular endothelium [35], and particulate flow based on continuum enrichment methods [90,91]. So far no mature theory has been available for the prediction of blood rheology and blood perfusion through micro-vessels and capillary networks. The different time and length scales as well as large motions and deformations of immersed solids pose tremendous challenges to the mathematical modeling of blood flow at that level [89,245].

In the following examples, we will demonstrate some preliminary successes in modeling red blood cells within viscous fluids, using immersed methods. The role of red blood cells in the human body is to pick up oxygen as the blood passes through the lungs and to release it to the cells over the entire body. RBC mem-

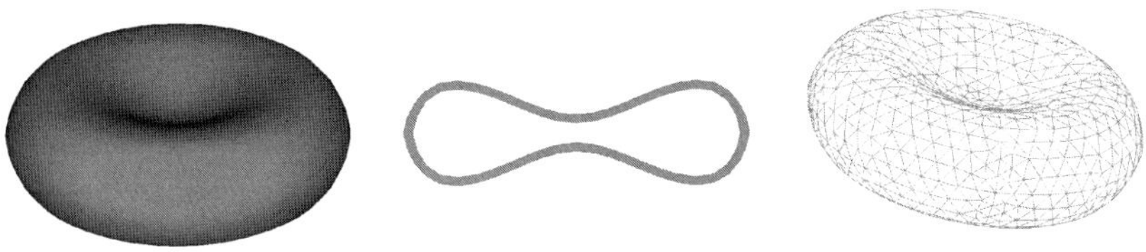

(a) 3D RBC Model (b) RBC Cross Section (c) RBC Mesh

Figure 7.62. A three-dimensional finite element mesh of a single RBC model.

brane contains a lipid bilayer structure. Unlike the white blood cell, in suspension culture, a typical RBC assumes a biconcave disc shape which permits its passage through capillaries. The biconcave disc shape enables the surface to volume ratio to be significantly higher than that of a sphere formed by a tensioned membrane. In addition, the biconcave disc shape suggests that the membrane cytoskeleton has both bending and membrane rigidities.

In general, RBC membrane has a shear modulus of $4.2 \times 10^{-3}$ dyn/cm$^2$, a dilation modulus of 500 dyn/cm$^2$, and a bending modulus of $1.8 \times 10^{-12}$ dyn/cm$^2$. Since the shear modulus of RBC membrane is much smaller than the dilation modulus, RBC membrane is susceptible to deformation by shear with very little area change. Furthermore, the cytoplasm contained within a typical RBC is incompressible, and even with the large deformation experienced by the cell membrane, there is no excessive membrane area change. Employing immersed methods, we concentrate on the rheological aspects of flow systems of arteriole, capillary, and venule which involve deformable cells, cell–cell interactions, and compliant vessels. As shown in Figure 7.62, to account for both bending and membrane rigidities, a typical RBC membrane is modeled with a three-dimensional finite element formulation with 1043 nodes and 4567 tetrahedra elements [65]. For simplicity, we also represent a typical RBC as a flexible three-dimensional thin structure enclosing an incompressible fluid with the same viscosity, i.e., 0.01 dyn s/cm$^2$. To account for different viscosity of the internal fluid, implicit and compressible immersed continuum methods must be used. Moreover, a function is used to describe the $x$- and $y$-coordinates of the cross-sectional profile:

$$\bar{y} = 0.5\left[1 - \bar{x}^2\right]^{1/2}\left(a_o + a_1\bar{x}^2 + a_2\bar{x}^4\right), \quad -1 \leqslant \bar{x} \leqslant 1, \tag{3.199}$$

with $a_o = 0.207$, $a_1 = 2.002$, and $a_2 = 1.122$, and the nondimensional coordinates $\bar{x}$ and $\bar{y}$ are scaled as $x/5$ and $y/5$ μm, respectively.

To further the discussion and include the viscoelastic material properties, we introduce a general Kelvin viscoelastic model as depicted in Figure 7.63. It is clear that the limiting cases of this model correspond to the familiar Maxwell, Voigt, and Kelvin models. Designating the shear strains at the spring $G_o$ and the

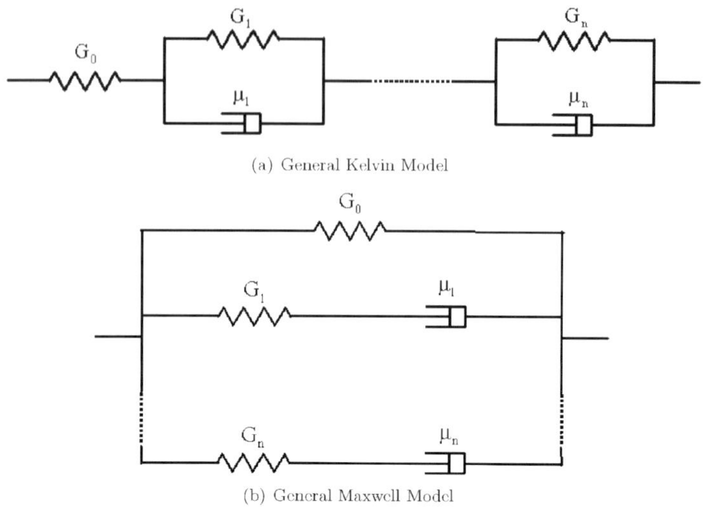


Figure 7.63. General viscoelastic models.

$i$th spring-dashpot combination $(G_i, \mu_i)$ as $\gamma_o$ and $\gamma_i$, respectively, and the stress within the series as $\sigma$, we obtain the following governing equations:

$$\sigma(t) = G_o \gamma_o(t), \tag{3.200}$$

$$\sigma(t) = G_i \gamma_i(t) + \mu_i \dot{\gamma}_i(t), \quad 1 \leqslant i \leqslant n. \tag{3.201}$$

Thus, from Eq. (3.201), we obtain

$$\gamma_i(t) = \int_{-\infty}^{t} \frac{1}{\mu_i} e^{-(t-s)/\alpha_i} \sigma(s)\, ds, \tag{3.202}$$

where the relaxation time $\alpha_i$ with respect to the $i$th spring-dashpot combination is defined as $\mu_i / G_i$.

Finally, the total shear strain $\gamma(t)$ can be evaluated as

$$\gamma(t) = \frac{\sigma(t)}{G_o} + \sum_{i=1}^{n} \gamma_i(t) = \int_{-\infty}^{t} C(t-s)\, d\sigma, \tag{3.203}$$

where the creep or compliance function $C(t)$ is defined as:

$$C(t) = \frac{1}{G_o} + \sum_{i=1}^{n} \frac{1}{G_i}\left(1 - e^{-t/\alpha_i}\right). \tag{3.204}$$

In a typical creep experiment, with $\sigma(s) = 0$ for $s < 0$ and $\sigma(s) = \sigma$ for $s \geqslant 0$, the shear strain is simply expressed as $\gamma(t) = \sigma C(t)$, which implies that the creep function can be determined experimentally. Naturally, a curve fitting will yield the physical parameters $G_o$, $G_i$, and $\mu_i$ in Eq. (3.204). Likewise, the governing equations for a general Maxwell viscoelastic model as depicted in Figure 7.63, can be expressed as

$$\sigma_o(t) = G_o \gamma(t), \tag{3.205}$$

$$\sigma_i(t) = \mu_i \dot{\gamma}(t) - \alpha_i \dot{\tau}_i(t), \quad 1 \leqslant i \leqslant n. \tag{3.206}$$

Thus, from Eq. (3.206), we obtain

$$\sigma_i(t) = \int_{-\infty}^{t} G_i e^{-(t-s)/\alpha_i} \dot{\gamma}(s)\, ds, \tag{3.207}$$

and the total stress $\sigma(t)$ can be evaluated as

$$\sigma(t) = G_o \gamma(t) + \sum_{i=1}^{n} \sigma_i(t) = \int_{-\infty}^{t} R(t-s)\, d\gamma, \tag{3.208}$$

where the relaxation or stiffness function $R(t)$ is defined as

$$R(t) = G_o + \sum_{i=1}^{n} G_i e^{-t/\alpha_i}. \tag{3.209}$$

In a typical relaxation experiment, with $\gamma(s) = 0$ for $s < 0$ and $\gamma(s) = \gamma$ for $s \geqslant 0$, the stress is simply expressed as $\sigma(t) = \gamma R(t)$, which implies that the relaxation function can also be determined experimentally. Naturally, a curve fitting will yield the physical parameters $G_o$, $G_i$, and $\mu_i$ in Eq. (3.209).

In our study of the behaviors of RBC aggregates and their effects on viscoelastic properties of blood flows, we must also consider cell–cell adhesion. The first attempt to study this phenomenon using three-dimensional continuum models of RBC aggregates with the effects of shear rates, RBC geometries, cell–cell interaction forces, and nonlinear material descriptions was published in Ref. [174]. Cell–cell adhesion plays an important role in various physiological phenomena including the recognition of foreign cells [322]. Although the exact physiological mechanisms of RBC coagulation and aggregation are still ambiguous, it has been found that both RBC surface structure and membrane proteins are key factors in producing adhesive/repulsive forces. In fact, RBC membranes held together by hydrophobic interactions are chemically and electrostatically active. Moreover, most membrane lipids and some proteins can drift laterally within the membrane. In particular, phospholipids move quickly ($\sim$2 μm/s) along the membrane while membrane proteins drift at a slower pace.

Conventional explanations in general fall into two categories: the bridging and adsorption models, with the adsorption and exclusion of the plasmatic macromolecules at RBC surfaces, respectively. The primary macromolecules that cause RBC aggregation are the so-called fibrinogen. The depletion layer results in a reduction of osmotic pressure in the cell gap, which consequently produces an attractive force. It can be assumed that the cell adhesion will occur only if cells are close enough. Nevertheless, when cells are too close to each other, a repulsive force will hinder contact. The repulsive forces include the steric forces due to the glycocalyx and the electrostatic repulsive force induced from the same negative fixed charges at cell surfaces. Of course, the level of RBC aggregation depends on the initial position of the cells and the strength of the adhesive force in comparison with other forces, such as hydrodynamic forces. The basic behavior of the interaction forces between two cells is simply illustrated as the weak attractive and strong repulsive forces at far and near distances. Due to the complexity of the aggregation process, we accumulate the intermolecular force, electrostatic force, and protein dynamics into a potential function, similar to an intermolecular potential. Here we adopt the Morse potential, found to be capable of quantitatively predicting the aggregation behaviors consistent with experimental observations:

$$\phi(r) = D_e\left[e^{2\kappa(r_o-r)} - 2e^{\kappa(r_o-r)}\right], \tag{3.210}$$

$$f(r) = -\frac{\partial\phi(r)}{\partial r} = 2D_e\kappa\left[e^{2\kappa(r_o-r)} - e^{\kappa(r_o-r)}\right], \tag{3.211}$$

where $r_o$ and $D_e$ stand for the zero force length and surface energy, respectively, and $\kappa$ is a scaling factor.

In practice, we often characterize blood flow with the strength of the shear flow, which is measured with the shear rate $\dot{\gamma}$ within the range from 0.1 to 100 $s^{-1}$. It is clear that as the shear rate increases the blood viscosity decreases initially, eventually reaching a plateau marking the plasma viscosity where the blood elasticity continues to decrease. In particular, Eqs. (3.204) and (3.209) demonstrate that the blood viscoelastic behaviors can be accurately predicted via actual or numerical experiments on the macroscopic level. On the microscopic level, deformable cells play an important role in the blood viscoelastic behaviors [317,318]. In the quiescent state, normal red blood cells tend to aggregate. Under low shear rate, aggregates are mainly influenced by the cell–cell interaction forces; in the mid-shear rate region, RBC aggregates start to disintegrate and the influence of the deformability gradually increases. Under high shear rate, red blood cells tend to stretch, align with the flow, and form layers. The illustration of these three different stages is shown in Figure 7.66.

With respect to cell–cell interaction forces, as shown in Figure 7.64, the cut-off length is chosen to be around 3 μm, beyond which the attractive force decays very fast. This means that nodes outside the influencing domain, essentially a sphere with the cut-off length as its radius, have very little effect. Moreover, the potential

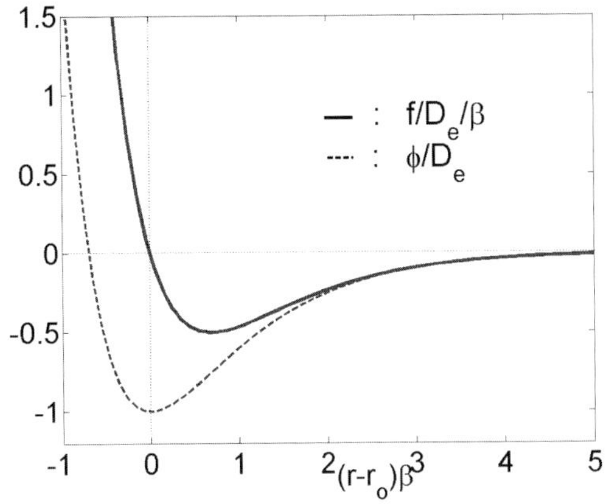


Figure 7.64. Nondimensionalized Morse potential and force.

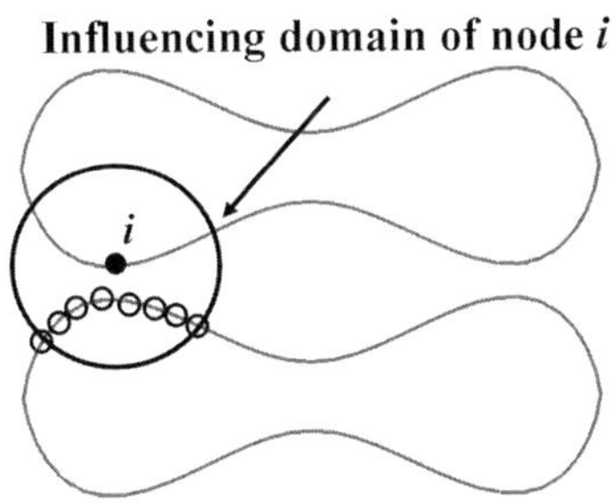


Figure 7.65. Illustration of the influencing domain.

function is chosen such that cells will de-aggregate at the shear rate above $0.5\ \mathrm{s}^{-1}$, according to the experimental observation [41]. For convenience, we set the zero force length as 250 nm. However, as we reduce the zero force length, local enrichment and multiscale coupling must also be introduced. It is important to recognize the fact that one microliter of normal human blood contains about 5,000,000 red blood cells, 7000 white blood cells, and 300,000 platelets. At this point, with current multiscale and multiphysics computational tools, it is impossible to handle such complex fluids as a whole.

In general, the cell–cell interaction forces are not sufficient to deform cell membranes. However, the ensuing aggregate could alter the surrounding fluid significantly. To illustrate the aggregation formation, we consider two red blood cells attracting each other in the quiescent fluid. In order to show clearly the effects of cell–cell interaction forces, we enlarge the magnitude of the interaction force by five times. As shown in Figure 7.67, under strong cell–cell interaction forces, two

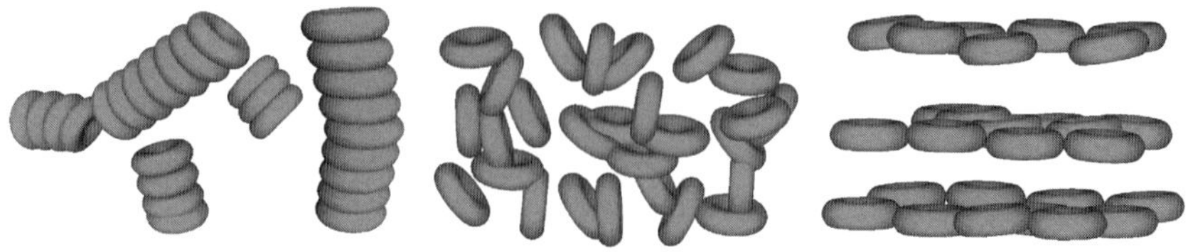

(a) Low shear (b) Mid-shear (c) High shear

Figure 7.66. Blood microscopic changes under different shear rates.

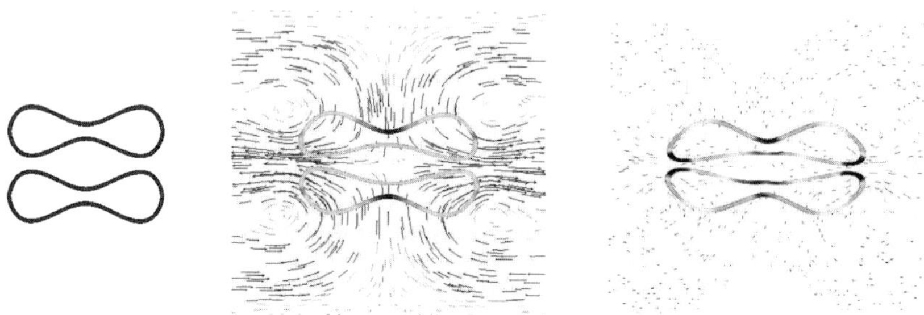

(a) (b) (c)

Figure 7.67. Cell configuration changes in quiescent fluid with cell–cell interactions.

disassociated cells gradually move together and form a doublet with moderate deformations.

RBC aggregation is one of the main causes of the nonNewtonian behaviors of blood flows. Due to the presence of cross-linking proteins fibrinogen on cell membranes and globulin in blood plasma, red blood cells tend to form aggregates called rouleaus, in which cells adhere loosely like a stack of coins. The presence of massive rouleaus can impair the blood flow through micro- and capillary vessels, causing fatigue and shortness of breath. The variation in the level of RBC aggregation may be an indication of thrombotic disease. In general, the lower the shear rate, e.g., the slower the blood velocity, the more prevalent, i.e., larger size and number density, are RBC aggregates. As the shear rate drops to zero, it is anticipated that human blood will become one big aggregate, which then behaves as a viscoelastic solid as illustrated earlier. On the contrary, as the shear rate increases, cell rouleaus tend to break up. Individual red cells also deform slightly, elongate, and line up with the streamlines. The deformability and the decrease of cell–cell interaction forces combine to reduce blood viscosity with the increase of the shear rate. As shear rate increases above certain level, the whole blood behaves almost like a Newtonian fluid with a nearly constant viscosity. It is important to

Figure 7.68. Illustrations of the four-RBC model and cell–cell interaction forces.

point out that for rigid particles including hardened red cells, normal leukocytes, etc., bulk viscosity is essentially independent of the shear rate.

To study RBC aggregates, we introduce a four-cell model with and without cell–cell interaction forces under different shear rates. The four-cell model is placed at the center of the fluid domain with vertical center distance of 3.96 μm. Due to the biconcave shape, the adhesive/repulsive forces mainly exist around RBC perimeters as shown in Figures 7.65 and 7.68. With the boundary shear velocity of 5 μm/s for both upper and bottom surfaces, we get a shear rate of $0.5\ s^{-1}$. In Figure 7.68, the four-cell model is subjected to three different shear rates. As expected, without cell–cell interaction forces, the cells will rotate and align with the flow separately while maintaining their vertical separations. On the other hand, RBC aggregates with cell–cell interaction forces rotate initially as a bulk with varying vertical distances. The comparison shows that cell–cell interaction forces restrain the disintegration of RBC aggregates and introduce elasticity in blood's macroscopic mechanical behaviors.

In a set of numerical experiments, we subject RBC aggregates and viscous fluid mixers to a low shear rate ($0.25\ s^{-1}$) and observe that RBC aggregates rotate as a bulk at $t = 0$, $t = 2$, and $t = 4$ s as shown in Figure 7.69. With an intermediate shear rate ($0.5\ s^{-1}$), our numerical simulation demonstrates that after initial rotations cells align with the shear direction and then de-aggregate at $t = 0$, $t = 0.5$, and $t = 1$ s, respectively, as shown in Figure 7.69. At a higher shear rate ($3.0\ s^{-1}$), RBC aggregates completely disintegrate and cells begin to orient themselves into parallel layers at $t = 0$, $t = 0.25$, and $t = 0.5$ s as shown in Figure 7.69. The disintegration of RBC aggregates with the increase of the shear rate is an indication of the decrease of the macroscopic viscosity, which is consistent with experimental observations.

In another experiment with a multiple-cell aggregate blocking the flow, a high shear rate induces cells to rotate first, then partially disintegrate, and finally align into parallel layers as shown in Figure 7.70. As evidence of excellent numerical resolution of the immersed finite element method and protein molecular dynamics, the vorticities surrounding deformable cells are clearly captured along with large structural motions and deformations.

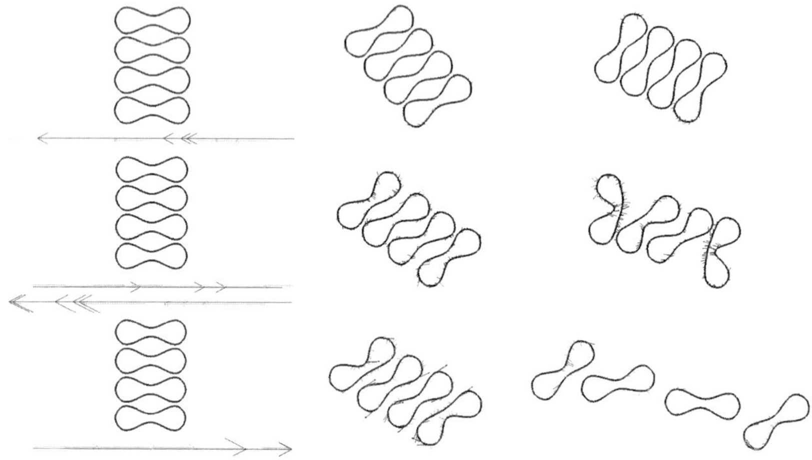

Figure 7.69. Four-cell model at different shear rates.

The typical diameter of the micro-vessel is 1.5–3 times larger than that of the cell. On the other hand, the capillary vessel diameter is about 2–4 μm, which is significantly smaller. The pressure gradient, which drives the flow, is usually around 3.2–3.5 KPa. For the chosen diameter and pressure, the Reynolds number in a typical capillary is around 0.01. In fact, in the process of squeezing through capillaries, large deformations of red blood cells not only slow down the blood flow, but also enable the exchange of oxygens through capillary vessel walls.

Sickle cell anemia is caused by genetic abnormalities in hemoglobin, which is a complex molecule constituting the most important component of red blood cells. When sickle hemoglobin loses oxygen, the deoxygenated molecule form rigid rods which distort the cell membrane into a sickle or crescent shape. The sickle-shaped cells are both rigid and sticky and tend to block capillary vessels and cause the blood flow blockage to the surrounding tissues and organs. To relate blood rheology to sickle cell anemia, we consider the normal and sickle red blood cells passing through a micro-vessel contraction. The strong viscous shear introduced by such a flow contraction leads to some interesting phenomena of RBC aggregation with respect to cell–cell interaction forces and cell deformability. Furthermore, the modeling of this complex fluid–solid system also demonstrates the capability of the coupling of the Navier–Stokes equations with protein molecular dynamics. We assign the sickle cell a crescent shape with higher rigidity and cell–cell adhesive/repulsive forces. Unlike the discrete-particle model presented in Refs. [64,267], in immersed methods, complex interaction forces between cells are modeled with a Morse potential and three-dimensional deformable biconcave cells are modeled as individual continuum objects. Thus, we can take advantage

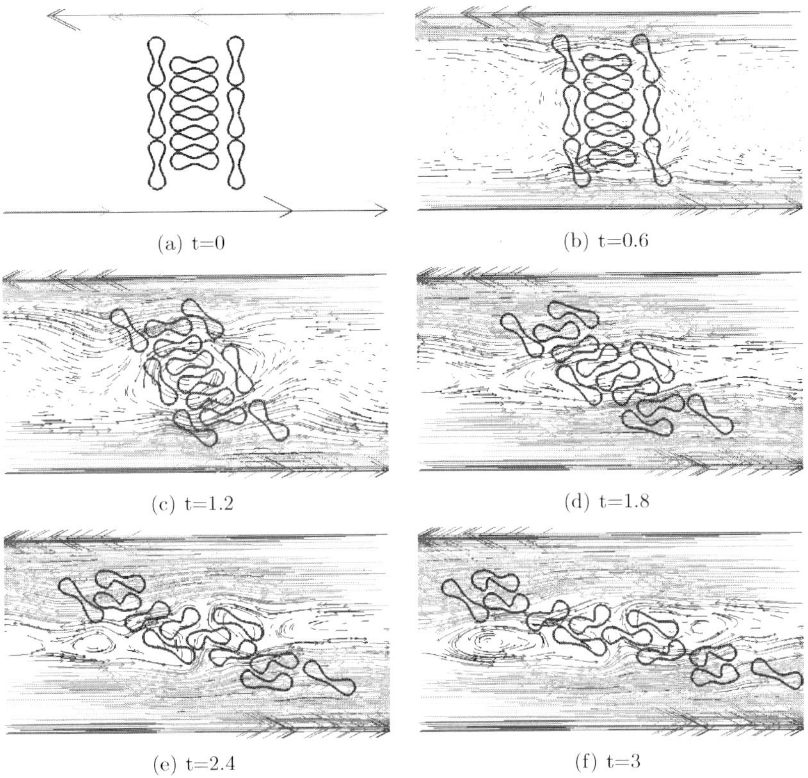


Figure 7.70. Twelve-cell model at different shear rates.

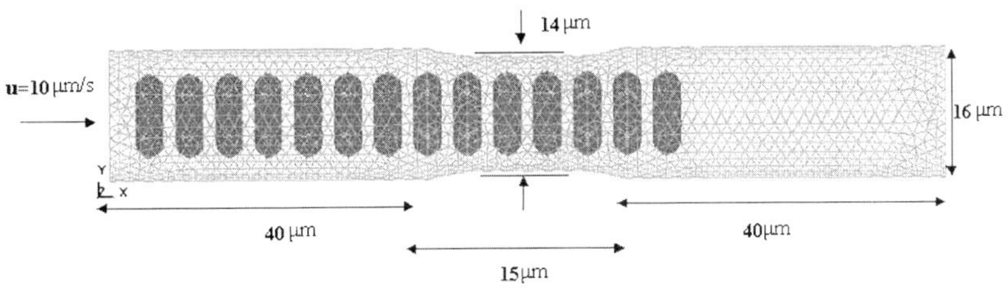


Figure 7.71. Finite element models of red blood cells passing through a vessel constriction.

of various mature finite element formulations for both fluid and solid domains. The initial configurations of the channel and cells are depicted in Figure 7.71. As shown in Figure 7.72, as cells pass the diffuser stage of the contraction with the inlet velocity of 10 μm/s, the deceleration of leading cells forms blockage for

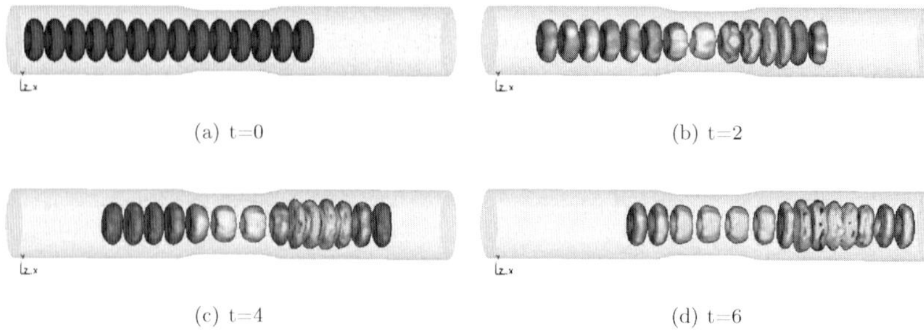


Figure 7.72. The normal red blood cell flow at different time steps.

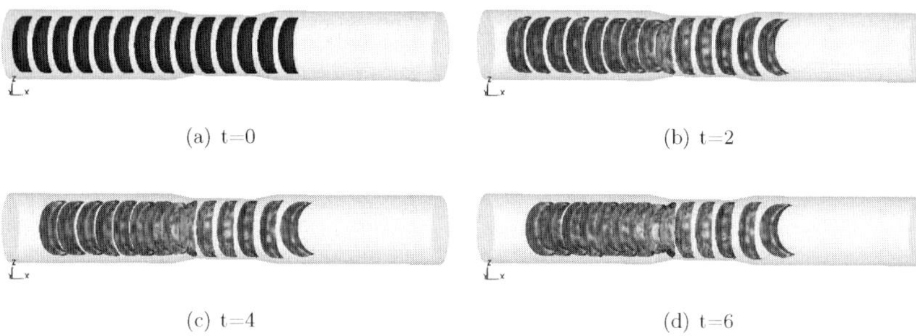


Figure 7.73. The sickle cell flow at different time steps.

incoming cells. Therefore, dilation of cells is coupled with pile-up of cells at the outlet of the vessel constriction. Also confirmed in Figure 7.73, under the similar flow conditions, rigid and sticky sickle cells eventually block the micro-vessel entrance, which will certainly result in de-oxygenation of surrounding tissues.

Consider a three-dimensional simulation of a single red blood cell squeezing through a capillary vessel. RBC diameter is 1.2 times larger than that of the capillary vessel, which leads to the divergence of the cytoplasm internal liquid to the two ends of the capsule by deforming into a slug during the squeezing process. During the exiting process, there is a radial expansion of the slug due to the convergence of the cytoplasm, which deforms the capsule into an acaleph or jellyfish shape. In Figure 7.74, four snapshots illustrate various stages of the red blood cell with respect to the capillary vessel.

In another example, a chain of three deformable objects with the similar material properties are released and move towards to an elastic bifurcation. Initially, these objects are perfectly centered and aligned with the bifurcation point. The slight difference between the upper and lower branches of the bifurcation breaks the symmetry. As shown in Figure 7.75, objects impact, deform, and conform with the viscous flow within the lower branch of the bifurcation. This type of study is

(a) t=0

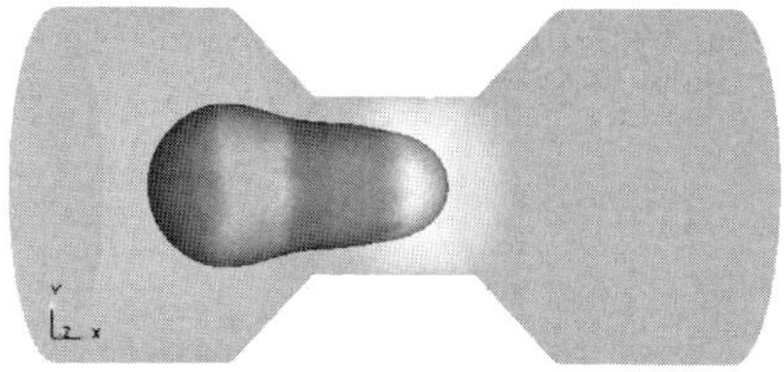

(b) t=0.48

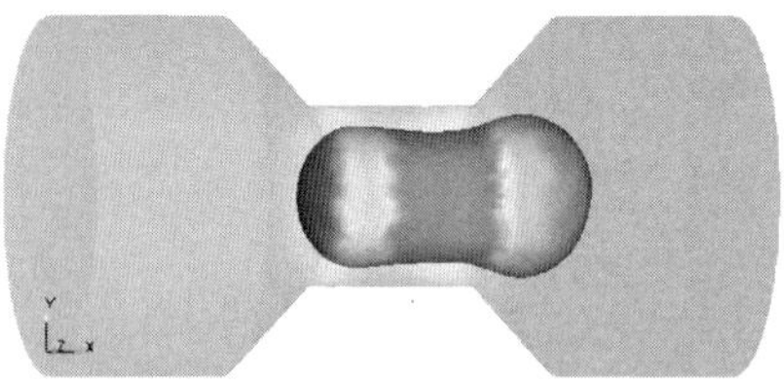

(c) t=0.96

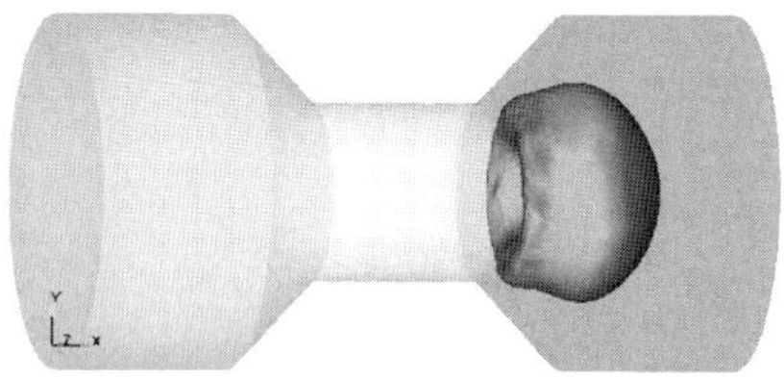

(d) t=1.44

Figure 7.74. A single white cell squeezing through a capillary vessel.

very important for the understanding of the adverse effects of artificial devices exerted on red blood cells.

*Mapping and kernel*

In the immersed boundary method, all discretized delta functions are continuous and have finite support points. Consider first a one-dimensional case, in accordance with the translation invariance [48], for all $r$, where $r$ is the parameter representing the position of the submerged boundary point and is scaled with respect to the grid size $h$, and the discretized delta function satisfies $\sum_j \phi^2(r-j) = C$, where $C$ is a numerical constant. In addition, to uniquely define the discretized delta function for all $r$, we also have $\sum_j (r-j)^m \phi(r-j) = 0$, where the selection of the $m$th moment depends on the number of support points. For instance, the discretized delta function with four support points is uniquely defined by the following:

(1) $\phi$ is a continuous function, with $\phi(r) = 0$ for $|r| \geqslant 2$;
(2) $\forall r, \sum_{j \text{ even}} \phi(r-j) = \sum_{j \text{ odd}} \phi(r-j) = 1/2$;
(3) $\forall r, \sum_j (r-j)\phi(r-j) = 0$; and
(4) $\forall r, \sum_j \phi^2(r-j) = C$, where $C$ is a numerical constant.

In general, for $0 < r < 1$, the discretized delta function $\phi(r-j)$ covers four nonzero support points. However, for the degenerate case of the 4-point discretized delta function centered at $r = 0$, we have five support points, namely,

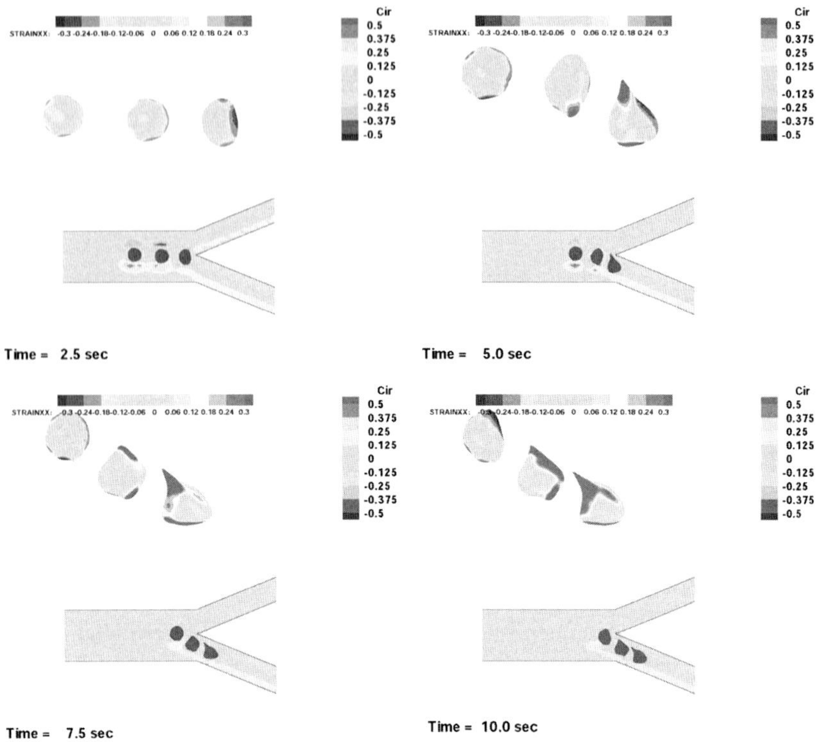


Figure 7.75. Deformable objects impact elastic bifurcation point and conform to flow condition.

$r - j = -2, -1, 0, 1, 2$. From the degenerate case, we can easily derive the constant $C$. Hence, we obtain the following four admissible branches of solutions for $0 < r < 1$:

$$\begin{aligned} &\phi(r) = \left[-(2r-3) + \sqrt{4r - 4r^2 + 1}\right]/8, &&\phi(r-2) = 1/2 - \phi(r), \\ &\phi(r-1) = -1/4 + r/2 + \phi(r), &&\phi(r+1) = 3/4 - r/2 - \phi(r). \end{aligned} \tag{3.212}$$

Therefore, in the three-dimensional case, one of the smoothed approximations to the delta function for uniform background fluid mesh with a mesh size $h$ is given by

$$\delta_h(\mathbf{x}) = \frac{1}{h^3}\phi\left(\frac{x_1}{h}\right)\phi\left(\frac{x_2}{h}\right)\phi\left(\frac{x_3}{h}\right). \tag{3.213}$$

It is interesting to note that the discretized delta function in Eq. (3.212) is very close to $(1 + \cos \pi r/2)/4$, with $r \in [-2, 2]$. Moreover, it is easy to confirm that the discretized delta function $\phi(r)$, with $r \in [-2, 2]$, defined in Eq. (3.212),

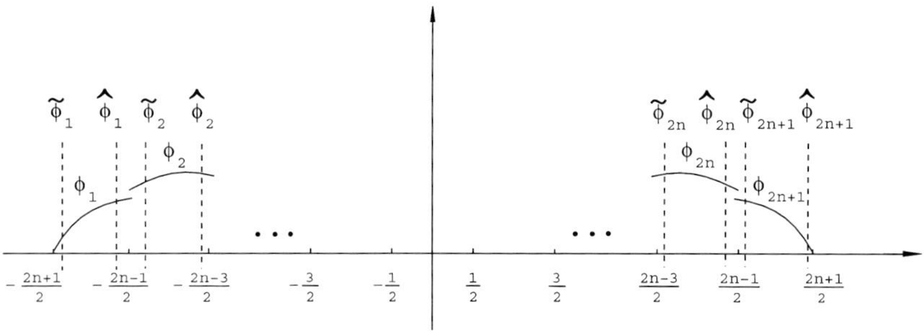


Figure 7.76. A discretized delta function with an odd number of support points.

has $C^1$ continuity [294]. In this section, we discuss in detail the construction of the discretized delta function employed in the immersed boundary method and prove mathematically that such a function is $C^1$ and not $C^2$ continuous. A general discretized delta function with an odd number of support points is shown in Figure 7.76. Assume $\phi_i$, $1 \leqslant i \leqslant 2n+1$, are piecewise $C^\infty$, and the global $\delta_h$ is written as $\phi(r)$ with $-(2n+1)/2 \leqslant r \leqslant (2n+1)/2$. For convenience, as illustrated in Figure 7.76, assign $\hat{\phi}_k^m$ and $\tilde{\phi}_k^m$ as the limit values before the right station and after the left station of the $k$th branch of the discretized delta function, denoted as $\phi_k(r)$. For the global discretized delta function $\phi(r)$ to be $C^m$ continuous, we must have

$$\hat{\phi}_k^{(m)} = \tilde{\phi}_{k+1}^{(m)}, \quad \text{for } 1 \leqslant k \leqslant 2n. \tag{3.214}$$

At this point, in order to preserve the properties of the delta function, we would like to enforce some essential properties for the corresponding discretized delta function.

THEOREM 7.3.1. *For a nonnegative integer m, the mth moment of the delta function is written as*

$$\int x^m \delta(x)\, dx = \begin{cases} 1, & m = 0, \\ 0, & m \geqslant 1. \end{cases} \tag{3.215}$$

*With a uniform grid, the discretized mth moment can be expressed either*

$$\sum_i (r-i)^m \phi(r-i) = 0, \tag{3.216}$$

*or*

$$\sum_i i^m \phi(r-i) = r^m. \tag{3.217}$$

PROOF. Referring to Chapter 2, we can easily derive $\int \delta(x)\,dx = 1$ and $\int x^m \delta(x)\,dx = 0^m = 0$ for $m \geqslant 1$. Eq. (3.216) is the direct application of Eq. (3.215) to a uniform grid. Suppose for any integer k, $0 \leqslant k \leqslant n-1$, Eq. (3.217) holds, i.e., $\sum_i i^k \phi(r-i) = r^k$. Employ the polynomial expansion

$$(r-i)^n = \sum_{j=0}^{n} r^j i^{n-j} (-1)^j C_n^j,$$

we can easily derive, from $\sum_i (r-i)^n \phi(r-i) = 0$,

$$\begin{aligned}
&\sum_i (r-i)^n \phi(r-i) \\
&\quad = \sum_i i^n \phi(r-i) + \sum_i \sum_{j=1}^{n} r^j i^{n-j} (-1)^j C_n^j \phi(r-i) = 0.
\end{aligned} \tag{3.218}$$

Moreover, based on the premise, for $1 \leqslant j \leqslant n$, we also have $\sum_i i^{n-j}\phi(r-i) = r^{n-j}$. As a consequence, Eq. (3.218) is equivalent to

$$\begin{aligned}
&\sum_i i^n \phi(r-i) + \sum_{j=1}^{n} r^j r^{n-j} (-1)^j C_n^j \\
&\quad = \sum_i i^n \phi(r-i) - r^n + (1-1)^n = 0,
\end{aligned} \tag{3.219}$$

from which we obtain $\sum_i i^n \phi(r-i) = r^n$, namely, Eq. (3.217) also holds for $k = n$.

Therefore, based on mathematical induction, Eq. (3.217) is valid for any non-negative integer $m$.

Consider $m = 1$, we have $2n+2$ choices of implementing Eq. (3.217) which corresponds to $2n+2$ stations as shown in Figure 7.76. Moreover, denote the station values as unknown vectors $\tilde{\boldsymbol{\phi}} = \langle \tilde{\phi}_1, \tilde{\phi}_2, \ldots, \tilde{\phi}_{2n+1} \rangle$ and $\hat{\boldsymbol{\phi}} = \langle \hat{\phi}_1, \hat{\phi}_2, \ldots, \hat{\phi}_{2n+1} \rangle$, we have

$$\tilde{\mathbf{A}}\tilde{\boldsymbol{\phi}} = \mathbf{r} \quad \text{and} \quad \hat{\mathbf{A}}\hat{\boldsymbol{\phi}} = \mathbf{r}, \tag{3.220}$$

with $\mathbf{r} = \langle -(2n+1)/2, -(2n-1)/2, \ldots, (2n-1)/2, (2n+1)/2 \rangle$, and the $(2n+2) \times (2n+1)$ matrices $\tilde{\mathbf{A}}$ and $\hat{\mathbf{A}}$ are expressed as

$$\tilde{\mathbf{A}} = \begin{bmatrix} 0 & -1 & -2 & \cdots & -2n \\ 1 & 0 & -1 & \cdots & -2n+1 \\ 2 & 1 & 0 & \cdots & -2n+2 \\ \vdots & \vdots & \ddots & \vdots & \vdots \\ 2n+1 & 2n & 2n-1 & \cdots & 1 \end{bmatrix} \quad \text{and}$$

$$\hat{\mathbf{A}} = \begin{bmatrix} -1 & -2 & -3 & \cdots & -(2n+1) \\ 0 & -1 & -2 & \cdots & -2n \\ 1 & 0 & -1 & \cdots & -(2n-1) \\ \vdots & \vdots & \ddots & \vdots & \vdots \\ 2n & 2n-1 & 2n-2 & \cdots & 0 \end{bmatrix}.$$

It is obvious that both $\tilde{\mathbf{A}}$ and $\hat{\mathbf{A}}$ are rank-two matrices and the following conditions are dependent

$$\sum_{i=1}^{2n+1} \hat{\phi}_i = \sum_{i=1}^{2n+1} \tilde{\phi}_i = 1. \tag{3.221}$$

In fact, Eq. (3.221) can be easily derived by subtracting one row of the matrices $\tilde{\mathbf{A}}$ or $\hat{\mathbf{A}}$ from the other. In order to prove $C^1$ and not $C^2$ continuity of the discretized delta function, we must prove that Eq. (3.214) holds for $m = 1$ and not for $m = 2$. The condition (3.217), which is used for the derivation of the discretized delta function, suggests that properties of the Vandermonde matrix can be introduced. Similar to the derivation of the matrices $\tilde{\mathbf{A}}$ and $\hat{\mathbf{A}}$, we denote the station values as unknown vectors $\tilde{\mathbf{x}} = \langle \tilde{\phi}_2, \tilde{\phi}_3, \ldots, \tilde{\phi}_{2n+1} \rangle$ and $\hat{\mathbf{x}} = \langle \hat{\phi}_1, \hat{\phi}_2, \ldots, \hat{\phi}_{2n} \rangle$, and employing the condition (3.221), we derive

$$\mathbf{A}\tilde{\mathbf{x}} = \tilde{\mathbf{b}} \quad \text{and} \quad \mathbf{A}\hat{\mathbf{x}} = \hat{\mathbf{b}}, \tag{3.222}$$

with

$$\tilde{\mathbf{b}} = \mathbf{b}_o + \tilde{\mathbf{b}}_f \tilde{\phi}_1 \quad \text{and} \quad \hat{\mathbf{b}} = \mathbf{b}_o + \hat{\mathbf{b}}_f \hat{\phi}_{2n+1}, \tag{3.223}$$

$$\mathbf{A} = \begin{bmatrix} 1 & 1 & 1 & \cdots & 1 \\ 0 & 1 & 2 & \cdots & 2n-1 \\ 0 & 1^2 & 2^2 & \cdots & (2n-1)^2 \\ \vdots & \vdots & \ddots & \vdots & \\ 0 & 1^{2n-1} & 2^{2n-1} & \cdots & (2n-1)^{2n-1} \end{bmatrix},$$

and

$$\mathbf{b}_o = \begin{bmatrix} 1 \\ \frac{2n-1}{2} \\ (\frac{2n-1}{2})^2 \\ \vdots \\ (\frac{2n-1}{2})^{2n-1} \end{bmatrix}, \quad \tilde{\mathbf{b}}_f = \begin{bmatrix} -1 \\ 1 \\ -1 \\ \vdots \\ 1 \end{bmatrix}, \quad \text{and} \quad \hat{\mathbf{b}}_f = -\begin{bmatrix} 1 \\ 2n \\ (2n)^2 \\ \vdots \\ (2n)^{2n-1} \end{bmatrix}.$$

As a key observation, we identify that $\mathbf{b}_o$ can be considered as $\mathbf{b}_r = \langle 1, r, r^2, \ldots, r^{2n-1} \rangle$ evaluated at $r = (2n-1)/2$, whereas the following vectors $\mathbf{b}_1$ and

$\mathbf{b}_2$ can be simply considered as the first and second derivatives of $\mathbf{b}_r$ evaluated at $r = (2n-1)/2$:

$$\mathbf{b}_1 = \left.\frac{\partial \mathbf{b}_r}{\partial r}\right|_{r=(2n-1)/2} \left\langle 0, 1, 2\left(\frac{2n-1}{2}\right), \ldots, (2n-1)\left(\frac{2n-1}{2}\right)^{2n-2}\right\rangle, \tag{3.224}$$

$$\mathbf{b}_2 = \left.\frac{\partial^2 \mathbf{b}_r}{\partial r^2}\right|_{r=(2n-1)/2} \left\langle 0, 0, 2, \ldots, (2n-1)(2n-2)\left(\frac{2n-1}{2}\right)^{2n-3}\right\rangle. \tag{3.225}$$

Therefore, the first and second derivatives of the vectors $\tilde{\mathbf{x}}$ and $\hat{\mathbf{x}}$ in Eqs. (3.222) can be evaluated from

$$\mathbf{A}\dot{\tilde{\mathbf{x}}} = \mathbf{b}_1 + \tilde{\mathbf{b}}_f \dot{\tilde{\phi}}_1 \quad \text{and} \quad \mathbf{A}\dot{\hat{\mathbf{x}}} = \mathbf{b}_1 + \hat{\mathbf{b}}_f \dot{\hat{\phi}}_{2n+1}, \tag{3.226}$$

$$\mathbf{A}\ddot{\tilde{\mathbf{x}}} = \mathbf{b}_2 + \tilde{\mathbf{b}}_f \ddot{\tilde{\phi}}_1 \quad \text{and} \quad \mathbf{A}\ddot{\hat{\mathbf{x}}} = \mathbf{b}_2 + \hat{\mathbf{b}}_f \ddot{\hat{\phi}}_{2n+1}. \tag{3.227}$$

Now, for the discretized delta function of the immersed boundary method, we also enforce the end support points, such that

$$\tilde{\phi}_1 = \hat{\phi}_{2n+1} = 0, \qquad \dot{\tilde{\phi}}_1 = \dot{\hat{\phi}}_{2n+1} = 0, \quad \text{and} \quad \ddot{\tilde{\phi}}_1 = \ddot{\hat{\phi}}_{2n+1} = 0. \tag{3.228}$$

As a consequence, derived from Eqs. (3.222), we obtain

$$\tilde{\mathbf{x}} = \hat{\mathbf{x}} = \mathbf{A}^{-1}\mathbf{b}_o. \tag{3.229}$$

It is clear that Eq. (3.229) is equivalent to Eq. (3.214) and the discretized delta function $\phi(r)$ is continuous. Notice that in the immersed boundary method, for the derivation of discretized delta functions, an additional constraint is introduced:

$$\sum_i \delta_h^2(r - ih) = C, \tag{3.230}$$

where $h$ stands for the grid size and $C$ is a constant independent of the position $r$.

Eq. (3.230) is provided based on the understanding of the translation invariance [48]. As a consequence, we also have the following additional constraint:

$$\sum_{i=1}^{2n+1} \tilde{\phi}_i^2 = \sum_{i=1}^{2n+1} \hat{\phi}_i^2 = C. \tag{3.231}$$

If the discretized delta function has $C^2$ continuity, by taking the first and second derivatives of Eq. (3.231), and employing the end conditions (3.228), we must have

$$\tilde{\mathbf{x}}^T \dot{\tilde{\mathbf{x}}} = \hat{\mathbf{x}}^T \dot{\hat{\mathbf{x}}} = 0 \quad \text{and} \quad \dot{\tilde{\mathbf{x}}}^T \dot{\tilde{\mathbf{x}}} + \tilde{\mathbf{x}}^T \ddot{\tilde{\mathbf{x}}} = \dot{\hat{\mathbf{x}}}^T \dot{\hat{\mathbf{x}}} + \hat{\mathbf{x}}^T \ddot{\hat{\mathbf{x}}} = 0. \tag{3.232}$$

It is therefore clear, based on Eqs. (3.226) to (3.229) and (3.232), that the $C^1$- and not $C^2$-continuity condition is equivalent to

$$\mathbf{b}_o^T\left(\mathbf{A}^{-1}\right)^T\left(\mathbf{A}^{-1}\right)\mathbf{b}_1 = 0, \tag{3.233}$$

$$\mathbf{b}_o^T\left(\mathbf{A}^{-1}\right)^T\left(\mathbf{A}^{-1}\right)\mathbf{b}_2 + \mathbf{b}_1^T\left(\mathbf{A}^{-1}\right)^T\left(\mathbf{A}^{-1}\right)\mathbf{b}_1 \neq 0. \tag{3.234}$$

Let us first prove the equality (3.233). Considering a general Vandermonde matrix, named after Alexandre-Théophile Vandermonde (1735–1796), we have

$$\det\begin{bmatrix} 1 & 1 & 1 & \cdots & 1 \\ x_1 & x_2 & x_3 & \cdots & x_{2n} \\ x_1^2 & x_2^2 & x_3^2 & \cdots & x_{2n}^2 \\ \vdots & \vdots & \vdots & \ddots & \vdots \\ x_1^{2n-1} & x_2^{2n-1} & x_3^{2n-1} & \cdots & x_{2n}^{2n-1} \end{bmatrix} = \prod_{1\leqslant i<j}^{2n} (x_j - x_i). \tag{3.235}$$

Obviously, the matrix $\mathbf{A}$ is a Vandermonde matrix with $x_1 = 0, x_2 = 1, \ldots, x_j = j-1, \ldots, x_{2n} = 2n-1$, and from Eq. (3.235), we have

$$\det\mathbf{A} = \prod_{0\leqslant k<j}^{2n-1} (j-k). \tag{3.236}$$

Supposing we construct another Vandermonde matrix $\mathbf{A}_i$ by replacing the $i$th column of the matrix $\mathbf{A}$ with $\langle 1, r, r^2, \ldots, r^{2n-1}\rangle$, we obtain

$$\det\mathbf{A}_i = \frac{\prod_{0\leqslant k<j}^{2n-1}(j-k)}{\prod_{k\neq i-1}^{0,2n-1}(i-1-k)} \prod_{k\neq i-1}^{0,2n-1} (r-k). \tag{3.237}$$

Employing Cramer's rule, we then have

$$y_i = \frac{\det\mathbf{A}_i}{\det\mathbf{A}} = \frac{\prod_0^{2n-1}(r-k)}{\prod_{k\neq i-1}^{0,2n-1}(i-1-k)}\frac{1}{r-i+1}. \tag{3.238}$$

As a consequence, we also have

$$\frac{\partial y_i}{\partial r} = \sum_{j\neq i-1}^{0,2n-1} \frac{\prod_0^{2n-1}(r-k)}{\prod_{k\neq i-1}^{0,2n-1}(i-1-k)}\frac{1}{(r-i+1)(r-j)}, \tag{3.239}$$

$$\frac{\partial^2 y_i}{\partial r^2} = \sum_{j\neq i-1}^{0,2n-1}\sum_{l\neq i-1,j}^{0,2n-1} \frac{\prod_0^{2n-1}(r-k)}{\prod_{k\neq i-1}^{0,2n-1}(i-1-k)}\frac{1}{(r-i+1)(r-j)(r-l)}. \tag{3.240}$$

Therefore, employing Eqs. (3.238), (3.239), and (3.240), we obtain

$$y_i\frac{\partial y_i}{\partial r} = \frac{\left[\prod_0^{2n-1}(r-k)\right]^2}{\left[\prod_{k\neq i-1}^{0,2n-1}(i-1-k)\right]^2}\frac{1}{(r-i+1)^2}\sum_{j\neq i-1}^{0,2n-1}\frac{1}{(r-j)}, \tag{3.241}$$

$$y_i \frac{\partial^2 y_i}{\partial r^2} = \frac{\left[\prod_0^{2n-1}(r-k)\right]^2}{\left[\prod_{k\neq i-1}^{0,2n-1}(i-1-k)\right]^2} \frac{1}{(r-i+1)^2} \times \sum_{j\neq i-1}^{0,2n-1} \sum_{l\neq i-1,j}^{0,2n-1} \frac{1}{(r-j)(r-l)}, \tag{3.242}$$

$$\frac{\partial y_i}{\partial r}\frac{\partial y_i}{\partial r} = \frac{\left[\prod_0^{2n-1}(r-k)\right]^2}{\left[\prod_{k\neq i-1}^{0,2n-1}(i-1-k)\right]^2} \frac{1}{(r-i+1)^2} \left[\sum_{j\neq i-1}^{0,2n-1} \frac{1}{(r-j)}\right]^2. \tag{3.243}$$

It is clear that the equality (3.233) is equivalent to

$$\sum_{i=1}^{2n} y_i \frac{\partial y_i}{\partial r} = 0. \tag{3.244}$$

Moreover, by replacing $i$ with $2n - i + 1$, we can easily confirm that

$$\prod_{k\neq i-1}^{0,2n-1} (i-1-k)^2 = \prod_{k\neq 2n-i}^{0,2n-1} (2n-i-k)^2. \tag{3.245}$$

Notice that at $r = (2n-1)/2$, we have the following anti-symmetrical properties for $i$ series, with $1 \leqslant i \leqslant 2n$ and $j$ series, with $0 \leqslant j < 2n-1$,

$$\begin{aligned} &r - i + 1 = (2n - 2i + 1)/2, \\ &r - (2n - i + 1) + 1 = (-2n + 2i - 1)/2, \end{aligned} \tag{3.246}$$

and

$$\begin{aligned} &r - j = (2n - 2j - 1)/2, \\ &r - (2n - j - 1) = -(2n - 2j - 1)/2. \end{aligned} \tag{3.247}$$

Hence, using the anti-symmetrical properties of (3.247), we can easily confirm that at $r = (2n-1)/2$, we have

$$\sum_{j=0}^{2n-1} \frac{1}{r-j} = 0, \tag{3.248}$$

$$\sum_{j\neq i-1}^{0,2n-1} \frac{1}{r-j} = -\frac{1}{r-i+1}. \tag{3.249}$$

Furthermore, by employing Eqs. (3.245) and (3.246), we derive

$$
\begin{aligned}
&-\frac{1}{\prod_{k\neq i-1}^{0,2n-1}(i-1-k)^2}\frac{1}{(r-i+1)^3}\\
&\quad-\frac{1}{\prod_{k\neq 2n-i}^{0,2n-1}(2n-i-k)^2}\frac{1}{(r-2n+i)^3}=0.
\end{aligned} \tag{3.250}
$$

Based on Eq. (3.250), it is then straightforward to prove Eq. (3.244) by grouping terms $i$ with $2n-i+1$, and consequently, Eq. (3.233). Employing similar procedures, we also derive for Eq. (3.242),

$$
\sum_{j\neq i-1}^{0,2n-1}\sum_{l\neq i-1,j}^{0,2n-1}\frac{1}{(r-j)(r-l)}=\frac{2}{(r-i+1)^2}-\sum_{j=0}^{2n-1}\frac{1}{(r-j)^2}. \tag{3.251}
$$

However, the anti-symmetrical properties will no longer hold for the cancellation of terms involving even powers of $(r-i+1)$ and $(r-j)$. Without much difficulty, we then also prove the inequality (3.234). Therefore, the discretized delta function in the immersed boundary method is $C^1$- and not $C^2$-continuous.

In immersed methods, the nonlinear mapping from the fluid mesh to the solid mesh is essential. At a typical solid node $j$, with a finite support domain $\Omega_j$, the discretized form of the constraint of the velocities of the immersed solid and the corresponding fluid occupying the same solid domain can be expressed as

$$
\mathbf{v}_j^s=\sum_i \mathbf{v}_i\phi_i\left(\mathbf{x}_i-\mathbf{x}_j^s\right),\quad \forall \mathbf{x}_i\in\Omega_j, \tag{3.252}
$$

where $\phi_i(\mathbf{x}_i-\mathbf{x}_j)$ is the kernel function centered at the solid node $j$, represented with $\mathbf{x}_j^s$.

It is very important to realize that the material points of the submerged solid domain will move in the entire domain, therefore, even if we do not adjust the size of the support domain attached to these material points, Eq. (3.252) represents a nonlinear mapping simply denoted as $\mathcal{M}$. Note that for the second view of immersed incompressible solids, namely dry, impermeable, and completely separated from the surrounding fluid, within the solid domain, we have the stress difference $\sigma_{ij}^s-\sigma_{ij}^f$ or $-(p^s-p^f)\delta_{ij}+(\tau_{ij}^s-\tau_{ij}^f)$. In order to use the definition of $\sigma_{ij}^f$, we must also map the unknown pressure from the fluid mesh denoted with node $i$ to the solid mesh denoted with node $j$. In this case, it is beneficial to use the continuous pressure mixed finite element formulation for both fluid and solid domains [300,13]. Consequently, like Eq. (3.252), we have

$$
p_j^f=\sum_i p_i\phi_i(\mathbf{x}_i-\mathbf{x}_j),\quad \forall \mathbf{x}_i\in\Omega_j. \tag{3.253}
$$

It turns out that such discretized mapping using various kernel functions has been studied recently in the meshless finite element methods. For example, reproducing kernel particle method (RKPM) was proposed as an alternative or enhancement to various numerical procedures including finite element methods [307,309]. Unlike the discretized delta function originally proposed in the immersed boundary method [49], the kernel functions in the meshless methods can handle nonuniform meshing, marking an important improvement for the increase of local resolutions near the interfaces. In the immersed boundary method, all discretized delta functions are continuous and have finite support points within a uniform background mesh. For nonuniform meshing in the fluid domain, different kernel functions must be introduced. As illustrated in Ref. [308], both wavelet and smooth particle hydrodynamics (SPH) methods belong to a class of reproducing kernel methods where the reproduced function $u^R(x)$ is derived as

$$u^R(x) = \int_{-\infty}^{+\infty} u(y)\phi(x-y)\,dy, \tag{3.254}$$

with a projection operator or a window function $\phi(x)$.

The reproducing condition requires that up to $n$th order polynomial be reproduced:

$$x^n = \int_{-\infty}^{+\infty} y^n\phi(x-y)\,dy. \tag{3.255}$$

From the convolution theorem in the Fourier transform domain, Eq. (3.254) can be expressed as

$$\hat{u}^R(\omega) = \hat{u}(\omega)\hat{\phi}(\omega), \tag{3.256}$$

where $\omega$ is the natural frequency or wave number of the functions $u(x)$ and $\phi(x)$.

Supposing $\hat{\phi}(\omega)$ is a perfect rectangular window function, a so-called ideal low-pass filter function, which corresponds to the sinc function in the function domain, spatial or temporal, $\hat{u}(\omega)$ contains signals within $|\omega| \leqslant \omega_c$, where $\omega_c$ is the cut-off or band-limited frequency or wave number defined by the rectangular window function. Nevertheless, it has been recognized that the ideal low-pass filter does not have a compact support in the function domain and is therefore not useful as an interpolation function in computational mechanics. Therefore, as presented in Ref. [308], a correction function $C(x; x-y)$ is introduced in the finite computation domain $\Omega$, i.e., the domain of influence or support:

$$C(x; x-y) = \sum_{k=0}^{n} \beta_k(x)(x-y)^k, \tag{3.257}$$

and the consequent modified window function $\phi$ is constructed as

$$\bar{\phi}(x; x-y) = \sum_{k=0}^{n} \beta_k(x)(x-y)^k\phi(x-y). \tag{3.258}$$

Define $m_k(x)$ as the $k$th moment of the window function $\phi(x)$, with $i = \sqrt{-1}$, and we have

$$m_k(x) = \int_{\Omega} (x-y)^k \phi(x-y)\, dy = i^k \hat{\phi}^{(k)}(0), \quad k = 0, 1, 2, \ldots, n. \tag{3.259}$$

Similarly, define $\bar{m}_k(x)$ as the $k$th moment of $\bar{\phi}(x)$, and we have

$$\bar{m}_k(x) = \int_{\Omega} (x-y)^k \bar{\phi}(x-y)\, dy = i^k \hat{\bar{\phi}}^{(k)}(0), \quad k = 0, 1, 2, \ldots, n. \tag{3.260}$$

In general, with a proper construction of the correction function with respect to the selected window functions, such as the scaling function, wavelet, or spline family, we will be able to enforce

$$\hat{\bar{\phi}}(0) = 1, \quad \hat{\bar{\phi}}^{(1)}(0) = 0, \quad \ldots, \quad \hat{\bar{\phi}}^{(n)}(0) = 0. \tag{3.261}$$

In essence, the window function is required to be flatter at $\omega = 0$ in the Fourier domain as the order $n$ of reproducing gets higher in Eq. (3.255). This implies that as $n \to \infty$, $\hat{\bar{\phi}}(\omega)$ becomes flatter at $\omega = 0$ and approaches an ideal filter. Therefore, based on Eqs. (3.259), (3.260), and (3.261), we can set up the moment equations to solve for $\beta_k$,

$$\begin{bmatrix} m_0(x) & m_1(x) & \cdots & m_n(x) \\ m_1(x) & m_2(x) & \cdots & m_{n+1}(x) \\ \vdots & \vdots & \ddots & \vdots \\ m_n(x) & m_{n+1}(x) & \cdots & m_{2n}(x) \end{bmatrix} \begin{bmatrix} \beta_o(x) \\ \beta_1(x) \\ \vdots \\ \beta_n(x) \end{bmatrix} = \begin{bmatrix} 1 \\ 0 \\ \vdots \\ 0 \end{bmatrix}. \tag{3.262}$$

Furthermore, we can also define a dilation parameter $a$ and introduce

$$\bar{\phi}_a(x-y) = \frac{1}{a}\bar{\phi}\left(\frac{x-y}{a}\right). \tag{3.263}$$

Thus, by changing the value of the dilation parameter $a$, usually by a factor of 2, a sequence of low-pass filters is defined. Note that the dilation parameter $a$ is directly linked to mesh density. The difference between two such low-pass filters defines a high-pass filter. Therefore, we can perform a multi-resolution analysis with the projection operator $P_a$ expressed as

$$P_a u(x) = \int_{\Omega} u(y)\bar{\phi}_a(x-y)\, dy. \tag{3.264}$$

Based on the resolution of the projection operator $P_{a/2^n}$, a hierarchical representation of a function $u(x)$ is defined as

$$u(x) = \lim_{n\to\infty} P_{a/2^n}u(x), \quad \ldots, \quad P_a u(x), \quad \ldots, \quad \lim_{n\to\infty} P_{2^n a}u(x) = \emptyset.$$

Furthermore, we also define a complementary projection operator $Q_{2a}$, such that the higher scale projected solution $P_a u$ can be represented by the following sum of low and high scale projections:

$$P_a u(x) = P_{2a} u(x) + Q_{2a} u(x), \tag{3.265}$$

where the $Q_{2a}$ projection, which can be viewed as a peeled off scale or the rate of variation of $P_a u(x)$, is simply given as a wavelet projection:

$$Q_{2a} u(x) = \int_{\Omega} u(y) \bar{\psi}_{2a}(x - y)\, dy, \tag{3.266}$$

with $\bar{\psi}_{2a}(x - y) = \bar{\phi}_a(x - y) - \bar{\phi}_{2a}(x - y)$.

In accordance with the reproducing property, we have

$$\int_{\Omega} y^k \bar{\psi}_{2a}(x - y)\, dy = 0, \quad \text{with } k = 0, 1, \ldots, n, \tag{3.267}$$

where $n$ can be considered as the order of the polynomial based wavelet.

The discretized reconstruction of the delta function for nonuniform spacing can be written as

$$\bar{\phi}_{aJ}(x) = C(x; x - x_J) a^{-1} \phi\left(\frac{x - x_J}{a}\right) \Delta x_J. \tag{3.268}$$

For uniform spacing, if the basis functions are chosen to be $1$, $x$, $x^2$, and the cubic spline as the window function, the modified window function satisfying the moment reproducing conditions is given as

$$\begin{aligned} \delta_h(r) &= \bar{\phi}_h(x) = \frac{1}{h} \bar{\phi}(r) \\ &= \frac{1}{h}\left(\frac{27}{17} - \frac{30}{17} r^2\right) \begin{cases} \frac{1}{6}(r + 2)^3, & -2 \leqslant r < -1, \\ \frac{2}{3} - r^2(1 + \frac{r}{2}), & -1 \leqslant r < 0, \\ \frac{2}{3} - r^2(1 - \frac{r}{2}), & 0 \leqslant r < 1, \\ -\frac{1}{6}(r - 2)^3, & 1 \leqslant r < 2, \\ 0, & \text{otherwise}. \end{cases} \end{aligned} \tag{3.269}$$

Likewise, to reproduce $1, x, x^2, x^3, x^4$ terms, the following correction function can be introduced:

$$\frac{170010}{80347} - \frac{429450}{80347} r^2 + \frac{178290}{80347} r^4.$$

In Figures 7.77 and 7.78, we provide a comparison of the function and Fourier domains of the modified window functions with the original discretized delta

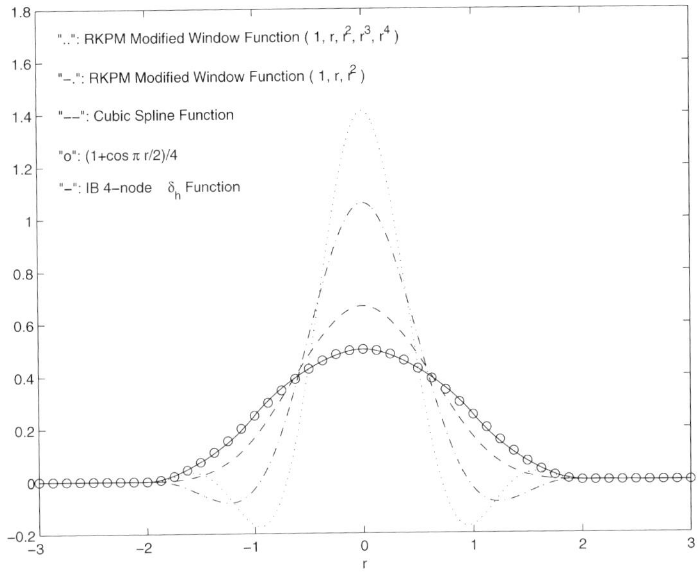


Figure 7.77. Various discretized delta functions.

function. It is clear, as we increase the reproducing order $n$, with the same finite support region, that the kernel functions approach more the ideal low pass filter. In the immersed continuum method, we employ the RKPM kernel functions. Instead of using the discretized delta function, it is possible to interpolate the solid point velocity within a single finite element [23]. The mathematical implication, with respect to the convergence behaviors and accuracy, is still to be discovered. In addition, as shown in Figure 7.79 the translation invariance is no longer satisfied for RKPM kernel functions. Numerical evidence also suggests that a restriction based on the translation invariance could be relaxed.

### 7.3.3. *Fictitious domain method; immersed continuum method*

#### *Fictitious domain method*

Suppose there exists a rigid cylinder (for two-dimensional cases) or a rigid sphere (for three-dimensional cases) occupying a volume $\Omega_s$ in the total domain $\Omega$. Again, on the FSI interface $\Gamma_s$, the unit normal vector of the solid is $\mathbf{n}^s$ or $\mathbf{n}$, which points outward to the flow region, and the unit normal vector of the fluid is denoted as $\mathbf{n}^f$ or $-\mathbf{n}$, which points inward to the submerged solid. Following the nonslip boundary condition on the interface $\Gamma_s$, we have

$$\mathbf{v}(\mathbf{x}, t) = \bar{\mathbf{v}}(t) + \bar{\boldsymbol{\omega}}(t) \times \big(\mathbf{x} - \bar{\mathbf{x}}(t)\big), \quad \forall \mathbf{x} \in \Gamma_s, \tag{3.270}$$

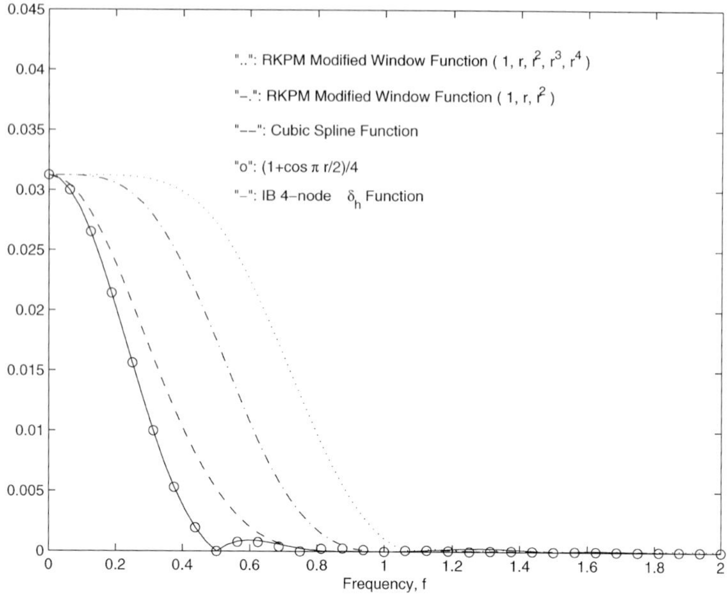


Figure 7.78. Spectra of various discretized delta functions.

where $\bar{\mathbf{x}}$, $\bar{\mathbf{v}}$, $\bar{\boldsymbol{\omega}}$, and $\mathbf{x}$ stand for the current position of the mass center, velocity, and angular velocity, and position on the interface of the rigid body.

Because the solid occupying $\Omega_s$ is a rigid body, Eq. (3.270) can be rewritten as

$$\mathbf{v}(\mathbf{x}^s, t) = \bar{\mathbf{v}}(t) + \bar{\boldsymbol{\omega}}(t) \times (\mathbf{x}^s - \bar{\mathbf{x}}(t)), \quad \forall \mathbf{x}^s \in \Omega_s. \tag{3.271}$$

Of course, on the FSI interface $\Gamma_s$, Eq. (3.271) is manifested as Eq. (3.270). Furthermore, the governing equations (strong form) of the FSI system can be depicted as

$$\rho_f \dot{v}_i = \sigma_{ij,j} + \rho_f g_i, \quad \text{in } \Omega_f (\text{or } \Omega \setminus \Omega_s), \tag{3.272}$$

$$M \dot{\bar{v}}_i = M g_i + F_i^s, \quad \text{for the rigid body } \Omega_s, \tag{3.273}$$

$$\mathbf{I}\dot{\bar{\boldsymbol{\omega}}} + \bar{\boldsymbol{\omega}} \times \mathbf{I}\bar{\boldsymbol{\omega}} = \mathbf{T}^s, \tag{3.274}$$

where $\mathbf{I}$ and $M$ are the rotational inertia tensor (or matrix) and the mass of the rigid body, respectively, and the resultant torque $\mathbf{T}^s$ and force $\mathbf{F}^s$, due to the fluid traction around the rigid body, are expressed as

$$\mathbf{T}^s = -\int_{\Gamma_s} (\mathbf{x}^s - \bar{\mathbf{x}}) \times \boldsymbol{\sigma}\mathbf{n}\, d\Gamma, \tag{3.275}$$

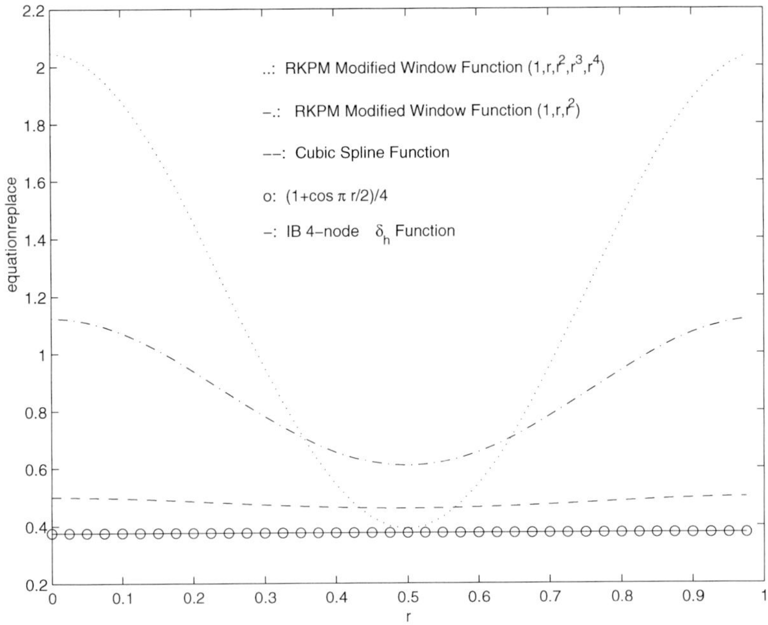


Figure 7.79. Various discretized delta functions with respect to translation invariance.

$$\mathbf{F}^s = -\int_{\Gamma_s} \boldsymbol{\sigma}\,\mathbf{n}\,d\Gamma. \tag{3.276}$$

In the fictitious domain method, an imaginary fluid is introduced to occupy the submerged solid domain $\Omega_s$ and to synchronize with the solid within $\Omega_s$. With the fluid velocity variation $\mathbf{w} \in (\mathcal{H}^1_{0,\Gamma_v}(\Omega))^d$, the rigid body velocity variation $\bar{\mathbf{w}} \in \mathcal{R}^d$, and the rigid body angular velocity variation $\bar{\boldsymbol{\theta}} \in \mathcal{R}^d$ combining the solid domain with the fluid domain, employing integration by parts, the divergence theorem, and Eqs. (3.271), (3.275), and (3.276), we can convert the governing equations (strong form) in Eqs. (3.272) to (3.274) into the variational equations (weak form):

$$\int_{\Omega} \left[\rho w_i(\dot{v}_i - g_i) + w_{i,j}\sigma_{ij}\right] d\Omega$$
$$+ r\left[M(\dot{\bar{v}}_i - g_i)\bar{w}_i + (\mathbf{I}\dot{\bar{\boldsymbol{\omega}}} + \bar{\boldsymbol{\omega}} \times \mathbf{I}\bar{\boldsymbol{\omega}}) \cdot \bar{\boldsymbol{\theta}}\right] = 0, \tag{3.277}$$

with $r = 1 - \rho/\rho_s$.

The key treatment in the immersed boundary method and its extensions is the introduction of the delta function to synchronize the fluid motion with the solid motion within the immersed structure/solid domain $\Omega_s$, namely:

$$\mathbf{v}^s = \mathbf{v}^f. \tag{3.278}$$

In fact, the constraint of Eq. (3.278) introduces the (distributed) Lagrangian multiplier, which acts as the equivalent body force. In the fictitious domain method, a similar distributed Lagrangian multiplier $\boldsymbol{\lambda}$ is introduced, along with the following traditional mixed formulation, and we obtain, $\forall \mathbf{w} \in (\mathcal{H}^1_{0,\Gamma_v}(\Omega))^d$ and $\boldsymbol{\lambda} \in (\mathcal{H}^1(\Omega_s))^d$,

$$\int_\Omega \left[\rho_f w_i(\dot{v}_i - g_i) + w_{i,j}\sigma_{ij}\right] d\Omega + r\left[M(\dot{\bar{v}}_i - g_i)\bar{w}_i + (\mathbf{I}\dot{\bar{\boldsymbol{\omega}}} + \bar{\boldsymbol{\omega}} \times \mathbf{I}\bar{\boldsymbol{\omega}}) \cdot \bar{\boldsymbol{\theta}}\right]$$
$$- \left(\boldsymbol{\lambda}, \mathbf{w} - \bar{\mathbf{w}} - \bar{\boldsymbol{\theta}} \times \left(\mathbf{x}^s - \bar{\mathbf{x}}\right)\right) = 0, \tag{3.279}$$

and

$$\left(\boldsymbol{\mu}, \mathbf{v} - \bar{\mathbf{v}} - \bar{\boldsymbol{\omega}} \times \left(\mathbf{x}^s - \bar{\mathbf{x}}\right)\right) = 0, \quad \forall \boldsymbol{\mu} \in \left(\mathcal{H}^1(\Omega_s)\right)^d, \tag{3.280}$$

where the inner product is defined as

$$(\boldsymbol{\mu}, \boldsymbol{\lambda}) = \int_{\Omega_s} \left(\mu_i \lambda_i + l^2 \mu_{i,j} \lambda_{i,j}\right) d\Omega, \tag{3.281}$$

with a scaling factor $l$ dependent of the characteristic length of $\Omega_s$.

Similar to the immersed boundary method, a clear advantage of the fictitious domain method is the use of the same background fluid mesh [94]. Nevertheless, such a formulation is currently limited to immersed rigid bodies. Since any rigid body is an incompressible body, the direct link between the fictitious incompressible fluid domain can be established. In addition, for incompressible viscous fluids, velocity/pressure formulation must also be used along with the distributed Lagrangian multiplier. A wealth of theoretical studies on the inf-sup conditions, in particular those for three-field mixed finite element formulations, are available in the context of the treatment of incompressible solids and fluids [13].

*Immersed continuum method*

In the case of the incompressible solid immersed in the incompressible fluid, the traditional explicit scheme can be employed, if there is no excessively stiff boundary springs connecting with the tether points [142] and the additional elasticity moduli of the immersed solid are reasonable. This is addressed in the extended immersed boundary method or immersed finite element method [303,321]. In this section, we mainly address the issue of implicit formulations for excessively stiff boundary springs and high elastic moduli. In particular, we use a general case of a compressible solid immersed in compressible fluid to cover one seemingly very different scenario, namely, compressible solid immersed in incompressible fluid. In the immersed continuum method, we use a physical argument such that the bulk modulus of the solid can be much higher than the bulk modulus of the fluid. With the assumption that the acoustic wave speed within the fluid is constant, a so-called pseudo compressible fluid model is often used in numerics to mimic incompressible fluid behaviors. Why do we use this type of fluid model coupled with a compressible solid? To use the concept of immersed methods, we need to replace the immersed solid with a fluid similar to the surrounding fluid. If the volume of the immersed solid changes ever slightly, the volume of the corresponding fluid must also change, hence, the fluid model of the surrounding fluid must also be slightly compressible.

It is important to note that here the immersed solid is viewed as impermeable, thus such a volume of fluid does not exist physically. To account for the correct effect of the submerged solid exerting on the surrounding fluid, we must subtract the inertial force, the external body force, and the internal stress effects of such an imaginary fluid volume $\Omega_s(t)$. This idea is very similar to the fictitious domain method, which was proposed to handle immersed rigid bodies [94]. The advantage of the fictitious domain method is the use of a fixed background fluid mesh, whereas the immersed rigid bodies are modeled with the Lagrangian description. Recently, such an approach has been extended to immersed flexible bodies [282, 320]. It is this author's opinion that, in full-fledged implicit forms, there must be some hidden mathematical equivalence of these two sets of approaches: namely, the immersed boundary/continuum methods and the fictitious domain method. Maybe some day, an inf-sup condition similar to the modeling of almost incompressible materials can be derived in the context of the treatment of compressible solids immersed in compressible fluids [13].

To clearly present this new form of immersed methods dealing with compressible immersed continuum, we study similar variational forms. Considering the domain $\Omega$, as depicted in Figure 7.80, and supposing there exists a submerged solid domain $\Omega_s$ enclosed by a sufficiently smooth boundary $\Gamma_s$ (a line for two-dimensional cases and a surface for three-dimensional cases), the entire domain $\Omega$ is subdivided into two regions: the solid region $\Omega_s$ and the fluid region $\Omega_f$. Therefore, the boundaries of fluid and solid regions can be simply expressed as $\partial\Omega_s = \Gamma_s$ and $\partial\Omega_f = \Gamma_s \cup \Gamma_v \cup \Gamma_f$. Denoting $\boldsymbol{\sigma}$ as the stress tensor, and $\mathbf{v}$ as the velocity vector, we establish the following set of governing equations (strong form):

$$\rho_s \dot{v}_i^s = \sigma_{ij,j}^s + \rho_s g_i, \quad \text{in } \Omega_s, \tag{3.282}$$

$$\rho_f \dot{v}_i^f = \sigma_{ij,j}^f + \rho_f g_i, \quad \text{in } \Omega_f, \tag{3.283}$$

$$[v_i] = 0, \quad \text{on } \Gamma_s, \text{ kinematic matching}, \tag{3.284}$$

$$[\sigma_{ij} n_j] = 0, \quad \text{on } \Gamma_s, \text{ dynamic matching}, \tag{3.285}$$

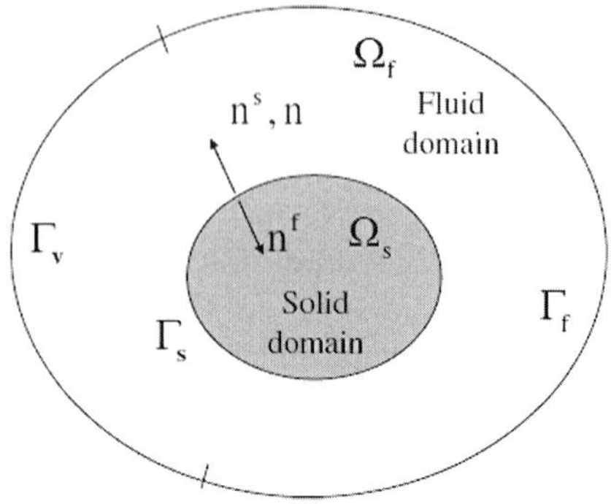


Figure 7.80. Immersed continuum illustration.

where the surface normal vector $\mathbf{n}$ is aligned with that of the solid domain $\mathbf{n}^s$ and opposite to that of the fluid domain $\mathbf{n}^f$.

Note that we use the subscript or the superscript $f$ and $s$ represent fluid and solid domains, respectively. Defining the same Sobolev space $(\mathcal{H}^1_{0,\Gamma_v}(\Omega))^d$, we express Eqs. (3.282) to (3.285) in the variational form (weak form):

$$\forall \mathbf{w} \in \left(\mathcal{H}^1_{0,\Gamma_v}(\Omega)\right)^d$$

$$\int_{\Omega_s} w_i\left[\rho_s\left(\dot{v}_i^s - g_i\right) - \sigma^s_{ij,j}\right] d\Omega + \int_{\Omega_f} w_i\left[\rho_f\left(\dot{v}_i^f - g_i\right) - \sigma^f_{ij,j}\right] d\Omega = 0. \tag{3.286}$$

REMARK 7.3.6. $\mathbf{w} \in (\mathcal{H}^1_{0,\Gamma_v}(\Omega))^d$ also implies that the kinematic matching of Eq. (3.284) is satisfied.

Again, using integration by parts and the divergence theorem, introducing dynamic matching at the interface $\Gamma_s$, and combining fluid and solid domains with $\Omega_f \cup \Omega_s = \Omega$, Eq. (3.286) can be rewritten as:

$$\forall \mathbf{w} \in \left(\mathcal{H}^1_{0,\Gamma_v}(\Omega)\right)^d$$

$$\int_{\Omega} \left[w_i \rho_f(\dot{v}_i - g_i) + w_{i,j}\sigma_{ij}\right] d\Omega - \int_{\Gamma_f} w_i f_i^{f,p}\, d\Gamma - \int_{\Omega_s} w_i^s f_i^s\, d\Omega = 0, \tag{3.287}$$

with

$$\int_{\Omega_s} w_i^s f_i^s\, d\Omega = -\int_{\Omega_s} \left[w_i(\rho_s - \rho_f)(\dot{v}_i - g_i) + w_{i,j}\left(\sigma^s_{ij} - \sigma^f_{ij}\right)\right] d\Omega. \tag{3.288}$$

REMARK 7.3.7. In Eq. (3.285), the unit surface normal vectors at the FSI interface $\Gamma_s$ are assigned as $\mathbf{n}^s = -\mathbf{n}^f = \mathbf{n}$; whereas in Eq. (3.288) the term involving the given surface traction $\mathbf{f}^{f,p}$ will remain the same if the variational forms are carried out for the entire domain $\Omega$ instead of the solid and fluid parts. Thus the focus will be on the submerged solid $\Omega_s$ and its interface with the fluid $\Gamma_s$.

REMARK 7.3.8. In Eq. (3.287), within the entire domain $\Omega$, the external work comes from the external body force $\rho_f \mathbf{g}$, and the surface traction $\mathbf{f}^{f,p}$ at the Neumann boundary $\Gamma_f$; whereas the power input within the submerged solid domain $\Omega_s$ includes the contribution from the inertial force difference, buoyancy force, and internal energy difference. Again, the kinematic matching at the submerged interface $\Gamma_s$ also implies that the submerged interface will move at the same velocity as that of the fluid particles in the immediate vicinity. In Eq. (3.287), we again do not stipulate the material derivative $d\mathbf{v}/dt$ and the stress $\boldsymbol{\sigma}$. Hence the turbulent and the nonNewtonian fluid models can eventually be incorporated.

In the explicit implementation, just as in the immersed boundary method, in recent finite element extensions [303,321], we introduce the following two key equations to synchronize the fluid occupying the submerged solid domain $\Omega_s$ with the solid and distribute the solid force $\mathbf{f}^s$:

$$f_i^{fsi} = \int_{\Omega_s} f_i^s \delta(\mathbf{x} - \mathbf{x}^s)\, d\Omega, \tag{3.289}$$

$$v_i^s = \int_{\Omega} v_i \delta(\mathbf{x} - \mathbf{x}^s)\, d\Omega, \tag{3.290}$$

where $\mathbf{f}^{fsi}$ represents the same equivalent body force as in the immersed boundary method.

It is very important to recognize that $\mathbf{f}^s$ is the force density within the solid domain $\Omega_s$, whereas $\mathbf{f}^{fsi}$ is the equivalent body force over the entire domain $\Omega$. The physical significance of $\mathbf{f}^s$ and $\mathbf{f}^{fsi}$ is quite different. As a consequence, the entire FSI system is represented with the same set of governing equations (3.184) (strong form) and the corresponding variational form (3.185) (weak form).

REMARK 7.3.9. Eqs. (3.289) and (3.290) are comparable to Eqs. (3.181) and (3.182) in the IB method. In Eq. (3.288), $\mathbf{f}^s$ can be considered as the equivalent force density within the submerged solid domain $\Omega_s$. In fact, this force density directly corresponds to the rigid link between the fluid occupying the submerged solid domain $\Omega_s$ and the solid. In other words, the force density $\mathbf{f}^s$ stands for the Lagrange multiplier corresponding to the constraint in Eq. (3.290). Of course, the definition in Eq. (3.288) also matches the virtual power input from the submerged solid domain $\Omega_s$.

Note that the inertial force difference $-\int_{\Omega_s} w_i(\rho_s - \rho_f)\dot{v}_i \, d\Omega$ of the submerged solid continuum corresponds to the inertial force $-\int_{\Gamma_s} w_i^s m \ddot{u}_i^s \, d\Gamma$ of the submerged elastic boundary, whereas the internal energy difference $-\int_{\Omega_s} w_{i,j}(\sigma_{ij}^s - \sigma_{ij}^f) \, d\Omega$ of the submerged solid continuum corresponds to the elastic force $-\int_{\Gamma_s} w_i^s f_i^s \, d\Gamma$ of the submerged elastic boundary. In comparison with the submerged elastic boundary, the contribution of the submerged solid includes an additional term to account for the external body force difference (the so-called buoyancy) $\int_{\Omega_s} w_i(\rho_s - \rho_f) g_i \, d\Omega$. This buoyancy force is the direct manifestation of the submerged solid which occupies a finite volume. Likewise, the inertial effect of the submerged solid includes the difference between fluid and solid densities, assuming the fluid occupying the submerged solid domain is forced to have the same motions as those of the solid.

We present here the velocity/pressure formulation for the compressible viscous fluid and the displacement/pressure formulation for the compressible solid with a hyperelastic material model. For simplicity in this section, we omit the superscript or subscript $f$ for fluid variables. The detailed formulations are discussed in Section 7.2.1. Unlike the fluid domain, since we use the Lagrangian description for the submerged solid, the treatments of the continuity equation and the Cauchy stress in nonlinear solid mechanics are not as straightforward. As a special case, if the submerged solid is a flexible structure with a linear elastic material law, we will only have the geometrical nonlinearity to deal with. In this case, we suppose the Young's modulus and the Poisson ratio are $E$ and $\nu$, respectively, and the bulk modulus for the solid can be simply expressed as $\kappa^s = E/3(1-2\nu)$. However, Cauchy stress must still be depicted on the current configuration, which itself is unknown. Define the Sobolev spaces so the weak form of governing equations can be modified as: $\forall q \in \mathcal{L}^2(\Omega)$, $q^s \in \mathcal{L}^2(\Omega_s)$, $\mathbf{w} \in (\mathcal{H}^1_{0,\Gamma_v}(\Omega))^d$, which includes $\forall \mathbf{w}^s \in (\mathcal{H}^1(\Omega_s))^d$, and we find $\mathbf{v}$ and $p$ in $\Omega$, $p^s$ in $\Omega_s$, such that

$$\int_{\Omega} w_i \rho(\dot{v}_i - g_i) \, d\Omega + \int_{\Omega} (w_{i,j}\tau_{ij} - p w_{i,i}) \, d\Omega - \int_{\Gamma_f} w_i f_i^{f,p} \, d\Gamma$$
$$+ \int_{\Omega_s} \left[ w_i^s(\rho_s - \rho)(\dot{v}_i - g_i) + w_{i,j}^s\left(\tau_{ij}^s - \tau_{ij}^f\right) - \left(p^s - p\right) w_{i,i}^s \right] d\Omega$$
$$+ \int_{\Omega} q\left(v_{j,j} + \frac{p_{,t}}{\kappa}\right) d\Omega + \int_{\Omega_s} q^s\left(J_3 - 1 + \frac{p^s}{\kappa^s}\right) d\Omega = 0. \tag{3.291}$$

We recognize that there are two sets of discretizations, namely, one for the Lagrangian solid mesh and one for the Eulerian fluid mesh. The communications between these two sets of nodes are illustrated in Figure 7.81. In essence, two methods can be employed. First, traditional finite element interpolation functions can be used to interpolate the solid nodal velocity once the particular host fluid

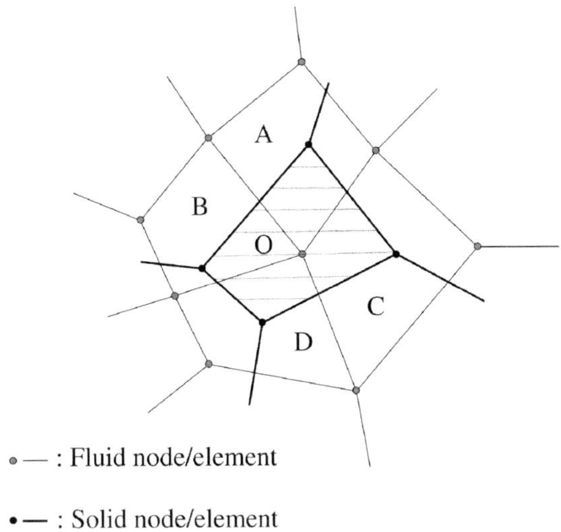


Figure 7.81. Communication between Lagrangian solid nodes and Eulerian fluid nodes.

element is identified through a search algorithm. Second, a technique similar to meshless finite element methods is introduced. In this procedure, a particular influence domain is attached to the solid node and the corresponding fluid nodes are identified within the influence domain. The stabilization finite element scheme is similar to what has been published by Wang [293]. Note that the convective terms are hidden in $\dot{v}_i^h$ and the detailed expressions of other stabilized Galerkin formulation for the Navier–Stokes equations can be found in Section 7.1 or Refs. [273, 274,278]. In general the interpolation functions for the velocity vector and the unknown pressures are different. Therefore, we retain the superscripts $v$ and $p$ to denote such differences. For the fluid domain $\Omega$, the following interpolations are used:

$$\mathbf{v}^h = h_i^v \mathbf{v}_i, \qquad \mathbf{w}^h = h_i^v \mathbf{w}_i, \qquad p^h = h_i^p p_i, \; q^h = h_i^p q_i, \tag{3.292}$$

where $h_i^v$ and $h_i^p$ stand for the interpolation functions at node $i$ for the velocity vector and the pressure, and $\mathbf{v}_i$, $\mathbf{w}_i$, $p_i$, and $q_i$ are the nodal values of the discretized velocity vector, admissible velocity variation, pressure, and pressure variation, respectively.

For the solid domain $\Omega_s$, the discretization is based on the following:

$$\begin{aligned} \mathbf{u}^{s,h} &= h_j^u \mathbf{u}_j^s, \qquad & \mathbf{w}^{s,h} &= h_j^u \mathbf{w}_j^s, \\ p^{s,h} &= h_j^{p^s} p_j^s, \qquad & q^{s,h} &= h_j^{p^s} q_j^s, \end{aligned} \tag{3.293}$$

where $h_j^u$ and $h_j^{p^s}$ stand for the interpolation functions at node $j$ for the displacement vector and the unknown pressures, and $\mathbf{u}_j^{s,h}$, $\mathbf{w}_j^{s,h}$, $p_j^{s,h}$, and $q_j^{s,h}$ are the nodal values of the discretized displacement vector, admissible velocity variation, pressure, and pressure variation, respectively.

Substituting both discretizations (3.292) and (3.293) into Eq. (3.291), we obtain the following discretization of the weak form: $\forall q^h \in \mathcal{L}^2(\Omega^h)$, $q^{s,h} \in \mathcal{L}^2(\Omega_s^h)$, $\mathbf{w}^h \in (\mathcal{H}^{1,h}_{0,\Gamma_v^h}(\Omega^h))^d$, which includes $\mathbf{w}^{s,h} \in (\mathcal{H}^{1,h}(\Omega_s^h))^d$,

$$\begin{aligned}
&\int_{\Omega^h} w_{ik} h_i^v \rho \dot{v}_k^h \, d\Omega - \int_{\Gamma_f^h} w_{ik} h_i^{v,f} f_k^{f,p} \, d\Gamma + \int_{\Omega^h} \left(w_{ik} h_{i,l}^v \tau_{kl} - p^h w_{ik} h_{i,k}^v\right) d\Omega \\
&\quad + \int_{\Omega_s^h} \left[w_{jk}^s h_j^u (\rho_s - \rho)\left(\dot{v}_k^h - g_k\right) + w_{jk}^s h_{j,l}^u \left(\sigma_{kl}^s - \sigma_{kl}^f\right)\right] d\Omega \\
&\quad - \int_{\Omega^h} w_{ik} h_i^v \rho g_k \, d\Omega + \int_{\Omega^h} q_i h_i^p \left(v_{l,l}^h + \frac{p_{,t}^h}{\kappa}\right) d\Omega \\
&\quad + \int_{\Omega_s^h} q_j^s h_j^{p^s} \left(J_3 - 1 + \frac{p^{s,h}}{\kappa^s}\right) d\Omega = 0.
\end{aligned} \tag{3.294}$$

For clarity, we introduce a displacement nodal unknown vector $\mathbf{U}$, although it is only evaluated in the solid domain $\Omega_s$ in which a Lagrangian description is prescribed. In fact, within the solid domain, $\mathbf{U}$ is denoted as $\mathbf{U}^s$ and evolves according to $\dot{\mathbf{U}}^s$ and $\ddot{\mathbf{U}}^s$, which are mapped from the velocity nodal unknown vector $\mathbf{V}$ and acceleration nodal unknown vectors $\dot{\mathbf{V}}$ for the fluid domain. Mathematically, we could say that $\mathbf{v}^s$ is $\mathbf{v}$ directly evaluated at the material point $\mathbf{x}^s$. Likewise, the pressure nodal unknown vectors $\mathbf{P}$ and $\mathbf{P}^s$ are introduced for fluid and solid domains, respectively. Moreover, in the discussion of numerical procedures, we denote the time derivative of a variable $a$ as $\dot{a}$. For simplicity, we employ the Newton–Raphson iteration, and apply the Newmark time integration schemes as illustrated in Eqs. (2.149), (2.150), (2.156), and (2.157). The details of this time integration scheme for linear and nonlinear systems have been elaborated upon in Chapter 6 as well as in Refs. [152,271,291,292]. Finally, for the entire domain $\Omega$, due to the arbitrariness of the variations $w_{ik}$, $q_i$, and $q_j^s$, we have four equations at each fluid node $i$ and one equation at each solid node $j$:

$$r_{ik}^v = 0, \qquad r_i^p = 0, \qquad r_j^{p^s} = 0, \tag{3.295}$$

or

$$\mathbf{r}\left(\mathbf{V}, \mathbf{P}, \mathbf{P}^s\right) = \mathbf{0}, \tag{3.296}$$

where the residuals are defined as

$$
\begin{aligned}
r_{ik}^{v} = {} & \int\limits_{\Omega^h} h_i^v \rho \dot{v}_k^h \, d\Omega + \int\limits_{\Omega^h} \big[h_{i,l}^v \tau_{kl} - p^h h_{i,k}^v\big] \, d\Omega - \int\limits_{\Gamma_f^h} h_i^{v,f} f_k^{f,p} \, d\Gamma \\
& + \int\limits_{\Omega_s^h} \mathcal{M}\big[h_j^u (\rho_s - \rho)\big(\dot{v}_k^h - g_k\big) + h_{j,l}^u \big(\sigma_{kl}^s - \sigma_{kl}^f\big)\big] \, d\Omega \\
& - \int\limits_{\Omega^h} h_i^v \rho g_k \, d\Omega, \\
r_i^p = {} & \int\limits_{\Omega^h} h_i^p \left( v_{l,l}^h + \frac{p_{,t}^h}{\kappa} \right) d\Omega, \\
r_j^{p^s} = {} & \int\limits_{\Omega_s^h} h_j^{p^s} \left( J_3 - 1 + \frac{p^{s,h}}{\kappa^s} \right) d\Omega.
\end{aligned}
$$

### 7.3.4. Implicit/compressible solver; FSI systems with immersed solids

#### Implicit/compressible solver

We adopt the same matrix-free Newton–Krylov iterative solver for the nonlinear system of Eqs. (3.295). In the $k$th Newton–Raphson iteration at time step $m+1$ of the nonlinear residual equation (3.295), from $\mathcal{R}^N$ to $\mathcal{R}^N$, with $N$ as the number of the total unknowns, we start with a first guess of the incremental unknowns $\Delta\boldsymbol{\Theta}^{k,0}$, namely, $\Delta\mathbf{V}^0$, $\Delta\mathbf{P}^0$, and $\Delta\mathbf{P}^{s,0}$, which often are zero vectors. Then the residual of the linearized systems of equations at the $k$th Newton–Raphson iteration is evaluated as

$$
\begin{aligned}
\mathbf{p} = {} & -\mathbf{r}^{m+1,k-1} - \mathbf{r}_{,v}^{m+1,k-1} \Delta\mathbf{V}^0 - \mathbf{r}_{,p}^{m+1,k-1} \Delta\mathbf{P}^0 \\
& - \mathbf{r}_{,p^s}^{m+1,k-1} \Delta\mathbf{P}^{s,0}.
\end{aligned} \tag{3.297}
$$

This error vector is used to construct the $n$-dimensional Krylov subspace $\mathcal{K}^n = \operatorname{span}\{\mathbf{p}, \mathbf{J}\mathbf{p}, \dots, \mathbf{J}^{n-1}\mathbf{p}\}$ where $\mathbf{J}$ is the $N \times N$ Jacobian matrix evaluated at time step $m+1$ and the $k$th Newton–Raphson iteration of the nonlinear residual equation (3.296) and can be rewritten as

$$
\mathbf{J} = \big(\mathbf{r}_{,v}^{m+1,k-1}, \mathbf{r}_{,p}^{m+1,k-1}, \mathbf{r}_{,p^s}^{m+1,k-1}\big). \tag{3.298}
$$

Employing the procedures illustrated in Section 7.2.3, the solution vector $\Delta\boldsymbol{\Theta}$, or rather $\Delta\mathbf{V}$, $\Delta\mathbf{P}$, and $\Delta\mathbf{P}^s$ is expressed as

$$
\Delta\boldsymbol{\Theta}^{k,n} = \Delta\boldsymbol{\Theta}^{k,0} + \sum_{i=1}^{n} y_i \mathbf{q}^i. \tag{3.299}
$$

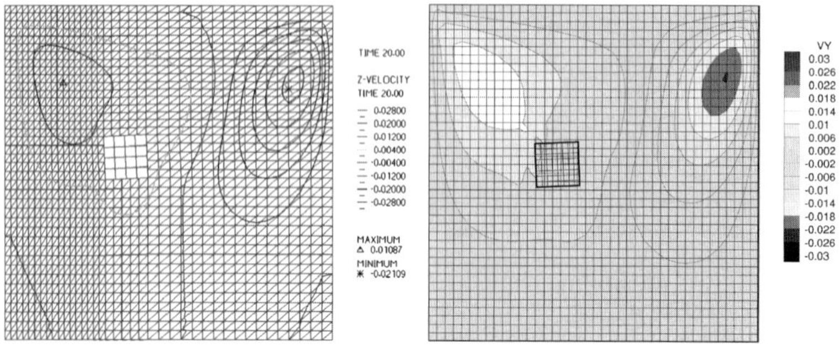


(a) Traditional ALE Formulation (b) Immersed Continuum Method

Figure 7.82. Horizontal velocity comparison.

*FSI systems with immersed solids*

Recently, some breakthroughs have been made in the development of finite element formulations for immersed boundary/continuum methods [303,321]. In addition, a preliminary formulation for beams with both bending and torsional moments has been presented [81,173]. In this section, we focus on a series of driven cavity problems with immersed solids. For the case with one immersed square solid, it is still possible to solve this FSI system with traditional modeling techniques such as the ALE formulation. We will compare the solution derived from the immersed continuum method with those from the traditional ALE approaches.

The driven cavity problem considered here has the dimension of $1 \times 1$ m. A typical immersed solid has the dimension of $0.13 \times 0.13$ m. In the immersed continuum method, the background fluid mesh for the entire cavity, which includes the space occupied by the immersed solids, consists of $20 \times 20$ $9/4c$ mixed finite elements. A typical immersed solid is represented by $4 \times 4$ $9/4c$ mixed finite elements. In the plotting stage, we split each 9-node element into four 4-node elements. As a result, the velocity mesh will be two times denser than the pressure mesh. As shown in Figures 7.82 and 7.83, the mesh construction clearly demonstrates the philosophical difference between the immersed boundary/continuum methods and the traditional ALE approaches. In the immersed continuum method, the solid mesh is right on top of the background fluid mesh, whereas in the ALE formulation, the solid mesh is surrounded by the fluid mesh with a different mesh density.

To compare the dynamical behaviors, the top shear velocity of the cavity is set as $0.1 \sin(2\pi/40t)$ m/s. The deformable solid is situated initially in the center of the cavity with zero velocity. The submerged solid is made of a compressible rubber material with the material constant $\kappa_s = 1 \times 10^7$ N/m$^2$ and the density $\rho_s = 1000$ kg/m$^3$. For hard solids we use $C_1 = 400$ and $C_2 = 200$ N/m$^2$,

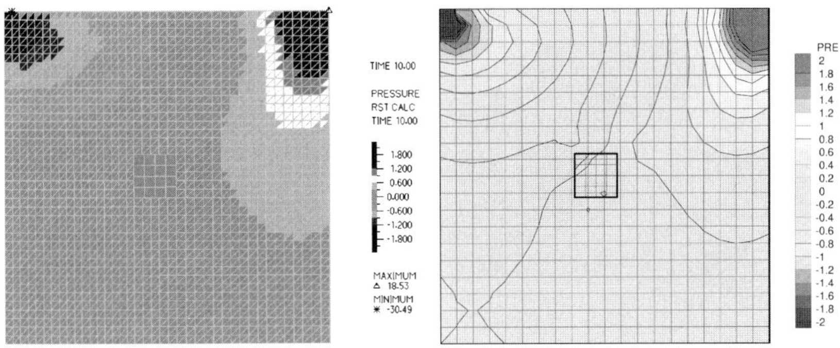


(a) Traditional ALE Formulation (b) Immersed Continuum Method

Figure 7.83. Pressure comparison.

whereas for soft solids we use $C_1 = 0.4$ and $C_2 = 0.2\,\mathrm{N/m^2}$. The case with hard solids resembles the driven cavity problem with immersed rigid bodies. In addition, the viscous fluid is represented with compressible model with a constant wave speed. In this case, we have the dynamic viscosity $\mu = 1$ Pa s, the bulk modulus $\kappa_f = 2.1 \times 10^6\ \mathrm{dyne/cm^2}$, and the density $\rho_f = 1000\ \mathrm{kg/m^3}$. Even with the coarse mesh used, the developed formulation with high-order elements provides reasonable results comparable to a reference solution. In addition, no spatial oscillation and checkerboard pressure bands are observed, as demonstrated in Figure 7.83. Of course, the proposed description can be used for refined meshes. As one of the main messages, we must point out that in engineering practice, before a large number of finite elements are used, it is always beneficial to employ coarse meshes with high-order elements to obtain a reasonable estimation of complicated problems.

As shown in Figure 7.86, even with the coarse mesh used, the implicit immersed continuum method with high-order mixed elements provides reasonable results comparable to reference solutions. It is very important to note that the discretized delta function provides the possibility of linking the immersed solid node with the surrounding fluid nodes, as done in meshless finite element methods. However, as pointed out in Ref. [23], the immersed solid node can also be restricted to communicate with very finite elements for the background fluid domain. As long as the virtual power input to the surrounding fluid is preserved, the concept of immersed methods will stay the same. In fact, this procedure is the same as the standard finite element procedure to distribute concentrated forces and to interpolate displacement/velocity within the element. Whether or not this procedure is as efficient as the meshless type of communication depends very much on the search algorithm for the purpose of locating the immersed solid node.

The pressure and vertical velocity distributions at different time stations are depicted in Figures 7.84 and 7.85. For soft immersed solids, the initial shear force of

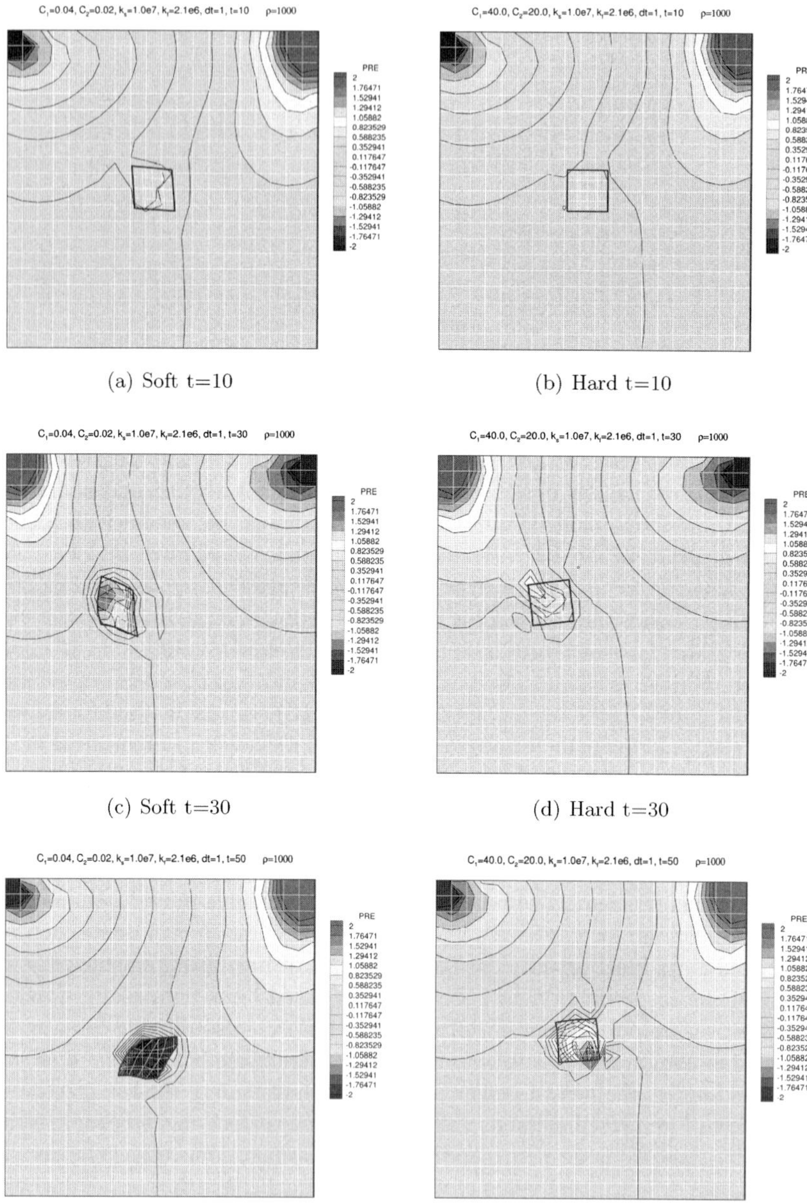


(a) Soft t=10 (b) Hard t=10

(c) Soft t=30 (d) Hard t=30

(e) Soft t=50 (f) Hard t=50

Figure 7.84. Pressure snapshots of a driven cavity with one immersed deformable solid.

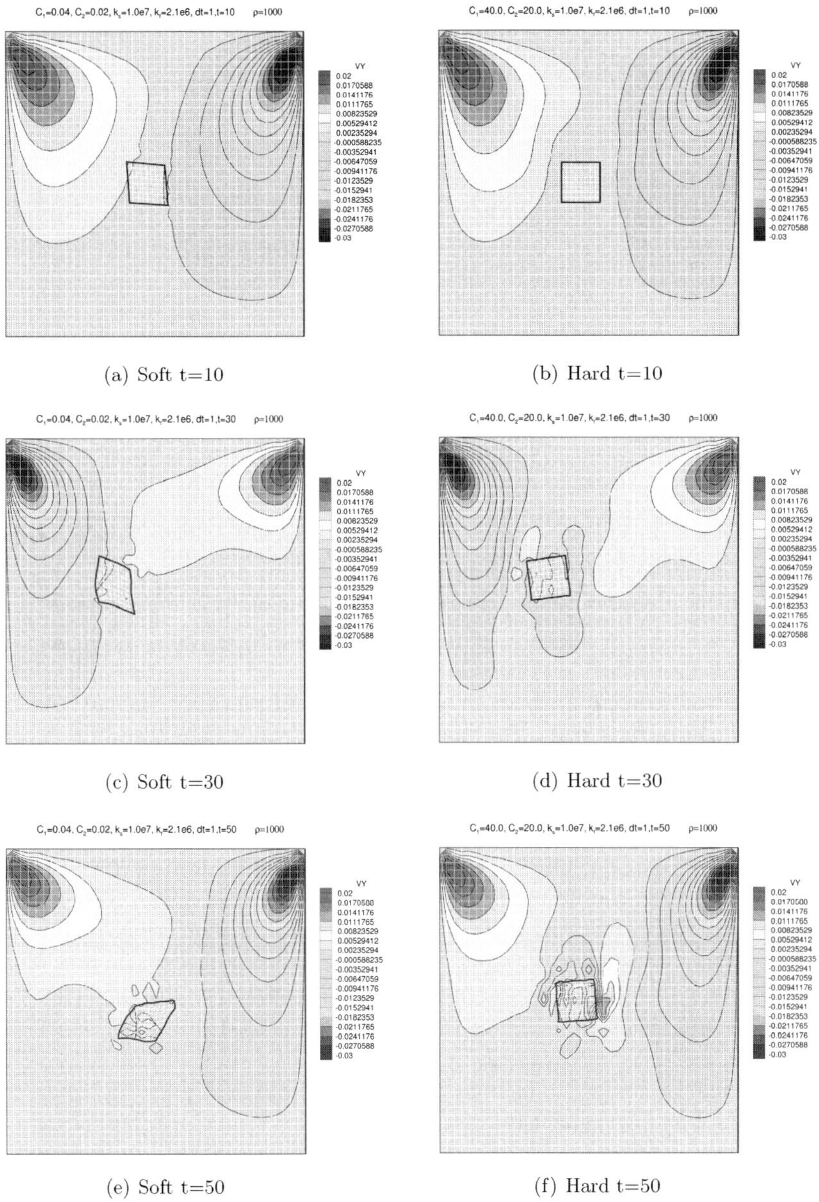


(a) Soft t=10 (b) Hard t=10

(c) Soft t=30 (d) Hard t=30

(e) Soft t=50 (f) Hard t=50

Figure 7.85. Vertical snapshots of a driven cavity with one immersed deformable solid.

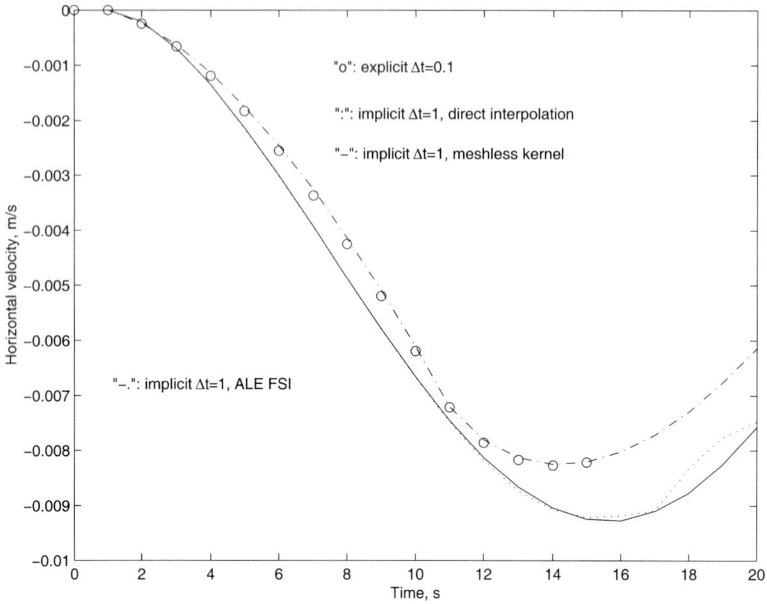


Figure 7.86. A comparison between the traditional method and the immersed methods.

the driven cavity problem causes an obvious shear deformation, whereas for hard immersed solids, rigid body like translations and rotations are observed. A side benefit of this implicit compressible immersed continuum method is the volume preservation of the immersed solid. In this case, for both soft and hard solids, the material is almost incompressible with a high bulk modulus. Although this implicit matrix-free Newton–Krylov solution scheme is not fast, the procedure is not extremely time consuming or prohibitive to carry out for a typical laptop. By choosing the proper preconditioner, the inner GMRES iteration can be minimized to two to three steps while the outer Newton iteration still depends on how close to the solution the initial guess is. In addition, no spatial oscillation and checkerboard pressure bands are observed. Of course, the proposed description can be used for refined meshes.

The benefit of immersed methods is clearly depicted in Figures 7.87 and 7.88 for the case with five immersed solids. For this case, it is no longer feasible to use the ALE formulation, whereas it is a simple task to add a few more deformable solids in the immersed continuum method. Note that in the immersed continuum method the contact between two immersed solids is accomplished for free just as in the material point method and the immersed boundary method. The immersed continuum method was implemented in a regular laptop and the documented run-

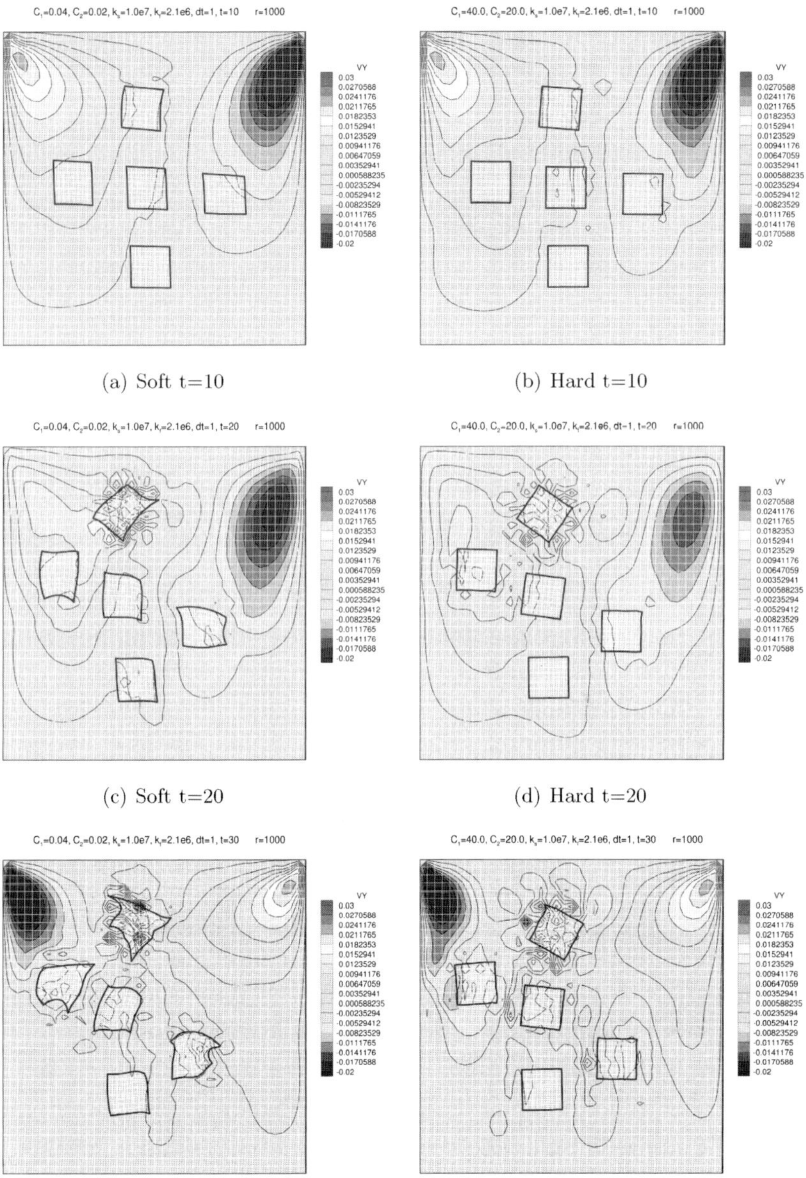


(a) Soft t=10 (b) Hard t=10

(c) Soft t=20 (d) Hard t=20

(e) Soft t=30 (f) Hard t=30

Figure 7.87. Vertical snapshots of a driven cavity with five immersed deformable solid.

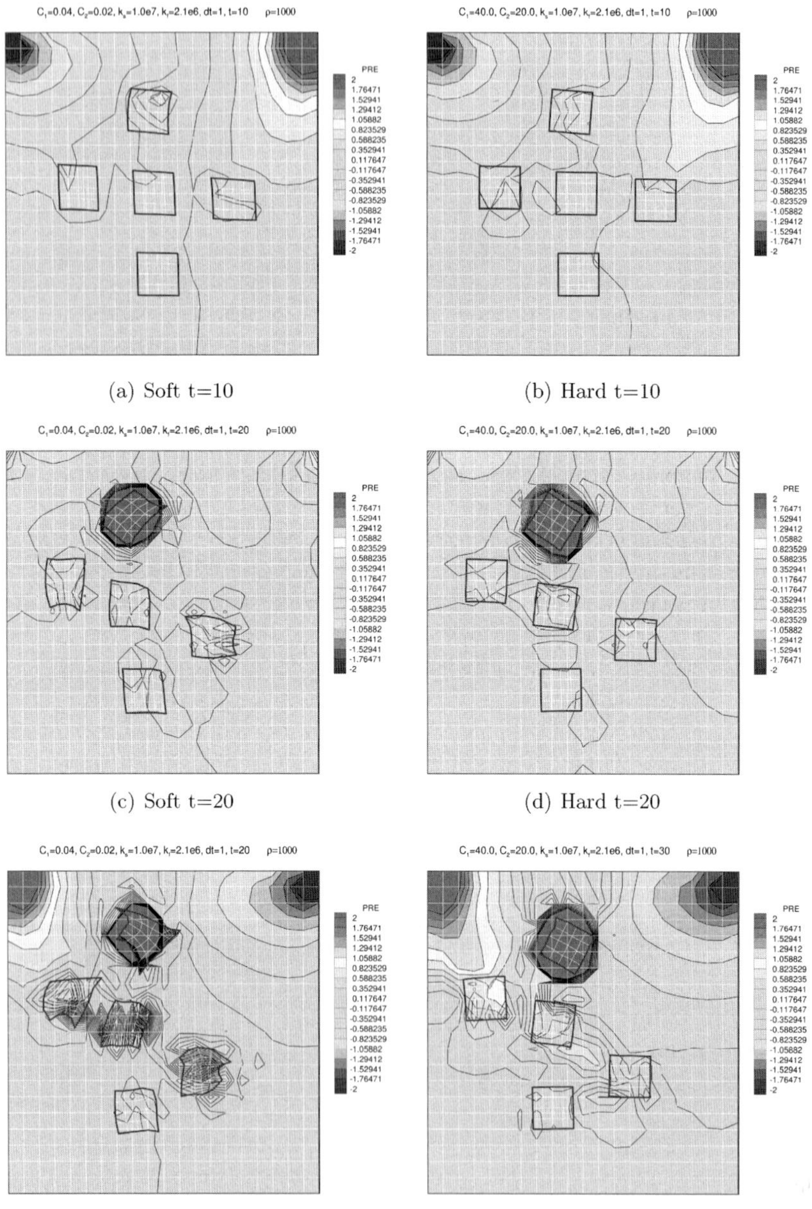


(a) Soft t=10 (b) Hard t=10

(c) Soft t=20 (d) Hard t=20

(e) Soft t=30 (f) Hard t=30

Figure 7.88. Pressure snapshots of a driven cavity with five immersed deformable solid.

Table 7.1 Comparison of the matrix-free Newton–Krylov iterative procedures

| 20 time steps step size 0.07 s | Preconditioner | |
|---|---|---|
| | With | Without |
| Newton iterations | 5 | 12 |
| GMRES iterations | 2 | 10 |
| CPU time (s) | 2168 | 10721 |

ning time for the implicit matrix-free Newton–Krylov iterative procedure is as follows:

To quantify the effectiveness of the preconditioner, we compare a typical FSI problem using the matrix-free Newton–Krylov iteration procedures with and without the preconditioner. The total number of time steps is 40. Three criteria are used: the number of Newton–Raphson iterations per time step, the number of GMRES iterations per Newton–Raphson iteration, and the total CPU time in seconds. A detailed digit by digit comparison also confirms that the velocity and pressure results differ only after the 5 or 6 decimal points, which is consistent with the error criteria set for the Newton–Krylov iteration. As demonstrated in Table 7.1, it is evident that the preconditioner is very effective in all three categories. Notice that with the preconditioner the maximum number of GMRES iteration stays at two for all time steps and Newton–Raphson iterations. Finally, the results of the solutions with and without the preconditioner are within the iterative error bounds, in this case, set to be as $10^{-5}$, less than the square root of the machine roundoff error $10^{-15}$.

In the immersed boundary/continuum methods, in order to satisfy energy conservation, namely, that the energy input to the fluid domain from the elastic boundary/solid is the same as that from the equivalent body force, the same Dirac delta function must be used in both the distribution of the resultant nodal force and the interpolation of the velocity based on the surrounding fluid grid point velocities. The key treatment in the immersed boundary/continuum methods is the synchronization of fluid motion with solid motion within the immersed solid domain $\Omega_s$, namely,

$$\mathbf{v}^s = \mathbf{v}^f. \tag{3.300}$$

In fact, the constraint of Eq. (3.300) introduces the distributed Lagrange multiplier as the equivalent body force. It is conceivable that full-fledged augmented Lagrange multipliers can be introduced as primary unknowns. In this case, the equivalent body forces can be directly calculated along with independent fluid and solid velocity vectors. Nevertheless, such a procedure will introduce a set of new unknowns equal to the number of velocity unknowns for solids. We conclude that mixed finite element formulations, namely, velocity/pressure for fluids and displacement/pressure for solids, can be successfully adopted to the immersed continuum method. The coupling of fluids and solids is the central feature in the study of the mechanics of the heart, arteries, veins, microcirculation, and pulmonary blood flow. Currently, the modeling of strong hemodynamic interaction with flexible structures is limited by severe fluid mesh distortions around flexible structures with large deformations and displacements. Recent breakthroughs have been made in immersed methods which prove to be a viable platform for multiscale and multiphysics modeling of biological systems. These new numerical algorithms for large FSI systems are essential to instability analysis of linear nonautonomous systems; the computation of limit sets, which include stationary points, periodic orbits, quasi-periodic orbits, and strange attractors; and the computation of turning points, which include super- and sub-critical saddle-node, pitchfork, and Hopf bifurcations in parameter space. Of course, it is only the beginning for this type of modeling treatment for complex FSI systems. Further mathematical understanding and study are urgently needed.

# References

[1] A.A. Kalinske, C.L. Pien, Eddy diffusion, International Chemical Engineering 36 (1944) 220–223.

[2] A.C.J. Luo, R.P.S. Han, Quantitative stability and bifurcation analyses of the generalized duffing oscillator with strong nonlinearity, Journal of the Franklin Institute 334B (1997) 447–459.

[3] A.D. Pierce, Acoustics—An Introduction to its Physical Principles and Applications, McGraw-Hill, 1981.

[4] A.H. Nayfeh, B. Balachandran, Applied Nonlinear Dynamics—Analytical, Computational, and Experimental Methods, John Wiley & Sons, 1995.

[5] A.J. Reynolds, Turbulent Flows in Engineering, John Wiley & Sons, 1974.

[6] A.L. Fogelson, C.S. Peskin, A fast numerical method for solving the three-dimensional stokes equations in the presence of suspended particles, Journal of Computational Physics 79 (1988) 50–69.

[7] A.N. Brooks, T.J.R. Hughes, Streamline upwind/Petrov–Galerkin formulations for convection dominated flows with particular emphasis on the incompressible Navier–Stokes equations, Computer Methods in Applied Mechanics and Engineering 32 (1982) 199–259.

[8] A.R. Teel, L. Praly, Tools for semi-global stabilization by partial state and output feedback, SIAM Journal of Control and Optimization 33 (1995) 1443–1488.

[9] H. Ashley, G. Haviland, Bending vibrations of a pipe line containing flowing fluid, Journal of Applied Mechanics 17 (1950) 229–232.

[10] O. Axelsson, Iterative Solution Methods, Cambridge University Press, 1996.

[11] K. Azer, C. Peskin, A one-dimensional model of blood flow in arteries with friction and convection based on the Womersley velocity profile, Cardiovascular Engineering: An International Journal 7 (2007) 51–73.

[12] I. Babuška, M. Suri, The $p$ and $h$-$p$ versions of the finite element method, basic principles and properties, SIAM Review 36 (1994) 578–632.

[13] W. Bao, X. Wang, K.J. Bathe, On the Inf-Sup condition of mixed finite element formulations for acoustic fluids, Mathematical Models and Methods in Applied Sciences 11 (2001) 883–901.

[14] J. Bear, Dynamics of Fluids in Porous Media, Dover, 1988.

[15] R. Bellman, Stability Theory of Differential Equations, McGraw-Hill, 1953.

[16] H. Benard, Formation de centres de giration à l'Arriere d'un obstacle en mouvement, Comptes Rendus 146 (1908).
[17] T.B. Benjamin, Dynamics of a system of articulated pipes conveying fluid. I. Theory, Proceedings of the Royal Society, Series A 261 (1961) 457–486.
[18] T.B. Benjamin, Dynamics of a system of articulated pipes conveying fluid. II. Experiments, Proceedings of the Royal Society, Series A 261 (1961) 487–499.
[19] A. Bermúdez, R. Rodríguez, Finite element computation of the vibration modes of a fluid–solid system, Computer Methods in Applied Mechanics and Engineering 119 (1994) 355–370.
[20] B.J. Briscoe, S.S. Panesar, The effect of surface topography on the adhesion of poly(urethane)-metal contacts, in: International Conference on Frontiers of Tribology, April 1991, pp. A20–A27.
[21] B.J. Hamrock, Fundamentals of Fluid Film Lubrication, McGraw-Hill, Inc., 1994.
[22] B.M. Irons, Role of part inversion in fluid-structure problems with mixed variables, AIAA Journal 8 (1970) 568.
[23] D. Boffi, L. Gastaldi, L. Heltai, On the CFL condition for the finite element immersed boundary method, Computers and Structures 85 (2007) 775–783.
[24] D. Boffi, L. Gastaldi, A finite element approach for the immersed boundary method, Computers and Structures 81 (2003) 491–501.
[25] T. Bohr, M.H. Jensen, G. Paladin, A. Vulpiani, Dynamical Systems Approach to Turbulence, Cambridge University Press, 1998.
[26] F. Brezzi, M. Bristeau, L.P. Franca, M. Mallet, G. Rogé, A relationship between stabilized finite element methods and the Galerkin method with bubble functions, Computer Methods in Applied Mechanics and Engineering 96 (1992) 117–129.
[27] F. Brezzi, M. Fortin, Mixed and Hybrid Finite Element Methods, Springer-Verlag, 1991.
[28] F. Brezzi, K.J. Bathe, A discourse on the stability conditions for mixed finite element formulations, Computer Methods in Applied Mechanics and Engineering 82 (1990) 27–57.
[29] F. Brezzi, K.J. Bathe, M. Fortin, Mixed-interpolated elements for Reissner–Mindlin plates, International Journal for Numerical Methods in Engineering 28 (1989) 1787–1801.
[30] M. Briscolini, P. Santangelo, Development of the mask method for incompressible unsteady flows, Journal of Computational Physics 84 (1989) 57–75.
[31] C.A. Felippa, R. Ohayon, Mixed variational formulation of finite element analysis of acoustoelastic/slosh fluid-structure interaction, Journal of Fluids and Structures 4 (1990) 35–57.
[32] C.A. Lacey, F.E. Talke, A tightly coupled numerical foil bearing solution, IEEE Transactions on Magnetics 26 (1990) 3039–3043.

[33] C.A. Tan, C.D. Mote Jr., Analysis of a hydrodynamic bearing under transverse vibration of an axially moving band, Journal of Tribology 112 (1990) 514–523.

[34] C.A. Taylor, T.J.R. Hughes, C.K. Zarins, Computational investigations in vascular disease, Computers in Physics 10 (1996) 224–232.

[35] J. Cao, B. Donell, D.R. Deaver, M.B. Lawrence, C. Dong, In vitro side-view imaging technique and analysis of human T-leukemic cell adhesion to ICAM-1 in shear flow, Microvascular Research 55 (1998) 124–137.

[36] P. Carbonell, X. Wang, Z.P. Jiang, L. Praly, On the suppression of flow-induced vibration with a simple control algorithm, International Journal of Nonlinear Science and Applications 8 (2003) 49–64.

[37] C.D. Bertram, Unstable equilibrium behaviour in collapsible tubes, Journal of Biomechanics 19 (1986) 61–69.

[38] C.D. Bertram, Simple resistive models for steady flow through uniform and tapered-stiffness collapsible tubes, Journal of Fluids and Structures 14 (2000) 593–612.

[39] D. Chapelle, K.J. Bathe, The inf-sup test, Computers and Structures 47 (4–5) (1993) 537–545.

[40] S. Chien, Electrochemical interactions between erythrocyte surfaces, Thrombosis Research 8 (1976) 189–202.

[41] I. Christie, D.F. Griffiths, A.R. Mitchell, O.C. Zienkiewicz, Finite element methods for second order differential equations with significant first derivatives, International Journal for Numerical Methods in Engineering 10 (1976) 1389–1396.

[42] P. Clément, Approximation by finite element functions using local regularization, RAIRO Analyse Numérique 2 (1975) 77–84.

[43] R. Clift, J.R. Grace, M.E. Weber, Bubbles, Drops, and Particles, Academic Press, 1978.

[44] R. Cortez, M. Minion, The blob projection method for immersed boundary problems, Journal of Computational Physics 161 (2000) 428–453.

[45] R. Courant, E. Isaacson, M. Rees, On the solution of nonlinear hyperbolic differential equations by finite differences, Communications on Pure and Applied Mathematics 5 (1952) 243–255.

[46] L. Cremer, M. Heckl, Structure-Borne Sound, Springer-Verlag, 1973 (translated and revised by E.E. Ungar).

[47] C.S. Peskin, Numerical analysis of blood flow in the heart, Journal of Computational Physics 25 (1977) 220–252.

[48] C.S. Peskin, The immersed boundary method, Acta Numerica 11 (2002) 479–517.

[49] C.S. Peskin, B.F. Printz, Improved volume conservation in the computation of flows with immersed elastic boundaries, Journal of Computational Physics 105 (1993) 33–46.

[50] D.A. Knoll, D.E. Keyes, Jacobian-free Newton–Krylov methods: A survey of approaches and applications, Journal of Computational Physics 193 (2004) 357–397.
[51] D.B. Spalding, A novel finite difference formulation for differential expressions involving both first and second derivatives, International Journal for Numerical Methods in Engineering 4 (1972) 551–559.
[52] D.C. Wilcox, Turbulence Modeling for CFD, 2nd ed., DCW Industries, Inc., 1994.
[53] R. Dillon, L. Fauci, A. Fogelson, D. Gaver III, Modeling biofilm processes using the immersed boundary method, Journal of Computational Physics 129 (1996) 57–73.
[54] D.M. McQueen, C.S. Peskin, Computer-assisted design of butterfly bileaflet valves for the mitral position, Scand. J. Thor. Cardiovasc. Surg. 19 (1985) 139–148.
[55] D.M. McQueen, C.S. Peskin, Shared-memory parallel vector implementation of the immersed boundary method for the computation of blood flow in the beating mammalian heart, The Journal of Supercomputing 11 (1997) 213–236.
[56] D.N. Ku, Blood flow in arteries, Annual Review of Fluid Mechanics 29 (1997) 399–434.
[57] J. Donea, A Taylor–Galerkin method for convective transport problems, International Journal for Numerical Methods in Engineering 20 (1984) 101–119.
[58] J. Donea, S. Giuliani, J.P. Halleux, An arbitrary Lagrangian–Eulerian finite element method for transient dynamic fluid-structure interactions, Computer Methods in Applied Mechanics and Engineering 33 (1982) 689–723.
[59] D.R. Roisum, Runnability of paper. 1. Predicting runnability, TAPPI Journal 73 (1990) 97–101.
[60] D.R. Roisum, Runnability of paper. 2. Troubleshooting web breaks, TAPPI Journal 73 (1990) 101–106.
[61] D.S. Drumheller, Introduction to Wave Propagation in Nonlinear Fluids and Solids, Cambridge University Press, 1998.
[62] J. Dugundji, E. Dowell, B. Perkin, Subsonic flutter of panels on continuous elastic foundations, AIAA Journal 1 (1963) 1146–1154.
[63] W. Dzwinel, K. Boryczko, D.A. Yuen, A discrete-particle model of blood dynamics in capillary vessels, Journal of Colloid and Interface Science 258 (2002) 163–173.
[64] E.A. Evans, Y.C. Fung, Improved measurements of the erythrocyte geometry, Microvascular Research 4 (1972) 335.
[65] E.F. DeCrosta Jr., Evaporative drying theory, in: Handbook of Practical Aspects of Pressing and Drying, 1979, pp. 1–10.

[66] E.F. DeCrosta Jr., Sheet flutter and windage problems on paper machines. I. Air movement; II. Vibrations; III. The dryer pocket; IV. Causes, in: TAPPI Notes—Sheet Flutter and Windage Problems Seminar, 1991, pp. 1–9.

[67] E.H. Dowell, E.F. Crawley, H.C. Curtiss Jr., D.A. Perters, R.H. Scanlan, F. Sisto, A Modern Course in Aeroelasticity, 3rd ed., Kluwer Academic, 1995.

[68] E.H. Dowell, E.F. Crawley, H.C. Curtiss Jr., D.A. Peters, R.H. Scanlan, F. Sisto, A Modern Course in Aeroelasticity, Kluwer Academic, 1995.

[69] E.H. Dowell, M. Ilgamov, Studies in Nonlinear Aeroelasticity, Springer-Verlag, 1988.

[70] E.L. Wilson, M. Khalvati, Finite elements for the dynamic analysis of fluid–solid systems, International Journal for Numerical Methods in Engineering 19 (1983) 1657–1668.

[71] D. Elad, M. Sahar, J.M. Avidor, S. Einav, Steady flow through collapsible tubes: Measurements of flow and geometry, Journal of Biomechanical Engineering 114 (1985) 84–91.

[72] E.M. Saiki, S. Biringen, Numerical simulation of a cylinder in uniform flow: Application of a virtual boundary method, Journal of Computational Physics 123 (1996) 450–465.

[73] E.N. Dvorkin, K.J. Bathe, A continuum mechanics based four-node shell element for general nonlinear analysis, Engineering Computation 1 (1984) 77–88.

[74] E.P. Newren, A.L. Fogelson, R.D. Guy, R.M. Kirby, Unconditionally stable discretizations of the immersed boundary equations, Journal of Computational Physics 222 (2007) 702–719.

[75] C. Evangelinos, D. Lucor, G.E. Karniadakis, DNS-derived force distribution on flexible cylinders subject to vortex-induced vibration, Journal of Fluids and Structures 14 (2000) 429–440.

[76] E.A. Fadlun, R. Verzicco, P. Orlandi, J. Mohd-Yusof, Combined immersed-boundary finite-difference methods for three-dimensional complex flow simulations, Journal of Computational Physics 161 (2000) 35–60.

[77] L.J. Fauci, C.S. Peskin, A computational model of aquatic animal locomotion, Journal of Computational Physics 77 (1988) 85–108.

[78] F.B. Hildebrand, Advanced Calculus for Applications, Prentice-Hall, 1976.

[79] F.C. Moon, Chaotic and Fractal Dynamics—An Introduction for Applied Scientists and Engineers, John Wiley & Sons, 1992.

[80] A. Ferent, C.S. Peskin, X. Wang, Modeling a bent twisted filament immersed in fluid, in: 7th US National Congress on Computational Mechanics, 2003.

[81] F.M. White, Viscous Fluid Flow, McGraw-Hill, 1991.

[82] G.A. Smook, Handbook for Pulp and Paper Technologists, 2nd ed., Angus Wilde Publications, 1992.

[83] L. Gastaldi, Mixed finite element methods in fluid structure systems, Numerische Mathematik 74 (1996) 153–176.
[84] G.C. Everstine, A symmetric potential formulation for fluid-structure interaction, Journal of Sound and Vibration 79 (1981) 157–160.
[85] G.E. Schneider, M.J. Raw, A skewed, positive influence coefficient upwinding procedure for control-volume-based finite-element convection–diffusion computation, Numerical Heat Transfer 9 (1986) 1–26.
[86] G.F. Roach, Green's Functions, Cambridge University Press, 1970.
[87] V. Girault, P.A. Raviart, Finite Element Methods for Navier–Stokes Equation: Theory and Algorithms, Springer-Verlag, 1986.
[88] G.J. Wagner, A numerical investigation of particulate channel flow, PhD thesis, Northwestern University, 2145 Sheridan Road, Evanston, IL 60208, June 2001.
[89] G.J. Wagner, N. Moës, W.K. Liu, T.B. Belytschko, The extended finite element method for rigid particles in Stokes flow, International Journal for Numerical Methods in Engineering 51 (2001) 293–313.
[90] G.J. Wagner, S. Ghosal, W.K. Liu, Particulate flow simulations using lubrication theory solution enrichment, International Journal for Numerical Methods in Engineering 56 (2003) 1261–1289.
[91] J. Gleick, Chaos—Making a New Science, Viking, 1987.
[92] P. Glendinning, Stability, Instability and Chaos—An Introduction to the Theory of Nonlinear Differential Equations, Cambridge University Press, 1994.
[93] R. Glowinski, T.W. Pan, T.I. Hesla, D.D. Joseph, J. Périaux, A fictitious domain approach to the direct numerical simulation of incompressible viscous flow past moving rigid bodies: Application to particulate flow, Journal of Computational Physics 169 (2001) 363–426.
[94] D. Goldstein, R. Handler, L. Sirovich, Modeling a no-slip flow boundary with an external force field, Journal of Computational Physics 105 (1993) 354–366.
[95] R. Haberman, Applied Partial Differential Equations with Fourier Series and Boundary Value Problems, Prentice-Hall, 2004.
[96] P. Hagedorn, Non-linear Oscillations, 2nd ed., Oxford University Press, 1988.
[97] B. Häggblad, K.J. Bathe, Specifications of boundary conditions for Reissner/Mindlin plate bending finite elements, International Journal for Numerical Methods in Engineering 30 (1990) 981–1011.
[98] M.J. Hannoyer, M.P. Paidoussis, Dynamics of slender tapered beams with internal or external axial flow. Part I: Theory, Journal of Applied Mechanics 46 (1979) 45–51.
[99] J. Happel, H. Brenner, Low Reynolds Number Hydrodynamics, Martinus Nijhoff Publishers, 1983.

[100] O. Haszpra, Modelling Hydroelastic Vibrations, Pitman, 1979.

[101] H.C. Chen, R.L. Taylor, Vibration analysis of fluid–solid systems using a finite element displacement formulation, International Journal for Numerical Methods in Engineering 29 (1990) 683–698.

[102] H. Hencky, Über die berücksichtigung der schubverzerrung in ebenen platten, Ingenieur Archiv 16 (1947) 72–76.

[103] D. Hendriana, On finite element and control volume upwinding methods for high Peclet number flows, PhD thesis, Massachusetts Institute of Technology, 77 Massachusetts Avenue, Cambridge, MA 02139, May 1994.

[104] H.J.-P. Morand, R. Ohayon, Fluid Structure Interaction—Applied Numerical Methods, John Wiley & Sons, 1995 (translated by C.A. James).

[105] H.J.P. Morand, R. Ohayon, Substructure variational analysis of the vibrations of coupled fluid-structure systems. Finite element results, International Journal for Numerical Methods in Engineering 14 (1979) 741–755.

[106] H.K. Khalil, Nonlinear Systems, Macmillan, 1992.

[107] H.M. Schey, Div, Grad, Curl, and All That, Norton, 1992.

[108] P. Holmes, J.L. Lumley, G. Berkooz, Turbulence, Coherent Structures, Dynamical Systems and Symmetry, Cambridge University Press, 1996.

[109] P. Holmes, J. Marsden, Bifurcation to divergence and flutter in flow-induced oscillations: An infinite dimensional analysis, Automatica 14 (1978) 367–384.

[110] P.J. Holmes, Bifurcations to divergence and flutter in flow induced oscillations: A finite-dimensional analysis, Journal of Sound and Vibration 53 (1977) 471–503.

[111] H.P. Lee, T.Y. Ng, Transverse vibration of a plate moving over multiple point supports, Applied Acoustics 47 (1996) 291–301.

[112] Y. Huang, C.M. Doerschuk, R.D. Kamm, Computational modeling of RBC and neutrophil transit through the pulmonary capillaries, Journal of Applied Physiology 90 (2001) 545–564.

[113] F. Ihlenburg, Finite Element Analysis of Acoustic Scattering, Springer-Verlag, 1998.

[114] U. Ingard, On the theory and design of acoustic resonators, Journal of Acoustic Society of America 25 (1953) 1037–1061.

[115] A. Iosilevich, K.J. Bathe, F. Brezzi, On evaluating the Inf-Sup condition for plate bending elements, International Journal for Numerical Methods in Engineering 40 (1997) 3639–3663.

[116] A. Isidori, Nonlinear Control Systems, vol. II, Springer-Verlag, 1999.

[117] J.A. Wickert, Non-linear vibration of a traveling tensioned beam, International Journal of Non-linear Mechanics 27 (1992) 503–517.

[118] J.A. Wickert, C.D. Mote Jr., Current research on the vibration and stability of axially moving materials, The Shock and Vibration Digest 20 (1988) 3–13.

[119] J.A. Wickert, C.D. Mote Jr., Classical vibration analysis of axially moving continua, Journal of Applied Mechanics 57 (1990) 738–744.

[120] J.C. Heinrich, D. Connolly, Three-dimensional finite element analysis of self-acting foil bearings, Computer Methods in Applied Mechanics and Engineering 100 (1992) 31–43.

[121] J.C. Heinrich, O.C. Zienkiewicz, Quadratic finite element schemes for two-dimensional convective-transport problems, International Journal for Numerical Methods in Engineering 11 (1977) 1831–1844.

[122] J.C. Heinrich, P.S. Huyakorn, O.C. Zienkiewicz, A.R. Mitchell, An 'upwind' finite element scheme for two-dimensional convective transport equation, International Journal for Numerical Methods in Engineering 11 (1977) 131–143.

[123] J.C. Strikwerda, Finite Difference Schemes and Partial Differential Equations, Wadsworth & Brooks/Cole Advanced Books & Software, 1989.

[124] J.G. Campbell, The in-plane elastic constants of paper, Australian Journal of Applied Science 12 (1961) 356–357.

[125] J.H. Ginsberg, The dynamic stability of a pipe conveying a pulsatile flow, International Journal of Engineering Science 11 (1973) 1013–1024.

[126] Z.P. Jiang, I. Mareels, Robust nonlinear integral control by partial-state and output feedback, in: Proceedings of the 39th IEEE Control Conference, Sydney, December 2000, pp. 2084–2089.

[127] Z.P. Jiang, L. Praly, Design of robust adaptive controllers for nonlinear systems with dynamic uncertainties, Automatica 34 (1998) 825–840.

[128] J.J. Slotine, W. Li, Applied Nonlinear Control, Prentice-Hall, 1991.

[129] J.K. Hale, Ordinary Differential Equations, 2nd ed., Krieger, 1980.

[130] J.K. Hale, Oscillations in Nonlinear Systems, Dover, 1992.

[131] J.K. Hale, H. Koçak, Dynamics and Bifurcations, Springer-Verlag, 1991.

[132] J.N. Newman, Maine Hydrodynamics, MIT Press, Cambridge, MA/London, 1979.

[133] C. Johnson, U. Nävert, J. Pitkäranta, Finite element methods for linear hyperbolic problems, Computer Methods in Applied Mechanics and Engineering 45 (1984) 285–312.

[134] J.R. Munkres, Analysis on Manifolds, Addison-Wesley Publishing, 1991.

[135] J.S. Hubert, E.S. Palencia, Vibration and Coupling of Continuous Systems, Springer-Verlag, 1989.

[136] M. Kawahara, T. Miwa, Finite element analysis of wave motion, International Journal for Numerical Methods in Engineering 20 (1984) 1193–1210.

[137] D. Kay, D. Loghin, A. Wathen, A preconditioner for the steady-state Navier–Stokes equations, SIAM Journal on Scientific Computing 24 (1) (2002) 237–256.

[138] K.E. Atkinson, The Numerical Solution of Integral Equations of the Second Kind, Cambridge University Press, 1997.

[139] L. Kiefling, G.C. Feng, Fluid-structure finite element vibrational analysis, AIAA Journal 14 (1976) 199–203.

[140] S. Kim, S.J. Karrila, Microhydrodynamics—Principles and Selected Applications, Butterworh–Heinemann, 1991.

[141] Y. Kim, L. Zhu, X. Wang, C.S. Peskin, On various techniques for computer simulation of boundaries with mass, in: 2nd MIT Conference on Computational Fluid and Solid Mechanics, 2003.

[142] K.J. Bathe, Finite Element Procedures, Prentice-Hall, Englewood Cliffs, NJ, 1996.

[143] K.J. Bathe, Simulation of structural and fluid flow response in engineering practice, Computer Modeling and Simulation in Engineering 1 (1996) 47–77.

[144] K.J. Bathe, Fluid-structure interactions, Journal of Mechanical Engineering 120 (1998) 66–68.

[145] K.J. Bathe, The Inf-Sup condition and its evaluation for mixed finite element methods, Computers and Structures 79 (2001) 243–252.

[146] K.J. Bathe, D. Chapelle, The Finite Element Analysis of Shells: Fundamentals, Springer-Verlag, 2003.

[147] K.J. Bathe, E.N. Dvorkin, A formulation of general shell elements—The use of mixed interpolation of tensorial components, International Journal for Numerical Methods in Engineering 22 (1986) 697–722.

[148] K.J. Bathe, D. Hendriana, F. Brezzi, G. Sangalli, Inf-Sup testing of upwind methods, International Journal for Numerical Methods in Engineering 48 (2000) 745–760.

[149] F. Brezzi, K.J. Bathe, M. Fortin, Mixed interpolated elements for Reissner–Mindlin plates, International Journal for Numerical Methods in Engineering 28 (1989) 1787–1801.

[150] K.J. Bathe, C. Nitikitpaiboon, X. Wang, A mixed displacement-based finite element formulation for acoustic fluid-structure interaction, Computers and Structures 56 (2–3) (1995) 225–237.

[151] K.J. Bathe, V. Sonnad, On effective implicit time integration in analysis of fluid-structure problems, International Journal for Numerical Methods in Engineering 15 (1980) 943–948.

[152] K.J. Bathe, J. Walczak, H. Zhang, Some recent advances for practical finite element analysis, Computers and Structures 47 (4–5) (1993) 511–521.

[153] K.J. Bathe, W.F. Hahn, On transient analysis of fluid-structure systems, Computers and Structures 10 (1979) 383–391.

[154] K.J. Bathe, H. Zhang, M.H. Wang, Finite element analysis of incompressible and compressible fluid flows with free surfaces and structural interactions, Computers and Structures 56 (2–3) (1995) 193–213.

[155] K.L. Koai, T. Yang, J. Chen, The application of finite element method in analyzing the noise reduction effect of Helmholtz resonator attachments,

Proceedings of the ASME Noise Control and Acoustics Division 22 (1996) 199–205.

[156] H. Kohno, K.J. Bathe, A flow-condition-based interpolation finite element procedure for triangular grids, International Journal for Numerical Methods in Fluids 51 (2006) 673–699.

[157] R. Kress, Linear Integral Equations, Springer-Verlag, 1999.

[158] M. Krstić, I. Kanellakopoulos, P.V. Kokotović, Nonlinear and Adaptive Control Design, John Wiley & Sons, 1995.

[159] K.U. Ingard, A. Akay, On the vibration damping of a plate by means of a viscous fluid layer, Journal of Vibration, Acoustics, Stress, and Reliability in Design 109 (1987) 178–184.

[160] L.A. Monclova, L.J. Forney, Numerical simulation of a pipeline tee mixer, Industrial Engineering Chemistry Research 34 (1995) 1488–1493.

[161] M.C. Lai, C.S. Peskin, An immersed boundary method with formal second-order accuracy and reduced numerical viscosity, Journal of Computational Physics 160 (2000) 705–719.

[162] H. Lamb, Hydrodynamics, 6th ed., Dover Publications, 1945.

[163] J. LaSalle, S. Lefschetz, Stability by Lyapunov's Direct Method with Applications, Academic Press, 1961.

[164] B.E. Launder, D.B. Spalding, The numerical computation of turbulent flows, Computer Methods in Applied Mechanics and Engineering 3 (1974) 269–289.

[165] L.B. Felsen, N. Marcuvitz, Radiation and Scattering of Waves, IEEE Press, 1994.

[166] L.C. Wellford Jr., T.H. Ganaba, A finite element method with a hybrid Lagrange line for fluid mechanics problems involving large free surface motion, International Journal for Numerical Methods in Engineering 17 (1981) 1201–1231.

[167] L.E. Kinsler, A.R. Frey, A.B. Coppens, J.V. Sanders, Fundamentals of Acoustics, 3rd ed., McGraw-Hill Publishing Company, 1982.

[168] L.G. Olson, Finite element analysis of fluid-structure interactions, PhD thesis, Massachusetts Institute of Technology, 77 Massachusetts Avenue, Cambridge, MA 02139, June 1985.

[169] L.G. Olson, K.J. Bathe, A study of displacement-based fluid finite elements for calculating frequencies of fluid and fluid-structure systems, Nuclear Engineering and Design 76 (1983) 137–151.

[170] L.G. Olson, K.J. Bathe, Analysis of fluid-structure interactions. A direct symmetric coupled formulation based on the fluid velocity potential, Computers and Structures 21 (1–2) (1985) 21–32.

[171] S. Li, W.K. Liu, Reproducing kernel hierarchical partition of unity. Part I: Formulation and theory, International Journal for Numerical Methods in Engineering 45 (1999) 251–288.

[172] S. Lim, A. Ferent, X. Wang, C.S. Peskin, Dynamics of a closed rod with twist and bend in fluid, SIAM Journal on Scientific Computing (2008), in press.
[173] Y. Liu, L. Zhang, X. Wang, W.K. Liu, Coupling of Navier–Stokes equations with protein molecular dynamics and its application to hemodynamics, International Journal for Numerical Methods in Fluids 46 (2004) 1237–1252.
[174] M.A. Hamdi, Y. Ousset, G. Verchery, A displacement method for the analysis of vibrations of coupled fluid-structure systems, International Journal for Numerical Methods in Engineering 13 (1978) 139–150.
[175] MacNeal, Citerley, Chargin, A symmetric modal formulation of fluid-structure interaction, including a static approximation to higher order fluid modes, in: ASME Pressure Vessels and Piping Conference, San Francisco, April 1980.
[176] R. Marino, P. Tomei, Nonlinear Control Design: Geometric, Adaptive and Robust, Prentice-Hall, 1995.
[177] Y. Matsuzaki, Y.C. Fung, Stability analysis of straight and buckled two-dimensional channels conveying an incompressible flow, Journal of Applied Mechanics 44 (1977) 548–552.
[178] M.C. Junger, D. Feit, Sound, Structures, and their Interaction, MIT Press, 1972.
[179] M.D. Tidriri, Preconditioning techniques for the Newton–Krylov solution of compressible flows, Journal of Computational Physics 132 (1997) 51–61.
[180] M.E. Rosar, C.S. Peskin, Fluid flow in collapsible elastic tubes: A three-dimensional numerical model, New York Journal of Mathematics 7 (2001) 281–302.
[181] L. Meirovitch, Methods of Analytical Dynamics, McGraw-Hill, 1988.
[182] B. Mercier, J. Osborn, J. Rappaz, P.A. Raviart, Eigenvalue approximation by mixed and hybrid methods, Mathematics of Computation 36 (1981) 353–427.
[183] R.D. Mindlin, Influence of rotatory inertia and shear on flexural motions of isotropic, elastic plates, Journal of Applied Mechanics 18 (1951) 31–38.
[184] M.K. Hakala, Application of the finite element method to fluid-structure interaction in ship vibration, Technical Research Center of Finland, Research Report, 1986, p. 433.
[185] M.M. Zdrvavkovich, Different modes of vortex shedding: An overview, Journal of Fluids and Structures 10 (1996) 427–437.
[186] J. Mohd-Yusof, Development of immersed boundary methods for complex geometries, Center for Turbulence Research Annual Research Briefs, NASA Ames/Stanford University, 1998, pp. 325–336.
[187] C.D. Mote, A study of band saw vibrations, Journal of the Franklin Institute 279 (1965) 430–444.

[188] M.P. Paidoussis, Dynamics of flexible slender cylinders in axial flow. Part 1. Theory, Journal of Fluid Mechanics 26 (1966) 717–736.

[189] M.P. Paidoussis, Dynamics of flexible slender cylinders in axial flow. Part 2. Experiments, Journal of Fluid Mechanics 26 (1966) 737–751.

[190] M.P. Paidoussis, Fluid-Structure Interactions—Slender Structures and Axial Flow, vol. I, Academic Press, 1998.

[191] M.P. Paidoussis, Fluid-Structure Interactions—Slender Structures and Axial Flow, vol. II, Academic Press, 2004.

[192] M.P. Paidoussis, N.T. Issid, Dynamic stability of pipes conveying fluid, Journal of Sound and Vibration 33 (1974) 267–294.

[193] M.P. Paidoussis, G.X. Li, Pipes conveying fluid: A model dynamical problem, Journal of Fluids and Structures 7 (1993) 137–204.

[194] M.P. Paidoussis, M.J. Pettigrew, Dynamics of flexible cylinders in axisymetrically confined axial flow, Journal of Applied Mechanics 46 (1979) 37–44.

[195] M.P. Paidoussis, C. Sundararajan, Parametric and combination resonances of a pipe conveying pulsating fluid, Journal of Applied Mechanics 42 (1975) 780–783.

[196] M.S. Howe, Acoustics of Fluid-Structure Interactions, Cambridge University Press, 1998.

[197] M.S. Marcus, A finite-element method applied to the vibration of submerged plates, Journal of Ship Research 22 (1978) 94–99.

[198] S. Müftü, R.C. Benson, Modelling the transport of paper webs including the paper permeability effects, in: Proceedings of ISPS at the ASME-International Congress and Exposition in San Fransisco, CA, November 1995.

[199] W.C. Müller, Simplified analysis of linear fluid-structure interaction, International Journal for Numerical Methods in Engineering 17 (1981) 113–121.

[200] Y.D. Murray, C.D. Mote Jr., Analysis of a plane inclined guide bearing under transverse vibration and translation of a plate, Journal of Lubrication Technology 105 (1983) 335–341.

[201] G. Muthuveerappan, N. Ganesan, M.A. Veluswami, Vibration of square cantilever plate immersed in water, Journal of Sound and Vibration 61 (3) (1978) 467–470.

[202] M.W. Hirsch, S. Smale, Differential Equations, Dynamical Systems, and Linear Algebra, Academic Press, 1974.

[203] D. Newman, G.E. Karniadakis, Simulations of flow over a flexible cable: A comparison of forced and flow-induced vibration, Journal of Fluids and Structures 10 (1996) 439–453.

[204] C. Nitikitpaiboon, K.J. Bathe, An arbitrary Lagrangian–Eulerian velocity potential formulation for fluid-structure interaction, Computers and Structures 47 (4–5) (1993) 871–891.

[205] T. Nomura, T.J.R. Hughes, An arbitrary Lagrangian–Eulerian finite element method for interaction of fluid and a rigid body, Computer Methods in Applied Mechanics and Engineering 95 (1992) 115–138.
[206] O.C. Zienkiewicz, The Finite Element Method, 3rd ed., McGraw-Hill Publishing Company, 1977.
[207] O.C. Zienkiewicz, P. Bettess, Fluid-structure dynamics interaction and wave forces. An introduction to numerical treatment, International Journal for Numerical Methods in Engineering 13 (1978) 1–16.
[208] O.C. Zienkiewicz, R.E. Newton, Coupled vibrations of a structure submerged in a compressible fluid, in: Proc. Symp. Finite Element Techniques, University of Stuttgart, Germany, 1969, pp. 359–379.
[209] L. Olson, T. Vandini, Eigenproblems from finite element analysis of fluid-structure interactions, Computers and Structures 33 (1989) 679–687.
[210] O.M. Faltinsen, Sea Loads on Ships and Offshore Structures, Cambridge University Press, 1990.
[211] P. LeTallec, V. Ruas, On the convergence of the bilinear-velocity constant pressure finite element method in viscous flow, Computer Methods in Applied Mechanics and Engineering 54 (1986) 235–243.
[212] D. Pantuso, K.J. Bathe, A four-node quadrilateral mixed-interpolated element for solids and fluids, Mathematical Models and Methods in Applied Sciences 5 (1995) 1113–1128.
[213] P.G. Ciarlet, The Finite Element Method for Elliptic Problems, North-Holland, 1978.
[214] P.J. Roache, Verification and Validation in Computational Science and Engineering, Hermosa Publishers, 1998.
[215] P.M. Morse, Vibration and Sound, 2nd ed., McGraw-Hill Publishing Company, 1948.
[216] P.M. Morse, K.U. Ingard, Theoretical Acoustics, Princeton University Press, 1986.
[217] D. Porter, D.S.G. Stirling, Integral Equations—A Practical Treatment, from Spectral Theory to Applications, Cambridge University Press, 1990.
[218] C. Pozrikidis, Boundary Integral and Singularity Methods for Linearized Viscous Flow, Cambridge University Press, 1992.
[219] A. Pramila, Sheet flutter and the interaction between sheet and air, TAPPI Journal 69 (1986) 70–74.
[220] A. Pramila, Natural frequencies of a submerged axially moving band, Journal of Sound and Vibration 113 (1987) 198–203.
[221] A. Quarteroni, Modeling the cardiovascular system—A mathematical adventure: Part I, SIAM News (2001) 34.
[222] E. Race, J.B. Wheeldon, D. Fraser-Clark, Air movement induced by felts and fabrics and its ventilation effect on dryer pockets, in: TAPPI Notes—Sheet Flutter and Windage Problems Seminar, 1991, pp. 26–31.

[223] B. Ramaswamy, M. Kawahara, Arbitrary Lagrangian–Eulerian finite element method for unsteady, convective, incompressible viscous free surface fluid flow, International Journal for Numerical Methods in Fluids 7 (1987) 1053–1075.

[224] B. Ramaswamy, M. Kawahara, T. Nakayama, Lagrangian finite element method for the analysis of two-dimensional slosh problems, International Journal for Numerical Methods in Fluids 6 (1986) 659–670.

[225] J.W.S. Rayleigh, Acoustical observations. II, Philosophical Magazine 7 (1889) 149–162.

[226] R.D. Blevins, Formulas for Natural Frequency and Mode Shape, Van Nostrand Reinhold Company, 1979.

[227] R.D. Blevins, Flow-Induced Vibration, 2nd ed., Krieger, 1994.

[228] R.E. Page, Method of suppressing paper web flutter, United States Patent, 4321107, March 1982.

[229] E. Reissner, Note on the effect of transverse-shear deformation in laminated anisotropic plates, Computer Methods in Applied Mechanics and Engineering 20 (1979) 203–209.

[230] R.H. Sabersky, A.J. Acosta, E.G. Hauptmann, Fluid Flow—A First Course in Fluid Mechanics, 3rd ed., Macmillan Publishing Company, 1989.

[231] R.J. LeVeque, Z.L. Li, Immersed interface methods for Stokes flow with elastic boundaries or surface tension, SIAM Journal on Scientific Computing 18 (1997) 709–735.

[232] R.L. Bisplinghoff, H. Ashley, Principles of Aeroelasticity, Dover, 1962.

[233] R.L. Bisplinghoff, H. Ashley, R.L. Halfman, Aeroelasticity, Dover, 1983.

[234] A.M. Roma, C.S. Peskin, M.J. Berger, An adaptive version of the immersed boundary method, Journal of Computational Physics 153 (1999) 509–534.

[235] A. Roshko, Experiments on the flow past a circular cylinder at very high Reynolds number, Journal of Fluid Mechanics 10 (1961) 345–356.

[236] N. Rouche, P. Habets, M. Laloy, Stability Theory by Lyapunov's Direct Method, Springer-Verlag, 1977.

[237] R.P. Beyer Jr., A computational model of the cochlea using the immersed boundary method, Journal of Computational Physics 98 (1992) 145–162.

[238] R.P. Beyer Jr., R.J. LeVeque, Analysis of a one-dimensional model for the immersed boundary method, SIAM Journal of Numerical Analysis 29 (1992) 332–364.

[239] S. Rugonyi, K.J. Bathe, On finite element analysis of fluid flows fully coupled with structural interactions, Computer Modeling in Engineering and Sciences 2 (2001) 195–212.

[240] A. Selamet, N.S. Dickey, J.M. Novak, Theoretical computational and experimental investigation of Helmholtz resonators with fixed volume: Lumped vs. distributed analysis, Journal of Sound and Vibration 187 (1995) 358–367.

[241] S.F. McCormick, Multigrid Methods, SIAM, 1987.
[242] A. Shanley, Pushing the limits of CFD, in: Chemical Engineering, December 1996, pp. 66–67.
[243] M. Shub, Global Stability of Dynamical Systems, Springer-Verlag, 1987.
[244] W. Shyy, M. Francois, H.S. Udaykumar, N. N'dri, R. Tran-Son-Tay, Moving boundaries in micro-scale biofluid dynamics, Applied Mechanics Review 54 (2001) 405–453.
[245] S.L. Diamond, Reaction complexity of flowing human blood, Biophysical Journal 80 (2001) 1031–1032.
[246] S.M. Dickinson, The buckling and frequency of flexural vibration of rectangular isotropic and orthotropic plates using Rayleigh's method, Journal of Sound and Vibration 61 (1) (1978) 1–8.
[247] S.M. Soedel, W. Soedel, On the free and forced vibration of a plate supporting a freely slosh surface liquid, Journal of Sound and Vibration 171 (2) (1994) 159–171.
[248] S.P. Timoshenko, On the transverse vibrations of bars of uniform cross section, Philosophical Magazine 43A (1922) 125–131.
[249] S.P. Timoshenko, J.M. Gere, Theory of Elastic Stability, McGraw-Hill, 1988.
[250] S.P. Timoshenko, S. Woinowsky-Krieger, Theory of Plates and Shells, 2nd ed., McGraw-Hill Publishing Company, 1987.
[251] S.S. Chen, Dynamic stability of tube conveying fluid, Journal of the Engineering Mechanics Division ASCE 97 (1971) 1469–1485.
[252] S.T. Ariaratnam, N.S. Namachchivaya, Dynamic stability of pipes conveying pulsating fluid, Journal of Sound and Vibration 107 (1986) 215–230.
[253] S.T. Noah, G.R. Hopkins, Dynamic stability of elastically supported pipes conveying pulsating fluid, Journal of Sound and Vibration 71 (1980) 103–116.
[254] I. Stakgold, Green's Functions and Boundary Value Problems, John Wiley & Sons, 1998.
[255] R. Stenberg, Analysis of mixed finite element methods for the Stokes problem: A unified approach, SIAM Journal of Numerical Analysis 42 (1984) 9–23.
[256] R. Stenberg, Error analysis of some finite element methods for the Stokes problem, Mathematics of Computation 54 (1990) 495–508.
[257] R. Stenberg, A technique for analysing finite element methods for viscous incompressible flow, International Journal for Numerical Methods in Fluids 11 (1990) 935–948.
[258] R. Stenberg, M. Suri, An *hp* error analysis of MITC plate elements, SIAM Journal of Numerical Analysis 34 (1997) 544–568.
[259] J.M. Stockie, B.R. Wetton, Analysis of stiffness in the immersed boundary method and implications for time-stepping schemes, Journal of Computational Physics 154 (1999) 41–64.

[260] V. Strouhal, Ueber eine besondere art der tonenegung, Annalen der Physik and Chemie, series 5 (Leüpzig) 5 (1878) 216–251.

[261] D. Sulsky, J.U. Brackbill, A numerical method for suspension flow, Journal of Computational Physics 96 (1991) 339–368.

[262] J. Sundqvist, An application of ADINA to the solution of fluid-structure interaction problems, Computers and Structures 17 (5–6) (1983) 793–807.

[263] M. Suri, I. Babuška, C. Schwab, Locking effects in the finite element approximation of plate models, Mathematics of Computation 64 (210) (1995) 461–482.

[264] T. Sussman, K.J. Bathe, A finite element formulation for nonlinear incompressible elastic and inelastic analysis, Computers and Structures 26 (1–2) (1987) 357–409.

[265] S.V. Patankar, Numerical Heat Transfer and Fluid Flow, Hemisphere Publishing, 1980.

[266] H. Tanaka, T. Araki, Simulation method of colloidal suspensions with hydrodynamic interactions: Fluid particle dynamics, Physical Review Letters 85 (2000) 1338–1341.

[267] D. Tang, J. Yang, A free moving boundary model and boundary iteration method for unsteady viscous flow in stenotic elastic tubes, SIAM Journal on Scientific Computing 21 (2000) 1370–1386.

[268] T.B. Belytschko, Fluid-structure interaction, Computers and Structures 12 (1980) 459–469.

[269] T.B. Belytschko, J.M. Kennedy, A fluid-structure finite element method for the analysis of reactor safety problems, Nuclear Engineering and Design 38 (1976) 71–81.

[270] T.B. Belytschko, W.K. Liu, B. Moran, Nonlinear Finite Elements for Continua and Structures, John Wiley & Sons, 2000.

[271] H. Tennekes, J.L. Lumley, A First Course in Turbulence, MIT Press, 1972.

[272] T.E. Tezduyar, Stabilized finite element formulations for incompressible-flow computations, Advances in Applied Mechanics 28 (1992) 1–44.

[273] T.E. Tezduyar, Finite element methods for flow problems with moving boundaries and interfaces, Archives of Computational Methods in Engineering 8 (2001) 83–130.

[274] T.J. Pedley, The Fluid Mechanics of Large Blood Vessels, Cambridge University Press, 1980.

[275] T.J. Pedley, X.Y. Luo, Modeling flow and oscillations in collapsible tubes, Theoretical and Computational Fluid Dynamics 10 (1998) 277–294.

[276] T.J.R. Hughes, The Finite Element Method: Linear Static and Dynamic Analysis, Prentice-Hall, 1987.

[277] T.J.R. Hughes, L.P. Franca, M. Balestra, A new finite element formulation for computational fluid dynamics: V. Circumventing the Babuška–Brezzi condition: A stable Petrov–Galerkin formulation of the Stokes problem

accommodating equal-order interpolations, Computer Methods in Applied Mechanics and Engineering 59 (1986) 85–99.
[278] T.J.R. Hughes, L.P. Franca, G.M. Hulbert, A new finite element formulation for computational fluid dynamics: VIII. The Galerkin/least squares method for advective–diffusive equations, Computer Methods in Applied Mechanics and Engineering 73 (1989) 173–189.
[279] T.J.R. Hughes, M. Mallet, A new finite element formulation for computational fluid dynamics. III. The generalized streamline operator for multidimensional advective–diffusive systems, Computer Methods in Applied Mechanics and Engineering 58 (1986) 305–328.
[280] T.J.R. Hughes, T.E. Tezduyar, Finite elements based upon Mindlin plate theory with particular reference to the four-node bilinear isoparametric element, Journal of Applied Mechanics 48 (1981) 587–596.
[281] R. Van Loon, P.D. Anderson, F.N. Van de Vosse, S.J. Sherwin, Comparison of various fluid-structure interaction methods for deformable bodies, Computers and Structures 85 (2006) 833–843.
[282] F. Verhulst, Nonlinear Differential Equations and Dynamical Systems, Springer, 1996.
[283] T. von Karman, Über den mechanismuss des widersstandes den ein bewegter korper in einen flussigkeit erfahrt, in: Nachrichten der K. Gesellschaft der Wissenschaften zu Gottingen, 1912, pp. 547–556.
[284] V.V. Bolotin, The Dynamic Stability of Elastic Systems, Holden-Day, 1964.
[285] W.A. Conrad, Pressure-flow relationship in collapsible tubes, IEEE Transactions on Biomedical Engineering 16 (1969) 284–295.
[286] W.A. Conrad, D.M. McQueen, E.L. Yellin, Steady pressure flow relations in compressed arteries: possible origin of Korotkoff sounds, Medical and Biological Engineering and Computing 18 (1980) 419–426.
[287] W.A. Conrad, M.L. Cohen, D.M. McQueen, Note on the oscillations of collapsible tubes, Medical and Biological Engineering and Computing 16 (1978) 211–214.
[288] D. Wahren, Wet webs in open draws—An engineering analysis, TAPPI Journal 64 (1981) 89–93.
[289] X. Wang, Finite element analysis of air-sheet interactions and flutter suppression devices, Computers and Structures 64 (5–6) (1997) 983–994.
[290] X. Wang, Numerical analysis of moving orthotropic thin plates, Computers and Structures 70 (1999) 467–486.
[291] X. Wang, Velocity/pressure mixed finite element and finite volume formulation with ALE descriptions for nonlinear fluid-structure interaction problems, Advances in Engineering Software 31 (2000) 35–44.
[292] X. Wang, On the discretized delta function and force calculation in the immersed boundary method, in: K.J. Bathe (Ed.), Computational Fluid and Solid Mechanics, Elsevier, 2003, pp. 2164–2169.

[293] X. Wang, A.G. Giorges, C. Park, Simulation of a deformable ball passing through a step diffuser, Computers and Structures 72 (1999) 435–456.
[294] X. Wang, F. Bloom, Dynamics of a submerged and inclined concentric pipe system with internal and external flows, Journal of Fluids and Structures 13 (1999) 443–460.
[295] X. Wang, F. Bloom, Stability issues of concentric pipes conveying steady and pulsatile fluids, Journal of Fluids and Structures 15 (2001) 1137–1152.
[296] X. Wang, Z. Feng, L.J. Forney, Computational simulation of turbulent mixing with mass transfer, Computers and Structures 70 (1999) 447–465.
[297] X. Wang, J.K. Hale, On monodromy matrix computation, Computer Methods in Applied Mechanics and Engineering 190 (2001) 2263–2275.
[298] X. Wang, K.J. Bathe, Displacement/pressure based finite element formulations for acoustic fluid-structure interaction problems, International Journal for Numerical Methods in Engineering 40 (1997) 2001–2017.
[299] X. Wang, K.J. Bathe, On mixed elements for acoustic fluid-structure interactions, Mathematical Models and Methods in Applied Sciences 7 (1997) 329–343.
[300] X. Wang, G. Renner, Computational simulation of three-dimensional pretensioned moving paper webs, in: Nineteenth Southeastern Conference on Theoretical and Applied Mechanics, May 1998.
[301] X. Wang, W.K. Liu, Extended immersed boundary method using FEM and RKPM, Computer Methods in Applied Mechanics and Engineering 193 (2004) 1305–1321.
[302] S. Wiggins, Global Bifurcations and Chaos, Springer-Verlag, 1988.
[303] W.J. Minkowycz, E.M. Sparrow, G.E. Schneider, R.H. Pletcher, Handbook of Numerical Heat Transfer, John Wiley & Sons, 1988.
[304] W.K. Blake, Mechanics of Flow-Induced Sound and Vibration, Academic Press, 1986.
[305] W.K. Liu, Y. Chen, C.T. Chang, T.B. Belytschko, Advances in multiple scale kernel particle methods, Computational Mechanics 18 (1996) 73–111.
[306] W.K. Liu, Y. Chen, R.A. Uras, C.T. Chang, Generalized multiple scale reproducing kernel particle methods, Computer Methods in Applied Mechanics and Engineering 139 (1996) 91–157.
[307] W.K. Liu, S. Jun, Y.F. Zhang, Reproducing kernel particle methods, International Journal for Numerical Methods in Fluids 20 (1995) 1081–1106.
[308] W.K. Liu, Y.J. Chen, Wavelet and multiple scale reproducing kernel methods, International Journal for Numerical Methods in Fluids 21 (1995) 901–932.
[309] W.McC. Siebert, Circuits, Signals, and Systems, 4th ed., The MIT Press and McGraw-Hill Book Company, 1989.
[310] W.W. Nichols, M.F. O'Rourke, McDonald's Blood Flow in Arteries—Theoretical, Experimental and Clinical Principles, 4th ed., Arnold, 1998.

[311] X.E. Guo, E. Takai, K. Liu, X. Wang, An exploration of cell stress and deformation under shear flow, in: ASME Winter Bioengineering Conference BED-51, 2001.
[312] T. Yabe, T. Ishikawa, P.Y. Wang, T. Aoki, Y. Kadota, F. Ikeda, A universal solver for hyperbolic-equations by cubic-polynomial interpolation. Part II. 2-dimensional and 3-dimensional solvers, Computer Physics Communications 66 (1991) 233–242.
[313] B. Yang, C.D. Mote Jr., Active vibration control of the axially moving string in the *s* domain, Journal of Applied Mechanics 58 (1991) 189–196.
[314] Y.C. Fung, An Introduction to the Theory of Aeroelasticity, John Wiley & Sons, 1955.
[315] Y.C. Fung, Biomechanics: Circulation, Springer-Verlag, 1996.
[316] Y.C. Fung, Biomechanics: Motion, Flow, Stress, and Growth, Springer-Verlag, 2000.
[317] Y.S. Uflyand, The propagation of waves in the transverse vibrations of bars and plates, Prikl. Mat. Meh. 12 (1948) 287–300.
[318] Z. Yu, A DLM/FD method for fluid/flexible-body interactions, Journal of Computational Physics 207 (2005) 1–27.
[319] L. Zhang, A. Gerstenberger, X. Wang, W.K. Liu, Immersed finite element method, Computer Methods in Applied Mechanics and Engineering 193 (2004) 2051–2067.
[320] C. Zhu, G. Bao, N. Wang, Cell mechanics: Mechanical response, cell adhesion, and molecular deformation, Annual Review of Biomedical Engineering 2 (2000) 189–226.

# Subject Index